THE CRAFT READER

THE CRAFT

READER

EDITED BY GLENN ADAMSON

BLOOMSBURY VISUAL ARTS
LONDON • NEW YORK • OXFORD • NEW DELHI • SYDNEY

BLOOMSBURY VISUAL ARTS
Bloomsbury Publishing Plc
50 Bedford Square, London, WC1B 3DP, UK
1385 Broadway, New York, NY 10018, USA
29 Earlsfort Terrace, Dublin 2, Ireland

BLOOMSBURY, BLOOMSBURY VISUAL ARTS and the Diana logo are trademarks
of Bloomsbury Publishing Plc

First published in Great Britain by Berg Publishers 2010
Reprinted by Bloomsbury Academic 2016 (twice), 2017
This edition published by Bloomsbury Visual Arts 2018
Reprinted 2019 (twice), 2020 (twice), 2021

Cover design: William Joseph
Cover image © (Front cover) Otto Hagel, Marguerite Wildenhain Showing The Motion of Hands
Making a Pot, ca. 1945. Marguerite Wildenhain papers. Archives of American Art. (Spine) W. A. S.
Benson, Electric table lamp, ca. 1900. Brass. Victoria and Albert Museum.

A catalogue record for this book is available from the British Library.

A catalog record for this book is available from the Library of Congress.

ISBN: HB: 978-1-8478-8304-9
PB: 978-1-3500-9264-8

Typeset by Apex CoVantage, LLC, Madison, WI, USA

To find out more about our authors and books visit www.bloomsbury.com
and sign up for our newsletters.

CONTENTS

SECTION 3: MODERN CRAFT: IDEALISM AND REFORM

SECTION 4: THE PERSISTENCE OF CRAFT IN THE AGE OF MASS PRODUCTION

SECTION 5: CRAFT IN THEORY: AESTHETICS, ESSENCE, STATUS

SECTION 6: CRAFT IN ACTION: LIFE, ART, DESIGN

ACKNOWLEDGEMENTS

The seeds of this anthology were planted over a decade ago in a graduate seminar titled 'The American Craftsman' at Yale that was led by Edward S. (Ned) Cooke, Jr. Ned's teaching has been much in my mind as I have shaped this anthology. Several of the readings in this book were on that course syllabus, and many more have been added as a result of conversations that he and I have had in the years since.

More recently, I have had the equally great fortune of working alongside Ned and the British craft historian Tanya Harrod as editors of the *Journal of Modern Craft*. That experience, too, has deeply informed this anthology. As we began to publish new studies of craft and its multivalent encounters with modernity, it struck me that a reference work on the subject was needed: a book that would indicate the overall shape of the field as it stands now, and collect the available bibliography into a single useful volume. Ned and Tanya have been the most significant influences on this project. They are here offered my heartfelt thanks.

Like the *Journal*, this anthology ranges as widely as possible, covering subjects from the industrial revolution forward to the present and without geographical limitation. This obviously meant stretching far outside of my own area of research (Anglo-American decorative art, design and fine art). Partly as a result I have incurred several other debts to friends and colleagues who offered valuable assistance, suggestions and thoughts: Elissa Auther, Jeremy Aynsley, Pennina Barnett, Richard Barbrook, Luke Beckerdite, Chris Breward, Alison Britton, Julia Bryan-Wilson, David Byron, Garth Clark, Alan Crawford, David Crowley, Edmund de Waal, David Doris, Johanna Drucker, Sabrina Gschwandtner, Christine Guth, Janis Jeffries, Louise Mazanti, Angela McShane, Kevin Murray, Jane Pavitt, Andrew Perchuk, Patricia Riboult (who directed my attention to Gilbert Simondon, and also kindly looked over my translation of his writing); Giorgio Riello, Catharine Rossi, Merryll Saylan, Jenni Sorkin, Penny Sparke, Sarah Teasley, Jorunn Veiteberg and Kristina Wilson. I also thank the anonymous peer reviewers who have looked over aspects of the book for their useful and positive comments.

Rafael Cardoso, Zandra Ahl, Anthea Black and Nicole Burisch, and Julia Bryan-Wilson contributed new or reedited texts that have not appeared in print before, allowing me to bring the book right up to date.

Tristan Palmer, my editor at Berg, has been an enthusiastic advocate for this project from the beginning and has seen it through the various stages of production with efficiency and aplomb.

Invaluable logistical support was provided by a team of interns: Keelin Burrows, Elizabeth Bisley and Tom Bisley did much of the hard graft of digitizing and correcting texts and securing reproduction rights. The Center for Craft, Creativity and Design generously funded this work; I am indebted to Dian Magie and her team at CCCD for their visionary support of craft scholarship.

A range of people and institutions graciously extended reproduction rights for both texts and images. I would here mention particularly Zandra Ahl; the American Craft Council; British Craft Council; British Museum; the Crafts Study Centre, Farnham; David Cripps; Liz Collins; Cranbrook Art Museum; David Doris; Haimi Fenichel; Sabrina Gschwandtner; Martin Bodilsen Kaldahl; Tami Katz-Freiberg; Kristian Kozul; Lisson Gallery; Frank Lloyd Gallery; the Museum of Contemporary Craft in Portland, Oregon; Sam Maloof; George Nakashima Woodworking; Jonathan Pollock; Science Museum, London; the Smithsonian Archives of American Art; Carole Tulloch; and the Victoria and Albert Museum.

My most heartfelt thanks go my family, particularly my grandfather Arthur and my parents, David and Joyce; my twin brother, Peter, whose experience and wisdom made him the perfect interlocutor (and translator of one text); and my partner, Alicia Volk, who contributed at every level to this book, as she has to everything I have done over the last decade.

I would also like to pay tribute to the ingenious, passionate and humane authors whose writings are collected in this volume. Any anthology can only be as good as the available material, and despite craft's reputation as an understudied subject I felt spoiled for choice. I hope that this book does justice to the women and men who went before and that it will serve as an inspiration to those who come after.

NOTE ON THE TEXTS

Except in the case of typographical errors, all texts have been transcribed with their original spelling, with some standardization of punctuation. Unless otherwise noted, referencing systems and footnotes have been retained, though their format and numbering has sometimes been changed for consistency. Short deletions within a text are marked with an ellipsis; longer deletions are marked with an ellipsis in brackets. Editorial additions for clarity (e.g., identification of individuals mentioned only by last name) are in brackets.

Formatting, illustrations and references to illustrations have been removed throughout. All images that appear in the book have been selected by the editor; only where noted are the images drawn from the original text.

Citations give only the original date and place of publication, and sometimes the date and place of first publication in English.

Translations, newly composed or amended texts and editorial matter are the copyright of Berg Publishers and the respective authors.

INTRODUCTION

In the introduction to one of the earliest books about craft published in English, the 1677 instructional guide *Mechanick Exercises*, Joseph Moxon wrote as follows: 'I thought to have given these *Exercises*, the Title of *The Doctrine of Handy-Crafts*; but when I better considered the true meaning of the Word *Handy-Crafts*, I found that *Doctrine* would not bear it; because *Hand-Craft* signifies *Cunning*, or *Sleight*, or *Craft* of the Hand, which cannot be taught by Words, but is only gained by *Practice* and *Exercise*.' It is a curious and telling sentence. Despite the fact that he was writing the introduction to a book, Moxon argued that craft is something beyond words: something learned with the body rather than the mind.

We might say that Moxon's contradictory state of mind—his notion that he was writing discourse about something fundamentally nondiscursive—perfectly captures the character of craft writing. The idea that making is its own particular sort of thinking is an appealing one. But it also constitutes a major challenge for anyone who wants to do justice to making through the seemingly inadequate tools of words and ideas.

Perhaps this is why one so often hears the complaint that craft suffers from a lack of intelligent writing. This may be true in a relative sense (an 'art reader' or 'politics reader' would have to be a lot longer than this one), but plenty has been written on the subject, and much of it is worth reading. The purpose of this book, *The Craft Reader*, is to gather some of the best of this literature together. In my role as editor, I have chosen what I thought to be the most representative examples, within certain parameters. For the most part practical considerations limited me to texts that were originally written in English, or had already been translated. Despite this, I have tried to cover as wide a geographical span as possible, an objective which entailed its own priorities. The process of selection was by no means inevitable or obvious, and many excellent writings had to be left out. For this reason each selection is contextualized in a short introduction and accompanied by a list of further readings. I have also included a recommended bibliography of important books that could not be easily excerpted.

There is currently a rapid expansion in writing about craft; perhaps in the near future, new anthologies will be produced to complement this one. In the meantime, it is my hope that almost anyone interested in craft, from any perspective, will find something here to interest them. Those who care deeply will want to read the book cover to cover and treat it as an indispensable reference. For those new to the subject, it should be valuable as a sort of introductory guide. (The reader is organized both thematically and roughly chronologically, with classroom use in mind.) And readers who care mainly about other things but are curious about craft will find surprises here. Ideally, the book will help to build new and unexpected connections, both within craft scholarship and beyond.

READING CRAFT HISTORY

So this book is offered mainly in the spirit of helpfulness. But it is also intended as a provocation. This is an anthology, not a manifesto, and so the introductory texts and the selected extracts themselves are intended to map various intellectual terrains, not to advance my own point of view. It would be disingenuous, however, to pretend that my choices in putting together this reader were studiously neutral.

Though it may not be immediately obvious, the book's structure is meant to challenge a commonly held set of intuitions about craft history, which can briefly be summarized as follows. With the onset of the industrial revolution in the late eighteenth century, craft began to suffer an irreversible decline—a process of deskilling and workplace alienation. In response, reformers and preservationists, most notably those associated with the Arts and Crafts Movement, emerged to rescue it. Though they were not able to maintain craft's economic value, they did raise awareness about its aesthetic importance and thus paved the way towards its rebirth as a distinct art form. Today more than ever, the artisan's place in culture is threatened by new technologies, from the internet to rapid prototyping. But a few institutions and individuals have been able to maintain a viable position for craft, partly by building new bridges to the worlds of contemporary art and design.

This narrative tacitly underpins the logic of many craft institutions: magazines, councils, museums, educational establishments and practitioners' organizations. It doubtless contains a great deal of historical truth. However, it can also be debated on almost every particular. First, craft skill was (and is, since the process is ongoing) not simply eroded as a result of industrialization. Rather, it has been continually transformed, and displaced into new types of activity. Second, the Arts and Crafts movement was not just a benevolent form of aestheticism, devoted to a backwards-looking idealism. It was modern and political in nature. It involved more invention than preservation, and it could sometimes be a corrosive and distorting cultural force. Third, the much-discussed transformation of craft into art could be seen more or less as a category error, and a pernicious one at that. Describing craft as an art form, or even as a fixed set of disciplines, disguises the otherwise obvious fact that craft is involved in an enormous range of cultural practices that have nothing to do with aesthetics or museums. It also blinds us to the potential radicality of craft's nonart status. Finally, the idea that bridges can or should be built between craft and art or design is from a certain perspective quite strange, since craft has always been a crucial aspect of art and design. The objective of 'crossing boundaries' serves only to produce boundaries that never existed in the first place.

What links all of these points of dissent from the so-called standard story is a conviction that craft should be seen in fluid and relative, rather than limiting and categorical, terms. Therefore, this reader is not restricted to the disciplines for which 'craft' is often reserved: ceramics; metalwork, including enameling and jewelry; the various textile arts; glass-making; and woodworking. (Occasionally boatbuilding, bookbinding, and paper-making might get a look in.) This use of craft to designate a short list of particular trades is culturally specific to English-speaking cultures. It doesn't exist in other European languages, much less in Asia or Africa. And it is far from self-evident as a way of dividing up the world of production.

One advantage of defining craft in a simple but open-ended manner—let us say, as the application of skill and material-based knowledge to relatively small-scale production—is that it allows us to draw connections across a much wider range of activities than the so-called 'crafts' themselves. Among these we might list architecture, painting, printmaking and sculpture; the creation of design prototypes, including digital rendering; routines of maintenance and repair; couture; gardening and cookery; factory, dockyard and construction work and so on. The concept is employed conventionally in all these arenas—think of phrases like 'hand-built house' or 'artisanal cheese'. Accepting the cultural logic behind such common usages would also mean accepting that craft is not a movement or a field, but rather a set of concerns that is implicated across many types of cultural production.

OUTLINE

With these principles in mind, a few further words about this book's structure are in order. It begins with 'how-to', or instructional literature, and ends with contemporary statements on curatorial practice—thus framing the contents between two genres that are unusually richly represented in writings on craft. These two bookends are also meant to imply a gradual broadening of the book's contents, which begin from the particularities of making and finally end with reflections about representation through exhibition.

In between, there is a similarly expansive movement which commences with relatively concrete questions of economy, progresses into ethics and politics and then shifts to abstract theorization and finally the messy contingency of experience. This main substance of the book is divided into five sections. These chart a rough chronology from about 1820 to the present but are organized thematically rather than in date order. Primary and secondary texts are mixed together throughout.

Section 2 is devoted to craft during the period of industrial revolution, when 'modern craft' could be said to have begun (more on that below). Readings in this section seem almost to debate one another. Some authors support the model of the industrial revolution described above, in which craft skill is thought to suffer a general decline, while others critique or complicate that notion. This section also incorporates *fin de siècle* arguments in favor of integrating the handmade with machine production.

We proceed with a section about the reformist and idealist tendencies that sprang up beginning in the middle of the nineteenth century, usually in conscious opposition to industrialization. Here we encounter John Ruskin and William Morris, who are for most English-speaking readers the best known of all craft theorists. While much of the rest of this anthology departs from their ideas, the other readings in this third section are largely consonant with them—whether that means the preservationist aesthetics of Soetsu Yanagi and Kamaladevi Chattopadhyay, or the emphasis on spiritual gratification seen in the writings of René Guénon and M. C. Richards.

As if in response to these statements of idealism, Section 4 focuses on various manifestations of real-world craft persistence. We have now moved well into the twentieth century, so these readings tend to deal either explicitly or implicitly with the conditions of craft within late capitalism. The geography is broad here, and the economic situations range equally widely—from the luxury trades of Italy and France to adaptive craftsmanship employed in Nigeria and India, or the disastrous industrial disarray of the former Soviet Union. There is also an engagement with issues of technology, developed primarily from the cybernetic theories of Norbert Wiener.

Next, in Section 5, we turn to the realm of theory. Aside from a brief foray into African aesthetics, the perspectives here are mostly Euro-American, though I have tried to be ecumenical in terms of intellectual position. Thus we have the close examination of skill in the work of David Pye; the formalist ideas of Henri Focillon and Elsie Fogerty; the Marxist cultural theorists Walter Benjamin and Theodor Adorno; the phenomenological theories of Martin Heidegger, and the application of his ideas within the field of architecture by Kenneth Frampton; and to close the section, voices on several sides of the debate over art and craft.

The extracts in Section 6, titled 'Craft in Action,' are divided into three parts, grouped under the headings of life, art and design. This allowed for the juxtaposition of readings which, taken collectively, engage with craft in depth, but in unexpected ways. All of the authors in this section look at craft as it is actually practiced in real circumstances, by real people, not in the notional realms of how it could be, or should be. The tone is often personal, or at least passionate. Of particular importance are the ideas of Feminist authors from the 1970s and later—women such as Lucy Lippard and Rozsika Parker, who claimed marginalized craft as a powerfully symbolic 'Other' to masculine-coded art and industry. In their diversity and inventiveness, the writings in this section will probably strike most readers as the most likely models for future thinking and research.

Feminism resonates strongly in the concluding section of the book, which surveys contemporary approaches to curating craft. It is no coincidence that while the preponderance of nineteenth- and early-twentieth-century authors in this anthology are men, the majority of today's essential writers on the subject are women. But the importance of Feminism to craft discourse is much more than a matter of gender. The repoliticization of craft that occurred in the 1970s—an infusion of urgency and ideas that had little to do with the Arts and Craft lineage—is the greatest single influence on the contemporary DIY or 'crafter' scene, which combines the expression of subcultural identities with an attempt to create anticorporate commercial opportunities. Equally, Feminist theory has been important in its contention that craft is best seen as a pervasive, 'everyday' activity, implicated in the contingent flux of modern life. For many practitioners today, craft need not be seen as a subgenre of fine art, nor as necessarily rooted in tradition. This open situation, in which craft is where you find it and what you make of it, is indeed the context in which this anthology was assembled.

MODERN CRAFT

This leads to a discussion of one final word, or rather two: 'modern craft'. I approached this anthology as a counterpart to my 2007 book *Thinking through Craft*, which discusses the role that craft has played in modern and contemporary art practice; and also to the *Journal of Modern Craft*, which I co-edit with Edward S. Cooke, Jr. and Tanya Harrod and which is intended as an open space for scholars and practitioners of every persuasion to write about the subject. As is obvious from the use of the two words 'modern craft' in the title of the journal, I am fascinated by their conjunction. This is an anthology about the complex meanings that these words give to each other: craft's transformations within the process of modernization; the way that Modernists and Postmodernists have used and viewed craft; and the insights we can gain about modernity itself, by paying attention to the way it variously structures, hides and celebrates craft activity.

Of course, 'modern craft' also designates a chronology. One can easily conceive of a 'craft reader' that included texts earlier than 1820. That book might begin in the eighteenth century in order to include the first closely argued statements about craft in relation to economy and aesthetics (by authors such as Bernard Mandeville, William Hogarth and Denis Diderot). Or it could start with the seventeenth century, when printing technology made instructional literature like Moxon's book for 'mechanicks' commonplace, and guilds were beginning to draft extensive legal documents. Or it might go back to the Renaissance, when probably the first and certainly the most self-aggrandizing of all artists' autobiographies was written by a metalsmith, Benvenuto Cellini. Or all the way back to antiquity—Plato certainly had some interesting things to say on the matter!

This anthology, however, concentrates on modernity—the phase of history that is coincident, more or less, with the emergence of fast-moving industrialization—because this is when craft becomes a problem, an issue that was widely considered to be worth worrying about. The role of artisans within culture has never been straightforward. But it is only when artisanal labor is placed in explicit contrast with other means of production (chiefly mechanization, fine art and technological mediation) that craft itself becomes a locus for discourse. Indeed, it could be argued that until its modern separation from these other possibilities, 'craft' itself did not exist, at least in our sense of the term. It is a term established and defined through difference.

Situating the emergence of modern craft in space and time is difficult, to say the least. Scholars have detected the stirrings of the industrial and consumer revolutions at least as early as the seventeenth century (and some dispute that there were any such revolutions at all) and have located key economic transformations in Germany rather than Britain, or indeed in India rather than Europe. The conscious distinction between the fine and the minor arts can be detected even earlier, in the Italian Renaissance (though this was not as pronounced or as uniform as many commentators claim). What we can say for sure is that by the early nineteenth century craft became a dialectical term, held consciously in opposition with other terms and therefore susceptible to widespread ethical, aesthetic and economic analysis.

This reader tracks the discursive formations that developed during and after this moment. If there is a single lesson to be learned, it is that craft is not simply antimodern. It is rather a strain of activity that responds to and conditions the putatively normative experience of modernity, in many and unpredictable ways. It is understandable that craft is often seen in simpler terms that that—as oppositional rather than adaptive. Modernity, after all, seems hard to stand up to. It is notionally defined by 'one size fits all' structures that are temporally and geographically transcendent: rationality, science, capitalism, mechanization, International Style architecture, autonomous artworks and secularism, to name just a few. Craft could be seen as diametrically opposing all of these. It entails irregularity, tacit knowledge, inefficiency, handwork, vernacular building, functional objects and mysticism. Further, craft's association with gendered, ethnic and local identities could be seen as inherently resistant to (or, potentially, critical of) modernity's homogenous transcendentalism.

Yet, as this reader hopefully demonstrates, it would be a mistake to cast the matter in such starkly oppositional terms. The point of the phrase 'modern craft' is that it contains within it *both* sides of these cultural conflicts. Modern craft would best be seen not as a paradox, or an anachronism, or a set of symptoms, but as a means of articulation. It is not a way of thinking outside of modernity, but a modern way of thinking otherwise.

SECTION 1

HOW-TO

SECTION INTRODUCTION

It is possible to write about craft from a bewildering number of perspectives. Art history, anthropology, poetry, sociology, philosophy, ergonomics: these are only some of the disciplinary frameworks which have laid claim to the subject. But only one type of literature is particular to craft—a genre that craft can call its own. This is the 'how-to' text: instructional writing that attempts to convey in words that which can only be done by hand—to describe the specifics of process and material that actually constitute craft.

How-to texts are often cited by craft professionals as proof of the subject's lamentable anti-intellectual qualities, and it's true that they can sometimes make for dull reading. But they can also be fascinating, and they have served a number of crucial roles. First and most obviously, they are a means of knowledge transfer—a way of codifying and distributing the norms of practice. This function has become more, not less, important with the onset of modernity. Indeed, it is only with the breakdown of the guild system that how-to writing comes into its own, for the obvious reason that training in earlier periods was conducted mainly orally. Certainly, there are medieval and renaissance period texts that describe the means of making (the twelfth-century *On Divers Arts* by the German monk Theophilus, and Cipriano di Picolpasso's *Three Books of the Potter's Art* of about 1548 are famous examples).[1] But it is only in the seventeenth and, especially, in the eighteenth centuries that manuals on various craft disciplines become a publishing phenomenon in their own right. During the period of the industrial revolution the genre proliferated. Diderot's *Encyclopedia* described the luxury trades of France to a readership of aristocrats; builders' manuals taught members of the working class not only craftsmanship but also mathematics and mechanical drawing; and chemical treatises on ceramic and glass contributed to the growing literature of the scientific community.

Even today, most publications about craft fall into the how-to category, whether they are books, magazines or Web sites. Indeed, if craft kits and hobby supplies are added to the calculation, then instructional products (rather than finished hand-crafted goods themselves) account for the vast majority of the total economic value of craft. This fact reflects a dynamic we will encounter often in this book: when it comes to modern craft, value is often located not in the act of making, but at one remove. Thus how-to writings are not only a means of establishing and disseminating norms of practice, but also philosophical claims about craft.

The selections in this part of the reader amply attest to this partiality. They are shot through with both overt and unstated assumptions about the social purposes of the crafts that they describe. Otto Salomon and W.A.S. Benson's didactic literature of improvement, George Sturt's elegiac description of traditional wagon wheel–making, Anni Albers's precise description of weaving procedures and Hal Riegger's no-nonsense advice on getting back to basics in pottery: these writings capture as well as any texts could the shifting terrain of modern craft from design reform to nostalgia to Modernism to subculture. How-to writings like these often pretend to be devoid of ideology, but this of course only makes them more effective as a delivery system for it.

This holds true for the massive flood of instructional literature that has accompanied the current fashion for knitting, crochet, embroidery and other textile arts. In recent years, a set of skills that once was associated with kindly grandmothers, in the public imagination, anyway, has been embraced by hobbyists young and old, contemporary artists and even political activists. Many who are attracted to such activity cite its association with community, its Feminist overtones and the satisfaction that can be derived from tactile experience in an increasingly abstract, artificial world. None of these motivations are new, of course, but the means by which the amateur 'crafter' craze has spread are unprecedented. Knitting circles, one of the oldest of social technologies, are promoted via blogs, one of the newest.

A related development is that of the craft media personality, for whom how-to writing serves as the basis for entrepreneurship. The first (and still the most successful) of these was arguably the American tastemaker Martha Stewart, but in recent years less-restrained figures have taken up the mantle of explaining craft to the masses. Debbie Stoller, editor of *Bust* magazine (which, according to its Web site, 'tells the truth about women's lives and presents a female perspective on pop culture'), is the most well known of this new wave of DIY celebrities. Packed with Feminist slogans and unintimidating step-by-step directions, Stoller's books are reminiscent of the 1970s Punk zines that advised readers, 'Here's one chord, here's another, now start your own band.' But if Punk was expressly anticapitalist, the 'Stitch'n Bitch nation' is simultaneously a viable subculture and a successful trademark in its own right. In the context of the astonishingly high economic value of the hobby craft as a whole (estimated by one organization at $30 billion in 2006 in the United States alone), even Stoller might be considered a fringe phenomenon—she stands at the hip end of a huge industry.[2] Nonetheless, her particular brand of how-to writing seems to capture the present moment, when, as she puts it, 'It's time to get your knit on.'[3]

This leads us to the consideration of a final role played by how-to literature: entertainment. As is obvious from the sheer volume of instructional publications produced annually, most are never put to direct use. Books are given as gifts or bought on impulse, paged through and left on the shelf. Magazine subscriptions lead inexorably to piles mounting up in the basement or next to the sewing machine. This apparent neglect is important, as it indicates that how-to texts play a largely aspirational part in their readers' lives. As Steven Gelber has argued, even when DIY craft activity pretends to be instrumental, it tends to serve mainly incidental ends, such as constructing a masculine identity within the increasingly automatized home.[4] It is no coincidence that in the modern era, modern craft itself has moved steadily towards a similar state of nonfunctionality; like ceramics magazines, handmade pots now are likelier to occupy a display shelf than a cupboard. As this anthology will go on to demonstrate in more detail, one of modern craft's distinctive characteristics is its lack of basis in raw necessity. If artisanal labor was once primarily economic, in the past century and a half it has been chiefly idealistic and aesthetic. There are many exceptions to this general rule; but the fact that instructional literature is avidly consumed but mainly goes unheeded might be viewed as emblematic of craft's situation within modernity in general.

1

'INTRODUCTORY REMARKS', FROM *THE TEACHER'S HANDBOOK OF SLÖJD*

Otto Salomon

A vital, but often overlooked, body of writings concerning modern craft is to be found in educational literature. From the middle of the nineteenth century onwards, the reach of formal schooling was dramatically extended to the middling and working classes. In this process craft (or 'shop') classes were a field of experimentation and a point of contention. Should working-class children and young adults be trained to join the swelling ranks of the industrial workforce? Or, rather, should they be given a broadly liberal education in the hopes of making them good citizens? The latter perspective, usually seen as more socially progressive, had its origins in the work of German educators such as Friedrich Froebel (who introduced the Kindergarten *in the 1840s) and in the Scandinavian* slöjd *system, as developed by Otto Salomon.* Slöjd *is a word with no precise English equivalent; it implies skill in work rather than craft as a productive activity. Salomon was quite explicit that his pioneering efforts were intended not to train manual laborers or artisans, but rather to inculcate in every student well-regulated habits of mind and body. Beginning in the 1870s he put this principle into practice at Nääs, a vocational school that he founded on his wealthy uncle's estate near Gothenburg, Sweden. Each student was put through a course of making 'models' based on everyday, functional objects, along the way learning basic woodworking techniques and (more importantly) a range*

of organizational skills and even lessons in good posture. Salomon proved to be as effective a promoter as he was as an educator, and through various means (including publications such as the following, as well as displays at international expositions and lecturing abroad) he made slöjd *an international phenomenon. His influence was felt not only in the shop classes of his own day, but also in the architecture, design and crafts of the generation that grew up in the late nineteenth century—those who would become the pioneers of modernism in the 1920s and 30s.*

Otto Salomon, 'Introductory Remarks', from *The Teacher's Handbook of Slöjd* (Boston: Silver, Burdett & Co., 1891), excerpted.

Slöjd is not to be confounded with the work of the artisan—a mistake which may easily happen if the distinction is not sufficiently strongly emphasized. Speaking generally, the slöjder does not practise his art as a trade, but merely as a change from some other employment; and in the nature of the articles produced, in the tools used in their production, in the manner of executing the work, etc., slöjd and the work of the artisan differ very decidedly the one from the other. Slöjd is much better adapted to be a means of education, because purely economical considerations do not come forward so prominently as must be the case with work undertaken as a means of livelihood.

Educational slöjd differs from so-called practical slöjd, inasmuch as in the latter, importance is attached to the work; in the former, on the contrary, to the worker. It must, however, be strongly emphasized that the two terms, educational and practical, ought in no way to be considered antagonistic to each other, as frequently happens in popular language; for, from the strictly educational point of view, whatever is educationally right must also be practical, and vice versa. When the educational and the practical come into conflict, the cause is always to be found in the pressure of adventitious circumstances, e.g., the number of pupils, the nature of the premises, and, above all, pecuniary resources, etc. To make educational theory and practice coincide is an ideal towards which every teacher must strive. One man, perhaps, may be able to come nearer to this common ideal than another, but everyone, as he runs his course, must have this goal clearly in view, and in every unavoidable compromise he must endeavour to make what ought to be done and what can be done come as close together as possible.

What, then, is the aim of educational slöjd? To utilise, as is suggested above, the educative force which lies in rightly directed bodily labour, as a means of developing in the pupils' physical and mental powers which will be a sure and evident gain to them for life. Views may differ as to what is to be understood by a "cultured" or an "educated" man, but however far apart in other respects these views may lie, they all have at least one thing in common, i.e., that this much disputed culture always appears in its possessors in the form of certain faculties, and that therefore the development of faculty, so far as this can be directed for good, must enter into all educational efforts. This being the case, the influence of slöjd is cultivating and educative, just in the same degree as by its means certain faculties of true value for life reach a development which could not be attained otherwise, or, at least, not in the same degree. Educational slöjd, accordingly, seeks to work on lines which shall insure, during and by means of the exercise it affords, the development of the pupil in certain definite directions. These are of various kinds. As the more important, it is usual to bring forward: pleasure in bodily labour, and respect for it, habits of independence, order, accuracy, attention and industry, increase of physical strength, development of the power of observation in the eye, and of execution in the hand. Educational slöjd has also in view the development of mental power, or, in other words, is disciplinary in its aim. [. . .]

THE SPECIAL KIND OF SLÖJD RECOMMENDED

Various materials, e.g., wax, clay, paper, pasteboard, wood, metal, &c., may be used in educational slöjd. Wood, however, is for several reasons the most suitable material; hence wood-slöjd has been the most popular of all, both in schools and for private instruction. As the name implies, wood-slöjd means slöjding in wood. This, again, includes several different kinds of work. Amongst these, however, it is the so-called slöjd-carpentry which best fulfils the conditions required when instruction in slöjd is given with educational ends in view. It is adapted to the mental and physical powers of children. By enabling them to make a number of generally useful articles, it awakens and sustains genuine interest. It encourages order and accuracy, and it is compatible with cleanliness and tidiness. Further, it cultivates the sense of form more completely than instruction in drawing does, and, like gymnastics and free play, it has a good influence upon the health of the body, and consequently upon that of the mind. Additional advantages are, that it is excellently adapted for methodical arrangement, comprising as it does a great number of exercises of varying degrees of difficulty, some of which are very easy; and that it gives a considerable degree of general dexterity by means of the many different tools and manual operations which it introduces. [. . .]

It must be borne in mind that although slöjd-carpentry and ordinary carpentry have something in common, inasmuch as the same raw material (wood) is employed, and to some extent the same or similar tools are used, yet they differ one from the other in several very important respects. For example, the articles made in slöjd-carpentry are in many cases quite different from those which fall within the province of the carpenter. The articles made in slöjd-carpentry are differentiated partly by their smaller size, for the articles made in workshops are generally much larger; partly by their form, for they are often bounded by variously curved outlines, whilst articles made by the carpenter are generally rectangular or cylindrical. This is especially shown in the case of the many different kinds of spoons, ladles, scoops, handles, etc., which form such an important element in slöjd-carpentry.

Further, though many tools are common to both kinds of work, there are also considerable differences in this respect. Several tools which are seldom or never used in the carpenter's workshop, e.g., the axe, the draw-knife, and the spoon-iron, occupy an important place in slöjd-carpentry.

The most characteristic tool in slöjd-carpentry is, however, the knife, and by the use of this, his chief instrument, the slöjder may always be distinguished from the carpenter, whose favourite tool is the chisel, and who, as seldom as possible, and never willingly, takes the knife in his hand. In carpentry, on the other hand, use is made of a number of tools more or less necessary, which are quite unknown to the slöjder, who works for the most part under more primitive conditions. Distinct differences can also be pointed out in the manner of executing the work (for while division of labour is practised in carpentry, it is not permitted in slöjd) and in the manner of using the tools. It will be seen from the foregoing that much may pass under the name of instruction in slöjd which, properly speaking, ought simply to be called instruction in carpentry. It is most important that this distinction should be maintained, because otherwise educational slöjd will by degrees be lost in instruction in carpentry as a trade.

In some schools where slöjd is taught we find turning and wood-carving as well as slöjd-carpentry. This, however, is not so common now as it was a few years ago. People seem to be coming more and more to the conclusion that both occupations are more suitable for the home than for the school. Neither of them is to be commended from the hygienic point of view. As regards turning, the difficulty of procuring suitable turning-lathes presents in many schools a serious obstacle to its general use; whilst the necessity of performing preliminary exercises, apart from the actual objects made (a proceeding of very doubtful educational value) places turning quite in the shade as compared with slöjd-carpentry. Wood carving, on the other hand, does not involve that energetic bodily labour which is of such great importance in connection with educational slöjd. Again, wood-carving, classed as it is with the so-called "finer" kinds of manual work, has a tendency to intensify in the child that contempt for rough bodily labour which has already unfortunately done so much social harm. The danger of this is however greatest when the children are imprudently permitted to ornament objects which they have not made. When wood-carving is used, not as a separate kind of slöjd, but in order to complete slöjd-carpentry, and when ornamentation is only allowed after the children are able in a satisfactory way to execute the articles to be embellished by its means, the disadvantages are minimised.

[. . .]

METHOD

It is an essential condition of any method of instruction in educational slöjd, that the work of the pupils shall be independently and accurately executed, for only thus can habits of self-reliance, order, and accuracy, so important in the formation of character, be developed. In order that self-reliance may be developed, the teacher must guard himself against giving more help than is absolutely necessary, whether this help consists in explaining the best way of doing the work, or in doing the work instead of the pupil. As regards the latter, the teacher will do well to lay down, as a general rule, that he never should touch the pupil's work, for only by this means can he avoid the temptation, to which unfortunately many teachers have succumbed, to execute the most important parts of the work instead of the pupil. At the same time he must remember that it is also hurtful to the pupil, and that it deprives his instruction of considerable educational value, if by unnecessary explanations he hinders the pupil from using his own judgment to discover the right way. The teacher's art in educational slöjd consists essentially in being as passive and unobtrusive as possible, while the pupil is actively exercising both head and hand. Only in this way can the feeling of self-reliance arise and gain strength. Let the teacher content himself with pointing out the way, and watching that the pupil walks in it. Let him as much as possible refrain from leading where this is unnecessary and, it may be, hurtful.

[. . .]

SOME RULES FOR THE SLÖJD TEACHER

In all teaching, and not least in slöjd teaching, the maintenance of order must be laid down as an indispensable condition. The following simple directions may serve for guidance to the teacher.

Every pupil should have a fixed place at a bench. When circumstances permit, it is advisable to have at disposal as many benches (or when benches intended for two are used, half as many benches) as there are pupils taking part simultaneously in a lesson.

The benches and tools should be furnished with numbers, so that they can easily be distinguished from one another. The following tools should, if possible, belong to each bench, and be marked with its number: knife, trying-plane, smoothing-plane, jack-plane, square, marking-gauge, compasses, rule or metre measure, and scraper. Other tools may serve the whole class in common.

All tools should have fixed places. Those belonging to the bench may be allowed to lie upon it until the close of the lesson, but all tools in common use should be laid by or hung up immediately after use, in order that they may be easily found.

The teacher must take care that all the edge tools in use are well sharpened, and that any tool which gets out of order, or is broken, is repaired as soon as possible. If practicable, the pupils should do their own repairs.

At the beginning of the lesson the pupils should, in an orderly way, get out their tools and work. The latter, if begun in a previous lesson, should be kept in boxes specially provided for the purpose, and should be marked with the pupils' names.

In order to teach and superintend in the full meaning of these terms, the teacher must not stand still in one place. He must go from one pupil to another with advice and criticism. The pupils, on the contrary, must, as far as possible, remain at their benches. If they desire any advice from the teacher, they must not attract his attention by calling out, but by some signal, e.g., holding up one hand, standing in front of the bench and looking towards him, etc. All unnecessary talking must be carefully avoided.

The pupil himself, guided by the teacher, must select suitable wood. Waste must be avoided as far as possible.

The pupil must not be allowed to polish with sand-paper until the teacher has examined the work and found that sufficient use has been made of cutting tools. The sand-paper is to be kept by the teacher and given out by him as required. About 6 sq. in. is calculated for each model. The calculation is founded on the supposition that though the models become larger as the course proceeds, the greater facility of the pupil diminishes in about the same degree his need of sand-paper.

At the end of the lesson all the tools should be put back in their places, care being taken that all the saws are loosened. The tools should be counted by the "captain," or monitor, appointed for the class, after which the teacher sees that everything is in its right place. The wood and the pieces of work are put away tidily. The benches are brushed and made clean with a brush which should hang by the side of each bench, and the floor is swept. The shavings, however, need not be carried away oftener than once or twice a week.

When the finished pieces of work have been "passed" by the teacher, a label should be stuck on, and on this label should be stated the number of the model and its name, the name and age of the pupil, and the number of hours spent in making it. If it is considered desirable to give every piece of work a value, this also may be mentioned on the label.

Although from the educational point of view it is advisable that the pupils should at once take home their work, it is generally for other reasons more expedient that it should remain in the school in the care of the teacher until it can be exhibited publicly at an examination or terminal breaking-up. After this has taken place, the articles are to be regarded as the property of the makers. The sale of work for the benefit of the school should never be thought of.

FURTHER READING

Norman Brosterman, *Inventing Kindergarten* (New York: Abrams, 1997).

John Dewey, *Art as Experience* (New York: Minton, Balch & Co., 1934).

June E. Eyestone, 'The Influence of Swedish Sloyd and Its Interpreters on American Art Education', *Studies in Art Education* 34/1 (Autumn, 1992), pp. 28–38.

2

ELEMENTS OF HANDICRAFT AND DESIGN

W. A. S. Benson

The influences of Otto Salomon's slöjd *system of craft education are clear in the ensuing excerpt by the British Arts and Crafts Movement metalsmith W.A.S. Benson, who was inspired to take up the hammer in 1880 after meeting William Morris. Benson went on to manufacture lighting, tablewares and other metalwares to be sold through Morris & Co. and became the chairman of that company upon the death of its leader in 1916. Benson's slim pedagogical volume* Elements of Handicraft and Design *was intended as a primer for schoolteachers and students alike and suggests a variety of simple woodworking and other craft activities. Despite its sometimes pedantic tone, the text is notable for its interweaving of progressive European ideas in education and clear instructional language. What distinguishes it from Salomon's text is a greater investment in skill and 'taste'. If* slöjd *employed craft exclusively as a means of imparting physical and mental discipline, Benson saw the instruction of craft to children as aesthetic training. From his perspective, an early encounter with rudimentary carpentry was a foundation for the broader project of design reform, as it would instill in the student not only orderly habits, but an understanding of beauty and a lifelong respect for craftsmanship.*

William Arthur Smith Benson, *Elements of Handicraft and Design* (London: Macmillan, 1893), excerpts.

Given a workshop fitted with carpenters' benches and a supply more or less adequate of the usual tools for working in wood, how to utilise them for the right education of boyhood, what should be the aim, and what the means to its attainment? This is a problem which confronts a large and increasing number of school-managers and teachers. Nor in saying boyhood would I exclude girlhood, knowing no reason why ready skill of hand and the practical knowledge that comes therewith should be less generally valuable to women than to men, while it is certain that many women possess a high natural aptitude for handicraft, and therefore the grammatical masculine in these pages must be taken inclusively; all questions of specialised employment of boys or girls being outside our present scope, which is occupied with the foundations for an intelligent education of the hand and eye.

As in the case of foundations for a material edifice, there is usually some soil and untrustworthy substance the removal of which is the first item in the builder's specification, so do we find the ground for our educational foundation cumbered with misconceptions contrary to edification which we can by no means disregard. Perhaps the most serious of these is the idea that it can ever be the function of a general school course wholly or in part to teach a trade. Though a school workshop will usually bear a superficial resemblance to a job-bing carpenter's shop, you cannot at school teach boys to be carpenters or cabinet-makers; and if you could it would be eminently unde-sirable to select one or more of the countless industries of life for such wholesale artificial cultivation. Special schools there may be, in which workers in wood, as well as others, can supplement the experience of their apprentice-ship and gain broader views of their craft and the history and theory of it, but our present concern is with another matter, that is, the teaching of handicraft as an element in general education—what Ruskin speaks of as "a lib-eral education founded on right handicraft." And as the aim, so will the methods differ from those of a trade shop. The purpose of a shop is to turn out goods, saleable wares of one kind or another; whereas a school aims at turning out better and more capable men and women. The themes and exercises which are the means to a literary education have no immediate marketable value; commercially, they are but spilt ink and spoilt paper. And so with the exercises of the handicraft school; the endeavour to make a saleable product of them is likely to prove a great hindrance to the true advantage of the scholars, which lies in pursuing a course designed to train mind and eye and hand to work together, and accustom-ing them to accurate and patient industry. And the wider the interest of such a course can be made the better; it may open out glimpses of the foundations of physical science and of the economics of industry; encouraging hab-its of observation, the love of nature, and the perception of form and colour; developing the latent sense of beauty, in which we have been supposed to fall short of other nations, though enough remains of the work of our forefathers to show that no innate defect in the race hin-ders the production and enjoyment of beautiful

things. For we must not forget in matters of education to regard life as well as livelihood; we have to rear up a race capable of taking a brave part in the competitive industry of the world, and at the same time of enjoying the rewards thereof as good citizens of a prosper-ous state, and to this end we are encouraged by the opinions of the strictest economists to cultivate the higher faculties and emotions in due proportion.

[. . .]

In education, as in after life, the incidental advantages of any course are often among the most valuable of its results; for instance, the pursuit of excellence in sports and games is usu-ally found to be a better road to physical health than any course of gymnastics specifically de-vised to that end, for indeed healthy pleasure is the best of stimulants, the safest of medi-cines. It is well that a man should find pleasure in his work, and when the men sing over it the manager knows that all goes smoothly; and it is just the sense of rightness, or well-employed dexterity, the nicety of perfection, and the delicate finish of detail, making the fabric seemly to the eye and touch, that give the sense of pleasure in the work and make song possible. But the workshop may be made an introduction to a less popular virtue than tune-fulness, and that without naming it by name: well-ordered thrift. A place for everything, and everything in its place, a use for everything, and everything ready for its use, should about sum up the whole economic philosophy of the workshop teacher. What eye, accustomed to the tidiness of a well-ordered factory, has not been grieved, in the enjoyment of his coun-try holiday, by the sight of machines wrecked for want of a bolt, or rotting for want of a coat of paint or a shed to cover them, and the fertilising liquors of the farmyard wasted in enriching the nettles of the roadside ditch?

Nor would like instances be far to seek in an urban district.

Now the practical workshop cannot be run at all without practical and orderly thrift, not the "spoil the ship for a ha'porth of tar" sort of thrift, but the "stitch in time, everything in its place, ready for use" sort of thrift, that grudges nothing to use but anything to waste; and the teacher that turns out pupils with the habit of careful use and tidy storage will assuredly deserve well of his country.

When one forsakes generalities and comes to consider in detail what work we are to set our pupils to, we are met by a serious difficulty. By the nature of things the material we must work in will be chiefly wood, and as in popular parlance workers in wood are called carpenters, a handicraft school is pretty sure to be often mistaken for a carpentry school, and criticised, either because it teaches carpentry or because it cannot teach carpentry, according to the prejudices of the critic. The answer to all this is that there is a multitude of trades all working in wood, as shipwrights, cartwrights, turners, cabinet-makers, and the rest, none of which can be taught at school, but that the rudimentary use of the hammer, saw, and file, and a few edge tools, can be taught along with a deal of common-sense science, and the understanding of drawings or diagrams, all of which go to make up a certain handiness which lies at the root of excellence in all trades, and is of incalculable value in any walk of life. And the better to carry out this view it is most desirable to add to the ordinary list of carpenters' tools, not merely the knife and rasp, as in the Swedish Slöjd, but a few tools for carving and for working sheet-metal, together with materials for painting, staining, and varnishing, and by thus doing we shall add immensely to the educational value of our course, and at the same time the outlay for establishing a workshop for a given number of pupils will be less on this extended system than where the course is severely confined to bare rudimentary exercises in joinery.

[. . .]

The beginnings of handicraft now make an excellent start in the kindergarten, with its numerous interesting operations with scissors and paste, needles and thread, plaiting with paper or straw or rush, all which, with the pretty craft of basket-weaving, we may safely leave to be developed by the ingenious hands which have already done so much to improve educational methods. But when we come to a more advanced stage, and have to deal with an age capable of handling the lighter sort of carpenters' tools, such guidance fails us. There are, it is true, a sufficiency of books giving an adequate account of the nature and use of the several tools, which might prove excellent introductions to the joiners' trade, and though this is not what we want, some such books should be in the possession of the teacher of handicraft. The Swedish Slöjd system, however, comes much nearer what is required, and though it by no means covers the whole ground, it supplies many exercises which can be used to supplement those given here, while we can scarcely do better than begin by taking a hint from their course in proposing for a first exercise the preparation of sticks for the flower-garden, or at least for the flower-pot. Those commonly sold are split from straight-grained wood and roughly rounded, but from the point of view of handicraft training we regard the task as a first exercise in the use of the saw and the plane,—the problem being, to cut a strip off the edge of a thin board and plane it up smooth and foursquare. When we can do this with some approach to accuracy we shall be able to get further instruction out of such stripes, but our first attempts ought to make capital flower-sticks. The board should be about 3/8 inch thick, and having planed one edge we take a gauge and mark off a strip 3/8 inch wide and say 16 inches long, then saw it off and plane the three rough sides as smooth and square as we can—the result, if we are fairly successful, being a stick about 1/4 inch square. If, then, we point one end and round off the edges with a paring chisel, or perhaps more easily with a rasp and coarse glass-paper, we shall have a very serviceable stick for a pot of hyacinths, the appearance of which we may greatly improve with a coat of green paint, at the same time making it much more durable. But I have seen much prettier sticks than this, the recollection of which suggests a first lesson in ornament. Before rounding off the angles mark off about 3 inches of one end and leave that square; then shave off the angles, reducing the square to an octagon, and as you approach the other or pointed end, reduce the octagon in turn to a rounded form; then with a sharp knife whittle the square end,—a process which should give the keenest pleasure to a properly constituted mind.

[. . .]

But perhaps it is now time to try our hand at making a box,—not a very difficult thing it might seem—one consecrated by all tradition as the first essay of the beginner; and yet in the perfect and accurate making of boxes lies the larger part of the craft and mystery of cabinet-making.

Our first box, however, must really be a very simple affair and by no means a piece of cabinet-work, but only a common nailed box, a baby packing-case, in fact, though we shall plane the wood as smoothly as we can. Almost any book on woodwork will give you ample instructions for making such a box; to which I need only add that it is much easier to nail the box together, and makes a stronger job when

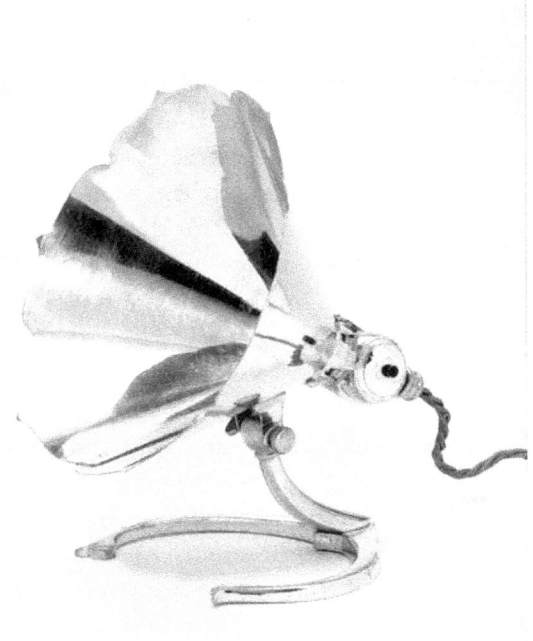

Figure 1 W.A.S. Benson, *Electric Table Lamp*, ca. 1900.

done, if you cut the ends or short sides out of rather stouter wood than the rest, for it gives you more stuff to drive the points of the nails into and a better bearing for the side pieces. And it is worth while dwelling a little on this point, and considering the why and wherefore of it with the aid of some simple experiments. Take two pieces of plank, one thicker than the other, say 3/4 inch and 1/2 inch respectively, and having squared off the ends fix them upright in the vice as if ready to nail to them, and try how much easier it is to balance a third piece on top of the thick than of the thin piece; and when a piece is placed in position for nailing, notice how much more steadily you can hold it in place on the thicker piece, and then reflect that what applies to the force of your hand applies also to the holding power of the nails, which is exerted to much better advantage in the case of the wider bearing. It is by the habit of noticing such matters that the mechanical instinct is cultivated and workmanship is elevated from mere rule of thumb to manly reasonableness. Though to make a box tidily is by no means the easiest possible task, yet it is one that is rightly set before the beginner at an early stage, for the reason that, however rough his achievement, so long as it will hold together it is of some utility; if it will not serve as a nail-box, at least it will do to fill with earth and grow seeds in.

[. . .]

Although we have already spent some time in studying the construction of boxes, and have even given some attention to embellishing them with colour, with metal, or with carving, we have not yet mentioned the typical workmanlike method of making a box—the dovetail jointing of its four sides, which we may regard as the first step towards cabinet-making . . . [T]he interest for us of this, as of other joints, lies rather in understanding the reason of its general excellence as a workmanlike joint, which we shall best apprehend by experimenting with various forms of construction, rather than by drawing our examples from the practice of a single trade. In so doing we shall but be following the lead of teachers of arithmetic who by no means confine their choice of sums to those that actually occur in casting the ledgers of a single business.

Now the dovetail is a joint which is principally of use when the ends of pieces of wood have to be joined at right angles to their lengths; practical instructions for making it will be found in any book which treats of the actual handling of carpenters' tools. For joining the end of one piece of plank to the side of another it is of no practical use, but for end joints it is a very pretty and effective joint, for the following reason:—wood, as you remember, swells with damp and contracts on drying in the width and thickness but not in the length of the plank; now as the grain is all one way with this joint, no swelling or shrinking will affect its tightness as both parts will work together. In the nailed box you will remember that the strength of the corners depended chiefly upon the thickness of the piece into which the nails were driven, and similarly it is obvious that the strength of a dovetail depends finally on the thickness of the sides; while for any given thickness accuracy of fitting determines the strength.

Now we shall find as we go on that the main structure of very many works of wood is just that of a box, four pieces of wood dovetailed together at the corners, their keeping in shape being dependent upon the perfection of the work at the joint.

But supposing that we want to make a box somewhat larger than those we have hitherto discussed, and that we distrust the adequacy of our skill in dovetailing large pieces of not very thick wood, what method lies open to us? I think that in that case we may find a large measure of practical science in the make of certain rough-looking and perhaps despised pieces of carpentry, packing-cases in fact. Of course a great many packing-cases are simply plain nailed boxes, such as we have already considered, but most of any size are constructed of rather thin boards nailed side by side on to narrow slips at either end called ledges; and very considerable skill may be shown in the disposition of these ledges, which serve at once to unite the narrow plank into a wider board, and to give a sufficient bearing for the adjacent sides to be nailed to. Having prepared one pair of sides you can either nail the plank for the other pair direct to the ledges of the first pair, or you may fit them with their own independent ledges, and then nail the separately completed sides together. We have the further choice between putting the ledges inside or outside, or one in and one out at each corner; the sides also may have additional ledges at mid-length. We may cut them projecting so as to receive the thickness of the lid and bottom; and you may be sure that a good workman has excellent reasons for his choice in any case, though perhaps he would not be able to explain them in very elegantly scientific phraseology. But however that may be, I hope that hereafter you will look at a packing-case with a trifle more respect than you have hitherto felt, and realise that very superior people need not be above learning various lessons from these humble constructions.

[. . .]

RECAPITULATION

I trust that any one who has followed me thus far will have obtained at least an inkling of the simpler aspects of the processes of design; and that those who in addition to following the letter, have put their hands to the plough (among other tools), practically treading the furrow set forth for their edification, will close this little volume with a higher idea of the dignity of handicraft than is common with the scholar: will realise that the mechanics of industry are not mechanical in the baser sense, but full of fine philosophy: that, as a study, the articulation of joinery is not so far below that of speech as it seemed; while a day's work upon it may yield a solid satisfaction often denied to the best efforts of a verbal logician. Here we have been dealing with the science of handicraft in its cradle; but hereafter I hope to meet, as readers of more specialised dissertations, some of those who, with the aid of saw and chisel, have puzzled through this elementary logic of box and chair and table.

FURTHER READING

W.A.S. Benson, *Elements of Handicraft and Design* (London: Macmillan, 1893).

Tage Frid, *Tage Frid Teaches Woodworking* (Newtown, CT: Taunton Press, 1979).

John Sedding, *Art and Handicraft* (London: Kegan Paul, Trench, Trübner & Co., 1893).

3

THE WHEELWRIGHT'S SHOP

George Sturt

'He had a steely and everlasting hatred of all sentimentality.'[1] This was how one of his closest friends, the novelist Arnold Bennett, summarized George Sturt's attitude to the people in the rural area of Farnham, about whom he wrote almost ten books. Sturt lived in this part of England for his entire life. Though he inherited the family wheelwright shop in 1884, and also counted a local potter among his older relations, he was first and foremost a writer. In addition to his novels and descriptive books he left extensive journals and correspondence, many of which center on the challenge of getting the realities of country life into prose. Like many skilled writers who turn their attention to the close recounting of craft process, Sturt had a basic conviction that it was only through direct, physical experience that one could understand workmanship, or even raw materials: 'My own eyes know because my hands have felt, but I cannot teach an outsider, the difference between ash that is "touch as whipcord", and ash that is "frow as a carrot", or "doaty", or "biscuity"'.[2] In the following excerpt from The Wheelwright's Shop, Sturt's best-known book, he describes the making of the three key parts of a wagon wheel: the spokes, the felloes (the curved parts that make up the wheel rim) and the central stock (or hub). Even this relatively technical description has an elegiac, awestruck tone. It is impossible to forget that the craft know-how he is describing had disappeared even by the time Sturt was writing in the 1920s, wooden wheels having been displaced by rubber tires. And yet he never presents the rural artisan as quaint. From the removal of unnecessary weight from every part of the wheel to the driving of the spokes into the hub, each step in the process is seen as exacting and efficient. For the men who practiced these skills, one senses the experience of the shop was something like the opposite of romance.

George Sturt, 'Spokes and Felloes' and 'Stocks', from The Wheelwright's Shop (Cambridge: Cambridge University Press, 1923).

SPOKES AND FELLOES

The device of lightening the horse-load in a vehicle by shaving away every superfluous hair's thickness of timber was carried to its highest in wheels, and especially in their spokes. Here, least of all, was any concession made to ornament. It is true I once saw, perhaps at South Kensington Museum, a pair of Chinese wheels the spokes of which had been turned in a lathe, to look as pretty as stair banisters, but this would not have done for English waggons and dung-carts on English farms. In the wheels for a use so rough, often in rough weather, it was needful to save every possible grain of strength, while shaving ruthlessly away every grain of mere weight.

The draw-shave by itself was not enough. A more delicate tool followed it, significantly called a spokeshave. In later years spokes were even finished off with sandpaper. But this was rather to please the painters, who by then were specialising in wheelwright's work as they had long done in coach-building—it was to please the painters rather than to lighten the wheel. Sandpaper was used also by clumsy boys, who had not learnt how to finish with a spoke-shave. Of course it was not the spokes alone that came in for this attention. The felloes had their share. A well-finished wheel, intelligently shaved up, showed where the expert aimed at strength, and where he knew material could be spared. It was a case for experts. Machine-made wheels were wont to be unnecessarily heavy; and in army wheels too I have seen material left that should have been shaved away, if there had been more intelligence at head-quarters. But old village wheelwrights knew better. Even I knew.

As soon as I could distinguish the heart of oak from the sap (it was not at first, for they had not taught me such accomplishments at school) I took to looking out, myself, the spokes to be used for every wheel that went from my shop, and at once the question of strength arose. At the back of the spoke was where strength was chiefly wanted; the front, "the face," did not matter half so much. In fact, a little sap was liked all up the face, because it would take a slightly better surface for painting, where painting showed. For all that it was the back of the spoke that was of chief importance. A little knot there—a little black knot no bigger than a pea—damned the whole thing. What I looked for was a length without flaw, which after being rounded up could be left at the full dimensions decided on. Nothing might be shaved away from the back. It was heart of oak, fit to last for ever. After thirty years or so, when hard wear had worn the wheel out, the spokes were still sound, and were taken out carefully. I sold thousands of them, at a penny a piece, for ladder rounds, but the better of them were set aside to be worked up again for smaller spokes or in barrow wheels. Then they might start another career of usefulness, for another twenty years or so. And still the oak continued good; only, the spoke could not be used again, for a reason soon to be shown.

When Cook had carried away the chosen spokes (others made wheels, but the most of that work was given to George Cook, so sure he was)—when Cook had carried the spokes to his bench and planed them true and, with compasses and bevel, marked out the "shoulders," his first job was to saw out the shoulders and cut out the "foot"—the tenon to be driven down into the nave as far as the shoulders would let it go. I think he used generally a fine tenon-saw—like a butcher's meat saw—for the whole of this work; yet in some cases, if the grain of the oak was favourable, he may have split out the foot with wide sharp chisel. I cannot remember now. Next, the foot being prepared for the nave, came the job of shaping up the parts of the spoke that would be exposed.

This was begun with axe and draw-shave, the front edge or "face" being roughly chopped-out and all the four corners deftly splintered away. A "smoothing plane" then made the sides of the spoke straight, and after that they were further cleaned up with a "jarvis"—a sort of hollow-bladed plane used for rounding spokes and for nothing else. But, down towards the shoulders of the spoke, near the nave, there were curves into which no other tool would go, and the spoke-shave came into play.

Here, I remember, was a nicety I never fathomed. Again and again I have carried a spoke to Cook, finished, for all that my eyes could see; but I think he never once let it go untouched. With a tolerant smile on his pursed lips he would take the spoke from my hands—as if recognising that I had tried; and then he would put his own spoke-shave to it, making some slight difference I never properly grasped. Yet I always believed that he knew best. There may indeed have been an element of craftsman's "swank" in his behaviour—a willingness to show that he could do better than his employer; but he never so much as hurt my feelings in this way. Moreover, he would always do just the same thing with the "turned spokes" sometimes supplied him. Heavy they were, these machine-made commodities originating from America—heavy, and clumsily finished, even to my eyes; but they seemed to offend George Cook still more, and save on emergency he unwillingly used them. He was too much of an artist in spokes.

After the spokes, the felloes—ash, elm, or beech—in looking out which, from the felloe-stacks, I used a "pattern," so as to get something sure to yield the required curve and the right length too. For of course felloes differed in curve according to the height (the diameter) of the wheel, and in length according to the number of spokes used.

For instance, for a front waggon wheel four feet high, felloes about two feet long would suffice if there were six of them; but if there were only five they needed to be almost six inches longer to make up the full circumference.

As may be remembered the felloes had been roughly hewn when green, then stacked away to season; but now they had to be shaped out exactly. The years that had dried them had made them very hard too—all the better under very sharp tools, but trying to an inexpert man. Moreover they had probably become a little "cast" (twisted or warped) in the drying, so that there was always something more to be cut away. A good deal of measuring and marking therefore had to be gone through before the felloes could even be bored.

The first step was to get a plane surface to work to—an apprentice's job, a boy's job—the beginning of all things in the wheelwright's craft. It is no exaggeration to say I hated it. Probably the plane was ill-sharpened and ill-set, and anything would have gone wrong; but a resentment took hold of me against that innocent curve of the surface of the felloe under my plane. If only the disgusting thing could have been straight! The felloe lay on one side, jabbed hard against the spiked bench-iron and with its rounded top surface towards me. It looked tractable enough. Yet when I tried to plane it, too often it jumped up over the bench-iron, or proved cross-grained and would not let the plane "shoot" comfortably and smoothly from end to end. I have been persuading myself that "trying-sticks" were brought in at last, to verify the workmanship, but probably this is wrong. "Trying-sticks" were two little straight bits of wood, about seven inches long by an inch square, and painted black. If you laid them in the same direction one at each end of a newly planed surface, and bent down and squinted across them from one to the other, it was easy to see if they lay level—("true out o' wind")—and so to judge whether or no the surface they lay on was truly plane. But I think this was for surfaces farther apart than the two ends of a felloe. But however that may have been, certain it is that a planing of the "face" or front side of every felloe was essential.

Figure 2 Peter Henry Emerson, *In the Haysel,* 1888.

By the time this was done the face was not only plane, but smooth and pale-coloured. The exact pattern could accordingly be pencilled out on it; and that done the wheelwright got to work with axe and adze, chopping down to the pencil line. The axe was for the convex outside; to get the inner line (the concave "belly") it was necessary to wedge the felloe into the felloehorse and go at it with the adze. Cook used to work this out so neatly that the belly shone, looking fit for polishing, it was so smooth. Lastly,

with the axe, the felloe was chopped down ("taken-down") to its required thickness, the fourth side of it having been marked out with a "scratch-bit"—a gauge held close against the face of the felloe and scratching into the newly-chopped "back," the newly-adzed belly. Thus finished, the felloe was a block of wood exactly squared—some three-and-a-half inches square, say, by thirty inches long—curved to be one segment of a wheel-rim. When the felloes for a new wheel had been brought to this stage they were piled up together ready, as soon as the "stock" and spokes should be ready for them.

STOCKS

Of the stock (the nave or hub) I hardly dare speak, such a fine product it was, and so ignorant about it do I feel. It is true I learnt to buy stocks with confidence in my own judgment: I seasoned them, chopped them into shape, chose them at last even to satisfy Cook. Nay, he occasionally asked my opinion, if anything dubious was discovered in working. But, as I had never enough skill of hand and eye myself, I always entrusted the actual turning and mortising of stocks to a trusty man—Cook as long as he lived, and after him preferably Hole. These men, I knew, would sooner have been discharged than work badly, against their own conscience. So I left the stocks to them, only liking to look at each stock when it was brought from the lathe, and to "weight" it (poise it) in my arms and hear the wheelwright say "rare stock that." His enthusiasm was catching. I felt a glow of pride in having ministered, however humbly, to so noble a tradition. Then I left the stock again to the workman.

A lumpish cylinder in shape—eleven or twelve inches in diameter and twelve or thirteen inches from end to end—a newly-turned stock was a lovely thing—to the eyes, I thought, but more truly to sentiment, for the associations it hinted at. Elm from hedge-row or park, it spoke of open country. Well seasoned, it was a product of winter labour, of summer care in my own loft under my own hands. Long quiet afternoons it had lain there, where I could glance from the stocks across the town to the fields and the wooded hills. I had turned it over and over, had chopped the bark away, had brushed off the mildew while the quiet winter darkness had stolen through the shed, and at last I had chosen the stock for use, and put it into Cook's hands.

And now it lay, butter-coloured, smooth, slightly fragrant, soon to begin years of field-work, after much more skill—the skill of ancient England—had been bestowed on it, though already telling of that skill in every curve. Certainly we did not consciously remember all these matters at the time: rather we concerned ourselves with the utility this block of elm would have, with its grip for many years of the oak spokes to be driven into it by and by. But, without thinking, we felt the glamour of the strong associations; and the skilled craftsmen must have felt it more than I, because they lived in that glamour as fishes live in water. They knew, better than any other may do, the answer of the elm when the keen blade goes searching between its molecules. This was, this is, for ever out of my reach. Only, I used to get some fellow-feeling about it, looking at a newly-turned stock. I understood its parts—the shallow hollows at back and front where the blacksmith would presently put on the bonds, the sloping "nose," the clean chisel-cut of the "breast stroke." This last was cut in all round the stock to mark where the face of the spokes was to be.

So, when I had had my look, the wheel-maker—Cook or another—carried the stock to his bench, there to mark on it with straddling compasses the place for the first auger-holes, preliminary to mortising it for the spokes. A tricky job, this. One young man, I remember, marking out his stock, prepared for an odd number of spokes—eleven or thirteen; though, every felloe requiring two, the spokes were always in even numbers; which error he did not detect until he had bored his stock and spoilt it. Too big for the fire, and too cross-grained to be easily split and thrown away, it lay about for months, an eyesore to the luckless youth who had spoilt it and a plain indication that it is not quite easy to mark a stock correctly.

Likewise was it not altogether a simple thing, though the skilled man seemed 'to find it easy enough, to fix the wobbly stock down for working upon. It was laid across a "wheel-pit"—a narrow trench with sills, about three feet deep—where iron clamps, themselves tightly wedged into the sills, held the stock steady back and front. Then the mortices were started, with auger-holes. How easy it looked! In my childhood I had heard the keen auger biting into the elm, had delighted in the springy spiral borings taken out; but now I learnt that only a strong and able man could make them.

The holes being bored, and before the actual mortising could begin, a gauge was attached to the front end of the stock, to be a guide for the coming operations. This gauge was a slender bar of wood—almost a lath—swinging round like one hand of a clock, but extending three feet or so beyond the stock. At the outer end of it a thin sliver of whalebone projected just so far as the front of the spokes would come if they had the right "dish." Note that. The spokes would have to lean forward a little bit; and the gauge was set so that this might be attended to even in mortising the stock. Before ever a spoke was actually put in the wheelwright tested the place for it, shutting one eye and squinting down with the other to see that the front edge of the mortice was properly in line with the whalebone sticking out from the gauge. The principle was very much like a marksman's taking his aim by foresight and backsight. One mortice having been cut, the stock was levered round with an iron bar so that the opposite mortice could be cut, and thus it was done all round, splinters or borings often dropping clear, right through

the stock from one side to the other into the wheel-pit. The uncut ribs of wood left between the mortices were called "meshes"—a word that will be wanted again. I do not think we shall want again the word "buzz"—the name for the strange three-cornered chisel used for cleaning out the mortices of a stock and, to the best of my belief, used for nothing else, unless for enlarging the central hole in the stock.

And now,—how dare I go on to describe that swinging drive of the wheelwright's action, fixing the spokes into the stock? Prose has no rhythm for it—no spring, no smashing blow recurrent at just the right time and place. The stock is to be imagined, ready at last, clamped down across the wheel-pit. From the front of it the gauge slants up; the dozen or fourteen spokes are near at hand, each with its tenon or "foot" numbered (in scribbled pencilling) to match the number scribbled against its own place in the stock. For although uniformity has been aimed at throughout, still every mortice has been chiselled to receive its own special spoke, lest the latter should by chance have had any small splinter broken away after all. The true wheelwright would not take that chance. He intended that every spoke should really fit tight; and there he has the spokes all numbered, to his hand.

He picks up one in one hand, and, with sledge-hammer in the other, lightly taps the spoke into its own mortice. Then he steps back, glancing behind him belike to see that the coast is clear; and, testing the distance with another light tap (a two-handed tap this time) suddenly, with a leap, he swings the sledge round full circle with both hands, and brings it down right on the top of the spoke—bang. Another blow or so, and the spoke is far enough into the mortice to be gauged. Is it leaning forward a little too much, or not quite enough? It can be corrected, with batterings properly planted on front or back of top, and accordingly the wheelwright aims his sledge, swinging it round tremendously again and again, until the spoke is indeed "driven" into the stock. It is battered over on the top, but the oak stands firm in the mortice, to stay for years.

For an hour or so, until all the spokes had been driven into a wheel, this sledge-hammer work went on, tremendous. I have seen nothing else like it. Road-menders greatly smite an iron wedge into the road they are breaking up; blacksmiths' mates use a ponderous sledge at some of their work; foresters, cleaving, make great play with beetle and wedges; but so far as I have noticed, these men (like the "Try-Your-Strength" men at a country fair) do not really know how to use sledge or beetle. They raise it up above their heads and bring it down, thump, with all the force of strong arms; but a wheelwright driving spokes, though not necessarily a very strong man, was able, with knack, to strike more powerful blows, and many of them too, in succession. With one hand close under the head he gave the sledge a great fling, then slipped the same hand down the handle, to help the other hand hold it in and guide it truly round its circle. By the time it reached the spoke the sledge had got an impetus. With the momentum of a stone from a sling, it was so to speak hurled down on its mark, terrific.

This way of driving spokes was probably very antique, and, being laborious and costly, it had died out from my shop before I had to retire myself. Hoop-tyres, superseding strakes, had indeed made such strenuous arm-work

less necessary; and the lighter wheels for spring-vans and carts, besides being more rapidly worn out on the harder roads and at the quicker pace, did not otherwise need putting together so strongly. But a dung-cart or a waggon was meant to last a life-time: the wheels were heavy; "strakes" of old could not pull them together as more modern tyres did; the wheels might have to lie on their face in a meadow all the summer for "stepping" a rick-pole and then be put to their proper use again, and if they could not stand all this the wheelwright was sure to hear of it from the farmer.

So, in my first five or six years at my shop, Cook (and perhaps others) made wheels in the right provincial style—wheels to stand hard work until they fairly wore out. As I saw it practised the art must have been time-honoured indeed. Village shops had carried it on for generations. I like to think that the twelve-spoke wheel, the cart wheel in one of the Canterbury Tales, was the work of men using the sledge as I saw George Cook using it.

NOTES

1. Arnold Bennett, introduction to George Sturt, *A Small Boy in the Sixties* (Horsham: Caliban Books, 1982; orig. pub. 1927), p. xi.
2. Sturt, *Wheelwright's Shop*, p. 24.

FURTHER READING

George Sturt, *William Smith: Potter and Farmer* (London: Catto and Windus, 1920).

E. D. Mackerness, *The Journals of George Sturt* (Cambridge: Cambridge University Press, 1967).

4

ON WEAVING

Anni Albers

Founded in Weimar in 1919 and closed by the Nazis in 1933 after two subsequent moves to Dessau and Berlin, the Bauhaus was the most influential modern art and design school. This was partly thanks to the emigration of many of the school's leading members prior to World War II. One of these was the weaver Anni Albers, who with her husband, Josef, came to the experimental Black Mountain College in North Carolina and subsequently moved on to Connecticut (Josef having been hired to teach in the art department at Yale University). Like most Bauhaus products Albers thought of herself principally as a designer rather than as a craftswoman who happened to design her own wares. And yet she saw the two activities of designing and making as inextricable. She faithfully preached the modernist precept that design should be simple and self-aware. Designers, she argued, should pay careful attention to the inherent quality of materials and work in a manner appropriate to their tools and processes. Thus weaving should be thought about, first and foremost, in terms of the grid imposed by the interweaving of warp and weft threads. Albers was also an early and adventurous investigator into the world's indigenous craft traditions. Particularly taken with Mexican and other Latin American textiles, she introduced her students at Black Mountain to the use of a simple backstrap loom and often used rough

fibers and natural dyes in her own work. Albers thus encapsulates the dichotomy that characterizes much modern craft, between progressive theorization and an attraction to long-established authenticity.

Anni Albers, 'The Fundament Constructions', from *On Weaving* (Middletown, CT: Wesleyan University Press, 1965).

The structure of a fabric or its weave—that is, the fastening of its elements of threads to each other—is as much a determining factor in its function as is the choice of the raw material. In fact, the interrelation of the two, the subtle play between them in supporting, impeding, or modifying each other's characteristics, is the essence of weaving.

The fundamental constructions, in common with all fundamental processes, have a universal character and are used today, as they were in our early history, here and everywhere. They show the principle of textile construction clearly. With only a few exceptions, all other constructions are elaborations or combinations of the basic three: the plain weave, the twill, and the satin weave. Of these three it is the plain weave that embodies the sum total of weaving and therewith reaches back the furthest.

All weaving is the interlacing of two distinct groups of threads at right angles. Wherever a fabric is formed in a different manner we are not dealing with a weaving. Where, for instance, the threads intersect diagonally in relation to the edge of the fabric, or radially from a center, we have a braided material; where only one thread is used to build up the material, we have a knitted or crocheted one; where threads intertwine or loop around each other, we have a lace or a net fabric. The horizontal-vertical intersecting of the two separate systems of thread is of great consequence for the formative side of weaving. The more clearly this original formation is preserved or stressed in the design, the stronger the weaving will be in those characteristics that set it apart from other techniques. Just as a sculpture of stone that contents itself to live within the limits of its stone nature is superior in formal quality to one that transgresses these limits, so also a weaving that exhibits the origin of its rectangular thread-interlacing will be better than one which conceals its structure and tries, for instance, to resemble a painting. Acceptance of limitations, as a framework rather than as a hindrance, is always proof of a productive mind.

The threads grouped vertically or lengthwise in the fabric are the warp threads; those running horizontally or crosswise are the filling threads. By collective names they are the warp and the weft or filling or woof or pick. The warp threads are stationary in the process of weaving while the filling threads are in motion, which indicates that the weaver for the most part deals with the filling threads and which may explain the greater number of terms for them.

In the plain weave this intersecting of warp and weft takes place in the simplest possible manner. A weft thread moves alternately over and under each warp thread it meets on its horizontal course from one side of the warp to the other; returning, it reverses the order and crosses over those threads under which it moved before and under those over which it crossed. This is the quintessence of weaving. The result is a very firm structure which, since it is comparatively unelastic, is strong under tension and also easily preserves its rectangular shape. It has an even, uniform surface, with warp and weft appearing in equal measure and producing the same effect on the front and the back of the fabric. It has a tendency to be stiff and, since the threads here cannot be pushed together very closely, it appears perforated when held against the light. Not more than two warp and two weft threads are necessary for its basic construction, and therefore only the simplest type of equipment is required. It is also a weave that demands less material for its construction and can be produced faster than any other. The usefulness of these characteristics is evident. There is probably no weave produced in more millions of yards the world over, now as in former times, than this plain weave. We recognize it in Egyptian mummy cloth and in our sheets, in unbleached muslin, potato sacks, and sail cloth—in short, wherever strength and a solid surface that does not permit threads to be caught accidentally are required.

It is interesting to note that this most practical of all thread constructions is at the same time also the one most conducive to aesthetic elaborations. The fact that warp and weft appear on the surface in equal amounts and intersect visibly leads to the use of contrasting materials and colors for them, thereby underlining the original structure of the weave. Emphasizing this structure still further are stripes in either warp or filling and, one step further, checked effects, another of the most typical designs of weaving in a plain weave. But beyond these elemental formative additions, the condensed quality of this weave, its use of only essential components, predisposes it also to be the construction used in work of a pictorial character, that is, in tapestries. Its shortcoming for such a purpose—the necessity of having to deal with a mixture of warp and weft—is overcome by deviating from the balanced proportion of warp and filling and using disproportionately more filling. By spacing the warp so widely that the weft can be beaten together closely, it is possible to cover the warp up entirely; the filling thereby becomes the sole agent of the surface. Gothic tapestries, those of the Renaissance, Aubusson tapestries all are executed in this simplest of constructions. The old truth applies here again—a process reduced to just the essential allows for the broadest application.

Another construction, also fundamental in its simplicity though already one step nearer complexity, is the twill weave. Whereas the plain weave is essentially a balanced weave—that is, warp and weft take an equal part in it and consequently produce the same appearance on the face of the fabric as on the back—the twill can be either a balanced or an unbalanced weave. It is unbalanced when either warp or filling is predominant, and in that case the face and back of the cloth are the reverse of each other. For where the filling covers most of the surface, the back naturally shows for the most part warp, and vice versa. A twill in which the warp prevails on the surface is called accordingly a warp twill, and the one that shows on the face more filling than warp, a filling twill.

The principle of construction in a twill is that the successive filling threads move over one warp thread or over a group of warp threads, progressively placing this thread or group of threads one warp thread to the right or left of the preceding one. Thus, in the smallest filling twill, which covers three warp and three filling threads, the first warp thread is raised over the first filling thread, which floats over the second and third warp threads; the second warp thread is raised over the second filling thread, which covers the first and third warp threads; the third warp thread runs over the third filling thread, which now floats over the first and second warp threads. This manner of intersecting warp and weft produces distinct diagonal lines, the characteristic twill lines. In a warp twill of the same size, the proportion of warp and filling on the face of the fabric will be reversed. The first and second warp threads will be raised over the first filling thread; the second and third warp threads over the second filling thread; and finally the first and third warp threads over the third one.

The diagonal twill line can, of course, run to either the right or the left. This is of consequence only in regard to the direction of the twist in the yarn used; for a slant to the right, for instance, will increase the relief effect of the ridge formed by a left twist warp, while a left slant would decrease it. The angle of the slant varies with the relationship of warp to weft in regard to the size of the threads and the closeness of the setting. If these are equal, the slant will be at an angle of 45°; if the warp is thicker or more closely set than the filling, the incline will be steeper; if it is thinner or more loosely set than the filling, it will be more gradual.

Innumerable twills can be designed: either balanced or uneven; either simple, with just one twill line, or compound, with a number of lines. Twills are often written in the form of numbers indicating the warp threads raised or lowered. For instance,

$$\frac{1 \quad\quad 3}{2 \quad\quad\quad 1}$$

would specify an uneven 7-leaf warp twill, in which one warp thread is raised, two are lowered, three are raised, and one is lowered. A balanced twill would read

$$\frac{2 \quad\quad 3}{\quad 2 \quad\quad 3} \text{ , etc}$$

Twill weaves, as a result of longer floating threads, are softer and can be woven more closely than plain weaves. They also are more pliable and inclined to give way more easily to diagonal pull, which makes them eminently suited for tailoring and thus for clothing purposes. We know them in the form of denims and either cotton materials for our work clothes and in countless wool tweeds. They were also known in ancient times, and in this hemisphere twills have been unearthed which date back to the Peruvian Mochica period.

The satin weave, the third of the fundamental constructions, is believed to have been invented by the Chinese.[1] In some ways it is the opposite of the plain weave. For, if the plain weave is essentially a construction that can only be balanced—that is, can only produce a fabric that is the same in front and back—the satin weave can only be unbalanced, can only produce a fabric different on either side, can show only either warp or filling. In contrast also the plain weave, where the closest intersection of warp and weft is sought, the farthest intersection within a given unit is chosen for a satin weave. The long, floating threads cover the points of intersection of warp and weft and permit the threads to be beaten together closely, so that a uniform, smooth surface is achieved, lacking any obviously visible structural effects.

We have found that the plain weave requires two warp and two filling threads for its construction, and the twill weave at least three. The satin weave calls for a minimum of five warp and weft threads.

To discover the best position for the points of intersection of warp and filling, technically termed "stitchers," the unit of threads that is to form the satin is divided into two groups of different size that are larger than one thread, that are not divisible one into the other, and that are not divisible by a common third. A unit of five threads, for example, is divided into one group of two and one of three. After interlacing the first warp thread with the first filling thread, the places for further intersection will be, for every following weft thread, either always two or always three warp threads removed from the preceding intersection. Thus the stitchers for a 5-leaf satin will be in the order 1, 3, 5, 2, 4; that is, the first warp thread intersects with the first weft thread, the third warp thread with the second, and so on. Progressing in the other possible order, the stitchers will be placed in the following arrangement: 1, 4, 2, 5, 3. Every warp thread has to be attached once within the unit to every weft thread, in a position that allows for the widest possible separation of the stitchers. Many satins can be formed by this method. The unit of six threads forms an exception, since it cannot be divided into any groups that comply with the requirements. Advancing in the order 1, 3, 5, 2, 4, 6 seems possible at first glance, when only the first unit is considered. But the repeat will reveal the defect that the first and the sixth stitchers come to be side by side. By exchanging the last two stitchers, a workable order can be given. Thus, instead of 1, 3, 5, 2, 4, 6, the progression will now read 1, 3, 5, 2, 6, 4. In larger units, more than two numbers of progression can be found. For instance, the unit of sixteen threads can be divided into groups of three and thirteen, five and eleven, seven and nine, all equally suited to our purpose here.

This wide separation between the points of interlacing in the satin weave makes for a very pliable, soft fabric which, in addition, can be highly glossy when executed in a lustrous material because of the homogeneous surface of either warp or weft. The contrast to the plain weave becomes apparent again when we compare the possible functions of the two; for, whereas we considered the plain weave to be the most serviceable construction, the satin weave is a luxurious one. The soft drape, the gloss that usually goes with the weave, and, on the negative side, the long floating threads that preclude hard wear, predispose it for an extravagant existence. It is a weave made for splendor. We know it in the form of silk satin, used in decorous draperies or, equally decorously, in our clothes of leisure.

The innumerable deviations from these three basic weaves show in varying degrees the main characteristics of their lineage, depending on how close or how distant their relationship is.

NOTE

1. Luther Hooper, *Hand-Loom Weaving* (New York, Pitman, 1920), p. 168.

FURTHER READING

Anni Albers, *On Designing* (Middletown, CT: Wesleyan University Press, 1962).

Nicholas Fox Weber, *Josef and Anni Albers: Designs for Living* (London: Merrell, 2004).

Virginia Gardner Troy, *Anni Albers and Ancient American Textiles: From Bauhaus to Black Mountain* (Aldershot: Ashgate, 2002).

5

PRIMITIVE POTTERY

Hal Riegger

The potter Harold Eaton (Hal) Riegger was trained in America's burgeoning academic ceramic programs—at Alfred University in New York and Ohio State University—but he made his name by rejecting everything that institutionalization stood for. A conscientious objector during World War II, he became an early proponent of raku, a Japanese technique of rapid firing that caught the imagination of many potters (notably Paul Soldner) in the immediate postwar period. In succeeding decades, influenced by the countercultural tenor of the times, he sought out more primal, communal techniques. His 'Experiment A' workshops brought a group of potters out into the austere desert of Panamint Valley, California. There, Riegger taught raku and the construction of anagama (hill-climbing channel kilns), as well as the use of clay and glaze ingredients scavenged from the surrounding hills. The strategy was to take potters away from the ready-made materials and electric kilns used in university clay studios and throw them back upon their own resources.[1] In the following passages taken from his how-to book Primitive Pottery, *he explains his reasons for rejecting the comforts and advance techniques of the modern ceramic studio and instructs readers how to fire pots out in the open, using nothing but materials that are ready to hand. Two points are worth highlighting: first, his insistence on the potter's 'responsibility' for the whole process, which has an ethical as well as a practical aspect; and second, his argument that apparently simple techniques can often require more rather than less skill, because they lack the safeguards of more technically complex processes. These ideas lift Riegger's text from the merely nostalgic. His 'primitivism' is idealistic, but he is also deeply respectful and knowledgeable about the tacit knowledge of Native American and African potters, and he is individualistic in a very modern sense, as his interest in using rubber tires for fuel suggests.*

Hal Riegger, *Primitive Pottery* (New York: Van Nostrand Reinhold, 1972).

First you have to find a place to make a fort. Then you outline it. After that you start digging until you get it deep enough. Then you put boards on and while there's still a little hole in the roof you make a fireplace. After that you make a door. Then you get in and seel off the cracks. Then you get out and get a bucket or something and fill it about to inches up with water and put dirt in it. After that you get some old newspapers and put them on the boards and put the mud over it and it is all finished. I guess.

—Theme written by Brad Johnson, age 10, Gridley, California, 1967

PREFACE

Some of my readers will wonder, "Why go backward in your craft and write about primitive ways of making pottery?"

Ceramics is one of the oldest crafts and its technology may not keep pace with that in other areas. But recently it has made significant advances for a craft that seems tenaciously to have hung on to the past—in maintaining centuries-old mechanical methods and being very slow to research and adapt new materials and methods. Many things are now taken for granted that did not exist sixty, or even fifteen, years ago. I think of two, for example: Pyrex glass and space capsule nose cones. Entertaining the thought of relinquishing these advances would be insane!

A few American Indians work in the primitive way, I suspect, like a few well-known Mexican potters, for the tourist trade. People in other parts of the world still work this way because they have not learned other methods, and their mode of living does not require a change. Some of us, like myself, do it even though our experience and knowledge have progressed much further.

For twelve years I have been working this way, for the most part, in preference to the more generally known ways. Why do I find primitive pottery interesting to the extent I do? And why do I feel it is so pertinent to teaching?

Without explanation, I could reply, 'I like to make pots this way,' and there could be no challenge. Yet, while no defense is necessary, an explanation helps. Justification is found in the effect upon the resulting pots. In its simplicity and therefore its demands upon the skills of the potter (which are greater because of the technical simplicity) a manner of clear, logical thinking is brought about. We are, in a sense, taken back to the uncluttered thinking of children. Not only is this refreshing in so complex an existence, it is good training for the mental processes.

INTRODUCTION

Throughout the world, wherever man found clay and discovered he could shape it with his hands and harden the resulting forms with fire, he experienced, in its most elemental and basic form, what we now call ceramics. Clay is a plastic mineral that can be permanently hardened in red heat.

Countless techniques and a vast store of knowledge have evolved from these two important facts about clay. No industrialized country is without its society of professional ceramists who represent an economically important industry: hardly a person today is unaware of ceramics, and in North America universities offering ceramic courses number into the hundreds, let alone the thousands of elementary and secondary schools that offer classes in pottery.

Yet ceramics can still be defined fundamentally as the craft and industry of forming objects out of plastic clay and firing these objects so as to harden them permanently. In this book on primitive ways of making pottery these fundamental, direct concepts must be remembered and practiced.

"Primitive" is a difficult word to use here and needs some explanation. Archaeologists are not at all agreed on how to use the word. Perhaps we should dream up a new one.

In a discussion once of my pottery activities with a member of one of the southwest American Indian tribes, the word 'primitive' was rejected in favor of 'traditional' to describe them. Yet this is not correct either. But it does point out that primitive, as applied to art and to the craft or making of objects, in no way implies crudity. If anything, craftsmanship was on a higher level than we are apt to see nowadays. The materials were not refined or man-made and were taken from nature as they were found, wherever and whenever man chose to express himself through art. I have no doubt that man then observed the qualities and character of natural materials around him far more accurately than we do now, and used them quite sensitively and selectively for what they were and what they could do and convey. Hard and soft woods, feathers, bark, animal furs and hides, obsidian, jade, walrus whiskers, nacre—these were the materials nature provided, not to mention clay, the material of our concern in this book.

So I would like to look at it this way: "primitive" defines the way we will observe our materials (awareness) and the way we will go about using them (sensitivity) or, in other words, our approach towards pottery fashioned and fired in a primitive manner without the tools of modern technology.

Much early pottery around the world is similar, whether it was made in what is now England, or in Africa, the South Pacific or Japan. All of this was happening before these lands had their present identity. Apparently man's needs for existence and his intelligence were pretty much the same the world over at this stage of his development. Of course this did not all happen at one time, some areas in the world saw the first development of the craft of pottery while in others it came later.

Figure 3 Hal Riegger, *Planet Pot,* ca. 1974.

In some places (Africa, southwestern United States, and Fiji, for example) people are still working in a primitive fashion; in some places this occurs alongside of the most advanced technology we know.

[. . .]

FIRING WITHOUT KILNS

Firing pottery without a kiln is a simple and direct operation that brings the potter into intimate contact with his pots and the elements of firing in a way other, more sophisticated firings cannot. But, as with all simple processes not assisted by technical aids, considerable skill and experience are necessary for good results. Every part of the operation is the direct responsibility of the person doing it. Because the process is simple does not mean its execution is without its demands.

For the first time some rarely considered aspects of fire become important: heat rises, small pieces of fuel burn faster and hotter than large pieces, wind can make the heat of a fire lopsided. These and other things must be known and brought under control for successful firing.

Following the same train of thought, one realizes that the pots must be so placed that heat will entirely surround them, that there should be no sudden rise in temperature, and that the heat around the pots must be as even as possible. Pots are not placed in a pit and a fire built over them. Fuel is selected and used in a way that brings the temperature up gradually, and firings are avoided on windy days if possible.

Let us examine first the manner of setting the pots for firing. American Indians originally put their raw pieces on rocks. Out of experience they chose rocks that would not explode in the firing. If we do not know already, we must test large rocks in a fire before using them . . . Nigerian potters make a bed of twigs and branches upon which their pots are put; the effect is the same as with rocks in that heat will reach the bottoms of the ware as well as the sides and top. Potters also raise their pots off the ground with an old metal grating, or any old metal that can stand the heat . . .

It is apparent that pots can be fired at almost any location. The potter will first survey the surroundings to determine what objects are at his disposal, then proceed to set up for firing on that basis.

The second element of firing concerns fuel and how it is controlled. The potter will naturally choose a location providing more or less natural, and freely available, fuel, and such a site will likely be outside an urban area.

Oil and gas, which need supplementary equipment, are not considered and have no place in this kind of firing; instead, fuels like wood, grass, coal, lignite, leaves, and animal dung are used. It matters not so much which of these is available as long as the potter knows how to use them.

Both the type of fuel and its form will affect the way the pottery setting is made. These two factors must be logically related to one another . . . Knowing how fuels burn is helpful: heavy, large chunks of wood burn more slowly than finely split wood; green wood burns more slowly than well-seasoned wood; dry grass burns hotter and faster than wood, but for a shorter time. American Indians in the past have used coal, lignite and animal dung as well as wood, and are always sure to gather sufficient fuel before starting the fire. The Nigerians must amass great stacks of fuel because it is mostly grasses tied into bundles. It is far better to have gathered more than enough fuel than to have to scrounge frantically during the firing.

The fuels about which I can speak with some authority are wood, grass, cow dung, and rubber tires. Wood is the most common, yet I prefer it the least. Both dung and grass are easier to handle, and with them I can avoid a too-sudden heat rise in the beginning. Of all the fuels I prefer dung; it catches fire easily and burns with a gentle, steadily increasing heat. Twigs or finely split wood are put under the pottery setting and lit, and dung is piled around and on top of the pots. Dung seems always to reach about the optimum temperature without getting too hot. Placed as it is, one or two layers thick around the pottery, it forms a protective barrier that holds in the heat during firing and keeps the pottery from cooling too rapidly after firing is completed. It is not necessary to slow the cooling by burying the pots and fire under dirt. This is true also of grass which, when burned, will leave a thick protective layer of fine ash all over the pots.

[. . .]

One contemporary version of primitive firing with no kiln uses automobile tires that provided a concentrated and easily handled source of fuel. Whether one's ethics allow the use of tires is another matter, for they do give off a thick black smoke. The same considerations are given to firing preparations for tires as were described earlier for other fuels. Physical dimensions of the tires dictate the size and number of pots that can be fired at one time. After the pots have been thoroughly dried, bricks, stones, cans, or a grate are arranged over the coals to support the pots within the inside diameter of the tires. Pieces are put in, then three, four or five bricks or stones are placed in a circle around the setting to support the tires. Air entering underneath provides for better combustion inside. Then three or four tires are placed on these supports, and within a few minutes they will be ignited by the burning coals. A stack of tires like this actually forms a kind of kiln, or wall, protecting pots from the cooler outside air as well as containing the fire's heat. Rubber is such a concentrated fuel that it will be impossible to avoid reducing the pots although sometimes they will reoxidize when cooling.

The main bulk of the tires will burn out in about one-half to three-quarters of an hour longer.

The idea of using strips of inner tube and blown out tires found along desert highways came to some of us during a workshop in southeastern California in 1964. One particularly observant student, quite aware of the necessity to use what the area provided, came back to camp one day with his truck full of torn tires and rubber scraps, along with a small amount of scarce wood.

I have worked many times since with rubber in different forms, in open fires as well as for simple kiln firing. When used to fire pottery in an enclosure such as a kiln, rubber will burn more efficiently, but not without some smoke. This leads me to mention that tire retreading shops around the country accumulate enormous masses of scrap rubber shavings similar in size to sawdust. Whereas its use may not have a legitimate place in firing primitively, it seems to me potters might do some experiments on how to fire their kilns with this fuel efficiently and smokelessly. The rubber is free. Tire retreading shops have no use for it and take it to our overloaded dumps for disposal.

[. . .]

A final bit of advice to a potter attempting projects like these is that he be where he is. By this I mean that his mind and thinking, as well as his body, must be where he is working and not elsewhere. The reality of the situation is all around him and visible in his surroundings; it is not at his home or workshop. Knowledge and experience out of his past are helpful, if not essential, but he is not in the past. He is in and of today, where he is, at the moment. Then his mind will be open and receptive.

It is good to approach these experiences as explorations and discoveries. The potter will then be in a mood to plan his activities rationally, and to accept results that have previously been outside his experience.

NOTE

1. Gillian Hodge, 'Hal Riegger's "Experiment A" Workshops', *Craft Horizons* 30/3 (May/June 1970), pp. 32–33.

FURTHER READING

Eliot Wigginton, ed., *The Foxfire Book* (New York: Anchor Press/Doubleday, 1972).

Hal Riegger, *Raku: Art and Technique* (New York: Van Nostrand Reinhold, 1970).

Stewart Brand, ed., *The Whole Earth Catalogue* (Santa Cruz: Portola Institute, 1969).

SECTION 2

CRAFT AND THE INDUSTRIAL REVOLUTION

SECTION INTRODUCTION

'On every hand', wrote Thomas Carlyle in 1829, 'the living artisan is driven from his workshop to make room for the speedier, inanimate one. The shuttle drops from the fingers of the weaver and falls into iron fingers that ply it faster.'[1] This is the picture of the industrial revolution that the nineteenth century has bequeathed to us: a decline of traditional skills and a replacement of the living, breathing, thinking craftsman by the inhuman machine. Nearly all twentieth-century craft discourse takes this narrative as a starting point, leading to a situation in which artisanal cultures tend to be viewed as either anachronistic survivals or conscious revivals. But should craftsmanship and the industrial revolution really be opposed in this stark manner? Certainly Carlyle was not alone in his assumption that the artisan was disappearing from the earth, but there were also many in the nineteenth century who saw the craft of their time in terms of continuity, or even ascendancy. It is an ironic fact: the moment when modern craft began is perhaps the least well-understood chapter in its history.

INDUSTRIAL ARTISANS

There are several ways in which one might rewrite craft history during the industrial revolution. First, and most obviously, the machines in question had themselves to be made. Metalworkers, in particular, had never been so in demand, so various in their skills (working to widely divergent degrees of tolerance in many materials), or as crucial to the economy as a whole. Breakthroughs in design and engineering were often premised not on the elimination of hand tools, but rather their improvement. Indeed, one could say during the industrial revolution, the 'traditional' value structure of the metals trade was upended in the craftsperson's favor: in previous centuries, metal had been very expensive and labor was very cheap, so meticulously crafted objects were routinely melted down for the value of their material. By 1800, by contrast, the trade focused more and more on base metals instead of silver, and there was an unprecedented level of respect and autonomy afforded to the invention and skill of the 'industrial artisans' who made such things as bridges, steam engines, and machine tools.[2] As the industrial booster Samuel Smiles noted in the excerpt included here, 'It is one thing to invent, and another thing to make the invention work.' In this sense, the modern industrial artisan may have had more control and autonomy than any craftsman (especially the vast majority working outside of elite patronage structures) had previously enjoyed.

A second, less clear-cut reason for the artisan's continuing importance within industrialism is the fact that different materials dictated different solutions. One example is the hand-finishing of metal, which was difficult to make absolutely smooth using an automated cutting tool. This meant that even one of the most celebrated cases of early mass production, the making of interchangeable components for guns, was made possible only through craft skills. (A *Report on Small Arms* delivered to the British Parliament in 1853 noted, 'Whenever we want great perfection of parts we must do it by hand labour.'[3]) As the historian Maxine Berg has pointed out, in 1851 the British economy was still principally based on agriculture and crafts untouched by automation (shoe-making, tailoring, building and the like), not mechanized industry.[4] There was a vivid contrast between industries, such as cotton spinning and weaving, that lent themselves very well to mechanization and others, such as furniture making, that did not. Such differences were determined by materials and processes, but they had dramatic social consequences. While the massive mills in the north of England inspired the early writings of Karl Marx, chairs continued to be made by artisanal 'jobbers' in the east end of London. Both of these populations were exploited, but only the first seemed to mark a change in the nature of work itself.

Finally, there is the matter of aesthetics. Again, this is usually seen as a simple story: the hand versus the machine. Automated production, it is assumed, led to the indiscriminate production of poorly designed goods, from overcrowded textile patterns to coarsely molded and printed ceramics. The design reform movements that dictated discussion of craft in the late nineteenth century, particularly in England in the wake of the Great Exhibition of 1851, were intended to correct this problem. But it is worth remembering that the Arts and Crafts Movement taste for 'irregularity' would have perplexed most eighteenth-century artisans, who strove to achieve the regulated effects that later became associated with machines.[5] Furthermore, the British case, while influential, was also exceptional. There was no real parallel in places like France, Russia and Italy, where the production of luxury goods by hand remained critical to the economy; and in Germany, mass production never led to widespread antipathy towards a 'machine aesthetic' (this was to be crucial in the developmental phases of modernism in the early twentieth century).[6] Even in the case of Britain, one must distinguish between rhetoric and reality. Some of the exhibition showpieces that came in for most strenuous criticism, often on the grounds of their unimaginative mining of previous historical styles, were in fact prodigious works of hand craftsmanship, untouched by the machine.

The Division of Labor

This leads us to an obvious question: if craft skill was still viable in the period of industrialism in all these ways, why were writers like Carlyle moved to eloquence on the behalf of the disappearing artisan? This is difficult to answer, but it seems to be largely a matter of politics. The phrase 'division of labor', so often used as a stand-in for the far-reaching process of industrialization itself, is often cited in craft discourse but rarely interrogated. Commentators have tended to focus on the nature of labor—the sense that division leads inevitably to deskilling. This itself is debatable; the writings of Harry Braverman and Michael Ettema, included here, exemplify opposing positions on the question. Those who, like Braverman, wish to argue for a general trend towards deskilling and specialization must contend with the inconvenient fact that certain trades experienced division of labor long before machines were in use. In eighteenth-century furniture shops, which of course were completely dependent on hand processes, divided labor was an important strategy to lower costs and increase production. It could even be a means of producing a superior product, as it allowed individuals to specialize in particularly difficult skills like carving.[7] The figure of the omni-competent yeoman artisan has some basis in truth, but it is mostly a myth.

Yet it is equally clear that craftspeople's lives and livelihoods were often radically disrupted by the division of labor, and that this process was already under way at the beginning of the century. An

1830 cabinetmaker's guide noted that despite the fact that 'the various trades of the Cabinet Maker, Chair Maker, Japanner, Gilder, and Lackerer [sic], are so intimately connected, that there is scarce a handsome piece of furniture where the combination of their joint efforts is not necessary . . . it is almost universally the case, that a workman in one branch is entirely ignorant of the methods used by another'.[8] This is a complaint about not the skills of the individual workman, but the lack of integration across the trade. While many craft disciplines continued to be practiced with little interruption to methods and materials, few of them were immune to another process of modernization: reorganization. Berg makes the point succinctly. Industries like printmaking, tobacco harvesting, linen and wool production, and iron manufacture were all trades practiced mainly in the countryside, 'outside the context of the urban craft structure'. Though these trades were slow to mechanize, workers were not organized into guilds and thus lacked the social and symbolic traditions associated with city-based artisanal communities. It was these rural workshops that 'became the pioneering factories. As such they constituted exercise in the organization of labor rather than any change in actual mechanical processes'.[9]

In Britain, then, even the relative continuity of craft practice on the workshop floor did not prevent great upheaval in the politics of labor. Karl Marx saw this point clearly. Unlike the Arts and Crafts thinkers that he influenced—preeminently William Morris—he was well aware of the continuing importance of craft within factory contexts. He was worried not about deskilling per se, but rather about questions of power within the factory system. One key insight of his treatise *Capital*, excerpted here, is that mass manufacture does not necessarily require the presence of machines. Even in an artisanal context, divided labor permits greater control over individual workmen, a more precise calculation of investment-to-profit ratios, and flexibility of workforce (for example, specialized workers can be more easily let go and then rehired to minimize losses during a slow sales period). This suggests that historians of nineteenth-century craft should pay less attention to questions of automation and more to questions of organization. This would include the redesign of workspaces and the displacement of systems of training from apprenticeships to formalized institutions.[10] It would also take into account changing politics within craft communities themselves. One key development in this regard was an emergence of a 'labor aristocracy' of highly skilled workers, protective of their position and inclined to resist the upward mobility of less skilled groups such as immigrants and women.[11] Such division within the artisanal workforce may have been just as important a factor in producing discontent in the workplace as the oppression of factory bosses.

Modern Craft at the Fin de Siècle

It might be argued, in contradiction to the preceding, that these complexities within the process of industrialization are all footnotes to the main story: the steady decline of craft under the conditions of modernity. By the end of the nineteenth century, certainly, traditionally trained metalsmiths and woodworkers were no longer in charge of machine tool production, and industries that had initially proved difficult to mechanize (such as glassmaking) had become fully automated. From this perspective, craft and industry may not have definitively parted company in the early days of the industrial revolution, but it was only a matter of time. The Modernist turn, towards machines and away from the hand, might be seen as the aesthetic and expressive proof of this divorce.

And yet, developments in Central Europe and Scandinavia suggest that leading-edge thought at the fin de siècle envisioned new forms of continuity between industry and craft. The goal of the Vienna Secessionist movement to create *Gesamtkunstwerke*, or total works of art, might itself be seen as an attack on the principle of 'division of labor'. Here the objective was to unite objects and their environment into a single stylistic gesture. This was partly an aesthetic goal, of course—and could be applied in completely apolitical ways, as is seen in the work and writings of the brilliant, egomaniacal American architect Frank Lloyd Wright or the Glasgow school of architects and designers led by Charles Rennie Mackintosh. But in Central Europe, the *gesamtkunstwerk* was also linked to the principle of equal collaboration between artists, craftsmen and architects. It thus had overtones of social reform and was meant as a challenge to the atomization of typical capitalist enterprise.

This alliance of design to a Marxist-inflected utopianism would be widely shared across Europe in the first half of the twentieth century.[12] The Swedish craft reformer Gregor Paulsson, for example, argued that 'the fact that art-craftsmen, as is now mostly the case, sit and make gew-gaws for an interested or rich minority, while industrial products are designed by poor draftsmen, can hardly be considered right. It ought rather to be that the special skill of the former should be used to benefit as great a part of the community as possible; hence, cooperation between industry and art'.[13] Even the Viennese architect Adolf Loos, whose polemical essay 'Ornament and Crime' is often mistakenly seen as an anticraft diatribe, was both in his writings and his practice highly attuned to the importance of appropriate workmanship. As we will see in Section 6, this 'designer-craftsman' ideal persisted into the postwar period. But the ambition to reinscribe artisanal knowledge into industry was already well under way at the turn of the century. Modernism was in this sense not (as its supporters often claimed) a clean break from the past. It was, rather, just one in a series of displacements in craft's complicated relations to industry.

NOTES

1. Thomas Carlyle, 'The Mechanical Age', in *Signs of the Times* (1829).
2. On the transition of artisans to machine making see L. A. Clarkson, *Proto-Industrialization: The First Phase of Industrialization?* (London: Macmillan/Economic History Society, 1985).
3. David A. Hounshell, *From the American System to Mass Production, 1800–1932* (Baltimore: Johns Hopkins University Press, 1984), p. 24.
4. Maxine Berg, *The Machinery Question and the Making of Political Economy* (Cambridge: Cambridge University Press, 1980), p. 21.
5. Peter Betjemann, 'Craft and the Limits of Skill: Handicrafts Revivalism and the Problem of Technique', *Journal of Design History* 21/2 (2008), pp. 183–93.
6. Nancy Troy, *Modernism and the Decorative Arts in France: Art Nouveau to Le Corbusier* (New Haven: Yale University Press, 1991).
7. Edward S. Cooke, Jr., *Making Furniture in Preindustrial America: The Social Economy of Newtown and Woodbury, Connecticut* (Baltimore: Johns Hopkins University Press, 1996).
8. George A. Siddons (attributed), *The Cabinet-Maker's Guide* (London: Sherwood, Gilbert and Piper, 1830), pp. viii–ix.
9. Maxine Berg, ed., *Technology and Toil in Nineteenth Century Britain* (London: CSE Books, 1979), p. 7.
10. The key theoretical text on this point is Michael Foucault, *Discipline and Punish: The Birth of the Prison* (New York: Pantheon, 1978; orig. pub. 1975). For more empirically grounded histories see Alfred D.

Chandler, Jr., *The Visible Hand: The Managerial Revolution in American Business* (Cambridge, MA: Harvard University Press, 1977); Philip Scranton, *Endless Novelty: Specialty Production and American Industrialization, 1865–1925* (Princeton: Princeton University Press, 1997); Lara Kriegel, *Grand Designs: Labor, Empire, and the Museum in Victorian Culture* (Durham, NC: Duke University Press, 2007).

11. E. J. Hobsbawm, 'The Labour Aristocracy in 19th Century Britain', in *Labouring Men* (London: Weidenfeld and Nicolson, 1964), pp. 278–84; John Rule, 'The Property of Skill in the Period of Manufacture', in Patrick Joyce, ed., *The Historical Meanings of Work* (Cambridge: Cambridge University Press, 1987); Bruce Laurie, *Artisans into Workers: Labor in Nineteenth-Century America* (New York: Hill and Wang, 1989); John Rule.

12. Christopher Wilk, *Modernism: Designing a New World 1914–1939* (London: V&A Publications, 2006).

13. Gregor Paulsson, 'The Artist and the Industrial Arts', *Metropolitan Museum of Art Bulletin* 22/2 (Feb. 1927), pp. 42–44.

6

ON THE ECONOMY OF MACHINES AND MANUFACTURES

Charles Babbage

For those who are looking for a villain—a spokes-man of the machine against the craftsman—Charles Babbage might seem a likely candidate. As the historian Maxine Berg wrote in her over-view of early British theories of mass manufac-ture, 'Babbage regarded the machine as a great corrective of the indiscipline of labor: it could function as a check against the inattention, idle-ness, and dishonesty of human labor.'[1] In this respect he could be seen as a precursor of Fred-erick Taylor and other rationalizers who sought to quantify the exact amount of labor necessary for a given task, and to hold workers to this stan-dard regardless of the consequences. This view of Babbage as a somewhat inhuman, calculat-ing figure is also underpinned by his activities as a mathematician and engineer. His famous 'differential engine', built in the 1820s—which he described as a 'machine for the manufacture of number'—is often considered the world's first computer.[2] And yet it is clear that Babbage was deeply respectful of workers' skills. Unlike the other economists of his era who wrote about the advantages to be gained from machine manufac-ture, he based his comments on direct observation on the factory floor. His calculating engine was based on the operation of jacquard looms and was itself a marvel of fine metalwork. Indeed, Bab-bage could not have missed the importance of craft to his undertaking, for it was not the theory or the mathematics that resulted in difficulties for him but rather the actual fabrication of his engines' many subtle parts.[3] In the following ex-cerpt, Babbage both enumerates the benefits of the division of labor and carefully makes a place in his system for craft skill—which the efficient capitalist should purchase in 'exactly that precise quantity' which is necessary.

Charles Babbage, *On the Economy of Machinery and Manufactures* (1832), excerpts from Preface, Chapters 10, 19.

The present volume may be considered as one of the consequences that have resulted from the calculating engine, the construction of which I have been so long superintending. Having been induced, during the last ten years, to visit a considerable number of workshops and fac-tories, both in England and on the Continent, for the purpose of endeavouring to make my-self acquainted with the various resources of mechanical art, I was insensibly led to apply to them those principles of generalization to which my other pursuits had naturally given rise. The increased number of curious pro-cesses and interesting facts which thus came under my attention, as well as of the reflec-tions which they suggested, induced me to believe that the publication of some of them might be of use to persons who propose to be-stow their attention on those enquiries which I have only incidentally considered . . .

There exists, perhaps, no single circumstance which distinguishes our country more remarkably from all others, than the vast extent and perfection to which we have carried the contrivance of tools and machines for forming those conveniences of which so large a quantity is consumed by almost every class of the community. The amount of patient thought, of repeated experiment, of happy exertion of genius, by which our manufactures have been created and carried to their present excellence, is scarcely to be imagined. If we look around the rooms we inhabit, or through those storehouses of every convenience, of every luxury that man can desire, which deck the crowded streets of our larger cities, we shall find in the history of each article, of every fabric, a series of failures which have gradually led the way to excellence; and we shall notice, in the art of making even the most insignificant of them, processes calculated to excite our admiration by their simplicity, or to rivet our attention by their unlooked-for results.

The accumulation of skill and science which has been directed to diminish the difficulty of producing manufactured goods, has not been beneficial to that country alone in which it is concentrated; distant kingdoms have participated in its advantages. The luxurious natives of the East, and the ruder inhabitants of the African desert are alike indebted to our looms. The produce of our factories has preceded even our most enterprising travellers. The cotton of India is conveyed by British ships round half our planet, to be woven by British skill in the factories of Lancashire: it is again set in motion by British capital; and, transported to the very plains whereon it grew, is repurchased by the lords of the soil which gave it birth, at a cheaper price than that at which their coarser machinery enables them to manufacture it themselves.

The advantages which are derived from machinery and manufactures seem to arise principally from three sources: The addition which they make to human power. The economy they produce of human time. The conversion of substances apparently common and worthless into valuable products . . .

At each increase of knowledge, as well as on the contrivance of every new tool, human labour becomes abridged. The man who contrived rollers, invented a tool by which his power was quintupled. The workman who first suggested the employment of soap or grease, was immediately enabled to move, without exerting a greater effort, more than three times the weight he could before.

The economy of human time is the next advantage of machinery in manufactures. So extensive and important is this effect, that we might, if we were inclined to generalize, embrace almost all the advantages under this single head: but the elucidation of principles of less extent will contribute more readily to a knowledge of the subject; and, as numerous examples will be presented to the reader in the ensuing pages, we shall restrict our illustrations upon this point.

[. . .]

OF THE IDENTITY OF THE WORK WHEN IT IS OF THE SAME KIND, AND ITS ACCURACY WHEN OF DIFFERENT KINDS

Nothing is more remarkable, and yet less unexpected, than the perfect identity of things manufactured by the same tool. If the top of a circular box is to be made to fit over the lower part, it may be done in the lathe by gradually advancing the tool of the sliding-rest; the proper degree of tightness between the box and its lid being found by trial. After this

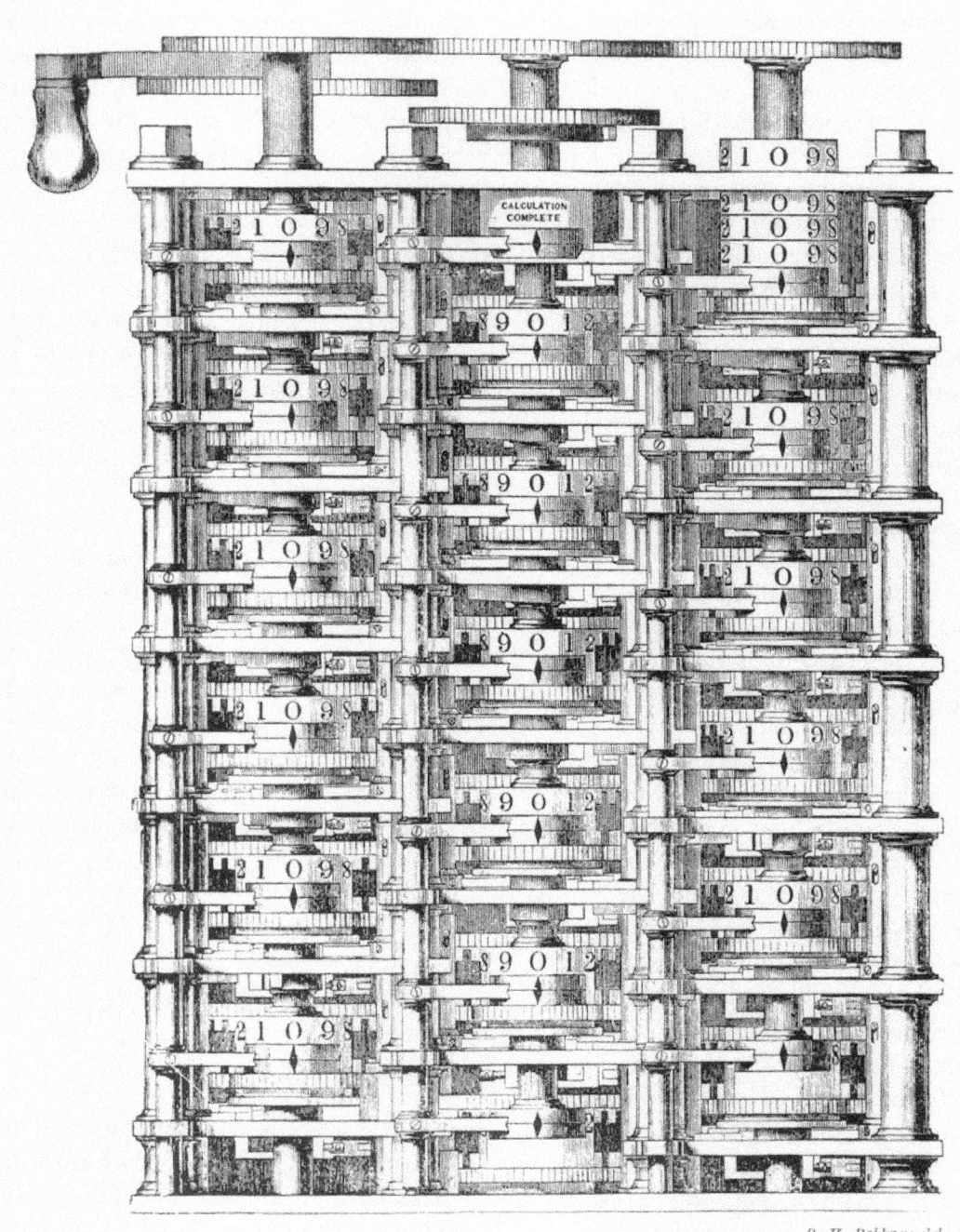

Figure 4 *Charles Babbage's Difference Engine No. 1,* ca. 1830s.

adjustment, if a thousand boxes are made, no additional care is required; the tool is always carried up to the stop, and each box will be equally adapted to every lid. The same identity pervades all the arts of printing; the impressions from the same block, or the same copperplate, have a similarity which no labour could produce by hand. The minutest traces are transferred to all the impressions, and no omission can arise from the inattention or unskilfulness of the operator. The steel punch, with which the cardwadding for a fowling-piece is cut, if it once perform its office with accuracy, constantly reproduces the same exact circle.

The accuracy with which machinery executes its work is, perhaps, one of its most important advantages: it may, however, be contended, that a considerable portion of this advantage may be resolved into saving of time; for it generally happens, that any improvement in tools increases the quantity of work done in a given time. Without tools, that is, by the mere efforts of the human hand, there are, undoubtedly, multitudes of things which it would be impossible to make. Add to the human hand the rudest cutting instrument, and its powers are enlarged: the fabrication of many things then becomes easy, and that of others possible with great labour. Add the saw to the knife or the hatchet, and other works become possible, and a new course of difficult operations is brought into view, whilst many of the former are rendered easy. This observation is applicable even to the most perfect tools or machines. It would be possible for a very skilful workman, with files and polishing substances, to form a cylinder out of a piece of steel; but the time which this would require would be so considerable, and the number of failures would probably be so great, that for all practical purposes such a mode of producing a

steel cylinder might be said to be impossible. The same process by the aid of the lathe and the sliding-rest is the everyday employment of hundreds of workmen.

[. . .]

ON THE DIVISION OF LABOUR

Perhaps the most important principle on which the economy of a manufacture depends, is the division of labour amongst the persons who perform the work. The first application of this principle must have been made in a very early stage of society, for it must soon have been apparent, that a larger number of comforts and conveniences could be acquired by each individual, if one man restricted his occupation to the art of making bows, another to that of building houses, a third boats, and so on. This division of labour into trades was not, however, the result of an opinion that the general riches of the community would be increased by such an arrangement; but it must have arisen from the circumstance of each individual so employed discovering that he himself could thus make a greater profit of his labour than by pursuing more varied occupations. Society must have made considerable advances before this principle could have been carried into the workshop; for it is only in countries which have attained a high degree of civilization, and in articles in which there is a great competition amongst the producers, that the most perfect system of the division of labour is to be observed. The various principles on which the advantages of this system depend, have been much the subject of discussion amongst writers on political economy; but the relative importance of their influence does not appear, in all cases, to have been estimated with sufficient precision. It is my intention, in the first instance, to state shortly those principles, and then to point out what appears to me to have been omitted by those who have previously treated the subject.

It will readily be admitted, that the portion of time occupied in the acquisition of any art will depend on the difficulty of its execution; and that the greater the number of distinct processes, the longer will be the time which the apprentice must employ in acquiring it. Five or seven years have been adopted, in a great many trades, as the time considered requisite for a lad to acquire a sufficient knowledge of his art, and to enable him to repay by his labour, during the latter portion of his time, the expense incurred by his master at its commencement. If, however, instead of learning all the different processes for making a needle, for instance, his attention be confined to one operation, the portion of time consumed unprofitably at the commencement of his apprenticeship will be small, and all the rest of it will be beneficial to his master: and, consequently, if there be any competition amongst the masters, the apprentice will be able to make better terms, and diminish the period of his servitude. Again, the facility of acquiring skill in a single process, and the early period of life at which it can be made a source of profit, will induce a greater number of parents to bring up their children to it; and from this circumstance also, the number of workmen being increased, the wages will soon fall.

A certain quantity of material will, in all cases, be consumed unprofitably, or spoiled by every person who learns an art; and as he applies himself to each new process, he will waste some of the raw material, or of the partly manufactured commodity. But if each man commit this waste in acquiring successively every process, the quantity of waste will be much greater than if each person confine his attention to one process; in this view of the

subject, therefore, the division of labour will diminish the price of production.

Another advantage resulting from the division of labour is, the saving of that portion of time which is always lost in changing from one occupation to another. When the human hand, or the human head, has been for some time occupied in any kind of work, it cannot instantly change its employment with full effect. The muscles of the limbs employed have acquired a flexibility during their exertion, and those not in action a stiffness during rest, which renders every change slow and unequal in the commencement. Long habit also produces in the muscles exercised a capacity for enduring fatigue to a much greater degree than they could support under other circumstances. A similar result seems to take place in any change of mental exertion; the attention bestowed on the new subject not being so perfect at first as it becomes after some exercise.

The employment of different tools in the successive processes is another cause of the loss of time in changing from one operation to another. If these tools are simple, and the change is not frequent, the loss of time is not considerable; but in many processes of the arts the tools are of great delicacy, requiring accurate adjustment every time they are used; and in many cases the time employed in adjusting bears a large proportion to that employed in using the tool. The sliding-rest, the dividing and the drilling-engine, are of this kind; and hence, in manufactories of sufficient extent, it is found to be good economy to keep one machine constantly employed in one kind of work: one lathe, for example, having a screw motion to its sliding-rest along the whole length of its bed, is kept constantly making cylinders; another, having a motion for equalizing the velocity of the work at the point at which it passes the tool, is kept for facing surfaces; whilst a third is constantly employed in cutting wheels.

The constant repetition of the same process necessarily produces in the workman a degree of excellence and rapidity in his particular department, which is never possessed by a person who is obliged to execute many different processes. This rapidity is still further increased from the circumstance that most of the operations in factories, where the division of labour is carried to a considerable extent, are paid for as piece-work. It is difficult to estimate in numbers the effect of this cause upon production. In nail-making, Adam Smith has stated, that it is almost three to one; for, he observes, that a smith accustomed to make nails, but whose whole business has not been that of a nailer, can make only from eight hundred to a thousand per day; whilst a lad who had never exercised any other trade, can make upwards of two thousand three hundred a day.

In different trades, the economy of production arising from the last-mentioned cause will necessarily be different. The case of nail-making is, perhaps, rather an extreme one. It must, however, be observed, that, in one sense, this is not a permanent source of advantage; for, though it acts at the commencement of an establishment, yet every month adds to the skill of the workmen; and at the end of three or four years they will not be very far behind those who have never practised any other branch of their art. Upon an occasion when a large issue of bank-notes was required, a clerk at the Bank of England signed his name, consisting of seven letters, including the initial of his Christian name, five thousand three hundred times during eleven working hours, besides arranging the notes he had signed in parcels of fifty each.

The division of labour suggests the contrivance of tools and machinery to execute its processes. When each process, by which any article is produced, is the sole occupation of one individual, his whole attention being devoted to a very limited and simple operation, improvements in the form of his tools, or in the mode of using them, are much more likely to occur to his mind, than if it were distracted by a greater variety of circumstances. Such an improvement in the tool is generally the first step towards a machine. If a piece of metal is to be cut in a lathe, for example, there is one particular angle at which the cutting-tool must be held to insure the cleanest cut; and it is quite natural that the idea of fixing the tool at that angle should present itself to an intelligent workman. The necessity of moving the tool slowly, and in a direction parallel to itself, would suggest the use of a screw, and thus arises the sliding-rest. It was probably the idea of mounting a chisel in a frame, to prevent its cutting too deeply, which gave rise to the common carpenter's plane. In cases where a blow from a hammer is employed, experience teaches the proper force required. The transition from the hammer held in the hand to one mounted upon an axis, and lifted regularly to a certain height by some mechanical contrivance, requires perhaps a greater degree of invention than those just instanced; yet it is not difficult to perceive, that, if the hammer always falls from the same height, its effect must be always the same.

When each process has been reduced to the use of some simple tool, the union of all these tools, actuated by one moving power, constitutes a machine. In contriving tools and simplifying processes, the operative workmen are, perhaps, most successful; but it requires far other habits to combine into one machine these scattered arts. A previous education as a workman in the peculiar trade, is undoubtedly a valuable preliminary; but in order to make such combinations with any reasonable expectation of success, an extensive knowledge of machinery, and the power of making mechanical drawings, are essentially requisite. These accomplishments are now much more common than they were formerly, and their absence was, perhaps, one of the causes of the multitude of failures in the early history of many of our manufactures.

Such are the principles usually assigned as the causes of the advantage resulting from the division of labour. As in the view I have taken of the question, the most important and influential cause has been altogether unnoticed, I shall restate those principles in the words of Adam Smith:

"The great increase in the quantity of work, which, in consequence of the division of labour, the same number of people are capable of performing, is owing to three different circumstances: first, to the increase of dexterity in every particular workman; secondly, to the saving of time, which is commonly lost in passing from one species of work to another; and, lastly, to the invention of a great number of machines which facilitate and abridge labour, and enable one man to do the work of many."

Now, although all these are important causes, and each has its influence on the result; yet it appears to me, that any explanation of the cheapness of manufactured articles, as consequent upon the division of labour, would be incomplete if the following principle were omitted to be stated: that the master manufacturer, by dividing the work to be executed into different processes, each requiring different degrees of skill or of force, can purchase exactly that precise quantity of both which is necessary for each process; whereas, if the whole work were executed by one workman, that person must possess sufficient skill to perform the most difficult, and sufficient strength to execute the most laborious, of the operations into which the art is divided.

NOTES

1. Maxine Berg, *The Machinery Question and the Making of Political Economy* (Cambridge: Cambridge University Press, 1980), p. 186.
2. Quoted in Anthony Hyman, *Charles Babbage: Pioneer of the Computer* (Oxford: Oxford University Press, 1982), p. 1.
3. Hyman, 124–5.

FURTHER READING

Maxine Berg, *The Machinery Question and the Making of Political Economy* (Cambridge: Cambridge University Press, 1980).

Anthony Hyman, *Charles Babbage: Pioneer of the Computer* (Oxford: Oxford University Press, 1982).

Sadie Plant, Zeroes + Ones: Digital Women and the New Counterculture (New York: Doubleday, 1997).

7

ARTISANS AND MACHINERY

Peter Gaskell

Long before the Arts and Crafts Movement, alarm bells began to sound about the effects of industrialization on the English working population. Most commentators on the 'machinery question' held fast to vigorous optimism, and voices of opposition, though in the minority, were hardly unified. The artisanal politics of the moment were complex. While traditional guilds had largely given way to labor collectives of various kinds, these were divided by geography, trade, and skill level. There was a world of difference between skilled urban craftsmen, workers in the textile mills of the North, and the agitators of the agricultural Swing Riot of 1830 (touched off by the introduction of mechanized threshing machines). Ideas about mechanization, even among the working class, were equally diverse. On the whole, both popular and erudite opposition to industrialization focused not on the erosion of skill, but on the more fundamental question of unemployment. Only a few writers emphasized the cultural and psychological harm wrought by factory labor. One was Peter Gaskell, whose 1836 Artisans and Machinery *was a revision and expansion of* The Manufacturing Population of England, *published three years earlier. Nominally intended as a medical study of factory workers in Lancashire, the book cited a bewildering variety of telltale signs of degradation, from diet and dress to speech and body shape. Gaskell thus anticipated the sense of broad cultural crisis*

more often associated with the writings of Karl Marx and Friedrich Engels. There is no call here for political revolution—nothing analogous to Engels's confident claim that workers are radicalized 'just in proportion as their handicraft has been invaded by the progress of machinery'—but there is no doubt that concerns about the 'dark satanic mills', and hence the energies that would eventually coalesce in socialism, were already well established.[1]

Peter Gaskell, *Artisans and Machinery: The Moral and Physical Condition of the Manufacturing Population Considered with Reference to Mechanical Substitutes for Human Labour* (London: J. W. Parker, 1836), excerpts.

THE EARLY MASTER MANUFACTURERS

It may be laid down as a maxim, that whenever numerous bodies of men—whatsoever their rank, and whatsoever the cause which has led to their congregation—are brought together, a deterioration, more or less marked, in the moral condition of some portions of the community, is the inevitable result. Large cities and populous districts have, in all ages, been the foci from whence have emanated, if not great, at least numerous crimes.

One principal effect of the steam engine has been, to crowd workmen together, collecting them from parts in which they had hitherto formed portions of a scattered population.

Example appears to be one of the most powerful agents in the production of the common actions of life. The various grades of society, from the most elevated to the most debased, are led equally away by it.

Many of the first successful manufacturers, both in town and country, were men who had their origin in the rank of mere operatives, or who sprung from the extinct class of yeomen. It has been already explained that this class had been driven, by the pressure of circumstances, to the adoption of spinning, at the period when trade was undergoing that series of changes which ended in the introduction of steam.

The celerity with which some of these individuals accumulated wealth in the early times of machine spinning and weaving, are proofs—if any such were wanting—that they were men of quick views, and great energy of character, possessing no small share of sagacity, and by these means were able to avail themselves to the utmost of the golden advantages which were presented to their grasp, at a time when they supplied the whole universe with the products of manufacture.

But they were men of very limited general information—men who saw and knew little of any thing beyond the demand for their twist or cloth, and the speediest and best modes for their production. They were, however, from their acquired station, men who exercised very considerable influence upon the multitudes of workmen who became dependent upon them.

The acquisition of wealth, unfortunately for the interests of all parties, was not, in the first instance, attended by a corresponding improvement in their moral and social character; on the contrary, all who had an opportunity of watching its effects, can only deplore and condemn the evil purposes to which, for many years, some portions of it were applied.

The extreme rapidity with which the returns were made for a considerable period—and this too with an immense profit—might well dazzle them. The animal enjoyments, the sensual indulgences, which were witnessed at the orgies of these parties, totally unchecked by any intercourse with polished society, should have had the veil of oblivion drawn over them, were it not that, to some degree, they tend to explain the depravity which in a few years spread, like a moral plague, over the factory artisans.

The sprinkling of men of more refined habits amongst the early successful cotton manufacturers, was extremely scanty. Very few who brought large capital into the trade, were fortunate, or even made satisfactory progress. Neither will this fact be considered singular, when it is remembered with whom the battle had to be fought. They had to oppose men who had a practical acquaintance with machinery, and who laboured themselves, assiduously and diligently; whereas the previous pursuits and education of the capitalist, had unfitted him, in some respects, for that rapidity of action and quickness of calculation, which were essentially necessary, if he must keep pace with the daily improvements projected and carried on around him.

Master manufacturers then, at the commencement of this important epoch, were in many instances men sprung from the ranks of the labourers, or from a grade just removed above these—uneducated, of coarse habits, sensual in their enjoyments, partaking of the rude revelry of their dependants—overwhelmed by success, but yet, paradoxical as it may sound, industrious men, and active and far-sighted tradesmen.

Wealth brought with it some of its usual accessories. Cottages were exchanged for mansions erected purposely for them, larger, more commodious, and furnished in a style of shew and expense, if not of taste, sufficiently indicative of the state of the owner's purse and prospects; and to these were transferred the manners which had unhappily disgraced their late more humble residences.

Destitute of every thing intellectual, and condemning every thing savouring of refinement, whether in manner or thought, they were in some measure driven to the indulgence of their animal sensations. This was generally sought for in the use of ardent spirits, which roused them for a time into furious excitement, and rendered them unconscious of all that was due to decency or propriety. Thus wallowing in intemperance, little wonder can be excited that other passions were stimulated into active operation; and from their situation, unbounded facilities were offered for their display.

The almost entire extinction of sexual decency, which has been one of the darkest stains upon the character of the mill artisans, the laxity in all the moral obligations which ought to exist between the sexes, and the consequent loss of this most important influence in the formation of social manners, may be traced, to some extent, to this period of their history.

Condemning, as every man must, the conduct of these parties, it may be remarked, that the mischief lay in no small degree with the particular juncture in which they were brought so conspicuously forward. Their want of education—the animal life they had previously led—the sudden accession of wealth—the contempt in some cases generated for refinement, by the discovery they soon made, that wealth, although burdened with blunt and coarse manners, was still an all-powerful agent for procuring worldly respect—the vanity which leads men to ascribe results to causes personal to themselves. Yes, keeping up their original vulgarity, in which they took a strange pride—the facilities for lascivious indulgence afforded them by the number of females brought under their immediate control—the herding together of workmen, the result of the factory system, more especially multitudes of boys and girls from ten to sixteen years of age, freed from domestic discipline—the separation of man and wife during the hours of labour—the dependence which naturally grew up on the part of the labourers—all these are matters which will serve to explain at least, the immorality which marked the bearing of many, though by no means the whole of the early mechanical manufacturers.

[. . .]

THE INFLUENCE OF MACHINERY ON THE VALUE OF HUMAN LABOUR— SUBSTITUTION OF AUTOMATA FOR HUMAN AGENTS—ITS EXTENT AND ULTIMATE CONSEQUENCES

It has been remarked by Mr. Babbage, that "if the competition between machinery and human labour is perceived to be perfectly hopeless, the workman will at once set himself to learn a new department of his art." Were this possible, the necessary consequences of mechanical improvement would signify nothing; but it is impossible, and a reference to his own table on hand-loom weavers will sufficiently show that there are at present insurmountable difficulties in the way of the conversion of a great body of operatives from one industrial condition to another.

Whoever is in the habit of visiting the workshops of the machine-makers, and the mills of the great cotton manufacturers, from time to time, cannot fail to be struck with the incessant improvements in the application of machinery. These improvements, though they may not enable the master to dismiss any of his hands, prevent the necessity for engaging fresh ones, though he doubles the productive powers of his mill.

The rapid growth of the staple branch of manufacture—the cotton trade, has caused vast immigrations into those districts in which it is principally carried on. The depression in the agricultural counties has pushed these immigrations beyond the demand; the repeated turnouts have brought sudden accessions of new hands in great numbers—thousands of Irish have deserted their native and miserable homes, in search of employment at the loom; these circumstances, one and all, have brought into the trade a surplus quantity of men, and that at a period when the necessity for them is daily lessening . . .

Whenever the pressure of foreign or domestic competition becomes more severe, the masters will be necessitated to avail themselves to the utmost of every thing which can assist in lowering the price of their products, and human labour must and will be pushed to the wall. Many great changes will, of course, take place before this inevitable result is gained, and reductions in wages for quantity will be constantly progressing; but the ultimatum is less remote than those interested in it are aware of; for let it be remembered that all mechanical applications, and the moving power derived from the expansive nature of steam, have as yet but arrived at one point in their career, and this point says nothing as to what may be done. There can be no question whatever that many processes, for which the human hand is at present indispensable, will very shortly have machines adapted to them; these, if they will not quite displace the workman, will render one man capable of producing, or rather of superintending, the production of quantity now requiring ten or twenty labourers. This is no theoretical opinion—the whole history of the cotton manufacture attests its truth, and collateral proofs are abundant in other branches of manufacture.

It does not follow that improvements in existing machinery, or every new machine, should at once throw out a number of hands. Those, however, who argue that machinery never has that effect, and never will have it, either wilfully delude themselves, or take a very limited and imperfect view of the subject. It must have one of two effects, the objects of every change, improvement, and addition, being to lessen the amount of labour required for production; these effects must be either to render fewer workmen necessary to produce a given quantity of manufacture, or so far lower the price of the manufactured article as at once to increase the demand for it so considerably, as to absorb the same number of men as are already engaged in it. In many instances, in fact generally, the latter has been the case hitherto, and would, perhaps, continue to be so, were Great Britain entirely to monopolize manufactures. But this cannot be; and, as it has been before stated, the maximum must be attained. All these improvements having therefore one end, all tending to the same point, namely, the cheapening of labour, the time must come when its value will be so small as to make it nearly worthless to the possessor.

[. . .]

The effects of mechanical production, as far as we have traced them, are, in the first place, to lower the value of human labour, and, in the next, to destroy it altogether, except in so far as the hands engaged in machine making are concerned: and even these are being encroached upon—machines making machines. The intermediate step between the two just mentioned, is its effects upon the higher qualities of the operative, namely, his skill, emulative pride, and respect for his own position.

It is singular to observe how widely apart are the opinions of those who contend that every mechanical improvement must of necessity benefit the workman; and in nothing more is this discrepancy visible than on this point. In a paper on the cotton manufacture, in the *Edinburgh Review*, No. 91, written by Mr. [John Ramsay] McCulloch, the following words occur. To the truth of some of these we have borne ample testimony; from others we entirely dissent.

"Our master manufacturers, engineers and artisans, are more intelligent, skilful, and enterprising, than those of other countries, and the extraordinary inventions they have already made, and their familiarity with all the principles and details of the business, will not only enable them to perfect the processes already in use, but can hardly fail to lead to the discovery of others. Our establishments for spinning, weaving, printing, bleaching, &c., are infinitely more complete and perfect than any that exist elsewhere, *the division of labour in them is carried to an incomparably greater extent, the workmen are trained from infancy to industrious habits,* and have attained that peculiar dexterity and sleight of hand in the performance of their separate tasks, that can only be acquired by long and unremitting application to the same employment."

This is the language of Mr. McCulloch, a leading authority in that particular school of political economy to which he belongs.

Its variance with the fact is extraordinary; and here Dr. [Andrew] Ure, in his "Philosophy of Manufactures," being a skilful *mechanician,* sets right the abstractions of the theorist.

"It is in fact the constant aim and tendency of every improvement in machinery to supersede human labour altogether, or to diminish its cost, by substituting the industry of women and children for that of men, *or that of ordinary labourers for trained artisans.*" —p. 23.

"This tendency to employ merely children with watchful eyes and nimble fingers, *instead of journeymen of long experience,* shows how the scholastic dogma of *the division of labour* into degrees of skill has been exploded by our enlightened manufacturers." —Ibid.

"Improvements in machinery effect a substitution of labour comparatively unskilled, for that which is more skilful." —p. 30.

"The principle of the factory system is, to substitute mechanical science for hand-skill, and the partition of a process into its essential constituents, for the division or gradation of labour among artisans. On the handicraft plan, labour, more or less skilled, was usually the most expensive element of production; but on the automatic plan, *skilled* labour gets progressively superseded, and will eventually be *replaced by mere overlookers of machines.*" —p. 20.

"Mr. Anthony Strutt, who conducts the mechanical department of the great cotton factories of Helper and Milford, has so thoroughly departed from the old routine of the schools, *that he wilt employ no man who has learned his craft by regular apprenticeship.*" —p. 21.

"An eminent mechanician of Manchester told me, that he does not choose to make any steam-engines at present, because, with his existing means, he would be obliged to resort to the old principle of the division of labour, so fruitful of jealousies and strikes among workmen; but he intends to prosecute that branch of business whenever he has prepared suitable arrangements on the equalisation of labour, or automatic plan." —p. 21.

We might multiply these extracts; but they are sufficient to show what is the truth. It is, as Dr. Ure justly remarks, the great aim of machinery to make skill or strength on the part of the workman valueless, and to reduce him to a mere watcher of, and waiter upon, *automata*. The term artisan will shortly be a misnomer as applied to the operative; he will no longer be a man proud of his skill and ingenuity, and conscious that he is a valuable member of society; he will have lost all free agency, and will be as much a part of the machines around him as the wheels or cranks which communicate motion.

NOTE

1. Friedrich Engels, *The Condition of the Working Class in England* (Leipzig, 1845). 'Dark satanic mills', coined in an 1804 poem by William Blake, is often used as a shorthand for the inhuman conditions of early industry.

FURTHER READING

William J. Ashworth, 'England and the Machinery of Reason', in Iwan Rhys Morris, *Bodies/Machines* (Oxford: Berg, 2002).

Michael Burawoy, 'Karl Marx and the Satanic Mills: Factory Politics Under Early Capitalism in England, the United States, and Russia', *American Journal of Sociology* 90/2 (Sept. 1984), pp. 247–82.

Robert Gray, *The Factory Question and Industrial England, 1830–1860* (Cambridge: Cambridge University Press, 1996).

Robert Southey, *Colloquies on the Progress and Prospects of Society* (1829).

E. P. Thompson, *The Making of the English Working Class* (London: Victor Gollanz, 1963).

8

'HOW AN ARISTOCRACY MAY EMERGE FROM INDUSTRY', FROM *DEMOCRACY IN AMERICA*

Alexis de Tocqueville

One of the abiding myths about nineteenth-century America is that it was a new kind of nation, where mobility, individualism, and the pursuit of the almighty dollar broke fundamentally from European social order. Historians continue to debate the extent to which this myth was based on fact. Those who dissent must deal, first and foremost, with the observations of Alexis de Tocqueville, the French political theorist. Tocqueville's Democracy in America, published in two volumes in 1835 and 1840, was based on a tour he had made in 1831–2, originally with the intention of studying prison reform. A combination of anecdotal detail and perceptive theorization made the book (especially the first volume) one of the runaway publishing successes of the era. For a French public conditioned to think in terms of class revolution, the picture Tocqueville painted was doubtless a fascinating one. 'Men in America, as with us, are arranged according to certain categories in the course of social life,' he had noted in his travel journal. 'Common habits, education, and above all wealth establish these classifications. But these rules are neither absolute, not inflexible, nor permanent . . . even though two individuals never meet in the same salons, if they meet on the public square, one looks at the other without pride, and in return is regarded without envy. At bottom they feel themselves equal, and are.'¹ Yet Tocqueville was by no means inclined to see America as a paradise of equality. In the following passage, Tocqueville muses on the possibility that a new hierarchy might be generated within America's capitalist manufacturing system. His interest is in the politics of skilled labor. 'Dexterity' may well result from specialization, he wrote, but only at the cost of dependency: 'The art advances, the artisan recedes.'

Alexis de Tocqueville, 'How an Aristocracy May Be Created by Manufactures', in *Democracy in America*, vol. 2 (1840).

I have shown how democracy favors the growth of manufactures and increases without limit the numbers of the manufacturing classes; we shall now see by what side-road manufacturers may possibly, in their turn, bring men back to aristocracy.

It is acknowledged that when a workman is engaged every day upon the same details, the whole commodity is produced with greater ease, speed, and economy. It is likewise acknowledged that the cost of production of manufactured goods is diminished by the extent of the establishment in which they are made and by the amount of capital employed or of credit. These truths had long been imperfectly discerned, but in our time they have been demonstrated. They have been already applied to many very important kinds of manufactures, and the humblest will gradually be governed by them. I know of nothing in politics that deserves to fix the attention of the legislator more closely than these two new axioms of the science of manufactures.

When a workman is unceasingly and exclusively engaged in the fabrication of one thing, he ultimately does his work with singular dexterity; but at the same time he loses the general faculty of applying his mind to the direction of the work. He every day becomes more adroit and less industrious; so that it may be said of him that in proportion as the workman improves, the man is degraded. What can be expected of a man who has spent twenty years of his life in making heads for pins? And to what can that mighty human intelligence which has so often stirred the world be applied in him except it be to investigate the best method of making pins' heads? When a workman has spent a considerable portion of his existence in this manner, his thoughts are forever set upon the object of his daily toil; his body has contracted certain fixed habits, which it can never shake off; in a word, he no longer belongs to himself, but to the calling that he has chosen.

It is in vain that laws and manners have been at pains to level all the barriers round such a man and to open to him on every side a thousand different paths to fortune; a theory of manufactures more powerful than customs and laws binds him to a craft, and frequently to a spot, which he cannot leave; it assigns to him a certain place in society, beyond which he cannot go; in the midst of universal movement it has rendered him stationary.

In proportion as the principle of the division of labor is more extensively applied, the workman becomes more weak, more narrow-minded, and more dependent. The art advances, the artisan recedes. On the other hand, in proportion as it becomes more manifest that the productions of manufactures are by so much the cheaper and better as the manufacture is larger and the amount of capital employed more considerable, wealthy and educated men come forward to embark in manufactures, which were heretofore abandoned to poor or ignorant handicraftsmen. The magnitude of the efforts required and the importance of the results to be obtained attract them. Thus at the very time at which the science of manufactures lowers the class of workmen, it raises the class of masters.

While the workman concentrates his faculties more and more upon the study of a single detail, the master surveys an extensive whole, and the mind of the latter is enlarged in proportion as that of the former is narrowed. In a short time the one will require nothing but physical strength without intelligence; the other stands in need of science, and almost of genius, to ensure success. This man resembles more and more the administrator of a vast empire; that man, a brute.

The master and the workman have then here no similarity, and their differences increase every day. They are connected only like the two rings at the extremities of a long chain. Each of them fills the station which is made for him, and which he does not leave; the one is continually, closely, and necessarily dependent upon the other and seems as much born to obey as that other is to command. What is this but aristocracy?

As the conditions of men constituting the nation become more and more equal, the demand for manufactured commodities becomes more general and extensive, and the cheapness that places these objects within the reach of slender fortunes becomes a great element of success. Hence there are every day more men of great opulence and education who devote their wealth and knowledge to manufactures and who seek, by opening large establishments and by a strict division of labor, to meet the fresh demands which are made on all sides. Thus, in proportion as the mass of the nation turns to democracy, that particular class which is engaged in manufactures becomes more aristocratic. Men grow more alike in the one, more different in the other; and inequality increases in the less numerous class in the same ratio in which it decreases in the community. Hence it would appear, on searching to the bottom, that aristocracy should naturally spring out of the bosom of democracy.

But this kind of democracy by no means resembles those kinds which preceded it. It will be observed at once that, as it applies exclusively to manufactures and to some manufacturing callings, it is a monstrous exception in the general aspect of society. The small aristocratic societies that are formed by some manufacturers in the midst of the immense democracy of our age contain, like the great aristocratic societies of former ages, some men who are very opulent and a multitude who are wretchedly poor. The poor have few means of escaping from their condition and becoming rich, but the rich are constantly becoming poor, or they give up business when they have realized a fortune. Thus the elements of which the class of poor is composed are fixed, but the elements of which the class of the rich is composed are not so. To tell the truth, though there are rich men, the class of rich men does not exist; for these rich individuals have no feelings or purposes, no traditions or hopes, in common; there are individuals, therefore, but no definite class.

Not only are the rich not compactly united among themselves, but there is no real bond between them and the poor. Their relative position is not a permanent one; they are constantly drawn together or separated by their interests. The workman is generally dependent on the master, but not on any particular master; these two men meet in the factory, but do not know each other elsewhere; and while they come into contact on one point, they stand very far apart on all others. The manufacturer asks nothing of the workman but his labor; the workman expects nothing from him but his wages. The one contracts no obligation to protect nor the other to defend, and they are not permanently connected either by habit or by duty. The aristocracy created by business rarely settles in the midst of the manufacturing population which it directs; the object is not to govern that population, but to use it. An aristocracy thus constituted can have no great hold upon those whom it employs, and even if it succeeds in retaining them at one moment, they escape the next; it knows not how to will, and it cannot act.

The territorial aristocracy of former ages was either bound by law, or thought itself bound by usage, to come to the relief of its serving-men and to relieve their distress. But the manufacturing aristocracy of our age first impoverishes and debases the men who serve it and then abandons them to be supported by the charity of the public. This is a natural consequence of what has been said before. Between the workman and the master there are frequent relations, but no real association.

I am of the opinion, on the whole, that the manufacturing aristocracy which is growing up under our eyes is one of the harshest that ever existed in the world; but at the same time it is one of the most confined and least dangerous. Nevertheless, the friends of democracy should keep their eyes anxiously fixed in this direction; for if ever a permanent inequality of conditions and aristocracy again penetrates into the world, it may be predicted that this is the gate by which they will enter.

NOTE

1. Quoted in George Wilson Pierson, *Tocqueville and Beaumont in America* (New York: Oxford University Press, 1938), p. 551.

FURTHER READING

Jean-Claude Lamberti, *Tocqueville and the Two Democracies* (Cambridge, MA: Harvard University Press, 1989; orig. pub. in French 1983).

Leo Marx, *The Machine in the Garden* (Oxford: Oxford University Press, 1964).

Walter Nugent, 'Tocqueville, Marx, and American Class Structure', *Social Science History* 12/4 (Winter, 1988), pp. 327–47.

Cheryl B. Welch, *The Cambridge Companion to Tocqueville* (Cambridge: Cambridge University Press, 2006).

9

INDUSTRIAL BIOGRAPHY:
IRON WORKERS AND TOOL MAKERS

Samuel Smiles

Best known as the author of the wildly success-ful book Self-Help, *Samuel Smiles seems in retrospect to have been the most optimistic of Victorians—against some pretty tough compe-tition. His comparatively little known volume* Industrial Biography *(a sequel of sorts to Smiles's earlier* Lives of the Engineers*) delivers hagio-graphic, summary accounts of the lives of great machine tool innovators such as Joseph Bramah, Henry Maudslay, and William Fairbairn. Such men were in the forefront of the industrial revolu-tion in that they developed the capital goods that lay at the heart of factory production: milling machines, replicating lathes, power looms, steam hammers and 'engines' of many other kinds. On the other hand, they and their direct associates were among the great artisans of the nineteenth century. The development, refinement and ongo-ing repair of such tools called for a combination of flexible skills and precise execution. As Smiles suggests in this excerpt, particularly in his descrip-tion of the trials and tribulations that James Watt and Matthew Boulton experienced while developing their famous steam engine, such craft aptitude was highly prized but hard to come by.*

Samuel Smiles, *Industrial Biography: Iron Workers and Tool Makers* (London: John Murray, 1863), excerpts.

"The true Epic of our time is, not *Arms and the Man*, but *Tools and the Man*—an infinitely wider kind of Epic."—Thomas Carlyle

While commemorating the labours and honouring the names of those who have striven to elevate man above the material and the mechanical, the labours of the important industrial class to whom society owes so much of its comfort and well-being are also entitled to consideration. Without derogating from the biographic claims of those who minister to intellect and taste, those who minister to util-ity need not be overlooked. When a French-man was praising Sir John Sinclair the artist who invented ruffles, the Baronet shrewdly remarked that some merit was also due to the man who added the shirt.

A distinguished living mechanic thus ex-presses himself to the Author on this point: "Kings, warriors and statesmen have heretofore monopolied not only the pages of history, but almost those of biography. Surely some niche ought to be found for the Mechanic, without whose skill and labour society, as it is, could not exist. I do not begrudge destructive heroes their fame, but the constructive ones ought not to be forgotten; and there *is* a heroism of skill and toil belonging to the latter class, worthy of as grateful record, less perilous and romantic, it may be than that of the other, but not less full of the results of human energy, bravery, and character. The lot of labour is indeed often a dull one; and it is doing a public service to endeavour to lighten it up by records of the struggles and triumphs of our more illustrious workers, and the results of their labours in the cause of human advancement."

[. . .]

It is always difficult to apportion the due share of merit which belongs to mechanical inventors, who are accustomed to work upon each other's hints and suggestions, as well as by their own experience. Some idea of this difficulty may be formed from the fact that, in the course of our investigations as to the origin of the planing machine—one of the most useful of modern tools—we have found that it has been claimed on behalf of six inventors . . . "There is nothing," says Mr. Hawkshaw, "really worth having that man has obtained, that has not been the result of a combined and gradual process of investigation. A gifted individual comes across some old footmark, stumbles on a chain of previous research and inquiry. He meets, for instance, with a machine, the result of much previous labour; he modifies it, pulls it to pieces, constructs and reconstructs it, and by further trial and experiment he arrives at the long sought-for result."

But the making of the invention is not the sole difficulty. It is one thing to invent, said Sir Marc Brunel, and another thing to make the invention work. Thus when Watt, after long labour and study, had brought his invention to completion, he encountered an obstacle which had stood in the way of other inventors, and for a time prevented the introduction of their improvements, if not led to their being laid aside and abandoned. This was the circumstance that the machine projected was so much in advance of the mechanical capability of the age that it was with the greatest difficulty it could be executed. When labouring upon his invention at Glasgow, Watt was baffled and thrown into despair by the clumsiness and the incompetency of his workmen. Writing to Dr. Roebuck on one occasion, he said, "You ask what is the principal hindrance in erecting machines? It is always the smith-work."

His first cylinder was made by a whitesmith, of hammered iron soldered together, but having used quicksilver to keep the cylinder airtight, it dropped through the inequalities into the interior, and "played the devil with the solder." Yet, inefficient though the whitesmith was, Watt could ill spare him, and we find him writing to Dr. Roebuck almost in despair, saying, "My old white-iron man is dead!" feeling his loss to be almost irreparable. His next cylinder was cast and bored at Carron, but it was so untrue that it proved to be useless. The piston could not be kept steam tight, notwithstanding the various expedients which were adopted of stuffing it with paper, cork, putty, pasteboard, and old hat. Even after Watt had removed to Birmingham, and he had the assistance of Boulton's best workmen, Smeaton expressed the opinion, when he saw the engine at work, that notwithstanding the excellence of the invention, it could never be brought into general use because of the difficulty of getting its various parts manufactured with sufficient precision. For a long time we find Watt, in his letters, complaining to his partner of the failure of the engines through "villainous bad workmanship." Sometimes the cylinders, when cast, were found to be more than an eighth of an inch wider at one end than the other; and under such circumstances it was impossible the engine could act with precision. Yet better work could not be had. First-rate workmen in machinery did not as yet exist; they were only in process of education. Nearly everything had to be done by hand. The tools used were of a very imperfect kind. A few ill-constructed lathes, with some drills and boring-machines of a rude sort, constituted the principal furniture of the workshop. Years later, when Brunel invented his block-machines, considerable time elapsed before he could find competent mechanics to construct them, and even after they had been constructed he had equal difficulty in finding competent hands to work them.

Watt endeavoured to remedy the defect by keeping certain sets of workmen to special classes of work, allowing them to do nothing else. Fathers were induced to bring up their sons at the same bench with themselves, and initiate them in the dexterity which they had acquired by experience; and at Soho it was not unusual for the same precise line of work to be followed by members of the same family in three generations. In this way as great a degree of accuracy of a mechanical kind was arrived at [as] was practicable under the circumstances. But notwithstanding all this care, accuracy of fitting could not be secured so long as the manufacture of steam-engines was conducted mainly by hand. There was usually a considerable waste of steam, which the expedients of chewed paper and greased hat packed outside the piston were insufficient to remedy; and it was not until the invention of automatic machine-tools by the mechanical engineers about to be mentioned, that the manufacture of the steam-engine became a matter of comparative ease and certainty. Watt was compelled to rest satisfied with imperfect results, arising from imperfect workmanship. Thus, writing to Dr. Small respecting a cylinder 18 inches in diameter, he said, "at the worst place the long diameter exceeded the short by only three-eighths of an inch." How different from the state of things at this day, when a cylinder five feet wide will be rejected as a piece of imperfect workmanship if it is found to vary in any part more than the 80th part of an inch in diameter!

Not fifty years since it was a matter of the utmost difficulty to set an engine to work, and sometimes of equal difficulty to keep it going. Though fitted by competent workmen, it often would not go at all. Then the foreman of the factory at which it was made was sent for, and he would almost live beside the engine for a month or more; and after easing her here and screwing her up there, putting in a new part and altering an old one, packing the piston and tightening the valves, the machine would at length be got to work.[1] Now the case is altogether different. The perfection of modern machine-tools is such that the utmost possible precision is secured, and the mechanical engineer can calculate on a degree of exactitude that does not admit of a deviation beyond the thousandth part of an inch. When the powerful oscillating engines of the *Warrior* were put on board that ship, the parts, consisting of some five thousand separate pieces, were brought from the different workshops of the Messrs. Penn and Sons, where they had been made by the workmen who knew not the places they were to occupy, and fitted together with such prevision that so soon as the steam was raised and let into the cylinders, the immense machine began as if to breathe and move like a living creature, stretching its huge arms like a new-born giant, and then, after practicing its strength a little and proving its soundness in body and limb, it started off with the power of above a thousand horses to try its strength in breasting the billows of the north sea.

Such are among the triumphs of modern mechanical engineering, due in a great measure to the perfection of the tools by means of which all works in metal are now fashioned. These tools are themselves among the most striking results of the mechanical invention of the day. They are automata of the most perfect kind, rendering the engine and machine-maker in a great measure independent of inferior workmen. For the machine tools have no unsteady hand, are not careless or clumsy, do not work by rule of thumb, and cannot make mistakes. They will repeat their operations a thousand times without tiring, or varying one hair's breadth in their action; and will turn out, without complaining, any quantity of work, all of a like accuracy and finish. Exercising as they do so remarkable an influence on the development of modern industry, we now propose so far as the materials at our disposal will admit, to give an account of their principal inventors.

NOTE

1. There was the same kind of clumsiness in all kinds of mill-work before the introduction of machine-tools. We have heard of a piece of machinery of the old school, the wheels of which, when set to work, made such a clatter that the owner feared the engine would fall to pieces. The foreman who set it agoing, after working at it until he was almost in despair, at last gave it up, saying, "I think we had better leave the cogs to settle their differences with one another: they will grind themselves right in time!"

FURTHER READING

Adrian Jarvis, *Samuel Smiles and the Construction of Victorian Values* (Stroud: Sutton, 1997).

Samuel Smiles, *Self-Help, with Illustrations of Character and Conduct* (1859).

Timothy Travers, *Samuel Smiles and the Victorian Work Ethic* (New York: Garland, 1987).

10

CAPITAL

Karl Marx

'In handicrafts and manufacture, the workman makes use of a tool, in the factory, the machine makes use of him.' This sentence would seem to sum up the Marxist understanding of craft history. And indeed Karl Marx's enormous and enormously influential treatise on industry and commodities Capital *included a powerful indictment of the factory for dehumanizing the worker. In the following excerpts, Marx conforms closely to what we might expect, and his grim descriptions of the alienating effects of industrial labor still have power and relevance today. But we also find here a surprisingly subtle inquiry into the variable texture of work in the modern and premodern periods. Rather than envisioning the bygone age of handicraft in simple terms, Marx subjects the past to the same close analysis as the contemporary mass manufacture that is his real subject. He also accounts for the continuing relevance of hand skills in some branches of factory work and points out some counterintuitive facts about the machine (for example, that it often* reduces *the division of labor by performing multiple operations that used to be accomplished repetitively by many separate hands). It must be remembered that Marx was not in any sense 'against' the industrial revolution. Nor did he look back fondly on feudal peasant society, as William Morris did. Rather, he thought that the tragic upheavals of modernity were necessary to bring about the proletarian revolution to follow.*

Karl Marx, *Capital* (London, 1887; orig. pub. in German, 1867). Vol. 1, Chapter 15 ('Machinery and Modern Industry'), Sections 1, 4, excerpted.

THE DEVELOPMENT OF MACHINERY

John Stuart Mill says in his "Principles of Political Economy":

"It is questionable if all the mechanical inventions yet made have lightened the day's toil of any human being."[1]

That is, however, by no means the aim of the capitalistic application of machinery. Like every other increase in the productiveness of labour, machinery is intended to cheapen commodities, and, by shortening that portion of the working-day, in which the labourer works for himself, to lengthen the other portion that he gives, without an equivalent, to the capitalist. In short, it is a means for producing surplus-value.

In manufacture, the revolution in the mode of production begins with the labour-power, in modern industry it begins with the instruments of labour. Our first inquiry then is, how the instruments of labour are converted from tools into machines, or what is the difference between a machine and the implements of a handicraft? We are only concerned here with striking and general characteristics; for epochs in the history of society are no more separated from each other by hard and fast lines of demarcation, than are geological epochs.

Mathematicians and mechanicians, and in this they are followed by a few English economists, call a tool a simple machine, and a machine a complex tool. They see no essential difference between them, and even give the name of machine to the simple mechanical powers, the lever, the inclined plane, the screw, the wedge, &c. As a matter of fact, every machine is a combination of those simple powers, no matter how they may be disguised. From the economic standpoint this explanation is worth nothing, because the historical element is wanting. Another explanation of the difference between tool and machine is that in the case of a tool, man is the motive power, while the motive power of a machine is something different from man, as, for instance, an animal, water, wind, and so on. According to this, a plough drawn by oxen, which is a contrivance common to the most different epochs, would be a machine, while Claussen's circular loom, which, worked by a single labourer, weaves 96,000 picks per minute, would be a mere tool. Nay, this very loom, though a tool when worked by hand, would, if worked by steam, be a machine. And since the application of animal power is one of man's earliest inventions, production by machinery would have preceded production by handicrafts. When in 1735, John Wyatt brought out his spinning machine, and began the industrial revolution of the 18th century, not a word did he say about an ass driving it instead of a man, and yet this part fell to the ass. He described it as a machine "to spin without fingers."

All fully developed machinery consists of three essentially different parts, the motor mechanism, the transmitting mechanism, and finally the tool or working machine. The motor mechanism is that which puts the whole in motion. It either generates its own motive power, like the steam-engine, the caloric engine, the electromagnetic machine, &c., or it receives its impulse from some already existing natural force, like the water-wheel from a head of water, the wind-mill from wind, &c. The transmitting mechanism, composed of fly-wheels, shafting, toothed wheels, pulleys, straps, ropes, bands, pinions, and gearing of the most varied kinds, regulates the motion, changes its form, where necessary, as for instance, from linear to circular, and divides and distributes it among the working machines. These two first parts of the whole mechanism are there, solely for putting the working machines in motion, by means of which motion the subject of labour is seized upon and modified as desired. The tool or working machine is that part of the machinery with which the industrial revolution of the 18th century started. And to this day it constantly serves as such a starting-point, whenever a handicraft, or a manufacture, is turned into an industry carried on by machinery.

On a closer examination of the working machine proper, we find in it, as a general rule, though often, no doubt, under very altered forms, the apparatus and tools used by the handicraftsman or manufacturing workman; with this difference, that instead of being human implements, they are the implements of a mechanism, or mechanical implements. Either the entire machine is only a more or less altered mechanical edition of the old handicraft tool, as, for instance, the power-loom,[2] or the working parts fitted in the frame of the machine are old acquaintances, as spindles are in a mule, needles in a stocking-loom, saws in a sawing-machine, and knives in a chopping machine. The distinction between these tools and the body proper of the machine, exists from their very birth; for they continue for the most part to be produced by handicraft, or by manufacture, and are afterwards fitted into the body of the machine, which is the product of machinery.[3] The machine proper is therefore a mechanism that, after being set in motion, performs with its tools the same operations that were formerly done by the workman with similar tools. Whether the motive power is derived from man, or from some other machine, makes no difference in this respect. From the moment that the tool proper is taken from man, and fitted into a mechanism, a machine takes the place of a mere implement. The difference strikes one at once, even in those cases where man himself continues to be the prime mover. The number of implements that he himself can use simultaneously, is limited by the number of his own natural instruments of production, by the number of his bodily organs. In Germany, they tried at first to make one spinner work two spinning-wheels, that is, to work simultaneously with both hands and both feet. This was too difficult. Later, a treadle spinning-wheel with two spindles was invented, but adepts in spinning, who could spin two threads at once, were almost as scarce as two-headed men. The Jenny, on the other hand, even at its very birth, spun with 12–18 spindles, and the stocking-loom knits with many thousand needles at once. The number of tools that a machine can bring into play simultaneously, is from the very first emancipated from the organic limits that hedge in the tools of a handicraftsman.

In many manual implements the distinction between man as mere motive power, and man as the workman or operator properly so called, is brought into striking contrast. For instance, the foot is merely the prime mover of the spinning-wheel, while the hand, working with the spindle, and drawing and twisting, performs the real operation of spinning. It is this last part of the handicraftsman's implement that is first seized upon by the industrial revolution, leaving to the workman, in addition to his new labour of watching the machine with his eyes and correcting its mistakes with his hands, the merely mechanical part of being the moving power.

[. . .]

As soon as tools had been converted from being manual implements of man into implements of a mechanical apparatus, of a machine, the motive mechanism also acquired an independent form, entirely emancipated from the restraints of human strength. Thereupon the individual machine, that we have hitherto been considering, sinks into a mere factor in production by machinery. One motive mechanism was now able to drive many machines at once. The motive mechanism grows with the number of the machines that are turned simultaneously, and the transmitting mechanism becomes a wide-spreading apparatus.

We now proceed to distinguish the co-operation of a number of machines of one kind from a complex system of machinery.

In the one case, the product is entirely made by a single machine, which performs all the various operations previously done by one handicraftsman with his tool; as, for instance, by a weaver with his loom; or by several handicraftsman successively, either separately or as members of a system of Manufacture. For example, in the manufacture of envelopes, one man folded the paper with the folder, another laid on the gum, a third turned the flap over, on which the device is impressed, a fourth embossed the device, and so on; and for each of these operations the envelope had to change hands. One single envelope machine now performs all these operations at once, and makes more than 3,000 envelopes in an hour. In the London exhibition of 1862, there was an American machine for making paper cornets. It cut the paper, pasted, folded, and finished 300 in a minute. Here, the whole process, which, when carried on as Manufacture, was split up into, and carried out by, a series of operations, is completed by a single machine, working a combination of various tools. Now, whether such a machine be merely a reproduction of a complicated manual implement, or a combination of various simple implements specialised by Manufacture, in either case, in the factory, i.e., in the workshop in which machinery alone is used, we meet again with simple co-operation; and, leaving the workman out of consideration for the moment, this co-operation presents itself to us, in the first instance, as the conglomeration in one place of similar and simultaneously acting machines. Thus, a weaving factory is constituted of a number of power-looms, working side by side, and a sewing factory of a number of sewing-machines all in the same building. But there is here a technical oneness in the whole system, owing to all the machines receiving their impulse simultaneously, and in an equal degree, from the pulsations of the common prime mover, by the intermediary of the transmitting mechanism; and this mechanism, to a certain extent, is also common to them all, since only particular ramifications of it branch off to each machine. Just as a number of tools, then, form the organs of a machine, so a number of machines of one kind constitute the organs of the motive mechanism.

A real machinery system, however, does not take the place of these independent machines, until the subject of labour goes through a connected series of detail processes, that are carried out by a chain of machines of various kinds, the one supplementing the other. Here we have again the co-operation by division of labour that characterises Manufacture; only now, it is a combination of detail machines. The special tools of the various detail workmen, such as those of the beaters, cambers, spinners, &c., in the woollen manufacture, are now transformed into the tools of specialised machines, each machine constituting a special organ, with a special function, in the system. In those branches of industry in which the machinery system is first introduced, Manufacture itself furnishes, in a general way, the natural basis for the division, and consequent organisation, of the process of production. Nevertheless an essential difference at once manifests itself. In Manufacture it is the workmen who, with their manual implements, must, either singly or in groups, carry on each particular detail process. If, on the one hand, the workman becomes adapted to the process, on the other, the process was previously made suitable to the workman. This subjective principle of the division of labour no longer exists in production by machinery. Here, the process as a whole is examined objectively, in itself, that is to say, without regard to the question of its execution by human hands, it is analysed into its constituent phases; and the problem, how to execute each detail process, and bind them all into a whole, is solved by the aid of machines, chemistry, &c.[4] But, of course, in this case also, theory must be perfected by accumulated experience on a large scale. Each detail machine supplies raw material to the machine next in order; and since they are all working at the same time, the product is always going through the various stages of its fabrication, and is also constantly in a state of transition, from one phase

Figure 5 Cotton spinning mill, South Carolina, 1903.

to another. Just as in Manufacture, the direct co-operation of the detail labourers establishes a numerical proportion between the special groups, so in an organised system of machinery, where one detail machine is constantly kept employed by another, a fixed relation is established between their numbers, their size, and their speed. The collective machine, now an organised system of various kinds of single machines, and of groups of single machines, becomes more and more perfect, the more the process as a whole becomes a continuous one, i.e., the less the raw material is interrupted in its passage from its first phase to its last; in other words, the more its passage from one phase to another is effected, not by the hand of man, but by the machinery itself. In Manufacture the isolation of each detail process is a condition imposed by the nature of division of labour, but in the fully developed factory the continuity of those processes is, on the contrary, imperative.

[. . .]

THE FACTORY

At the commencement of this chapter we considered that which we may call the body of the factory, i.e., machinery organised into a system. We there saw how machinery, by annexing the labour of women and children, augments the number of human beings who form the material for capitalistic exploitation, how it confiscates the whole of the workman's disposable time, by immoderate extension of the hours of labour, and how finally its progress, which allows of enormous increase of production in shorter and shorter periods, serves as a means of systematically getting more work done in a shorter time, or of exploiting labour-power more intensely. We now turn to the factory as a whole, and that in its most perfect form.

Dr. Ure, the Pindar of the automatic factory, describes it, on the one hand, as:

"Combined co-operation of many orders of workpeople, adult and young, in tending with assiduous skill, a system of productive machines, continuously impelled by a central power" (the prime mover); on the other hand, as "a vast automaton, composed of various mechanical and intellectual organs, acting in uninterrupted concert for the production of a common object, all of them being subordinate to a self-regulated moving force."

These two descriptions are far from being identical. In one, the collective labourer, or social body of labour, appears as the dominant subject, and the mechanical automaton as the object; in the other, the automaton itself is the subject, and the workmen are merely conscious organs, co-ordinate with the unconscious organs of the automaton, and together with them, subordinated to the central moving-power. The first description is applicable to every possible employment of machinery on a large scale, the second is characteristic of its use by capital, and therefore of the modern factory system. Ure prefers therefore, to describe the central machine, from which the motion comes, not only as an automaton, but as an autocrat. "In these spacious halls the benignant power of steam summons around him his myriads of willing menials."[5]

Along with the tool, the skill of the workman in handling it passes over to the machine. The capabilities of the tool are emancipated from the restraints that are inseparable from human labour-power. Thereby the technical foundation on which is based the division of labour in Manufacture, is swept away. Hence, in the place of the hierarchy of specialised workmen that characterises manufacture, there steps, in the automatic factory, a tendency to equalise and reduce to one and the same level every kind of work that has to be done by the minders of the machines; in the place of the artificially produced differentiations of the detail workmen, step the natural differences of age and sex.

So far as division of labour re-appears in the factory, it is primarily a distribution of the workmen among the specialised machines; and of masses of workmen, not however organised into groups, among the various departments of the factory, in each of which they work at a number of similar machines placed together; their co-operation, therefore, is only simple. The organised group, peculiar to manufacture, is replaced by the connexion between the head workman and his few assistants. The essential division is, into workmen who are actually employed on the machines (among whom are included a few who look after the engine), and into mere attendants (almost exclusively children) of these workmen. Among the attendants are reckoned more or less all "Feeders" who supply the machines with the material to be worked. In addition to these two principal classes, there is a numerically unimportant class of persons, whose occupation it is to look after the whole of the machinery and repair it from time to time; such as engineers, mechanics, joiners, &c. This is a superior class of workmen, some of them scientifically educated, others brought up to a trade; it is distinct from the factory operative class, and merely aggregated to it. This division of labour is purely technical.

To work at a machine, the workman should be taught from childhood, in order that he may learn to adapt his own movements to the uniform and unceasing motion of an automaton. When the machinery, as a whole, forms a system of manifold machines, working simultaneously and in concert, the co-operation based upon it, requires the distribution of various groups of workmen among the different kinds of machines. But the employment of machinery does away with the necessity of crystallising this distribution after the manner of Manufacture, by the constant annexation of a particular man to a particular function.[6] Since the motion of the whole system does not proceed from the workman, but from the machinery, a change of persons can take place at any time without an interruption of the work . . .

Although then, technically speaking, the old system of division of labour is thrown overboard by machinery, it hangs on in the factory, as a traditional habit handed down from Manufacture, and is afterwards systematically re-moulded and established in a more hideous form by capital, as a means of exploiting labour-power. The life-long speciality of handling one and the same tool, now becomes the life-long speciality of serving one and the same machine. Machinery is put to a wrong use, with the object of transforming the workman, from his very childhood, into a part of a detail-machine. In this way, not only are the expenses of his reproduction considerably lessened, but at the same time his helpless dependence upon the factory as a whole, and therefore upon the capitalist, is rendered complete. Here as everywhere else, we must distinguish between the increased productiveness due to the development of the social process of production, and that due to the capitalist exploitation of that process. In handicrafts and manufacture, the workman makes use of a tool, in the factory, the machine makes use of him. There the movements of the instrument of labour proceed from him, here it is the movements of the machine that he must follow. In manufacture the workmen are parts of a living mechanism. In the factory we have a lifeless mechanism independent of the workman, who becomes its mere living appendage.

"The miserable routine of endless drudgery and toil in which the same mechanical process is gone through over and over again, is like the labour of Sisyphus. The burden of labour, like the rock, keeps ever falling back on the worn-out labourer."[7]

At the same time that factory work exhausts the nervous system to the uttermost, it does away with the many-sided play of the muscles, and confiscates every atom of freedom, both in bodily and intellectual activity. The lightening of the labour, even, becomes a sort of torture, since the machine does not free the labourer from work, but deprives the work of all interest. Every kind of capitalist production, in so far as it is not only a labour-process, but also a process of creating surplus-value, has this in common, that it is not the workman that employs the instruments of labour, but the instruments of labour that employ the workman. But it is only in the factory system that this inversion for the first time acquires technical and palpable reality. By means of its conversion into an automaton, the instrument of labour confronts the labourer, during the labour-process, in the shape of capital, of dead labour, that dominates, and pumps dry, living labour-power. The separation of the intellectual powers of production from the manual labour, and the conversion of those powers into the might of capital over labour, is, as we have already shown, finally completed by modern industry erected on the foundation of machinery. The special skill of each individual insignificant factory operative vanishes as an infinitesimal quantity before the science, the gigantic physical forces, and the mass of labour that are embodied in the factory mechanism and, together with that mechanism, constitute the power of the "master." This "master," therefore, in whose brain the machinery and his monopoly of it are inseparably united, whenever he falls out with his "hands," contemptuously tells them:

> "The factory operatives should keep in wholesome remembrance the fact that theirs is really a low species of skilled labour; and that there is none which is more easily acquired, or of its quality more amply remunerated, or which by a short training of the least expert can be more quickly, as well as abundantly, acquired . . . The master's machinery really plays a far more important part in the business of production than the labour and the skill of the operative, which six months' education can teach, and a common labourer can learn."[8]

The technical subordination of the workman to the uniform motion of the instruments of labour, and the peculiar composition of the body of workpeople, consisting as it does of individuals of both sexes and of all ages, give rise to a barrack discipline, which is elaborated into a complete system in the factory, and which fully develops the before mentioned labour of overlooking, thereby dividing the workpeople into operatives and overlookers, into private soldiers and sergeants of an industrial army.

NOTES

1. Mill should have said, "of any human being not fed by other people's labour," for, without doubt, machinery has greatly increased the number of well-to-do idlers.
2. Especially in the original form of the power-loom, we recognise, at the first glance, the ancient loom. In its modern form, the power-loom has undergone essential alterations.
3. It is only during the last 15 years (i.e., since about 1850), that a constantly increasing portion of these machine tools have been made in England by machinery, and that not by the same manufacturers who make the machines. [. . .]
4. "The principle of the factory system, then, is to substitute . . . the partition of a process into its essential constituents, for the division or graduation of labour among artisans." Andrew Ure, *The Philosophy of Manufactures* (London, 1835), p. 20.
5. Ure, p. 18.
6. Ure grants this. He says, "in case of need," the workmen can be moved at the will of the manager from one machine to another, and

he triumphantly exclaims: "Such a change is in flat contradiction with the old routine, that divides the labour, and to one workman assigns the task of fashioning the head of a needle, to another the sharpening of the point." He had much better have asked himself, why this "old routine" is departed from in the automatic factory, only "in case of need."

7. Friedrich Engels, *The Condition of the Working Class in England* (London, 1887; orig. pub. in German, 1845), p. 217.

8. "The Master Spinners' and Manufacturers' Defence Fund. Report of the Committee." Manchester, 1854, p. 17.

FURTHER READING

Marshall Berman, *All That Is Solid Melts Into Air: The Experience of Modernity* (New York: Penguin Books, 1982).

Friedrich Engels, *The Condition of the Working Class in England* (London, 1887; orig. pub. in German, 1845), p. 217.

Karl Marx and Friedrich Engels, *The Communist Manifesto* (London, 1850; orig. pub. in German 1848).

John Rule, "The Property of Skill in the Period of Manufacture," in Patrick Joyce, ed., *the Historical Meanings of Work* (Cambridge: Cambridge University Press, 1987).

11

'THE PRIMARY EFFECTS OF SCIENTIFIC MANAGEMENT', FROM *LABOR AND MONOPOLY CAPITALISM*

Harry Braverman

According to his wife, Harry Braverman 'once said he didn't think he could write another book like Labor and Monopoly Capital. *What he meant was that it represented a melding of his political and intellectual life with his experience as a factory worker'.[1] Before going on to a brilliant career as an editor (among his achievements was the publication of Alex Haley's* Autobiography of Malcolm X*), Braverman had worked as a coppersmith in the Brooklyn Navy Yards, and later as a sheet metal fabricator and pipefitter in the Midwest. From the age of seventeen, he was also a Communist. His view of industry was therefore shaped by a genuine allegiance with the skilled working class and a deep hostility to modern capitalism. These opinions combined powerfully in* Labor and Monopoly Capital, *an updating of Marx's writings to the conditions of the late twentieth century. Like other Marxists of the postwar period, such as Herbert Marcuse and Louis Althusser, Braverman focused on 'alienation', the process by which workers are disconnected from the meaning and value of their own labor. What made his writing on this theme distinctive was its evocation of work as lived experience, rather than as an economic variable. The 'degradation of work' became the key theme in this analysis. In the following passage, he attacks the principle of 'scientific management', the attempt to maximize shop floor efficiency through exact measurement of time, energy and output. (This pseudo-discipline had been pioneered by the early twentieth-century business consultant Frederick Winslow Taylor, whose villainy is comprehensively catalogued in Braverman's book.) Though he has subsequently been criticized by economists and sociologists who find his descriptions of deskilling misleading—chiefly because he underestimates the persistence of craft skill across the diversity of industrial and postindustrial contexts—Braverman's passionate advocacy on the part of craft workers cannot be questioned.*

Harry Braverman, 'The Primary Effects of Scientific Management', from *Labor and Monopoly Capitalism: The Degradation of Work in the Twentieth Century* (New York: Monthly Review Press, 1974).

The generalized practice of scientific management, as has been noted, coincides with the scientific-technical revolution. It coincides as well with a number of fundamental changes in the structure and functioning of capitalism and in the composition of the working class. In this chapter, we will discuss, in a preliminary way, some of the effects of scientific management upon the working class; later chapters will return to this discussion after the necessary conditions for understanding it more fully have been established.

The separation of mental work from manual work reduces, at any given level of production, the need for workers engaged directly in production, since it divests them of time-consuming mental functions and assigns these functions elsewhere. This is true regardless of any increase in productivity resulting from the separation. Should productivity increase as well, the need for manual workers to produce a given output is further reduced.

A necessary consequence of the separation of conception and execution is that the labor process is now divided between separate sites and separate bodies of workers. In one location, the physical processes of production are executed. In another are concentrated the design, planning, calculation, and record-keeping. The preconception of the process before it is set in motion, the visualization of each worker's activities before they have actually begun, the definition of each function along with the manner of its performance and the time it will consume, the control and checking of the ongoing process once it is under way, and the assessment of results upon completion of each stage of the process—all of these aspects of production have been removed from the shop floor to the management office. The physical processes of production are now carried out more or less blindly, not only by the workers who perform them, but often by lower ranks of supervisory employees as well. The production units operate like a hand, watched, corrected, and controlled by a distant brain.

The concept of control adopted by modern management requires that every activity in production have its several parallel activities in the management center: each must be devised, precalculated, tested, laid out, assigned and ordered, checked and inspected, and recorded throughout its duration and upon completion. The result is that the process of production is replicated in paper form before, as, and after

it takes place in physical form. Just as labor in human beings requires that the labor process take place in the brain of the worker as well as in the worker's physical activity, so now the image of the process, removed from production to a separate location and a separate group, controls the process itself. The novelty of this development during the past century lies not in the separate existence of hand and brain, conception and execution, but the rigor with which they are divided from one another, and then increasingly subdivided, so that conception is concentrated, insofar as possible, in ever more limited groups within management or closely associated with it. Thus, in the setting of antagonistic social relations, of alienated labor, hand and brain become not just separated, but divided and hostile, and the human unity of hand and brain turns into its opposite, something less than human.

This paper replica of production, the shadow form which corresponds to the physical, calls into existence a variety of new occupations, the hallmark of which is that they are found not in the flow of things but in the flow of paper. Production has now been split in two and depends upon the activities of both groups. Inasmuch as the mode of production has been driven by capitalism to this divided condition, it has separated the two aspects of labor; *but both remain necessary to production, and in this the labor process retains its unity.*

The separation of hand and brain is the most decisive single step in the division of labor taken by the capitalist mode of production. It is inherent in that mode of production from its beginnings, and it develops, under capitalist management, throughout the history of capitalism, but it is only during the past century that the scale of production, the resources made available to the modern corporation by the rapid accumulation of capital, and the conceptual apparatus and trained personnel have become available to institutionalize this separation in a systematic and formal fashion.

The vast industrial engineering and record-keeping divisions of modern corporations have their origins in the planning, estimating, and layout departments, which grew in the wake of the scientific management movement. These early departments had to make their way against the fears of cost-conscious managers, whom [Frederick Winslow] Taylor sought to persuade with the following argument: "At first view, the running of a planning department, together with the other innovations, would appear to involve a large amount of additional work and expense, and the most natural question would be is [sic] whether the increased efficiency of the shop more than offsets this outlay? It must be borne in mind, however, that, with the exception of the study of unit times, there is hardly a single item of work done in the planning department which is not already being done in the shop. Establishing a planning department merely concentrates the planning and much other brainwork in a few men especially fitted for their task and trained in their especial lines, instead of having it done, as heretofore, in most cases by high priced mechanics, well fitted to work at their trades, but poorly trained for work more or less clerical in its nature."[2] But to this he added the following caution: "There is no question that the cost of production is lowered by separating the work of planning and the brain work as much as possible from the manual labor. Where this is done, however, it is evident that the brain workers must be given sufficient work to keep them fully busy all the time. They must not be allowed to stand around for a considerable part of their time waiting for their particular kind of work to come along, as is so frequently the case."[3] This is by way of serving notice that no part of capitalist employment is exempt from the methods which were first applied on the shop floor.

At first glance, the organization of labor according to simplified tasks, conceived and controlled elsewhere, in place of the previous craft forms of labor, have a clearly degrading effect upon the technical capacity of the worker. In its effects upon the working population as a whole, however, this matter is complicated by the rapid growth of specialized administrative and technical staff work, as well as by the rapid growth of production and the shifting of masses to new industries and within industrial processes to new occupations.

In the discussion of this issue in Taylor's day, a pattern was set which has been followed since. "There are many people who will disapprove of the whole scheme of a planning department to do the thinking for the men, as well as a number of foremen to assist and lead each man in his work, on the ground that this does not tend to promote independence, self-reliance, and originality in the individual," he wrote in *Shop Management.* "Those holding this view, however, must take exception to the whole trend of modern industrial development."[4] And in *The Principles of Scientific Management:* "Now, when through all of this teaching and this minute instruction the work is apparently made so smooth and easy for the workman, the first impression is that this all tends to make him a mere automaton, a wooden man. As the workmen frequently say when they first come under this system, 'Why, I am not allowed to think or move without someone interfering or doing it for me!' The same criticism and objection, however, can be raised against all other modern subdivision of labor."[5]

These responses, however, clearly did not satisfy Taylor, particularly since they seemed to throw the blame on his own beloved "modern subdivision of labor." And so in both books he went on to further arguments, which in *Shop Management* took this form:

It is true, for instance, that the planning room, and functional foremanship, render it possible for an intelligent laborer or helper in time to do much of the work now done by a machinist. Is not this a good thing for the laborer and helper? He is given a higher class of work, which tends to develop him and gives him better wages. In the sympathy for the machinist the case of the laborer is overlooked. This sympathy for the machinist is, however, wasted, since the machinist, with the aid of the new system, will rise to a higher class of work which he was unable to do in the past, and in addition, divided or functional foremanship will call for a larger number of men in this class, so that men, who must otherwise have remained machinists all their lives, will have the opportunity of rising to a foremanship.

The demand for men of originality and brains was never so great as it is now, and the modern subdivision of labor, instead of dwarfing men, enables them all along the line to rise to a higher plane of efficiency, involving at the same time more brain work and less monotony. The type of man who was formerly a day laborer and digging dirt is now for instance making shoes in a shoe factory. The dirt handling is done by Italians or Hungarians.[6]

This argument gains force in a period of growth, of the rapid accumulation of capital through production on an ever larger scale, and of the constant opening of new fields of capital accumulation in new industries or the conquest of pre-capitalist production forms by capital. In this context, new drafts of workers are brought into jobs that have already been degraded in comparison with the craft processes of before; but inasmuch as they come from outside the existing working class, chiefly from ruined and dispersed farming and peasant populations, they enter a process unknown to them from previous experience and they take the organization of work as given. Meanwhile, opportunities open up for the advancement of some workers into planning, layout, estimating, or drafting departments, or into foremanships (especially two or three generations ago, when such jobs were customarily still staffed from the shop floors). In this manner, short-term trends opening the way for the advancement of some workers in rapidly growing industries, together with the ever lower skill requirements characteristic at the entry level where large masses of workers are being put to work in industrial, office, and marketing processes for the first time, simply mask the secular trend toward the incessant lowering of the working class as a whole below its previous conditions of skill and labor. As this continues over several generations, the very standards by which the trend is judged become imperceptibly altered, and the meaning of "skill" itself becomes degraded.

[. . .]

The destruction of craftsmanship during the period of the rise of scientific management did not go unnoticed by workers. Indeed, as a rule workers are far more conscious of such a loss while it is being effected than after it has taken place and the new conditions of production have become generalized. Taylorism raised a storm of opposition among the trade unions during the early part of this century; what is most noteworthy about this early opposition is that it was concentrated not upon the trappings of the Taylor system, such as the stopwatch and motion study, but upon its essential effort to strip the workers of craft knowledge and autonomous control and confront them with a fully thought-out labor process in which they function as cogs and levers. In an editorial which appeared in the *International Molders Journal*, we read:

> The one great asset of the wage worker has been his craftsmanship. We think of craftsmanship ordinarily as the ability to manipulate skillfully the tools and materials of a craft or trade. But true craftsmanship is much more than this. The really essential element in it is not manual skill and dexterity but something stored up in the mind of the worker. This something is partly the intimate knowledge of the character and uses of the tools, materials and processes of the craft which tradition and experience have given the worker. But beyond this and above this, it is the knowledge which enables him to understand and overcome the constantly arising difficulties that grow out of variations not only in the tools and materials, but in the conditions under which the work must be done.

The editorial goes on to point to the separation of "craft knowledge" from "craft skill" in "an ever-widening area and with an ever-increasing acceleration," and describes as the most dangerous form of this separation:

> the gathering up of all this scattered craft knowledge, systematizing it and concentrating it in the hands of the employer and then doling it out again only in the form of minute instructions, giving to each worker only the knowledge needed for the performance of a particular relatively minute task. This process, it is evident, separates skill and knowledge even in their narrow relationship. When it is completed, the worker is no longer a craftsman in any sense, but is an animated tool of the management.[7]

A half-century of commentary on scientific management has not succeeded in producing a better formulation of the matter.

NOTES

1. Miriam Braverman, quoted by Michael G. Livingston, "Harry Braverman: Marxist Activist and Theorist," a talk presented at the "Explorations in the History of U.S. Trotskyism Conference," New York University, Sept. 29 to Oct. 1, 2000.
2. L. and Barbara Hammond, *The Rise of Modern Industry* (London, 1925; reprint ed., New York, 1969), p. 119.
3. Frederick W. Taylor, *Shop Management*, in *Scientific Management* (New York and London, 1947), pp. 65–66.
4. Ibid., p. 121.
5. Ibid., p. 146.
6. Frederick W. Taylor, *The Principles of Scientific Management* (New York, 1967), p. 125.
7. E. P. Thompson, *The Making of the English Working Class* (New York, 1964), pp. 291–92.

FURTHER READING

Daniel Bell, *The Coming of Post-Industrial Society* (New York: Harper, 1973).

Carl Bridenbaugh, *The Colonial Craftsman* (Chicago: University of Chicago Press, 1966).

Herbert Marcuse, *One-Dimensional Man* (Boston: Beacon, 1964).

Frederick Winslow Taylor, *The Principles of Scientific Management* (New York: Harper and Brothers, 1911).

12

THE WORKSHOP OF THE WORLD: STEAM POWER AND HAND TECHNOLOGY IN MID-VICTORIAN BRITAIN

Raphael Samuel

Raphael Samuel's interest in the experience of the proletariat was not merely historical. He was the founder of the History Workshop movement, based (appropriately enough) at Ruskin College, an independent school in Oxford where he began teaching in 1962. The aim of this populist experiment in education was to bring the ideas of historians to a broad audience, outside of the traditional universities; and to teach research skills to adults with non-academic backgrounds, often preparing them for further study. (Such students, Samuel noted, 'were well peculiarly placed to write about many facets of industrial and working class history'.[1]) Expressly socialist, the History Workshop movement was a counterpart to cultural studies, as it was developed in Birmingham by historians like Raymond Williams and Stuart Hall; and material culture and folklore studies in America. While never a prolific scholar, Samuel created a space for others through his editing and was an extraordinary researcher in his own right. His long essay 'The Workshop of the World'—excerpted here—remains, over three decades later, the best single overview of handwork in the Victorian period. Its message is best summarized in the line 'Nineteenth century capitalism created many more skills than it destroyed, though they were different in kind from those of the all-round craftsmen, and subject to a wholly new level of exploitation.' Ranging widely from the maritime industry to textiles to furniture making, Samuel makes no attempt to deliver a final judgment on craft's fate during the period. Rather, he shows just how complex the situation was. It is easy to get lost in the details, perhaps, but Samuel's careful research is as powerful a counternarrative as we have to presumptions about the vanishing of craft in the nineteenth century.

Raphael Samuel, 'The Workshop of the World: Steam Power and Hand Technology in Mid-Victorian Britain', *History Workshop Journal* 3 (Spring 1977), excerpts.

[. . .]

THE MACHINERY QUESTION

Whatever their disagreements about the origins of the industrial revolution, economic historians are in little doubt about its effects. Steam power and machinery transformed the labour process and acted on society as an independent or quasi-independent force, demonic or beneficent according to the point of view, but in any event inescapable. Commodities were cheapened and new markets opened up for them; labour was made enormously more productive at the same time as the physical burden of toil was eased; mechanical ingenuity took the place of handicraft skill. David Landes' summary in *The Unbound Prometheus* is both influential and representative:

In the eighteenth century, a series of inventions transformed the manufacture of cotton in England and gave rise to a new mode of production—the factory system. During these years, other branches of industry effected comparable advances, and all these together, mutually reinforcing one another, made possible further gains on an ever-widening front. The abundance and variety of these innovations almost defy compilation, but they may be subsumed under three principles: the substitution of machines—rapid, regular, precise, tireless—for human skill and effort; the substitution of inanimate for animate sources of power . . . thereby opening to man a new and almost unlimited supply of energy; the use of new and far more abundant raw materials, in particular, the substitution of mineral for vegetable or animal substances.[2]

This account has the merit of symmetry, but the notion of substitution is problematic, since in many cases there are no real equivalents to compare. The fireman raising steam in an engine cab, or the boilermaker flanging plates in a furnace, were engaged in wholly new occupations which had no real analogy in previous times. So too, if one thinks of the operations they were called upon to perform, rather than the nature of the finished product, were the mill-hands of Lancashire and the West Riding. And if one looks at technology from the point of view of labour rather than that of capital, it is a cruel caricature to represent machinery as dispensing with toil. High-pressure engines had their counterpart in high-pressure work, endless chain mechanisms in non-stop jobs. And quite apart from the demands which machinery itself imposed there was a huge army of labour engaged in supplying it with raw materials, from the slave labourers on the cotton plantations of the United States to the tinners and copper miners of Cornwall. The industrial revolution, so far from abridging human labour, created a whole

new world of labour-intensive jobs: railway navvying is a prime example, but one could consider too the puddlers and shinglers in the rolling mills, turning pig-iron into bar, the alkali workers stirring vats of caustic soda, and a whole spectrum of occupations in what the Factory legislation of the 1890s was belatedly to recognise as 'dangerous' trades. Working pace was transformed in old industries as well as new, with slow and cumbersome methods of production giving way, under the pressure of competition, to overwork and sweating.

Nor is it possible to equate the new mode of production with the factory system. Capitalist enterprise took quite different forms in, for instance, cabinetmaking and the clothing trades, where rising demand was met by a proliferation of small producers. In agriculture and the fisheries it depended upon an increase in numbers rather than the concentration of production under one roof. In metalwork and engineering—at least until the 1880s—it was the workshop rather than the factory which prevailed, in boot and shoemaking, cottage industry. The distributive trades rested on the broad shoulders of carmen and dockers, the electric telegraph on the juvenile runner's nimble feet. Capitalist growth was rooted in a sub-soil of small-scale enterprise. It depended not on one technology but on many, and made use, too, of a promiscuous variety of profit-making devices, from the adulteration of soot (in which there was an international trade with the West Indies, as well as a local one with farmers for manure) to the artificial colouring of smoked haddocks.[3] Bread was dosed with liberal sprinklings of alum to disguise inferior wheats; low-grade cloths were camouflaged with 'size'.[4] In domestic house-building scamped workmanship kept the speculative builder afloat, while in the East End furniture trade orange boxes provided the raw materials for piano stools and Louis Quatorze cabinets. The 'Golden Dustman', immortalized by Charles Dickens in *Our Mutual Friend,* is as representative a figure of mid-Victorian capitalism as the Bradford millionaires pilloried by John Ruskin for their taste. So too—from the same novel—are the Veneerings, whose provincial counterparts rose to affluence by cotton 'corners' on Liverpool or Manchester Exchange. One thousand needlewomen made the fortunes of Nicoll, the Regent Street sweater,[5] while the railway speculations of the 1840s rested on the muscle power of three hundred thousand navvies.

Economic historians have had remarkably little to say about either labour process or the relationship of technology to work. They are much more concerned with business cycles and measuring rates of growth. Commercial achievement excites them, and whole histories will be written to celebrate the achievements of individual firms. Railways are discussed as a source of investment, and their comparative contribution to economic growth is a subject of hot debate; nothing at all is said about how the rolling stock was made or the engine cabins staffed or merchandise unloaded. Bricks, too, are treated as an index of investment, without so much as a word being said of the primitive conditions in which they were made, or the ferocious toil imposed on the men, women and children who made them.[6] Production is seen at second or third remove, in terms of inventory cycles and aggregate profitability: we do not learn how the furnaces were de-clinkered, or the iron steam ships coaled.

Except in the 'heroic' age of invention, economic historians have very little to say about machinery. They may tell us what it did for production, but not what it meant for the producers, and their preoccupation in recent years with 'take-off'—'that decisive interval in the history of a society when growth becomes its normal condition'—means that they give far more attention to the progress of mechanisation, and the constellation of circumstances favouring it, than to measuring its human costs. The plight of the hand-loom weavers in the 1830s is admitted, even insisted upon, but since they are regarded as a solitary and to some extent exceptional case, they do not seriously obstruct the linear march of improvement, and once they have been disposed of the historian passes quickly to the problems of a 'mature' economy, and the triumphs of Free Trade.

For labour historians, the machinery question attracts attention chiefly in the 1820s and 1830s, when Cartwright's loom was throwing thousands out of work, and when the rival merits of an agrarian and an industrial society ('past and present') were being vigorously canvassed on all sides. The scenario is arresting, with midnight raiding parties, rickyard incendiaries and factories besieged. But the drama is short-lived, and once the protagonists have performed their parts they are quickly shuffled off-stage. Opposition to machinery is assigned to the pre-history of socialism, when it was 'utopian' rather than 'scientific'; and the machine-breakers, despite Eric Hobsbawn's pioneering attempt to interpret their action in the light of modern collective bargaining (machine-breaking as a form of strike)[7] take their place in the gallery of 'primitive', pre-industrial rebels, along with such other early 19th century martyrs to oppression as Jeremiah Brandreth and Dic Penderyn. Luddism appears as a doomed, if heroic, resistance to the ineluctable forces of change—a fight against the inevitable—the Swing Riots of 1830 as 'the last labourers' revolt'. Yet in industry after industry the machinery question was still being fought out in mid-Victorian times, and there was a whole spectrum of occupations where mechanisation was still being resisted, or its scope drastically curtailed, in the 1890s: the last great machinery strike in the boot and shoe trade did not take place until 1895; while as late as 1898 a steam saw mill was blown up in the Forest of Dean.[8] There were also striking regional variations in the application of invention and progress of the machine, and in some cases at least the strength or otherwise of the workers' opposition seems to have been the deciding factor. In carpet weaving, for instance, the 'extra speeded' Moxon (an improved power loom of the 1870s) was kept out of Kidderminster entirely, where the Weavers' organisation was strong, but installed with apparent ease in Rochdale, Halifax and Durham, the northern centres of the trade.[9] In printing, the Hattersley, an early mechanical typesetter, was widely employed by provincial newspapers (the first was installed in the offices of the *Bradford Times* in 1868) but the London Society of Compositors was successful in keeping it at bay.[10] Similarly in boot and shoe making, the 'stabbing machine'—an application of the sewing machine to waxed threads—was excluded from Northampton, the metropolis of the wholesale trade, after three general strikes against it, fought between 1857 and 1859; but it was widely employed at Leicester, Norwich and Bristol.[11] In metalworking, the treadle-worked 'Oliver', a semi-mechanical stamp which had been common in Staffordshire for 'generations', was still apparently unknown in Manchester in 1865, and when in that year a local manufacturer attempted to introduce it, the nut and bolt makers (or at any rate the anonymous correspondent who wrote on their behalf) threatened to kill him.[12]

Even when machinery was eventually installed, the struggle to control it remained unresolved, and one of the most common complaints of employers in the late 19th century was that tools were not run at their proper speeds, but were being sabotaged by worker lethargy or resistance. In a cotton mill every spindle was potentially a battleground as mules increased in size: in an ironworks every attempted economy in fuel or alteration to the 'heat'. Often the machine proved disappointing to its patentees and promoters, either for want of precision, or because of the recalcitrance of the raw material, or because of the irreplaceability of handicraft skill. Patent could follow patent without anything like profitability being achieved, and the employer's dream of a 'self-acting' mechanism—equal to the best hand labour, but driven by itself—remained elusive. Mechanisation, in short, was a process rather than an event. It did not begin with the great inventions of the 18th and early 19th centuries; nor did it end with their application. The process itself was neither linear nor smooth but, on the contrary, discontinuous and subject to a whole complex of competing claims, pulling in opposite directions. For the most part it advanced by small increments rather than by leaps, and forward movements were often followed by retreat, as workers reasserted their claims. In the study of which this article forms a part, I want to argue that the machinery question, so far from being settled by the defeat of the Luddites, is in some sense coterminous with capitalism itself; that resistance to machinery, though often opaque and only intermittently recorded in the documents, was an endemic feature of 19th century industrial life. I also want to look at the repercussions of machinery on skill, and at the ways in which the labour process was reconstructed both from above and below, under the impact of technical change. Finally I want to look at machinery in relationship to the 'reserve army of labour' and the demographic changes of the early and middle years of the nineteenth century, and to consider the relationship of factory industry to capitalism in the countryside, domestic outwork and the workshop trades.

Readers of *Capital* will know that such a discussion inevitably bears on Marx's 'stages' of capitalist development. In chapters XIII to XVI of *Capital* Vol. I he proposes three great epochs of capitalist development, which are both chronologically and analytically distinct.

1. The handicraft stage, or that of petty commodity production—the chrysalis from which later capitalism grew.
2. Capitalist 'manufacture'—the concentration of artisan and handicraft production under the control of a single capitalist, and the systematic extension of the division of labour.
3. 'Modern' industry—the epoch inaugurated by the coming of machine tools and the factory system.

In Marx's discussion each of these epochs appears to supersede its predecessor and in the case of 'manufacture' and 'modern industry' at least a clear chronology is suggested, the first being assigned to the period from the middle of the sixteenth to about the middle of the eighteenth century, the second to the age of invention. But as Marx's lengthy chapter on 'modern industry' unfolds—it takes up fully 150 pages of the book—it becomes clear that modern industry incorporates older systems of production rather than superseding them, and that it is in fact a mixed development, in which 'modern' domestic industry and 'modern' manufacture play no less distinctive a part than the machine-based factories.[13] Here, as elsewhere in *Capital*, there are plainly shifts of emphasis in Marx's discussion, and one way of elucidating them—as well as of determining

their theoretical status—would be to consider the historical phenomena to which they were addressed. The discussion of such questions has in recent years been left to the philosophers and the economists, each of them concerned, in their own way, with the theoretical consistency of Marx's texts rather than the industrial reality which he was attempting to dissect. The historian may be ill-equipped to undertake a work of epistemological clarification, or to explore the more problematical reaches of the law of value. But that does not or should not mean that he or she has no contribution to make to theoretical discussion. The territory of *Capital* Vol. I is, after all, a historian's territory, one whose landmarks are in many cases familiar, and whose signposts the historian will sometimes be better placed than an economist or a philosopher to read.

[. . .]

COMBINED AND UNEVEN DEVELOPMENT

Steam power and hand technology may represent different principles of industrial organisation, and to the historian they may well appear as belonging to different epochs, the one innovatory, the other 'traditional' and unchanging in its ways. But from the point of view of 19th century capitalist development they were two sides of the same coin, and it is fitting that the Great Exhibition of 1851—'the authentic voice of British capitalism in the hour of its greatest triumph'[14]—should have given symbolic representation to them both. 'Steam power', an admiring commentator noted, 'wholly turned the mahogany which runs round the galleries of the Crystal Palace'.[15] But the 300,000 panes of glass which covered it were blown by hand,[16] and so was the Crystal Fountain which formed the centre-piece of the transept, 'glittering in all the colours of the rainbow'.[17] The promoters were intoxicated with the idea of 'self-acting machinery', and the technological miracles it might perform. But they devoted a great deal of their space to—among other things—needlework; and in demonstrating the competitive capabilities of British industry they were heavily dependent on artisan skills. Most of the manufactures on display were handicraft products, and even in the Machinery Court many of the exhibits were assembled from hand-made components. 'Few objects' excited more attention among foreigners than the displays of Sheffield cutlery and edge-tools (the Sheffield Court was one of the most extensive in the building),[18] while domestic visitors, it seems, were no less enraptured by the impenetrable locks, 'myriopermutation' keys and incombustible safes of Messrs. Chubb, Bramah and Mordan. Superimposed on the idea of mechanical progress there was also a nascent commercial aesthetic which the Exhibition's promoters rather grandly labelled 'the marriage of industry and art'. In subsequent years it was to make 'taste' a very principle of production and the marginal differentiation of products a primary axis of growth.

The balance of advantage between steam power and hand technology was, in mid-Victorian times, very far from settled, and many manufacturers though experimental in making new products and multiplying novelties of design, remained wedded to conservative production routines. Human beings, the main alternative to machinery, were, from a commercial point of view, often a much more attractive proposition. They were a great deal cheaper to install than a power house, and much more adaptable in their action than a self-acting stamp or press. When they broke down, the master did not have to pay for repairs; when they made a mistake, he could fine them; when there was no work for them to do he could give them the sack. Skills too were cheaper than machinery to come by. A steam sawing machine, in 1850, cost £700 to install;[19] a pair of travelling sawyers could be hired to do a job for five shillings, while a circular saw—such as the one used at Joseph Severn's shop in Codnor—could be ginned by a horse for free. Machinery was thus often adopted as a last resort, when every alternative means of extracting surplus value had failed to yield an adequate return, and it is no accident that manufacturers—like the Sheffield file makers of 1866—so often turned their eyes to it when they were faced with demands for higher wages.

The orthodox account of the industrial revolution concentrates on the rise of steam power and machinery, and the spread of the factory system. It has much less to say about alternative forms of capitalist enterprise (such as those to be found in mining and quarrying), about the rise of sweating, or the spread of back-yard industries and trades. Nor does it tell us much about the repercussions of technology on work. Landes' picture has the compelling power of paradigm, with mechanisation on an 'ever-widening front' and steam power—'rapid, regular, precise'—effortlessly performing labour's tasks. But if one looks at the economy as a whole rather than at its most novel and striking features, a less orderly canvas might be drawn—one bearing more resemblance to a Brueghel or even a Hieronymus Bosch than to the geometrical regularities of a modern abstract. The industrial landscape would be seen to be full of diggings and pits as well as of tall factory chimneys. Smithies would sprout in the shadows of the furnaces, sweatshops in those of the looms. Agricultural labourers might take up the foreground, armed with sickle or scythe, while behind them troops of women and children would be bent double over the ripening crops in the field, pulling charlock, hoeing nettles, or cleaning the furrows of stones. In the middle distance there might be navvies digging sewers and paviours laying flags. On the building sites there would be a bustle of man-powered activity, with housepainters on ladders, and slaters nailing roofs. Carters would be loading and unloading horses, market women carrying baskets of produce on their heads; dockers balancing weights. The factories would be hot and steamy, with men stripped to the singlet, and juvenile runners in bare feet. At the lead works women would be carrying pots of poisonous metal on their heads, in the bleachers' shed they would be stitching yards of chlorined cloth, at a shoddy mill sorting rags. Instead of calling his picture 'machinery' the artist might prefer to name it 'toil'.

Skill was as important as toil (the two often went hand in hand) and in mid-Victorian times it was plentifully available. The domestic housebuilder could draw on a vast substratum of carpentering skills: so could such booming industries as Kentish Town pianos and High Wycombe chairs. The new iron shipyards were quickly filled with artisans and mechanics drawn from a dozen different trades; by the 1870s they were already a very cockpit of sectarian craft rivalries. Engineering employers recruited their labour from those who had served their apprenticeships in the 'country' branches of the trade, with wheelwrights, blacksmiths, and in the small town foundries; and it was a matter of real anxiety in the industry when, towards the end of the century, this source of recruitment began to dry up. 'There is no evidence that labour supply impeded any of the machine tool firms', writes Roderick

Floud in his recent book. 'Even as early as the 1830s, Nasmyth was able to break a strike in his works, aimed at forcing him to employ only men who had served an apprenticeship, by importing sixty-four Scottish mechanics and he remarked that "we might easily have obtained three times the number . . . " No other machine tool maker appears to have had difficulties in securing labour, or, indeed, . . . in dismissing it when times were bad, in the confident expectation that the men could be re-employed if trade improved'.[20]

It was not only in craft industry that capitalism drew on reservoirs of skill, but in every branch of economic activity where a mainly hand technology prevailed. Tunnel bricklayers on the railway works were in their own way as skilled as stonemasons; so were the coal heavers and timber porters in the docks, the carters and wagoners in road haulage, the ploughmen and rick-builders on the farms, the shot-firers and hewers in the pits. As well as the 'aristocracy of labour', on whom British historians have lavished such continuous attention, there was also a whole army of rural mechanics and small town artisans, like the 'ragged trousered philanthropists' of Mugsborough, who still await their chronicler. So do the poor artisans of Shoreditch and Bethnal Green, in East London, who at the time of the 1891 census constituted no less than 60 percent of the local working population.

Nineteenth century capitalism created many more skills than it destroyed, though they were different in kind from those of the all-round craftsmen, and subject to a wholly new level of exploitation. The change from sail to steam in shipping led to the rise of a whole number of new industrial crafts, as well as providing a wider arena for the exercise of old ones. The same may be said of the shift from wood to iron in vehicle building, and of horse to steam in transport. In the woodworking trades a comparatively small amount of machinery supported a vast proliferation of handicraft activities, while in metallurgy the cheapening of manufacturing raw materials led to a multiplication of journeymen-masters. The mid-Victorian engineer was a tool-bearer rather than a machine minder, the boilermaker was an artisan rather than a factory hand. In coal mining activity increased by the recruitment of a vast new class of workers who were neither exactly labourers, nor yet artisans, but who very soon laid claim to hereditary craft skills. Much the same was true of workers in the tinplate mills and ironworks. The number of craftsmen in the building trade increased by leaps and bounds, though the rise of new specialities led to a narrowing of all-round skills.

In juxtaposing hand and steam-powered technologies one is speaking of a *combined* as well as of an *uneven* development. In mid-Victorian times, as earlier in the 19th century, they represented *concurrent* phases of capitalist growth, feeding on one another's achievements, endorsing one another's effects. Both were exposed to the same market forces; both depended for their progress upon the mobilisation of wage labour on a hitherto unprecedented scale, and both were equally subject to the new work discipline, though it affected them in different ways. The industrial revolution rested on a broad handicraft basis, which was at once a condition of its development and a restraint on its further growth. In mid-Victorian times—as I shall attempt to show in a second article—the handicraft sector of the economy was quite as dynamic as high technology industry, and just as much subject to technical development and change. It was indeed in the first rather than the second that mass production methods in many cases were pioneered; that new classes of commodity were

created; and that modern capitalist methods of exploitation—both of producers and consumers—were most clearly prefigured and explored.

NOTES

1. Quoted in Vic Gammon, "In Memoriam: Raphael Samuel, 1935–1996," *Folklore* 108 (1997), p. 105.
2. David Landes, *The Unbound Prometheus*, Cambridge 1969, p. 41.
3. For an autobiographical account by one who traded in adulterated soot, George Elson, *The Last of the Climbing Boys*, London 1900, pp. 78–81; [. . .] *Liverpool Mercury*, 10 Oct. 1845, p. 400; Henry Mayhew, *London Labour and the London Poor*, London 1861.
4. For the effect of heavy sizing on the health of those who had to work with it in the mills, Parliamentary Papers 1872 (203) LIV and 1884 (c. 3861) LXXII. For a moving autobiographical account, Alice Foley, *A Bolton Childhood*, Manchester 1973, p. 51.
5. Goldsmith's Coll., Henry Mayhew, *Labour and the London Poor*, London 1851, p. 40.
6. H. A. Shannon, 'Bricks, a trade index', *Economica* 1934.
7. Hobsbawm, 'The Machine Breakers', in *Labouring Men*.
8. [National Archive] For. 3/559.3/263. The explosion, like others in the district, seems to have been the work of Walter Virgo and the Blakeney gang.
9. Howell Collection, Bishopsgate Institute, *The Moxon Loom Arbitration Proceedings*, Kidderminster 1879. Cf. also J. N. Bartlett, 'The Mechanization of the Kidderminster Carpet Industry', *Business History*, 1967, IX.
10. J. Reynolds, *The Letterpress Printers of Bradford*, Bradford 1971, pp. 5–6; Sid Wills, 'The Compositor's Frame' in *The Workshop Trades*.
11. John Ball, 'Account of the Northamptonshire Boot and Show Makers Strike in 1857/8/9', *Trade Societies and Strikes*, London 1860.

There are very full reports of these strikes in the *Northampton Mercury* and the *Northamptonshire Free Press*.

12. 'Murderous Threats to Workmen in the Screw Bolt Trade', *The Ironmonger*, 30 Dec. 1865, VII 180–1.

13. Karl Marx, *Capital* XV 464–96; XVI 519.

14. Francis Klingender, *Art and the Industrial Revolution*, Paladin ed., London 1975, p. 144. Klingender uses this excellently apposite phrase to describe Sir Matthew Digby Wyatt's official catalogue of the Exhibition.

15. George Dodd, *Curiosities of Industry*, London 1852, 'Wood,' p. 18.

16. Gustav Strauss, *England's Workshops*, 1864, p. 186.

17. *Illustrated Exhibitor*, 7 June 1851.

18. *Illustrated Exhibitor*, 23 August 1851.

19. *Morning Chronicle*, 4 July 1850.

20. Roderick Floud, *The British Machine Tool Industry, 1850–1914*, Cambridge 1976, p. 49.

FURTHER READING

Royden Harrison and Jonathan Zeitlin, eds, *Divisions of Labour: Skilled Workers and Technological Change in Nineteenth Century England* (Sussex/Chicago: Harvester Press/University of Illinois Press, 1985).

Raphael Samuel, ed., *People's History and Socialist Theory* (London: Routledge Kegan & Paul, 1981).

Raphael Samuel, ed., *Village Life and Labour* (London: Routledge Kegan & Paul, 1975).

Martin J. Wiener, *English Culture and the Decline of Industrial Spirit, 1850–1980* (Cambridge: Cambridge University Press, 1981).

13

TECHNOLOGICAL INNOVATION AND DESIGN ECONOMICS IN FURNITURE MANUFACTURE

Michael Ettema

Like Raphael Samuel, the historian of technology Michael Ettema argues for a nuanced view of craft's role in late nineteenth-century manufacture. Influenced by a few exceptional cases such as the textile industry, historians have tended to overestimate the degree to which mechanization penetrated and transformed craft trades. He focuses on furniture making, which was particularly difficult to automate due to rapid changes in fashion, and the difficulty of achieving complex ornament using machines. In fact, it is likely that stylistic preferences exerted a brake on industrialization—the reverse of the commonly held assumption that Victorian furniture was overdecorated because machines had made it cheap to do so. Ettema's article is also important in its insistence on the specifics of the making process. Just as Samuel examines separate trades, Ettema looks at different tools within the furniture industry and emphasizes the limitations imposed by certain processes and economic considerations. As he puts it, furniture makers in a competitive environment were always looking to provide 'the most chair for the price'. In the nineteenth century, more often than not, that meant relying on skilled craftsmanship rather than abandoning it.

Michael Ettema, 'Technological Innovation and Design Economics in Furniture Manufacture', *Winterthur Portfolio* 16, nos. 2/3 (Summer/Autumn 1981), excerpted.

Critics of late nineteenth-century furniture design often comment on the effects of technological innovation. In an age that takes for granted the omnipotence of technology, it seems reasonable to assume that the industrial revolution physically altered the nineteenth-century built environment. The correlation between frantic mechanization and effervescent Victorian design appears to be more than coincidence, and scholars have long assumed that some type of causal relationship must exist.

Early twentieth-century commentators linked woodworking machinery with furniture design in order to repudiate the stylistic preferences of their parents. Horrified by Victorian hyperbole, the generation of designers and connoisseurs who embraced the myth of the early American artist-craftsman found a convenient and compelling rationale for their rejection of the late nineteenth-century aesthetic. By 1850, they said, machines had destroyed the traditional bond between art and industry, eliminating skilled workmen and undermining the small-shop system of manufacture. Without the artistic sensitivity and control of the craftsman, machines capable of mass-producing expensive-looking ornament lured manufacturers into grotesque design excesses and seduced them into aesthetic compromises for the sake of profit. In this view, technology directly caused elaborate and degraded styles.[1]

In more recent years, the idea of cultural relativism has softened reactions to Victorian design. Lacking a stylistic crusade, contemporary historians have more cautiously evaluated the effect of machinery on design, while remaining enthusiastic about the degree of impact. In her contribution to *Technological Innovation and the Decorative Arts* (1974), Polly Anne Earl deflated some post-Victorian criticisms by pointing out that the adoption of steam- or water-powered machinery in the furniture industry was a slow and uneven process. Especially in the eastern cities, relatively small, unpowered shops produced a significant proportion of America's furniture as late as the 1870s. Earl quickly speculated, however, that many small producers may have used foot- or hand-powered machinery to gain some of the advantages of the new technologies. Thus, while she concluded that large-scale, powered machinery was not responsible for Victorian styles, nor styles responsible for machines, she still assumed that technological innovations led to increased furniture production at lower prices, bringing more decoration into the average home.[2] Although Earl, like other recent historians, transformed the manipulative and destructive power of machinery into a democratizing force, mechanization remained for her a central issue in the study of Victorian design.[3]

The willingness of historians to emphasize the efficacy of machinery reflects a lingering confusion between the impact of technological innovations and broader changes effected by the industrial revolution. Furniture historians have taken for granted the extraordinary capabilities of woodworking machines without actually explaining their operation, capacity, or advantage over hand tools. Moreover,

historians assume that machinery necessarily had some pervasive influence on the furniture industry, but they ignore the factors that governed the application of these tools to production.[4] The result has been an overestimation of the capabilities of machinery. Decorative arts historians have unconsciously created the concept of a technological monolith that must be dismantled and examined to find new information and new perspectives on the relationship between process and product. Undeniably, machinery helped to restructure the furniture industry and to alter product design, but it had the least influence on expensive, trend-setting goods. Not only did the adoption of machinery diffuse slowly through the industry, but the history of woodworking machinery reveals a surprising lack of automation, especially for more complex operations. In general, machinery allowed an increase in furniture production but failed to democratize style because machines could not produce inexpensive copies of expensive-looking ornament. Proliferation, not elaboration, was the legacy of technological innovation in the nineteenth-century furniture industry.

Much of the confusion about the way technology was employed in furniture manufacture stems from the persistent desire to view furniture as a "decorative art." It is essential to discard the romantic notion of eighteenth-century artist-craftsmen or nineteenth-century manufacturers using tools to create whatever their aesthetic sensibilities or insensibilities dictated. Furniture production in this country is and always has been a business subject to the same basic economic laws that govern the manufacture of other consumer durable goods. Manufacturers who wish to stay in business cannot produce indiscriminately and without considering the consumers from whom they draw their livelihood. They have free rein to express themselves only if there is a market for their expression. In spite of devices such as advertising which attempt to manipulate the market, manufacturers in a traditionally competitive field such as furniture cannot afford to make what the customer will not or, more importantly, cannot buy. Manufacturers learn the market and equip themselves to fill a niche in it.

[. . .]

Combining style, materials, and technology in a desirable product at a price the market will accept has always been the basic problem facing furniture makers. Since costs are, in part, dependent on the labor intensity of their technologies, manufacturers must design pieces with the capabilities of their tools in mind, constantly compromising between cost and style. This system of give and take is the economic interface between technology and style. It is the economics of design.

The assumption implicit in most writing on nineteenth-century furniture has been that technological innovations profoundly disrupted the economics of furniture production, altering the traditional relationship between design and cost, and making high-end products available at low-end prices. The first part of this statement is correct; machinery did change the manner of production at most levels. If the last part logically follows, however, it would have been necessary for machines to disrupt the cost hierarchy of the various woodworking processes, upsetting the system of techno-economic compromises. To test the accuracy of this supposition it is necessary to understand in general terms what makes one woodworking process more or less efficient than another, and then to look for those qualities in the machines themselves.

Woodworking machinery tends to maximize efficiency by replacing skilled labor with cheaper semiskilled or unskilled labor, and speeding the operation to allow greater output per man-hour of labor. In *The Nature and Art of Workmanship*, designer David Pye offers two incisive concepts which point out how these goals are often achieved. Workmanship, he states, can be divided into the "workmanship of risk" and the "workmanship of certainty." The former is "workmanship using any kind of technique or apparatus in which the quality of the work is not pre-determined but depends on the judgment, dexterity and care which the maker exercises as he works." In the latter, "the quality of the result is exactly predetermined before a single saleable thing is made."[5] Cutting a molding with a chisel is risk because only the workman controls the depth and direction of the cut. Running a molding with a molding plane, however, approaches certainty because the contoured plane iron predetermines the shape of the cut, and the plane block prevents the iron from entering the work too deeply. By changing risk to certainty, machinery can increase productivity by reducing the care and dexterity required to form the product. In effect, technological innovations transfer the workmanship of risk from the manufacture of the product to the manufacture of the tools.

Jigs, such as plane blocks, which help control the movement of the tool through the work, are often used to reduce the workmanship of risk. On many machines, the workpiece can also be jigged in order to insure accurate cutting and save time in production. Since furniture is made in multiple units called cuttings, jigs or guides are set up and premeasured to hold the work pieces and allow them to be introduced to the cutter at the correct point. This eliminates the necessity of measuring or marking off the work to be done on each piece. Thus, identical parts

Figure 6 Henry Eyles, *Chair*, 1851.

can be fed in succession, almost as fast as the machine can work them.

In *A Treatise on the Construction and Operation of Wood-Working Machines* (1872), English engineer John Richards stated that "the operations in which machinery effects the greatest saving, are those where much power can be used, where long cutting edges can be applied, and where few adjustments are needed along the progress of work."[6] The use of power is a prime factor in most machinery. In effect, it decreases the resistance of the material to the action of the tool, speeding its movement through the work. In certain cases, the use of power also allows the employment of broad cutting edges that, self-evidently, work more of the wood at one time. Planing machines,

for example, can be built to work a board of almost any size along its entire width, unlike hand planes which take cuts only a few inches wide. Stopping the action of the tool to adjust the cut is less efficient than continuous operation. Since the workmanship of certainty and flexibility in operation usually run at odds, however, machines built for more complex processes cannot always avoid this drawback.

Indeed, because machines built for different purposes had very different requirements, they could not employ the same cost-reducing qualities. The significant factor in furniture manufacture is the extent to which technological innovations changed the relative costs of various woodworking operations, upsetting the economic hierarchy. If the new machines dramatically reduced costs in all aspects of manufacture, the techno-economic system would have collapsed, resulting in the democratization of style or the inexpensive production of expensive-looking goods. The machines themselves, however, reveal that the greatest changes came in those operations that were already inexpensive, while the more costly technologies remained labor intensive.

Carving, perhaps the most important ostentatious technology, is an excellent case in point. Early twentieth-century furniture historian Thomas Ormsbee dated the beginning of full mechanization of the furniture industry very conveniently at 1850 when the most heavily carved of Victorian objects were becoming popular in this country. At a glance, this might seem plausible since an Englishman named Jordan patented, built, and used a carving machine in 1845.[7] The coincidence is enticing, but in fact no practical carving machines were widely used in this country until the last quarter of the nineteenth century—after the most elaborate styles began to pass from high fashion. Richards stated: "As to the history of carving machines thus far, leaving out special cases and taking the result generally, it has been an even race against hand labour, to say the best, and gives no great promise of gain in the future. In this assumption we are guided by the only fact that is entirely reliable in the matter, which is, that carving in both England and America, as well as on the continent is mainly done by hand."[8] Few machines were built on the Jordan principle until, in the 1880s, a small number of firms brought out machines purporting to duplicate hand carving. The only other carving machine to see widespread use in this country was the spindle carver, a simple machine requiring great skill in operation because the work was hand held against the cutters.

Although both types of carving machines sped the labor-intensive process, neither was capable of producing high-style ornament without the aid of skilled hand carvers. Using contoured, rotating cutters, they carved many simplified forms with ease, but for intricate work, the knives had to be repeatedly changed for cuts of varying sizes and shapes. If the carving was extremely complex, the number of adjustments was so great that conceivably it could take more time to use a machine than to use hand tools. In any event, the diversity of cuts produced by hand carving could not be imitated using even a large number of jigged, rotating cutter heads. Without hand finishing, the products of carving machines were conspicuously unrefined.

As a result, carving machines merely roughed out the work on more expensive productions. A Grand Rapids observer reported in 1906 that "the automatic carving machines supplanted the laborious process of removing by hand superfluous wood preparatory to the final artistic touch of the hand tool, which infuses life into each upturned leaf as guided by the skilled carver."[9] Manufacturers' advertising

as well as articles in trade magazines continually stressed the compatibility of hand and machine work.[10] Indeed, the furniture industry employed large numbers of hand carvers well into the twentieth century to do finish work or to carve decoration of which machines were incapable. For example, Sligh Furniture Company of Grand Rapids, a producer of middle- and low-end furniture, still required a sizeable staff of skilled workers fifty years after Ormsbee's time of "full mechanization".

[. . .]

To summarize, the degree to which machinery was capable of reducing labor costs in furniture manufacture was inversely proportional to the total cost of the product. Machines could produce furniture more quickly than ever before, and processes such as shaping and embossing increased the variety and availability of inexpensive ornament. But owing to the technical limitations of the machines that made it, cheap decoration was visually inappropriate for most expensive goods. Maximum use of laborsaving machinery produced low-end furniture. This should not be surprising since the nineteenth century saw the explosion of a middle-class market in addition to the maintenance of a large working-class sector. Certainly it was no coincidence that the furniture industry employed machinery to produce precisely what was in greatest demand. If critics observed an increase in second-rate ornament on furniture in the second half of the nineteenth century, it undoubtedly occurred. However, it resulted not so much from the evils of technology as from the increase in the number of people who could afford to buy only second-rate goods.

A cautious view of the revolutionary potential of woodworking machinery can still recognize significant change within the furniture industry in the nineteenth century because machines did not constitute the entire system of manufacture. Other crucial variables included shop organization, price and availability of labor, cost of materials, and efficiency of marketing and transportation systems. Although the combination of these factors permanently altered the structure of the industry, their effect on the shape of the product was subject to physical, economic, and cultural restraints. Changes in machine efficiency and factory size were balanced by a continuity in the economic relationship between process and product. Like the hand tools they were meant to replace, woodworking machines reinforced the hierarchical structure of design economics in furniture manufacture.

Understanding the design economics of nineteenth-century furniture requires at least a general acquaintance with the physical and economic capabilities of the tools of production. The following pages list the major classes of woodworking machines available to the Victorian furniture industry, outlining their use, operation, limitations, and approximate date of introduction. Rather than provide a complete history of woodworking machinery, the purpose here is to contribute to an understanding of process where it most directly intersects product design.

[. . .]

NOTES

1. For example, see Thomas Ormsbee, *Early American Furniture Makers: A Social and Biographical Study* (New York: Tudor, 1930); Carl Drepperd, *The Primer of American Antiques* (Garden City, N.J.: Doubleday, 1944); Ethel Hall Bjerkoe, *The Cabinetmakers of America* (Garden City, N.J.: Doubleday, 1926); and Charles Cornelius, *Early American Furniture* (New York: Century, 1926).

2. Polly Anne Earl, "Craftsmen and Machines: The Nineteenth-Century Furniture Industry," in *Technological Innovation and the Decorative Arts*, ed. Ian M. G. Quimby and Polly Anne Earl (Charlottesville: University Press of Virginia, 1974), pp. 316–18, 324.

3. See also Mary Jean Smith Madigan, *Eastlake Influenced American Furniture, 1870–1890* (Yonkers, N.Y.: Hudson River Museum, 1973), and Donald C. Peirce, "Mitchell and Rammelsberg: Cincinnati Furniture Manufacturers, 1847–1881," in *American Furniture and Its Makers: Winterthur Portfolio 13* ed. Ian M. G. Quimby (Chicago: University of Chicago Press, 1979), pp. 209–29.

4. For example, see Jan Seidler, "The Furniture Industry in Victorian Boston," *Nineteenth Century 3*, no. 2 (Summer 1977): 65–69.

5. David Pye. *The Nature and Art of Workmanship* (New York: Van Nostrand Reinhold, 1968), p. 7

6. John Richards, *A Treatise on the Construction and Operation of Wood-Working Machinery* (London: E. and F. N. Spon, 1872), p. 275.

7. Ormsbee, *Furniture Makers*, p. 89; Charles Knight, Knight's *Cyclopedia of the Industry of All Nations* (London: By the author, 1851), p. 546.

8. Richards, *Treatise*, p. 44.

9. Dwight Goss, *History of Grand Rapids and Its Industries* (Chicago: C. F. Cooper, 1906), pp. 1037–38. Both Robert Hunt and Charles Tomlinson stressed the point that variations of the Jordan carver were intended only to speed production by roughing out the carving before the handwork was done (Hunt, ed., Ure's *Dictionary of Arts, Manufacture, and Mines*, 3 vols. [5th ed., London: Longman, Green, Longman, and Roberts, 1860], 1: 627; Tomlinson, *Cyclopaedia of Useful Arts and Manufactures*, 2 vols. [London and New York: George Virtue, 1854], 1: 334–35).

10. For example, see Charles Cist, *Sketches and Statistics of Cincinnati in 1851* (Cincinnati: W. H. Moore, 1951), p. 203; and "Industries

of Michigan—I: Furniture Making," *Michigan Technic* 17, no. 30 (January 1927): 32.

FURTHER READING

Judy Attfield, "'Give 'em Something Dark and Heavy": The Role of Design in the Material Culture of Popular British Furniture, 1939–1965', *Journal of Design History* 9/3(1996), pp. 185–201.

Sharon Darling, *Chicago Furniture: Art, Craft, and Industry, 1833–1983* (New York: W. W. Norton, 1984).

Clive Edwards, *Victorian Furniture: Technology and Design* (Manchester: Manchester University Press, 1993).

Adrian Forty, *Objects of Desire: Design and Society, 1750–1980* (London: Thames and Hudson, 1986).

Pat Kirkham, *The London Furniture Trade, 1700–1870* (London: Furniture History Society, 1989).

14

ARTISTIC AMERICA

Siegfried Bing

The Art Nouveau style, with its whiplash lines, natural forms and giddy allegorical subject matter, is much beloved today. Its proper interpretation, however, is a matter of some difficulty. Art Nouveau was a conscious return to the old days of Parisian luxury manufacture, when hand-carved and inlaid furniture with gilt bronze mounts had dictated la mode *in interiors across the Western hemisphere. But it has also been seen as a forerunner of modernism, and as a design reform movement with strong connections to the Arts and Crafts movement in Britain. It looked to Japan for stylistic influences, too, drawing on* ukiyo-e *woodblock prints and handmade pottery for ideas. With the founding of his Parisian shop* L'Art Nouveau *in 1895, Bing set himself up as the most exciting purveyor of this new design idiom—a position that was confirmed in grand style when his installations at the World's Fair of 1900 were an international sensation. Given that most of Bing's merchandise was both handmade and expensive, it is surprising to see him arguing that 'our decorative arts have suffered from the exclusive prestige accorded to what we pompously call Fine Art'. But this was a conviction that America had confirmed in him. When praising Louis Comfort Tiffany, the glass and metalwork designer, he made it clear that what he admired most was Tiffany's ability to maintain 'artistic' quality while producing at large scale (though not necessarily using*
heavy machinery). As Nancy Troy has argued, Bing thought that 'in order to maintain the high quality of French design that had been achieved before the breakdown of traditional craft production, it was necessary to establish an equivalent production system that would respond to contemporary industrial conditions'—a blend of past and present that was the obverse of William Morris's, but no less paradoxical.*

Siegfried Bing, excerpts from *Artistic America*, 1895. As translated by Bettina Eisler and reprinted in Robert Koch, ed., *Artistic America: Tiffany Glass and Art Nouveau* (Cambridge, MA: MIT Press, 1970).

[. . .]

Many of those amateurs whose interest extends to all areas of art will still remember their surprise on seeing, at the Exposition of 1878, several examples of metalwork of the most extraordinary quality. Although not intrinsically original in concept—their decorative principles were taken directly from the Japanese—the borrowed elements were so ingeniously transposed to serve their new function as to become the equivalent of new discoveries. In any case, these useful objects were attractive, not least because they had ceased to embody the constant reincarnation of our own traditional forms, however charming, whose interest had long palled with repetition . . . And suddenly, America, which only shortly before had experienced its first artistic stirrings, was to bear proof of singular powers of initiative and youthful vigor, in sharp contrast with the thinning blood that progressively has weakened the industrial arts throughout Europe and made impotent our most precious hereditary gifts.

By what surprising phenomenon were the roles reversed? We French, tireless teachers of all nations, who, for centuries have continuously sowed throughout the world the seed of our artistic knowledge, should be especially awed by this sudden flowering in a barren, far-off land. And it would be still more appropriate to translate our wonder into discovery of how America can enlighten us in our present weakness.

As we proceed to a serious examination of this kind, we should first recognize that our decorative arts have suffered for too long from the exclusive prestige accorded to what we pompously call Fine Art. This truth has already begun to be felt by our artists themselves. Many have finally stopped considering the products of industrial art as unworthy of their talents. For the last several years we have observed them happily accept the pariahs of yesterday into their public exhibitions; sometimes the artists themselves have even conceived the design, made ingenious models and molds of every kind, no longer feeling their genius demeaned by working with their own hands, in materials once thought too commonplace. But nothing produced by all of this sincere effort seems clothed in the specific nature of its function; none of it has the really practical appearance imprinted upon objects by real craftsmen. The latter, conversely, when they are experts in a particular technique, and thoroughly familiar with the organic structure of each object, are too much the prisoners of laboriously learned doctrines, with no inspiration in their ideas. Indeed we have our schools of the decorative and industrial arts, invaluable in the training of skilled hands, capable of obtaining every subtlety from the most polished execution. But just as schools of Fine Art fail in their own realm, so schools of Applied Art are unable to galvanize the imagination or ignite a spark of brilliance. For many years, what has been lacking is the artist—of born talent, to be sure—who will commit himself wholeheartedly to an artisan's work.

We sometimes come across a laborer-artist; what we lack is the *artist-laborer.*

Aspiring young artists pursue an unending search for the Ideal; but they can perceive it only in its abstract manifestations; things will be ever thus as long as the world limits its honors to those for whom the supreme dream of Beauty takes certain privileged forms. Further, as soon as anyone feels vibrate within him burgeoning creative gifts, he wonders why he should expend it working on a thousand anonymous projects, when an inspired brush or skillful chisel promises the immediate possibility of sudden deification?

In America, things happen differently. The same democracy that serves as a basis for the entire social structure of the country has to the same extent penetrated the world of art. Neither accident of birth nor choice of one career over another confers any aristocracy. No caste system could long endure in an environment where all roads can lead to honor and fame. When an American artist holds an honored place in public esteem, it is in no way due to his choice of painting or sculpture; but rather because he has given shape to a new concept of Beauty—and any tool may have been used, with equal brilliance, to serve this distinguished cause—it makes no difference whether it is called brush, chisel, or something else.

Three men were in the forefront of this movement: Samuel Colman, John La Farge, Louis C. Tiffany, son of the founder of Tiffany & Co., and all three had started their careers as painters. It is safe to assume, however, that their contribution as painters would have been limited. Their painting is primarily distinguished for its warm and harmonious coloring, and this fine tonal harmony would be of inestimable help to them in their new field of endeavor.

Now, is this to say that the ardent conviction and sound judgment of a few original minds must forcibly suffice to change suddenly a given level of art? And should we assume that if similar initiative were to appear in our own country, it could suddenly revive our talents, dulled for too long?

Indeed no. In America this task has been facilitated, first of all, by the fact that the American mind is not haunted by so many memories. Her youthful imagination can have free rein, and, when it comes to the making of things, her hand is not restricted to a circumscribed number of movements, ever and predictably the same. Not that the American people are anything other than an offshoot of our own earliest roots and, thus, of our traditions as well. What has given them a different destiny is that they do not, as we do, make a *religion* of these same traditions. It is their rare privilege to make use of our aged maturity, adding to it the bursting energy of youth's prime.

From the happy fusion of these two elements, America today has formulated her own theory, as applied to industrial arts, a theory that may be summarized in the following formula.

First, try to enrich the arsenal of usable materials with every element in nature, down to the lowliest, which the blinders of old habits have until now ignored. To manipulate these materials, master every known process and method, in their most diverse applications. Then, after we have learned and analyzed everything, acquired every secret technique, every trick of the trade as taught by the experience of centuries . . . then, completely forget the way these have been used in the past, banish from memory any lingering obsession with inherited forms; in a word, place old and tried knowledge in the service of an entirely new spirit, with no guidelines other than those of intuitive taste and natural laws of logic.

Since this method has already brought forth sufficiently tangible results to augur for its practical value, it should lead to serious reflection on our part. Could we not try, through a virile force of will, to escape the suffocating bonds of past memories? We do not suggest the renunciation of our glorious heritage. Quite the reverse: justifiably proud of our earlier artistic triumphs, we should let their brilliance shine inviolate in the distance of history. If we think about it, we do an injustice to the great influence of these works of art, to their very reason for existence, if we use them as instruments of imitation, in the sole aim of absolving ourselves from the exhausting labor of creation. Why, instead of continuing to reproduce the forms of earlier art, not try to equal the creative genius that gave them birth?

For the last few years, such problems have once again begun to preoccupy our most brilliant thinkers. But for the moment at least, their efforts have not been commensurate with the magnitude of the task. Success in this realm will require something more than witty and imaginative creations, not without quality in their own way, but most often remaining on the level of refined knickknacks, for the exclusive delectation of a few avid collectors.

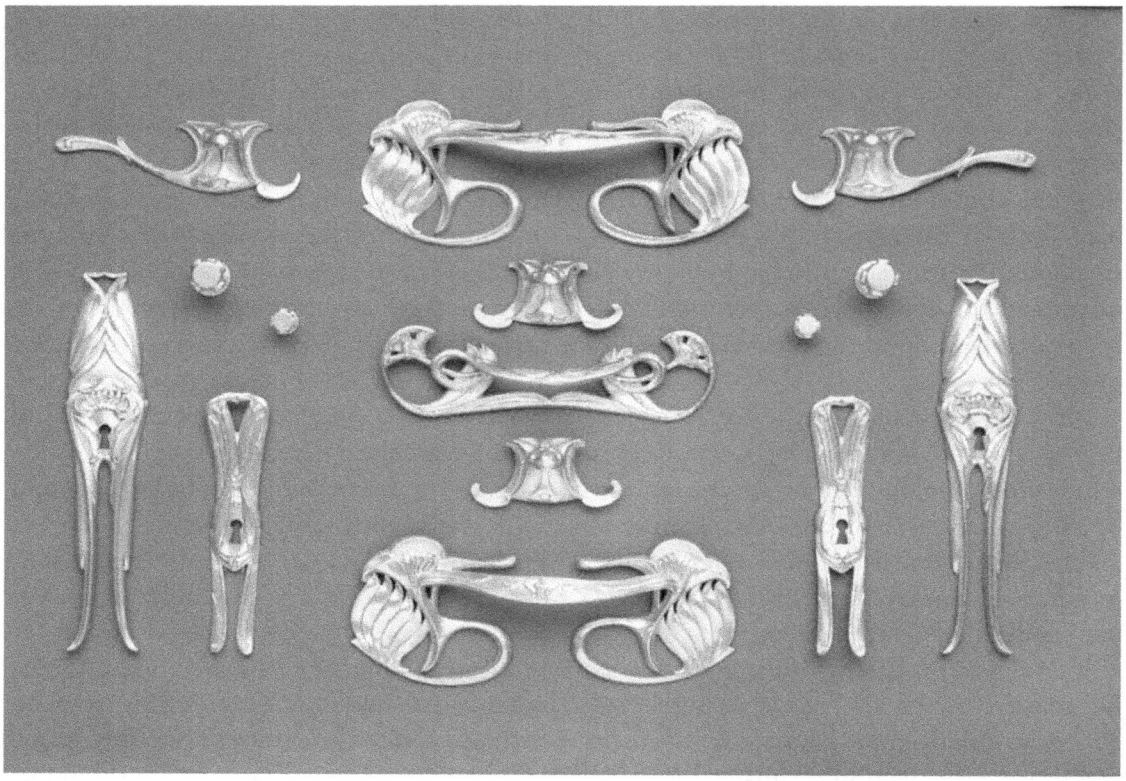

Figure 7 Georges de Feure, *Furniture Mounts*, ca. 1900.

We are well aware of our objective but hesitate over the best way of reaching it.

Let me describe the American method: Instead of exhausting her strength in sporadic efforts, whose outcome is neither definitive nor conducive to continuity, the American leaders of the new movement act with the same resolve that gives force to every American undertaking. After they have thoroughly planned the projects which they want to introduce, after they have laid the groundwork and decided upon the course to follow, together they consolidate the results of previous random efforts, thus establishing close ties and a real sense of solidarity among the most diverse art forms—major or minor—summoning the most humble individual techniques to join the most lofty concepts. Harmony emanates from these variegated elements focused upon a common goal, unity within complexity. Having achieved this result, we should expect a single style to emerge. But this is not the case, if by style we mean our own habitual aping of previous art forms, these undisguised confessions of impotence; the Americans, to the contrary, are engaged in a search for the subtle and mysterious rhythm which constitutes *style* in the noble sense of the word.

These are the new doctrines; but when it comes to putting them into practice, each artist has his individual way of seeing and proceeds according to his particular abilities and temperament.

[. . .]

Tiffany saw only one means of effecting this perfect union between the various branches of industry: the establishment of a large factory, a vast central workshop that would consolidate under one roof an army of craftsmen representing every relevant technique: glassmakers and stone setters, silversmiths, embroiderers and weavers, casemakers and carvers, gilders, jewelers, cabinetmakers—all working to give shape to the carefully planned concepts of a group of directing artists, themselves united by a common current of ideas.

Through the boldness of such corporate enterprises America may well insure a glorious future to her industrial art. But it would be an excessive optimism to harbor an absolute confidence in this regard. The path chosen, an excellent one in itself, could easily become a dead end if the nobility of the goal is not constantly kept in mind. The danger is that no single man, least of all an artist, can usually provide, from his own resources, the means necessary for so large an undertaking. Circumstances require, consequently, finding capitalization and founding a large company whose principal concern must be the material prosperity of their business. Above all, orders must keep coming in. However, the time is not yet come when the majority of American art lovers are sufficiently discerning to follow the lead of the few talented artists among them. Unsure of themselves, this great majority copies European examples, which bear the guarantee of long traditions. In spite of a proclaimed disdain for vulgar prejudice, what constitutes Society remains in perpetual thrall to European fashions. And this is especially true of external glitter, when the issue is one of social status and impressing others. For the moment, large companies of this kind cannot survive solely through the patronage of collectors of taste. They are constantly forced by

circumstance to submit to the crass demands of local tradition in order to subsist.

In contrast to our own procedures, where efforts have not yet rallied to the task at hand, Americans feel compelled to go faster than reason dictates. We must all sincerely hope, in the interest of our common cause, that Americans will come to a more balanced understanding of circumstances. For example, it would suffice to start with a few companies, launched on a modest scale, which could, without immediate financial worries, depend upon the limited clientele in whose patronage lies all hope for the future of art in America.

Another point which should be noted is the fact that all of these diverse experiments are in the realm of a sumptuary art which—intelligent or not—is intended only for certain classes of society. Most prosperous families have simply added a larger element of comfort to the simple habits of their ancestors. In this middle class is preserved intact the uncomplicated mentality, the practical and upright nature of the first settlers, whose strength tempered the nation, and whose presence is never more clearly visible than in the habits of domestic life. It was in ever-growing response to the needs of these patriarchal families that American industry made such extraordinary strides forward. Without being contaminated by any self-seeking ambition, its development took place in a rational way, following the improvement in taste and ever-new possibilities of technical resources. In this particular area of art, the manufacturing processes played an important role. Never before had there been such close association between art and labor. In terms of execution, strictly speaking, art played no role whatsoever. Nor was it even the worker's hand that intervened; with unremitting regularity, the machine did all the work, piece by piece, cutting, grinding, and polishing the thousands of models of exactly identical things . . . But well before this entry into the fray of the mechanical unconscious, art had already completed its own work, giving shape to the task.

The result was an art very different from our own, but an art nonetheless. What we sometimes call "giving everyday objects an artistic appearance" consists of applying to them some form of ornament. This is the system of additions, but additions which add nothing whatsoever to the utility of the object but which almost invariably render them less practical to handle, less harmonious of line, more difficult to maintain; above all, additions which rob them of the real character of their intended use. Furthermore, this intemperate passion for ornament rarely emerges from any spontaneous concept of an original motif, but is almost invariably a travesty of some earlier design, borrowed from things entirely different in nature. In all fairness, it should be added that our own period is not the only one to be at fault in this respect. The Renaissance itself has left us some disastrous examples; furnishings built to resemble vast monuments, where even the smallest utensils, such as tiny saltcellars, were transformed into palaces a few inches high, upon which cornices, entablatures, and caryatides abound. Even the fifteenth century, generally characterized by an art of great clarity, occasionally went similarly astray, particularly in the area of metalwork. The first instincts of a people have always produced a perfect sense of proportion, and we find nowhere greater synthesis of construction than in the products of primitive peoples; or by chance those found in remote corners of civilized countries where tradition has perpetuated them.

It is always in the name of art that everything is spoiled.

America, without being a nation any longer in its infancy, has nonetheless been able to seize the essence of a great many things by returning to their original basis, thus spurring our generation to react finally against earlier mistakes. Americans have never understood why a utilitarian object should be embellished with a load of ornament, as opposed to more perfect finish in its workmanship, or the simplicity of a more graceful, practical improvement. The beauty embodied by the appropriate form of things is not imposed through the efforts of reason alone; even when judged purely from an aesthetic point of view, sensibly built objects attract us, so strongly are the laws of logic linked to those of beauty.

[. . .]

One axiom has it that the machine is the predestined enemy of art. The hour has finally come to discredit such ready-made ideas. The machine can propagate beautiful designs, intelligently thought-out and logically conditioned to facilitate multiplication. It will become an important factor in raising the level of public taste. Through the machine, a unique concept can, when sufficiently inspired, popularize endlessly the joy of pure form, while preventing the distribution of a multitude of inept creations whose sole claim to being works of art stems from the presumable difficulty or skill involved in making them by hand.

It cannot be said, however, that America was guided by this kind of reasoning when she put these doctrines into practice. Rather, it was through the hidden force of existing circumstances that mechanization took place in the United States, through a sort of inherent intimacy with her own latent needs,

Figure 8 Louis Comfort Tiffany, *Decanter and Stopper*, 1902.

needs for which new industries had to provide equally new methods.

But if this same demonstration was nowhere as conclusive as in the lighting industry, culminating in the above-mentioned results, it is equally true that all of the industrial arts developed more or less along the same lines, all profited from having appeared at just that period when the conditions of life took on an unexpected aspect, thus formulating new requirements with which they could cope without being shackled by a multitude of time-honored customs. Every day we encumber our ancient intellectual structure with all sorts of additions; *by using every modern discovery, the New World can build its own structure with a single effort.*

And if, in the impassioned fervor of this work, errors are made constantly while we look on with a skeptical smile, let us be careful. We should be fully aware that these mistakes stem from the naive inexperience of a youthful people in the full force of their power, and whose every try, even those which miss the mark, is made with all possible strength. Which is why when America is wrong, she is never wrong by halves.

If in the same field of industrial design, we analyze instead the excellence of the results obtained, we can distinguish the following prevalent conditions:

The complete and total rehabilitation of that category of art which, renouncing the glories long restricted to painting and lofty sculpture, is concerned with enhancing the prestige of everyday objects, those whose perfection is a thousand times more important than any other.

The establishment of huge factories to concentrate the most diverse branches of decorative art in a search for a determined goal, prescribed by the powerful will of a single directing spirit.

A moral bond and tacit collaboration uniting scattered efforts; taking a multitude of external forms, identical tendencies, unthwarted by unreliable recollections of a past centuries old, and adapted to the particular conditions of time and place.

The strict subordination of questions of ornament to those of organic structure; the inner conviction that every useful object should draw its beauty from the rhythmic ordering of lines, which, before all other considerations, is subject to the practical function to be fulfilled.

The enthusiastic adoption, even at enormous financial sacrifice, of the most advanced

methods; the organization of model factories and workshops, a constant readiness
to relinquish old machinery as soon as
new, improved models are available. In a
word, the rule is: be ever and always equal
to the perpetual metamorphoses of the
times, the day and the present hour—in
every branch of human activity ripe for
development.

FURTHER READING

Paul Greenhalgh, *Art Nouveau 1890–1914* (New York: Abrams, 2000).

Nancy J. Troy, *Modernism and the Decorative Arts in France: Art Nouveau to Le Corbusier* (New Haven: Yale University Press, 2001).

Gabriel P. Weisberg, et al., *The Origins of L'Art Nouveau: The Bing Empire* (Amsterdam: Van Gogh Museum/ Cornell University Press, 2004).

15

IN THE CAUSE OF ARCHITECTURE: THE ARCHITECT AND THE MACHINE

Frank Lloyd Wright

Often described as the greatest architect of the twentieth century, Frank Lloyd Wright was also an important figure for the crafts. He designed furniture, metalwork and stained glass and gathered around him a 'fellowship' of architect/builders at Taliesin, his Wisconsin estate. While his visionary ideas sometimes outpaced his practicality—he is infamous for leaky roof problems—his buildings nonetheless exemplified sensitivity to materials and building processes. Wright wrote and lectured widely, and craft was one of his regular themes. His most well-known talk on the subject is probably 'The Art and Craft of the Machine', written in 1901. It is a rather confusing piece of writing, filled with portentous proclamations such as 'the Machine is Intellect mastering the drudgery of the earth that the plastic art may live'.[1] But the basic thrust was clear; like Siegfried Bing, Wright saw the machine as a handmaiden to artistic creativity, not as a corrosive force. This marked a departure from the ideas of Ruskin and Morris ('In the field of art activity', he wrote, 'they have wrought much miserable mischief'). Though Wright's buildings and designs are often grouped stylistically with the Arts and Crafts movement, in fact his ideas were more akin to those of German design reformers of the time, who were similarly in favor of an integrated approach to production in which handwork and machines played distinct roles. The following essay, a more cogent and concise reprise of his arguments of 1901, frames that position in no uncertain terms: architects can either adapt themselves to the techniques of machine production or become irrelevant.

Frank Lloyd Wright, 'In the Cause of Architecture: The Architect and the Machine', *The Architectural Record* (May 1927).

The Machine is the architect's tool—whether he likes it or not. Unless he masters it, the Machine has mastered him.

The Machine? What is the machine?

It is a factor Man has created out of his brain, in his own image—to do highly specialized work, mechanically, automatically, tirelessly, and cheaper than human beings could do it. Sometimes better.

Perfected machines are startlingly like the mechanism of ourselves—anyone may make the analogy. Take any complete mechanistic system and compare it with the human process. It is new in the world, not as a principle but as a means. New but already triumphant.

Its success has deprived Man of his old ideals because those ideals were related to the personal functions of hands and arms and legs and feet.

For feet, we have wheels; for hands, intricate substitutes; for motive power, mechanized things of brass and steel working like limited hearts and brains.

For vital energy, explosives or expansives. A world of contrivance absorbs the inventive energy of the modern brain to a great extent and is gradually mastering the drudgery of the world.

The Machine is an engine of emancipation or enslavement, according to the human direction and control given it, for it is unable to control itself.

There is no initiative will in machinery. The man is still behind the monster he has created. The monster is helpless but for him.

I have said monster—why not savior?

Because the Machine is no better than the mind that drives it or puts it to work and stops it.

Greed may do with it what it did with slaves in "the glory that was Greece and the grandeur that was Rome"—only do it multiplied infinitely. Greed in human nature may now come near to enslaving all humanity by means of the Machine—so fast and far has progress gone with it.

This will be evident to anyone who stops to study the modern mechanistic Moloch and takes time to view it in its larger aspects.

Well—what of it! In all ages Man has endured the impositions of power, has been enslaved, exploited, and murdered by millions—by the initiative wills [sic] back of arms and legs, feet and hands!

But there is now this difference—the difference between a bow and arrow and gunpowder. A man with a machine may murder or enslave millions, whereas it used to take at least thousands to murder millions. And the man behind the machine has nothing on his conscience. He merely liberates an impersonal force.

What is true of the machine as a murderer is just as true of it as a servant.

Which shall it be? It is for the creative artist to decide—For no one else. The matter is sociological and scientific only in its minor aspects. It is primarily a matter of using the machine to conserve life, not destroy it. To enable human beings to have life more abundantly. The use of the machine can not conserve life in any true sense unless the mind that controls it understands life and its needs, as *life*—and understands the machine well enough to give it the work to do, that it can do well and uses it to that end.

Every age and period has had its technique. The technique of the age or period was always a matter of its industrial system and tools, or the systems and tools were a matter of its technique. It doesn't matter which. And this is just as true today.

This age has its own peculiar—and, unfortunately, unqualified technique. The system has changed. The Machine is our normal tool.

America (or let us say Usonia—meaning the United States—because Canada and Brazil are America too)—Usonia is committed to the Machine and is Machine-made to a terrifying degree. Now what has the mind behind and in control of the Machine done with it to justify its existence, so far? What work suited to its nature has been given it to do? What in the way of technique has been developed by its use that we can say really serves or conserves Life in our country outside mere acceleration of movement?

Quantity production?—Yes. We have ten for one of everything that earlier ages or periods had. And it is worth so far as the quality of life in it goes, less than one-tenth of one similar thing in those earlier days.

Outside graceless utility, creative life as reflected in "things" is dead. We are living in the past, irreverently mutilating it in attempting to modify it—creating nothing—except ten for one. Taking the soul of the thing in the process and trying to be content with the carcass or shell or husk—or whatever it may be, that we have.

All Man-made things are worthy life. They may live to the degree that they not only served utilitarian ends, in the life they served but expressed the nature of that service in the form they took as things. That was the beauty in them and the one proof of the quality of life in those who used them. To do this, love entered into the making of them. Only the joy of that love that gives life to the making of things proves or disproves the quality of the civilization that produced them.

See all the records of all the great civilizations that have risen and fallen in course of Time and you may see this evidence of love as joy in the making of their things. Creative artists—that is workmen in love with what they were making, for love of it made them live. And they remain living after the human beings whose love of life and their understanding of it was reflected in them, are thousands of years dead. We study them longingly and admire them lovingly and might learn from them—the secret of their beauty.

Do we?

What do we do with this sacred inheritance?

We feed it remorselessly into the maw of the Machine to get a hundred or a thousand for one as well as it can do it—a matter of ubiquity and ignorance—lacking all feeling, and call it progress.

Our "technique" may therefore be said to consist in reproduction, imitation, ubiquity. A form of prostitution other ages were saved from, partly because it was foolish to imitate by hand the work of another hand. The hand was not content. The machine is quite content. So are the millions who now have as imitations bearing no intimate relation to their human understanding, things that were once the very physiognomy of the hearts and minds—say the souls of those whose love of life they reflected.

We love life, we Usonians as much as any people. Is it that we are now willing to take it in quantity too—regardless of inferior quality and take all as something canned—long ago?

One may live on canned food quite well—But can a nation live a canned life in all but the rudimentary animal expressions of that life? Indefinitely?

Canned Poetry, Canned Music, Canned Architecture, Canned Recreation. All canned by the Machine.

I doubt it, although I see it going on around me. It has its limits.

We must have the technique to put our love of life, in our own way, into the things of our life, using for our tool the Machine, to our own best advantage—or we will have nothing living in it at all—soon.

How to do it?

Well! How does anyone master tools? By learning the nature of them and, by practice, finding out what and how they do, what they do best—for one thing.

Let architects first do that with the Machine. Architects are or must be masters of the industrial means of their era. They are, or must be—interpreters of the love of life in their era.

They must learn to give it expression in the background for that life—little by little, or betray their office. Either that or their power as normal high priests of civilization in a Democracy will never take its place where it is so badly needed. To be a mason, plasterer, carpenter, sculptor, or painter won't help architects much—now.

They may be passing from any integral relation to life as their architecture, a bad form of surface decoration superficially applied to engineering or buildings would seem to indicate and their function go to something other and else. An embarrassment of riches, in the antique, a deadly facility of the moment, a polyglot people—the necessity of "ready-made" architecture to clothe the nakedness of steel frames decently or fashionably, the poisonous taste of the period; these alibis have conspired with architects to land us where we all are at the mercy of the Machine. Architects point with pride to what has happened. I can not—I see in it nothing great—at least nothing noble. It is as sorry a waste as riches ever knew. We have every reason to feel ashamed what we have to show for our "selves" in any analysis that goes below the skin.

A kind of skin disease is what most architecture is now as we may view it today. At least it never is organic. It has no integrity except as a "composition." And modern artists, except architects, ceased to speak of "composition" long ago.

Fortunately, however, there is a growing conviction that architecture is something not in two dimensions—but with a third and that third dimension in a spiritual sense may be interpreted as the integral quality in the thing or that quality that makes it integral.

The quality of *life* in man-made "things" is as it is in trees and plants and animals, and the secret of character in them which is again "style" is the same. It is a materialization of spirit.

To put it baldly—Architecture shirks the machine to lie to itself about itself and in itself, and we have Architecture for Architecture's sake. A sentimental absurdity. Such "Architecture" being the buildings that were built when men were workmen—and materials and tools were otherwise—instead of recognizing Architecture as a great living Spirit behind all that—a living spirit that left those forms as noble records of a seed time and harvest other than ours, thrown up on the shores of Time, in passing. A Spirit living still only to be denied and belied by us by this academic assertion of ours that they are that spirit. Why make so foolish an assertion? I have asked the question in many forms of many architects, in many places, and always had to answer myself. For there is no philosophy back of the assertion other than a denial or a betrayal—that will hold together. Instead there is a doctrine of Expediency fit only for social opportunists and speculative builders or "schools." There is no other sense in it.

The Machine does not complain—it goes on eating it all up and crying continually for more.

Where is more coming from? We have already passed through nearly every discovered "period" several times forward and gone backward again, to please the "taste" of a shallow present.

It would seem, now, time to take the matter seriously as an organic matter and study its vitals—in a sensible way.

Why not find out what *Nature* is in this matter. And be guided by principles rather than Expedients? It is the young man in architecture who will do this. It is too late for most successful practitioners of today to recover from their success. These essays are addressed to that young man.

NOTE

1. Frank Lloyd Wright, 'The Art and Craft of the Machine', 1901; reprinted in B. B. Pfeiffer, *Frank Lloyd Wright: Collected Writings, Vol. 1* (New York: Rizzoli, 1992).

FURTHER READING

Anthony Alofsin, ed., *Frank Lloyd Wright: Europe and Beyond* (Berkeley: University of California Press, 1999).

Alan Crawford, 'Ten Letters from Frank Lloyd Wright to Charles Robert Ashbee', *Architectural History* 13 (1970), pp. 64–132.

Neil Levine, *The Architecture of Frank Lloyd Wright* (Princeton NJ: Princeton Architectural Press, 1996).

Thorstein Veblen, *The Instinct of Workmanship and the State of the Industrial Arts* (1914).

16

ART AND THE MACHINE

Hermann Muthesius

For British and American audiences, the German architect and design theorist Hermann Muthesius is best known today for his groundbreaking three-volume architectural study The English House *(1904–5). The book was written while Muthesius was posted at the German Embassy in London, from 1896 to 1902, with the specific purpose of studying English design and housing. To this task he brought an outsider's incisive analysis to the Arts and Crafts Movement, which he admired but also considered to be idealistic. The following short article, published in* Dekorative Kunst *(an influential journal edited by the art critic Julius Meier-Graefe) at the end of Muthesius's British sojourn, is arguably the most clear and concise statement of his thinking on the subject. Like other German designers such as Peter Behrens and Richard Riemerschmid, he felt strongly that reformers must work with the machine rather than against it and argued that the opposition of art work and factory production was a false one. Five years later in 1907, Muthesius would take the leading role in forming the Deutscher Werkbund, an organization of designers and artists based in Munich. Often considered the precursor to the Bauhaus, the Werkbund immediately became the focus for debates about craft and design in central Europe—similar to the Wiener Werkstätte in Vienna, with which it had close ties (particularly through the figure of Josef*

Hoffman, a co-founder of both organizations, who also espoused a machine aesthetic). The Werkbund shut down in the 1930s during the Nazi era but reopened in 1950 and continues to be active today.

Hermann Muthesius, 'Art and Machine', *Dekorative Kunst* (1902). Translated by Peter Adamson.

Amongst all the tasks that remain to be solved regarding the topic of modern art, the proper assessment of the work done by machines is the most difficult, but also the most wide-reaching and meaningful. From an "artistic" point of view, we are accustomed to dismiss, or even to lament, the products of machines. Many today claim that the commercial arts can be enlivened again only by means of handicraft. Yet our factories keep churning out their wares, sending them into the marketplace in huge quantities. No one claims that these have anything to do with art; only the products of the "art industry" may present themselves as such. Only these products are allowed into our art journals and our books about art history. We find ourselves, then, in a situation strikingly opposed to that of the culture which machines began to chew apart a century ago: then, everything belonged to (what we now consider as) art, whereas now, we distinguish between the legitimate children and the misbegotten.

But when we consider things from the economic point of view, rather than from an artistic point of view, things look rather different. The products that are economically natural are of course the things made by machines. They come about under conditions that have developed in natural circumstances, are produced upon the foundations of the marketplace, and supply the most useful benefits with the least amount of labor. These are, then, the products which are "modern" in the truest sense: they are appropriate to our current living conditions. Whereas the products of the "art industry" are more or less a matter of taste, available only to the well-off, that is, to the few, and could never provide for the daily needs of the people. If one takes handicraft as an ideal, one commits oneself to the economically unnatural. The immediate result is the bizarre cultural ideal championed by William Morris and the English socialist artists, who began from the idea of an "art by the people for the people" and ended up with such expensive products that only the wealthiest few thousand people could consider owning them.

There have been frequent attempts, up to the present day, to argue on any grounds possible that factory products must necessarily be non-artistic. But as the years pass, the building built upon these grounds (which derive from the aesthetic of our earlier culture) wobbles more and more. Nowadays many are willing to say they find a landau, a suspension bridge, a locomotive "beautiful"—all things that have no connection with art, but rather grow wild outside its preserve, so to speak. Hand in hand with this development comes an increasing aversion to ornamentation, to impractical elaboration of form, to decoration in general, things to which the old art had no objection. There is certainly a current towards recognizing the aesthetic legitimacy of the children

that are produced in a practical, "non-artistic" way, and to allow them into the ring of art. By contrast the things that were "artistic" ten, twenty or thirty years ago now seem to many "unmodern," if not "non-artistic," and precisely because of the "art" that was imposed upon them. As soon as these attitudes become more widespread, the old prejudice against machine products must disappear, and no one will designate these items as falling outside the realm of the aesthetic.

Aesthetic judgment is built upon prejudice; habit is its midwife. The basic, shared human susceptibility to beauty is quite primitive. It was the same basic susceptibility to beauty that gave rise to the Ionic temple in ancient Greece and the fantastic, extravagantly ornamented buildings of middle India, which are incomprehensible to us Europeans. Yet both developed from the same starting-point in the human endeavor to create beauty, and here, as there, they rank as the highpoint of art.

Even the most impossible is thus a possible development and reconstruction of aesthetic judgment. We get a glimpse of this truth even from ladies' fashion. So it is entirely possible that eventually, the common man will find our factory-made products beautiful. That this is possible is shown by a fabricated product which we have for centuries habitually, and thus unhesitatingly, called "artistic", namely the printed book. Gutenberg's achievement was nothing other than the introduction of machine-production into the printing of books, in place of handwork. Who would nowadays deny that books, treated in the right way, can be products of art in good standing?

In the story of human tectonics, we may observe the oddity that new developments must often go through an intermediate phase before reaching their final form, not unlike the metamorphosis of insects. Indeed this seems to be the norm. As soon as new conditions arise, they demand that we put our hands to tasks for which we have no point of reference. Thus in the nineteenth century, gas replaced candles, and electric light replaced gas. The right form to suit the means of illumination never appeared immediately; rather there was a step-wise development away from imitation candles. When linoleum came on to the scene, it was at first always given a pattern of granite, tiles, or parquet, and wallpaper was made to imitate fabric. In each case we see intermediate phases which imitate the earlier form.

Figure 9 Josef Hoffmann, *Vase*, 1905–10.

Our fabricated products have gone a similar path. Whenever we began to make some previously handmade item using mass production, great ingenuity was applied to imitating the peculiarities of the handmade form. So we had stamped tin ornamentation, imitation wood carvings in paper, and other forgeries which could only bring artistic discredit upon the work of machines. Because, from an artistic point of view, we had not previously paid any heed to machine-made things, we now began to disdain them. From the outset, the best prospects for appreciation lay with those machine products which arose entirely independently, without reminding us of any previous objects. To take a convenient example, the bicycle had no earlier form it could imitate: and so in its case we glimpse an appropriate, pure form of the machine-made. Here the form of the object is in keeping not only with its origin in mass production, but also—because so-called "artistic" issues did not even arise in its design—with the highest degree of functionality (*Zweckmässigkeit*). And yet the bicycle is quite pleasing. Perhaps because in this case, the conditions of existence for this thing, and the special form in which the thing appears, overlap completely. It embodies a certain genuineness; it has style. Whether one calls this style "functional style," "machine style," or whatever, there is no reason the bicycle should not be pleasing.

The opposition between this machine style and the so-called artistic style is no doubt a stage of aesthetic appreciation which the world will grow out of. It cannot be a matter of how a manmade object came to be; it is, rather, important that it wears its manner of coming to be on its sleeve, that it embody a clear style. If this is not the case, it presents itself in disguise and sooner or later repels us because of the false conviction that lies hidden within its form. For our entire human tectonic to be subject to the same laws, one and the same notion would have to cover all tectonic achievements; but nowadays we have no word for this concept. One ought to think carefully before using the resounding word "art." For we now have such particular associations with this term that utterly mistaken judgments will arise. As we have seen, it has come to the point that we call things made under the natural economic conditions of today un-artistic, while we call the things made in accordance under unnatural conditions artistic. Something is amiss here.

It would seem our minds are haunted by the fossils of an old culture. We tend to name a tectonic achievement as artistic only when it has no admixture of functionality whatever. This leads to a dichotomy, and to a current uncertainty, which express themselves not only in the arena of judgment, but also in the creation of tectonic achievement. A fluctuation between realism and the idyllic: that is the present situation in the so-called "new" art industry. The one who takes it in hand to create properly does not concern himself with the question whether he is setting out to make this object artistic, that object unrelated to art. He creates however that which the human condition of his day (*menschliche Bildungsgeist*) leads him to. And so long as this undertaking suffers from no learned prejudices, it is always by its very nature artistic (keeping this word, for lack of a better one). Regardless whether man's products emerge from a machine or from our hands, they will inevitably be artistic, so long as they are done properly. The machine is nothing but a more perfect instrument.

Nowadays one often hears that we should make our life artistic once more. But those who say this go astray, wishing to reach this goal by setting up an opposition to the work done by machines. It is from this false perspective that all modern English art, from Morris to the present day, has proceeded. The scope of its influence will continually shrink as machines march forwards towards victory. As soon as we instead look this new emergence in the eye, acknowledge it, give it our attention, expand upon it, lovingly develop it within the pure, clear conditions of its making, we are then not confined but enriched. It is a sign of old age to want to preserve a situation which has passed beyond its natural growth. A youthful view of art faces forwards. Any view of art today must take the machine-made under its wing, or abandon itself to the prospect of an imminent end.

FURTHER READING

Paul Betts, *The Authority of Everyday Objects: A Cultural History of West German Industrial Design* (Berkeley: University of California Press, 2004).

Le Corbusier, *A Study of the Decorative Art Movement in Germany*, edited by Mateo Kries and Alexander von Vegesack (Berlin: Vitra Art Museum, 2008).

17

BUILDING MATERIALS

Adolf Loos

Perhaps the most vilified and least well-understood theorist on the subject of craft is Adolf Loos, the Viennese architect who in 1908 penned the notorious essay 'Ornament and Crime'. This hysterical condemnation of the use of decoration was intended by Loos primarily as an attack on the richly embellished style of the Viennese Secession. However, it has become famous (or infamous) for its combination of a seeming prescience regarding the Modernist architecture of succeeding decades—much of which did indeed eschew ornament in favor of white walls and clear glass—and a horrifyingly racist attitude towards 'primitive' cultures such as that of Papua New Guinea. Loos characterized the practice of tattooing as 'degenerate'" and argued that cultural progress could be indexed by the degree to which useless decoration had been abandoned in favor of useful and simple design. He also extended this argument to the design of his own day, pointing to the fussy ornament on contemporary shoes—little holes and deckled edges—as an instance of misplaced, inefficient labor. There is no getting around the vile aspects of Loos's thinking, which are all too redolent of the later rhetoric of racial purity deployed by the Nazis. But it cannot be too strongly emphasized that his anti-ornament diatribe was an argument not against good craftsmanship, but rather in favor of it. Loos did not by any means want to do away with the traditional skills of the

shoemaker; on the contrary, he insisted on directing the craftsman's energies towards less arbitrary ends. This is clearer, perhaps, in the following less widely known, and less controversial, essay, in which Loos states his high regard for the mastery of materials and his 'awe for the human work' that this entails—an attitude amply evident in his buildings, which often exhibit great sensitivity to workmanship.

Adolf Loos, 'Building Materials', *Neue Freie Presse* (28 August 1898), reprinted in Loos, *Speaking Into the Void: Collected Essays 1897–1900* (Cambridge, MA: MIT Press, 1982).

Which is worth more, a kilogram of stone or a kilogram of gold? The question probably seems ridiculous. But only to the merchant. The artist will answer: All materials are equally valuable as far as I am concerned.

The Venus of Milo would be equally valuable whether it were made of the rubble which paved the streets—in Paros, the streets were paved with Parian marble—or gold. The Sistine Madonna would not be worth a penny more if Raphael had mixed a few pounds of gold into his colors. A merchant who has to consider melting down the golden Venus in case of need or scraping off the Sistine Madonna will, of course, calculate differently.

The artist has only one ambition: to master his material in such a way that his work is independent of the value of the raw material. Our architects, however, have not heard of this ambition. For them, a square meter of wall surface out of granite is more valuable than a square meter out of plaster.

But granite in and of itself is worthless. It lies all around outside in the fields; anyone can get hold of it. It forms whole mountains, whole mountain ranges, which one has only to dig up. The streets are topped with it, and the cities are paved with it. It is the most common stone, the most ordinary material that we know. And yet there are people who consider granite our most precious building material.

These people say "material" but they mean "work." Human labor, technical skill, and artistry. For granite demands much work to wrest it from the mountains, much work to bring it to the designated location, work to give it the correct form and to endow it with a pleasing appearance by cutting and polishing. Our hearts beat with reverential awe at the sight of the polished granite wall. Awe for the material? No, awe for the human work.

So might granite then be more valuable than plaster? We have still not said that. For a wall with a plaster decoration by the hand of Michelangelo would overshadow even the most highly polished granite wall. It is not just the quantity, but the quality of the work performed that determines the value of an object.

We live in a time that gives precedence to the quantity of work performed. For quantity is easily controlled; it is immediately obvious to anyone and demands no skilled eye or special knowledge. Thus there are no errors. So many workers have worked at a job for so many hours at such and such a wage. Anyone can calculate it. And we want to make the value of the things with which we surround ourselves easy to understand. Or else there

would be no point to them. Thus, those things that took a longer time to make must deserve more respect.

It was not always this way. Formerly one built with the materials that were the most easily obtainable. In some regions this was with brick, in some with stone; in some the walls were stuccoed. Did those who used stucco consider themselves somewhat inferior to those architects who built in stone? Of course not, why should they have? The idea did not occur to anyone. If there were quarries in the vicinity, one simply built out of stone. But to bring stone to a building from far away seemed more a matter of money than of art. And art, the quality of a work, meant more formerly than it does today.

Times like those brought out proud, strong natures in the field of architecture. Fischer von Erlach did not need granite to make himself understood. He created works out of clay, limestone, and sand, works that capture our attention as powerfully as the best buildings made out of materials that are the most difficult to handle. His spirit, his artistry mastered the most miserable materials. He was capable of bestowing the nobility of art on the most plebeian dust. He was a *king* in the realm of materials.

Today it is not the artist who rules, but rather the day laborer, not the creative idea, but the working hours. And the rulership is gradually being wrested even from the hands of the day laborer, for something has appeared that has a qualitatively better and cheaper work output: the machine.

But any amount of production time, whether of the machine or the coolie, costs money. And if one has no money? Then one begins to fake the working hours and to imitate materials.

Figure 10 Adolf Loos, *Chest of Drawers*, ca. 1900.

The reverence for the quantity of work done is the most fearsome enemy that the crafts profession has. For it results in imitation. And imitation has demoralized a large part of our crafts. All pride, all handicraft spirit have left it. "Book printer, what can you do?" "I can print books in such a way that they are taken for lithographs." "And lithographer, what can you do?" "I can make lithographs that are taken for prints." "Carpenter, what can you do?" "I can carve ornaments that look so easy you could mistake them for stuccowork." "And stucco worker, what can you do?" "I can imitate moldings and ornaments exactly and make hairline joints that appear so authentic that they look like the best stonemasonry." "But I can do that too!" cries the sheet-metal worker proudly. "When my ornaments are painted and sanded, no one would suspect that they are made out of tin." What a pitiful group!

A spirit of self-degradation pervades our crafts. It is no surprise that this profession is doing badly. Such people cannot help but do badly. Carpenter, be proud that you are a carpenter! It is the stucco worker who makes ornaments. You should pass him by without jealousy or envy. And you, stucco worker, what have you to do with the stonemason? The stonemason makes joints, unfortunately has to make joints, since little stones are cheaper to come by than big ones. Be proud of the fact that your work does not exhibit the paltry joints that cut the stonemason's columns, ornaments, and walls into sections. Be proud of your profession, be happy that you are not a stonemason!

But I am talking to the wind. The public does not want a proud craftsman. For the better the craftsman can imitate, the more the public will support him. Reverence for expensive materials—the surest sign of the parvenu stage in which our nation currently finds itself—will have it no other way. The parvenu considers it disgraceful not to be able to adorn himself with diamonds, disgraceful not to be able to wear furs, disgraceful not to be able to live in a stone palace—ever since he has learned that diamonds, furs, and stone palaces cost a great deal of money. He does not know that the lack of diamonds, furs, and stone facades has no effect on elegance. Therefore, since he is short of money, he grasps for surrogates. A ridiculous enterprise. For those people whom he wants to deceive, those, that is, endowed with the means to surround themselves with diamonds, furs, and stone facades, cannot be fooled. They find his efforts laughable. And his efforts are further unnecessary vis-à-vis those of a lower standing than his if he is conscious of his own superiority anyway.

In the last decade imitation has dominated the entire building industry. Wall coverings are made out of paper, but this they may by no means show. They must retain the patterns of damask silk or Gobelin tapestries. Doors and windows are made out of softwood. But since hardwood is more expensive, the softwood must be painted to look like it. Iron must be painted to look like bronze or copper. But against poured cement, an achievement of this century, we are entirely helpless. Since cement is in and of itself a splendid material, we have just one thought whenever we use it, the same thought that we have upon first confronting any new material: what can we imitate with it? We used it as a surrogate for stone. And since poured cement is so extremely inexpensive, like the parvenus that we are, we indulged in the most thoroughgoing wastefulness. A true cement epidemic gripped the century. "Oh, my dear Herr Architect, couldn't you put just a little more art on the facade for another five gulden?" the vain contractor probably said. And the architect tacked as many gulden worth of art onto the facade as were demanded of him, and sometimes a little more.

Nowadays poured cement is being utilized for the imitation of stuccowork. It is characteristic of our Viennese situation that I who am against the violation of materials, who have combated imitation energetically, am dismissed as being a "materialist." Just look at the sophistry: these are the people who attribute such a value to materials that they have no fear of their becoming characterless and who freely resort to surrogates.

The English have exported their wallpaper to us. Unfortunately they cannot send over entire houses as well. But we can see from their wallpaper just what the English are aiming for. This is wallpaper that is not ashamed to be made of paper. And why should it be? There are certain wall coverings that cost more. But the Englishman is not a parvenu. In his home, it could never occur to anyone that the money had run out. Likewise, his clothes are made of sheep's wool, and they display this

honestly. If the leadership in clothing were left to the Viennese, sheep's wool would be woven to look like velvet and satin. Even though it is only made out of wool, English clothing material, and thus our clothing material, never manifests the Viennese "I'd really like it, but I can't afford it."

And that should bring us to a chapter that plays the most important role in architecture, to a principle that should form the ABC of every architect—namely the principle of cladding. But I will reserve discussion of this principle for my next article.

FURTHER READING

Hal Foster, *Design and Crime (And Other Diatribes)* (London: Verso, 2003).

Kenneth Frampton, *Adolf Loos: Architecture 1903–1932* (New York: Monacelli Press, 1996).

Benedetto Gravagnuolo, *Adolf Loos: Theory and Works* (London: Art Data, 1995).

Janet Stewart, *Fashioning Vienna: Adolf Loos's Cultural Criticism* (London: Routledge, 2000).

18

HANDWERK/KUNSTHANDWERK

Stefan Muthesius

The words that we use to describe craft have a dramatic, and often unnoticed, effect on the way that crafted objects themselves are understood. The Japanese term kurafto *(a direct adaption of the English term), for example, was invented in the middle of the twentieth century to describe industrial products that refer to traditional materials and techniques (which are designated by the older word* kôgei*). The following essay by Stefan Muthesius (an art historian at the University of East Anglia and relative of Hermann Muthesius) employs the tools of etymology as a way of understanding the relation between craft reform and industry in Germany. Muthesius argues that in the late nineteenth century the wide compass of the term 'handwerk' (hand work) was a linguistic analogue of—and perhaps even a contributor to—the higher tolerance for mass production in German craft reform than in English-speaking nations. It also helps us to trace the continuity of guild structures in Germany, which arguably did not suffer the same rate or degree of decline as Britain did when it came to such customs as apprenticeship. Finally, Muthesius uses terminology as a way of tracking the emerging polarization of debates about design reform around the turn of the century, with some reformers arguing for* Kunsthandwerk *(artistic crafts) and pro—mass production designers such as Richard Riemerschmid*

wielding neologisms like Maschinenmöbel *(machine furniture). Subsequently, Modernists would try to create a master theory of craft and design, which centered on a simpler term:* form. *Taken as a whole, Muthesius's terminological account provides an unexpected map of the complex terrain of craft's evolving relations with industry.*

Stefan Muthesius, 'Handwerk/Kunsthandwerk', *Journal of Design History* 11/1(1998).

One way of trying to understand the complexities of nineteenth- and twentieth-century Western crafts and design is to enquire into their basic terminology and to reflect on their terminologies in different languages. It is essential to state, at the outset, that for this exercise it is not enough simply to present English 'translations of foreign words', certainly not in the case of the major terms. Rather, what is needed are definitions, using a wide vocabulary of adjacent words as well as their etymology, which means going backwards and forwards between languages. We need to be aware, throughout, of the common roots of Germanic, or Romance words, which, of course, simply reflect the shared Western social and cultural history of all major concerns in the fields of art and techniques. An English speaker does not need to know much German in order to get at the basic meaning of 'Hand Werk' or 'hand work'; on that general level English and German are still simply the same language. And yet we must be cautious at precisely this point. To translate literally Handwerk as handiwork ('hand work', curiously, lacks meaning) could be misleading, because the latter, if retranslated into twentieth-century German as 'Handarbeit', has the very much restricted meaning of ladies' needlework (useful as well as ornamental).[1]

Handwerk, in all German-speaking lands, is the chief twentieth-century umbrella term for anything which does not come under 'Industrieproduktion' (industrial production). A better twentieth-century English equivalent is thus another old Germanic word, craft, or 'the crafts' (its modern German derivative, 'Kraft', however, cannot be used in our context because it means, very generally, force, strength or energy). Like craft, Handwerk can be seen in opposition to the products of 'industry', i.e. it is perceived to possess values which are different from, and better than those of industry—although that opposition is not nearly as strong as in the case of crafts. Like crafts, Handwerk can, furthermore, be understood in opposition to design, although, because of the vagueness of the term design, the juxtaposition of Handwerk with Design is often nebulous, too. We shall come back to this when we mention some of the German twentieth-century equivalents and variants of 'design'.

On a basic level, Handwerk can also be opposed to Kunst.[2] However, as one is acutely aware of the perennial European confusion reigning within the semantic field of arts, of the fine, applied and technical 'arts', it is not surprising to note that the Germans could create the seemingly paradoxical combination 'Kunst Handwerk'. But this happened only during the later nineteenth century and hence our second term is very much easier to understand in English-speaking countries. The German Kunsthandwerk movement of the late nineteenth or certainly the twentieth century was pretty much the equivalent of our Arts and Crafts movement, and of today's studio crafts. Kunsthandwerk will be discussed further below and we first turn to Handwerk as such, although we have to be aware that some Arts and Crafts values were instilled into plain Handwerk, too. A further English term comes to mind—'artisan'—together with the French 'artisanat'. The way it refers to the skilled worker, in contrast to the unskilled one, indeed corresponds to the notion of skill in Handwerk; but as it was the designation of the step on the social status ladder that was the chief purpose of the term artisan, and as it was mainly used in the nineteenth century, its comparison with the broadly used Handwerk is of limited value.

At the outset we must stress that Handwerk comprises a vastly greater sphere of activities than twentieth-century English 'crafts'. Handwerk is a term that has a firm position in the realm of economics and statistics. A Handwerker is the generic term for the person who repairs one's plumbing, who cuts one's hair; the baker belongs to the 'Backerhandwerk', the bricklayer to the 'Bauhandwerk'. Bauhandwerk is opposed to 'Bauindustrie', although here, in particular, the borderlines between Handwerk and Industrie are increasingly difficult to draw. Another English term needs to be introduced here—'trade'. Handwerk comprises the 'trades', such as in the 'building trades'; it shares some of the imprecisions of the term 'trade', including (and contrary to what has been said above) the blurred borderline towards industry. On the whole, though, modern German

Handwerk usually tries to play down any purely commercial element. Thus, at its very briefest, Handwerk comprises both crafts and trades. One could branch out here into general perceptions of German products, the ethos and myth of 'Made in Germany', by reflecting on the ways in which a properly trained, 'professional' Handwerker combines the best of the 'trader' and of the 'craftsman', being the conductor of a small business with his or her feet on the ground, but also practising individualistic, contemplative or arty ways of making, designing and inventing.[3]

If one goes back 150 or 200 years, the situation in both German- and English-speaking countries was much simpler. There is actually a German equivalent to trade, and that is 'Gewerbe'. An old synonym to Gewerbe is 'Gewerke', its nucleus simply being 'Werk' (work). An old equivalent to modern German 'arbeiten', and the correspondent for 'to work', is, indeed, 'werken'. We shall come below to the way in which nineteenth- and twentieth-century applied art and design ideologues put special emphasis on certain terms and made them sound powerful: 'Werk' was certainly one of those. Up to the last decades of the nineteenth century Gewerbe/Gewerke, and, for that matter, Industrie, covered everything, from the roughest kinds of large-scale production and the finest kinds of manufacture ('Manufaktur' is a German term, too, but its use has been restricted to 'Porzellanmanufaktur') down to the small jobbing craftsman or trader.[4] The big change came, as everywhere else, with industrialization. Its main phase, in Germany, is witnessed later than in Britain and its impact was more sudden. Full 'Gewerbefreiheit'—the complete freedom to set up any kind of business, anywhere—was only introduced in 1869, in anticipation of the complete economic and political unification of Germany. Modern industry had finally swept away the remnants of the old guild restrictions. It soon, however, appeared that this might spell disaster for the future of all those manufacturing branches which had not acquired, or saw no prospect of acquiring, large-scale workforce or machinery. There would seem to be nothing left to do for the smaller trades, everything was to be made by machine; only repair work would remain for the impoverished jobbing artisan. Large branches of trade, even the whole of Handwerk could be defined negatively, as that which was left behind by modernization. It was now that Handwerk began to reflect on its state and status and rapidly built up its modern ethos, terminology and complex organization.

German social stratification models, being somewhat different from those which are normally used in Britain, speak of the 'class of the Handwerker', or the 'class of the peasant farmer'. The class of the Handwerker (Handwerkerstand) was held to be very largely part of the middle class (Mittelstand), of the bourgeoisie. It was during the later nineteenth century that a large section of the Buergerstand, and with it most of the smaller trades, took a political turn to the right; they became opposed to both liberal internationalism and internationalist socialism; more importantly, many of them, by 1900, had embarked on an ideological stance of cultural pessimism or scepticism, an ideology of anti-Modernism and anti-progress as well as a widespread nationalism.[5] A reference to the seemingly intact world of the 'medieval craftsman'—a notion that had first been voiced around 1800—became de rigueur.

At the same time, Handwerk gave itself a modern organizational framework with nationwide associations, conventions, cleverly organized publicity and political lobbying. Each town or district established a 'Handwerks-kammer' (equivalent to the Chamber of Commerce and Industry, the Industrie- und Handelskammer) or a 'Handwerksverband/Handwerksverein', i.e. associations or societies. At times, the old term 'innung', Handwerks-innung, was revived, though on the whole the modern organization stayed clear of the other old German term for guild, 'Zunit' (to add to the richness, the Germans also use Gilde). What was modern about this setup was the carefully orchestrated network at regional and national level; 'Professionalization' is another English term (without a precise German equivalent) which could be applied at this point. There were numerous pieces of legislation, especially during the 1880s, which attempted to continue some of the old rules, or at least the old nomenclature of the guilds, for instance the old system of apprentices and journeymen, and the protection of the term 'Meister'. As with brand names, this was and is a legal protection of the terminology, a re-inforcement of its ethos, but not an absolute protection of production monopolies. To this day, Handwerk represents a seemingly firmly defined group of crafts, trades or professions, and at the same time an officially defined social group within German society.[6]

Another crucial way of helping Handwerk in its 'competition with industry' was to lean towards the art side of manufacturing. As elsewhere, Germans in the second half of the nineteenth century were preoccupied with 'applying art to industry' in as many branches of manufacture as possible. Applied arts is trans-lated as 'Kunstgewerbe', or 'angewandte Kue-nste' and sometimes as 'Kunsthandwerk'—all terms, as well as Kunstindustrie, were used interchangeably until the 1890s. Kunstgewerbe meant the belief that artistic values in manu-facture could be inculcated through educa-tion. In the early decades, that is until about 1870–80, the Kunstgewerbe movement still be-lieved in a comprehensive way of applying art to Gewerbe; i.e. art could be applied to all produc-tion processes, to machine and mechanized processes, as well as to hand processes. The art in applied arts was largely understood as 'applied' ornament.[7] There was a belief, which was, for instance, still shared by Alois Riegl, namely that good, ornamented products could be created without much additional physi-cal effort, without much added cost, simply through a better understanding of ornament and by acquiring good taste ('Geschmack').[8]

But by the late 1870s, certainly by the mid-1880s, German critics began to divide industry up. The twentieth-century attempt of Handwerk to define itself as a distinct kind of activity had begun. It was now believed that machine-produced ornament was, or had lately developed a tendency towards the 'cheap' and bad. It was held, furthermore, that it was German products, in particular, which had succumbed to this new trend. French products appeared consistently superior, and, to a growing extent, English ones, too. We now enter the familiar Arts and Crafts–Modernist trajectory. It is not the purpose of this contribution to rehearse its main tenets and those of subsequent Modern design; on the other hand, it is impossible to understand the meaning and aims of twentieth-century Handwerk and Kunsthandwerk without bear-ing Modernism in mind. From the 1860s–70s onwards it was the organizers of the Applied Art museums, the 'Kunstgewerbemuseen', who provided the tone of the discussion and who pushed arguments forward. In their wake, the writers on the applied arts, merging into what today would be called design criticism, began their powerful discourses,[9] appearing mainly in a new type of publication, the arty kind of applied arts journal. The key term was 'Reform'. Reform meant, of course, aiming for the new, but it also aimed at the assurance, or reassurance of quality, artistic merit and a high cultural ranking.

The term Kunstgewerbe continued to be used; after 1900 the term Kunsthandwerk slowly gained prominence, though, as we shall see, it was not until about 1930 that it acquired its full present-day use. The well-known key date for all German-speaking countries was 1897. In 1897 Jugendstil members of the Munich Secession, the breakaway artists' group founded in 1893, initiated the 'Münchner Werkstätten'. This represented a two-pronged attack on the older and the late nineteenth-century kinds of manufacture of the applied arts. The choice of the term Werkstätten—as such a perfectly good German term was very probably influenced by the English Arts and Crafts new use of 'workshops', signifying small-scale work organizations. More important in the early years, from 1897 to about 1905, was the insistence on pure artistic input. All products of the movement were designed by named designers, by artists, as was common within the British Arts and Crafts groups.

Returning to our term Handwerk, the situation grew more and more complex. A typical Handwerker, say a maker of reasonably up-market furniture in a small or medium-sized firm, now saw himself squeezed not only by industry, or the large manufacturing firms, but also from the other side, so to speak, from a new kind of artist-designer. The Meister had to watch, but was not able to understand, the meteoric rise of Jugendstil and Secession furnishings which dominated the new smart journals, where those designers—and the firms who made the pieces—received free advertising, and attracted the highest patronage.

It seemed all the more paradoxical, as in many ways the smaller class of Handwerker and the new smart designers were sharing the same platform of anti-machine and anti-industry. What the Handwerker did not want to understand was an actual need for a designer; had he not always produced useful and beautiful cabinets whereby the act of designing was integral of the whole process?

The second complicating factor was the way in which our new group of critics rediscovered, as in England, what they saw as the old traditions of Handwerk. When we say in German 'das traditionelle Handwerk', we come close to the English term 'traditional crafts'. What this meant, in actual fact, was to project on to old work (as distinct from contemporary work) certain values: a sympathy for the more 'basic' kinds of craft practices and 'simple', 'traditional' ways of life. This was done under the banner of 'Volkskunst'. Folk art and folk crafts, for instance 'peasant furniture', suddenly, from around 1890 onwards, appeared eminently attractive in terms of outline and colour and its quaint, often sparse, motifs of decoration. Furthermore, this furniture was described as 'practical', rationally constructed, it was held to possess a 'sense of the material', honesty. Compared with this, the ordinary Handwerk of the day appeared stuck in the depths of 'meretricious' late nineteenth-century ornamentation and copyism. Thus, paradoxically, the new art movement, especially in interior design, would throw in the works of the contemporary, the 'ordinary' cabinetmaker-Handwerker with 'bad taste' and the unoriginality of late nineteenth-century mass-produced items.[10]

By about 1905 a strong polarization can be observed. There was avant-garde Modernism in design, and there were the forces of the old-fashioned, the remnants of the nineteenth

Figure 11   Richard Riemerschmid, *Table and Chair,* 1898.

century. This was the line taken by Hermann Muthesius who gathered most of the avant-garde designers behind him. As a critic put it in the 1920s: 'The Handwerker is stubborn [ein Dickkopf]; he wants to make everything beautiful, and that means he adds ornament . . . [through] imitation.'[11] By 1907–10 the situation was further complicated by the way in which many of the new artist-designers united with the very newest ideas in design reform, namely that machine work could be good, if it were appropriately designed, such as with the 'künstlerische Maschinenmöbel' designed by the top Munich artist-designer Richard Riemerschmid in 1905. The next catchword of this group was 'Typenmöbel'-type furniture.[12] The new group of critics, artists and also some manufacturers then began to use an evocative term, by combining the ring of the powerful 'traditional' word Werk with the even more archaizing term 'Bund', which means a close-knit, brotherly group. Like the term Bauhaus a decade or so later, these artificially constructed but seemingly natural names stick in everybody's mind, although, taken literally, they say little and may even be misleading. The

Deutsche Werkbund's platform was that the vast majority of German products, whether Handwerk or industry, were bad from the point of view of art or taste and even in their practical functionality.

It must be noted that all these controversies were conducted, on both sides, among a small elite; there was a broad spectrum of producers up and down the country who created work which was never dealt with in the critical press. On the other hand, there were the numerous professional associations, mentioned earlier, who saw it as their task to discuss publicly the new problems. Some of those who did not adopt Modern styles of design and did not belong to the Werkbund united and protested. For the Handwerker, a redefinition of his or her role and values was now imperative. Under the banner of Kunst, here, too, a process of stratification took place. We are henceforth concerned with the somewhat separate group of those trades which concerned themselves with interior design. The fact was—and this needs much further investigation—that at least the domestic furnishing branch of Handwerk consolidated itself to a large extent during the years 1910–25. This demonstrated a new openness in two directions: firstly, there was no need to condemn outright the use of machines; in woodwork, for instance, their use for the roughest kinds of work would not harm the image of quality. Secondly, some of the larger firms had begun to employ some of the new designers on a freelance basis. Furthermore, many of the ideas of the once assertive early Jugendstil phase were now being rejected by new avant-garde trends, and their designers had had to shed some of their pride.

Above all, Handwerk managed to consolidate its image by catering for the higher and the highest segments of the market. It gave itself an image of absolute quality. By the early 1920s, both Handwerk and much designer Kunstgewerbe demonstrated a certain reversal of the 1900 radical position—unornamented = artistic = high-class—and a return to a more traditional hierarchy of decoration. An expensive interior of 1920 would be praised for its 'nobel' and restrained decor. This was then coupled with an emphasis on 'craftsmanship' or 'fine craftsmanship'; the Germans introduced a more official-sounding term, 'handwerkliche Qualität' (craftsman-like quality). 'Quality', pure and simple, had been one of the Werkbund's watchwords; but there seemed no absolute, indivisible quality; each group of producers could add the prefix that suited them. Occasionally, 'handwerkliche Qualität' could actually be used in the negative sense, namely by those who meant 'only handwerkliche Qualität', in contrast to top-class industrial design. Handwerklich meant high dexterity, finish and, especially during the 1920s, the demonstrative use of expensive materials—in other words, 1920s 'art deco', a term that had not yet arrived on the scene. Handwerk's aspiration to high culture was underlined by the term 'Handwerkskultur'. 'Nobel', 'vornehm' were the German equivalents of 'refined', a key value in France and England, too.[13]

There was, besides all this, much production of furnishings in the 'Volkskunst' style, based on some of the ideas of the Volkskunst revival of 1900, mentioned a while ago—or in the 'Heimatstil' (using the German evocative combination of home and homelands); but this now acquired, in conjunction with greater precision in folklore studies, a more specific and thus restrictive regional ethnic categorization. Looking at English-speaking countries, there was, and is, a vast amount of 'Crafts-Shop' crafts in Germany (for a long time, and confusingly, this branch of work continued to be referred to by the nineteenth-century term Kunstgewerbe).

To sum up: from 1920 until almost the present day there are a number of principal strands in the production of interior design and objects of daily use: 1) volume, or mass manufacture; 2) localized small-trade manufacture, called Handwerk, quasi-anonymously producing many of the same goods as 1); 3) furnishings by named designers from the Modernist art movements; 4) a more limited range of mostly smaller kinds of products under the banner of folklore; 5) the beginnings of the Arts and Crafts–minded workshops or individuals making a very limited range of goods; in other words, the beginnings of Kunsthandwerk in the narrower, studio-crafts sense of the word.

Kunsthandwerk slowly acquired today's meaning in complex debates with other strands of Modernism. One of the purposes of the noted Werkbund Exhibition of 1924 and the ensuing book of 1925 entitled *Die Form Ohne Ornament* (*Form Without Ornament*) was to please both the industrial design faction and the Handwerk faction.[14] 'Form' is a powerful word in German twentieth-century art debates and even a brief definition is difficult. It means, as the title of the 1924 undertaking indicates, that there is an artistically valuable element in the 'body' of a work, of an object, with the ornament left off. Furthermore, form—'the deepest expression of inner forces . . . inescapable necessity and the best proof for a liveliness and the health of the times'—also implied the choice of the right style. But what matters most to the authors is that the term form, and especially 'simple form', can be applied and can serve as a system of value for both industrial form, here called 'technische Form' and for its opposite, here called 'primitive' or 'natural' form. It is

the latter we are most interested in; we read of the 'warmen bilden aus der Hand . . . wachstuemliche Form—warmly creating from hand . . . the feeling-for-growth form'. There is a further characterization of technical form as being simple, but 'raffiniert'—here the German word is much closer to the French 'raffiné' than the English 'refined'. Crafts' form, on the other hand, is 'primitive', simple; what is more, examples of 'primitive form' are almost all by women, whereas the best examples of 'technical form' are by men. Predictably, we are told that we need both types of form and that they complement each other.[15]

It is, again, important that at this juncture both industrial design and crafts were seen as equally valid spheres of art. The debates went on for many years and are best known from accounts of the Bauhaus. On the one hand, there was the central position of Modernism around 1930 which for various reasons ('Zeitgeist', 'education of the masses') went for all-out industrial design. On the other hand, the values of Handwerk, of 'traditional' Handwerk were also reiterated more and more frequently. A Werkbund publication of 1931 by the art historian Georg Friedrich Hartlaub was entitled *Das Ewige Handwerk im Kunstgewerbe der Gegenwart: Beispiele modernen kunsthandwerklichen Gestaltens*.[16] Its catch-all phrases are indicative of the wobbliness of all those terms. There is 'the present', but also 'the eternal'; there is the by then, strictly speaking, meaningless term Kunstgewerbe—'applied' art—at any rate, something that Hartlaub does not deal with; there is the most important new term, Kunsthandwerk; there is, for good measure, Handwerk itself.

And there is a newish term—'Gestalten/Gestaltung'—which means literally, to give something a 'Gestalt', Gestalt (a fairly common German word) being synonymous essentially with shape. Another term of the same period is 'Formgebung' (giving). The meaning of these terms is, of course, nothing other than that of the international Romance term 'designing'. It is only since 1970 or so that Germans have commonly used 'design' and pronounced it in the English way, with the German words having disappeared. Perhaps the terms 'form-giving' and 'gestalten' were too comprehensive, signifying, as they do, both making and designing. There was, incidentally, yet another paraphrase of Kunsthandwerk, the more seldomly used 'Werkkunst'. Turning the German term for work of art, Kunstwerk, on its head, it illustrates again the well-known German proclivity for producing endless combinations of words, as well as the desire to sound important by sounding basic.

The book by Hartlaub again admits to the then 'popular technoid form' but its chief aim is to preach the preservation of the 'ewige (eternal) Handwerk' and to pinpoint those forms and values which are tied to the hand or to Handwerk kinds of production. This now leads to a severe reduction of activities; only a tiny fraction of the sphere of 'traditional' Handwerk is admitted into Kunsthandwerk. It means, first of all, a production of individualized pieces for individual consumers. These individual producers can no longer be part of the general run of anonymous commercial, local Handwerk producers, but the Kunsthandwerker is usually a specially trained artist. Only a few materials are under consideration: glass, enamelling, stone-cutting, ceramics, batik, certain textiles, silversmithing. A crucial exclusion at this point is furniture. Clearly, the definition of Kunsthandwerk now approaches the narrow 'studio crafts'—a term without direct German equivalent. Thus by now the essential set of definitions of the twentieth century—Handwerk, Kunsthandwerk and in-

dustrial design (although the terms for the latter were only in their infancy)—had been arrived at.

To close with a few remarks about later decades: with regard to the later 1930s it has been emphasized by Joan Campbell that the Nazi dictatorship did not mean a very significant break with previous German design policies. Under the Third Reich, writers continued to rail against kitsch, against anything termed dishonest or pretentious. The main characteristic of Nazi pronouncements was simplification; the question of quality seemed solved by simply stating that production by Germans was German 'Wertarbeit' (quality work). Quality work was, furthermore, linked with the concept of the 'enjoyment of work'. There was less concern now for designer names, consultancies, etc. Debates of 'style', about modern vs. revivalist, internationalist/technoid style were largely eliminated; there were no competing groups, no Handwerk vs. International Modern industrial design, etc. All design policies, whether industrial or Handwerk, were pronounced by state or party agencies. At the same time, an unproblematic, seemingly natural hierarchy from ornate state representation downwards was upheld. The term design was taken care of, as well as Handwerk, by the formulation 'gestaltendes (form-creating) Handwerk'—the latter not an invention of the Nazis. Theorems returned to the eminently simple: 'form, function, materials', or the 'unconscious feeling for rules and scale'.[17]

After the Second World War the debates of the inter-war period were reopened, international Modernism had a voice again; on the other hand, divisions between various branches of producers appeared less severe, or controversial, than in the early decades of the century. Handwerk survived as previously defined, or, as most would say, in its 'traditional' structures. Kunsthandwerk, the full notion of studio crafts, now became finally established and institutionalized: 'A Kunsthandwerk is taking shape which neither wants to serve as model to industrial design, nor does it want to serve as Handwerk's alternative to industrial design; Kunsthandwerk means a lebensnotwendiges [life-necessary] supplement' (to the other spheres).[18] This is followed by the well-rehearsed juxtapositions: organic/technoid, etc. We now witness the foundation of pressure groups, the organization of prizes and exhibitions for Kunsthandwerk, some by the state, others as cooperatives. In some cases, such as at the renowned Munich Handwerker Messe, this support was, and is still organized under the umbrella of the powerful bodies of Handwerk itself. To 'learn a Handwerk', and that means going through the age-old process of becoming a Meister, is considered the basis for all activities, but does not suffice for the Kunsthandwerker; there has to be subsequent training at an art academy. The Kunsthandwerker sees himself or herself primarily as a 'freischaffender' (free-creating), a freelance artist.[19] Lastly, it has to be emphasized again that party politics did not play much of a role: the development in the GDR was not very different from that in West Germany with regard to Kunsthandwerk and even Handwerk. Much of the organizational structure of Handwerk was preserved, although the central state now had a far greater say. West Germans remarked that this was due to the fact that the still privately run old Handwerk provided efficient services which communist state-run industries were unable to deliver. Apart from a short phase in the mid-1950s, when a return to ornament was demanded and certain mildly folksy styles were revived, Arts and Crafts/Modernist notions of 'form', and, latter, 'design' prevailed in the GDR, too.[20]

There is thus little problem in translating present-day German Kunsthandwerk into studio crafts, while the divergencies and contrasts, as well as the similarities, of 'das Handwerk' remain. Most important, however, seems the ring of romantic and Modernist mysticism which is contained in both. The role of the Arts and Crafts movements was crucial in both countries, in fact in all Germanic-language countries. Some of its origins lay with early nineteenth-century German and British Romanticism. In the early twentieth century the various interest groups took these mysticisms into the market place. In support of their claims of quality they concluded discourses in which they tried to maximize the impact of powerful basic words, especially of Werk, in all its combinations. Werk's, or work's impact lies in the way it denotes both the process of devoted working and the satisfying results of working. In the end, two kinds of questions arise from the concerns of this article on terminology. Firstly: what will be the future of the terms work/craft *vis à vis* design and art; secondly: what fate awaits rich national and regional terminologies within the increasing globalization of languages?

NOTES

1. Cf. C. Muller, *Das Grosse Fachwörterbuch für Kunst und Antiquitäten, Deutsch, Englisch, Franzosisch*, Weltkunst Verlag, Munich, 1982.
2. Cf. W. Tatarkiewicz, *History of Aesthetics*, The Hague, 1970.
3. J. Campbell, *Joy in Work: German Work: The National Debate 1800–1945*, Princeton, 1989.
4. e.g. Kohlenbergwerk/ coalmine.
5. S. Volkov, *The Rise of Popular Antimodernism In Germany: The Urban Master Artisans 1873–1896*, Princeton, 1978.

6. A. Zelle, *Das Handwerk in Deutschland*, under the editorship of the Zentralverband (central association) des deutschen Handwerks, published by the Presse und Informationsdienst der Bundesregierung, Bonn, 1953. [. . .]

7. B. Mundt, *Das deutsche Kunstgewerbemuseum im 19. Jahrhundert*, Munich, 1974; H. M. Wingler (ed.), *Kunstschulreform 1900–1933*, Berlin, 1977; for the development of German design ideas and policies and for the English influence, see S. Muthesius, *Das Englische Vorbild, Die deutschen Reformbewegungen in Architektur, Wohnbau und Kunstgewerbe im späteren 19. Jahrhundert*, Munich, 1974; J. Heskett, *Design in Germany, 1870–1918*, London, 1986; M. Schwartzer, *German Architectural Theory and the Search for Modern Identity*, Cambridge, 1995.

8. A. Riegl, 'Kunsthandwerk und kunst-handwerkliche Massenproduction', *Zeitschrift des Kunstgewerbevereins*, Munich, 1895, p. 6. See S. Muthesius, 'Riegl and the folk art revival', in R. Woodfield (ed.), *Alois Riegl* (Series: *Critical Voices in Art, Theory and Culture*), G+B Arts International, 1998.

9. e.g. Julius Lessing (Director of the Berlin Kunstgewerbemuseum), Handarbeit, Berlin, 1887.

10. B. Deneke, 'Die Beziehungen zwischen Kunstharidwerk und Volkskunst um 1900', Anzeiger des Germanischen Nationalmuseums Nürnberg, 1968, pp. 140–61; S. Muthesius, *Das englische Vorbild*; R. Mielke, *Volkskunst*, Magdeburg, 1896. There was, again, some English influence at play here, this time directly through the writings of Morris, Ruskin and Walter Crane and their utopian socialist views. The term Volkskunst was chosen partly because the term 'social' was politically taboo for most the whole of the German middle classes. Walter Crane's chapter 'Art and Social Democracy' (in his book *The Claims of Decorative Art*, 1892, translated into German literally as *Die Forderungen der Dekorativen Kunst* in 1896) was rendered as 'Kunst und Volkstum'.

11. J. Campbell, *The German Werkbund: The Politics of Reform in the Applied Arts*, Princeton, 1978; F. J. Schwartz, *The Werkbund*, Yale University Press, 1996; H. Ghrist, 'Der "Fall Mutheius" und die Künstler', *Die Kunst*, vol. 18 (i.e. vol. XI, Dekorative Kunst), 1908, pp. 42–4; See S. Hubrich, *Hermann Muthesius: Die Schriften zu Architektur, Kunstgewerbe und Industrie in der 'neuen Bewegung'*, Berlin, 1981. 'Dickkopf . . .' Dr Lotz-Hanau, 'Kunsthandwerk und Kunstindustrie', *Deutsche Kunst und Dekoration*, vol. 55, October 1924–March 1925, p. 73.

12. 'Künstlerische Maschinenmobel' (produced by the Dresdner Werkstätten fur Handwerkskunst), *Deutsche Kunst und Deknration*, vol. XVII, October 1905–March 1906, pp. 247–64; 'Typenmobel', *Dekorative Kunst*, November 1908, pp. 86–95; March 1909, pp. 258–64.

13. The Werkbund's main 'enemy' was one of the major trades associations, comprising the 'older' (from the Werkbund's perspective 'uncritical') forces of Handwerk, industry and some nineteenth-century Kunstgewerbe organizations, the Verband für wirtschaftliche Interessen des Kunstgewerbes (Association for the economic interest of the Applied Arts); see K. Junghans, *Der Werkbund: Sein erstes Jahrzehnt*, Berlin, 1982, p. 2; A. Koch, *Das neue Kunsthandwerk in Deutschland und Oesterreich unter Berücksichtigung der Deutschen Gewerbeschau in München 1922*, Darmstadt, 1923 (some of the exhibits were also published in Deutsche Kunst und Dekoration, 1922–3, (see [3]); E. Redslob, 'Handwerkskultur', Deutsche *Kunst und Dekoration*, vol. 59, October 1926-March 1927, p. 234; see Campbell, *The German Werkbund*.

14. Series: *Bucher der Form*, vol. 1, Stuttgart, Berlin, Leipzig, 1925; c.f. G. Naylor, *The Bauhaus Re-assessed*, London, 1985.

15. *Die Form Ohne Ornament (Series: Bucher der Form)*, pp. 6, 9. The most important material at this point for 'free forming' appears

ceramics, e.g. the extremely rough work by the otherwise unknown Dorkas Harlin, Stuttgart.

16. *The Eternal Crafts in the Applied Arts of the Present: Examples of Modern Arts and Crafts Design, Werkbund Buch*, Berlin, 1931. Included are textiles by the Handweberei Sigmund von Weech, Munich and Metalwork by Waldemar Ramisch, Berlin.

17. Campbell, *The German Werkbund*, pp. 243ff.; see also B. Siepen, *Deutsche Wertarbeit, Veröffentlichungen Vortragsreihe Württembergisches Landesgewerbemuseum*, Stuttgart, 1938; W. Passarge, Deutsche Werkkunst der Gegenwart, Berlin, 1937. The organizing body of all production was the Deutsche Arbeitsfront, for design its Sektion Schonheit der Arbeit (German work brigade(s), Section beauty of work).

18. F. Kampfer & K. W. Beyer, *Kunsthandwerk in Wandel*, Berlin (East), 1984, pp. 8–9.

19. The most important regular shows are held in conjunction with the Internationale Handwerksmesse in Munich and with the Frankfurter Herbst Messe (Autumn Fair). [. . .]

20. See Zelle, *Das Handbuch in Deutschland*, pp. 35–6; see above Kampfer & Beyer, *Kunsthandwerk in Wandel*; *Deutsches Kunsthandwerk*; *Veröffentlichungen Institut für Angewandte Kunst*, Dresden, 1956.

FURTHER READING

John V. Maciuika, *Before the Bauhaus: Architecture, Politics, and the German State, 1890–1920* (Cambridge: Cambridge University Press, 2006).

Frederic J. Schwartz, *The Werkbund: Design Theory and Mass Culture Before the First World War* (New Haven: Yale University Press, 2005).

SECTION 3

MODERN CRAFT: IDEALISM AND REFORM

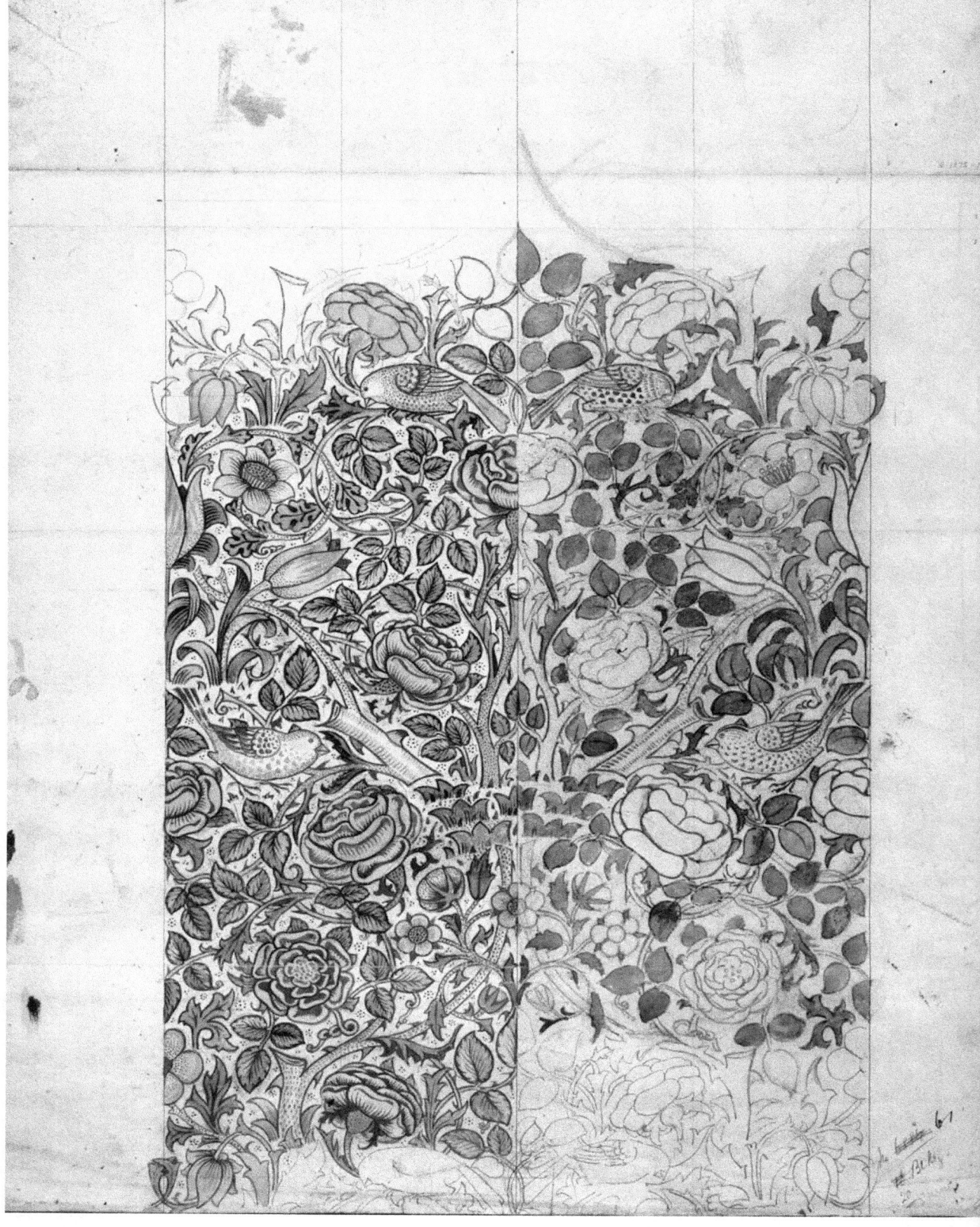

SECTION INTRODUCTION

When the book is closed on modern craft—if it ever is—the largest number of entries in the index will be found under the heading 'idealism'. That principle can be detected throughout the history covered in this book, from the earliest texts of the Arts and Crafts Movement to tomorrow's blogs. Idealism has animated a diversity of crafted objects, from humble baskets to handwritten calligraphy to elegant glass sculptures, and the goal of social reform according to such ideals can take an equally diverse range of forms. There is craft in its utopian mode, in which case the ambition is to build a new community, or perhaps a new world, around the principle of honest work. There is a more personal, spiritual form of idealism in which craft practice is directed towards the improvement of the self. Then there is countercultural craft, explicitly antagonistic to the mainstream. And finally, there is aesthetic idealism in all its varieties, dedicated not to social change but to the cause of beauty for its own sake.

A TRAGIC PROJECT

Idealism, then, is no simple matter, especially given that its various forms blend together and overlap. What all craft idealism seems to have in common, however, is a fundamental disconnection with the capitalist marketplace. No Arts and Crafts project was financially viable for long, unless it transgressed the principles of the movement by resorting to cheap labor or mass production. Similarly, craft countercultures have a way of dropping out of fashion—or worse, becoming all too fashionable, their motifs co-opted by the very marketplace they tried to critique. Even craftspeople content with the seemingly uncontroversial goal of making the world a more beautiful place have had to contend with the marginality of their enterprise: very few people have seen fit to allow handmade pots and homespun cloth to completely displace plastic bowls and T-shirts in their lives. The more extreme forms of artisanal opposition to the profit motive have tended to be even more remote from political reality. This is evident in the writings of A. J. Penty, the founder of the 'guild socialism' movement, which sought to revive medieval guild ownership of the means of production. Penty was convinced that 'our industrial system is doomed' and that a 'new social order' was just around the corner.[1]

More worrying is the fact that the annals of modern history are littered with examples in which craft idealism has been turned to sinister purposes. Folk revivals, often grouped under the heading of Romantic Nationalism, have involved a great deal of stereotype and cultural chauvinism, but they rarely turn violent.[2] The same cannot be said, of course, for the fascist regimes of the 1930s and 1940s, all of which placed a great emphasis on craft traditions as authentic expressions of the homeland, or as means to cleanse a visual culture tainted by foreign influence. Forthright, hand-carved stone was the architectural idiom of choice for Mussolini and Hitler—an appropriate backdrop for their pageants of military order.[3] Craft-inflected domestic goods were promoted through various organs of the Nazi political machine as a means of advancing the moral righteousness of the home.[4] Sôetsu Yanagi and the other proponents of the *mingei* or folk craft movement in Japan were hardly fascists, but they have been seen as complicit with imperialism and the authorities' attempts to achieve a 'new order of daily life'.[5] In the postwar period, the imagery of agrarian and industrial craft has been central to the propaganda of repressive states as diverse as Franco's Spain and the Soviet Union.[6] More recently, in the 1980s, revolutionaries in Nicaragua employed an all-too-familiar rhetoric of ethnic purity and the preservation of artisanal culture.[7] Those who assume that craft is an inherently liberatory affair must contend with histories that suggest otherwise.

All this said, if the reform of culture through craft remains an impossible and sometimes a tragic project, it is also still an attractive one. The act of failing can bring its own rewards. This can take the form of martyrdom, as in the case of Mahatma Gandhi, who did not live to see the Indian craft revival he inspired flourish in the post–World War II era. But more often, the message of craft idealism ends up having an impact precisely through its own undoing. This process might be given the Hegelian name of sublation (aufhebung in German), which has been used to describe the partial absorption of seemingly indigestible cultural energies into the mainstream.[8] From this perspective it could be argued, counterintuitively, that it is mainly by 'selling out' that the craft movement brings about widespread cultural transformation. Yes, the earnestly democratic chairs of Gustav Stickley have become high-priced collectibles today, and the craftsy attire of 1960s communes is adopted by kids in prefab suburbia. But though such reversals of fortune may be depressing to contemplate, we should remember that they are also testaments to the meaning and value that was opened up by these moments of craft idealism. Reformist impulses do not need to be completely realized in order to have an effect.

THINKING OTHERWISE

One might go further and suggest that perhaps we should judge craft's idealists not by their success in shaping the future, but rather the acuity which they bring to their own present-day concerns. Craft affords an opportunity to 'think otherwise', a framework for reflection and critique. This is, first and foremost, a question of temporality. For nearly all of those who view craft through an idealist lens, it is a good thing because it stands outside of the hurried routines of modern efficiency. Craft is slow and anachronistic, and this is exactly what gives it value. Or at least this is the message conveyed by the selections in this section of this book, from the medievalism of John Ruskin to George Nakashima's eloquent description of the drawn-out rhythms of nature.

The idea that craft is temporally out of joint has also dictated the way that craft reformers have situated themselves in space. From the late nineteenth century onwards, geographies that were supposedly untouched by the processes of modernization have been identified, studied and exploited in an effort to retain contact with an authentic past. Examples of such regions are too numerous to count: Appalachia in the American South; the Cotswolds in England; the Gaeltacht (Irish-speaking counties) of Ireland; Karelia in Finland; the peasant provinces of Russia, such as Smolensk; the Podhale region of Poland; and Okinawa, the island to the south of Japan are just a few examples.[9] Each of these sites has been subjected to that particular combination of care and condescension that is the hallmark of the idealist impulse. The places are real, but within modern craft discourse they often function as a purer, ungraspable Other—something not unlike nature itself, which we can admire but never reproduce, since our attempts to do so would be, by definition, artificial.

FLOWERS IN RIO

Given the subject of idealism, perhaps it would be appropriate to conclude with a metaphor. The scholar Odina Leal has researched the habits of Brazilian peasants who have recently moved from the countryside to new, urban and suburban homes. She noticed that her subjects tended to prefer artificial plastic flowers to real ones, which reminded them of the home that they had recently left behind. They also had rapidly acquired a connoisseurship of fake flowers and were readily able to tell high-quality from shoddy (and, even when poor, were often willing to pay for the former). This contrasts with the general taste of the Brazilian urban middle class, who universally favor fresh-cut flowers—widely available in the countryside but expensive in the city.[10]

Leal's example captures much that is at stake in modern craft idealism. Notably, it involves issues of space, time, cultural resources and social class. In moving from their rural homes to the city, the peasants have traveled not only spatially, but (figuratively speaking) temporally as well: they have undergone a process that we might call modernization, or alternatively 'detraditionalization'.[11] In the process they cultivate new tastes, which gravitate towards the artificiality of their new environment. Conversely, middle-class taste, already modernized, prefers signs of rustic authenticity for its decoration. Both parties are willing to pay for the privileges taken for granted—indeed, actively rejected—by the other. (As the French theorist Jacques Rancière has observed, 'Those who exalt or denounce the "tradition of the new" usually forget that this tradition has as its strict complement the "newness of the tradition".'[12]) In each case, the grass literally seems greener on the other side of the fence.

There are many ways to complicate this little allegory. A Marxist might want to look past questions of symbolism and think about all the flowers, both real and fake, as commodities—whose values are interchangeable, and determined not by peasants and bourgeois but by a system of production and distribution. A Postmodernist would argue that the distinction between 'real' and 'fake' in the anecdote is a false one. (After all, the fresh flowers are probably grown on an industrialized farm, an environment just as artificial as a plastics factory.)[13] A scholar of material culture might be struck by the fact that the story centers on flowers, which are ornamental rather than functional: for it is precisely in such seemingly inessential matters that unspoken cultural desires may be most clearly expressed. An aesthetician might focus attention on the act of cutting the flower and putting it into a vase, which makes a valueless plant into an object of visual pleasure. An anthropologist would want to know what else is going on in these people's lives besides flower collecting. Indeed, Leal herself (as an anthropologist) points out that the peasants in her study would often decorate their televisions not only with plastic flowers, but also with photographs of loved ones they had left behind in their villages—suggesting that a narrative of radical detraditionalization goes only so far.

What seems clear, however, is that craft is a means of navigating all these issues. Handmade pots and homespun cloths made under the conditions of modernity, in this sense, are quite similar to fresh-cut flowers decorating bourgeois apartments in Rio. Like those flowers, pots and textiles contain worlds of complexity in their own right. They can send mixed messages. They can be either unselfconscious or pointedly reformist. They are, in themselves, cultural texts that require decoding. For most of the authors gathered in this section, whether they recognize it or not, writing about craft is a way of doing just this: addressing the relationship between the traditional and the modern, between the genuine and the artificial. This is never a simple matter. Modern craft signifies authenticity, but authenticity at (at least) one remove. And if we adopt a broad conception of craft, including art, design, industry and ritual, then we begin to sense that what we have on our hands is not a well-kept garden, but a rich and varied landscape.

NOTES

1. Arthur J. Penty, *Post-Industrialism* (London: Macmillan, 1922), p. 117. See also Penty, *The Restoration of the Gild System* (1906). See also Christopher Bailey, 'Progress and Preservation: The Role of Rural Industries in the Making of the Modern Image of the Countryside', *Journal of Design History* 9/1(1996), pp. 35–53.

2. Wendy Kaplan, ed., *The Arts and Crafts Movement in Europe and America, 1880–1920: Design for the Modern World* (Los Angeles: Los Angeles County Museum of Art, 2004); Benedict Anderson, *Imagined Communities: Reflections on the Origin and Spread of Nationalism* (London: Verso, 1983).

3. Simonetta Falasca-Zamponi, *Fascist Spectacle: The Aesthetics of Power in Mussolini's Italy* (Berkeley: University of California Press, 2000).

4. Paul Betts, *The Authority of Everyday Objects: A Cultural History of West German Industrial Design* (Berkeley: University of California Press, 2004).

5. Kim Brandt, *Kingdom of Beauty: Mingei and the Politics of Folk Art in Japan* (Durham, NC: Duke University Press, 2007).

6. David Crowley, "Stalinism and Modernist Craft in Poland," *Journal of Design History* 11/1 (1998), pp. 71–83; David Crowley and Jane Pavitt, *Cold War Modern: Design 1945–1970* (London: V&A Publications, 2008).

7. Les W. Field, 'Constructing Local Identities in a Revolutionary Nation: The Cultural Politics of the Artisan Class in Nicaragua, 1979–90', *American Ethnologist* 22/4 (Nov. 1995), pp. 786–806.

8. Dick Hebdige, *Subculture: The Meaning of Style* (London: Methuen, 1979); Fredric Jameson, "On 'Cultural Studies'," *Social Text* 34 (1993), pp. 17–52.

9. Focused studies include Julia S. Ardery, *The Temptation: Edgar Tolson and the Genesis of Twentieth-Century Folk Art* (Chapel Hill: University of North Carolina Press, 1998); Sam Smiles, ed., *Going Modern and Being British: Art, Architecture and Design in Devon c. 1910–1960* (Exeter: Intellect Books, 1998); Charlotte Ashby, 'Looking Backwards and Forwards: Fennomane Furniture Design in Finland Around 1900', *Journal of Modern Craft* 1/2 (July 2008), pp. 181–96; Wendy Salmond, 'A Matter of Give and Take: Peasant Crafts and Their Revival in Late Imperial Russia', *Design Issues* 13/1 (Spring 1997), pp. 5–14; Andrzej Szczerski, 'Sources of Modernity: The Interpretations of Vernacular Crafts in Polish Design Around 1900', *Journal of Modern Craft* 1/1 (March 2008), pp. 55–76; Brandt, *Kingdom of Beauty*.

10. Odina Leal, 'Popular Taste and Erudite Repertoire', *Cultural Studies* 4/1 (1990), pp. 19–29. See also John Fiske, 'Cultural Studies and the Culture of Everyday Life', in Lawrence Grossberg et al., *Cultural Studies* (New York: Routledge, 1992).

11. Paul Heelas, Scott Lash and Paul Morris, eds, *Detraditionalization: Critical Reflections on Authority and Identity* (Oxford: Blackwell, 1996).

12. Jacques Rancière, *The Politics of Aesthetics: The Distribution of the Sensible*, trans. Gabriel Rockhill (London: Continuum, 2004 [orig. pub. 2000]), p. 25.

13. Catherine Ziegler has recently analyzed the flower trade as a way to understand the movement of commodities in contemporary global culture. *Favored Flowers: Culture and Economy in a Global System* (Chapel Hill, NC: Duke University Press, 2007).

19

'THE NATURE OF GOTHIC', FROM *THE STONES OF VENICE*

John Ruskin

John Ruskin would likely have been bemused by his posthumous reputation as a revolutionary craft theorist. Known in his own day primarily as an art historian, critic and aesthete, Ruskin actually wrote little about making—especially in comparison to the voluminous amounts he published on the subjects of painting and ornament, as well as many subjects further afield (such as geology, botany, religion and literature). His politics were complicated. He called himself a 'Tory of the old school', but his sympathy for the working class helped to inspire the Arts and Crafts Movement. Ironically, it was Ruskin's deep love of the past that made him so electrifying for his contemporaries. The craftspeople who most caught his imagination were the anonymous masons who fashioned Gothic buildings, to which he devoted his first major publication, The Stones of Venice. *William Morris said of this work: 'To some of us when we first read it, now many years ago, it seemed to point out a new road on which the world should travel.'[1] It is a strange response to a book of medieval architectural history. Ruskin saw Venice as an organic synthesis of respect for nature, religious spirit, and joyous craftsmanship, all of which he saw as neglected in modern culture. Many commentators, notably David Pye, have argued that Ruskin's analysis was both romantic and illogical—a fantasy quite disconnected from the realities of the medieval building trade.[2] Morris and many others, though, found this antimodernist vision to be both beautiful and generative. Ruskin was also inspirational for later reformers because of his dedication to architectural preservation, and his belief in the virtues of labor (famously, in 1874 he induced his undergraduate followers at Oxford University to rebuild a road in a nearby village, a truly shocking transgression of British norms of class). Most of all, though, it was Ruskin's tendency to suggest ways forwards by looking backwards that made him the grandfather-figure of the modern craft movement.*

John Ruskin, 'The Nature of Gothic', from *The Stones of Venice*, Vol. II (1851–1853), excerpted.

We all have some notion, most of us a very determined one, of the meaning of the term Gothic; but I know that many persons have this idea in their minds without being able to define it: that is to say understanding generally that Westminster Abbey is Gothic and St. Paul's is not, that Strasburg Cathedral is Gothic, and St. Peter's is not, they have nevertheless, no clear notion of what it is that they recognize in the one or miss in the other such as would enable them to say how far the work at Westminster or Strasburg is good and pure of its kind; still less to say of any nondescript building, like St. James Palace or Windsor Castle, how much right Gothic element there is in and how much wanting. And I believe this inquiry to be a pleasant and profitable one; and that there will be found something more than usually interesting in tracing out this grey shadow many-pinnacled image of the Gothic spirit in us; and discerning what fellowship there is between it and our Northern hearts. And if, at any point of the inquiry, I should interfere with any of the reader's previously formed conceptions, and use the term Gothic in any sense which he would not willingly attach to it, I do not ask to accept, but only to examine and understand, my interpretation, as necessary to the intelligibility of what follows in the rest of the work.

We have, then, the Gothic character submitted to our analysis, just as the rough mineral is submitted to that of the chemist, entangled with many other foreign substances, itself perhaps in no place pure, or ever to be obtained or seen in purity for more than an instant; but nevertheless a thing of definite and separate nature; however inextricable or confused in appearance. Now observe: the chemist defines his mineral by two separate kinds of character; one external, its crystalline form, hardness, lustre, etc., the other internal, the proportions and nature of its constituent atoms. Exactly in the same manner, we shall find that Gothic architecture has external forms and internal elements. Its elements are certain mental tendencies of the builders, legibly expressed in it; as fancifulness, love of variety, love of richness, and such others. Its external forms are pointed arches, vaulted roofs, etc. And unless both the elements and the forms are there, we have no right to call the style Gothic. It is not enough that it has the Form, if it have not also the power and life. It is not enough that it has the Power, if it have not the form.

[...]

What characters, we have to discover, did the Gothic builders love, or instinctively express in their work, as distinguished from all other builders? Let us go back for a moment to our chemistry, and note that, in defining a mineral by its constituent parts, it is not one nor another of them, that can make up the mineral, but the union of all: for instance, it is neither in charcoal nor in oxygen, not in lime, that there is the making of chalk, but in the combination of all three in certain measures; they are all found in very different things from chalk, and there is nothing like chalk either in charcoal or oxygen but they are nevertheless necessary to its existence.

So in the various mental characters which make up the soul of Gothic. It is not one nor another that produces it; but their union in certain measures. Each one of them is found in many other architectures beside Gothic; but Gothic cannot exist where they are not found or, at least, where their place is not in some way supplied. Only there is this great difference between the composition of the mineral and of the architectural style, that if we withdraw one of its elements from the stone, its form is utterly changed and its existence as such and such a mineral is destroyed; but if we withdraw one of its mental elements from the Gothic style it is only a little less Gothic than it was before, and the union of two or three of its elements is enough already to bestow a certain Gothicness of character, which gains in intensity as well as the others, and loses as we again withdraw them.

I believe, then, that the characteristics of Gothic are the following, placed in the order of their importance:

1. Savageness.
2. Changefulness.
3. Naturalism.
4. Grotesqueness.
5. Rigidity.
6. Redundance.

These characters are here expressed as belonging to the building; as belonging to the builder they would be expressed thus:

1. Savageness or Rudeness.
2. Love of Change.
3. Love of Nature.
4. Disturbed Imagination.
5. Obstinacy.
6. Generosity.

And I repeat that the withdrawal of any one, or any two will not at once destroy the Gothic character of a building, but the removal of a majority of them will. I shall proceed to examine them in their order.

Savageness. I am not sure when the word "Gothic" was first generically applied to the architecture of the North but I presume that, whatever the date of its original usage, it was intended to imply reproach, and express the barbaric character of the nations among whom that architecture arose. It never implied that they were literally of Gothic lineage, far less that their architecture had been originally invented by the Goths themselves; but it did imply that they and their buildings together exhibited a degree of sternness and rudeness, which, in contra-distinction to the character of Southern and Eastern nations, appeared like a perpetual reflection of the contrast between the Goth and the Roman in their first encounter. And when that fallen Roman, in the utmost impotence of his luxury, and insolence of his guilt, became the model for the imitation of civilized Europe, at the close of the so-called Dark Ages, the word Gothic became a term of unmitigated contempt, not unmixed with aversion. From that contempt, by the exertion of the antiquaries and architects of this century, Gothic architecture has been sufficiently vindicated; and perhaps some among us, in our admiration of the magnificent science of its structure, and sacredness of its expression, might desire that the term of ancient reproach should be withdrawn, and some other, of more apparent honourableness, adopted in its place. There is no chance, as there is no need, of such a substitution. As far as the epithet was used scornfully, it was used falsely; but there is no reproach in the word, rightly understood; on the contrary, there is a profound truth, which the instinct of mankind almost unconsciously recognizes. It is true, greatly and deeply true, that the architecture of the North is rude and wild; but it is not true, that, for this reason, we are to condemn it, or despise it. Far otherwise: I believe it is in this very character that it deserves our profoundest reverence.

[. . .]

The second mental element above named was Changefulness, or Variety.

I have already enforced the allowing independent operation to the inferior workman, simply as a duty to him, and as ennobling the architecture by rendering it more Christian. We have now to consider what reward we obtain for the performance of this duty, namely, the perpetual variety of every feature of the building.

Wherever the workman is utterly enslaved, the parts of the building must of course be absolutely like each other; for the perfection of his execution can only be reached by exercising him in doing one thing, and giving him nothing else to do. The degree in which the workman is degraded may be thus known at a glance, by observing whether the several parts of the building are similar or not; and if, as in Greek work, all the capitals are alike, and all the mouldings unvaried, then the degradation is complete; if, as in Egyptian or Ninevite work, though the manner of executing certain figures is always the same, the order of design

Figure 12 John Ruskin, *Sketch of Gothic Tracery in Venice*, 1845.

not impeach love of order: it is one of the most useful elements of the English mind; it helps us in our commerce and in all purely practical matters; and it is in many cases one of the foundation stones of morality. Only do not let us suppose that love of order is love of art. It is true that order, in its highest sense, is one of the necessities of art, just as time is a necessity of music; but love of order has no more to do with our right enjoyment of architecture or painting, than love of punctuality with the appreciation of an opera. Experience, I fear, teaches us that accurate and methodical habits in daily life are seldom characteristic of those who either quickly perceive or richly possess, the creative powers of art; there is, however, nothing inconsistent between the two instincts, and nothing to hinder us from retaining our business habits, and yet fully allowing and enjoying the noblest gifts of Invention. We already do so, in every other branch of art except architecture, and we only do not so there because we have been taught that it would be wrong. Our architects gravely inform us that, as there are four rules of arithmetic, there are five orders of architecture; we, in our simplicity, think that this sounds consistent, and believe them. They inform us also that there is one proper form for Corinthian capitals, another for Doric, and another for Ionic. We, considering that there is also a proper form for the letters A, B, and C, think that this also sounds consistent, and accept the proposition. Understanding, therefore, that one form of the capitals is proper and no other, and having a conscientious horror of an impropriety we allow the architect to provide us with the said capitals, of the proper form, in such and such a quantity, and in all other points to take care that the legal forms are observed; which having done, we rest in forced confidence that we are well housed.

is perpetually varied, the degradation less total; if, as in Gothic work, there is perpetual change both in design and execution, the workman must have been altogether set free.

How much the beholder gains from the liberty if the labourer may perhaps be questioned in England, where one of the strongest instincts in nearly every mind is that love of order which makes us desire that our house windows should pair like our carriage horses, and allows us to yield our faith unhesitatingly to architectural theories which fix a form for everything, and forbid variation from it. I would

But our higher instincts are not deceived. We take no pleasure in the building provided for us, resembling that which we take in a new book or a new picture. We may be proud of its size, complacent in its correctness, and happy in its convenience. We may take the same pleasure in its symmetry and workmanship as in a well-ordered room, or a skilful piece of manufacture. And this we suppose to be all the pleasure that architecture was ever intended to give us. The idea of reading a building as we would read Milton or Dante, and getting the same kind of delight out of the stones as out of the stanzas, never enters our mind for a moment. And for good reason; there is indeed rhythm in the verses, quite as strict as the symmetries or rhythm of the architecture, and a thousand times more beautiful; but there is something else than rhythm. The verses were neither made to order, nor to match, as the capitals were; and we have therefore a kind of pleasure in them other than a sense of propriety. But it requires a strong effort of common sense to shake ourselves quit of all that we have been taught for the last two centuries, and wake to the perception of a truth just as simple and certain as it is new: that great art, whether expressing itself in words, colours, or stones, does not say the same thing over and over again; that the merit of architectural, as of every other art, consists in its saying new and different things; that to repeat itself is no more a characteristic of genius in marble than it is of genius in print; and that we may without offending any laws of good taste, require of an architect, as we do of a novelist, that he should be not only correct, but entertaining. Yet all this is true, and self-evident; only hidden from us, as many other self-evident things are by false teaching. Nothing is a great work of art, for the production of which either rules or models can be given. Exactly so far as architecture works on known rules, and from given models, it is not an art, but a manufacture; and it is, of the procedures, rather less rational (because more easy) to copy capitals or mouldings from Phidias, and call ourselves architects, than to copy heads and hands from Titian, and call ourselves painters.

[. . .]

The third constituent element of the Gothic mind was stated to be Naturalism; that is to say, the love of natural objects for their own sake, and the effort to represent them frankly, unconstrained by artistical laws. This characteristic of the style partly follows in necessary connection with those named above. For, so soon as the workman is left free to represent what subjects he chooses, he must look to the nature that is round him for material, and will endeavour to represent it as he sees it, with more or less accuracy according to the skill he possesses, and with much play of fancy, but with small respect for law. There is, however, a marked distinction between the imaginations of the Western and Eastern races, even when both are left free; the Western, or Gothic, delighting most in the representation of facts, and the Eastern (Arabian, Persian, and Chinese) in the harmony of colours and forms . . . The Gothic builders were of that central class which unites fact with design; but the part of the work which was more especially their own was the truthfulness. Their power of artistical invention or arrangement was not greater than that of Romanesque and Byzantine workmen: by those workmen they were taught the principles, and from them received their models, of design. But to the ornamental feeling and rich fancy of the Byzantine the Gothic builder added a love of fact which is never found in the South.

[. . .]

The fourth essential element of the Gothic mind was above stated to be the sense of the

Grotesque; but I shall defer the endeavour to define this most curious and subtle character until we have occasion to examine one of the divisions of the Renaissance schools, which was morbidly influenced by it. It is the less necessary to insist upon it here, because every reader familiar with Gothic architecture must understand what I mean, and will, I believe, have no hesitation in admitting that the tendency to delight in fantastic and ludicrous, as well as in sublime, images, is a universal instinct of the Gothic imagination.

The fifth element above named was Rigidity; and this character I must endeavour carefully to define, for neither the word I have used, nor any other that I can think of, will express it accurately. For I mean, not merely stable, but active rigidity; the peculiar energy which gives tension to movement, and stiffness to resistance, which makes the fiercest lightning forked rather than curved, and the stoutest oak-branch angular rather than bending, and is as much seen in the quivering of the lance as in the glittering of the icicle.

[. . .]

Last, because the least essential, of the constituent elements of this noble school, was placed that of Redundance; the uncalculating bestowal of the wealth of its labour. There is, indeed, much Gothic, and that of the best period, in which this element is hardly traceable, and which depends for its effect almost exclusively on loveliness of simple design and grace of uninvolved proportion: still, in the most characteristic buildings, a certain portion of their effect depends upon accumulation of ornament; and many of those which have most influence on the minds of men, have attained it by means of this attribute alone. And although, by careful study of the school, it is possible to arrive at a condition of taste which shall be better contented by a few perfect lines than by a whole facade covered with fretwork,

the building which only satisfies such a taste is not to be considered the best. For the very first requirement of Gothic architecture being, as we saw above, that it shall both admit the aid, and appeal to the admiration, of the rudest as well as the most refined minds, the richness of the work is, paradoxical as the statement may appear, a part of its humility. No architecture is so haughty as that which is simple; which refuses to address the eye, except in a few clear and forceful lines; which implies, in offering so little to our regards, that all it has offered is perfect; and disdains, either by the complexity or the attractiveness of its features, to embarrass our investigation, or betray us into delight. That humility, which is the very life of the Gothic school, is shown not only in the imperfection, but in the accumulation, of ornament. The inferior rank of the workman is often shown as, much in the richness, as the roughness, of his work; and if the cooperation of every hand, and the sympathy of every heart, are to be received, we must be content to allow the redundance which disguises the failure of the feeble, and wins the regard of the inattentive. There are, however, far nobler interests mingling, in the Gothic heart, with the rude love of decorative accumulation: a magnificent enthusiasm, which feels as if it never could do enough to reach the fulness of its ideal; an unselfishness of sacrifice, which would rather cast fruitless labour before the altar than stand idle in the market; and, finally, a profound sympathy with the fulness and wealth of the material universe, rising out of that Naturalism whose operation we have already endeavoured to define. The sculptor who sought for his models among the forest leaves, could not but quickly and deeply feel that complexity need not involve the loss of grace, nor richness that of repose; and every hour which he spent in the study of the minute and various work of Nature, made him feel more forcibly the barrenness of what was best in that of man: nor is it to be wondered at, that, seeing her perfect and exquisite creations poured forth in a profusion which conception could not grasp nor calculation sum; he should think that it ill became him to be niggardly of his own rude craftsmanship; and where he saw throughout the universe a faultless beauty lavished on measureless spaces of broidered field and blooming mountain, to grudge his poor and imperfect labour to the few stones that he had raised one upon another, for habitation or memorial. The years of his life passed away before his task was accomplished; but generation succeeded generation with unwearied enthusiasm, and the cathedral front was at last lost in the tapestry of its traceries, like a rock among the thickets and herbage of spring.

NOTES

1. Quoted in Robert Hewison, "John Ruskin," *Oxford Dictionary of National Biography* (Oxford: Oxford University Press, 2004).
2. David Pye, "On the Nature of Gothic," in *The Nature and Art of Workmanship* (Cambridge: Cambridge University Press, 1968).

FURTHER READING

Tim Barringer, *Men at Work: Art and Labour in Victorian Britain* (New Haven: Yale University Press, 2005).

Dinah Birch, *Ruskin and the Dawn of the Modern* (Oxford: Oxford University Press, 1999).

Eileen Boris, *Art and Labor: Ruskin, Morris and the Craftsman Ideal in America* (Philadelphia: Temple University Press, 1986).

Rosemary Hill, *God's Architect: Pugin and the Building of Romantic Britain* (London: Penguin, 2007).

John Ruskin, *The Seven Lamps of Architecture* (1849).

20

THE REVIVAL OF HANDICRAFT

William Morris

If William Morris had not existed it would have been necessary to invent him, so completely does he embody the idealism that lies at the heart of modern craft. Morris was an astonishing polymath. He wrote some of the best-known prose and poetry of the Victorian era. Following his conversion to Socialism in 1883, he became an effective political operator and polemical essayist. Above all, though, he was an artist—an identity that for him involved being a craftsman, designer and entrepreneur all at once. After an apprenticeship in the office of Gothic revival architect G. E. Street, he co-founded the quintessential Arts and Crafts workshop, Morris, Marshall, Faulkner & Co., in 1861. Through the activities of "the Firm," as he called it, he taught himself innumerable crafts, ranging from manuscript illumination and calligraphy to hand-weaving and fabric dyeing. And yet Morris was in some senses, and in his own estimation, a failure. He was unable to put into practice his key theoretical principle, the unity of aesthetic and political reform. Despite his commitment to making art for the masses, his clients were almost invariably well-to-do people sympathetic with his aims. Perhaps it was only in his own homes, Red House and Kelmscott House, that he fully realized the aesthetic ideal that he wished for all people. Characteristically, Morris was alive to this sad irony. As he writes in the essay extracted here, I think 'we have already got in all branches of culture rather more geniuses that we can comfortably bear, and that we lack, so to say, audiences rather than preachers.'

Morris, then, embodies craft idealism in both a positive and a negative sense: both its possibilities and its frustrations. Much of this book, The Craft Reader, is an attempt to recontextualize his Arts and Crafts ideal and its attendant contradictions by emphasizing other approaches to the subject. But there is no denying Morris's centrality to craft discourse and the continuing value of his thought. Improbably, this Victorian designer of decorative wallpapers was hugely influential on modernist designers—who took to heart his most famous proclamation, 'Have nothing in your houses that you do not know to be useful or believe to be beautiful.'[1] His thundering condemnation of modern capitalism and his vivid evocations of authentic craftsmanship have resonated powerfully over the last century and continue to do so today. He was also a perceptive commentator on other anxieties we still feel today, ranging from the globalist economy ('the Indian or Javanese craftsman may no longer ply his craft leisurely, working a few hours a day, in producing a maze of strange beauty on a piece of cloth: a steam-engine is set a-going at Manchester, and . . . the Asiatic worker [is] driven himself into a factory to lower the wages of his Manchester brother worker'[2]) to the superficiality of fashion ('a strange monster born of the vacancy *of the lives of rich people'³). Most of all, no one has better articulated the wishful thinking that surrounds craftspeople than Morris. For him they were simply 'makers of things by their own free will'—an ideal that may be impossible to achieve in practice but should never be forgotten.⁴*

William Morris, 'The Revival of Handicraft'. Originally published in the *Fortnightly Review* (November 1888), excerpted.

For some time past there has been a good deal of interest shown in what is called in our modern slang Art Workmanship, and quite recently there has been a growing feeling that this art workmanship to be of any value must have some of the workman's individuality imparted to it beside whatever of art it may have got from the design of the artist who has planned, but not executed the work. This feeling has gone so far that there is growing up a fashion for demanding handmade goods even when they are not ornamented in any way, as, for instance, woollen and linen cloth spun by hand and woven without power, hand-knitted hosiery, and the like. Nay, it is not uncommon to hear regrets for the hand-labour in the fields, now fast disappearing from even backward districts of civilized countries. The scythe, the sickle, and even the flail are lamented over, and many are looking forward with drooping spirits to the time when the hand-plough will be as completely extinct as the quern, and the rattle of the steam-engine will take the place of the whistle of the curly-headed ploughboy through all the length and breadth of the land. People interested, or who suppose that they are interested, in the details of the arts of life feel a desire to revert to methods of handicraft for production in general; and it may therefore be worth considering how far this is a mere reactionary sentiment incapable of realization, and how far it may foreshadow a real coming

change in our habits of life as irresistible as the former change which has produced the system of machine-production, the system against which revolt is now attempted.

In this paper I propose to confine the aforesaid consideration as much as I can to the effect of machinery versus handicraft upon the arts; using that latter word as widely as possible, so as to include all products of labour which have any claims to be considered beautiful. I say as far as possible: for as all roads lead to Rome, so the life, habits, and aspirations of all groups and classes of the community are founded on the economical conditions under which the mass of the people live, and it is impossible to exclude socio-political questions from the consideration of aesthetics. Also, although I must avow myself a sharer in the above-mentioned reactionary regrets, I must at the outset disclaim the mere aesthetic point of view which looks upon the ploughman and his bullocks and his plough, the reaper, his work, his wife, and his dinner, as so many elements which compose a pretty tapestry hanging, fit to adorn the study of a contemplative person of cultivation, but which it is not worth while differentiating from each other except in so far as they are related to the beauty and interest of the picture. On the contrary, what I wish for is that the reaper and his wife should have themselves a due share in all the fulness of life; and I can, without any great effort, perceive the justice of their forcing me to bear part of the burden of its deficiencies, so that we may together be forced to attempt to remedy them, and have no very heavy burden to carry between us.

To return to our aesthetics: though a certain part of the cultivated classes of to-day regret the disappearance of handicraft from production, they are quite vague as to how and why it is disappearing, and as to how and why it should or may reappear. For to begin with the general public is grossly ignorant of all the methods and processes of manufacture. This is of course one result of the machine-system we are considering. Almost all goods are made apart from the life of those who use them; we are not responsible for them, our will has had no part in their production, except so far as we form part of the market on which they can be forced for the profit of the capitalist whose money is employed in producing them. The market assumes that certain wares are wanted; it produces such wares, indeed, but their kind and quality are only adapted to the needs of the public in a very rough fashion, because the public needs are subordinated to the interest of the capitalist masters of the market, and they can force the public to put up with the less desirable article if they choose, as they generally do. The result is that in this direction our boasted individuality is a sham; and persons who wish for anything that deviates ever so little from the beaten path have either to wear away their lives in a wearisome and mostly futile contest with a stupendous organization which disregards their wishes, or to allow those wishes to be crushed out for the sake of a quiet life.

Let us take a few trivial but undeniable examples. You want a hat, say, like that you wore last year; you go to the hatter's, and find you cannot get it there, and you have no resource but in submission. Money by itself won't buy you the hat you want; it will cost you three months' hard labour and twenty pounds to have an inch added to the brim of your wideawake; for you will have to get hold of a small capitalist (of whom but few are left), and by a series of intrigues and resolute actions which would make material for a three-volume novel, get him to allow you to turn one of his hands into a handicraftsman for the occasion; and a

very poor handicraftsman he will be, when all is said. Again, I carry a walking-stick, and like all sensible persons like it to have a good heavy end that will swing out well before me. A year or two ago it became the fashion to pare away all walking-sticks to the shape of attenuated carrots, and I really believe I shortened my life in my attempts at getting a reasonable staff of the kind I was used to, so difficult it was. Again, you want a piece of furniture, which the trade (mark the word, Trade, not Craft!) turns out blotched over with idiotic sham ornament; you wish to dispense with this degradation, and propose it to your upholsterer, who grudgingly assents to it; and you find that you have to pay the price of two pieces of furniture for the privilege of indulging your whim of leaving out the trade finish (I decline to call it ornament) on the one you have got made for you. And this is because it has been made by handicraft instead of machinery. For most people, therefore, there is a prohibitive price put upon the acquirement or the knowledge of methods and processes. We do not know how a piece of goods is made, what the difficulties are that beset its manufacture, what it ought to look like, feel like, smell like, or what it ought to cost apart from the profit of the middleman. We have lost the art of marketing, and with it the due sympathy with the life of the workshop, which would, if it existed, be such a wholesome check on the humbug of party politics.

It is a natural consequence of this ignorance of the methods of making wares, that even those who are in revolt against the tyranny of the excess of division of labour in the occupations of life, and who wish to recur more or less to handicraft, should also be ignorant of what that life of handicraft was when all wares were made by handicraft. If their revolt is to carry any hope with it, it is necessary that

Figure 13   William Morris and Philip Webb, *'Trellis' Wallpaper*, designed 1862.

they should know something of this. I must assume that many or perhaps most of my readers are not acquainted with Socialist literature, and that few of them have read the admirable account of the different epochs of production given in Karl Marx's great work entitled *Capital*. I must ask to be excused, therefore, for stating very briefly what, chiefly owing to

Marx, has become a commonplace of Social-ism, but is not generally known outside it. There have been three great epochs of produc-tion since the beginning of the Middle Ages. During the first or medieval period all produc-tion was individualistic in method; for though the workmen were combined into great asso-ciations for production and the organization of labour, they were so associated as citizens, not as mere workmen. There was little or no division of labour, and what machinery was used was simply of the nature of a multiplied tool, a help to the workman's hand-labour and not a supplanter of it. The workman worked for himself and not for any capitalistic employer, and he was accordingly master of his work and his time; this was the period of pure handicraft. When in the latter half of the sixteenth century the capitalist employer and the so-called free workman began to appear, the workmen were collected into workshops, the old tool-machines were improved, and at last a new invention, the division of labour, found its way into the workshops. The division of labour went on growing throughout the sev-enteenth century, and was perfected in the eighteenth, when the unit of labour became a group and not a single man; or in other words the workman became a mere part of a machine composed sometimes wholly of human beings and sometimes of human beings plus labour-saving machines, which towards the end of this period were being copiously invented; the fly-shuttle may be taken for an example of these. The latter half of the eighteenth cen-tury saw the beginning of the last epoch of production that the world has known, that of the automatic machine which supersedes hand-labour, and turns the workman who was once a handicraftsman helped by tools, and next a part of a machine, into a tender of machines.

Figure 14 William Morris, *Sussex Chair*, designed ca. 1860.

[. . .]

This is very briefly the history of the evolu-tion of industry during the last five hundred years; and the question now comes: Are we justified in wishing that handicraft may in its turn supplant machinery? Or it would per-haps be better to put the question in another way. Will the period of machinery evolve itself into a fresh period of machinery more inde-pendent of human labour than anything we can conceive of now, or will it develop its con-tradictory in the shape of a new and improved period of production by handicraft? The sec-ond form of the question is the preferable one, because it helps us to give a reasonable answer to what people who have any interest in external beauty will certainly ask: Is the change from handicraft to machinery good or bad? And the answer to that question is to my mind that, as my friend Belfort Bax has put it, statically it is bad, dynamically it is good. As a condition of life, production by machinery is altogether an evil; as an instrument for forcing on us better conditions of life it has been, and for some time yet will be, indispensable.

Having thus tried to clear myself of mere reactionary pessimism, let me attempt to show why statically handicraft is to my mind desirable, and its destruction a degradation of life. Well, first I shall not shrink from saying bluntly that production by machinery necessarily results in utilitarian ugliness in everything which the labour of man deals with, and that this is a serious evil and a degradation of human life. So clearly is this the fact that though few people will venture to deny the latter part of the proposition, yet in their hearts the greater part of cultivated civilized persons do not regard it as an evil, because their degradation has already gone so far that they cannot, in what concerns the sense of seeing, discriminate between beauty and ugliness: their languid assent to the desirableness of beauty is with them only a convention, a superstitious survival from the times when beauty was a necessity to all men. The first part of the proposition (that machine-industry produces ugliness) I cannot argue with these persons, because they neither know, nor care for, the difference between beauty and ugliness; and with those who do understand what beauty means I need not argue it, as they are but too familiar with the fact that the produce of all modern industrialism is ugly, and that whenever anything which is old disappears, its place is taken by something inferior to it in beauty; and that even out in the very fields and open country. The art of making beautifully all kinds of ordinary things, carts, gates, fences, boats, bowls, and so forth, let alone houses and public buildings, unconsciously and without effort, has gone; when anything has to be renewed among these simple things the only question asked is how little it can be done for, so as to tide us over our responsibility and shift its mending on to the next generation.

It may be said, and indeed I have heard it said, that since there is some beauty still left in the world and some people who admire it, there is a certain gain in the acknowledged eclecticism of the present day, since the ugliness which is so common affords a contrast whereby beauty, which is so rare, may be appreciated. This I suspect to be only another form of the maxim which is the sheet-anchor of the laziest and most cowardly group of our cultivated classes, that it is good for the many to suffer for the few; but if any one puts forward in good faith the fear that we may be too happy in the possession of pleasant surroundings, so that we shall not be able to enjoy them, I must answer that this seems to me a very remote terror. Even when the tide at last turns in the direction of sweeping away modern squalor and vulgarity, we shall have, I doubt, many generations of effort in perfecting the transformation, and when it is at last complete, there will be first the triumph of our success to exalt us, and next the history of the long wade through the putrid sea of ugliness which we shall have at last escaped from. But furthermore, the proper answer to this objection lies deeper than this. It is to my mind that very consciousness of the production of beauty for beauty's sake which we want to avoid; it is just what is apt to produce affectation and effeminacy amongst the artists and their following. In the great times of art conscious effort was used to produce great works for the glory of the City, the triumph of the Church, the exaltation of the citizens, the quickening of the devotion of the faithful; even in the higher art, the record of history, the instruction of men alive and to live hereafter, was the aim rather than beauty; and the lesser art was unconscious and spontaneous, and did not in any way interfere with the rougher business of life, while it enabled men in general to understand and sympathize with the nobler forms of art. But unconscious as these producers of ordinary beauty may be, they will not and cannot fail to receive pleasure from the exercise of their work under these conditions, and this above all things is that which influences me most in my hope for the recovery of handicraft. I have said it often enough, but I must say it once again, since it is so much a part of my case for handicraft, that so long as man allows his daily work to be mere unrelieved drudgery he will seek happiness in vain. I say further that the worst tyrants of the days of violence were but feeble tormentors compared with those Captains of Industry who have taken the pleasure of work away from the workmen. Furthermore I feel absolutely certain that handicraft joined to certain other conditions, of which more presently, would produce the beauty and the pleasure in work above mentioned; and if that be so, and this double pleasure of lovely surroundings and happy work could take the place of the double torment of squalid surroundings and wretched drudgery, have we not good reason for wishing, if it might be, that handicraft should once more step into the place of machine-production?

I am not blind to the tremendous change which this revolution would mean. The maxim of modern civilization to a well-to-do man is, Avoid taking trouble! Get as many of the functions of your life as you can performed by others for you! Vicarious life is the watchword of our civilization, and we well-to-do and cultivated people live smoothly enough while it lasts. But, in the first place, how about the vicars, who do more for us than the singing of mass for our behoof for a scanty stipend? Will they go on with it for ever? For indeed the shuffling off of responsibilities from one to the other has to stop at last, and somebody has to bear the burden in the end. But let that pass, since I am not writing politics, and let us consider another aspect of the matter. What wretched lop-sided creatures we are being made by the excess of the division of labour in the occupations of life! What on earth are we going to do with our time when we have brought the art of vicarious life to perfection, having first complicated the question by the ceaseless creation of artificial wants which we refuse to supply for ourselves? Are all of us (we of the great middle class I mean) going to turn philosophers, poets, essayists—men of genius, in a word, when we have come to look down on the ordinary functions of life with the same kind of contempt wherewith persons of good breeding look down upon a good dinner, eating it sedulously however? I shudder when I think of how we shall bore each other when we have reached that perfection. Nay, I think we have already got in all branches of culture rather more geniuses that we can comfortably bear, and that we lack, so to say, audiences rather than preachers. I must ask pardon of my readers, but our case is at once so grievous and so absurd that one can scarcely help laughing out of bitterness of soul. In the very midst of our pessimism we are boastful of our wisdom, yet we are helpless in the face of the necessities we have created, and which, in spite of our anxiety about art, are at present driving us into luxury unredeemed by beauty on the one hand, and squalor unrelieved by incident or romance on the other, and will one day drive us into mere ruin.

Yes, we do sorely need a system of production which will give us beautiful surroundings and pleasant occupation, and which will tend to make us good human animals, able to do something for ourselves, so that we may be generally intelligent instead of dividing ourselves into dull drudges or duller pleasure-seekers according to our class, on the one hand, or hapless pessimistic intellectual personages, and pretenders to that dignity, on the other. We do most certainly need happiness in our daily work, content in our daily rest; and all this cannot be if we hand over the whole responsibility of the details of our daily life to machines and their drivers. We are right to long for intelligent handicraft to come back to the world which it once made tolerable amidst war and turmoil and uncertainty of life, and which it should, one would think, make happy now we have grown so peaceful, so considerate of each other's temporal welfare.

Then comes the question, How can the change be made? And here at once we are met by the difficulty that the sickness and death of handicraft is, it seems, a natural expression of the tendency of the age. We willed the end, and therefore the means also. Since the last days of the Middle Ages the creation of an intellectual aristocracy has been, so to say, the spiritual purpose of civilization side by side with its material purpose of supplanting the aristocracy of status by the aristocracy of wealth. Part of the price it has had to pay for its success in that purpose (and some would say it is comparatively an insignificant part) is that this new aristocracy of intellect has been compelled to forgo the lively interest in the beauty and romance of life, which was once the portion of every artificer at least, if not of every workman, and to live surrounded by an ugly vulgarity which the world amidst all its changes has not known till modern times. It is

not strange that until recently it has not been conscious of this degradation; but it may seem strange to many that it has now grown partially conscious of it. It is common now to hear people say of such and such a piece of country or suburb: 'Ah! it was so beautiful a year or so ago, but it has been quite spoilt by the building.' Forty years back the building would have been looked on as a vast improvement; now we have grown conscious of the hideousness we are creating, and we go on creating it. We see the price we have paid for our aristocracy of intellect, and even that aristocracy itself is more than half regretful of the bargain, and would be glad if it could keep the gain and not pay the full price for it. Hence not only the empty grumbling about the continuous march of machinery over dying handicraft, but also various elegant little schemes for trying to withdraw ourselves, some of us, from the consequences (in this direction) of our being superior persons; none of which can have more than a temporary and very limited success. The great wave of commercial necessity will sweep away all these well-meant attempts to stem it, and think little of what it has done, or whither it is going.

Yet after all even these feeble manifestations of discontent with the tyranny of commerce are tokens of a revolutionary epoch, and to me it is inconceivable that machine-production will develop into mere infinity of machinery, or life wholly lapse into a disregard of life as it passes. It is true indeed that powerful as the cultivated middle class is, it has not the power of recreating the beauty and romance of life; but that will be the work of the new society which the blind progress of commercialism will create, nay, is creating. The cultivated middle class is a class of slave-holders, and its power of living according to its choice is limited by the necessity of finding constant livelihood and

Figure 15   William Morris, *'Iris' Furnishing Fabric,* 1876. Retailed by Morris & Co.

employment for the slaves who keep it alive. It is only a society of equals which can choose the life it will live, which can choose to forgo gross luxury and base utilitarianism in return for the unwearying pleasure of tasting the fulness of life. It is my firm belief that we shall in the end realize this society of equals, and also that when it is realized it will not endure a vicarious life by means of machinery; that it will in short be the master of its machinery and not the servant, as our age is.

Meantime, since we shall have to go through a long series of social and political events before we shall be free to choose how we shall live, we should welcome even the feeble protest which is now being made against the vulgarization of all life: first because it is one token amongst others of the sickness of modern civilization; and next, because it may help to keep alive memories of the past which are necessary elements of the life of the future, and methods of work which no society could afford to lose. In short, it may be said that though the movement towards the revival of handicraft is contemptible on the surface in face of the gigantic fabric of commercialism, yet, taken in conjunction with the general movement towards freedom of life for all, on which we are now surely embraced, as a protest against intellectual tyranny, and a token of the change which is transforming civilization into socialism, it is both noteworthy and encouraging.

NOTES

1. William Morris, 'The Beauty of Life', lecture first delivered 1880. Subsequently published in *Hopes and Fears for Art* (1882).
2. William Morris, 'How We Live and How We Might Live', lecture first delivered 1884. First published in *Commonweal* (1887) and subsequently in *Signs of Change* (1888).
3. William Morris, 'Art and Socialism', lecture first delivered 1884.
4. William Morris, 'The Aims of Art', lecture first delivered 1886. Subsequently published in *Signs of Change* (1888).

FURTHER READING

Caroline Arscott, 'William Morris: Decoration and Materialism', in Andrew Hemingway, ed., *Marxism and the History of Art: From William Morris to the new Left* (London: Pluto Press, 2006).

Charles Harvey and Jon Press, *William Morris: Design and Enterprise in Victorian Britain* (Manchester: Manchester University Press, 1991).

Fiona MacCarthy, *William Morris: A Life for our Time* (London: Faber and Faber, 1994).

Gillian Naylor, ed., *William Morris by Himself: Designs and Writings* (London: Macdonald/Orbis, 1988).

E. P. Thompson, *William Morris: Romantic to Revolutionary* (London: Lawrence and Wishart, 1955).

21

ART AND LABOR

Ellen Gates Starr

It was not only design that was being 'reformed' in the late nineteenth century. Just as the 1880s and 90s were the height of the Arts and Crafts Movement in Europe and America, they were also decades in which attitudes to class and gender were undergoing rapid transformation. These reform movements, aimed at improving the lives of workers (especially immigrants, women and children), were not unprecedented, but they did raise social awareness of exploitation and inequity in ways that still resonate today. This was also the period in which the first large-scale feminist movements began, organized initially around voting rights, or suffrage. Chicago was a hotbed for activity in all these areas. The most progressive institution in the city was Hull House, co-founded by Jane Addams and Ellen Gates Starr. Their efforts were inspired by Toynbee Hall, a 'settlement house' in the East End of London that offered temporary residence, food and social services to the urban poor. Addams and Starr founded Hull House in 1889 on this model, focusing their efforts particularly on women. One of the key offerings was a course in bookbinding, a craft that Starr had learned from T. J. Cobden-Sanderson in London, as well as other skills like drawing, embroidery and clay modeling. A 'Labor Museum' was also set up in 1900 to provide a history of craft and industry through displays and live demonstrations (an early example of this practice: 'so trivial a thing as a girl cleaning gloves, or a man polishing metal', one observer noted, 'will almost inevitably attract a crowd, who look on with absorbed interest').[1] Like C. R. Ashbee, who founded his Guild of Handicraft while a resident at Toynbee Hall, Starr hoped that the practice of craft would lead to not only well-designed objects but also, more importantly, beneficial social effects. In this essay she outlines these ideas and explicitly connects the aesthetic goals of the Arts and Crafts Movement with the broader objectives of social reform.

Ellen Gates Starr, 'Art and Labor', *Hull-House Maps and Papers* (1895), excerpted.

To anyone living in a working-class district of a great city today, the question must arise whether it be at all worth the cost to try to perpetuate art under conditions so hopeless, or whether it be not the only rational or even possible course to give up the struggle from that point, and devote every energy to "the purification of the nation's heart and the chastisement of its life." Only by recreation of the source of art can it be restored as a living force. But one must always remember the hungering individual soul which, without it, will have passed unsolaced and unfed, followed by other souls who lack the impulse his should have given. And when one sees how almost miraculously the young mind often responds to what is beautiful in its environment, and rejects what is ugly, it renews courage to set the leaven of the beautiful in the midst of the ugly, instead of waiting for the ugly to be first cleared away.

A child of two drunken parents one day brought to Hull-House kindergarten and presented to her teacher a wretched print, with the explanation, "See the Lady Moon." The Lady Moon, so named in one of the songs the children sing, was dimly visible in an extreme corner of the print otherwise devoted to murder and sudden death; but it was the only thing the child really saw.

The nourishment to life of one good picture to supplant in interest vicious story-papers and posters; of one good song to take the place of vulgar street jingles, cannot, I believe, be estimated or guessed. A good picture for every household seems unattainable until households can produce, or at least select, their own; but certainly a good one in every schoolroom would not be unattainable, if the public should come to regard it as a matter of moment that the rooms in which the children of the land spend their most impressionable days be made beautiful and suggestive, instead of barren and repellent.

Mr. T. C. Horsfall, of Manchester, England, who has developed a system of circulating collections of pictures in the schools of that unhappy city, says that the decision as to whether art shall be used in education is, to modern communities, a decision as to whether the mass of the people shall be barbarian or civilized. Assuredly it has a direct bearing upon the art-producing possibilities of the communities in question.

Let us consider what is the prospect for an "art of the people" in our great cities. And first let us admit that art must be of the people if it is to be at all. We must admit this whether we look into the life of the past or into our own

life. If we look to any great national art, that of Athens or of Venice or of Florence, we see that it has not been produced by a few, living apart, fed upon conditions different from the Common life; but that it has been, in great part, the expression of that Common life. If it has reached higher than the Common life, it has done so only by rising through it, never by springing up outside it and apart from it. When Florence decked herself with reliefs of the Madonna and the Infant, the life of Florence was a devotion to these shrines. Giotto and Donatello only expressed with a power and grace concentred [sic] in them what all the people felt; and more than that, had not the people felt thus, there could have been no medium for that grace and power.

If we are to have a national art at all, it must be art of the people; and art can only come to a free people. The great prophet of art in our day, John Ruskin, has said that "all great art is praise," showing man's pleasure in God's work; and his disciple, William Morris, expresses another side of the same truth when he says that "to each man is due the solace of art in his labor, and the opportunity of expressing his thoughts to his fellows through that labor." Now, only a free man can express himself in his work. If he is doing slave's work, under slavish conditions, it is doubtful whether he will ultimately have many thoughts worth the name; and if he have, his work can in no wise be their vehicle. It is only when a man is doing work which he wishes done, and delights in doing, and which he is free to do as he likes, that his work becomes a language to him. As soon as it does so become it is artistic. Every man working in the joy of his heart is, in some measure, an artist. Everything wrought with delight in the work itself is, in some measure, lovely. The destructive force of the ugly is its heartlessness. The peasant's cottage in the Tyrol, built with its owner's hands, decorated with his taste, and propounding his morals and religion in inlaid sentences under its broad eaves, blesses the memory with a beauty but half obliterated by daily sight of dreary parallelograms and triangles, joylessly united, which make up the streets of our working-people. The streets of Venice, of Verona, of Rouen, were built by men working in freedom, at liberty to vary a device or to invent one. They were not built by lawlessness or caprice, but under a willing service, which alone is perfect freedom.

The same men who built so nobly the cathedrals and council-halls of Rouen and Venice, built as harmoniously, though more simply and modestly, as was fit, their own dwellings. Had they been capable of making their own houses ugly, they would have been incapable of housing beautifully the rulers of their city or the King of kings.

This is the fatal mistake of our modern civilization, which is causing it to undo itself and become barbarous in its unloveliness and discord. We have believed that we could force men to live without beauty in their own lives, and still compel them to make for us the beautiful things in which we have denied them any part. We have supposed that we could teach men, in schools, to produce a grace and harmony which they never see, and which the life that we force them to live utterly precludes. Or else we have thought—a still more hopeless error—that they, the workers, the makers, need not know what grace and beauty and harmony are; that artists and architects may keep the secrets, and the builders and makers, not knowing them, can slavishly and mechanically execute what the wise in these mysteries plan.

The results should long ago have taught us our mistake. But only now are we learning, partly from dismal experience of life barren of beauty and variety, and partly from severe but timely teaching from such prophets as Ruskin and Morris, that no man can execute artistically what another man plans, unless the workman's freedom has been part of the plan. The product of a machine may be useful, and may serve some purposes of information, but can never be artistic. As soon as a machine intervenes between the mind and its product, a hard, impassable barrier—a non-conductor of thought and emotion—is raised between the speaking and the listening mind. If a man is made a machine, if his part is merely that of reproducing, with mechanical exactness, the design of somebody else, the effect is the same. The more exact the reproduction, the less of the personality of the man who does the work is in the product, the more uninteresting will the product be. A demonstration of how uninteresting this slavish machine-work can become may be found in the carved and upholstered ornamentation of any drawing-room car—one might also say of any drawing-room one enters.

I have never seen in a city anything in the way of decoration upon the house of an American citizen which he had himself designed and wrought for pleasure in it. In the house of an Italian peasant immigrant in our own neighborhood, I have seen wall and ceiling decorations of his own design, and done by his own hand in colors. The designs were very rude, the colors coarse; but there was nothing of the vulgar in it, and there was something of hope. The peasant immigrant's surroundings begin to be vulgar precisely at the point where he begins to buy and adorn his dwelling with the products of American manufacture. What he brings with him in the way of carven bed, wrought kerchief, enamel inlaid picture of saint or angel, has its charm of human touch, and is graceful, however childish.

The peasants themselves secretly prefer their old possessions, but are sustained by a proud and virtuous consciousness of having secured what other people have and what the world approves. A dear old peasant friend of Hull House once conceived the notion that the dignity of his wife—whom he called "my lady"—required that she have a dress in the American mode. Many were the mediatorial struggles which we enacted before this "American dress" was fitted and done. And then, by the mercy of Heaven, her courage gave out, and she never wore it. She found it too uncomfortable, and I know that in her inmost heart she found it too ugly.

Could men build their own houses, could they carve or fresco upon casing, door, or ceiling any decoration which pleased them, it is inconceivable that, under conditions of freedom and happiness, they should refrain from doing so. It is inconceivable that, adorning their own dwellings in the gladness of their hearts, they should not develop something of grace, of beauty, of meaning, in what their hands wrought; impossible that their hands should work on unprompted by heart or brain; impossible then, as inevitable now, that most men's houses should express nothing of themselves save a dull acceptance of things commercially and industrially thrust upon them.

A workingman must accept his house as he finds it. He not only cannot build it, he cannot buy it; and is usually not at liberty to alter it materially, even had he the motive to do so, being likely to leave it at any time. The frescoed ceiling to which I have referred, as the only example within my experience of any attempt at original decoration, was in a cottage tenement. If the author had any affection for the work of

his hands, he could not take it away with him. He would probably not be permitted, were he inclined, to carve the doorposts; and the uncertainty of tenure would deter him from yielding to any artistic prompting to do so. It would be disheartening to find one's belongings set into the street, and be obliged to leave one's brave device half finished.

A man's happiness, as well as his freedom, is a necessary condition of his being artistic. Ruskin lays it down as a law that neither vice nor pain can enter into the entirely highest art. How far art can be at all co-existent with pain, ugliness; gloom, sorrow, and slavery concerns very vitally the question of an art of the people.

No civilized and happy people has ever been able to express itself without art. The prophet expands his "All great art is praise," into "The art of man is the expression of his rational and disciplined delight in the forms and laws of the creation of which he forms a part." A rational and disciplined delight in the forms and laws of the creation of which a denizen of an industrial district in one of our great cities forms a conscious part, is inconceivable. Some of the laws which govern its conscious life may be traced in their resultant forms.

Its most clearly manifested law is "the iron law of wages." [. . .] Of the law of love manifested in the harmonious life of the universe, these little toilers know nothing. Of the laws of healthy growth of mind and body by air, sunlight, and wholesome work, neither they nor their children can know anything. Of the laws of heredity they know bitterly, and of the law of arrested development.

It is needlessly painful to say here in what forms these laws have made themselves known to them, and to all who look upon them. It is equally needless to say that they can have no delight in these forms, no wish to reflect and perpetuate them. Need it be said that they can have no art?

[. . .]

There is one hope for us all—a new life, a freed life. He who hopes to help art survive on earth till the new life dawn, must indeed feed the hungry with good things. This must he do, but not neglect for this the more compassionate and far-reaching aim, the freeing of the art-power of the whole nation and race by enabling them to work in gladness and not in woe. It is a feeble and narrow imagination which holds out to chained hands fair things which they cannot grasp—things which they could fashion for themselves were they but free.

The soul of man in the commercial and industrial struggle is in a state of siege. He is fighting for his life. It is merciful and necessary to pass in to him the things which sustain his courage and keep him alive, but the effectual thing is to raise the siege.

A settlement, if it is true to its ideal, must stand equally for both aims. It must work with all energy and courage toward the rescue of those bound under the slavery of commerce and the wage-law; with all abstinence it must discountenance wasting human life in the making of valueless things; with all faith it must urge forward the building up of a state in which cruel contrasts of surfeit and want, of idleness and overwork, shall not be found. By holding art and all good fruit of life to be the right of all; by urging all, because of this their common need, to demand time and means, for supplying it; by reasonableness in the doing, with others, of useful, wholesome, beneficent work, and the enjoyment, with others, of rightful and sharable pleasure, a settlement should make toward a social state which shall finally supplant this incredible and impious warfare of the children of God.

Whatever joy is to us ennobling; whatever things seem to us made for blessing, and not for weariness and woe; whatever knowledge lifts us out of things paltry and narrowing, and exalts and expands our life; whatever life itself is real and worthy to endure, as there is measure of faith in us, and hope and love and patience, let us live this life. And let us think on our brothers, that they may live it too; for without them we cannot live it if we would; and when we and they shall have this joy of life, then we shall speak from within it, and our speech shall be sweet, and men will listen and be glad. What we do with our hands will be fair, and men shall have pleasure therein. This will be art. Otherwise we cannot all have it; and until all have it in some measure, none can have it in great measure. And if gladness ceases upon the earth, and we turn the fair earth into a prison house for men with hard and loveless labor, art will die.

NOTE

1. Jessie Luther, 'The Labor Museum at Hull House', *The Commons* 70/7 (May 1902): 1–13.

FURTHER READING

Jane Addams, *Twenty Years at Hull-House* (1910).

Alan Crawford, *C. R. Ashbee: Architect, Designer & Romantic Socialist* (New Haven: Yale University Press, 1986).

Rivka Lissak, *Pluralism and Progressives: Hull House and the New Immigrants, 1890–1919* (Chicago: University of Chicago Press, 1989).

Lisa Tickner, *The Spectacle of Women: Imagery of the Suffrage Campaign* (London: Chatto and Windus, 1987).

22

ART AND WORKMANSHIP

W. R. Lethaby

The British Arts and Crafts Movement was a combination of progressive and conservative impulses. A case in point is the career of the architect William Richard Lethaby, who today seems the most attractive of the movement's leaders. Born in rural Devon, the son of a carver and gilder, Lethaby was strongly influenced by the hand-built ships and farmhouses that surrounded him as a child. He got his start in the profession after taking up a position in the office of the influential architect Norman Shaw (himself a master of historicist and vernacular details). When Lethaby joined William Morris's 'anti-scrape' Society for the Preservation of Ancient Buildings, he became a friend and ally of the great reform theorist. As was the case with Morris's interest in the medieval, it was Lethaby's interest in the past that brought him to some of his most original ideas. His first book, Architecture, Mysticism, and Myth, *drew on Gothic architecture in forwarding an argument for building as a symbolic art form; in his later career, he would spend two decades as the surveyor of Westminster Abbey. Though Lethaby was hardly prolific as an independent architect—he completed only six buildings—his work and writings served as a powerful example of Morris's principles in application. As a founding director of the Central School of Arts and Crafts, he also promoted the movement's ideals in an educational context.*

W. R. Lethaby, 'Art and Workmanship', *Imprint* (January 1913). As reprinted in Lethaby, *Form in Civilization: Collected Papers on Art and Labour* (London: Oxford University Press, 1922).

We have been in the habit of writing so lyrically of art and of the temperament of the artist that the average man who lives in the street, sometimes a very mean street, is likely to think of it as remote and luxurious, not 'for the like of him.' There is the danger in habitual excess of language that the plain man is likely to be frightened by it and it may be feared that much current exposition of the place and purpose of art only widens the gap between it and common lives.

A proper function of criticism should be to foster our national arts and not to frighten timid people off with high-pitched definitions and far-fetched metaphors mixed with a flood of (as Morris said) 'sham technical twaddle'. It is a pity to make a mystery of what should most easily be understood. There is nothing occult about the thought that all things may be made well or made ill. A work of art is first of all a well-made thing. It may be a well-made statue or a well-made chair, or a well-made book. Art is not a special sauce applied to ordinary cooking; it is the cooking itself if it is good. Most simply and generally art may be thought of as *the well-doing of what needs doing.* If the thing is not worth doing it can hardly be a work of art, however well it may be done. A thing worth doing which is ill done is hardly a thing at all.

Fortunately people are artists who know it not—bootmakers (the few left), gardeners and basket-makers, and all players of games. We do not allow shoddy in cricket or football, but reserve it for serious things like houses and books, furniture and funerals.

If it is necessary that everything must be translated into words, our art critics might occupy quite a useful place if they would be good enough to realize that behind the picture-shows of the moment is the vast and important art of the country, the arts of the builder, furniture maker, printer, and the rest, which are matters of national well-being.

It is doubtful if we have it in us to form a leading school of painting at the present time; indeed, we seem to be occupied in trying to catch up with Europe at the wrong moment. It cannot be doubted, however, that we might lead in the domestic arts. And this is shown by the great interest which foreign observers take in the English Arts and Crafts movement. The Germans, indeed, who know the history of this development in England better than we do ourselves, realizing its importance from an economic point of view, have gone so far as to constitute a special branch of political economy which shall deal with the subject. One university, I believe, has established a professor's chair in the economics of arts and crafts. English study of fine lettering has in Germany been put into types which English printers are hastening to buy. We have now many highly trained men among us who might make books as notable as those of the finest presses if there were a steady demand for fine modern work.

During the last thirty years many English designers have set themselves to learn the crafts as artists; that is, so that they may have complete mastery of both design and workmanship. I may remark here that a characteristic of a work of art is that the design interpenetrates workmanship as in a painting, so that one may hardly know where one ends and the other begins. The master-workman, further, must have complete control from first to last to shape and finish as he will. If I were asked for some simple test by which we might hope to know a work of art when we saw one I should suggest something like this: *Every work of art shows that it was made by a human being for a human being.* Art is the humanity put into workmanship, the rest is slavery. The difference between a man-made work and a commercially-made work is like the difference between a gem and paste. We may not be able to tell the difference at first, but, when we find out, the intrinsic worth of the one is self-evident. Still it is highly important that commercial work shall be properly done after its own kind.

Although a machine-made thing can never be a work of art in the proper sense, there is no reason why it should not be good in a secondary order—shapely, smooth, strong, well fitting, useful; in fact, like a machine itself. Machine-work should show quite frankly that it is the child of the machine; it is the pretence and subterfuge of most machine-made things which make them disgusting.

In the reaction from the dull monotony of early Victorian days it must be admitted that many workers fell into the affectation of over-designing their things. Rightly understood, 'design' is not an agony of contortion but an effort to arrive at what will be obviously fit and true. The best design is one which, cost apart, should become a commonplace. A fine piece of furniture or a fine book-binding should be shaped as inevitably as a fiddle.

Usually the best method of designing has been to improve on an existing model by bettering it a point at a time; a perfect table or chair or book has to be very well bred.

Another phase of the reaction from modern ways has been an excessive regard for old things, so that original workers have not had a fair chance of maintaining the full traditions of their arts. For instance, the social results of 'collecting old furniture' of course were not foreseen, but they certainly inflicted great injury on an essentially noble craft. At the present moment people who would like to do things in the best way would be well advised to have what they require made by capable men in modern forms. Now that we know all about it there is something pawnshoppy about gatherings from auctions, and the highly misdirected skill of the imitator has often made it next to impossible for even the expert to tell the difference between an original work and a copy.

Of course the scarcity, value, and historical interest of old pictures, and of books printed by [William] Caxton, made it inevitable that they should be sought for and bought at great prices, but undoubtedly such collecting of antiques has had a most injurious effect on all kinds of modern production. One of the great phenomena of recent time has been a drift away from production towards dealing. We have to re-establish doing.

Of many problems this one of bringing back art to workmanship is not the least serious, or the most hopeful. It is a tremendous thing that whereas a century or so ago the great mass of the people exercised arts, such as bootmaking, bookbinding, chair-making, smithing, and the rest, now a great wedge has been driven in between the craftsman of every kind and his customers by the method of large production by machinery. 'We cannot go back'—true; and it is as true that we cannot stay where we are.

Once more let me try to make it clear that by art, instructed thinkers do not only mean pictures or quaint and curious things, or necessarily costly ones, certainly not luxurious ones. They mean worthy and complete workmanship by competent workmen.

FURTHER READING

Sylvia Backemeyer and Theresa Gronberg, *W. R. Lethaby: Architecture, Design and Education* (London: Lund Humphries, 1984).

W. R. Lethaby, *Architecture, Mysticism, and Myth* (London: Percival, 1891).

W. R. Lethaby, *Home and Country Arts* (London: Home and Country, 1923).

Godfry Rubens, *W. R. Lethaby: His Life and Work* (London: Architectural Press, 1986).

23

'SLOGANS', 'THE WORK AHEAD OF US' AND 'THE PROBLEM OF THE RELATIONSHIP BETWEEN MAN AND OBJECT'

Vladimir Tatlin

Like William Morris, Vladimir Tatlin was a socialist who believed in the goal of unifying art and everyday life. But in the context of the Communist Revolution in Russia, he had the opportunity to put this precept into practice in ways that Morris never could have imagined. Tatlin viewed artists like himself as aesthetician/engineers, who would reshape all of society through the construction of new forms. Despite the grand scale of this ambition, he always spoke of his work as a form of 'fine craft' (izobrazitel'noe delo); perhaps this is explained by the fact that Tatlin had spent his youth as a sailor, making his own clothes and learning the many skills necessary to life on board. Along with other key Constructivists, such as the photographer and graphic designer Alexander Rodchenko, the painters El Lissitzky and Kazimir Malevich and the textile designer Varvara Stepanova, Tatlin taught at the VKhUTEMAS (an acronym for 'Higher Artistic and Technical Studios' in Russian), an experimental design school founded in 1920 that bears close comparison to the Bauhaus. There Tatlin produced his best-known work, the astonishing design for the Monument to the Third International. *This massive spiraling tower with rotating glass structures inside it would, if it had been built, have been the ultimate statement of radical utopian architecture. To this day, this project and the other unrealized dreams of the Constructivists are seen as defining exemplars of the modern avant-garde. Though their story is rarely discussed in terms of craft history, it is clear that—for Tatlin, at least—integrating art, craft and life was the very definition of a better future.*

Selected writings as translated and reprinted in Larisa Zhadova, *Tatlin* (London: Thames and Hudson, 1988).

SLOGANS (1920–23)

Engineers and bridge-builders, make calculations for an invented new form.

> Material culture. Down with Tatlinism.
> Painting + engineering – architecture = construction of materials.
> Organized material is a utilitarian form.
> Let us place the eye under the control of touch.
> Through the discovery of material to the creation of a new object.
> Move not to the left, nor to the right, but to the necessary.
> Not the old, not the new, but the necessary.

Remove from our environment people who have displayed empty slogans not confirmed by their craft.

THE WORK AHEAD OF US (1920)

The principles on which our fine art—our craft—stood were discredited, and any connection among painting, sculpture and architecture was lost, as a result of which individualism i.e. the expression of merely personal habits and tastes, and artists in their treatment of material reduced it to the level of being distorted in relation to one of the branches of art. So, at best, the artist decorated the walls of private dwellings (individual nests) left us a series of 'Iaroslavl' stations . . .

What happened in '17 in a social sense had been carried out in our fine craft in 1914, when 'material, volume and construction' were established as a principle.

Distrusting the eye, we place it under the control of touch.

1916 in Moscow there was an exhibition of laboratory models made of real materials of reliefs and counter-reliefs.

An exhibition in 1917 gave a number of examples of materials selected on the basis of more complex research into the material itself as well as its resultant movement, tension and between the two.

This research into material, volume and construction allowed us in 1918 to begin creating an artistic form of a selection of materials like iron and glass, as materials belonging to modern classicism, equal to marble in the past in their austerity.

In such a way it becomes possible to combine purely artistic forms with utilitarian goals. For example: the project of a Monument to the Third Communist International (Exhibited at the Eighth Congress).

The fruits of this are models which give rise to discoveries serving the creation of a new world and which call upon producers to control the forms of the new life around us.

THE PROBLEM OF THE RELATIONSHIP BETWEEN MAN AND OBJECT (1930)

Let us declare war on chests of drawers and sideboards.

We are now waging war for a collective way of life.

Socialist cities, 'green cities,' communal residences, palaces of culture are being built. In this construction there arises before us in all its breadth the problem of *man and object*.

The object in our conception must become not a sign of social distinction but that unit which is called on to realize specific functions allotted to it. At moments this object may disintegrate, become only a part of the whole, but continue to fulfil functions.

Against the old artistic thinking it is necessary to set the new form: material culture.

Working in this area since 1914, first alone and then with a group of students, I became convinced that our industry will be able to produce objects of high quality only when the artist-production worker takes a direct part in the organization of the object.

A way of thinking based on the culture of material makes it possible to take account both of the properties of individual materials and of the most advantageous features of their interrelationships. In such a way the artist, in creating an object, furnishes himself with a palette of different materials which he uses on the basis of their properties. Taken into account here are colour, texture, density, elasticity, weight, strength, etc.

With the task of creating a concrete everyday object with determined functions, the artist of material culture takes account of all properties of suitable materials and their interrelationships, the organic form (man) for which a given object is created, and finally the social side: this man is a worker and will use the object in question in the working life he leads.

Here must be considered the maximum functionality of the object which can be achieved when there is a great understanding of the properties of materials. This factor creates the possibility for an intelligent selection of materials for a functional object, and for the introduction of completely new and hitherto unexplored materials. This in turn gives a completely exceptional result: an object which is original and radically different from objects in the West or in America. This last fact is very important inasmuch as our everyday life is being built on completely new principles.

The demands we make of an object which has to serve us are considerably greater given the conditions of everyday life here than the demands made in capitalist countries.

Our everyday life is built on healthy and natural principles and an object from the West cannot satisfy us. We must search for completely different points of departure for creating our object. It is for this reason that I show such a great interest in organic form as a point of departure for the creation of the new object. I came to the unalterable view that studying organic form will give the richest material for the creation of a new object.

All our life, and production too, is overburdened by things, and mainly things which contain other things. We are also striving to eliminate these, to take from them only certain parts and introduce those parts into a building's architecture (shelves into the recess of a wall and so on). What do we use in constructing one object or another? Modern technology is working on those questions first and foremost. But that is not enough. Besides 'what', 'how' is very important, the organic form is important. For this we take and analyse existing objects, we use technical constructions as models for the forms of everyday objects, and finally, we also use as models the phenomena of living nature. Such are our principal tasks in working on the organization of the new object in the new collective way of life.

FURTHER READING

John E. Bowlt, ed., *Russian Art of the Avant Garde: Theory and Criticism* (New York: Thames and Hudson, rev. ed., 1988).

Rosalind Krauss, *Passages in Modern Sculpture* (Cambridge, MA: MIT Press, 1977), ch. 2.

Christina Lodder, *Russian Constructivism* (New Haven, CT: Yale University Press, 1983).

John Milner, *Vladimir Tatlin and the Russian Avant-Garde* (New Haven, CT: Yale University Press, 1983).

24

THE WAY OF CRAFTSMANSHIP

Sôetsu Yanagi

The gentleman-scholar Sôetsu Yanagi was the foremost theorist of Japan's mingei *movement, the twentieth century's greatest success story in the revival and promotion of traditional craft. A shorthand for 'folk craft'* (minshuteki kogei), *the term* mingei *was coined in 1925 or 1926. Yanagi was an editor of the avant-garde journal* Shirakaba, *placing him at the center of discussion about progressive art movements in Japan. He was also strongly influenced by the British Arts and Crafts Movement, especially William Morris, but focused on the revaluation of antique folk artifacts rather than the transformation of economy and labor. This meant that* mingei *was initially defined by collecting, often in a mode akin to the Japanese tea ceremony* (chanoyu). *Korean objects, long treasured among tea enthusiasts, were central to Yanagi's aesthetic. He saw in white Choson dynasty ceramics a 'beauty of sadness'—a controversial attitude, given that Japan had formally annexed Korea as a colony in 1910 and was often brutal in its suppression of independence movements there. Subsequently* mingei *would be further drawn into the tragic political trajectory of modernizing Japan, as it became a tool of wartime 'life culture' training both domestically and in occupied China.*

Like tea ceremony, the acquisition and display of mingei *objects was at first an urbane pursuit. But by the 1930s the movement broadened, and the production of so-called 'new* mingei' *was in full swing. First there were studio craftspeople influenced by tradition, such as the potters Shôji Hamada, Tomimoto Kenkichi and Kawai Kanjirô; the printmaker Munakata Shikô; and the textile artist Serizawa Kisuke. There were also cases in which Yanagi and others acted as self-appointed consultants to traditional pottery kilns, woodworking shops and basket makers, encouraging a return to premodern ways and then promoting the results through expositions and department store sales. This was the peak of* mingei *as a middle-class taste in Japan. Yanagi continued to be influential in the postwar period, however. Partly this was due to the efforts of his British colleague Bernard Leach, who published (with the assistance of a translator) the widely read anthology of Yanagi's writings,* The Unknown Craftsman, *from which the following essay is taken.*

Note: The characters of Yanagi's given name can be read either 'Sôetsu' or 'Muneyoshi'. Yanagi used both, but the former has become standard in English-language publications.

Sôetsu Yanagi, 'The Way of Craftsmanship' (1927), in *The Unknown Craftsman: A Japanese Insight into Beauty,* adapted by Bernard Leach (Tokyo: Kodansha, 1972).

I have been writing for a long time about crafts, digging into almost virgin soil, and what I say may seem strange to unaccustomed ears, dubious, and difficult to accept because it is contrary to prevalent thought. I have continuously received a flow of doubting enquiries from friends and strangers alike, so I decided to gather my ideas together in the form of a series of questions and answers reviewing the bone structure of my arguments.

Q. What are crafts?

A. Things made to be used by people in daily life, such as clothes and furniture. Something different from fine arts, such as pictures made to look at.

Q. What is the particular kind of beauty in crafts?

A. Beauty that is identified with use. It is beauty born of use. Apart from use there is no beauty of craft. Therefore, things made that do not stand up to use or that ignore utility can barely be expected to contain that kind of beauty.

Q. What is the meaning you attach to the word "use"?

A. The word is not to be understood merely in its materialistic sense. The reason for this is that mind and matter must not be thought of as separate. Use therefore covers both. Such objects are to be looked at and touched with the responsive feeling of pleasure in use. If crafts are only judged from a utilitarian point of view, then pattern, for example, is uncalled for. But good pattern adds to the function of that utensil. It becomes an indispensable part of use. On the other hand, however useful an artifact may be, if it causes in the mind a feeling of ugliness, it detracts from total service. The issue becomes clear in the province of food. Satisfying the demand of hunger is not the sole object of good cooking. We need good presentation and good flavors— that helps our appetite. Again, use that fulfils the mind alone is meaningless, like a wax replica of food. By use, then, I intend the indivisibility of mind and matter.

Q. What is the special quality of beauty in crafts?

A. The special quality of beauty in crafts is that it is a beauty of intimacy. Since the articles are to be lived with every day, this quality of intimacy is a natural requirement. Such beauty establishes a world of grace and feeling. It is significant that in speaking of craft objects, people use terms such as savour and style. The beauty of such objects is not so much of the noble, the huge, or the lofty as a beauty of the warm and familiar. Here one may detect a striking difference between the crafts and the arts. People hang their pictures high up on walls, but they place their objects for everyday use close to them and take them in their hands.

Q. How many types of craft are there?

A. CRAFT

FOLKCRAFTS	*ARTIST CRAFTS*
Guild Crafts	Aristocratic Crafts
Industrial Crafts	Individual Crafts

Folkcrafts—unself-consciously handmade and unsigned for the people by the people, cheaply and in quantity, as for example, the Gothic crafts, the best work being done under the Medieval guild system.

Individual or artist crafts—made by a few, for a few, at a high price. Consciously made and signed. Examples, [Aoki] Mokubei or [William] Staite Murray.

Industrial crafts—such as aluminum saucepans, etc., made under the industrial system by mechanical means.

Aristocratic crafts—examples, Nabeshima ware in Japan under the patronage of a feudal lord, or Stanley Gibbons in England.

Broadly, such are the divisions.

Q. Out of all these, which has most craft character?

A. Folkcraft, especially things made by a community of craftsmen, for that is where you find the purest form of craft. The reason for this is artist-craftsmanship places utility second and tends to pursue beauty for its own sake, thereby breaking the laws of craftsmanship. Artist-craftsmen separate themselves from the real nature of crafts and approach the fine arts. From the point of view of pure craftsmanship, folk-craft carries the rightful lineage. I do not wish to enter into the discussion at this time of the poverty of material and beauty in most industrial machine-made goods. The *Mishima* wares of Korea are genuine folkcraft, but the individualistic pots by [Nin'ami] Dôhachi in the *Mishima* style, by approaching fine arts, divide from the main stream.

Q. Which contain greater beauty, folkcrafts or artist crafts?

A. If we place them side by side, strangely enough the artist crafts cannot be said to be better, for they depend upon the personality of the artist rather than the character of the craft. If the names of the artists were unknown, could they have stood the contest? There are people who buy the name of the maker rather than quality. As to aristocratic crafts, in their attention to technique and over-refinement, they, too, are separated from the main stream. It is truly strange that folkcrafts should be better than the work of artists in pursuit of beauty. The works of artist craftsmen are not primarily intended to be just good pots so much as to display the fine sensibility or strength of personality of the maker—the flavour of himself rather than the flavour of mankind, which crafts exude.

Q. Why is the product of the artist craftsman defeated by the folk craftsman?

A. I would like to answer this by saying that "individualistic beauty" is lower that transcends the individual. To the latter type folkcraft belongs, whereas the individual artist is often so wrapped up in himself and his expression that he goes against the law of nature. This can also be explained by the fact that the power of the individual is weaker than that of tradition. Personality, however great, is nothing compared with nature. Surprisingly enough, the history of art is full of examples of the products of humble craftsmen that are far finer than the work of clever individuals. This is because their work contains no signs of egotism. It is like looking for true belief in a world infested with self-centredness. Only when egotism diminishes does true belief make an appearance. Just as it is rare to find a sincere man among Pharisees, so it is rare to find good work in signed crafts. What artist woodworker has produced furniture to compare with the Gothic? If we were to select a hundred examples of the most beautiful crafts out of the past and present, ninety-nine percent, no possibly one hundred percent, would be unsigned.

Q. Are you denying the importance of personality in the crafts?

A. To negate personality is an error; however to remain satisfied with personality is yet a greater error. There is a beauty that emerges from individual art, but it is not purest. In the case of a really great individual, the greatness lies in his having gone beyond his individualism. The reason that the products of artist-craftsmen are found lacking from the standpoint of beauty is because they rarely rise above their individualism. Furthermore, things that are highly individualistic are unsuitable for daily use. The assertion of one individuality is almost sure to produce a clash with another individuality. It is rare to find restfulness in beauty of an individual kind. If in such a way a craft object becomes unsuitable for our daily living, it fails of its purpose. Craftsmanship must not be impeded by individualism.

Stress upon individualism is totally unsatisfactory; on the other hand, where do we find beauty without individualism? Having no individuality and transcending it—these two issues must not be confused.

The virtue of folkcrafts is that one feels no obtruding personality in them. The thing shines, not the maker. Consider Persian rugs; one feels their beauty before any question arises as to who made them. Actually, almost any Persian could have made them. The work was subdivided, it was certainly not done by one pair of hands, nor conceived by one mind. Moreover, of these rugs, can any one be called ugly? Again, let me reiterate that that craftsmen must go beyond individualism.

Q. How does the unlettered craftsman produce beauty?

A. He may be unlettered, uneducated, and lacking any particular force of personality, but it is not from these causes that beauty is produced. He rests in the protecting hand of nature. The beauty of folkcraft is the kind that comes from dependence on the Other Power. Natural material, natural process, and an accepting heart—these are the ingredients necessary at the birth of folkcrafts. Hence it is the kind of beauty that saves us. The craftsman has not the power to save himself. It is nature that does the saving, and therefore whatever is made is lovely. Can we find any ugly or false work amongst folkcrafts? By contrast, if everything depended upon the worker on his own, just think how many mistakes would result. Whereas, left to nature, every piece is saved.

One would be hard pressed to find amongst the myriad artifacts of Gothic craftsmen downright bad work. Likewise in Japanese textiles of the eighth century it would be difficult to find bad colours or patterns.

Q. Is it not possible for the artist-craftsman to make beautiful things?

A. I am not saying that it is absolutely impossible, but it is well to realize that the artist-craftsman's solo path is fraught with difficulty. As long as he lingers in the stage of individualism he can never arrive at the beauty of "no-thought" of folkcraft. To fund pure and simple faith in the ranks of intellect is a rarity of rarities. If one wishes to travel by reliance on one's own

power, one must pass through great inner discipline akin to that of the Zen monks. Attachment to individualism guarantees no beauty, nor does it even provide the requisite ease in technique. If the way of the individual should become the main stream of craftsmanship, the crafts of the people will suffer. Why? Because the people possess neither a real individuality, nor a real intellect. And yet surprisingly enough it is the crafts of the people that have produced the greatest blossoming. Implicit in this point is the story of how great the difficulty is for the individual. Only a rare genius is able to produce something extraordinary. Today we have individual craftsmen galore. But who can claim that they are all geniuses? A genius may appear once in a generation. The world is already flooded with the works of unenlightened craftsmen.

Q. What, then, is the value of an artist-craftsman?

A. If the object of a piece of work is the expression of individualist beauty, then we must admit that the way of craftsmen is limited, for the road of fine arts is better suited for that end. That which ends with individuality does not agree with the nature of craft. In these days of deterioration of the art of the people nobody else is available who can set the standards of beauty other than the artist-craftsman. Today, having our way, we need the capacity of those who can show us how to properly appreciate beauty in work. In the world of crafts we hunger for this leadership. If the artist-craftsman does not rise to this task, our horizons will darken. This phenomenon is required of this age of consciousness. The presence of artist-craftsmen is to serve as a bridge between this period and the next flowering of the art of the people. Their value, therefore, lies in their ability to understand beauty rather than in their expression of it. Consequently, their work takes on a great significance as a gift to the world of thought. Unfortunately, so few know clearly what the target is, and the number of those who have the

Figure 16 *Ainu Robe,* mid-nineteenth century.

genius to express true beauty in their work is so limited. Rather, as things are, there are too many who are poisoning the crafts.

In actuality, the artist-craftsman's function is to point the way as a compass does, rather than as a maker. For example, take the case of an artist-potter who makes a pot and puts on it a drawing of a landscape, which is then copied in thousands by many other artisans, as was the case in Ming dynasty China. Now the curious thing is that, at the point when awareness of the original dies away, a new beauty far greater than the original comes into being. The object now no longer belongs to the work of the individual but to the craft world of tradition. The work of an artist is thus less than the expression of the people. The value, then, of the individual is principally in his contribution to the world of intellectual thought.

Q. Which is more significant for the future—folk- or artist craft?

A. The work of the artist-craftsman is to clear the way ahead by pointing in the right direction for the eventual return of craftsmanship to the hands of the people themselves. The ultimate aim is not the expression of the self but to see true beauty, specifically in "people's art." The intent is not to save one individual and his work but to save craftsmen and their work for the future. From the spiritual and from society's point of view, the art of the people as a whole is much more important than the art of any one individual. The decline of folkcraft almost means the death of craftsmanship; such an event would make the Kingdom of Beauty impossible. Confining beauty in the hands of a very few artists is cold comfort. In this situation the artist-craftsman pursuing his lone path of personal integrity is like the hermit of former days. Nevertheless, should not self-purification mean purification of others as well? Hiding away from this world is certainly not the objective. To make a move from the fine arts toward art of the people implies a change from individual

salvation to the salvation of society. It is not the task of crafts proper to foster individualism and the individualist approach to work.

Q. What is to be expected most of the artist-craftsman?

A. It is to be hoped that the artist-craftsman will awaken to his obligation to contribute beauty to the community. Today this is his greatest failure. As things stand, whatever he does he pursues it with the consciousness of beauty in mind. But the artist of the future needs to be concerned with the requirements of people around him. The extent to which he contributes to society determines his value. Whilst he holds aloof, leaving the people to their fate, the horrible lampshades remain, benefitting neither him nor society, and so forth. It is far more significant to be saved with others than alone. The trend of social evolution causes one to anticipate the time when artist craftsmen, too, will begin to think of themselves as part of a whole. And the time will come when they will no longer be satisfied to pursue beauty by itself and will cease to be able to turn a cold shoulder upon the art of the people.

Q. What's lacking in the artist-craftsman?

A. His products are so few and so expensive. They are more decorative than useful. Even if they are made for use they are expensive and are therefore not employed in daily life, thus becoming luxury items. From the very beginning they are made for art collectors, and become disconnected from the life of the people. The only person who benefits is the favored purchaser. The artist-craftsman separates himself from need, and thereby divorces himself from the people around him. Is this not a mortal wound to craftsmanship? Apart from use and the people there is no meaning in either craftsmanship or beauty. If the artist-craftsman continues isolating himself from society, he has a responsibility to admit with humility [out of his own experience] that his position of self-expression is one of insufficiency. And in view of the achievement of the arts of the people, he needs to feel an awakened respect for them and pave the way towards the re-expression of that congregate power. At that moment when the work of the artist-craftsman ceases to be individual and he thus joins the ranks of all men, let him place his work next to the old work that he used to do. And he may see truth for the first time, for his old work will not stand up in service or in beauty.

Q. What are the strong points in folkcrafts?

A. They are never made for other than use; they are inexpensive; they are made in quantity sufficient to serves masses of people daily. Their quantity production means repeated practise in their technique, thereby freeing them from ailments arising from artfulness. They are made without obsessive consciousness of beauty; thus we catch a glimpse of what is meant by "no-mindedness," whereby all things become simplified, natural, and without contrivance. These are the qualities that provide a permeance [sic] of strength throughout the social and aesthetic edifice. There are so few evidences of disease in the arts of the people (*getemono*). Rarely are there cases of ugliness to be found in them. The people and their crafts are harmoniously interrelated. How little fine work has come out of intellect, technique, and individuality. By contrast, how little evidence of ugliness there is to be found in those ordinary articles of folk life of the past? This is parallel with Buddhist experience, in which but few Zen monks, relying on their own endeavours, reach true Enlightenment. Whereas amongst the ranks of unlettered, good, simple men and women of Buddhist dependence on Other Power (*tariki*) we find many of profound, humble faith.

Q. Why do you focus your attention to such an extent on folkcraft?

A. (1) My intuition has perceived a far richer beauty in folkcraft than in fine arts. (2) Hitherto in their discussion of crafts almost no one has taken up the contributions of these humble craftsmen and given them their due evaluation. (3) Art historians and collectors, on the contrary, have been biased in favour of individual artists. (4) The artist thereby has been kept locked up in his ivory tower of individualism and is out of touch with the people. (5) No one has as yet led the way toward communal expression for crafts. We must bring back the realization of values and those days when all things required in daily, ordinary life were beautiful. Only when we succeed in this can we speak of an epoch of craftsmanship. The beauty of craftsmanship lies with society rather than with the individual. There are many people wanting to be artist-craftsmen, but who is concerned about the improvement of nameless crafts? In fact, nobody believes that it is in this very namelessness that the deep roots of great craft find their sustenance. The only people who are on the increase are those who are fortifying themselves behind castle walls of self-enclosure. Even amongst purchasers, the habit of dependence upon name becomes ever more apparent. The more this habit predominates in the world, the more I feel the need of strengthening the voice of the humble artisan. Unless a wide public takes up this issue, the history of genuine craftsmanship will come to an end.

Q. Why are the folkcrafts in decline?

A. History clearly indicates that as industrial capitalism flourished, handcrafts declined in the East and the West alike. As we look backwards, suddenly in Japan about the year 1887 there was a sharp loss of beauty in all the crafts. Needless to say, this was around the time when the industrial system surged forward. In the Western world, with the spread of industry, the last of the glory attached to the Middle Ages and its craftsmanship came to an end.

Q. How does industrial capitalism destroy beauty of folkcraft?

A. Because the objective of production is profit alone, and even objects' utility is secondary. In front of the eyes of the capitalist is the word "profit." The quality, beauty, and health of an object are all secondary considerations. Greed for profit is destructive of both use and beauty. In addition, under capitalism, craftsmanship leans away from human hands towards machinery. As a consequence, beauty loses its sensibility more and more and tends towards hardness.

Crafts originally sprang from a person's making things for his own use. That was followed by selling for the use of others, and the next stage was the change from handwork to the machine. On the face of things this is a natural evolution, but seen from the other side this process means a change from what was at first healthy into insensibility, from freedom to cold expression, and from kindness to avarice. In this way an age ended that produced almost no ugliness, and a new period was launched where beauty in craft became very difficult. We are living in a time of severe change.

Q. Why are today's folkcrafts so impoverished?

A. Because the capitalist system launches us in a whirlpool of competition, we are forced to use sensational means to attract buyers. The immediate reflection of this is to be seen in bad colours and poor shapes. This bad influence unconsciously affects man's very heart, which is a grave problem. Crafts have lost their lasting values by exchanging them for passing fancy. It is unlikely that the clothes we wear today will ever be displayed in art galleries. This is because they are poor in material and design. Things go on being made that can only be described as bad, and interest lies only in the new and changing. Such an environment fails to deepen the sense of creative imagination, and even the taste of the educated becomes poor.

Q. What is the effect of industry on crafts?

A. At present exceedingly bad. As things are, the desire for a world of normal beauty, once more, is unlikely to arise. Although there can be a kind of beauty in things made mechanically, yet nothing so made has surpassed the beauties of the age of handiwork. The shape of things is at present hopeless. Since a tool is a kind of mechanical aid, one cannot say that hand and machine are utterly apart, and, for that matter the hand itself is a machine; why then are things made by hand both more beautiful and more lasting? Actually it is because it is a freer and more complex machine. However intricate the mechanics of a machine, they are nothing to those of the hand. Man's power is foolish in comparison with nature's.

[. . .]

Q. What is meant by getemono?

A. *Ge* means "ordinary" or "common," and *te* means "by nature." That is to say, *nami no mono,* "something of a quite practical nature." . . . I find it an astonishing providence that in these unsigned, cheap, abundant, quite ordinary articles there so often lies hidden a beauty that one could hardly expect to discover. The uncovering of this truth is a great affirmation of the common man. It bespeaks the total harmony between the concepts of economy and aesthetics.

Q. Are you stating that in getemono alone one discovers the beauty of craftsmanship?

A. No, I am not arriving at such a crude conclusion, for even in fine crafts what is beautiful is beautiful. But we must note that in fine crafts the examples of beauty are extremely rare, and even in them the expression or the state of mind from which they sprang stands upon the same basis as that of the *getemono:* there must be neither over-calculation nor complexity; the direct response to innate nature, naturalness, and simplicity are seen therein. But after all, are these not the very qualities of the beauty of the *getemono?* Here we may see how closely the perception of beauty in *getemono* and in crafts is connected. We have come all this way without clarification of this truth. In contrast to the habitual way of thinking that beauty in crafts is almost always dependent upon refinement, this view may bring about a reversal of values.

Q. Why do fine crafts so often fail?

A. To the extent to which they become separate from use, they are stripped of craftsmanlike content. The nearer to uselessness, the nearer to sickness. They seldom escape from the affliction of self-consciousness. They fall so easily into the pitfall of themselves. The craftsman is apt to become over-anxious about sheer skill. Thus the increase of complexity, additional décor, and self-conscious effort all become accentuated. Not only do fine crafts remove themselves from use by pursuing artistry, but they do not even fill demand, because their output becomes less. Consequently, the article becomes expensive, and economic problems arise. Surrounded by such sick conditions, the production of healthy articles diminishes. Fine wares are generally meant for admiration in a glass case, and, not being intended for use, they of necessity lack constructive strength. We cannot find in such things the main flow of craftsmanship.

Q. Why is healthy beauty more richly manifest in getemono than in fine crafts?

A. *Getemono* are things that work and serve us from day to day, not things kept in a glass case merely to be looked at. Their role is work, and therefore they do not lean towards frivolity. The worker must be sober in look and strong of body. If the body is weak, one cannot work: health is a natural requisite to perform work properly. In this obvious fact lies such objects' honest, simple, humble beauty. To reiterate, the principle of craftsmanship, where beauty and use are perfectly equated, may be found before one's eyes, like the thumb and fingers of one hand. Where else can we find greater beauty, in which naturalness, balance, and stability predominate?

[. . .]

Q. What is the meaning of placing such importance on nature?

A. First, nature must be freely at work in the mind when anything is well made. Though painstaking efforts may have their contribution to make in carrying out a work, more astonishing is the effect that "no-mindedness" has upon it. One gains greater insight into nature by open trust rather by attempts at intellectual understanding.

Secondly, procedures must be natural. Nature's simplicity hides a greater complexity than man's. Beauty requires neither indirectness nor intricacy. Try to add or contrive, and life vanishes. Great detail and high finish have to do with technique but have nothing to do directly with beauty. In fact, they interfere with it. Lovely things are almost always simply made.

Thirdly, the material provided by nature is nearly always best. Nothing is more precious than the unspoiled character of raw material. For it is always richer than the man-made. Man thinks that artificial material [such as glass] is pure, but from nature's side it is impure and forced. When we think back on great periods, we can almost say the material is synonymous with craftsmanship. One aspect of the beauty of crafts lies in the beauty of the materials. May we not accept crafts as generally being local? Crafts are born where the necessary raw materials are found. The closer we are to nature the safer we are; the further away, the more dangerous.

Q. What is the fundamental principle of the beauty of craft?

A. The principle of the beauty of craft is no different from the law that rules the spirit underlying all things. There is then no truer source than the words of the religious scriptures. A true example of craft is the same as a passage of a holy scripture. Only in the place of words, truth is conveyed through material, shape, colour, and pattern. Gothic crafts and Gothic religious spirit spoke with the same voice. It is also this same spiritual law that one sees expressed in the crafts of the Sung dynasty. Even in one single piece of good work, one finds expounded in material form the commandment to refrain from attaching oneself to the ego, the heart of Zen, which teaches "no-thought," the standpoint of the Other Power (*tariki*) school, which embraces and saves all beings without exception. Faith and beauty are but different aspects of the Absolute Reality.

I hope with these lines I have been able to express very nearly what I mean by craft.

FURTHER READING

Kim Brandt, *Kingdom of Beauty: Mingei and the Politics of Folk Art In Japan* (Durham, NC: Duke University Press, 2007).

Yuko Kikuchi, *Japanese Modernisation and Mingei Theory: Cultural Nationalism and Oriental Orientalism* (London: Routledge, 2004).

Kakuzo Okakura, *The Book of Tea* (Boston: Shambala, 2003; orig. pub. 1906).

Jun'ichirô Tanizaki, *In Praise of Shadows* (New Haven, CT: Leete's Island Books, 1977; orig. pub. 1933).

25

A POTTER'S BOOK

Bernard Leach

In 1954, Bernard Leach traveled through the United Kingdom and United States with his friends, the mingei *theorist Sôetsu Yanagi and the potter Shôji Hamada. Though he was nearly seventy years of age, it was arguably the peak moment of his influence. Born in Hong Kong, Leach had always been a cosmopolitan figure, traveling back and forth between 'east and west', as he put it. In the interwar period he had been (with Yanagi) part of the* Shirakaba *circle of aesthetes in Tokyo and had become intimately familiar with the history and practices of Japanese ceramics. In 1920 he returned to England with Hamada and set up a kiln and workshop at St. Ives, hoping to put his newfound respect for traditional ceramics into practice. To a great extent, he was successful. Leach became the most well-known of a small coterie of craftspeople who were patronized by (and to some degree part of) the well-to-do British intelligentsia. With the support of Dorothy and Leonard Elmhirst he set up a second pottery at their estate, Dartington Hall. At St. Ives he taught apprentices (such as Michael Cardew and Warren MacKenzie, both of whom went on to a prominence nearly matching Leach's), wrote books and essays and made pots that drew not only on Japanese traditions but also on the 'taproot' of English slipware and stoneware. By the 1950s, the Leach Pottery was churning out a rather uninspired line entitled 'standard ware' (seemingly without irony) and he was largely dependent on the* assistance of skilled assistants. But his ideas were in the ascendancy. The postwar rapprochement with Japan, combined with an emerging popular taste for Zen, helped his message of reverence for authentic tradition resonate throughout Britain and America. Leach would often verbally cross swords with those who dissented from this position, such as the Bauhaus-trained Marguerite Wildenhain, and to this day the 'brown pots' that he favored are seen as a symbol for conservatism in the crafts. Yet, precisely because his writings so fully exemplify the idealism of the studio pottery movement, it is unlikely that they will ever go completely out of fashion.*

Bernard Leach, excerpts from 'Towards a Standard', in *A Potter's Book* (London: Faber and Faber, 1940).

Very few people in this country think of the making of pottery as an art, and amongst those few the great majority have no criterion of aesthetic values which would enable them to distinguish between the genuinely good and the meretricious. Even more unfortunate is the position of the average potter, who without some standard of fitness and beauty derived from tradition cannot be expected to produce, not necessarily masterpieces, but even intrinsically sound work.

The potter is no longer a peasant or journeyman as in the past, nor can he be any longer described as an industrial worker: he is by force of circumstances an artist-craftsman, working for the most part alone or with a few assistants. Factories have practically driven folk-art out of England; it survives only in out of the way corners even in Europe, and the artist-craftsman, since the day of William Morris, has been the chief means of defence against the materialism of industry and its insensitivity to beauty.

Here at the very beginning it should be made clear that the work of the individual potter or potter-artist, who performs all or nearly all the processes of production with his own hands, belongs to one aesthetic category, and the finished result of the operations of industrialized manufacture, or mass-production, to another and quite different category. In the work of the potter-artist, who throws his own pots, there is a unity of design and execution, a co-operation of hand and undivided personality, for designer and craftsman are one, that has no counterpart in the work of the designer for mass-production, whose office is to make drawings or models of utensils, often to be cast or moulded in parts and subsequently assembled. The art of the craftsman, to use Herbert Read's terminology, is intuitive and humanistic (one hand one brain); that of the designer for reduplication, rational, abstract and tectonic, the work of the engineer or constructor rather than that of the 'artist'. Each method has its own aesthetic significance. Examples of both can be good or bad. The distinction between them lies in the relegation of the actual making not merely to other hands than those of the designer but to power driven machines. The products of the latter can never possess the same intimate qualities as the former, but to deny them the possibility of excellence of design in terms of what mechanical reproduction can do is both blind and obstinate. A motor car such as a Rolls Royce Phantom achieves a kind of perfection although its appeal is mainly

intellectual and material. There I think we come to the crux of the matter: good hand craftsmanship is directly subject to the prime source of human activity, whereas machine crafts, even at their best, are activated at one remove—by the intellect. No doubt the work of the intuitive craftsman would be considered by most people to be of a higher, more personal, order of beauty; nevertheless, industrial pottery at its best, done from the drawings of a constructor who is an artist, can certainly have an intuitive element.[1]

The trouble, however, is that at a conservative estimate about nine-tenths of the industrial pottery produced in England no less than in other countries is hopelessly bad in both form and decoration. With the exception of a few traditional shapes and patterns for table-ware, and others designed by the best designers available today and painted by the best available artists (none of whom is a potter), turned out notably by Wedgwood and Royal Worcester and Minton factories and by the Makin and Gray firms in Hanley, and excluding also a few purely functional and utilitarian designs, some of which are also traditional, such as Doulton's acid-jars, we meet everywhere with bad forms and banal, debased, pretentious decoration—qualities that are perhaps most conspicuous in 'fancy vases', flower-pots and other ornamental pieces, in which we find a crudity of colour combined with cheapness and inappropriateness of decoration and tawdriness of form that must be seen to be believed. And although the mechanical processes are indeed marvellous, as for example the automatic glazing, cleaning, measuring and stamping of many millions per month of bathroom tiles, fired in a single nonstop tunnel kiln, the mere fact of their being mass-produced is no reason why these tiles should be as cheaply designed and as dull and miserable in colour as it is possible for tiles to be; nor in the case of hollow-ware is the casting of shapes so exactly and so quickly and with such perfect pastes an adequate excuse for dead shapes, dead clay, dead lithographed printing or the laboured painting of dead patterns. Indeed the more elaborate and expensive the decoration the more niggling and lifeless it is, and the nearer it approaches the long deceased fashion of naturalism of the nineteenth century, when close attention to detail and the careful painting of pictures upon porcelain in enamel colours was considered the summit of ceramic art—'applied' art with a vengeance! . . .

It is obvious that the standards of the world's best pottery, for example, those of the T'ang and Sung periods in China and the best of the Ming, Korean celadons and Ri-cho, early Japanese tea-master's wares,[2] early Persian, Syrian, Hispano-Moresque, German Bellarmines, some delft and English slipware, cannot well be applied to industrial work, for such pottery was a completely unified human expression. It had not been mechanized. Yet there is no doubt that much can be learned by the industrial potter or designer from the wares especially of the Sung and early Ming dynasties. The Chinese potters' use of natural colours and textures in clays, the quality of their glazes (e.g. the Ying-ching and T'zu-chow families), the beauty and vitality of their well-balanced and proportioned forms, could be a constant source of inspiration to the designer for mass-production no less than to the craftsman.

It is no discredit to the scientific and utilitarian advances of the English pottery industry to say that the beauty to which the Sung potters attained was far beyond the highest that from its beginnings in Josiah Wedgwood the English factories ever aimed at. The two traditions and methods of production are radically different, and the intuitive, organic qualities of Sung pottery can never be completely expressed by the rational and tectonic methods of big industry. Concentration upon mechanical production and utilitarian and functional qualities is today necessary and justified, and as already said there is no reason to suppose that factory-made utilitarian wares may not by reason of their precision, their pleasing lines and perfection of technique, added to complete adaptation to use, have a great beauty of their own. Even during the course of the last two centuries moulded English tea ware of admirable design has been made, and often its decoration, especially the 'Japan' and other conventionalized set patterns of the late eighteenth and early nineteenth centuries, has been, if not great art, at least possessed of much charm. It would be surprising if equally good patterns could not be turned out by able designers today.

It is quite otherwise with the studio potter. He is indeed constrained to look to the best of the earlier periods for inspiration and may, so far as stoneware and porcelain are concerned, accept the Sung standard without hesitation. As it is, there are a few English craftsman potters today who do accept it, and their work is incomparably the best that is now being turned out.[3] Others go back to an outmoded 'arts and crafts' tradition, which seems to have had its origin in France in the last quarter of the nineteenth century and to have been largely influenced by modern Japanese designs, which became fashionable soon after the Paris Exhibition of 1867. Its characteristic features are weakness of form, especially of lip and foot, and, except in the case of the salt-glazed wares of the Martin Brothers (much of which was influenced by the same school of design), crudely coloured glazes in which all aesthetic quality is lost in technique, as always happens when the means are mistaken for the end . . .

In the absence of some agreement, however inarticulate, as to a common standard, one may hope to find an occasional work of genius in the free, or so-called fine arts (frequently then only the outcome of pain and poverty and lifelong obscurity); but in applied art, which depends upon collaboration in the workshop and constant sales to a public, there is even less hope. Indeed, amongst some at least of the free arts there does exist what one may call a classic standard, according to which the work of today, especially in literature and music, is compared with the great work of the past. That the criterion of beauty is a living thing and constantly in flux, is true, but here at least there is a continuous if ever changing consensus of opinion as to what may be called great achievement. In regard to pottery such a criterion can hardly be said ever to have entered the consciousness of Western man. In the East it has long been in existence, especially in Japan, where the aesthetic sensibility of educated people has been stimulated by the ablest of critics for some three hundred or more years.

[. . .]

A potter's traditions are part of a nation's cultural inheritance and in our time we are faced with the breakdown of the Christian inspiration in art. We live in dire need of a unifying culture out of which fresh traditions can grow. The potter's problem is at root the universal problem and it is difficult to see how any solution aiming at less than the full interplay of East and West can provide either humanity, or the individual potter, with a sound foundation for a world-wide culture. Liberal democracy, which served as a basis for the development of industrialism, provides us today with a vague humanism as insufficient to inspire art as either the economics of Karl Marx or the totalitarian conception of national life,

but at least it continues to supply an environment in which the individual is left comparatively free.

Our need of a criterion in pottery is apparent and seems to be provided by the work of the Tang and Sung potters which during the last twenty years has been widely accepted as the noblest achievement in ceramics. But the successful assimilation of strange stimuli requires a healthy organism, and it remains to be seen whether there is enough vitality in Europe to absorb from early Chinese pottery even more than we did during the eighteenth and nineteenth centuries from late Chinese porcelain. At the moment it is difficult to believe that the general arrogance of our materialism and the particular self-sufficiency of the pottery trade will permit the subtler scale of early oriental values to be perceived, except by artists and some sensitive people of leisure. Influences from alien cultures either upon art, or industry must pass through an organic assimilation before they can become part and parcel of our growth: This happens, moreover, only when they supply an inherent need, and is usually inaugurated by the enthusiasm and profound conviction of men who have themselves succeeded in making the synthesis. The superficial imitation of early Chinese shapes, patterns, colours and technique signify nothing unless new life emerges from the fresh combination. The temptation for the individual potter is to stand back with the paralysis of frustration in face of such a sea of change, but we cannot afford to wait until the tide of a new culture rises.

The necessity for a psychological and aesthetic common foundation in any workshop group of craftsmen cannot be exaggerated, if the resulting crafts are to have any vitality. That vitality is the expression of the spirit and culture of the workers. In factories the principal

Figure 17 Bernard Leach's work bench, showing pots, some of his working sketches, and his seals.

objectives are bound to be sales and dividends and aesthetic considerations must remain secondary. The class of goods may be high, and the management considerate and even humanitarian, but neither the creative side of the lives of the workers nor the character of their products as human expressions of perfection can be given the same degree of freedom which we rightly expect in hand work. The essential activity in a factory is the mass-production of the sheer necessities of life and the function of the hand worker on the other hand is more generally human.

The problem is made increasingly difficult for the reason that the people who are attracted today by the hand crafts are no longer the simple-minded peasantry, who from generation to generation worked on in the protective unconsciousness of tradition, but mainly self-conscious art students. They come to me year after year from the Royal College, or the Central School, or Camberwell, for longer or shorter, usually shorter, periods of apprenticeship. As soon as they have picked up enough knowledge, or what they think is enough, off they go to start potting on a studio scale for themselves. Very few have proved themselves to be artists. And what of the others, those thousands who pass through these schools and then either disappear from sight or continue to produce bad work. Again, in the past tradition would have developed and used their more moderate talents; in our own one cannot escape the sense of a great wastage.

In crafts the age-old traditions of hand work, which enabled humble English artisans to take their part in such truly human activities as the making of medieval tiles and pitchers and culminated in magnificent co-operations like Chartres Cathedral, have long since crumbled away. The small establishments of the Tofts and other slipware potters were succeeded by the factories of the Wedgwoods and the Spodes, and in a short space of time the standard of craftsmanship, which had been built up by the labour of centuries, the intimate feeling for material and form, and the common, homely, almost family workshop life had given way to specialization and the inevitable development of mass production. For that no individual can be praised or blamed: like many another institution it arose in response to a human need, moving parallel on the one hand with the slow progress of economic democracy, and on the other with an unprecedented rise in the population. But although we have now reached a point where for the first time in history we are able to produce enough and more than enough for all, the trouble from the artist's or craftsman's, or for that matter any sane person's point of view, is not only that the problem of equitable distribution is still unsolved, but that so many of the things we have thus contrived to make are inhuman.

In the field of ceramics the responsibility for the all-pervading bad taste of the last century and the very probable ninety per cent bad taste of today lies mainly with machine production and the accompanying indifference to aesthetic considerations of individual industrialists and their influence on the sensibility of the public.[4] Yet although industrialists will as time goes on become more and more conscious of the desirability of, if not the necessity for good form and decoration, it is also plain that during the last twenty-five years a far reaching change in aesthetic judgment has come about, not only in England, but literally all over the civilized world. A new type of craftsman, called individual, studio, or creative, has emerged, and a new idea of pottery is being worked out by him as a result of an immensely broadened outlook. Another wave of inspiration has come to us from the Far East, and out of the tomb-mounds of long dead Koreans and Chinese, looted and disturbed by the encroachment of Western commercialism, has arisen a new appreciation of ceramic beauty.

[. . .]

I can still remember vividly how twenty-five years ago I stood before the magnificent examples of the pottery of the Sung dynasty in the Tokyo Museum wondering how an individual potter of today could possibly appropriate to himself a beauty so impersonal, so inevitable— the patient unassuming outcome of centuries of tradition gradually developing through the experience of material and increasing complexity of need, and the sublimated emotion of a long succession of Chinese or Korean workers. I was abashed. I know now that it is a task beyond the power of any one man, and what makes the matter still worse, far from there being any unity of purpose and faith, at the present moment there is such an obsession with the individual point of view among English craftsmen, that one often hears them ridicule the very idea of a new communal standard. Independence once achieved is very precious, but an exaggerated pride in its possession stands bluntly in the way of concurrence in either aim or action, and the pride is only too often merely that of an artist on a dunghill. Since the Great War, however, there have been at least some signs of change, in science, in philosophy, in politics, even in the world-wide acceptance by the younger artists of a more or less common geometric abstract. But even this new common factor has been accompanied by a growing awareness of emptiness and sterility.

We craftsmen, who have been called artist, have the whole world to draw upon for incentive beauty. It is difficult enough to keep one's head in this maelstrom, to live truly and work sanely without that sustaining and steadying power of tradition, which guided all applied art in the past. In my own particular case the problem has been conditioned by my having been born in China and educated in England. I have had for this reason the two extremes of culture to draw upon, and it was this which caused me to return to Japan, where the synthesis of East and West has gone farthest. Living there among the younger men, I have with them learned to press forward in the hope of binding together those elements from the ends of the earth which are now giving form to the art of the coming age. I may tend to overstress the significance of East and West to one another, yet if we consider how much we owe to the East in the field of ceramics alone, and how recent a thing is Western recognition of the supreme beauty of the work of the early Chinese, perhaps I may be forgiven for the sake of the firsthand knowledge which I have been able to gather both of the spirit and manner in which that work was produced.

The manner, or technique, will be dealt with in the following chapters: here at the outset I am endeavouring to lay hold of a spirit and a standard which applies to both East and West. What we want to know is how to recognize the good or bad qualities in any given pot, and we are at least able to say that one should look first for the nature of the pot and know it for an expression of the potter in the background. He may be an unknown peasant or he may be a [William] Staite Murray. In the former case his period and its culture and his national characteristics will play a more important role than his personality; in the latter, the chances are that personality will predominate. In either case sincerity is what matters, and according to the degree in which the vital force of the potter and that of his culture behind him flow through the processes of making, the resulting pot will have life in it or not.

I have often sought for some method of suggesting to people who have not had the experience of making pottery a means of approach to the recognition of what is good, based upon common human experience rather than upon aesthetic hairsplitting. A distinguished Japanese potter, Mr. [Kanjirô] Kawai of Kyoto, when asked how people are to recognize good work, answered simply, 'With their bodies'; by which he meant, with the mind acting directly through the senses, taking in form, texture, pattern and colour, and referring the sharp immediate impressions to personal experience of use and beauty combined. But as pottery is made for uses with which we are all familiar, the difficulty probably lies less in one's ability to recognize proper adaptation of form to function than in other directions, primarily perhaps in unfamiliarity with the nature of the raw material, clay, and its natural possibilities and limitations, and also in uncertainty as to the more imponderable qualities of vitality and relative excellence of form, both of which are indispensable constituents of beauty. It must always be remembered that the dissociation of use and beauty is a purely arbitrary thing. It is true that pots exist which are useful and not beautiful, and others that are beautiful and impractical; but neither of these extremes can be considered normal: the normal is a balanced combination of the two. Thus in looking for the best approach to pottery it seems reasonable to expect that beauty will emerge from a fusion of the individual character and culture of the potter with the nature of his materials—clay, pigment, glaze—and his management of the fire, and that consequently we may hope to find in good pots those innate qualities which we most admire in people. It is for this reason that I consider the mood, or nature, of a pot to be of first importance. It represents our instinctive total reactions to either man or pot, and although there is no guarantee that our judgment is true for others, it is at least essentially honest and as likely to be true as any judgment we are capable of making at that particular phase of our development. It is far better to run the risk of making an occasional blunder than to attempt cold-blooded analyses based upon other people's theories. Judgment in art cannot be other than intuitive and founded upon sense experience, on what Kawai calls 'the body'. No process of reasoning can be a substitute for or widen the range of our intuitive knowledge.

This does not mean that we cannot use our common sense in examining the qualities in a pot which give us its character, such as form, texture, decoration and glaze, for analytic reasoning is important enough as a support to intuition. Beginning with the colour and texture of the clay, one must ask; apart from its technical suitability, whether it is well related to the thrown or moulded shape created by the potter and to the purpose for which the pot is intended—what, for example, is appropriate for a porous unglazed water jug is utterly unsuitable for an acid jar. Does its fired character give pleasure to the eye as well as to the touch; its texture contrast pleasingly with the glaze? Has it where exposed to the flame turned to a dull brick red which contrasts happily with the heavy jade green of a celadon? Does it show an interesting granular surface under an otherwise lifeless porcelain glaze? Has its plasticity been such as to encourage the thrower to his best efforts, for the form cannot be dissociated from its material. The shape of a pot cannot be dissociated from the way it has been made, one may throw fifty pots in an hour, on the same model, which only vary in fractions of an inch, and yet only half a dozen of them may possess that right relationship of parts which gives vitality—life flowing for a few moments perfectly through the hands of the potter.

[. . .]

It is interesting to see an Oriental pick up a pot for examination, and presently carefully turn it over to look at the clay and the form and cutting of the foot. He inspects it as carefully as a banker a doubtful signature—in fact, he is looking for the bona fides of the author. There in the most naked but hidden part of the work he expects to come into closest touch with the character and perception of its maker. He looks to see how far and how well the pot has been dipped, in what relation the texture and colour of the clay stand to the glaze, whether the foot has the right width, depth, angle, undercut, bevels and general feeling to carry and complete the form above it. Nothing can be concealed there, and much of his final pleasure lies in the satisfaction of knowing that this last examination and scrutiny has been passed with honour.

As for the shapes of pots and good proportions in different types, it is impossible to do more than offer a few general suggestions in the footnotes to the illustrations of particular examples. Artists of many races have believed that there are fundamental laws of proportion and composition, and I too believe it; for what we call laws are no more than generalizations founded on our sense experience, but when the attempt is made to reduce such generalizations to mathematical formulae, it is difficult to believe that they can be applied in practice without robbing the craftsman's work of its vitality. No formula, however accurate, can take the place of direct perception.

Here, for example, are a few of the constructional ideas that I have found useful:

1. The ends of lines are important; the middles take care of themselves.
2. Lines are forces, and the points at which they change or cross are significant and call for emphasis.
3. Vertical lines are of growth, horizontal lines are of rest, diagonal lines are of change.
4. Straight line and curve, square and circle, cube and sphere are the potter's polarities, which he works into a rhythm of form under one clear concept.
5. Curves for beauty, angles for strength.
6. A small foot for grace, a broad one for stability.
7. Enduring forms are full of quiet assurance. Overstatement is worse than understatement.
8. Technique is a means to an end. It is no end in itself.

NOTES

1. 'Whenever the final product of the machine is designed or determined by anyone sensitive to formal values, that product can and does become an abstract work of art in the subtler sense of the term.'—Read, *Art and Industry*, p. 37.
2. i.e. pottery approved of by the Japanese tea-masters, adepts in the *Cha-no-yu*, or tea-ceremony, who have for several centuries been the foremost art critics in Japan and have counted among their numbers many creative artists of the first rank. For an account of the spirit of *Cha-no-yu*, see *The Book of Tea* by Okakura Kakuzo, also A. L. Sadler, *Cha-no-yu*, London (1934).
3. There has never been a European stoneware tradition except that of the Rhenish salt-glazed wares. 'Accepting the Sung standard' is a very different thing from imitating particular Sung pieces. It means the use so far as possible of natural materials in the endeavour to obtain the best quality of body and glaze; in throwing and in a striving towards unity, spontaneity, and simplicity of form, and in general the subordination of all attempts at technical cleverness to straightforward, unselfconscious workmanship. A strict adherence to Chinese standards, howsoever fine, cannot be advocated, for no matter what the source and power of a stimulus, what we make of it is the only thing that counts. We are not the Chinese of a thousand years ago, and the underlying racial and social and economic conditions which produced the Sung traditions in art will never be repeated; but that is no reason why we should not draw all the inspiration we can from the Sung potters.
4. This is not to say that any better taste was shown in the work of the late nineteenth and early twentieth-century hand-potters in England up to fifteen or twenty years ago, or by many of them even now; but it is probable that the example set by industrialism and the strain of getting away from it was largely responsible even for their demoralization.

FURTHER READING

Emmanuel Cooper, *Bernard Leach: Life and Work* (London: Yale University Press, 2003).

Edmund De Waal, *Bernard Leach* (London: Tate Gallery, 1998).

Tanya Harrod, *The Crafts in Britain in the Twentieth Century* (London: Yale University Press, 1999).

26

INITIATION AND THE CRAFTS

René Guénon

Modernism is usually seen in terms of rationality: slogans such as 'form follows function' and 'truth to materials', and an investment in the rigors of abstraction. Yet it had another side, too: a world of ideas that emphasized such ineffable terms as expression, spirit and mysticism. This line of thinking, which had its roots in the turn-of-the-century aesthetician Henri Bergson and the theosophy of Helena Petrovna Blatavsky, leads eventually to the figure of René Guénon. A French metaphysician, Guénon (like many of the thinkers and artists associated with theosophy) turned to eastern religions as a source of universal spiritual values. Hinduism was his first and most important non-Western influence, but when he decided to leave Europe behind for good in 1930, he went to live in Cairo, adding Islam to his study of the world's religions. In the following text, Guénon considers the relationship between spirituality and craft, seizing on the metaphor of initiation as a link between the two. Of particular importance to him is the Indian notion of svadharma, *a Sanskrit term which might be translated 'one's own way'. For Guénon, initiation into craftsmanship was a means of following this individual destiny—a paradigm of truth to oneself, and one that runs counter to the fluidity that characterizes modern life.*

René Guénon, 'Initiation and the Crafts', *Journal of the Indian Society of Oriental Art* 6 (1938).

We have frequently said that the "profane" conception of the sciences and the arts, such as is now current in the West, is a very modern one and implies a degeneration with respect to a previous state in which both of them had an altogether different character. The same can be said about the crafts; the distinction, moreover, between arts and crafts or between "artist" and "craftsman" is also specifically modern, as if it were born of this profane deviation and had no meaning outside it. The *artifex* with the ancients is, without differentiating, a man who practises an art or a craft. He is neither an artist nor a craftsman in the sense these words have today, but something more than the one or the other, for his activity, in its origins at least, issues from principles of a far more profound order.

In all the traditional civilisations, in fact, every activity of man, whatever it be, is always considered as essentially derived from the principles; on account of that derivation it is as if "transformed" and, instead of being reduced to what it is simply in its exterior manifestation (this would be the profane point of view), it is integrated in the tradition and, for the one who performs it, it is a means of effectively participating in this tradition. Even from the simple exoteric point of view this is so: if one views, for example, a civilisation like that of Islam or the Christian civilisation of the Middle Ages, it is easy to see the "religious" character which the most ordinary acts of existence assume in it. Religion there is not a thing that holds a place apart and unconnected with everything else as in the case of the modern Westerners (those at least who still consent to acknowledge a religion); on the contrary, it pervades the whole existence of the human being; or, it would be better to say, all that constitutes this existence and the social life particularly, is as if included in its domain, so much so that under such conditions there cannot really be anything "profane," but for those who for one reason or another are outside the tradition and whose case is then a mere anomaly. In other civilisations, where there is nothing to which the name religion can be properly applied, there is none the less a traditional and "sacred" legislation which, while having different characteristics, exactly fulfils the same role; these considerations can therefore be applied without exception to all traditional civilisations. But there is something further still; if we pass from the exoteric to the esoteric (we use these words here for the sake of greater convenience, although they do not fit all the cases with equal rigour), we observe, generally, the existence of an initiation bound up with the crafts and taking them as its basis; these crafts then are still susceptible of a superior and more profound significance; we would like to indicate how they can effectively furnish a way of access to the domain of initiation.

Our understanding of it is made easier by the notion of what in Hindu doctrine is called *svadharma*, that is the performance by every being of an activity consistent with his own nature, and it is also by this notion, or rather by its absence, that the deficiency of the profane conception is most clearly marked. In the latter, a man can adopt any profession and he can even change it according to his will, as if this profession were something purely exterior to him, without any real connection with that which he really is and by virtue of which he is himself and not another. According to the traditional conception, on the contrary, everyone must normally fulfil the function for which he is destined by his very nature; and he cannot fulfil any other without a grave disorder resulting from it which will have its repercussion over the whole social organisation to which he belongs; more than that: if such a disorder becomes general, it will have its effects on the cosmical realm itself, all things being linked together according to strict correspondences. Without insisting any further on this last point, which, however, could easily be applied to the conditions of the present epoch, we may remark that the opposition of the two conceptions, in a certain connection at least, can be reduced to that of a qualitative and a quantitative point of view: in the traditional conception, the essential qualities of beings determine their activities; in the profane conception, the individuals are considered as mere units, interchangeable, and as if in themselves they were without any quality of their own. This last conception is closely connected with the modern ideas of equality and uniformity (the latter is contrary to true unity, for it implies the pure and inorganic multiplicity of a kind of social atomism) and can lead logically to the exercise of a purely mechanical activity only in which nothing properly human subsists; it is just this, in fact, that we can see today. It is thus well understood that the mechanical crafts of the modern age, being but a product of the profane deviation, cannot by any means offer the possibilities of which we intend to speak here; they even cannot in truth be considered as crafts, if one wishes to preserve the traditional meaning of the word, the only one with which we are concerned at present.

If the craft is something of the man himself and is, in a way, a manifestation or expansion of his own nature, it is easy to understand, as we have already said, that it can be used as a basis for an initiation and that generally even it is the fittest thing for this end. In fact, if initiation essentially has for its aim a surpassing of the possibilities of the human individual, it is equally true that only this individual such as he is in himself, can be taken as its point of departure; this accounts for the diversity of the ways of initiation, that is to say, of the means wrought up to act as "supports," in conformity with the difference of individual natures, a difference which subsequently intervenes less and less, as the being goes on advancing on his way. The means thus employed can be efficient only if they correspond to the very nature of the beings to whom they are applied, and as it is necessary to proceed from the more accessible to the less accessible, from the outer to the inner, it is normal to take these means from the activity by which the nature is manifested outwardly. It is evident, however, that this activity can play such a part only inasmuch as it really expresses the inner nature; here is truly a question of "qualification," in the initiatory sense of this term; in normal conditions this qualification should be a necessary condition for the exercise itself of the craft. This is at the same time related to the fundamental difference which separates the initiatory teaching from profane teaching: whatever is simply learnt from outside is here without any value; the question is to wake up the latent possibilities which the being has in himself (and this ultimately is the true significance of Platonic "reminiscences").

Following these last considerations, one can also understand that the initiation, taking the craft as its "support," will have at the same time, and inversely in some way, a repercussion in the practice of this craft. The being, in fact, having fully realised the possibilities of which his professional activity is but an external expression, and having thus an effective knowledge of the principle itself of this activity, will henceforth fulfil consciously what hitherto had been but an "instinctive" consequence of his nature; if thus the initiatory knowledge, for him, is born of the craft, the latter, in its turn, will be the field of application of this knowledge from which it can never be separated any more. There will be then a perfect correspondence of the interior and the exterior, and the work produced will be an expression, not only to some degree and more or less superficially, but a really adequate expression of the man who conceived and executed it; it will be a master-work in the true sense of this word.

This, one sees, is very far from the so-called "inspiration," unconscious or subconscious, in which modern people want to see the criterion of the real artist, who is nevertheless considered superior to the artisan or craftsman, according to the—more than contestable—distinction which they are in the habit of making. The artist or artisan, if he acts under such an inspiration, is in any case but a profane person; he shows, no doubt, by his inspiration that he carries within himself certain possibilities; as long however as he has not effectively become conscious of them, be it even that he attains to being what is generally called a "genius," this does not make any difference; unable as he is to control his possibilities, his success will be but accidental and this is granted as one commonly says that the inspiration is sometimes lacking. All one may concede so as to bring the present case nearer to the other where true knowledge intervenes, is, that the work which consciously or unconsciously flows from the nature of the

Figure 18 Architectural Model of the Jagannatha Temple in Puri, nineteenth century.

person who performs it, will never give the impression of a more or less painful effort; the effort always carries with it some imperfection, being anomalous, whereas such a work derives its perfection from its conformity with the nature; this conformity implies directly and necessarily that it is exactly suited to the end for which it is destined.

If now we intend to define more rigorously the domain of what may be called the initiations through the crafts, we have to say that they belong to the "lesser mysteries," referring as they do to the development of the possibilities which belong to the human state proper; this is not the last aim of initiation, but constitutes at least its first obligatory phase. It is necessary, in fact, that this development is accomplished in its integrity in order then to allow a surpassing of the human state; beyond this, however, it is evident that individual differences, in which these initiations through the crafts have their support, disappear completely and play no part any more. As we have explained elsewhere, the lesser mysteries lead to the restitution of the "primordial state" as it is called in traditional doctrines; yet, once the being has arrived at this state, which still belongs to the domain of human individuality (and which is the point of communication between it and the superior states), the differentiations which give birth to the diverse specialised functions have disappeared, although it is there that they all have equally their source, or rather on account of this very fact; to this common source one has to remount so as to possess in its plentitude all that is implied by the exercise of any function whatever.

If we view the history of humanity as taught by traditional doctrines, in conformity with cyclical laws, we must say that in the beginning man had the full possession of his state of existence and with it he naturally had the possibilities corresponding to all the functions prior to any distinction of these. The division of these functions came about in a subsequent phase, representing a state already inferior to the primordial state, in which however every human being, while having as yet only some definite possibilities, still spontaneously had the effective consciousness of them. It is only in a period of greater obscuration that this consciousness became lost; hence initiation became necessary so as to enable man to find once more along with consciousness, also the former state in which it inheres; this is, in fact, the first of its aims, and the one at which it aims immediately. In order to be possible, this implies a transmission going back by an uninterrupted chain to the state to be restored and thus step by step to the primordial state itself; still, the initiation does not stop there and the lesser mysteries being but the preparation for the great mysteries, that is for the taking possession of the superior states of the being, one has to go back even beyond the origins of humanity. In fact, there is no true initiation, even in the most inferior and elementary degree, without the intervention of a non-human element, which is the spiritual influence regularly communicated by the initiatory rite. If this is so, there is obviously no room for searching historically for the origin of initiation—a search which now appears bereft of sense—nor the origin of the crafts, arts and sciences, viewed according to their traditional and legitimate conception, for all these, through multiple, but secondary, differentiations and adaptations, derive similarly from the primordial state which contains them all in principle, and from there they link up with other orders of existence, even beyond humanity itself; this is necessary so that all and each, according to its rank and measure, can concur effectively in the realisation of the plan of the Great Architect of the Universe.

FURTHER READING

Ananda Coomaraswamy, *The Indian Craftsman* (London: 1909).

René Guénon, *Crisis of the Modern World* (1927).

Mark Sedgwick, *Against the Modern World: Traditionalism and the Secret Intellectual History of the Twentieth Century* (Oxford: Oxford University Press, 2004).

Robin Waterfield, *René Guénon and the Future of the West* (Wellingborough: Crucible, 1987).

27

INDIAN HANDICRAFTS

Kamaladevi Chattopadhyay

The iconic image of Indian craft idealism is a photograph of Mahatma Gandhi, dressed only in a loincloth, working at a spinning wheel to produce cotton thread for khadi *(homespun cloth): a picture of ascetic self-sufficiency.[1] When combined with Gandhi's theory of nonviolent resistance to British imperialism, this politicization of craft galvanized India—leading, finally, to independence in 1947—and resonated across the world. Gandhi's ideals and tactics have inspired protest movements ranging from African anticolonialism to the American civil rights movement to contemporary antiglobalization. But it was not Gandhi himself who directly led the Indian revival of handicraft. That credit goes to Kamaladevi Chattopadhyay. She was arguably the most successful of a long list of women who have done the real work of organizing craft movements in the twentieth century. Most, like her, were wellborn: Princess Maria Tenisheva in Russia; Ishbel, Countess of Aberdeen, in Ireland; Aileen Osborn Webb in the United States. Chattopadhyay was a liberated woman by the standards of her day, acting in films at one point in her career. She fell in with Gandhi in the 1920s, participating in various self-sufficiency schemes that he inspired. It was only after his death in 1948, however, that she became the leader of the Indian craft revival, writing many books on the subject and founding museums across the country, a national award system, and the All India Handicrafts Board. All*

the while she continued her charitable work with refugees and others who lived in poverty. In the following excerpt, taken from her most explicitly ideological book, Chattopadhyay *lays out her Gandhian vision in no uncertain terms, connecting it to traditional ways of life and belief. Of particular note is her discussion of international interventions into Indian craft economies, which anticipates many of the issues that are still discussed today within NGO-supported handicraft support schemes.*

Kamaladevi Chattopadhyay, excerpt from *Indian Handicrafts* (New Delhi: Allied Publishers, 1963).

Handicraft is rightly described as the craft of the people. In India it is not an industry as the word is commonly understood; for the produce is also a creation symbolising the inner desire and fulfilment of the community. The various pieces of handicrafts whether metalware, pottery, mats or woodwork, clearly indicate that while these are made to serve a positive need in the daily life of the people, they also act as a vehicle of self-expression for they reveal a conscious aesthetic approach. At the same time, they manifest in their structure the principles of Silpa Sastra, the ancient scientifically evolved formulae and regulations for manufacturing.

In the peace and quiet seclusion of the countryside the village community evolved a culture of its own out of the steady flow of its own life and of the nature around it. The community acted as a single personality because of the common integrated pattern of life, in responding to the common joys and burdens of life, to the common occasions and landmarks that stood out in the flux of time and the change of seasons. Out of a million coloured strands of tradition filled with song and verse, legends, myths, native romances and episodes, from the substance of the every day life of the community, and out of nature's own rich storehouse, was woven a rich, creative and forceful art.

The craftsman's position in the predominantly agricultural society was pivotal, for it made the village society self-contained, a characteristic of India through the long ages and which later inspired in Gandhi the dream of Sarvodaya—a self-supporting community which stood for the good of all. The social functioning was based on a code of personal relations and duties handed down from generation to generation instead of on contract and competition, with services being paid for in kind rather than cash, normally in grain at harvest time or a share in the communal land. A rigid adherence to the concept, that each man is born to his ordained work through which alone he can progress spiritually; and through the fulfilment of this Dharma or duty to final Moksha or liberation, provided the sanctions and the stability to this vast but well ordered system and ensured a high degree of perfection to the arts and crafts. The execution of the craft was not just an economic compulsion but a sacred duty. This largely explains the very meticulous care and devotion with which the humblest work was performed. The commonest of articles were endowed with beauty, for each task was a dedication.

The artisan was an important factor in the equation of the Indian society and culture. By performing valid and fruitful social functions for the community, he earned for himself a certain status and a responsible position in the society. He worked for those whom he knew and this gave a touch of personal intimacy to the work. He made things mainly for the use of the people around him and not so much for sale in a distant market-place. His work was evaluated not in mere terms of money but rather to entitle him to the necessities of life and leisure, and rest in sickness and old age. He was not at the mercy of the middle man or a changing clientele. He was an heir to the people's traditions and he wove them into his craft making it into an art. The bold local styles that the village artisans evolved operated as a great lever in the evolution of Indian art adding to the wealth and variety of colour and design.

To discover the sources of inspiration and to gather the full significance of the vast field of Indian handicrafts they have to be seen in the context of the background from which they have emerged, the dark toned bodies of the people for whom they were made and to fit into the serene flow of their lives.

Innumerable invasions of virile nomadic peoples who migrated over the length of India, the intermingling of vast civilisations, the impacts of alien myths, symbols and superstitions, the ancient gods and cults of the original inhabitants of this land, the geographical distribution of mountains, deserts and lush vegetation and the presence of minerals, salts and water were factors that moulded the aesthetic norms of a people nurtured on space-time concepts which emerged from chaos and formulated through millenniums.

It was against this background that hereditary groups of every type of artisans arose, organised within the rigid systems and protected by rigid laws, which ensured the high standard and continuity of these crafts. Tracing their origin to Visvakarma, the deity of crafts and the very source of the creative intellect, the craftsman combined within his being the functions of both conceiver and executor. He became in society the symbol of the outer manifestations of the creative purpose. The integration of creative endeavour for livelihood and the refusal to permit outer influence to loosely permeate and corrupt the unconscious process of renewal, lead to a great flowering of the craft tradition. The craftsman was the unbroken link in the tradition that embraced both the producer and the consumer within the social and religious fabric. Art and aesthetics were deeply rooted in function. Ornamentation and decoration was not divorced from utility.

On the organisational side, community of interests drew together the artisans who soon came to form guilds in India even as in Europe, Egypt and other regions. The guilds did not originally correspond to a sectarian or ethnical caste as is generally believed. As a matter of fact, the same trade was sometimes followed by men of different castes. Membership was normally hereditary but newcomers were admitted on payment of a fee. It was the guilds which regulated the hours of labour and the quantity of work for each through strictly enforced by-laws and fining defaulters. The guild also prevented undue competition between guilds and negotiated in cases of dispute. No overtime in a trade was allowed to any workman if there was unemployment in the same trade. The guilds not merely regulated wages but enforced the use of pure materials and a high excellence of workmanship. Each guild was managed by a court of Mahajans or kind of aldermen, with a special

position to the Seths or chiefs of the guilds. In the larger towns and cities, the guilds seemed to have wielded considerable influence, for the Seth of the guild became Nagar Seth, the titular head of all the guilds and the highest personage in the city and accepted as its representative by the government. The guilds built temples and also spent for welfare work from the guild funds which were augmented through special collection drives.

Based as the craft tradition was on a background of myth, symbol and fantastically rich imagery culled from the stories of the Puranic legends, there was no scope for stagnation. Although the forms in a manner repeated themselves, they were free from imitative intention, and each productive act was spontaneously linked with the stream of man's life and was a dynamic symbol of man's endeavour to express universal human emotions and interests.

Two main channels of craft expression developed. The one concerned with the treatment of surface as symbolised by inlay or enamel, reflecting as in a mirror the streams of people's lives and culture patterns that for a time commanded its patronage, responding to every sophistication and rarity of elegance. The other, structural in concept, rooted in the endless search, reflecting the familiar forms of the unchanging pattern of the village unit, the romance and emotional background of nomadic tribes, the rituals that bound man in invisible chains of a hoary past.

Indian handicrafts have thus been in a class by themselves. They express a great national heritage. While aesthetically fine, they were, nevertheless, essentially articles of utility. From the humble water-pot of clay to the curved knife to cut vegetables, from the cloth which covered the human form to the fabric flung on the bullock's back, every piece was a work of art, enriched by beautiful lines, vivid colours and alluring designs. Nothing was created to be kept as a dead piece in a glass case to be merely looked at or to trumpet the affluence of the owner. Beauty was not an isolated item, it was an integral part of one's intimate life. Whatever the article in use, no matter how mundane, it had to be beautiful. Decoration was not an end in itself. It had to serve a social purpose. In fact as in music, dance and painting, each creation had a mood or essence of its own—Rasas as they are called; and each was expressed in a recognised form like the Abhinaya or gesture symbols in dance or raga in music. Like acting in the classical Indian drama nothing was left to chance or the vagaries of the artist. Each move was worked out with care and precision.

Today people have an idea that beauty is the prerogative of the rich alone, for it is believed that beautiful things are expensive and beyond the reach of the ordinary man. Our tradition, however, is that an industrial object is also a work of art, and even though the Indian artisan seldom rose above the traditions, he was all the same an artist.

[. . .]

The role of cottage industries in the economy of the country that is building up its industrial structure anew after its freedom, like India and other Asian countries, has come to assume world-wide importance. Economists all over the globe have turned their vision and experience to this study. Several organisations in the United Nations like the UNESCO, FAO, [and] ILO have sections that deal with the promotion and development of handicrafts. There is fair unanimity on the conclusion that the industrialisation of the Asian countries has to be on a different pattern and largely through smaller industrial units and establishments.

In modern economy large scale industries supported by smaller ones represent prosperity for a country and a high standard of living for its people. But for regions whose natural potential has not been developed or deliberately retarded as in colonial and semi-colonial countries, causing vast unemployment and depressing the standards of living, at least one of the answers is many small scale industries. Moreover, where there is abundant labour but little easy capital for investment, particular care has to be taken to employ its limited funds in such a way as to obtain maximum productivity and profit. In small industries the capital cost per unit of production is generally low and the ratio of productivity per unit of capital higher. Small amounts for investment are easier to raise. Moreover, in a predominantly rural economy where capital formation is slow and laborious and savings is mostly sunk in agriculture and investment, and is still calculated in terms of land, people are generally not ready to risk their meagre hard earned savings in what seems a gamble, like remote industrial enterprises which they can neither see nor visualise. But they are more willing to join a local industry that they can see and understand and above all has an assured place in their everyday life and the village economy.

The number of unemployed is always larger in the less developed countries. Unfortunately improvement in agriculture methods instead of decreasing really adds to the number of unemployed. In addition agriculture being a seasonal occupation has to depend on supplementary industries to ensure the population even a nominal living standard.

Underdeveloped areas also present other problems. Transport facilities and general lines of communication are few and slow. These are different factors that aid in the growth of industry and commerce. Small industries

Figure 19 Margaret Bourke-White, *Mohandas K. Gandhi, India's Leader in the Struggle for Independence from Great Britain, Reading near a Spinning Wheel at Home*, 1946.

manufactured largely if not wholly out of raw material locally obtainable and the finished goods consumed easily in the industry's neighbourhood brighten the prospects of success for any small industrial enterprise. The high transportation costs as also the usual bottlenecks in free and quick movements are avoided, if the industry is closely integrated as possible to the local economy.

Within the last generation there has been a rapid transformation in the social fabric of the country. The building of roads, the introduction of machines, the breakdown of caste barriers, the bringing of an urban civilisation through the radio and the cinema to the door of the rural unit, have led to a rapid change of the norms that had evolved craft traditions. Today it is the town that is dictating the fashions and in some village fairs the clothes that are sold are no longer the resist or tie-dyed clothes produced by the local craftsmen, but the latest design woven by the nearby textile mill.

The snapping of the link between the creative impulse and livelihood, that is the inevitable outcome of mechanisation; and the introduction of an alien concept of designer as distinct from the craftsman has only destroyed further the craftsman's natural response to good form. This has led to increasing tensions in the craft tradition and a confusion in the unconscious background that is the very source of the creative process. That it has happened accidentally and not from a conscious awareness of the situation has only tended to produce greater chaos. The Indian craftsman is faced with a situation where on the one hand he hears the cry 'back to the past' or 'break with the past', produce something new, and on the other hand he is dazzled by the incomprehensible forms evolved by the West after decades of experimentation. To go wholly back to the past is impossible, for the past was a background of life that has less and less significance in terms of the new social order. Equally to absorb the Western forms has no meaning, for they are alien and have no link with the craftsman's comprehensions and concepts. What then is possible: The question has no easy solution. It may well be that the very laying bare of the problem with all its intricacies, conflicts and tensions will itself project the answer. No single human mind can mould the unconscious impulses of a craft tradition; what it can do is to help cleanse the eye of the craftsman of the corrupt forms that have blurred his vision and leave it to the unfailing

creative force that still lies deeply embedded within the craftsman's eyes and hands to dictate and create a new tradition.

As one looks upon the traditional craft products, one sees awe-inspiring beauty, expressing a vision and variety of force and feeling which has few parallels in the art treasures of the world. Here we gaze into the heart of our cultural soul, the well from which the creative spring has spurted. Here we see the long passage of history and the infinite moods of a people. As one looks at the Indian handicrafts, one instinctively senses the unity of all arts even as one is made aware of the unity of life in Indian philosophy. Yet nowhere probably could one find so great a diversity in form, shape and colour as in these crafts. At the same time, it is in this infinite variety that one sees the eternal search for unity. Whether it is in the ensemble of flowers and fruits, birds and animals, leaves and creepers, gods and human beings, whether it is in the *Phulkaris* of Punjab and *Kinkhwabs* of Banaras, or the *Patolas* of Pattan and the *Bandhanis* of Rajasthan, there is a sense of rhythm and harmony.

The sentiment of traditionalism alone cannot however take us very far in our effort to rehabilitate the Indian crafts. The modern demand is for beauty as a supplement to usefulness. Then again the concept of usefulness itself has changed because of the transformation in the mode of our thought, of living, habits and environment. Nor is there any longer the same fastidiousness for the purity of the material or the authenticity of the form. With the advent of cheap alloys for jewellery, artificial silk and synthetic stuff like plastic, the emphasis has definitely shifted to cheapness. Similarly the insistence on durability has been replaced by demand for greater variety. Modern taste is restless and prepared to renew and replace articles more easily and quickly.

One wonders if anyone has the heavy task such as rests on the craftsmen. With an artist, it is simple enough. He creates and sets up the standard of art forms and the public falls in with it. But with the craftsman, he has to meet the clientele more than half way. He has to combine beauty with utility and make the new product still embody the old symbol. He has to cater to a customer thousands of miles away and unknown to him, unlike the intimate community in which he lived in olden times. Yet at the same time he has to remain loyal to the traditions of his heritage. While he is expected to produce goods which can stand competition with machine products, he has to produce the same precision and finish with his fingers and hand.

The public should remember that development of handicrafts is entirely different from that of village industries or small scale industries. Not only is each craft highly complicated involving numerous processes, it is also very individualistic and local. Handprinting in Rajasthan is quite different from printing in Andhra. Metal inlay in Hyderabad is different from that in Uttar Pradesh. Every craft differs from region to region. Each has its own traditional ways of production and its own design, shape, colour etc. No one single plan can be applied wholesale all over as in the case of chakkis, cart wheels, charkhas or carpentry and smithy tools. The craftsmen are also scattered, some here and some there. In some cases the craft is reduced to a few individuals who are almost lost in the interior villages and have to be sought and villages combed to find them. Not only each craft, but each centre needs careful study and understanding before any measures for its treatment can be applied. It is, therefore, in the very nature of this country that it calls for great sensitivity and delicate handling, infinite patience and tireless service

before any results can be produced. The public must bear with this great heritage of ours and remember that its flowering belonged to another age, another atmosphere, a totally different pattern of living and tempo. In some ways handicrafts seem out of tune with our modern living and approach. There is nothing spectacular about them. You do not find them in imposing structures humming with life and lit by a million candle power lights. They have mostly to be unearthed in dark hovels with air pungent with stench. Even though millions are engaged in handicrafts all over the country, they are never found in large concentrations. The tools that produce these crafts are modest and unostentatious. Development of handicrafts cannot therefore be measured through spectacular structures or noisy machines. They speak an age when dignity lay in silence and beauty in subtlety.

NOTE

1. On Gandhi and craft see Arindam Dutta, *The Bureaucracy of Beauty: Design in the Age of its Global Reproducibility* (New York: Routledge, 2007), p. 258ff.

FURTHER READING

Patrick Brantlinger, 'A Postindustrial Prelude to Postcolonialism: John Ruskin, William Morris, and Gandhism', *Critical Inquiry* 22/3 (Spring 1996), pp. 466–85.

Kamaladevi Chattopadhyay, *The Awakening of Indian Women* (Madras: Everyman's Press, 1939).

Ajit K. Dasgupta, *Gandhi's Economic Thought* (London: Routledge, 1996).

Abigail McGowan, 'All That Is Rare, Characteristic or Beautiful: Design and the Defense of Tradition in Colonial India, 1851–1903', *Journal of Material Culture* 10/3, pp. 263–87.

Reena Nanda, *Kamaladevi Chattopadhyay: A Biography* (New York: Oxford University Press, 2002).

28

'THE RELATION OF THE PAST TO THE DEMANDS OF THE PRESENT', WORLD CRAFTS CONFERENCE PROCEEDINGS (1964)

The postwar American studio craft movement owes almost all of its shape and much of its success to Aileen Osborn Webb. Born into a wealthy New York family with strong commitments to the arts, Webb had her first experiences with craft development during the Depression. During and after World War II, she sponsored one initiative after another: a shop and gallery called America House in 1940; Craft Horizons *magazine (today called* American Craft*) in 1941; the American Craftsmen's Educational Council (today the American Craft Council) in 1943; the School for American Craftsmen in 1944 and the Museum of Contemporary Crafts (subsequently the American Craft Museum, and today the Museum of Arts and Design) in 1956, not to mention numerous conferences and exhibitions. All of these benefitted from her largesse as well as her strategic vision, open-mindedness and indefatigable energy. Perhaps her most ambitious undertaking of all was the World Crafts Council, founded in 1964 at a conference held on the campus of Columbia University. An extraordinary lineup of speakers was assembled, including the art theorists Rudolf Arnheim and Harold Rosenberg, novelist Ralph Ellison, designer Tapio Wirkkala, architect Louis Kahn, Museum of Modern Art director Rene d'Harnoncourt, anthropologist Frederick Dockstader and craft advocates such as Kamaladevi Chattopadhyay and Pupul Jayakar from India, Czeslaw Knothe*

from Poland, Dr. Rubin de la Borbolla from Mexico and Remy Alexander from Italy. As is clear from the following transcription of one panel discussion held at the event, the conference proved to be the occasion for lively debate. The egalitarian, democratic and sometimes utopian ambitions of Euro-American craft reform ran headlong into the concrete economic problems and strategies of craftspeople and advocates in Asia, Africa and Latin America. In subsequent years this diversity of opinion would become still more contested, with some members concerned mainly with preserving the traditional, and others desperate to see established craft economies modernized. The WCC still exists today, and its efforts are paralleled by many NGOs (nongovernmental organizations) that continue to promote craft as an economic, spiritual and cultural resource within an increasingly interconnected, technologically driven marketplace.

Excerpts from the panel 'The Relation of the Past to the Demands of the Present', World Crafts Council proceedings, 1964.

Dr. Rudolf Arnheim: Being the kind of psychologist who believes more in what people have in common than what distinguishes them from each other, I am convinced that to make things with their own hands for their own purposes is a continuing need of all human beings, regardless of where they live, what language they speak, and what their level of economic and technological development may be. This is what I would call the "first International of workmanship," the community of all men in their need and capacity to make fine objects with their hands. But, as you know, this "first International" is often put in an all too easy relationship with what I call the "second International of craftsmanship,' by which I mean mass production by machine, standardized for international trade. In terms of this relationship, craftsmanship is called old, and industry is called new. The question is then: Will the new replace the old, and should it do that? Viewed in this fashion, craftsmanship may seem to be fighting a purely defensive rear guard action. Instead, I believe, our thinking is based on the conviction that craftwork is here to stay, or, as Mr. Rubin de la Borbolla put it, that it is a continuing and eternal aspect of human nature. It is on this basis that we shall discuss the relationship of the past to the demands of the present. In other words, we are addressing ourselves to such questions as: To what extent can and should the traditional crafts preserve the ancient shapes? Can and should they change with the times?

[. . .]

Dr. Czeslaw Knothe: The great technical developments throughout the world have created too many possibilities in industrial and craft production. The enormous number of alternatives that face designers and craftsmen today amount to a surplus of possibilities and thus a devaluation in the product. The continuous search for novelty and economic gain produces an anxiety which is a characteristic of contemporary society. It allows no time for a contemplative approach to the phenomena of life. Subconsciously we feel a lack of subjective and aesthetic impressions of our environments. Recently a tendency to instill

aesthetic values in industrial production has become noticeable throughout the world; and this is not only true of advertising, which increases the market value of products. However, this process is a complicated one having many different aspects, and is responsible for spreading industrial products among the broad public when average taste must be taken into consideration.

Craftsmanship has another character. The relationship between the craftsman and his work is direct, often spontaneous, and has a human value. Craftsmanship can seldom compete on an economic level with industrial production, but it often surpasses it in cultural values. For that reason craftsmanship can be and often is a treasure house of creative thought. In general, an increase in craft production for economic reasons requires an increase in mechanical methods and leads to a lowering of the level of craftsmanship. Mass production and craftsmanship are primarily distinguished by the role the tool plays in each. For the craftsman it is a means of personal expression. In mass production the pervasive tool is the machine, and its role is that of an executor. This is antithetical to the handicrafts.

The role of craftsmanship varies in different countries. In underdeveloped countries where work is cheap, craft products may compete with factory production and still maintain their high cultural value. This is the normal state of things, but unfortunately matters are getting worse as a result of international industrialization. Inevitably, it will not be possible to preserve the old forms, and there is undisputed value in carefully preserving this inherited tradition. The general principle accepted in the conservation of relics can be expressed in one sentence: preserve the state of things as they are found. The most vital way of preserving and cultivating crafts-manship is to profit by the inspiration of past eras, enlivened by a contemporary view of life . . . We should foster the production of craft objects based on traditional forms. We should build numerous museums and pre-serve sites of ancient art, and finally, we must develop creativity in craftsmanship.

Schooling plays a prominent role in maintaining and spreading the practice of craftsmanship. In traditional schooling the education of individuals was usually limited to the teaching of general knowledge, as it is in present schooling—especially at the univer-sities. So the problem in schooling is partly separated from that of education. That is, by schooling we mean the teaching of ready-made knowledge which implies a rather passive role for the student. On the other hand, by educa-tion we mean the awakening and forming of the potential individual values of the student. We must admit, of course, that this division is partly theoretical because in most cases the stu-dent learns and is educated simultaneously.

With regard to the above, we believe that strict specialization in higher education is not advisable. In some countries, the universi-ties during the first years of teaching do not strictly enforce the division of students ac-cording to their specialties. The individual tendencies and abilities of the students are crystallized into specialization only during the later years. This enables the students to choose their specialty after studying a broad range of subjects. Knowledge of painting, architecture, sculpture, and so on, taught in the early years, enables them to understand the interpenetra-tion of the various art disciplines.

Finally, it is important that we develop international contacts such as this Congress. Also, international exhibitions should be held so that individuals, groups, and nations can present their achievements for the benefit of world cultures.

Dr. Frederick J. Dockstader: May I start off by saying that the term "traditional" needs definition. Tradition as we understand it in the United States is quite different from that in other parts of the world. We are a young coun-try. In many ways we are still establishing tra-ditions, many of which have function, many of which have lost their function. In my own field I have inherited a tradition going back at least ten thousand years, for the American Indian has been active that long in the area extending from north of the Canadian border down to the southern part of South America. I can draw from that venerable tradition, or from its recent manifestations—in both cases the tradition is identical.

I believe that the crafts as the Indian knew them did not die, although they were, con-trary to one of the statements we have heard, very definitely disrupted. The crafts lost their place among the Indians simply because the incoming migrant from Europe did not allow them to fill their original function. Neverthe-less, a tradition of this sort could, if allowed to, flourish and strengthen itself with the com-ing of a new order and exchange in the sense of giving and taking. Our great loss today is that very often we are so anxious to rush on to the future, we rarely look over our shoulder to learn from the past.

Today I was struck by our remarks about linking new developments in technical pro-cesses to the ways of the past. All of us seem to agree that human dignity can result from identification with a tradition. I think one of the greatest advantages that the underde-veloped countries—if you will forgive the term—have is such identification. I don't like the term "underdeveloped" because, frankly, I feel we are all underdeveloped in one way or another. For the term "underdeveloped" I would substitute the phrase, " a country which is developed in a different way." My country has television sets and bathtubs in the home, but we have much to learn about the other things that life could offer. I traveled in some countries where the people did not have television or bathtubs in their homes, yet from them I have learned things which meant a great deal more to me. Development is, like tradition, merely a relative matter.

When I visited the [New York] World's Fair I was again struck by this compulsion to show "development." Many of the foreign exhibits which I had looked forward to seeing disturbed me. Instead of the lovely textiles and magnificent pottery which I knew came from those countries and which is appreciated in museum collections, I saw motorcycles, bicycles, radios, things which were meaningless to me for identification with those countries. I found the lovely things only after I patiently climbed upstairs or downstairs, or went into the back rooms.

So I would like to say to all of you, why should you lose identity simply to live in today's world? Many of you will say economic necessity forces it. This is not always true. And if you lose your individuality simply to become one of the mass, you lose a great deal of the value of life.

The effort by a minority to find an identity in the majority culture has made our work with the American Indian craftsman difficult. The Indian very often finds it difficult to bridge the gap between the past and the current stream of life. He feels that the old ways are gone. There are no more buffalo, and he does not paint on buffalo hides as he once did. He must now paint on canvas or paper. "The old traditions have died," an Indian may say. It is not so much that old traditions have died, as that appreciation of their values has decreased.

I would also like to emphasize one other thing. After examining some of the contemporary shows, I wonder if some of you have not sacrificed utility for something which is eye-catching. One of the nice things about Indian art to me (and I think it is quite true of most traditional art) is that it is functional within its own terms. If you design pottery, textiles, or jewelry which does not serve a purpose, you have deserted the ranks of the creative artist. You simply are, as one of the speakers noted, a manufacturer of gimmicks. I think this sort of product should be sold in the curio or souvenir stores, rather than in a quality marketplace . . .

When the Indian arts and crafts board of the United States was established in 1934, one of the thoughts was of reviving some of those things which has been superseded by non-Indian motifs. It was an attempt to revive traditions and work which in some areas had died. The idea behind this was to develop and strengthen the Indian economy again. To accomplish this it was necessary to consider not only the needs of the Indian producers but also those of the consumer. Without the consumer, production means nothing, so there was an equal attempt to educate the buyer along with the seller.

Perhaps this is the one thing you have overlooked in your current activity. You must keep it in mind or you will not develop the respect with which your work should be viewed.

With the Indian development of the thirties and forties, many craftsmen "found themselves" for the first time. They gained an identity which they had not previously been able to establish. Much the same thing has been happening in Latin America. Dr. Rubin de la Borbolla has established a national museum of folk art in

Mexico City and regional museums throughout the country. These regional museums have the important function of showing the native craftsman what other craftsmen are doing in his area, and giving him an opportunity to show, in a dignified setting, the beauty which he can create with his hands. This movement is spreading in many of the Latin American countries, and in consequence throughout the Western Hemisphere there is a much greater degree of appreciation than there was, say, twenty-five years ago for what may be called traditional expression.

But all of this depends on the simultaneous education of both consumer and craftsman. And so, please, when you make your next exhibits, do not put your motorcycles, your radios and your bathtubs out front and hide your pottery in the back. We know you make motorcycles and radios and bathtubs but we esteem much more your basketry, your textiles and your pottery.

Mme. Pupul Jayakar (audience): I have been connected for the last few years in India with the development of handwoven textiles of which we are all very proud. The problem which we had about ten or fifteen years ago, really after independence, was a crisis in traditions that was still alive, brought about by a change in the nature of communications, the bringing of the village in contact with the town, a change in symbols, a change in producer-consumer relations. Tradition is not a static point. It is in constant movement bringing into its contours all that the craftsman perceives and experiences as a given moment along with the great sub-conscious storehouse he carries within him of history and knowledge of his craft. The problem, however, which any organizer is concerned with, is a change in the consumer need of function. A creative tradition undergoes

change because with changes in nature, function and relationships it is no longer within the hereditary craftsman's capacity to answer the kind of challenges with which he is faced.

In the handwoven textile field, we have three million handlooms with seven million people employed in the industry. The vastness of the problem makes it necessary to observe and carefully examine those focal points out of which solutions could emerge. In a country like India, or in any other country which is faced with problems arising out of the existence of a great traditional form of craft production which is totally different from craft production in countries where people take up crafts as a profession and where craft is not a matter of ancestry, the only solution possible is to create a milieu where the craftsman can contact in a precise and yet creative context new functions and needs. By bringing these elements together tensions are created out of which solutions may emerge which are at the highest possible quality level. This is important because it is out of such contact, and the emergence of new solutions, that a tradition can continue its vitality.

If we can create a situation where the challenge in terms of new needs is brought to the traditional craftsmen's consciousness and if they are given necessary creative stimulus and facilities of new techniques and technology, the new laboratory equipment, et cetera, then out of this tension responses will emerge which can carry the tradition forward without its being crystallized into imitative form. The moment a tradition becomes static and tends to imitate the past or present, it is dead. The whole attempt and emphasis of the craftsman must be to serve his environment and needs, in the present. This observation itself is the creation of a focal point of energy, which will propel and communicate its own answer.

[. . .]

Dr. Rubin de la Borbolla: A good craftsman has always solved the problems of function, and he has always given his product beauty and dignity. We should not be frightened by the problems of the present world where men need to change the functions of those things [they use] daily. When things were not made by machines, one would often go to a craftsman and request that he make a piece of furniture or something else. One would explain what one wanted, and the craftsman would have to solve the problems involved. A good craftsman can always solve the functional problems of his product. In the world today we only seem to face the problem of function. I shall give you an example. The people of the United States are buying the crafts of other people of the world for two reasons. One, the products are functional, and two, perhaps the craftspeople in the United States have not yet fulfilled the needs of their people.

Ramy Alexander (audience): I would like to go back very briefly to Mme. Jayakar's remarks. She touched upon a very concrete point when she said that any good craftsman can make the transition by himself. When I was in India I had the good fortune to witness this. My only observation in this case would be that not every traditional craftsman has enough strength, energy, imagination, enough creative capacity for understanding the new functions that a new society demands. When a craftsman is not accustomed to new functions and does not have the possibility of speaking to an individual customer who explains what he wants, he should be given a chance somehow to have contact with the new milieu. When this is done he begins to understand the new functions and then he comes up not only with adapted ideas, but sometimes with completely new ideas—as new as those of any sophisticated designer.

Glen Kaufman (audience): One of the problems in today's discussion is that the situation differs so vastly in each of our countries. The purposes of the craft movements are different in different cultures. In certain countries which are definitely concerned with making things that have a ready market, the purpose of the crafts is different from that in the United States. Many things that Dr. Dockstader said disturbed me, but I realize he spoke from an anthropologist's point of view. He looks at tradition from the point of view of use. I think that tradition may be narrowed down into three aspects: tradition of technique, tradition of form and design, and tradition of use. I think if we are going to consider crafts and their traditions as they affect us today, we must consider each one of these aspects differently. The tradition of use has changed greatly in the United States.

Borbolla: We should not be confused by a multitude of definitions and linguistic twists. We must agree upon several terms. Otherwise, we will not be able to reach successful and final conclusions, uplifting to the audience and to all of us who have an interest in crafts.

When we are talking about tradition, we are not talking about the past, and I must insist on that. In my paper I remarked that I did not want the audience to think of tradition as somehow dead. We are not discussing skeletons in our closets. We are talking about leading problems today, and tradition is one of them, because tradition is a functional part of culture.

A craftsman is more than an expert. He must be a man who carries the tradition of his culture, who has knowledge of this culture, who has the dexterity, creativeness, command of techniques and the sensibility to create beauty. Then he can solve the functional and aesthetic problems of any people at any given time in any part of the world. A man who does not have these qualifications is not a craftsman. He may copy what someone else has created, but he is not a craftsman. You can call him, instead, a laborer or a repeater or whatever you want. The man may have dexterity, may be able to work with his hands, and may have feeling, but he does not have creativity. In Latin America we have today perhaps more than seven million artisans, most of whom are not craftsmen but repeaters of tradition. We must not confuse them with the real creative craftsmen.

Arnheim: Isn't the question really how this living substance of tradition can be instilled into the products of our time? The problem of training and education is: how do you translate the artistic spirit of our time, the twentieth century, into those shapes which become the craft of our time? This is a challenge which those of you who are in education may want to discuss. What is it that makes a pot a product of the twentieth century, in spite of the fact that it still is enlivened by a tradition which may be 2,000 years old?

[. . .]

R. Vanhjah Richards (audience): We have been speaking of about tradition in art. I must say that in Africa, or some parts of it, the tradition of our art has suffered terribly. Why? Because of the tourists. Because of the so-called love of African art expressed by some people. As a result tradition has been corrupted; real art does not come out. Who is spoiling tradition? Is it the craftsmen, or those who buy the craft works? These are the things we must take into consideration. The craftsman might not always want to carry on the old tradition, but the market might demand it. As craftsmen we are supposed to be honest and true to our work, to create it as we feel, and we want to do this. But what about the purchaser? Perhaps he has only the ability to appreciate what he thinks is the "real tradition." And he says, "I didn't know it was this way. I wanted it this way, I wanted it that way." This helps to spoil our tradition.

Tradition, to be significant, must go along with the times. The way we expressed the sight of the Goddess of the Moon, or the way we expressed hunger in a piece of carving years ago, must be expressed differently today. We, the craftsmen, must express it in our own way, not the way the purchaser wants it. Tradition must come from us, not from the buying world. We have to live, of course, but let our conscience be clear in our work as we live with it.

FURTHER READING

Helen R. Lane, ed., *In Praise of Hands: Contemporary Crafts of the World* (New York: World Crafts Council, 1974), excerpted.

June Nash, ed., *Crafts in the World Market: The Impact of Global Exchange on Middle American Artisans* (Albany: State University of New York Press, 1993).

Rose Slivka, ed., *The Crafts of the Modern World* (New York: Horizon Press, 1968).

29

CENTERING

M. C. Richards

*The poet and amateur potter Mary Caroline Rich-
ards stood at the spiritual center of the postwar
American avant-garde. Though she had been pursuing an academic career as a professor of English since
the 1930s, it was her arrival at Black Mountain
College in 1945 that set the direction of her future
work. An experimental school in North Carolina,
Black Mountain was a crossroads at which radical
figures in multiple disciplines met and influenced
one another. Richards studied ceramics there under
the tutelage of Robert Turner and also encountered
such figures as composers David Tudor and John
Cage, choreographer Merce Cunningham, and the
potter Karen Karnes (with whom she would later
share a studio at a commune in Stony Point, New
York). She adopted pottery-making, particularly
the act of throwing on a wheel, as an apt symbol
for her developing life philosophy. Richards's book*
Centering *was the eventual result: a mixture of
popular psychology, Zen Buddhism, poetic verse
and thoughts about craft, which compares closely
with the contemporary activities of Cage in particular. That rare thing, a best-seller about craft,*
Centering *seems in retrospect to have anticipated
the tone of 'new age' or self-help books published
in the 1970s and thereafter, many of which also
appeal to the imagery of artisanal work as a metaphor for well-being.*

M. C. Richards, excerpts from *Centering* (Middletown,
CT: Wesleyan University Press, 1966).

Because I am a potter, I take my image, centering, from the potter's craft. A potter brings his
clay into center on the potter's wheel, and then
he gives it whatever shape he wishes. There are
wide correspondences to this process. Such extensions of meaning I want to call attention
to. For centering is my theme: how we may
seek to bring universe into a personal wholeness, and into act the rich life which moves
so mysteriously and decisively in our bodies,
manifesting in speech and gesture, material-
izing as force in the world the unifying energy
of our perceptions.

This book began when I accepted an invitation to give an "inspirational" speech to
craftsmen, who said they were dry for meaning in their efforts. They asked me because I
am teacher and poet as well as potter. They
wanted "the contemplations of the poet upon
the craft." I decided to share with my hosts
certain meanings that had inspired me: still-
ing my thirst, opening my eyes, freeing my
imagination and hearing, strengthening my
nerve, inspiriting my limbs. On the basis of
the response, Wesleyan University Press asked
for a book-length elaboration of the themes I
had broached. "We would like to have your
spirit between hard covers." Why? Because,
they implied, my experience builds bridges
between disciplines which are often considered separate if not antagonistic. This press

wishes to speak to the need for interdisciplinary participation.

I have written this book out of the feel of a process, and a feel of commitment to it. I speak from and to a diverse fellowship: in the arts and thought and research, poets and craftsmen and students and teachers, homemakers and community members and solitary citizens.

The imagery of centering is archetypal. To feel the whole in every part: The Mystery and Action and Being of the whole living organism of oneself and of that Self which all of us together make, and of that earth where we are humanly born, and of that sun-sphere that nourishes us too, and of all that universe that beats its way to us now through millions of trillions of light years, making our future its long past, and making the double-talk of mystics who drown time like a puppy in the flood of something else seem like a handbook to cosmic thinking, home style.

It is my hope to create a mood which will inspire and strengthen a confidence in man and his life earthwise and cosmic. A mood sympathetic to natural processes of forming and transforming. Human beings have many stories to tell, and this is one of them. I sense things that have happened to me as somehow characteristic of the human lot, transcending personality, bearing within them a form which can reveal to my consciousness and to others deeper meanings than those of private sensation. I sense structures everywhere at work, in realms to which sensations lead us but where they change into insight and compassion. The deeper we go into these realms, the more contact we make with another's reality. The sharper the sense of pain and bliss as they interweave through the heartbreak and luck of life, the more the line between self and other may dissolve. It is a physique-soul-alchemy: a transformation of inner and outer. This book is a story of transformation.

Its autobiographical aspect arose as I have said, and is emblematic. I claim that the center holds us all, and as we speak out of it, we speak in a common voice. It is as well my part of a common pledge: for I ask others how they have come to believe as they do. What we profess is spiritual autobiography, whether it be science or myth or religion or politics or art or educational philosophy. What I know about centering makes it impossible for me to pretend that truth is either objective or subjective; the practice of centering casts upon such dualisms another light. I very much hope that a relish for person and personal destiny will be conveyed from my breast to the reader's, that he experience himself in full depth, and experiencing himself so, confirm his capacity to experience his fellow man. There is this path forward, to One Another.

CENTERING AS TRANSFORMATION

There are two things which I have been thinking a lot about. One is the experience which in pottery we call Centering. And one is the experience which in nature we call Metamorphosis.

As human beings functioning as potters, we center ourselves and our clay. And we all know how necessary it is to be "on center" ourselves if we wish to bring our clay "into center" and not merely to agitate it or bully it. As organisms in the natural rhythms of birth, growth, and death, we experience metamorphosis throughout our lives, as our bodies grow and change from infancy to ripeness, as our capacities for inner experience enlarge and strengthen. As potters, we have an especially immediate and concrete daily experience of both these more-than-physical processes, For as potters we handle our medium in the full range of its transformations. We dig our clay out of its earth bed; or if we do not always dig it ourselves, we do know the experience of digging and preparing it. We experience the mud, we experience the forces of time and destiny that have transmuted rock into plastic dust. We experience the raw ware, the sudden spell of a mobile act brought into stillness. The newly thrown or constructed pot has a quality which is not to be found in any other phase of its life. Part of our craft may be to perpetuate that "life" and feeling of plasticity in the rigid stone. We experience all the colors and textures of the raw ware and its decoration. The double life of color in pottery, unfired and fired. The biscuit, the glaze, the oxides. The transformations in the kiln during the firing which we follow through the peepholes, seeing in our imaginations the physical changes: the elimination of chemical water, the clay "moving" into its stoneware form. The changing atmosphere in the kiln during the cooling period. That faint glow just before the darkness when we pull the damper out for the last drop in temperature. The fired pot. But even the fired pot stands in the long narrative of these transformations with only its own authenticity. For it too will disappear; it will be sold or given away. It will almost certainly be broken in time. The shards will then stand with their own special charm and symbolism. They may even be pounded up for grog and thus enter bodily into the process at another beginning point. Or they may be turned into mosaic for yet another experience of form.

And though shapes change, though each moment dies into the next, though no thing is being made to last, something is happening. Each moment bears life forward. It is as if the form that grows within our acts sheds each successive moment like a skin; it is as if the inner form which grows as a being within us is brought to maturity through the successive deaths of its material stages. It seems that the potter and his craft have had a special aura from the earliest times. Pottery is the ancient ur-craft, earth-derived, center-oriented, container for nourishment, water carrier. Experiences of centering and of personal metamorphosis grow within the craft.

Both of these experiences answer man's hunger for freedom—a state of being in which man's relatedness to life is unobstructed. Unobstructed either by concepts or by fear or by ignorance or by deformity. Freedom permits us to live into experience within and without. The outer shape of the clay is the extension of its center. We press out from the center and make the pot: the outside is the surface of the inside. We turn inward and outward with the same naturalness.

Man is hungry. The baby is born hungry. He is born yowling. Hunger is a built-in signal. Man seeks throughout his life to satisfy it. Hunger for food, hunger for love, hunger for sexual satisfaction, hunger for money, hunger for power, hunger for truth, hunger for pleasure and approval. Man's hunger keeps him always turning outward, turning toward nature and other people. He gets plenty to eat, he gets a mate, he gets financial security and professional recognition. He has leisure to enjoy himself and to explore the world. Still his hunger recurs to tell him that his quest is not ended.

All his satisfactions represent attempts to gain his freedom. And the freedom he wants consists at one level in the capacity to experience in a living way a dialogue with the presence of life in which his own self-center spins. As he brings himself into center, he brings into center all the knowledge and relatedness he has drawn from the larger life presence that surrounds him. He finds that food, for example, is a sacrament of the dialogue between him and the plant and animal and mineral world. He finds that sexuality is a sacrament of the yielding of one center to another, the sacrament of love. He finds that his will expresses his impulse to give himself back to the world. If he is attentive, his hunger can teach him the interconnections of surrender and satisfaction and feeling.

Man has many hungers. But they all seem to me to be versions of a twofold one: hunger for freedom, and hunger for union, a dance of each individuality with the world.

Now of course these hungers can be sick, or "fallen" as the theologians say. And all of us are in varying degrees sick or fallen. But we aspire to being well. We aspire to redeeming our energies so that they serve our highest consciousness. And we redeem them, not by wrestling with them and managing them, for we have not the wisdom nor the strength to do that, but by letting the light to shine upon them. And where does this light come from? It seems to shine in all created things, but in our sickness we are often opaque to it. It is our task to make ourselves permeable to light by yielding ourselves up to it.

To yield means both to lose and to gain. See how the paradox is wisely caught in the words we use. I yield, and my being increases and takes form by having been given up in this way. Love becomes easier and more natural and steadier as over and over again I practice this act of yielding, from the secret inner center, the quiet will. As I open myself to the presence that faces me, it enters. It is a union. It is communion.

Freedom is presence, not absence. Centering is an act of bringing in, not of leaving out. It is brought about not by force but by coordinations. It is difficult if not impossible for a potter to force his clay into center simply by exerted pressure. In order to take its new shape, the clay has to move. It is therefore advisable technically to press down and in and then to squeeze up, holding the rising cone broad across the top, and then down again, one hand pressing the clay against the other. Tensions in the fingers, in the arms and back, holding the breath—these things count. The potter has to prepare his body as he does that of the clay. Because the wheel is center-oriented, the ball of clay will take a centered position naturally if we create the necessary support and influence. Once it has become centered, it will remain so unless there is a flaw in the clay or unless it is knocked off center by some outside force. The path to freedom is itself a series of transformations.

A capacity to yield is strengthened in the potter who does not merely use his material to certain ends, but who yields up his soul as well as his hands and his intelligence to his love of the clay. Once his soul is yielded up, the transformations of the clay will speak to him as his own. The inner laws of life will seem to be simultaneously unique centers spinning in continuous relation to each other. Peripheries will seem to breathe in and out like silken scarves. The art of the dancer in his nakedness and in his draperies suggests this self-indwelling and union of beings within the flesh. It is as if one could see how the life-body slips from the corpus; or how the body of another person's feelings and thoughts enters one's own, like lovers no longer truly separated by membrane or epidermis. It is a marriage of forces. It is a continuous dialogue.

We are transformed, not by adopting attitudes toward ourselves but by bringing into center all the elements of our sensations and our thinking and our emotions and our will: all

Figure 20 Maija Grotell, *Self-Portrait Vase*, 1937.

the realities of our bodies and our souls. All the dark void in us of our undiscovered selves, all the small light of our discovered being, all the drive of our hungers, and our fairest and blackest dreams. All, all the elements come into center, into union with all other elements. And in such a state they become quite different in function than when they are separated and segregated and discriminated between or against. When we act out of an inner unity, when all of our selves is present in what we do, then we can be said to be "on center." Part of our skill as potters is to use all the clay on the wheel in any given form. Our wholeness as persons is expressed in using all of our selves in any given act. In this way the self integrates its capacities into a personal potency, as a being who serves life from his center at every instant. In this way knowledge can become a quality of consciousness and illumine our behavior spontaneously and truthfully. Personal transformation, or the art of becoming a human being, has a very special counterpart in the potter's craft.

[. . .]

Life is an art, and centering is a means. Art is a mode of being in which elements of form and content; style and meaning; feeling and rhythm—all the living perception may be imaged forth in a way that does not sacrifice the moving character of the world.

Every person is a special kind of artist and every activity is a special art. An artist creates out of the materials of the moment, never again to be duplicated. This is true of the painter, the musician, the dancer, the actor; the teacher; the scientist; the business man; the farmer—it is true of us all, whatever our work, that we are artists so long as we are alive to the concreteness of a moment and do not use it to some other purpose. Worshipers of happening, tender craftsmen in the full range and pull of substance, as faithful to God in the blown fuse and the disappointment and the difficulty as in the serene fulfillment.

The teacher works as an artist with the particular student or group, the particular situation, his own vision and his insight into the hungers of those in his charge. Every class becomes a composition, producing its unique revelation and tone. Simple or complex, harmonious or dissonant, galactic or linear, muted or brassy, teacher and students alike may awaken to the artistic processes at work.

We may develop a way of sensing each other, artistically, poetically. A person's smell, his hair, his skin, the tone of his voice, his teeth, his attitudes and gestures, his walk, the tempo of his breathing, his unspoken hungers, everything that emanates from him—all this emanation, as it were, bespeaks his wholeness: he breathing the world in and breathing it out again; a poetical understanding which we reap from experiencing each other in depth. This is not an act performed in the spirit of research, and may, indeed, occur from the briefest contact. A mere whiff. Sometimes one is unaccountably stirred by the presence of a person whom one hardly knows. But something knows something, that's certain. We give off, the poetic understanding takes in. It is a kind of knowledge that derives from the innerness of things. It is this innerness that kindles in the surfaces which so dazzle us. I find myself often in the plain gesture of shielding my eyes as I look into the glare of The Real Presence everywhere. One of the most stirring experiences of a teacher is to walk into a classroom for the first meeting and to sense within that room so much Life, so many hopes and fears and dreams and worldly innocencies. [sic] I bow my head before the power of the person. To speak to five or twenty or forty or two hundred persons with the continuous sensation of their unique individual realities humming like supersensible energy-systems in a room, with their lives at stake, shivers my timbers every time.

All the arts we practice are apprenticeship. The big art is our life. We must, as artists, perform the acts of life in alert relation to the materials present at any given instant. This is not a simple requirement. For each instant, as it ticks off, ticks off into the past; but the past is present in the forms we have taken. We stand between past and future, between the forces that have shaped us and those yet to lend their transforming powers to our growth. Eastern religions talk about karma—reaping what one has sown. We must think as well about transforming our karma through initiative. It is a matter for deep thought, to see wherein we are bound and wherein we are free. How best to grasp the paradoxes of obedience and originality. How best to educate our imagination, our initiative, and our will.

Moral initiative may be able to create an alternative to the death fantasy which seems to be a popular current mode of satisfying man's hunger for freedom. For the desire to kill and the desire to die are, I perceive, the other face of love. The other face of the hunger for union.

Death may tempt a hungry man, whom desperation has made stupid, with its false face of oblivion. We may think all our estrangements will be dissolved if we go back to the condition from which we emerged when we were born. But in our desperation and frustration, we forget that life is everywhere on the march, that time does not stand still, that the universe is evolving and men with it. We cannot go backward to a previous unconscious condition. We can go forward through the portal of Death, but we will step into that future carrying whatever spirit lives in us. This is what makes life so great: it is for keeps, it is on the level, absolutely serious. Everything we do makes a difference. Everything is important.

Since life is not dissolved by death, but again only changed in its form, how much more practical to heal the estrangements with all our present energy. (The Big Death is the Big Change: perhaps those who resort to it have lost all other resource; it is their desire for birth that compels them.)

If we look frankly about us, we can clearly see that there is no safety nor joy except in truth. And, as the most positivistic and pragmatic philosophers say, truth is what works. But surely nothing can be said to work which stands between a man and the fullness of his being.

Here the image of centering is useful. It is an act not of "sufficiency" but of "perfection." It calls upon an understanding of man's needs of every kind: physical, emotional, psychic, social, every need has its meaning in a man's life. Some needs are temporary and stand in the way of deeper needs. But we must honor all the needs as they arise and allow them to yield to ever deeper and more thorough fulfillments.

Art creates a bridge between being and embodiment. What are pigments and gestures, the ephemera of painting? Surely when we look at a painting, we are not seeing the paint merely. We are seeing something that is not there visibly, but which enters our perception through the eye. Paintings fade, peel, dirty, tear, rot. Pots break. Art in its material aspects is as impermanent as breath.

But meanwhile what has been its task? To perpetuate the supersensory awareness of man. To demonstrate over and over again how the joy of life is not locked within its tissues any more than the joy is locked within the smear of ink on a piece of Japanese paper. It somehow lives within it, and at the same time is freed by it. The power of a man's face, the supersensory impact of physiognomy, was somehow born into the world through Renaissance portraiture. The artists did not invent it; Rembrandt did not invent the faces he painted, he saw them. But they were there, and he was there. The artist in man performs this kind of function; he is geared somehow to stand at the frontier of perception, his soul pouring into his senses. As soul evolves, as times change, what he sees changes. He stands as a kind of prophet for his society. He sees space before science does. He hears simultaneity before technicians do. He experiences indeterminacy before theologians do.

Ordinary education and social training seem to impoverish the capacity for free initiative and artistic imagination. We talk independence, but we enact conformity. The hunger in many people for what is called self-expression is related to this unrealized intuitive resource. Brains are washed (when they are not dogged), wills are standardized, that is to say immobilized. Someone within cries for help. There must be more to life than all these learned acts, all this highly conditioned consumption. A person wants to do something of his own, to feel his own being alive and unique. He wants out of bondage. He wants into the promised land.

The artist and craftsman, however far he may be from an ultimate liberation, is continually willing his work. He devotes his life to acts which are a personal commitment to value. He is, to varying degrees, an example of a practicing initiative. A creative person. Initiating, enacting. Out of personal being. Using his lifetime to find his original face, to awaken his own voice, beyond all learning, habit, thought: to tap life at its source.

When the human community finally knows itself, it will discover that it lives at that center. Men will be artists and craftsmen in their life and labor. They will live as a community in moral autonomy, each man his own judge, with a minimum of external governing laws. The common laws will be found to operate within each man, so that when human beings become awake to their inner nature, they find that for the first time they know their neighbors. *Communitas* is built into the spirit of men. They have but to perceive it to create it.

This kind of society, where individuals live together in mutual service and fellowship, and in independence, feeling the separation between individual and community transformed into an organism which functions as both, is the society which lives life as an art. Man as artist is on the move. He is not an institution, but a moving pillar of light.

FURTHER READING

Mary Emma Harris, *The Arts at Black Mountain College* (Cambridge, MA: MIT Press, 1987).

Vincent Katz, ed., *Black Mountain College: Experiment in Art* (Cambridge, MA: MIT Press, 2003).

M. C. Richards, *The Crossing Point: Selected Talks and Writings* (Middletown, CT: Wesleyan University Press, 1973).

30

CRAFTSMAN LIFESTYLE:
THE GENTLE REVOLUTION

Eudorah Moore

The thirteen California Design exhibitions held from 1955 to 1976 were a unique context for craft. Woven hangings were put alongside plastic dishes, hand-thrown pots next to scuba gear and chainsaws. The shows projected a stereotype of the West Coast lifestyle that included not only fun in the sun (lawn furniture, pool design) but also an innovation-led infrastructure that had been built up during World War II. Small-batch potteries thrived alongside aircraft manufacturers, industrial designers like Charles and Ray Eames alongside 'designer-makers' like Sam Maloof. In the beginning, the series had been curated by Clifford Nelson in a fairly straightforward fashion—it was essentially a selection from the Los Angeles Furniture Mart, supplemented by contributions from local organizations like the Southern California Handweavers Guild. In 1962, though, Eudorah Moore took over the California Design exhibitions. A woman of startling energy and vision, she took the series in new directions, ringing the changes as craft entered an exploratory avant-garde phase. All jurying was done in person, requiring a massive logistical operation in which hundreds of objects were gathered in a warehouse for judging. Increasingly ambitious catalogues showed objects in natural surroundings: a Maloof chair in a grove of trees, a clutch of pots on a windswept beach. The focus on lifestyle remained, but as California became the center of the national

subculture in the late 1960s, Moore moved away from Nelson's focus on the market. In 1974, she helped to organize 'California Design 1910', the first survey of the West Coast version of the Arts and Crafts Movement—which, she noted, was a 'value statement, not a design style'.[1] Moore's perspective was perhaps best embodied in the book Craftsman Lifestyle: The Gentle Revolution, *published in 1976. The text was a series of interviews of craftspeople, in which they were asked not about their work (though that came up, of course) but about their homes, their shops, the rhythms of their average day, the satisfactions and difficulties they found in being craftspeople. California Design finally came to an end in 1976, as its home base, the Pasadena Art Museum, became a showcase for the private collection of Norton Simon. To the end Moore held fast to her sense that California was a sort of cultural laboratory, in which craft was the most important experiment: a solution to 'the dilemma of the existence of the whole person living within the machine society'.[2]*

Eudorah Moore, introduction to *Craftsman Lifestyle: The Gentle Revolution* (1976).

This is a book about people and attitudes and changing value judgments. It is not a book about crafts but about craftspeople—insights into what they regard as important, into how they view their life and work. It is a humble book, seeking in simple and informal terms to present a composite picture of lifestyles from which the reader may draw his own deductions. [. . .]

In 1974 we presented an exhibition titled "California Design 1910." The researches in conjunction with the preparation of that exhibition were father to the thought of this book, and led to that moment when a pattern seemed to emerge in the attitudes we had heard and observed in our dealings with contemporary craftsmen. Reading the literature of the Arts and Crafts movement and the thinking which generated it, reading the formulations of the ideal of living as expressed in publications of that time, the thought repeatedly and urgently recurred to us that many of today's craftspeople whose work had been in our shows, and whom we had come to know, were, in fact, now living that ideal articulated at the turn of the century. The life attitudes of contemporary craftspeople actually represent answers to the concerns of Carlyle, of man in relation to his labors; they reflect and activate Ruskin's feeling of the necessity for life pervaded with the consciousness of beauty and art, and Morris's ideals of the doing and the making—the process. It seemed more important than ever to record our observations of the craftsman value judgments and lifestyle for now they appeared to be part of a philosophical continuum, rooted in nineteenth-century ideas, and growing steadily and serenely to notable numbers. The idea of the book as vehicle to relate these observations began to form.

In the fall of 1976, in the raw unfinished rooms kindly loaned to us for receiving entries for the California Design '76 exhibition, we felt again the desire to record the phenomenon of the twentieth-century craftsman's movement: What it means in terms of attitude and value judgments—in a way of life as well as in finite

work. What extraordinary energy it represents, what excitement, what vitality! As hundreds of craftspeople came first in Los Angeles and then into the San Francisco warehouse, each carrying in his work—representing a very special extension of his ego, we knew we were privileged to be part of an electric scene. As we talked to these people we knew more. We found that most were highly educated, that each, in becoming a craftsperson, had made a significant and knowing choice; that being a craftsman implied a certain stance, and attitude, a way of life. We recognized that it was a path of personal exploration and commitment, of repudiation of expected roles, and, in many cases, of the material goods whose acquisition has so long been an aim in itself. We became intrigued to know more about these people who had left their work to be judged. When the jurying was over, we sent a simple questionnaire to those whose work was chosen to be in the exhibition. Essentially we said we knew they were interesting people, and we wondered if they'd tell us who they were, where they came from, where they were going, and what they thought was important. The replies made us race for the mail for weeks. Diverse, amusing, profound, infinitely interesting, they made us know we had to talk to as many of them as we could. We had to record their attitudes, their view of life, what they thought was important, for we were convinced that it *was* important, and a significant bellwether for the future . . .

First, and most important, is the fact that the role of craftsperson is universally a conscious and considered choice. All the people we talked to work creatively with their hands because it gives them joy, because it is fun, as they expressed it, because they "had to." The educational level of the group is high, and other career opportunities have obviously been

Figure 21 Alexandra Jacopetti, macramé playpen, erected at a sale of the Baulines Craftsmen's Guild in Bolinas, California, ca. 1973–4.

available, but being craftspeople is an unregretted choice, and results in an extraordinary degree of commitment and identification with their work. The desire for freedom is ubiquitous, even at material cost.

Second, in almost all cases the act of doing supersedes in importance the end result, the monument. In short, for these people, art and life are a single fabric, and the quality of living is the monument.

Third, manifestations of love of nature and identification with the unity of all things runs like a refrain through the interviews.

Fourth, one notes a thoughtful re-evaluation of standard American priorities. Quality supersedes quantity in importance. And quality is

concerned with the experiences of daily living for the individual rather than in appearance or image presented to outsiders. There is a standard disregard for long-held hierarchical distinctions within the social structure. Old ideas of "suffer now for ultimate rewards," are replaced by, "extract from every moment the joy it offers; whether pleasure in one's work, in visual perceptions, in good food, or in quiet repose. LIVE IT." The earliest and strongest influences on the Arts and Crafts movement stemmed from an intellectual and social elite whose concerns were political, or at least broadly social rather than pragmatically personal and perhaps were more theoretical than practical in living terms. Today's movement is deeply introspective and subjective, and is concerned with each individual's relationship to work, to living, to family, to nature and to himself. It is as if, in the words of T. S. Eliot, "We shall not cease from exploration/And the end of all our exploring/Will be to arrive where we started/And know the place for the first time."

Last, and a corollary to four, is the fact that the craftsman style is one of doing. He is a participator, not a spectator.

Although this thesis of changing value judgments is presented here through a group in which there is a broad occurrence of the shift, it is obvious that it is happening across the social structure. It is demonstrated in the craftsman lifestyle, but there are many proponents of the values other than craftsmen. Although the shift can be seen across the country the great numerical preponderance is in California and on the West Coast. It might be of interest to explore the reasons for this geographic centering of the "New Craftsmen's Movement."

The great wave of migrations into California in the first half of this century consisted largely of people who were coming for a new beginning, who were seeking freedom from old restraints, and the pleasures of a benign climate. They found a land with unpatterned social structure, with a zestful appetite for new ideas. Unpretentious, and caring little for the social mores of a more structured society, these people, when asked if they didn't feel distant and remote, would answer, "far away from what? It's here." A discreet hedonism began to infiltrate the puritan ethic. Nature, the land, the place, pervaded the consciousness. The odd brilliant light, so different from the softer glow of eastern days, gave a heightened vision.

Although using the benefits of the products of the industrialized east (indeed, in the case of the automobile, patterning its growth on its use) coastal productivity tended to develop in industries such as agriculture, tourism, movies, aircraft, oil, space, furniture, etc. which, by their nature allow people to function as individuals rather than assembly-line robots. This permitted a sense of "why not?", of personal possibility and identity to prevail.

Without the numbers of long-established private schools, here was developed an extraordinary, broad system of public education long before the east. California led in establishing the pattern of the multi-campus university and the second-level State University System emerging from the teacher's college structure, backed up in turn by the system of tuition-free community colleges. Education and climate have without doubt been contributing forces to the new movement. Educational facilities have provided experiences in numbers of well-equipped workshops, and learning has brought questioning and awareness of the quality of life. In a harsher climate more time is absorbed in the simple demands of living; a benign climate has made a more frugal and simple way of life, with free time for creativity a possibility.

Though these attitudes were formulated by European philosophers, the germination and groundswell has occurred here. We see these ideas being embraced by craftspeople and others across the country. We feel the attitudes herein delineated are significantly pointing a social direction. Where the traditional craftsman was an artisan because of material necessity, fashioning objects necessary to society, the new craftsman in the industrial society chooses the path of making the unessential necessity, fashioning his lifestyle to realize the creative impulse so vital to the whole person, providing those objects of the hand and mind so necessary to us all.

NOTES

1. *California Design 1910* (Pasadena: California Design Publications, 1974), p. 7.
2. *California Design 1976* (Pasadena: California Design Publications, 1976), p. 9.

FURTHER READING

Glenn Adamson, 'California Dreaming', *Furniture Studio 1: The Heart of the Functional Arts* (Free Union, VA: The Furniture Society, 1999).

Suzanne Baizerman, Jo Lauria, and Eudorah M. Moore, et al., *California Design: The Legacy of West Coast Craft and Style* (San Francisco: Chronicle Books, 2005).

31

THE SOUL OF A TREE

George Nakashima

The spiritualist dimension of craft found one of its greatest representatives in George Nakashima, a Japanese American woodworker, designer and architect. Though his career was long and his output varied, Nakashima is best known for tables and other furniture built around cross-sections of wood with a natural 'free edge.' The popularity of these works—much copied, especially in the heyday of the counterculture—probably has something to do with an association with Zen, though they have very little to do with traditional Japanese furniture and even less to do with Buddhism. In fact Nakashima's mysticism sprang from a very different source. In 1937, he had gone to Pondicherry, India, to work on an architectural project and had become deeply influenced by his client there, the Hindu spiritual leader Sri Aurobindo. This relationship kindled a mysticism in Nakashima that would last his whole life. The impulse found a focus during World War II, when (as a result of the government's policy of imprisoning Japanese Americans) he was interred in a camp in Idaho. There he met a woodworker named Gentaro Hikogawa, from whom he learned the use of traditional Japanese hand tools. After the war, he shifted his focus to furniture, making his own work and also designing for large production firms such as Knoll. Though his studio production was executed mainly by a team of helpers—his own handwork consisted mainly of selecting and preparing timber—he wrote eloquently of his reverence for the living wood from which his objects were made. The following passage, taken from Nakashima's widely read book The Soul of the Tree, exemplifies his particular brand of idealism, which is materialist and aesthetic in nature, more or less divorced from any agenda of social reform.

George Nakashima, excerpts from *The Soul of a Tree: A Master Woodworker's Reflections* (Tokyo/New York: Kodansha, 1981).

A THOUSAND SKILLS, A THOUSAND VOICES

Our skills have been sharpened, the designs made. The shed has been adequately stocked, the decision of solid wood versus veneers has been settled in favor of solid. The inventory of experience has been accumulated. The work commences.

The object is to make as fine a piece of furniture as is humanly possible. The purpose is usefulness, but with a lyric quality—this is the basis of all my designs.

The selection of timber is made in the shed, brought into the workshop and marked out for cutting. As far as possible, all elements are from the same tree. However, for some purposes, such as the need for strength, a different material may be used—for instance, hickory for spindles in a chair.

It is a stirring moment when out of an inert mass drawn from nature we set out to produce an object never before seen, an object to enhance man's world; above all, a tree will live again . . .

For millennia, the working of wood was almost entirely a hand operation. The lathe was often powered by the craftsman's assistants. It could also be activated by waterpower, and sometimes simply by feet.

The reality of the age, however, brings up the question of machinery. As much as man controls the end product, there is no disadvantage in the use of modern machinery and there is no need for embarrassment. Gandhi and his spinning wheel were more quixotic than realistic. A power plane can do in a few minutes what might require a day or more by hand. In a creative craft, it becomes a question of responsibility, whether it is man or the machine that controls the work's progress.

Woodworkers follow a long tradition dating certainly from the beginning of civilized man. The first wheel must have been wood. Possibly, man moved heavy objects on wood rollers. Woodworkers of quality must have existed in the Vedic age in India and, of course, among the Egyptians and the Greeks. Joseph and Jesus of Nazareth, as well as the shrine and temple carpenters of Japan, followed an honored craft. These skilled artisans from the fargone past speak with urgency and insistence. They are lighting the lamp for us to follow.

The selection of furniture parts is always most important. Of the roughly ten thousand boards available in my warehouse, the perfect choice must be made for each part of each board. Sometimes five or ten years pass before a board is selected for use. There must be a union between the spirit in wood and the spirit in man. The grain of the wood must relate closely to its function. The abutment of the edge of one board to an adjoining board can mean the success or failure of a piece. There must be harmony, grace and rhythm. It is so easy to place the wrong board, out of the ten thousand available, next to the wrong one, resulting in a mismatch. Sometimes a number of boards are shifted about until the right combination is found to make the happy whole.

There is so much individuality in these boards. Some are of great distinction and nobility, others plain and common, still others of such poor aspect that they must be relegated to the scrap pile. Each species of wood too has its own strong personality. The long fibers of the cypress contrast with the exuberance and beauty of its fine burls. The strong figuring vibrates with joy, at times through the whole bole, at other times only at the junction of several main limbs branching out. Roots, too, have strong personalities, especially where they meet the tree's trunk, producing fantastic richness of graining. Roots must be used in a precise and exact way. They may be cut round or square or oval. Or they may be left entirely natural, or "free."

Quite often the shape, size, texture and the extravagances of graining dictate the design and function of an object. Here the relationship of man to timber prevails as the two live comfortably together day after day, without tiring of each other.

Gradually a form evolves, much as nature produced the tree in the first place. The object created can live forever. The tree lives on in its new form. The object cannot follow a transitory "style," here for a moment, discarded the next. Its appeal must be universal. Cordial and receptive, it should invite a meeting with man.

The rough dressing down of a plank, the study of the contours, sizes, shapes, thicknesses; the rough chalk markings, the fine marking, the final cutting; the joinery and assembly, the finishing to bring out the depth of grain—all these steps follow one after the other, each with its own responsibility.

We must make as perfect an object as we know how. The final but essential requirement is to finish the top surface by hand. A good workman can achieve perfect surface work with a hand plane alone. To achieve a fine result, a carpenter may spend days surfacing the faces of a post in a Japanese house. For the best work, the bit is sharpened after each stroke, not because it is dull, but because the finest finish demands it.

What a wonder fine furniture can be—a chair to rest a human body, a table at which to work or to partake of food. A cabinet to organize and store the things for daily use. The parts are assembled, the joints designed and made ready to be put together. Solid wood moves, breathes and lives. The joints must be designed with this in mind.

It is in the making of the joints that skills count. The joints should not be too loose or so tight as to split a member. The shoulders of the joint should fit tightly and slightly compress the wood of the receiving member. In a house the joints should be a drive fit, that is, pounded in. A slop fit, even if held with a wedge, is not adequate. Compound tenons and mortices, shoulders for cinching and drawing up, compound shoulders for wedging, the thousand methods of joining several pieces of timber together—all of this joinery is designed so that a certain distension due to drying need not be serious and indeed often helps by tightening a bond.

The decline in quality of modern furniture is probably due in part to the use of the quick, easy and cheap dowel joint. The decline of modern domestic architecture can be traced to the popularity of the stud wall put together with hammer and nails, a type of construction calling for no joinery at all. By contrast, the early American house and barn with their excellent joinery still represent the best we have produced and will greatly outlast contemporary buildings.

Good joinery, whether in buildings or for furniture, is difficult to design and even more difficult to execute. It should be thought of as an investment, an unseen morality.

The precision, intricacy and sophistication of the *asa-no-ha* grille, in which twelve members come together, are indeed staggering.

Joinery was highly developed in Japan, originally the land of great evergreen forests where the fervor of the people drove them to construct great shrines and temples, some on the grand scale of the European churches and cathedrals. The work was done by carpenters, the *daiku,* master builders, without comparison any place.

Japanese joinery is a highly developed technique, formalized and exact. Each type of joint has a name and the procedures for each follow precisely in the proper order. The work was and still is done by hand with tools of an excellence that the West has not known, tools to meet every joinery requirement. The work goes extremely fast, the chips fly.

The greatest of these master builders who traditionally were also the designers and architects in wood were the *miya daiku,* the "shrine builders." These were the princes of construction, versed not only in building but also in structural proportions. Theirs was a skill acquired only by creative doing for generations. They were masters of the mystery of how four beams and two posts meet, the perfect pitch of a roof, the subtle relationship of one material to another.

Preserving the techniques of fine joinery can help save us from the onslaught of mediocrity in our furniture and housing. The great

Figure 22   George Nakashima, *Conoid Bench with Back*, 1961.

cypresses and cedars of the Orient are fast being depleted, the Lebanon cedar is almost extinct and even the fine American cypresses, the Port Orford cedar and the Alaska cedar, are in short supply. But new trees grow and the wood that can still be joined is luckily still adequate, though limited. We can still make joints to our hearts' content, joints that are honest, sound and enduring.

Hardwoods with beauty of grain and texture, even when accompanied by twisting, warping and other irregularities, can often be used for furniture. Joints for furniture are normally simpler than those used in buildings, since the structural demands are not so great. The craftsmanship, however, must be precise, since the stability of the entire piece of furniture depends on it.

In Japanese, *kodama*, the "spirit of a tree," refers to an experience known to almost all people of this island nation. It involves a feeling of special kinship with the heart of a tree. It is our deepest respect for the tree which impels us to master the difficult art of joinery, so that we may offer the tree a second life of dignity and strength.

Any joint can be made by hand, but where the machine can do the work more efficiently, I use the machine. Yet, there are many areas where only hand tools can be used. At such times the finely developed Japanese tools are effective and in many cases the only answer to tight problems.

The Japanese hand tools include the offset chisels, fan-shaped chisels, chisels ground hollow in the back with a slightly tempered carbon steel cutting edge and a soft steel backing; saws that can start cutting in the middle of a board (saws that cut on the pull), marking gauges with a knife for a point, an ax like nothing seen in the West. After using the Japanese wood block plane, there seems to be something sacrilegious about a plane with a steel block!

Toolmaking is still a great art, even in power machinery. The research on tools can be a study in itself. Familiarizing oneself with the many different types, the various toolmakers, the care and sharpening of tools and, of course, the techniques of using them—all this could alone fill a lifetime.

TABLES

We are faced with the problem of making a working surface, a table top. This surface can be composed of glued-together smaller parts, a book-matched pair, or a single slab. Often the characteristics of a single slab dictate the size and design of the piece. In book-matching the butterfly inlay is invaluable. It adds strength to the joinery and, just as important, it adds a creative design element. Where it is placed, its size, color and texture are all vital considerations. The butterflies should contrast with the boards, matched for attractiveness. Their grains should be perpendicular to each other for strength.

Originally the entire process was done with hand tools. Today we use a power tool for the preliminary work. First we hollow out the areas to be fitted with the butterfly-shaped piece with a router. Then we fit the joint with a flat chisel and clean out the corners with a fan-shaped Japanese chisel. Finally, the butterfly is placed in the area prepared to receive it and glued securely. The top, made to protrude

a bit, is planed smooth and sanded before final finishing.

Some of the happiest table forms are accidental or realized through imagination. Triangular pieces are extremely beautiful and useful in furniture making. The flares at the base of a tree usually produce the richest graining, nature's fantasies. These are unique and must be designed for one specific object, and that alone.

Then there are rectangular tables of various dimensions and woods—great long tables for conference and dining, all of solid wood. Contemporary international treaties might benefit from being written on a good honest table. How can one expect a sincere treaty signed on veneer, green felt, or on wood grained to look like marble?

Finally, to meticulous care, proper proportions, sound structure and honest creative design all working together, we add a fine finish. The finish stands for our faith, our integrity.

CHAIRS

What a personality a chair has! Chairs rest and restore the body, and should evolve from the material selected and the predetermined personal requirements which impose their restrictions on form, rather than the other way around. Some parts, such as spindles; are used primarily for strength, and aesthetics becomes a secondary consideration. These can be beautiful, however, and the error of just a sixteenth of an inch in the thickness of a spindle can mean the difference between an artistically pleasing chair and a failure.

Function, and beauty and simplicity of line are the main goals in the construction of a chair. If a chair's purpose is only to impress people or to show rank, then it becomes ostentatious and carved to death.

CABINETS

Cabinets are useful for the organization and storing of things; they may also be objects of beauty. Fine woodworkers are often called cabinetmakers. Making a fine cabinet for storing the warrior's armor, sweets for the child or any other purpose can be a truly rewarding accomplishment.

A large vertical storage piece in the grand tradition of Chinese cabinets or European armoires, chests of drawers, small bedside units, small chests of drawers for organizing small things, ladies' dressing boxes, filing cabinets, wall-hung pieces—all these help to make life orderly.

Traditionally in a Japanese family, a *kiri* (paulownia) tree is planted at the birth of a daughter. This tree is fast growing, beautiful of grain, light in weight and easy to work. When the daughter marries, the boards from this tree are made into a chest for her trousseau. The lightness of the wood makes it easy to transport. *Kiri* wood is never finished; in time, if it becomes soiled, it is resurfaced simply with a plane.

DESKS, BEDS AND LAMPS

Some of the most creative moments of the poet and prophet must have taken place at a desk, so the desk must be sincere. One must start with a great or modest slab of wood of rich or simple graining. Nothing much more is actually needed, except possibly a small cabinet to store writing necessities.

Beds and reclining pieces are important objects of furniture, and too often they become ostentatious. But they can be so simple as to be put away during the day so that the room may be used for other purposes . . .

The key to fine workmanship lies in the drive for perfection and the development of skills to achieve it. Perhaps as a backlash to industrialism and commercialism, a new concept seems to be taking hold. The large number of young people, many of them college graduates, who want to do truly fine work is astonishing. Even in my shop, where many questioned at first whether our work made sense, the reactions are now enthusiastic. There is a pride evident today in work well done. Many strive to create and to create well.

In my shop, each woodworker is an individual craftsman, free to work out his own *sadhana*, spiritual training to attain deep concentration resulting in union with the ultimate reality. Each person can do what he finds most suitable within certain guidelines. Our relationship and attitudes are based on the teachings of Sri Aurobindo and the ashram of Pondicherry. The will must aspire to produce as fine an object as is humanly possible. Each man must find his own personal truth. The endeavor must be to bring out the beauty and proportion, the textures and depth of the material used, to produce something that may last forever.

This may seem an anachronism in our age. But there is a battle to be won. For many years I have struggled to make my dream a functioning enterprise. At times, I felt it unjust that all the wild birds in my woods could sing all day long while I sweated in a craft the world didn't seem ready for. Now the points of light appear in the wilderness more frequently and afford additional opportunities for creativity.

To the advantage of craftsmen, the modern commercial system has produced its own built-in difficulties. The costs to a large manufacturer of mass-selling small objects are now so high, involving so many middlemen, that it is possible for craftsmen to build a better product and sell it for less by direct contact with the buyer. This helps in some measure to explain the resurgence of the craft movement.

The maker of fine wood furniture reaches out into hundreds of lives, listens to voices and shares in the lives of so many people, giving and receiving.

FURTHER READING

James Krenov, *A Cabinetmaker's Notebook* (New York: Van Nostrand Reinhold, 1975).

Mira Nakashima, *Nature, Form and Spirit: The Life and Legacy of George Nakashima* (New York: Harry N. Abrams, 2003).

Derek Ostergard, *George Nakashima: Full Circle* (New York: American Craft Museum, 1989).

32

THE LONG SHADOW OF WILLIAM MORRIS: PARADIGMATIC PROBLEMS OF TWENTIETH-CENTURY AMERICAN FURNITURE

Edward S. Cooke, Jr.

In this essay Edward S. Cooke, Jr., a scholar of American decorative arts at Yale University, examines the ideology of the American studio craft movement through the example of the Californian furniture maker Sam Maloof. In the 1960s, there were only a handful of studio furniture makers in America, as compared with hundreds of metalsmiths and perhaps thousands of potters. Nonetheless, it was Maloof who became the public face of American studio craft in these years. His personal modesty, work ethic and stylistic consistency projected an image of authenticity that lent him the stature of a national treasure. But what does it take to build such an image? Without accusing Maloof himself of any dishonesty, Cooke subjects his reputation to a critical examination, in three parts. First, he shows how the idealized notion of the craftsman that originated in the writings of William Morris serves as the basis for the rhetoric surrounding Maloof and his work. Second, he looks closely at Maloof's business strategies, ranging from pricing to the deployment of signatures. Finally, he examines the reception of Maloof's furniture in the market, pointing to the dramatic rise in prices of his signature form, the rocking chair, as an instance of a broader commodification in the studio craft field. Cooke's analysis is a fitting conclusion to this section on idealism: it provides both a summary of the tradition in craft that Morris initiated, and a call to think outside that framework.

Edward S. Cooke, Jr., 'The Long Shadow of William Morris: Paradigmatic Problems of Twentieth-Century American Furniture', *American Furniture 2003* (Milwaukee, WI: Chipstone Foundation/ University Press of New England, 2003), excerpted.

[. . .]

In its most public guises—museums and journals—the past century's material culture rarely exists as a coherent package. Many museums have departments of American decorative arts that focus on work made before 1920 (up through the Arts and Crafts Movement, celebrated as the last gasp of the individual craftsman) and assign responsibility for the twentieth century to different departments of design or contemporary art. In some other museums, departments of American decorative arts focus on one-off craft objects and luxury goods of the twentieth century, identifying them as the logical extensions of the historical collections, and thereby include only a small part of the period's material culture. Such institutions reject the industrial or commercial products of the past century, seeing them as commonplace kitsch. Few institutions collect and display a wide variety of objects from the twentieth century. The split is also seen in the mutually exclusive contents of design magazines such as *Metropolis* and *Design Issues* and decorative arts periodicals such as *Antiques* and *American Craft*. The origin of manufacture, the role of machinery, the numbers produced, and the market all play an important role in distinguishing an object's taxonomical category, exhibition relations, publication venue, and scholar profile. Separate, distinct discourses characterize the American field.

Compartmentalization can also be seen in much of the scholarship on American studio furniture produced in the past quarter of a century. A recent example, *The Furniture of Sam Maloof*, is the latest and largest consideration of that individual furniture maker. Much of the work on Maloof celebrates his singular genius, draws on a general context to provide background for hagiography, makes uncritical use of the maker's own words and philosophy to bestow meaning on the work, and relies on a descriptive consideration of the objects that focuses upon technique. Such an approach links current work on studio craft to the conservative decorative arts canon rather than examining this work within the context of various modes of furniture production or under the lens of material and visual culture theory. Why has this canon colored the focus of twentieth-century decorative arts, favoring the studio crafts and precluding design and theory? The investigation of this question leads back to William Morris, the English designer-craftsman, social critic, writer, and Socialist.

Morris was one of the first writers, and certainly the most prolific and influential, to use the term decorative arts in the manner in which we commonly understand it today, as "that great body of art, by means of which men have at all times more or less striven to beautify the familiar matters of everyday life." By focusing on "ornamental workmanship" Morris sought to elevate quotidian objects so that society ascribed value to them even though they might not equal the "higher" "arts of the intellect" (architecture, painting, and sculpture). In his call for serious consideration of this class of artistic production, Morris popularized the term decorative art, rejecting the other period terms such as "industrial arts" and "applied arts" because of their manufacturing and commercial connotations. Morris's extensive writings inspired many American cultural capitalists and early museum professionals, who then began to collect and institutionalize decorative arts at the turn of the century, thereby ensuring that Morris's terms and criteria became the foundation for the field. Therefore it is important to recognize Morris's particular construction of the decorative arts. Two themes stand out in his writings on the subject—his sense of history and his great esteem for the maker.[1]

Responding against the misery, alienation, and inequities of the contemporary British economy, Morris looked back to the fourteenth and fifteenth centuries as an idealized organic past. He wrote how craftsmen at that time had their own fields or lived adjacent to sites of agricultural production, were complete masters of their tools, enjoyed a rhythm of work that combined artisanal freedom and harmonious cooperation, and produced worthy objects. A certain communalism, born of the agrarian life and fostered by church fellowship, circumscribed the craftsman:

> The theory of industry among these communes was something like this. There is a certain demand for the goods which we can make, and a certain population to make them: if the goods are not thoroughly satisfactory we shall lose our market for them and be ruined: we must therefore keep up their quality to the utmost. Furthermore the work to be done must be shared amongst the whole of those who can do it, who must be sure of work always as long as they are well behaved and industrious.[2]

For Morris, we should learn from history and use it as a template for reform and restructuring. History is the supreme teacher that would guide and inspire those living in the present. Morris also believed that object making was ultimately a local activity, rooted in a specific context for an immediate audience.

The other tenet central to Morris' writings was his supreme regard for the individual craftsman as the heart of a successful society. Morris celebrated the craftsman for his ability to create a "new art of conscious intelligence" that was distinct from "mechanical toil," the useless work that provided the baubles of contemporary fashion. Deriving great pleasure in his activity, the true craftsman used dexterity and thoughtfulness within the comforts of the guild and the local community to improve the built environment. To Morris there was direct linear linkage from harmonious thoughts and processes to beautiful built environments and harmonious societies. In celebrating the medieval craftsman who designed his own products, Morris wrote:

> The medieval man sets to work at his own time, in his own house; probably makes his own tool, instrument, or simple machine himself, even before he gets on to his web, or his lump of clay, or what not. What ornament there shall be on his finished work he himself determines, and his mind and hand designs it and carries it out; tradition, that is to say the minds and thoughts of all workmen gone before, this, in its concrete form of the custom of his craft, does indeed guide and help him.[3]

The imaginative work of a fully skilled craftsman ensured pleasure in making and use.

Morris's idealized vision of production relations led ultimately to a critique of commercial capitalism, the products of which were "trivial, mechanical, unintelligent, incapable of resisting the changes pressed upon them by fashion or dishonesty." With the rise of the division of labor at the expense of the master designer-maker and a concern for profit rather than livelihood, the maker became "condemned for the whole of his life to make the insignificant portion of an insignificant article of the market." As the all-around craftsman gave way to the narrow specialist, items for use gave way to items for sale. Morris thus made selective use of Karl Marx to focus upon the process of manufacture as a determinant in the ultimate value of the object and paid little attention to the reception or social meaning of such objects outside of their processural origins. It is easy to see how Morris's ideology influenced decorative arts scholarship that celebrated the individual craftsman, identified the colonial and early national periods as the Golden Age of American craftsmanship, and linked the colonial craftsman to the myth of the self-sufficient American farmer.[4]

Such a privileging of the individual craftsman working directly with low technology to produce objects of exchange precludes both the notion of a professional industrial designer who works with a team to develop prototypes and then plans or orchestrates manufacture as well as the concept of a craft object existing as a commercial commodity with recursive meanings. Morris has thus cast a long shadow over the scholarship of the decorative arts, limiting the focus of study to the idealized shop floor, the time period of study to a preindustrial period, and the meaning of the product to decontextualized original use. Reliance on Morris's terms fails to engage with the twentieth-century discourse on new forms of production, the rise of batch and mass production, and the possibility of multiple simultaneous meanings . . .

Figure 23 Shop mark used by Sam Maloof from the late 1950s through 1971. Photograph by Jonathan Pollock.

To overcome the limits of the Morris paradigm, we should draw inspiration from English design historians and look at decorative arts and design as a whole, to examine production, reception, and theorization of a whole range of domestic material culture. To demonstrate the possibilities of this approach, the work of Sam Maloof and other pioneering studio furniture makers will be examined in a more theoretical manner.

Sam Maloof (b. 1916) has been making furniture in his own shop for the past fifty-five years. In an early article on the woodworker, art journalist Sherley Ashton described him in a manner that strikingly recalls Morris's idealized craftsman:

Working with disciplined hands and a free spirit, Maloof is rewarded with great warmth in his designs; but achieving warmth in his designs is a fetish with him. In his opinion the weakness in much contemporary American furniture is its coldness, a result of the fact that, in the United States, designer and maker are usually two people instead of one . . . His one enthusiastic concern is that every piece he turns out shall demonstrate usefulness, beauty, and craftsmanship . . . Maloof thrives on the freedom and demands of his one-man operation. In the course of the day he may be salesman, designer, craftsman, supply buyer, truck driver, but he is sublimely free to design and build, without interference from such commercial factors as cost accountants, advertising executives, sales managers, or shop superintendents whose foibles tend to destroy the subtleties of craftsmanship for the sake of profits.[5]

His happy lot was further linked to the pleasant lemon grove surrounding his shop. All subsequent writings on Maloof have repeated the importance of his pastoral utopia in the San Bernardino Valley and unwavering commitment to the designer-craftsman as distinct from and superior to industry.

[. . .]

In interviews, public lectures, or instructional workshops, Maloof talks modestly about how his work ethic enables him to survive without his wife ever taking a job outside the house, how he brings his skills to bear on every object that leaves his shop, how he has maintained close friendships with his clients, and how he has taken such pleasure in a lifetime of craft. Certainly the imprint of William Morris is clear, a connection implicitly noted by Maloof's followers. The ultimate result of his presentations is an awed audience that reverently approaches him and seeks more assurances about the values of the craftsman lifestyle. Viewers of his furniture also approach the objects with a similar sense of respectful deference . . .

But what lies below this placid surface of reverence and adulation? Essential to a more rigorous analysis of studio furniture practice is the recognition that pastoralism and an emphasis on the aura of craftsmanship often mask commercialism. While Maloof's gross production would not qualify him as an industry . . . he certainly demonstrates a very conscious interest in maintaining his public image. While he claims to have worked "not for recognition or for monetary reward," the evidence in the literature on him suggests otherwise.[6] His prominent leadership roles within local and national organizations such as the Southern California Designer-Craftsmen and the American Craft Council from the 1950s through the 1980s reveal not only that he felt responsibility to give something to the field but also suggests a willingness to help write craft history with himself as a major protagonist. When friends and publishers began to suggest in the 1970s that he write a book, he dismissed any solicitation of a mere how-to book but held to his conviction that he was worthy of a well-illustrated monograph that celebrated his contributions to woodworking and studio crafts. His talks have consistently dwelt on first-person stories, providing narratives that feature him, celebrate his work, and push other makers to secondary roles.

While priding himself as a full-time maker rather than as a teacher who made just a few pieces of furniture, Maloof has made it a priority to take time away from the shop to attend important conferences and keep his name in circulation. He began in the 1970s to devote increasing amounts of time to travel in order to lecture, teach workshops, and talk about his work in conjunction with the growing number of craft or furniture exhibitions. However, Maloof's use of public lectures and demonstrations/workshops—the staples of American craft marketing—to promote his work is hardly unique. The public seems to crave firsthand exposure to iconic figures like Maloof and [James] Krenov, and the makers simply respond to that demand. Yet each maker has their own particular spin: Maloof prides himself as the most successful full-time working craftsman, Krenov presents himself as the guardian of meaningful refined workmanship, and Wendell Castle (b. 1932) identifies himself as the leading art furniture maker. Each has taken up a facet of the Morris idealized craftsman.

Pricing is another area in which Maloof's activity can be interpreted in different ways. Throughout his autobiography he constantly recalls how people told him his work was underpriced or confesses his unfamiliarity with pricing. In 1971, he told one such story: "Most people tell me I don't charge enough. But my problem is setting a price myself. I'm hesitant. One friend said he wouldn't buy any more furniture from me unless I raised my prices. He'd contracted for a piece. I sent him the bill. And he sent me $200 more than I asked for."[7] Such posturing could be interpreted as either naiveté or a strategic form of self-deprecation intended to stir up additional commercial interest, but the frequency with which price is discussed in print or in conversation leads one to believe the latter.

[. . .]

Astute business practices can also be seen in Maloof's record keeping and awareness of milestones. He kept track of when he first produced a particular form, proudly recalled the number of firsts for which he was responsible (maker of the first piece of contemporary furniture accessioned by the Museum of Fine Arts, Boston; maker of first piece of contemporary furniture in the White House collection; first woodworker elected Fellow of the American Craft Council; first craftsman to receive a MacArthur Grant, etc), and adapted his practice of signing furniture. Initially he used a brand that linked him to more commercial practices. Beginning with the "Woodenworks" exhibition of 1972, he began to sign his work with an electric burning pen. Upon his election as an ACC Fellow in 1975, he burned in his signature, the number of that example in that year's work, and the initials "fACC" (Fellow of the American Crafts Council. After he received an honorary degree from the Rhode Island School of Design in 1992, he substituted "d.f.a. r.i.s.d." (doctorate of fine arts, Rhode Island School of Design) for "fACC." While such signatures surely embody his pride of accomplishment, they also provide a sense of commercial identity and authorship that are useful in the marketplace for custom furniture. Initially he found the stamp sufficient to link him to design community of Southern California, but his subsequent switch to his actual signature signals an interest in emphasizing the involvement of the human hand and the authenticity of the author/maker. His subsequent inclusion of a serial number and his honorific legitimization suggests his pursuit of a larger and different market—the national studio crafts market.

[. . .]

Closer scrutiny of Maloof's shop activities also provides some distance from the Morris ideal. Much of his ability to focus upon his shop was enabled by his wife and helpmate Freda, who selflessly served as business manager and salesperson, while also running the household, for fifty years. She may not have been able to take a job since she held a full time job in the house. He recognized her many contributions to his success, but it is important to see her not only as an enabler or inspiration, but rather as a full-time unpaid partner who oversaw the books, managed his time, entertained clients, and ran the showroom, which also happened to be their home. Employee relationships rather than pleasurable work characterize other aspects of Maloof's shop. For consistent production, he has hired a number of workers whose main tasks remain the tedious acts of sanding, finishing, and clean-up, freeing Maloof to focus on and control the riskier "signature" elements such as cutting out parts, assembling, and rough shaping. Maloof believes that good pay and benefits rather than providing opportunities for satisfying craftsmanship are the most important part of his relationship with these workers. Gendered and wage-based relationships seem more modern than those guild relationships favored by Morris and call into question the myth of the happy small shop. The intensification of a domestic type of production in order to accommodate aspects of market capitalism also links the Maloof enterprise to the rural New England craft shops of the early national period of American history, when decreased agricultural returns and increased markets for consumer goods spurred widespread familial exploitation and outwork.

[. . .]

While the preceding discussion offers important adjustments to the received history, it is in the reception of that furniture where more recent scholarship can shed dramatic new light. In the 1950s and 1960s most studio furniture embodied use value and consistently remained true to the function of furniture. Much of this early work continues to be used and enjoyed by the families who originally bought it from the maker. There remains a personal connection that exists outside the marketplace and a sense that those works were simply locally made versions of modern furniture. More complex to unwrap are those pieces bought from the 1970s on, when the changing context of the crafts world shifted the meaning of studio furniture. Maloof's furniture itself remained remarkably consistent in terms of form and process, with only slight refinements such as scooped wooden

seats, hard-lined edges, and stronger router joints where the legs met the seat. However what changed dramatically were the motives and expectations of the audience. Some clients might have bought his work at the shop but more began to order it based on seeing images of the furniture in print or seeing it on view in gallery shows or museum exhibitions. Some of these new buyers even made pilgrimages to Maloof's shop in Alta Loma . . .

In the late 1970s and 1980s, the changing marketplace, especially the emergence of a body of clients who could be considered collectors, elevated the work of the early studio furnituremakers, especially Maloof. Purchasing fine furniture because it might make them feel happier or more fulfilled, allow them to express their discerning individuality, or demonstrate their social power, this new type of client often followed trends and asked for a well publicized Maloof form. Maloof also began to offer his work in curly maple, a showier wood that appealed to the new clients, in addition to his familiar black walnut. The fetishizing of the craft object thus triggered the emergence of a well-understood Maloof style, making a recognized fashion line out of what had once been a decidedly anti-fashion, styleless pursuit.

One particular Maloof form underscores this transformation—the rocker. Maloof made his first spindle back rocker in about 1960, then began to think about its market potential when President John Kennedy's doctor endorsed the rocking chair as a relaxing therapeutic seat for those with lower back pain. However, Maloof's rocker remained a slow seller for much of that decade. He sold only one in 1963 or 1964, and had only been able to sell five in 1969. He then included a walnut rocker in "Woodenworks" and another one for the "Please Be Seated" program at the Museum of Fine Arts, Boston, in 1975.

Figure 24 Maloof, *Rocking Chair*, 1993. Walnut and ebony. Photograph by Jonathan Pollock.

Even though the latter example was only on display for a short time and was never used for public gallery seating, it had an enormous impact on the demand for Maloof's work. By 1980, Maloof had built about one hundred rockers, then priced at $2,500, and had orders for another sixty. Joan Mondale, the wife of Vice-President Walter Mondale, bought a rocker for the Vice President's house in 1979, and in 1981 a rocker purchased at the auction celebrating the twenty-fifth anniversary of the American Craft Museum was donated to President Ronald Reagan. The rocker became a celebrity item, purchased by entertainment figures such as Anthony Quinn, Gene Kelly, and Jim Henson; serious craft collectors such George and Dorothy Saxe of San Francisco, Sydney and Frances Lewis of Richmond, and Peter and Daphne Farago of Providence; and three Presidents (Reagan, Carter, and Clinton).[8]

In the 1980s Maloof averaged between twenty-five and thirty rockers a year, with prices in 1989 that ranged from $8,000 for walnut, to $12,000 for maple to $15,000 for rosewood. In two important 1980s exhibitions—"California Woodworking" (1980), and "Craft Today: Poetry of the Physical" (1986)—Maloof placed a rocker. But the real popularizer was a 1986 article on him in *People* magazine, titled "King of the Rockers," that followed his receipt of the MacArthur Grant in 1985. In the 1990s, rockers comprised more than half of his yearly production.[9]

Thus over the past quarter century, the rocker has come to represent or stand in for Sam Maloof. Maloof's furniture began to symbolize the maker; purchase of a Maloof rocker translated to acceptance of the myth of the American craftsman and control over the creative producer, or Morris's master of "ornamental workmanship." Distinguished from

elements of mass culture by the attention to workmanship and detail implied in the concept of craftsmanship, the rocker embodied the optimism, warmth, and masculine individuality of its maker. Yet there is a certain contradiction at play. The rocker, more limited in function in comparison to Maloof's other seating and more space intensive in terms of its action, became the signature object for the woodworker who presented himself as a designer-craftsman committed to functional straightforward furniture.

One can thus look at a Maloof rocker as a commodity, a tool or implement that commercial capitalism uses to lull its audience into passivity and acceptance, and an aesthetic object purchased more for financial or social investment than for functional need or as a local purchase. As the rustic furniture maker Dan Mack commented: "Nobody needs a Sam Maloof chair. A Maloof chair has become an attractive cultural icon of the 1950s, of an elder craftsman, of a noble savage, of conspicuous consumption, of museum-endorsed taste."[10] The reference to the happy craftsman toiling in his own shop thus reinforces the desired belief that handcraft remains an economically viable option even as many inequitable and exploitative forms of production persist to provide desired consumable goods. In this light, Maloof's conscious marketing strategies can also be read as symptoms of the larger socio-economic structure of commercial capitalism.

The foregoing reexamination of the furniture made by Sam Maloof and other early studio furniture makers reveals the complicated nature of production and reception. These artisans have produced beautiful objects for more than a half century, but one needs to situate this work outside of aesthetic appreciation and a Morrisian paradigm. More recent theoretical scholarship can shed light on a variety of meanings embodied in this furniture and ascribed to it. This is our task, to recognize the need to go beyond William Morris in the analysis and interpretation of late twentieth-century American furniture.

NOTES

1. William Morris, *The Decorative Arts: Their Relation to Modern Life and Progress* (London: Ellis and White, 1878), pp. 4, 25. In *Art and Its Producers* (1888; London: Longmans & Co., 1901), p. 3, Morris used the term architectural arts to describe "the addition to all necessary articles of use of a certain portion of beauty and interest, which the user desires to have and the maker to make."

2. William Morris, "Art and Industry in the Fourteenth Century" (1890) in *The Collected Works of William Morris,* edited by May Morris, 24 vols. (1910–15; reprint edition; New York: Russell and Russell, 1966), 22: 386. See also William Morris "Architecture and History" (1884) in Ibid, 22: 296–317; and *William Morris on History*, edited by Nicholas Salmon (Sheffield, Eng.: Sheffield Academic Press, 1996).

3. William Morris "The Prospects of Architecture" (1881) in Morris, ed., *The Collected Works of William Morris*, 22: 119–51. William Morris, "Architecture and History" (1884) in Ibid., 22: 312. See also "Textile Fabrics" (1884) and "Art and Its Producers" (1888) in Ibid., 22: 270–95, 342–355.

4. Morris, *The Decorative Arts*, p. 4. Morris, "Architecture and History," in Morris, ed., *The Collected Works of William Morris*, 22: 308. For an insightful exploration of this myth, see Laural Thatcher Ulrich, *The Age of Homespun* (New York: Alfred A. Knopf, 2001). See also Mary Douglas, "American Craft and the Frontier Myth," *New Art Examiner* 21 (September 1993): 22–26.

5. Sherley Ashton, "Maloof: Designer, Craftsman of Furniture," *Craft Horizons* 14/3 (May/June 1954), p. 15–19: 15, 18.

6. Sam Maloof, *Sam Maloof, Woodworker* (New York: Kodansha, 1983), p. 23.

7. Glenn Loney, "Sam Maloof," *Craft Horizons* 31, no. 4 (August 1971): 16–9, 70: 17. On pricing, see Maloof, *Sam Maloof, Woodworker*, esp. 31, 39, 41, and 45.

8. Rick Mastelli, "Sam Maloof," *Fine Woodworking* 25 (Nov./Dec. 1980), p. 48–55: 52; and Jeremy Adamson, *The Furniture of Sam Maloof* (New York: W. W. Norton, 2001), p. 114, 120–3, 175, 185–89, 203, and 216–7.

9. *California Woodworking* (Oakland, Ca.: Oakland Museum of Art, 1981); Paul Smith and Edward Lucie-Smith, *Craft Today: Poetry of the Physical* (New York: American Craft Museum, 1986). Barbara Manning, "Master Craftsman Sam Maloof is King of the Rockers Because He Rocks by the Seat of His Pants," *People Magazine*, January 6, 1986.

10. Dan Mack, "Thoughts on Chairs and Change and Creativity's Eternal Vitality," *Woodshop News* 12, no. 11 (October 1998), p. 12.

FURTHER READING

Jeremy Adamson, *The Furniture of Sam Maloof* (New York: W. W. Norton, 2001).

Edward S. Cooke, Jr., et al., *The Maker's Hand: American Studio Furniture* (Boston: MFA Publications, 2003).

Edward S. Cooke, Jr. *New American Furniture: The Second Generation of Studio Furnituremakers* (Boston: MFA Publications, 1989)

SECTION 4

THE PERSISTENCE OF CRAFT IN THE AGE OF MASS PRODUCTION

SECTION INTRODUCTION

The historical arc that began in Section 2 with narratives about industrialization and continued in Section 3 with countervailing idealistic or utopian pronouncements concludes here with a set of texts which discuss craft's continuing place within modern economies. Despite the many obituaries issued on its behalf, and the ever-proliferating forces of mass manufacture, modern craft is not just a symbolic matter. Its productive importance persists in many circumstances. In the heritage industry, on building sites, in the context of 'folk' production, in luxury markets, and even in the factory itself, craftsmanship is alive and well, if we know where to look.

Once again, it is a question of decline or displacement. Are ongoing craft practices best seen as the last vestiges of grand artisanal traditions, as compromised hybrids or rather as reimaginings of those traditions which expand their possibilities and relevance?[1] Perhaps it's a bit of both, and perhaps too, the answer varies from place to place. It is striking that the geography of craft's persistence is so very broad; this section includes examples from France, Italy, Russia, Africa, India and China and could have been even more wide-ranging.[2] The temptation is to think of craft economies as marking a receding frontier: quite simply, it is found wherever industrialism has yet to assert its forces completely, due to lack of economic development. But this fails to take into account the complexity of most of these cases, which see craft transforming alongside and in response to the effects of industrialization.[3] We might also choose to side with the author Arindam Dutta, who has recently argued: 'As a figure of difference, the artisan does not disappear with the advent of industrialism. Rather, it appears within it. The imago of the Oriental artisan is born and bred in the anthropological chrysalis of industrial capitalism . . . Had colonial officials found no artisanry in the vast territories under imperial control, it would have been necessary *to invent some*.'[4]

This section of *The Craft Reader* might be considered a generalized test of Dutta's thesis. Is the presence of craft during and after industrialism merely a matter of persistence? Or is 'traditional' craft, on the contrary, a fiction that emerges within modern ideology itself? A first set of answers is suggested by two case studies drawn from the world of luxury goods manufacture. The Italian shoemaker Salvatore Ferragamo and the Parisian chocolatiers described by anthropologist Susan Terrio both exhibit willful resistance to economic transformation. But insisting that 'old ways are best' is not enough. As both extracts show, it is only through the conscious manipulation of production and marketing that supposedly traditional standards of workmanship can be maintained. Ferragamo's story is particularly entrancing. Unable to sway traditional Italian cobblers to fabricate his modish designs, but also unwilling to submit to the mediocrity of machine manufacture, he ends up creating his own bespoke assembly line of specialist child laborers. Apparently there is always a price to pay for quality.

BRICOLAGE, THE AD HOC, AND THE POSTMODERN

David Doris's account of the fate of 'textile casualties' in Africa is a very different example of craft persisting in the face of economic change. He tracks the path of off-casts from Disney's factories in America, as they are bundled up and sent in bales to Nigeria, among other places. There they are refashioned into cloths of a distinctly local sensibility—a prime example of the creative recycling that Suzanne Seriff has identified as 'a kind of recycling in which yesterday's newspapers are transformed by hand into tin trunk liners; empty food cans become kerosene lanterns; and old tires are refashioned into spouted water vessels or bracelets for bodily adornment'.[5] This sort of adaptive response to large-scale economic forces, over which the individual craftsperson has no control, was celebrated in Nathan Silver and Charles Jencks's 1972 book *Adhocism*, a survey of informal practices that played on the old idea that necessity is the mother of invention.[6] While Silver and Jencks drew inspiration from situations like the one Doris describes, where poverty and geography combine to place artisans in a reactive relation to mass production, they also advocated improvisation to those inhabiting fully modernized contexts. If practiced widely, they felt, this would make for a more sustainable and diverse environment—a sentiment that was entirely in tune with the times, as other publications such as *The Whole Earth Catalog* (a sort of Sears, Roebuck catalogue for the environmentally hip whose motto was 'access to tools') and Victor Papanek's *Design for the Real World* attest.[7]

The rule-averse collage craft that adhocism entails seems in retrospect to have anticipated the strategies of Postmodernism—not least because Jencks himself would go on to define and popularize that term, especially in the context of architecture. Postmodern design and craft emphasizes the fragment over the whole and creativity over system. This predilection for the rough and ready is often interpreted as a kind of anticraft, but in fact most Postmodern design of the 1970s and 80s was made by hand. From Frank Gehry's architectural experiments with reused chain link fencing and salvaged lumber, to ceramics by the British potter Carol McNicoll, to the vividly clashing, strident furniture of Italian designers like Alessandro Mendini, do-it-yourself bricolage was the dominant mode.[8] (See Andrea Branzi's text, in Section 6, for more on this.) German designers in the 1980s spoke of having made their designs using rudimentary equipment *in den Hinterhof*—roughly, 'in the space out back'—just as Silver and Jencks had recommended. Even graphic designers working with computers for the first time, such as Neville Brody for *The Face* and David Carson for *Raygun*, drew on a handmade ad hoc aesthetic.

Although adhocism and Postmodernism had their playful side, both were meant to offer a genuine critique of mass-produced homogeneity—not too different from punk music, that other great expression of the bricoleur's aesthetic. Nonetheless, the conceits of Euro-American design rarely stir one's political conscience as much as front-line reporting from situations in which the persistence of craft is a matter of enforced (rather than willingly assumed) necessity. The debate over sweatshop labor in the so-called 'third world' is an unavoidable context for such reports. The question is whether sweatshops are a permanent system of exploitation, as antiglobalism activists contend, or a necessary stage on the road to development, as many economists argue.[9] Rather than adduce texts which address this dispute directly, this part of the book draws on the experiences of those who are responding to unequal economic relations through some novel form of craft. In addition to Doris's work on African textiles, there are two examples drawn from the outskirts of industry (ball bearing production in Russia, auto maintenance in Africa), as well as the astonishing case of a Chinese town where the main employment is the 'mass production' of paintings by hand. In none of these situations is critique an option—people are too busy maintaining their livelihoods for that—and what is most striking is the extraordinary lengths that people will go to in order to maintain established value systems in the face of economic change. We are at the other end of the economic spectrum here from the luxury goods manufacture with which the section begins, but it is a case of extremes meeting: as craft comes to grips with modernity, maintaining the old ways requires its own sort of innovation.

FEEDBACK: CRAFT AND CYBERNETICS

Another, and again a very different aspect of craft's persistence in the late twentieth century is in the field of advanced technology. What happens when the human body is no longer considered to be the sole instrument by which craft processes are executed? This may initially seem perverse. Surely once a process is divorced from the hand, it becomes something else—technology rather than craft? That intuition has some merit, but runs into immediate problems. As David Pye (see Section 5) pointed out some years back, 'Some things actually can be made without tools it is true, but the definition is going to be rather exclusive for it will take in baskets and coiled pottery, and that is about all!'[10] Craft is almost always a matter of triangulation between maker, tool and material (after all, even the naked hand might be considered a tool), and there is no obvious reason why any particular type of tool should be considered ineligible for this relation.

Here things get tricky. One might well argue that a bulldozer, because of its size, or a computer-driven laser cutter, because it is automated, or possibly even a big wooden club, because it is so difficult to use accurately, cannot possibly be considered craft tools. In each of these cases, some physical property of the tool prevents the user from applying fine motor control—and this, arguably, is usually what we expect from a craft process. (Pye's concepts of the workmanship of risk and the workmanship of certainty, as we will see, offer a good way of thinking through this distinction.) But this is a far cry from saying that advanced technology is incompatible with craft—for as we all know, technology can dramatically enhance our ability to perform fine motor operations rather than inhibit it, as in the case of medical surgery. How do we draw the lines?

Enter the figure of Norbert Wiener, the founder of the modern theory of cybernetics. Wiener's signature idea was that systems operate according to a principle of feedback, by which some aspect of the system's output reenters the system in such a way that it regulates future results. (This is often called a 'feedback loop.') The distinctiveness of this idea is that it can be applied with equal relevance to systems made up of motions, people, machines or even abstract economic values. The principle is the same, that is, whether used to analyze the operation of a steam engine or a stock market. Feedback is obviously rife in craft work. Many tools, like hand planes, are self-correcting to a degree, and more importantly, craftspeople operate by calibrating the motions of their work in direct response to the work that was just performed. Because it is equally applicable to a person's decision making, the performance of a manual tool, or the operation of a very complex system like a computer, cybernetic theory is a crucial tool for thinking about the dispersion of craft across technological systems of various kinds.

The readings selected here, beginning with Wiener's own extrapolation of cybernetics into a commentary about the ethics of work, follow the idea of feedback from the workbench out into the society as a whole. In each case, configurations of people and machines are seen not as inherently oppositional, but rather as an interwoven field in which the organic and the artificial combine. Malcolm MacCullough draws our attention to the specific routines by which craft is performed via a computer, while Rafael Cardoso discusses 'open source' design and its implications for craft theory. And finally, in the spirited (and partly satirical) manifesto for a fictional guild called the Digital Artisans, we arrive at the most counterintuitive claim of all: 'Skilled workers are best able to assert their autonomy precisely within the most technologically advanced industries.' In this declaration, the question seems settled in favor of displacement: a future for craft as contentious and uncertain as its past.

NOTES

1. Annie E. Coombes, 'The Recalcitrant Object: Culture Contact and the Question of Hybridity', in Francis Barker et al., *Colonial Discourse/Postcolonial Theory* (Manchester University Press, 1994), pp. 89–114.

2. For further studies of craft persistence see Rob Allen, 'The Myth of a Redundant Craft: Potters in Northern Nigeria', *The Journal of Modern African Studies* 21/1 (March 1983), pp. 159–66; Scott Cook, 'B. Traven and the Paradox of Artisanal Production in Capitalism', Mexican Studies 11/1 (Winter 1995), pp. 75–111; Eugene Cooper, 'Craft Development: Socialist and Capitalist', *The China Quarterly* 83 (Sept. 1980), pp. 447–60; Donald R. DeGlopper, 'Artisan Work and Life in Taiwan', *Modern China* 5/3 (July 1979), pp. 283–315; Deema Kaneff, *Who Owns the Past? The Politics of Time in a 'Model' Bulgarian Village* (Oxford: Berghahn Books, 2004); Barbro Klein, 'The Moral Content of Tradition: Homecraft, Ethnology, and Swedish Life in the Twentieth Century', *Western Folklore* 59/2 (Spring 2000), pp. 171–95; Brian Moeran, 'Materials, Skills, and Cultural Resources: Onta Folk Pottery Revisited', *Journal of Modern Craft* 1/1 (March 2008), pp. 35–54.

3. An influential thinker on the dynamic interrelation of modern and traditional economies is Karl Polanyi. See his *The Great Transformation: The Political and Economic Origins of Our Time* (Boston: Beacon Press, 1944); and George Dalton, ed., *Primitive, Archaic and Modern Economies: Essays by Karl Polanyi* (New York: Doubleday, 1968). For a study of the influence of informal craft activity on a mainstream technology industry see Yuzo Takahashi, 'A Network of Tinkerers: The Advent of the Radio and Television Receiver Industry in Japan', *Technology and Culture* 41/3 (2000), pp. 460–84.

4. Arindam Dutta, *The Bureaucracy of Beauty: Design in the Age of its Global Reproducibility* (New York: Routledge, 2007), pp. 31, 77.

5. Suzanne Seriff, 'Recycled, Re-Seen: Folk Art from the Global Scrap Heap', *African Arts* 29/4 (Autumn 1996), pp. 42–94: 46.

6. Charles Jencks and Nathan Silver, *Adhocism* (London: Secker and Warburg, 1972).

7. *The Whole Earth Catalogue* (Santa Cruz, CA: Portola Institute, 1969); Victor Papanek, *Design for the Real World* (New York: Thames and Hudson, 1972); Victor Papanek and Jim Hennessy, *Nomadic Furniture* (New York: Pantheon, 1973).

8. J. Fiona Regan, *Frank Gehry: Architect* (New York: Guggenheim Museum Publications, 2001); Mark DelVecchio, *Postmodern Ceramics* (New York: Thames and Hudson, 2001); Peter Weiss, *Alessandro Mendini: Design and Architecture* (Milan : Electaarchitecture, 2001); Barbara Radice, *Memphis: Research, Experiences, Results, Failures and Successes of New Design* (New York: Rizzoli, 1984).

9. For opposing statements in this argument, see Andrew Ross, *Low Pay, High Profile: The Global Push for Fair Labor* (London: New Press, 2004); and Paul Krugman, 'In Praise of Cheap Labor: Bad Jobs at Bad Wages are Better Than No Jobs at All', *Slate* (20 March 1997). See also Louise Amoore, *The Global Resistance Reader* (London: Routledge, 2005).

10. David Pye, *The Nature and Art of Workmanship* (Cambridge: Cambridge University Press, 1968), p. 9.

33

SHOEMAKER OF DREAMS

Salvatore Ferragamo

In the fashion industry, the handmade retains undoubted symbolic authority. According to the logic of couture, the unique skirt, hat or shoe is held to influence the course of mass-produced garments, in a sort of cultural trickle-down effect. The 'trendsetting' function of high fashion is inextricably bound to its craft manufacture, as it is in the specifics of cut and material that its special value is located. Of course, this rather alchemical ideal bears only a distant relationship to the realities of clothing production. The public image of a designer rarely reflects their actual business model. An early example of this kind of gamesmanship is Salvatore Ferragamo's breathlessly self-promotional autobiography, composed with the assistance of professional ghostwriters Douglas and Elizabeth Warner. In the following excerpt we see Ferragamo, who is in the process of becoming a luxury brand name through his bespoke work for Hollywood starlets, attempting to bring his shoes to a wider market. Initially repulsed by his encounter with American-style mass production, he returns to his artisanal roots in Italy, only to come into conflict with the craftsmen there. ('Not two pairs of shoes were alike', he fumes. 'One shoemaker would make them his way, the next in his way.') Eventually he finds a solution: he replicates in Florence the assembly-line methods that had so horrified him in America, seemingly without realizing that he has done so. Though it makes for delightful reading, there is an anticipation here of the third-world craft sweatshops that have been such a scandal for the fashion industry in more recent years. Ferragamo sweeps his own reservations aside, partly by consistently invoking the principle of 'comfort'—as if fine workmanship were primarily a pragmatic requirement rather than a way to sell fabulously expensive shoes.

Salvatore Ferragamo, excerpts from *Shoemaker of Dreams* (London: George G. Harrap, 1957).

So was I introduced to the United States. Two days later I was introduced to the world of miracles. Joseph [Covelli, Ferragamo's brother-in-law] arranged with one of the bosses of the Queen Quality Shoe Manufacturing Company—then, as now, one of the biggest and best shoe companies on the east coast of the United States—for me to be shown through the factory. When I had inspected the set-up, Joseph told me excitedly (he was anxious to keep me in Boston and eager to put me on the path to a good job) I could choose which section of the work I liked best and begin employment almost at once.

I went into the factory. Yes, it was exactly as [my brother] Alfonso had described it. Everything I could do the machines did in the twinkling of an eye. Yet I was not impressed—I was appalled. This was not shoemaking. This was an inferno, a bedlam of rattles and clatters and whizzing machines and hurrying, scurrying people. I stood dazed; I walked about dazed, watching the thousands of pieces of shoes going in at one end of the assembly line and pouring out at the other on endless belts, rows upon rows of finished shoes, hundreds of them, thousands of them, even—or so it seemed after an hour or so—millions of them. They were good shoes according to the standards of machine-made shoes; yet to me they were heavy, clumsy, and brutal, not to be compared even with the shoes I had seen in Naples and far, far below the standard I had set for myself. I stared and wandered, miserably. How could I choose a job in this labyrinth? This was not my home. I could not be happy here. I was a shoemaker, not a finisher or a trimmer or an edger or any other of the piecemeal jobs which went into these mass-produced shoes. There was no craftsmanship here, not an ounce.

Lunchtime came at last, and Joseph hurried over to me from his bench, smiling, eager to know what I thought of the factory. My reply took his smile away so quickly that I felt sorry afterwards that I had not been more tactful. I said vehemently: "No, no, no! I can't work here! I *won't* work here! This is not shoemaking. This is not craftsmanship. I am never going to have anything to do with machine-made shoes, never!"

[. . .]

Long before I moved to Hollywood I knew that to be a success there I would have to make special arrangements for shoemaking in quantity. I could not maintain an expensive establishment on the mere output of Taylor, Dietrich [English and German shoemakers that worked with Ferragamo], myself, and a few extra hands on the side. An organization concentrating on the hand-made article could do no more than satisfy a meagre number of orders—an absurd situation when I was the sole possessor of a discovery that would revolutionize shoemaking. Therefore I made up my mind from the beginning that I must enlist the aid of machines to cope with the larger section of my output. I was still opposed to the use of machines in my hand-made shoes—it is still true, and will always be true as long as I am in control of the work, that what are to-day known as "Ferragamo Originals" have never seen even the simplest machine—but that was no reason, I argued to myself in Santa Barbara, why the machine should not be brought in to accomplish the manufacture of shoes to my designs and patterns and, more particularly and essentially, to my special lasts, for stock sizes which could be retailed in the normal way. My discovery, I considered, would ensure that even machine-made shoes would not hurt the feet. They would not possess the custom-tooled perfection of the hand-made shoe fitted to the personal order—no shoe in the world can compare to that quality—but no single shoemaker can make all the shoes in the world; and all the shoemakers in the world, making by hand to my designs and lasts, could not make all the shoes that the world consumes as it walks. The custom-made shoe remains, and by the necessity of things must always remain, a narrow market which is the exclusive prerogative of those who can afford the price. This was not good enough for me. I wanted to put my shoes on the feet of as many people as possible; therefore I must enlist the aid of machines.

I sought out manufacturers throughout the United States—in Lynn, Massachusetts, in Brooklyn and New York City, in Philadelphia and Chicago—who would agree to manufacture shoes according to my specially prepared lasts and to my patterns. The arrangement called for me to make up the lasts in the appropriate range of fittings which my gathering experience of feet and my growing knowledge of the potential of my discovery had taught me was necessary for most accurate fittings 'off the peg,' and the makers would turn them out in the quantities I required, to my styles, patterns, and designs. The Hollywood shop I used not primarily as a salon for the fitting of custom-made shoes, but as an ordinary retail store. Little shoe-making was done on the premises; people came in and bought their shoes in any retail shop in the world.

Up to a point the plan worked extremely well. At the peak of production my manufacturers were turning out several hundreds of pairs a week, where I and my work-people could turn out only a handful. They sold readily because my system of fitting proved as workable in machine manufacture as it did my hand. The extras and small-part players who could not afford hand-made shoes could pay the price—though it was somewhat higher than that charged by the normal bootshop, because of the extra costs involved in the manufacture of limited lines exclusively for me—for footwear which they had discovered by wearing my shoes in the films gave them so much greater comfort than any others they could buy. They came, they bought, they raved about my machine-made shoes. Better still, the big stores in the major cities of the United States began to order my models in quantity.

Yet if it worked so far it did not work far enough for me. I was not completely happy with the quality right from the start, and, as time passed and my methods of last-construction and fittings were refined and improved with my ever-increasing knowledge of the structure and variations in feet, I became increasingly unhappy. What did it matter that my customers were content? I was not. I would send out my ideas, my instructions, my lasts, and my patterns and back the lines would come, an ungodly sight! To my mind they did not fit at all. The finish was poor; vamps were the wrong height, the height of the heels was incorrect, the finish on the back of the heels had been overlooked, the shanks were not according to the last because the manufacturers had not bothered to make special ones. There were many occasions when I was so disheartened that I had not the heart to charge the price I should have asked. To me they were a disgrace.

Yet how could I escape the impasse? If I ceased machine production I would be driven back to my rut as a maker of custom-made shoes with a strictly limited clientele. I could not go into every factory and personally supervise the manufacture of each new line. All I could do was send angry letters complaining that the work was not up to standard—and even in this the strength of my language was limited because some of the makers were half ready to cease making shoes to these new-fangled principles and get back to the normal production methods which were so much easier and less bothersome.

Yet my conscience and my pride would not allow me to continue selling these shoddy shoes indefinitely. They offended my every instinct, they offended every principle on which my standards and reputation were built.

[. . .]

The answer came one day, as the answer to a stalemate usually comes, in a wild simplicity. If I could maintain my output only by using the methods of mass production, and if the only way I could maintain my standards and my reputation was by the manufacture of hand-made shoes, why not a system of making hand-made shoes by mass production?

At first glance the idea seemed ridiculous, yet I asked myself: was it truly as absurd as it sounded? Was it really a paradox? I decided it was not. One man could make only so many shoes in a week; but multiply the number of shoemakers and you multiply the number of shoes. If enough shoemakers could be found and trained in my new principles I could provide them with patterns, models, and designs, and they would only have to execute the orders. And where else would I find the shoemakers but in Italy? My thoughts flew back to my childhood, to the days when I was wandering Naples and working in all the shoe shops among the makers of fine hand-made shoes. In Italy there was an inexhaustible supply of shoemakers; in Italy, the country of fine craftsmen, there would be scores of master shoemakers only too glad to make shoes to my instructions. It would be profitable, it would be secure, it would be magnificent!

I pondered the scheme, looking at it from all angles. Thee seemed to be no snags except finance. My personal resources were not sufficient to establish the organization as I imagined it, but the fact did not worry me. I knew that this was what I now had to do, and I would do it.

I began to cast around for sources of additional capital. My first thought was of my brothers, Secondino, Girolamo, and Alfonso, who were now helping me again. Their repair shop at Hermosa Beach was a doomed venture from the first . . . After two years they had closed up and had entered again into work with me, though not into partnership. Now I thought that if they would join in with my Italian venture we could resume our old happy association.

I went to them and outlined my scheme.

"In Italy," I told them eagerly, "I can find as many shoemakers as I need. I will do with them as I now do with the machines: supply the lasts, the designs, the patterns, and the styles, and they can ship them over to the States. We can use the Hollywood shop as a centre for distribution and in time we will open a chain of retail outlets throughout America, either with shops of our own or through the leading stores in the great cities. We might even establish our own shops in England and France—anywhere in the world. It will be terrific! It will be unique!"

I spoke of the economic problems: "Wages in Italy are lower than in America, which means that even with shipping costs we can produce as cheaply, if not cheaper, than we do now in the States. In any case, hand-made shoes pay a smaller import tariff than machine-made shoes."

I talked of my own part: "I will go over to Italy and arrange with the leading shoemakers over there—the ones with the best-equipped places and the finest workmen in their employ—to take over the manufacture, and then I will return to America to handle the distribution. I shall give up making custom shoes so that my mind and my time will be free to work on new designs, more wonderful designs, magnificent ideas which I cannot—I dare not—attempt to give to the machines."

It was no use. Just as in Santa Barbara they refused to listen. Their answer was the answer of several years before: "Better an egg to-day than a hen to-morrow." Why did I want to rush off with another crazy idea? Why didn't I stop where I was? Who was worrying about the shoes, anyway? Not your customers, look at your customers, more and more every day! Listen to what they say about your shoes, they cannot praise them

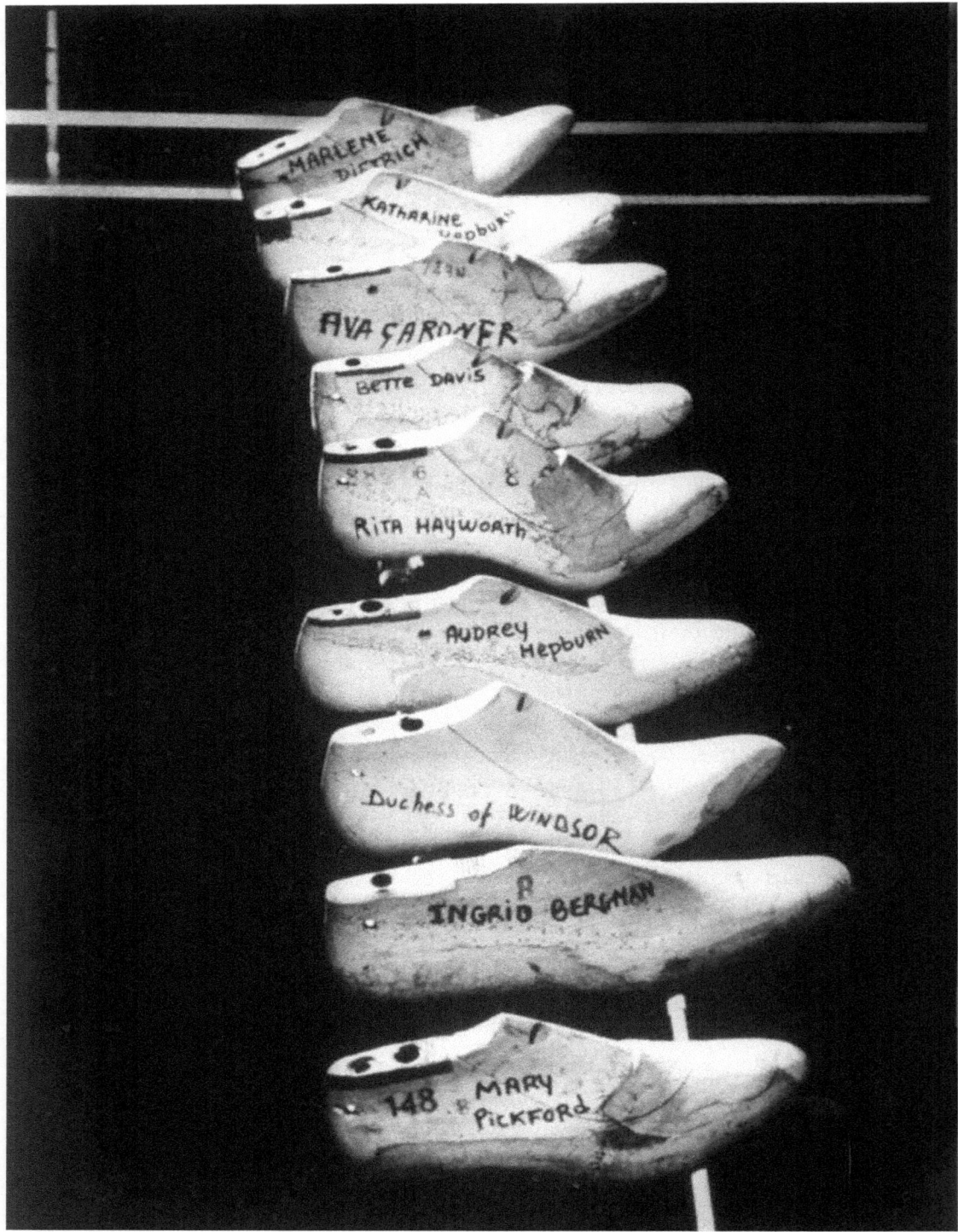

Figure 25 Wooden shoe forms of famous women at shoemaker Ferragamo.

highly enough. Besides, if you want to make hand-made shoes sure you can make hundreds of pairs if you will only be sensible and introduce a few elementary machines into the manufacture: a sewing machine, a perforating machine, a skiving machines. The sewing machine, for instance, does the work better, finer, and more accurately than the finest shoemaker in the world and in a fraction of the time. Why then are you so stubborn that you will not use any machines at all in your hand-made work?

Why indeed? I could only answer that, just as the ear is hurt if a concert pianist misses a note, so the shoe which is not entirely hand-caressed will, even if only in a small degree, hurt the feet.

They did not appreciate my point of view. I suppose that they considered that I was splitting futile hairs. I was disappointed but I was obsessed with my scheme. I felt that it was work I was called to do, and the call was stronger than my affection for my brothers. Once again, and this time finally, we parted in the way of business, and I took my thoughts and schemes among my friends in Hollywood and Los Angeles, talking, persuading, arguing. Gradually I aroused interest and at last, early in 1927, a corporation was formed, and a public announcement made that Salvatore Ferragamo, the shoemaker, was returning temporarily to Italy to establish a new production method which would bring Ferragamo shoes, hand-made in quantity by Italian master-craftsmen, to the feet of American women. The effect of the announcement was heartening and even astonishing. Press, radio, and magazines featured the idea lavishly. Orders poured in. Every woman of fashion, every sophisticated star of Hollywood, seemed determined to be the first to buy Ferragamo's Italian shoes. They came in a stream, full of

vivid ideas. They all wanted shoes that would be artistic, beautiful, rich, unthinkable.

I tied up the last loose ends in the States, making arrangements for the manufacturers of the machine shoes to continue during the short time that I expected to be away, and then I entrained for the east. On the way across the continent I dropped in on friends I had made during the past few years with my lines of machine-made shoes: the buyers and owners of big stores in the great cities. My progress was a triumphal procession. Every buyer wanted to see my new shoes and every buyer wanted to see them first and take them exclusively. In New York, for instance, George Miller, head of I. Miller and Sons' chain of shoe stores, put the matter briefly and succinctly. "When you get back, Salvatore," he said, "I want to see your new styles first. We have a lot of shops of beautiful shoes, and yours will fit into them to perfection. Your future will be assured out of the presentation we shall make of Ferragamo creations, made in Italy, through the Miller organization."

As I sailed away I felt that I was travelling in a legend. If I had possessed the slightest doubts about my success I might have thought that it was too good to be true.

The Italy I had not seen for thirteen years welcomed me with another great blast of publicity on radio, in the Press, and in periodicals. After two days' rest in Bonito to enjoy the homecoming to a mother who had feared that she would never see me again, strolling the streets renewing old acquaintances and looking with affectionate, nostalgic interest at the room in which I had opened my first shop—the village houses all looked like tiny boxes after the great buildings of America—I travelled to Naples, where I had dreamed of establishing my Italian headquarters, in search of shoemakers who would execute my grand plan.

My reception in Naples was like a blow in the face. Naples would have nothing to do with me. One by one the shoemakers of the city turned their back on the scheme. Many would not even listen. Those who consented to listen rejected me outright. It was impossible, it was wild, it was crazy. My lasts would not fit, my plan would not work. They made it clear that in their eyes I was nothing more than an Italo-American go-getter, a hustler, an expatriate with the usual quota of high-pressure ideas from across the Atlantic.

Hurt and upset, but refusing to be disheartened, I left Naples and went south, consoling myself with the thought that perhaps it was natural that the shoemakers of Naples would resent the approaches of the small boy from the tiny local village who had achieved such a reputation that he could command extensive publicity in the Italian Press. Things would be better in the big cities of Southern Italy, where there would be less antagonism to the 'local boy made good.' I was wrong. Southern Italy met me with the same blank refusals, the same indifference, the same scepticism, the same curled, disapproving lips, the same words: "Crazy, impossible, it couldn't work."

I left the south of Italy and went northward. In the more industrialized districts of central and northern Italy, I told myself, shoemakers would be more progressive, less hidebound. I went to Rome and when Rome failed me I went still further north, to Verona and Milan and Turin, to Venice and Padua. Everywhere the story was the same.

I was baffled and angry. The months and the money were slipping away, and these people would not listen to me. Why would they not listen? I was offering them comfortable business with pleasant profits and an assured market; the way they were treating me you would think I was trying to sell them a gold brick. I pleaded in vain. I used every argument I could think of—in vain. I was in despair. I had been in Italy for far too long and nothing had been accomplished, nothing. What could I do? I could not go back to America and confess failure. I could not return and cancel all those orders, saying, "I'm sorry, but it can't be done." I could not disappoint the hundreds of women who were waiting for the new designs that Ferragamo had promised. Besides, I knew it *could* be done. I knew it in my bones. I knew it in my heart.

I came at last to Florence, knowing that soon I should have to make a decision. I wandered the city. I knew no one and none knew me, but as I strolled through the soft summer night and felt the impact of its great beauty I thought that perhaps in Florence I might realize my dream. I wandered round the great cathedral and the slim, elegant Campanile; I peered through the dimness at the Gates of Paradise on the Baptistry and I stood, quiet and alone, in the Piazza Signoria, in the shadow of the Palazzo Vecchio, before which all the most stirring events in Florentine history have been enacted. I thought: Surely in this beautiful city, with its centuries of wealth in art and its long traditions of noble leatherwork, I can find the answer to my problem.

Next day I toured the city again, this time on business, meeting the master-craftsmen and their artisans. The answer was the same: it was impossible for one man to organize the workmen into the sort of production I needed. Nevertheless, I was not disappointed. It seemed to me as I talked with the workmen that they, if not their masters, were more interested than the men of any city I had yet visited.

At the end of the day I made up my mind. My original plan was impossible in one detail: there was no one to organize the output for me—therefore I must amend my scheme and do it myself. I must establish my own factory, with artisans under my own control, and leave the distribution arrangements in the hands of my colleagues in America.

To decide was to act. Within a few days I had found premises in the Via Mannelli large enough to house the number of men I needed. I stocked it with equipment and materials. I scoured Italy for the best shoemakers in the country, offering them better wages than they could obtain elsewhere. Soon I had nearly sixty men housed under one roof. I showed them my methods of shoemaking and refused to be dismayed by their cautious approach to ideas which were to them revolutionary. I insisted on my own processes, of which they were ignorant. Most important of all, out of the excitement and enthusiasm that my dream was at last coming true, I designed a series of shoes more beautiful, more astonishing, more extraordinary, than any I had ever designed in my life; shoes different from any that had ever been seen before; designs which my shoemakers had never even thought of doing.

Now I had to work quickly. Angry letters were reaching me from California, demanding to know then the promised shoes would arrive and warning me of financial difficulties. I needed no warnings. Money was short and time was running long.

When at last I sailed for New York I carried in my baggage the sum total of all my shoe production during those difficult months: eighteen shoes, all singles—there was not one pair among them and not one shoe looked in the least like any other. Yet I was not dismayed. In Florence I had left an organization of master craftsmen, their wages paid, doing the work I wanted. In my hands were the designs that would capture America. In a few days I

would show them to George Miller and set in motion the practical side of the business. I ignored all the difficulties. My workmen would learn my methods, my shoes would sell, my finances would improve. I needed only time, patience, and the assistance of my associates in California to make a tremendous success. Out of my new organization and the power of my new designs I would win all three.

[. . .]

I returned to Florence to find not a business but a shambles. Although my shoemakers had been fully paid during my absence, they had played ducks and drakes with my organization. Many had left. The others—fewer than thirty in number now—had done no work or had done so little, and done it so badly, that it was useless. As soon as I returned there began a series of arguments and yet more arguments until I felt sick. Many of my shoemakers were fifty and sixty years old and had been making shoes all their lives. Now they refused to make the shoes the way I wanted them made.

"We've always made shoes in our own way," they said. "Why do you want to change it? Your method is no good." They pointed to my lasts and said: "These are impossible. You can't make shoes on these."

My orders called for the manufacture of lines of thirty-six or fifty pairs of shoes, all identical. It was impossible to produce them. Not two pairs of shoes were alike. One shoemaker would make them his way, the next in *his* way. Despite all my efforts, despite my every moment of surveillance, despite my repeated and increasingly heated instructions, I could make no headway. The bolder ones walked out, saying angrily: "Nobody has ever made remarks like that to me! I know what I am doing! I have been a shoemaker for so many years. I taught such-and-such to make shoes, so why should you try to teach *me* how to do it? If you like the way I work, all right. If you don't like it I'll go."

They went. My working force dwindled steadily. My customers pestered me for delivery of their orders. Their representatives in Florence called day after day. They did not cancel—they wanted my shoes too much—but I just could not deliver. I could not tolerate the workmanship, and even if I had been willing to lower my standards the stores in America would never have accepted lines of shoes in which every pair, supposed to be identical, was different. Shipments were microscopic, and California started to worry me again, demanding shoes, shoes, and still more shoes. Money ran short, wasted by workmen who would not work as I wanted for the wages I was paying.

The position became intolerable. The Gordian knot had to be cut—and so I cut it. One day I dismissed the remaining workmen, looked round the empty factory with its litter of lasts and tools and leather, all idle, and went out in search of new men. This time I was determined to have nothing to do with any shoemaker who had learned the craft. I would have no more prejudice in my business. I would have people who knew and believed in my work.

I knew of a number of good, clever boys in Florence who were learning to be shoemakers but whose technical knowledge was incomplete. I knew that they could not make one single pair of shoes between them. Nevertheless, these were the lads I sought out. I established a school for them in Florence, and advertised for other boys who wished to learn the trade, offering to pay them while they learned.

The desperate plan worked. Within a few days I had gathered around me the nucleus of a working force in the persons of thirty eager youths. In another week the first shoes began to trickle off the assembly line—and it *was* an assembly line. I knew that inexperienced youths could never be taught to make a line of identical shoes, so I taught each boy how to do one job perfectly. When the job was finished he handed the shoes to the next boy for the next operation.

Because they were young and knew little, and because I was paying them well, they stuck to their tasks splendidly. Within a month of starting my school I had turned out a few complete lines, and shipments—minute shipments but they were a beginning—were going abroad, the first Italian shoes ever to be exported. There were, of course, slight mistakes in some of these first batches; there was workmanship which could have been bettered, though the errors were so slight that only I would have noticed them. Nevertheless, the shoes that went from my factory were perfect; the work that was not right I kept. But, best of all, I knew that the worst hurdle had been surmounted. All that remained was to keep on improving, adding to my labour force, expanding my output, and all would be well.

34

CRAFTING *GRAND CRU* CHOCOLATES IN CONTEMPORARY FRANCE

Susan J. Terrio

The ephemeral arts of cuisine are an understudied part of craft history and theory. Food writing often invokes craft as a cipher for quality; think of the term 'artisanal cheese', for example, or the way that distillery tours present us the lovingly painstaking process of making whiskey. Just as with 'indigenous' craft objects, authenticity and regional specificity are crucial to the ways that food is made, distributed and sold. Yet few studies have brought the tools of the craft historian to this subject. One notable exception is the following study by the Georgetown University anthropologist Susan J. Terrio. The essay focuses on France, a nation whose identity and international reputation is inseparable from its haute cuisine; and on chocolate, famous for inflaming the passions of makers and eaters alike. Given these high expectations, it is perhaps unsurprising that French confectioners are among the most intensely scrutinized of food professionals. Like all artisans, however, they are susceptible to the pressures of economic competition. Terrio describes the way that the small coterie of grand cru *chocolatiers have responded to unwelcome competition (principally from Belgium) by promoting their own 'premodern' fine craftsmanship. Many of the same dynamics that one would expect with reference to traditional pottery or textiles surface here: connoisseurship, nationalism and debates about standards and innovation. What is unusual, perhaps, is that this dark, sweet stuff becomes a test of cultural capital and 'powerfully illustrates how taste is produced and reproduced'. In this respect, her study serves as a model not only for the examination of other food crafts, but also of the complex role that craft continues to play in luxury trades in general.*

Susan J. Terrio, 'Crafting *Grand Cru* Chocolates in Contemporary France', *American Anthropologist* 98/1 (March 1996). Excerpted; some expository notes removed.

C'est un magasin où le chocolat règne en maître, traité par un maître. C'est du travail cent pour cent artisanal au sens "artist" du terme, qui sait firer de la sublime fève d'Amerique la substantifique splendeur.[1]

—Le guide des croqueurs de chocolat, 1988

I noted the display of Parisian master chocolatier Michel Chaudun in the window of his seventh arrondissement confectionery boutique when I arrived to interview him in late October 1990. It featured the lush tropical flora, tools, and raw materials associated with third-world cacao harvests. A framed text above assured customers that "notre chocolat provient des plus grands crus de cacaos du monde" (our chocolate comes from the best cacao bean growths in the world). Next to this was a basin of liquid dark chocolate, specialized handicraft tools, and *Le guide des croqueurs de chocolat (The guide of chocolate eaters)* listing the "170 best *chocolatiers* of France," including Michel Chaudun. A photocopy of the guide page devoted to Michel Chaudun revealed that his chocolates rated an 18 out of 20.

Michel Chaudun greeted me at the door and ushered me into his tiny, elegant boutique. Inside, dark chocolate candies with evocative names like Esmeralda and Véragua were invitingly displayed on an open central island. A small hand-printed sign indicated the price per kilo: 340F, or roughly $68. A stunning array of confectionery art, from baby bottles to life-size animals, was shelved alongside porcelain and crystal figurines, next to chic confectionery gift boxes. The boutique decor combined neutral earthen tones and rich woods with an abundant use of mirrors. Through a door separating the boutique from the adjacent workshop a young craftsman, Chaudun's only full-time worker and former apprentice, could be seen preparing a batch of house specialities. Next to him were newly coated rows of glossy, ebony-black chocolate bonbons. The intoxicating aroma of chocolate permeated the boutique whenever the workshop door opened.

Along its complex trajectory from cultivation and harvest in the third world to processing and consumption in the first world, chocolate is transformed and differentiated into many culturally relevant categories of food. In France these include breakfast breads, snacks, drink mixes, dessert cuisine, specialty candies which are sold as gifts, for personal consumption, and for ranking in connoisseur tastings, and finally, confectionery art.

In the 1980s Belgian producers of chocolate candies made a swift and successful incursion into the French market by specifically targeting the specialized niche dominated until then by French artisanal chocolatiers. Over the same period, European Community (EC)

representatives prepared for the Maastricht Treaty by proposing a set of European norms of chocolate production which threatened to undercut existing French legislation. Facing the intensified international competition of the 1980s and heightened fears of increasingly centralized regulation, French chocolatiers and cultural taste makers attempted to stimulate new demand for craft commodities by promoting "genuine," "grand cru," or "vintage" French chocolate.[2] Despite the publication of a plethora of works on the logic of consumption in late capitalist societies and a recent volume on the increasing demand for culturally authentic, handicraft goods from developing nations among first-world consumers, little is known about the economic and sociocultural dimensions of craft commodity production in advanced capitalism.[3] Few studies have examined the complex process whereby craft objects are culturally marked and endowed with social, aesthetic, and economic value as they are produced, exchanged, and consumed in postindustrial centers.

The exploration of the relationship between the elaboration of chocolate as a cultural commodity and the affirmation of national identity is important to consider in the wake of EC unification. The 1992 ratification of the Maastricht Treaty by a slim margin of French voters and the hostility it continues to generate among many British people are only two examples of the ambivalence engendered by the creation of a unified Europe. One of the strategies chosen by EC bureaucrats to forge a closer union among factious member nations has been to create a pan-nationalism grounded in a common European culture and shared cultural symbols (Shore and Black 1992).

Attempts in Brussels to build and impose a universal European culture threatened to undermine a notion of French culture defined in identical terms. A universalist notion of civilization still survives in France and is strongly linked to the view that French culture itself best embodies it (Rigby 1991). Many French people see their achievements in literature, philosophy, and the arts, both high and popular, as evidence of this. Moreover, the French state and its representatives take seriously the protection of their language and cultural forms from intrusive foreign influences. Current debates on the ubiquitous spread of English and the effect of European norms on traditional foods such as cheese illustrate this. Thus, even as France asserts her diplomatic, political, and economic presence in the "new" Europe, the arena of culture remains highly charged and contested.

On the eve of 1993, French chocolatiers and taste makers responded to repeated calls for European uniformity in various areas by invoking the uniqueness of their cultural products as exemplified in the specifically French "art" of chocolate making. This art was grounded in superior aesthetic standards and in the preeminence of French culinary arts and skilled artisanship, both constituent elements and potent emblems of French culture. Thus French chocolate, one of the commodities that connote the value of traditional craft production and the prestige of haute cuisine, provides a means of investigating the production of taste and its relation to key elements at the core of contemporary French culture.

ARTISANAL CHOCOLATE PRODUCTION: THE PAST AS PRESENT

It is perhaps wise to begin with a description of contemporary chocolate businesses and a brief discussion of the evolution of both the craft and French patterns of confectionery consumption.

Despite a continuous restructuring of the craft since chocolate was introduced to France in the late 16th century, the arrival of Belgian chocolate franchise outlets in France in the 1980s was reported as a unique event. It served as an important catalyst in the creative reinvention of chocolate candies as prestige cultural commodities. The organization of artisanal chocolate businesses like Chaudun's reveals the continuing salience of certain "traditional" work and social forms such as skilled craft production and independent entrepreneurship. Family members, both blood relations and in-laws, control daily business operations, which usually include two complementary and mutually reinforcing activities: sales and production. These businesses also adhere to a strictly gendered division of labor according to which men generally produce goods in the private space of the workshop and women sell them in the public sphere of the adjacent boutique. Skill is transmitted largely through experiential training and work is organized hierarchically, according to skill and experience, under the authority of the craftsman-owner in the workshop and his wife in the boutique.

Through their window displays and boutique interiors, French chocolatiers actively capitalize on the enduring association between contemporary artisanal production and the idealized, aestheticized image of a "traditional," premodern France.[4] This image evokes a "simpler," "better" time when family workshops provided the exclusive context within which a solidaristic community of uniformly skilled masters guaranteed the production of quality goods. French masters like Chaudun celebrate contemporary craftsmanship while linking it to a rich past of preindustrial guild traditions. Chaudun's elaborate pieces of confectionery art recall the masterpieces (chefs d'oeuvre) completed as a necessary rite of passage in French craft guilds and journeymen brotherhood associations (*compagnonnage*) (Coornaert 1966; Sewell 1980). The small size of Chaudun's boutique evokes the traditional artisanal shop and its place in a distinctively French national tradition of small-scale, skill-based family modes of entrepreneurship. The display of raw materials and artisanal tools reinforces, for the consumers' benefit, the human labor embodied in the goods. House candies are handmade on the premises by Michel Chaudun. The creation and prominent public presentation of individually named candies, as well as the culinary guide rating his chocolates, invoke a renowned French gastronomic heritage based on taste and aesthetics. Chaudun is not only a master craftsman but also a master chef.

At the same time, Chaudun's business is a testament to the changes that have transformed the craft of artisanal chocolate production. Progressive mechanization over the course of the 19th and 20th centuries provoked a two-stage restructuring of the craft. Initially, small- and medium-sized family chocolatiers who mechanized their workshops displaced craftsmen manually producing chocolate from cacao beans. These small-scale family producers were in turn definitively displaced by large-scale industrial manufacturers. By the 1950s the skills associated with the production of chocolate from cacao beans had shifted entirely to industrialized mass production. The craft of chocolate production was redefined and its skills came to center exclusively on the fabrication of dipped chocolate candies, molded chocolate figurines and, most recently, confectionery art. Currently, artisanal chocolatiers occupy a specialized niche within a fully industrialized sector; they purchase industrially manufactured blocks of chocolate and transform them into a personalized line of goods.[5]

In France, chocolate candies are purchased primarily as gifts and distributed to relatives, friends, and colleagues at significant social occasions. The purchase of artisanal candies is embedded within stylized gifting relations and remains closely linked to seasonal and ceremonial occasions such as private rite-of-passage observances and religious holidays such as Christmas and Easter. Until quite recently, French customers of family confectionery businesses purchased equal numbers of dark and milk chocolate candies as gifts, chose from fewer house specialities, saw virtually no confectionery art, and had no specialized culinary guides with which to rate the best French chocolates. A series of developments in the 1980s coalesced to effect considerable change.

During the 1970s and 1980s, competition increased and patterns of confectionery consumption changed. The purchase of artisanally produced candies for distribution as gifts increased modestly in the 1970s but stagnated at virtually the same level in the 1980s (Casella 1989). In contrast, the sale of mass-produced chocolate products registered a significant increase. Over the same period, foreign multinationals, including the American (Mars) and the Swiss (Lindt) companies, came to dominate the French market for mass-produced chocolate products.

In addition, from the early 1980s on, Belgian franchise outlets specifically targeted the market for confectionery gifts by selling mass-produced chocolate candies in store fronts that closely resembled French artisanal boutiques[6] . . . Mass-produced in Belgium for export, these candies were sold by franchise owners who had no training and little or no contact with the family entrepreneurs of the local craft community.

The French were dismayed by the increasing popularity and market share of candies they judged to be of inferior quality and taste. According to them, Belgian candies are too large (*gros*), too sweet (*sucré*), and too full of fillers (*gras*). They contrast French candies made from pure, dark, bittersweet chocolate with the larger milk and white chocolate products that predominate *chez les Belges* (in Belgian shops). In postindustrial societies such as France, cuisine defines a critically important area where economic power and cultural authority intersect. French cuisine has long enjoyed a preeminent reputation among the cuisines of the world; continuing dominance of the culinary world order is a matter of national pride. Yet in this context what counts as French taste and confectionery *savoir faire* is not at all clear. As Dorinne Kondo (1992:177) notes for Japanese fashion, "nation" and "culture" are problematized for French artisans when chocolates produced by foreign competitors gain French market share. How can one speak of a distinctive French chocolate when the French are just as likely to eat bars made by Mars or Lindt or to offer gifts of bonbons made by Belgian franchises as they are French candies?

Persistent concerns related to chocolate mirrored the tenor of wider debates on the central themes of French national identity. These themes include French competitiveness, economic power, political stature, and, especially, cultural autonomy in new European and world orders.

DEMAND, COMMODITIZATION, AND CRAFT

[. . .] The growing exchange of "traditional" craft commodities in global markets suggests that their purchase and consumption may be an essential feature of the present world economy (Nash 1993). Yet the mechanisms that underlie the demand for and consumption of craft commodities produced in postindustrial centers require further study. Craft commodities acquire and shed culturally specific meanings and symbolic value as they are circulated and consumed. While closely tied to local contexts, the exchange and consumption of craft commodities is also mediated by complex, shifting class and taste distinctions which are in turn shaped by global developments. Few studies address the question of how and to what extent the demand for craft objects is linked to taste-making processes such as rapid fashion shifts, direct political appeals, and the development of late capitalism itself.

If the globalization of markets and transnational consumerism characterize the continuing expansion of industrial capitalism, then this development also engenders a contradictory trend. This trend is manifest in the reassertion of local, culturally constituted identities, places, work practices, and commodities as a source of distinction and authenticity in the face of rapid change and the perceived homogeneity of transnationalism (Harvey 1989). Claims of cultural authenticity in advanced capitalism are often linked to an ideal, aestheticized premodern past as well as the groups, labor forms, and products associated with it.[7] Indeed it is the politics of cultural authenticity in the globalization of markets that enables "genuine," locally produced craft work and commodities to be maintained, revived, and/or reinvented precisely because they can be commoditized and sold as such.

What makes the chocolates sold in French boutiques "authentic" and those retailed in Belgian franchises "inauthentic"? How are these labels linked to changing habits of taste and the status struggles associated with them? In a cultural model of consumption where elite habits are disseminated downward and taste makers have heightened power to manipulate taste, chocolatiers and taste makers collaborated to codify and promote a new set of expert criteria for determining both the quality and the authenticity of "vintage" chocolates (Harvey 1989; Zukin 1991). The French differentiate and validate their chocolates through reference to a definitive taste standard adapted from wine connoisseurs. In the pursuit of social distinction, connoisseurship plays an important role. It drives demand for the prestige goods associated with it by reinforcing their rarity and conferring cultural capital on those who consume them. In this game of newly formulated rules of chocolate connoisseurship, consumers demonstrate that they are worthy of symbolically appropriating the objects they purchase through their mastery and display of esoteric taste protocols (Bourdieu 1984).

Moreover, in advanced capitalist societies where consumers have little if any direct experience with production, which itself is a symbol of alienation, Chaudun's chocolates are incarnated signs. Unlike mass-produced commodities, they do not require significant cultural work on the part of consumers to be moved symbolically from the realm of the standardized, impersonal commodity into the realm of personalized gift relations (Carrier 1990). Craft commodities do this cultural work for consumers; they make visible both a particular form of production (linking the conception of a product to its execution) and its attendant social relations. They are imbued with and are the bearers of the social identities of their makers and for this reason retain certain inalienable properties (Mauss 1990 [1925]; Weiner 1992). Produced in limited quantities, using traditional methods and/or materials, they evoke uninterrupted continuity with the past. The historicities of these goods, even if invented or altered, give them special value for both use and gift exchange. This is what

makes them "authentic" and distinguishes them from the "fake" or "inauthentic" chocolate made from identical materials. The silver jewelry made by Navajo Indians, the confections crafted by Japanese artisans, the pottery produced by Onta craftsmen, and the French candies crafted by master chocolatiers all have cultural authenticity in this sense.

[. . .]

Authentic Taste and Artisanal Savoir Faire

Their access to public media allows famous masters to play an important role in defining and celebrating the special skills that distinguish them as craftsmen from industrial producers. This differentiation has involved adroitly manipulating the knowledge of the transformations that a commodity like chocolate undergoes between cultivation and consumption. Knowledge about both the production and consumption of commodities has "technical, mythological and evaluative components and . . . [is] susceptible to mutual and dialectical interaction" especially as the complexity and distance of their flows increases (Appadurai 1986:41). As noted above, French artisans no longer select, blend, and process cacao beans. All purchase industrially produced blocks of chocolate. Yet few French consumers know this, and Parisian chocolatiers like Linxe and Chaudun make creative use of an oenological model, which gives the impression that their knowledge of and involvement in the productive process extends from the choice of the best vintages of beans to their transformation and presentation in the family boutique. Craftsmen on the local level have enthusiastically followed this lead.

As Linxe explained in a 1989 radio interview, "there are different types of cacao beans each from a different place, each with a climate and soil which endows it with particular properties" (Champs-Elysées, série 8, numéro 5). In asserting their skill as expert "alchemists," chocolatiers invoke a system for blending cacao beans that closely parallels that of the highest quality officially classified growths or estates (les grands crus) in the Bordeaux winegrowing region. While no such classification or regulation of cacao bean plantations exists, French chocolatiers nevertheless assure consumers that they select only the best vintages from renowned domains in South America.

They also remind consumers that industrially produced candies are "mummified" with preservatives and lack the "purity" and originality of handcrafted candies. The authenticity of their candies is linked to traditional methods passed down intergenerationally from father to son which privilege manual versus mechanized production and guarantee goods freshly made on the premises.

In postindustrial societies like France, craft can serve as a metaphor for an alternative set of cultural values and work practices in contrast to the dominant norm.[8] In these settings the persistence, reinvention, or creation of traditional craft cultural forms, work practices, and communities can be a means to reassert cultural distinctiveness and identity in response to rapidly changing circumstances (Harvey 1989). Master craftsmen can be celebrated as symbols of local and/or national cultural values. Craft commodities can be marketed on the basis of the nostalgia for an aestheticized, preindustrial work ethos. Here tradition serves as a model of the past that changes constantly because it is continually reinvented and reconstructed from the vantage point of the present. Indeed, the uses of the past outlined in the next section reveal it to be a social construction

strongly mediated and shaped by persistent contemporary concerns.

[. . .]

CONCLUSION

The recent conjuncture of the rise of a broader middle-class group of French consumers with the means to purchase expensive, handcrafted chocolates as gifts and for their own consumption, on the one hand, and the appearance of foreign franchises selling mass-produced candies in settings that replicate French artisanal boutiques, on the other, is a unique one in the history of the craft. The swift proliferation of these franchises changed the terms of the dialogue between French consumers and artisanal producers. The issue of exclusivity that had informed this dialogue in the past, when chocolate was a rare and costly luxury reserved for elite consumption, gave way to the issue of authenticity (Appadurai 1986: 44). Authenticity in this context is determined by culturally elaborated judgments involving connoisseurship, taste, and correctness.

In contemporary postindustrial economies like that of France, discriminating consumers want distinctive goods that are both culturally genuine and esoteric. Yet in these settings the only way to preserve or recreate the elite resonance of commodities that can be mass-produced is to elaborate the criteria of authenticity surrounding them. Through this elaboration and dissemination of an esoteric taste standard, French chocolatiers and cultural taste makers have differentiated authentic French chocolates handcrafted in French workshops by master craftsmen from foreign imitations. As bricoleurs they adapted a number of relevant elements of French culture in order to transform traditional craft candies into dessert cuisine with enhanced value and cachet for both individual consumption and gift exchange. Informing the cultural authenticity of these commodities are oenological criteria of connoisseurship in taste, a culinary discourse of freshness, purity, and aesthetics, and a French heritage of skilled craftsmanship and family entrepreneurship.

The craft commodities displayed in French boutiques like Chaudun's draw their power and value from their symbolic loading. Both craft and cuisine are potent, manipulable symbols of French culture on which numerous ideas can be projected and validated. In postindustrial economies marked by the "production of volatility" (Harvey 1989) handcrafted commodities satisfy the nostalgia for and appeal of the localized goods and modes of production associated with a traditional past. Chaudun's chocolates are both locally produced and distinctly French. The very persistence of skilled craftsmen and family modes of entrepreneurship in these economies means they can be absorbed within and designated as unique manifestations of a unified national culture. They can be enshrined as part of the nation's historic patrimony and redefined as genuine, living cultural forms.

The reconception of French chocolates as culturally genuine food occurred amid the uncertainty generated by the impending unification of the European community. Attempts to forge Europeanness in the name of a universal culture were especially problematic given the existence of a notion of French culture also defined as universal and embodied in French cultural achievements from literature to cuisine. Belgian candies, marketed as if they were freshly made, locally crafted French goods, were particularly threatening because they represented an incursion into sensitive cultural terrain.

The proposed implementation of European production norms for chocolate only heightened fears of increased cultural homogeneity in the name of Europe. The promotion of signature candy recipes and confectionery art were a reassertion of French cultural integrity as it is manifest in the culinary arts, master craftsmanship, and aesthetic standards.

The selective appropriation, reinvention, and exoticization of the historical origin and uses of chocolate in the New World also serve this purpose. By constructing a specifically French history of chocolate and celebrating its transformation from a primitive, foreign foodstuff to a refined French one, chocolatiers and connoisseurs reinforce received notions concerning French taste even as these notions are used to promote new confectionery criteria for determining it. In the skilled hands of French craftsmen, chocolate is sweetened but retains the powerful taste of its wild, natural origins. It is domesticated yet remains inextricably linked to the consuming habits of elites redefined as both cultured and hedonistic.

Preliminary French studies of confectionery consumption patterns (Casella 1989; Mathieu 1990) as well as surveys I conducted among producers and consumers during 1990–91 revealed that French consumers embraced the new standard by routinely specifying dark, semisweet chocolates in both personal and gift purchases. It remains to be seen if they will continue to indulge, even satiate, their appetite for the chocolates of masters like Linxe and Chaudun whose art, according to *The Guide*, fully reveals the "powerful subtleties of the Aztec cacao bean" (1988).

NOTES

1. Author's translation: This is a boutique where chocolate is master, crafted by a master. It is

100 percent artisanal work in the artistic sense of the term, which excels in drawing out the full-bodied splendor of the sublime American [cacao] bean.

2. Here taste maker refers to food critics, chefs, restaurateurs, journalists, social and artistic elites, and intellectuals such as social historians with access to visual and print media in France and the power to shape taste.

3. Some important works on consumption in advanced capitalist contexts include Appadurai 1986; Baudrillard 1981; Bourdieu 1984; Harvey 1989; Jameson 1984; Miller 1987; Sahlins 1976; Tobin 1992; and Zukin 1991. See Nash 1993 for an analysis of craft goods in the world market.

4. Historically France has been a preeminent nation of small manufacture, skilled artisanship, and craft associations such as guilds. France's reputation in luxury craft production was established through the worldwide export of French perfume, fashion, porcelain, sculpted furniture, wine, and cuisine. French artisanship also enjoys a positive resonance because small-scale, skill-based modes of family entrepreneurship dominated trade and industry well into the 20th century.

5. Attempts to depict the size of the French artisanal chocolate industry result in different statistics. The number of French businesses specializing in chocolate production and employing fewer than ten employees totals 720. If one considers small and medium-sized businesses (up to 50 employees), this number increases to 3,500 (Mathieu 1990).

6. It is important to distinguish among the Belgian franchises in question. These do not include the Godiva chocolate franchises firmly implanted in the French market. The new Belgian firms include Leonidas, Daskalides, Jeff de Bruges, and Neuville.

7. See Badone 1991; Bestor 1992; Moeran 1984; and Tobin 1992.

8. See Ennew 1982; Harevan 1992; Kondo 1990; and Moeran 1984.

FURTHER READING

Appadurai, Arjun, ed. 1986. *The Social Life of Things: Commodities in Cultural Perspective.* Cambridge: Cambridge University Press.

Badone, Ellen. 1991. *Ethnography, Fiction, and the Meanings of the Past in Brittany.* American Ethnologist 18(3):518–545.

Bestor, Theodore C. 1992. "The Raw, the Cooked, and the Industrial: Commoditization and Food Culture in a Japanese Commodities Market." Paper presented at the Department of Anthropology Colloquium at New York University. New York, NY.

Bourdieu, Pierre. 1984. *Distinction: A Social Critique of the Judgement of Taste.* R. Nice, trans. Cambridge, MA: Harvard University Press.

Carrier, James. 1990. *The Symbolism of Possession in Commodity Advertising.* Man 25:693–706.

Casella, Philippe. 1989. *La profession de chocolatier.* Paris: Agence Nationale pour le Développement de l'Éducation Permanente.

Coomaert, Emile. 1966. *Les compagnonnages en France du moyen âge à nos jours.* Paris: Éditions Ouvrières.

Ennew, Judith. 1982. "Harris Tweed: Construction, Retention and Representation of a Cottage Industry." In *From Craft to Industry: Ethnography of Proto-Industrial Cloth Production.* Esther Goody, ed. pp. 167–199. Cambridge: Cambridge University Press.

Le guide des croqueurs de chocolat (The guide of chocolate eaters). 1988. With a preface by Claude Lebey. Paris: Olivier Orban.

Harevan, Tamara. 1992. "The Festival's Work as Leisure: The Traditional Craftsmen of Gion Festival," in *Worker's Expressions.* J. Calagione, D. Francis, and D. Nugent, eds. pp. 98–128. Albany: State University of New York Press.

Harvey, David. 1989. *The Condition of Postmodernity: An Enquiry into the Origins of Cultural Change.* Cambridge: Basil Blackwell.

Kondo, Dorinne. 1990. *Crafting Selves: Power, Gender, and Discourses of Identity in a Japanese Workplace.* Chicago: University of Chicago Press.

———. 1992. "Aesthetics and Politics in Fashion. In Remade in Japan: Everyday Life and Consumer Taste in a Changing Society." In Tobin, ed. pp. 176–203. New Haven: Yale University Press.

Mathieu, Johann. 1990. *La Confiserie du Chocolat: Diagnostic de l'Univers et Recherche d'Axes de Développement.* Paris: L'Institut d'Observation et de Décision,

Direction de l'Artisanat du Ministère et l'Artisanat et du Commerce.

Mauss, Marcel. 1990 [1925]. *The Gift: The Form and Reason for Exchange in Archaic Societies*. W. D. Wells, trans. New York: W. W. Norton.

Moeran, Brian. 1984. *Lost Innocence: Folk Craft Potters of Onta, Japan*. Berkeley: University of California Press.

Nash, June, ed. 1993. *Crafts in the World Market*. Albany: State University of New York Press.

Rigby, Brian. 1991. *Popular Culture in Modern France: A Study of Cultural Discourse*. London: Routledge.

Sewell, William H., Jr. 1980. *Work and Revolution in France: The Language of Labor from the Old Regime to 1848*. Cambridge: Cambridge University Press.

Shore, Chris, and Annabel Black. 1992. *Anthropology Today* 8(3):10–11.

Tobin, Joseph J., ed. 1992 *Remade in Japan: Everyday Life and Consumer Taste in a Changing Society*. New Haven: Yale University Press.

Weiner, Annette. 1992. *Inalienable Possessions: The Paradox of Keeping-While-Giving*. Berkeley: University of California Press.

Zukin, Sharon. 1991. *Landscapes of Power: From Detroit to Disney World*. Berkeley: University of California Press.

35

FROM PEASANT TO ARTISAN: MOTOR MECHANICS IN A NIGERIAN TOWN

Sara Berry

Perhaps it is a measure of the seductions of capitalism that even its critics tend to think about production entirely in terms of new objects. The emphasis on novelty applies not only to factory-fresh commodities, but also to the compelling instances of remaking and hybridization that occur in undercapitalized economies. While craftsmanship in such contexts is no doubt of vital importance, the commonplace crafts of repair and maintenance have received comparatively slight attention from scholars. The following close examination of auto mechanics in Nigeria by anthropologist Sara Berry is one such study. As her fieldwork (carried out in 1979) shows, the questions that arise for these tradesmen are similar to those faced by any so-called 'traditional' artisans: how to structure an apprenticeship so that it will be mutually beneficial to master and learner; how to establish a trusting (and trusted) clientele; how to gain access to tools and raw materials; and how to balance time in the shop against other priorities in business and life. In the thirty years since Berry took this snapshot, the Nigerian economy has changed dramatically—in most respects for the worse—and the strategies of auto mechanics have doubtless moved on as well. When it comes to basic issues of craft organization, however, it remains the case that whether craftspeople are jerry-rigging old machines or
fashioning new ones sometimes makes very little difference.

Sara Berry, 'From Peasant to Artisan: Motor Mechanics in a Nigerian Town', from *Working Papers in African Studies*, no. 76 (December 1983), excerpts. A version of this essay later appeared in Catherine Coquery-Vidrovitch, ed., *Entreprises et Entrepreneurs en Afrique* (Paris: Harmattan, 1983).

MOTOR MECHANICS IN IFE

Motor vehicle repair is a product of the colonial economy in Nigeria, whose growth has been closely related to that of foreign trade and the state. Motor vehicles were introduced into Nigeria by the British and their use expanded along with the development of colonial administration and commerce. Road transport competed successfully with the railway and with water-borne transport for both freight and passenger traffic in the colonial period (Hawkins, 1958). Some of the earliest opportunities for Nigerians to engage in successful business expansion lay in road transport, and the rapid growth of the industry served, in turn, to generate demand for complementary services, such as motor repair, typically provided by small-scale self-employed artisans.

After 1950, the growth of the indigenous professional, bureaucratic and business classes further stimulated demand for vehicles and for mechanics' services, and this trend was greatly accelerated by the inflow of oil revenues. By the mid 1970's, domestic demand for motor vehicles was sufficient to warrant the construction of a domestic source of supply and the first Volkswagen assembly plant was opened. However, while petroleum exports have served to enrich the state bourgeoisie and its clients, so far they have done little to alter the structure of the Nigerian political economy. The Nigerian state remains dependent for most of its revenues on the export of primary products; growth has occurred primarily in the tertiary sector; and private enterprise remains heavily dependent on the state for access to both fixed and working capital and opportunities to employ it profitably. Access to state contracts, import licenses, loans and even technical assistance has in turn come to depend increasingly, if not exclusively, on education. Even in private commerce and service enterprises, although formal educational prerequisites for entry remain fairly low, educated people have an advantage in terms of accumulation and upward mobility—because of their superior ability to deal with the educated civil servants, military officers and (recently) elected officials who control the wealth and regulatory powers of the state. Conversely, lack of education tends to limit the scope and prospects for accumulation and upward mobility among private entrepreneurs. In small-scale service enterprise, such as trade or motor repair, low educational requirements serve both to facilitate entry and to limit the scope for individual advance—through increased market competition and reduced access to the means to escape its pressures. Under these circumstances, most small-scale enterprises are likely to remain small, not so much from choice as from necessity. While among my informants, mechanics were

likely to be earning more in 1979 than cocoa farmers, the social distance they had traveled from their fathers' farms was not yet very great.

GETTING ESTABLISHED

The mechanics I interviewed in Ife were mostly married men in their thirties, although individuals' ages ranged from 25 to over 60. None of the men were natives of Ife: all had come there in order to work as mechanics, and one third had served apprenticeships in Ife as well. Each of my informants specialized in some particular aspect of motor vehicle maintenance and repair: some worked only on motorcycle or lorry engines rather than on automobiles; others specialized in panel beating, welding or electrical work. In terms of assets and volume of business, they ranged from a couple of motorcycle mechanics who owned less than N100 worth of tools, employed only one or two non-paying apprentices and earned perhaps N100 a month, to one mechanic and one panel beater who each employed about 30 people (apprentices and paid workers), owned tools and buildings worth N5,000 or more, and earned about N2,000 a month net of business expenses.

Like most mechanics in western Nigeria (Aluko, 1972; Koll, 1969) all but one of my informants began their careers as apprentices. All but the eldest had attended primary school; sixteen had completed it and half of those had had some secondary schooling as well. Most of the latter group decided to become apprentices only because they were financially unable to continue in school. Once it had been decided (often by a boy's parents or other kinsmen) that he should learn mechanics' work, a master was found for him by his relatives. The master agreed to teach the boy his craft in exchange for a fee and—more important—the promise that the boy would remain with him for three to five years serving in effect as an unpaid laborer. The fees varied according to the status of the master, his relationship with the boy's sponsor—usually an elder agnatic [patrilineal] kinsman—and whether or not the master undertook to feed the boy during the apprenticeship period. In general, the fees were not very large compared to, say, school fees: N25 to N50 for three to four years was typical for most of my informants' own training in the 1950's and 1960's, and even the most successful charged their own apprentices no more than N50 per annum in 1979. Much more important from the master's point of view was the apprentice's labor. All but four of my informants had apprentices at the time of my interviews; the majority employed no other type of labor. While apprentices were not very skilled or efficient, the fact that they did not have to be paid was of crucial importance since mechanics' receipts tend to be irregular and working capital, therefore, implicitly expensive. My informants often knew little about the terms of their own apprenticeships, since these had invariably been arranged between their master and their senior relatives, but the terms on which they themselves engaged apprentices illustrate the importance of apprentice labor for reducing the master's financial risks. Several mechanics said, for example, that they required their apprentices' relatives to assume responsibility for any damage which the apprentice might do to the master's tools or to customers' vehicles, and some said that they charged *higher* fees to apprentices who wished to remain with them for only two years, instead of the customary three to five, because of the shorter period of service.

Apprentices are bound labor for the term of their training. They can be asked to run errands and perform household tasks (especially those who live with their masters); they are at their masters' beck and call at all times; they may be disciplined by flogging, though not all masters choose to use corporal punishment. Their status is expressed in rhetoric and ritual as well as in daily routine. When an apprentice completes his term of service he "becomes free." (Usually the English word is used, even in Yoruba conversation: "o to *free*.") To mark the occasion, the freed apprentice and his relatives are expected to give a party for the families and friends of both master and apprentice. Often the amount spent on such a "freedom ceremony" is two to four times the amount of the apprenticeship fee itself, marking the significant change in the young man's status which "freedom" implies.

[. . .]

We might expect initial capital requirements to be lower in Ife than in, say, Lagos or Ibadan, because the cost of living is lower there, but the difference is not likely to be on the order of four or six to one. Instead I think that aspiring mechanics in Ife derived a further advantage from the fact that they had previously worked as apprentices and journeymen, rather than as factory employees. To establish a viable enterprise, a mechanic (or any other artisan) needs customers as well as equipment. In the highly competitive environment of the urban "informal" sector, the average mechanic is not likely to attract enough customers to make a living simply by hanging out a signboard. He also needs to build up a clientele—a group of loyal customers who can be counted on to do business with him, time and again, rather than with his competitors. As one of my informants put it, a "regular customer is someone who sends for me when his car breaks down in Ibadan," rather than get it repaired by any of the hundreds of mechanics in Ibadan itself.

Working as a journeyman may not pay very well but, as a number of my informants pointed out, it affords the young mechanic an opportunity to begin to accumulate customers who will one day follow him when he sets up on his own. With such a clientele, a young man can establish his own business with less initial capital than would be required if he had to maintain himself for several months while trying to develop some business. Factory workers have no opportunity to form contacts with potential customers in their chosen field of business before becoming independent, and hence require a larger stock of initial working capital to tide them over the establishment period. Thus, initial employment in the "informal sector" may serve as a partial substitute for savings in the establishment of one's own firm.

[. . .]

GETTING AHEAD

Over a lifetime however, a person's use of resources is influenced by more than just those relationships he relied on in launching his career. Other ties, with individuals or institutions, may come to be important in the operation and expansion of the original enterprise, and of course opportunities for diversifying or even changing one's economic activities and assets are likely to change over time. The fact that Yoruba farmers, traders and artisans often used similar methods of getting established does not necessarily imply that they also employed similar strategies for getting ahead.

Although the market for their services has grown rapidly in recent years, the individual mechanic faces a number of difficulties in trying to develop a viable enterprise. Since barriers to entry are relatively low, the market is highly competitive. Customers are hard to find and to keep; demand, even from regular customers, is unpredictable in the short run; and it is difficult to enforce unwritten contracts with customers, employees, and suppliers of spare parts. Accordingly, mechanics tend to rely heavily on personal relationships to increase revenue and reduce some of the risks, or costs of risk-bearing, associated with doing business in a competitive and "informal" market. Cultivation of personal loyalties—with one's customers, suppliers, employees, partners and colleagues—is one of the principal strategies which mechanics employ both to expand their own firms and to try to regulate the market for their services.

One of the most important conditions of business success is, as we have seen, the accumulation of loyal customers. Building a clientele depends, in turn, not only on a mechanic's reputation for skill, honesty, fairness and dispatch, but also on whether or not people like to do business with him. The success of an enterprise can thus be augmented through the adroit use of culturally sanctioned modes of interpersonal behavior. As one customer put it, commenting on the popularity of a particular mechanic, "he doesn't know everything there is to know about Volkswagens, but he is a very respectful young man." Indeed, regular customers often find it to their own as well as their mechanic's advantage to remain fairly loyal. A vehicle owner in a moderate-sized town such as Ife has literally hundreds of mechanics, welders, battery chargers, vulcanisers, panel beaters, electricians and other specialized servicemen to choose from. Far from being a paradise of consumer sovereignty, however, such a market presents the vehicle owner with a formidable problem of information, for there is no institutional mechanism for guaranteeing or standardizing quality of service. Without some form of prior information, the customer does not know until afterward whether any particular mechanic can be counted on to do a good job. To obtain such information in advance, most vehicle owners rely on the advice of fellow consumers. Accordingly, the more loyal customers a mechanic has, the more people are likely to recommend him to their friends and the larger his volume of business is likely to be. Developing a sizeable clientele is especially important since individual customers' needs for repair services are inevitably irregular. In order to foster the loyalty on which their revenues depend, therefore, mechanics tend to offer more or better services to their regular customers. As one informant put it, "if someone is a regular customer, I have to repair his vehicle, even if he owes me money."

Mechanics also find it advantageous to cultivate special relationships with suppliers of spare parts. Such suppliers rarely extend credit—most mechanics avoid the expense of stocking parts by buying them only as needed with money supplied by the customer—but since most parts are imported, supplies are not always regular and prices are subject to sudden change. The mechanic who "knows" his suppliers may be able to get parts more quickly, and at a discount on black market prices, than one who does not. This in turn, enables him to offer better service to his customers.

In addition to cultivating good personal relationships with individual customers and suppliers, mechanics sometimes cooperate with one another to reduce the risks or raise the returns to their trade. Most mechanics specialize in one aspect of motor vehicle maintenance and repair—e.g., engines, electrical system, body

work, tires—so that a customer with more than one type of work to be done may have to go to several different shops. One strategy which a number of my informants used to attract customers was to form partnerships with men whose specialties were different from their own. Among my informants, seven were involved in partnerships at the time of my interviews and two others had had partners in the past. In all cases, each partner owned his own tools; in most, each employed his own apprentices and journeymen. Jobs were contracted for independently, and the partners' financial transactions were kept separate, although one partner might accept money on the other's behalf if their relationship was an especially good one. (An apprentice rarely accepts payment on the master's behalf; to do so would render him liable to accusations of theft.) I encountered only one case in which partners pooled receipts, and that partnership had broken up acrimoniously. In many respects, these partnerships resemble the form of "cooperation" which Koll found in Ibadan, where small groups of craftsman often

> associate ad hoc or permanently to share certain things or to do certain things together, without abandoning their economic independence . . . Since recruitment into a craft is no longer through kinship, such partnerships are apparently of a contractual nature; but in fact they have more in common with kinship relations than with typical contractual relations since prospective partners take years to judge each other's moral behavior and technical skills; once established, the partnership rest on mutual trust and understanding rather than on formalities. (Koll, 1969: 57)

Customer loyalty is not, of course, unwavering. In a relatively open market, price competition is inevitable; in addition there is the problem of managing credit relations. Most mechanics do not stock spare parts, but buy them as needed with money furnished by the customer, thus minimizing their own need for working capital. However, willingness to supply parts and/or services on credit can help to attract customers. The mechanic faces a continual choice between short-term and long-term gains: whether to risk immediate loss by cutting prices and/or accepting delayed payment in order to expand volume now and in the future. Like capitalists in other economies, mechanics in Ife have relied on combination, as well as individual initiative, to try to reduce risks and raise returns to their productive activities. In addition to partnerships among pairs or small groups of mechanics, there is an industry-wide body—the Ife Mechanics' Association—which seeks to regulate the market and to represent mechanics' interests in local affairs.

[. . .]

I mentioned earlier the fact that none of my informants employed hired labor, in the sense of workers paid on a time or piece-rate basis. The majority of their employees were apprentices; the rest were journeymen who received a share of the firm's proceeds. Employment of both apprentices and journeymen is financially advantageous to the artisan because apprentices are not paid at all and journeymen usually receive a share of the firm's proceeds. Thus, neither creates a fixed claim against the fluctuating and unpredictable cash receipts of the firm. By relieving the mechanic of the expense of maintaining a cash reserve, return on the such labor contracts effectively increase the internal rate of firm's assets. However, they also provide relatively weak incentives to employees to maximize their productivity—especially in the case of apprentices.

Other examples of managerial inefficiency abound. Most mechanics did not trust their apprentices or their journeymen to handle money, and hence could receive payment for services rendered only when they were actually in their shops. Few kept any form of written records—at most a journal of daily receipts—or felt that records were useful for any purpose other than that of obtaining credit from institutional sources. One man said that he had kept records for a couple of years, but quit when his application for a bank loan was turned down in spite of his bookkeeping efforts. Another informant's description of his (mental) accounting system is especially revealing. When I asked what he earned from his business, net of expenses, he replied that out of average gross weekly receipts of about N50, he spent N35 on maintaining his family, which left N15 "*for myself*. I use this money for rent, tools and entertainment" (my emphasis).[1] For this man, as for many others, capital accumulation was clearly a central personal goal, but cost accounting was not perceived as a means to further that end.

Similarly, labor productivity depended heavily on the master's physical presence in the shop. This was not only a matter of supervision but also of the delegation of authority. If a vehicle is brought to a shop when the master is not there, it must usually wait his return before work can be started at all, even if the apprentices are not already occupied on other work. Thus, time spent idle is often not the result of laziness or recalcitrance on the part of the apprentices, but simply of poor organization. To avoid damage to customers' vehicles, apprentices are schooled never to work on them without specific instructions; hence, they must wait to work until someone is there to tell them what to do. In firms which employ journeymen, apprentices may waste less time, but financial control may become less effective, if not a cause of outright dispute between journeyman and master.

The weaknesses of artisans' managerial performance are not, I think, simply a matter of inexperience (Harris, 1971) or of culture (Kilby, 1965), though both are involved to some extent. Ineffective management is also a direct consequence of the strategies which small-scale entrepreneurs use to increase their clientele and to reduce risk-bearing costs. Building loyalties with customers, suppliers and fellow mechanics requires that the mechanic often spend time away from his shop, either on ceremonial visits or, more often, negotiating and carrying out transactions in person. This is partly due to the underdevelopment of communications facilities in Nigeria and partly to the high value placed on personal contact and interchange in Yoruba culture (cf. Aronson, 1978; Eades, 1975). Similarly, diversification of the enterprise into technically complementary undertakings, such as dealing in spare parts or transport, may help to reduce costs or increase returns to the entrepreneur's technical skills but also tends to reduce the amount of time he spends supervising labor in the original enterprise.

Kinship and communal relations, which have often been used to solve problems of communication and financial management in African trading diasporas (Cohen, 1969; Baier, 1980; Curtin, 1975; Leighton 1979; Eades 1975), do not seem as well suited to those of labor supervision in service or manufacturing enterprise. Although nearly all of my informants had been assisted by their relatives in getting trained, and several had also received help in purchasing tools or constructing a shed, none of them had taken relatives as partners or paid employees, and most avoided having relatives as apprentices.

Several informants commented that they had been under some pressure to take junior kinsmen as apprentices in return for their elders' assistance in arranging their own training. Most felt, however, that once they had trained two or three junior brothers or cousins, they had discharged their obligation to the family.

In part, mechanics' reluctance to use kinship ties in managing their firms has to do with the norms which govern interpersonal relations in Yoruba society. Kinsmen's obligations to one another are determined in large part by seniority: elders are expected to protect and provide for their juniors and, in turn, to control them. The terms of their control are, however, circumscribed by custom. "Any senior has a right to unquestioned service, deference, and submissiveness from any junior" (Aronson, 1978: 94)—but not necessarily to efficiency. Any elder member of a descent group, or segment thereof, may discipline a junior member for disobedience or even laziness, but not for low productivity. Indeed, innovativeness may be interpreted as a sign of insubordination and is not therefore encouraged among junior people. Such norms are not, of course, confined to descent groups: they govern relations between superiors and subordinates throughout Yoruba society, and thus to some extent limit the authority of a master over his apprentices, whether or not they are related to him. His obligations to junior kinsmen are, however, greater than those to subordinates who are not members of his descent group. In taking relatives as apprentices, he therefore gains little in the way of effective control over their productivity, and may have to spend more in providing for them than he does for non-kin. Similarly, there is little to be gained by taking kinsmen as partners and, if a cousin absconds with money or tools from the firm, the family is likely to put pressure on the proprietor not to prosecute him, which he would do if the culprit were not a relative.

Thus, my observations tend to bear out the common conclusion that Yoruba artisans are not always effective managers, but do not support Kilby's suggestion that this is because their early socialization creates emotional inhibitions against mastering technical and organizational skills. Rather, in the case of mechanics, anyway, the strategies they have devised to solve problems of marketing and risk-aversion often exacerbate those of labor control, to which kinship and culture offer no readymade solutions.

In the absence of pre-existing institutional mechanisms for stimulating and controlling labor productivity, most of my informants either did the best they could through direct personal supervision, or devised some form of trusty system, whereby senior apprentices were delegated to supervise junior ones. Even in the latter case, however, the proprietor reserved to himself the right to negotiate with customers and to handle all cash flows, so that such transactions were delayed if he were absent from the shop. Similarly, even the strictest master might admonish, flog, or sack an apprentice for insubordination or theft, but not for incompetence. The result was often poor service and a good deal of wasted time.

In conclusion, it seems clear that inefficient management is not a symptom of indifference to profit. The mechanics described here are not target accumulators, whose capitalist propensities are blunted by cultural norms or social obligations. On the contrary, for most of my informants business expansion was a central personal goal—a point illustrated by the mechanic who described his income net of family expenses as income "for myself," to be reinvested in his firm. Their prospects for accumulation are limited, however, both by the structure of the market and by their own strategies for improving the terms on which they participate in it. The question then is how far they manage to transcend those limits through economic, social or political transactions outside the market.[2]

NOTES

1. Entertaining friends establishes one's reputation for generosity and sociability—qualities which, in turn, attract customers as well as friends and followers—and thus represents investment in the firm just as does the purchase of tools. (See Aronson, 1978).

2. In the larger study which this essay is based, I collected life histories and observed patterns of social organization among cocoa farmers, schoolteachers and professionals as well as mechanics. The majority of my informants were selected from descendants of a single community, although among the mechanics only two (of twenty) came from this town. One of the eighteen claimed to have four wives and twenty-four children. If he is excluded, the average number of children per wife drops to 2.6.

REFERENCES

E. O. Akeredolu-Ale, *The Underdevelopment of Nigeria* (Ibadan: Ibadan University Press, 1975).

S. A. Aluko et al., *Small-scale Industries, Western State of Nigeria* (He-Ife: University of He Press, 1972).

D. Aronson, *The City is Our Farm* (Boston, MA: Schenkman, 1978).

S. Baier, *An Economic History of Central Niger* (Oxford: Clarendon Press, 1980).

K. Baker and C.E.F. Beer, *The Politics of the Ibadan Peasantry* (Ibadan: University of Ibadan Press, 1976).

K. Barber, "Oriki in Lkuku: Relationships Between Verbal and Social Structures" (Ph.D. thesis, University of Ife, 1979).

S. S. Berry, *Cocoa, Custom and Socio-economic Change in Rural Western Nigeria* (Oxford: Clarendon Press, 1975). [1975a]

———, "Export Growth and Rural Class Formation in Western Nigeria," in R. Dumett and L. Brainerd, eds., *Problems of Rural Development* (Leiden: E. J. Brill, 1975).

A. Callaway, *Nigerian Enterprise and the Employment of Youth* (Ibadan: Nigerian Institute of Social and Economic Research, 1973).

A. Cohen, *Custom and Politics in Urban Africa* (Berkeley and Los Angeles: University of California Press, 1969).

R. Cohen and R. Sandbrook, eds., *The Development of an African Working Class* (London: Longman, 1975).

P. D. Curtin, *Economic Change in Precolonial Africa* (Madison, WI: University of Wisconsin Press, 1975).

J. S. Eades, "The Growth of a Migrant Community: The Yoruba in Northern Ghana," in J. Goody, *Changing Social Structure in Ghana* (London: International African Institute, 1975).

———, "Kinship and Entrepreneurship among Yoruba in Northern Ghana," in W. Shack and E. P. Skinner, eds., *Strangers in Africa* (Berkeley, CA: University of California Press, 1979).

J. R. Harris, "Nigerian Enterprise in the Printing Industry," *Nigerian Journal of Economic and Social Studies*, 10/ 2 (July 1968).

———, "Nigerian Entrepreneurship in Industry," in P. Kilby, ed., Entrepreneurship and Economic Development (New York: Free Press, 1971).

J. R. Harris and M. P. Rowe, "Entrepreneurial Patterns in the Nigerian Sawmilling Industry," Nigerian Journal of Economic and Social Studies, 8/1 (March 1966).

FURTHER READING

Vivek Bald, 'Appropriating Technology', in Alondra Nelson, Thuy Linh N. Tu, and Alicia Hines, eds, *Technicolor: Race, Technology, and Everyday Life* (New York: New York University Press, 2001).

Jamie Barber, 'Skill Upgrading within Informal Training: Lessons from the Indian Auto Mechanic', *International Journal of Training and Development* 8/2 (2004), pp. 128–39.

Sidney Kasfir, 'Apprentices and Entrepreneurs: The Workshop and Style Uniformity in Subsaharan Africa', in C. D. Roy, ed., *Iowa Studies in African Art: The Stanley Conferences at the University of Iowa* (Iowa City: University of Iowa, 1987).

36

DESTINY WORLD: TEXTILE CASUALTIES IN SOUTHERN NIGERIA

David T. Doris

Craft is usually seen as existing in opposition to globalization. In the following essay, however, David Doris—a professor of African art history and visual culture at the University of Michigan—provides a compelling narrative in which globalization has given rise to a new craft-based art form. In many respects it's not a pretty story. Doris trails the circulation of waste textiles, from the factories of multinational corporations to Nigerian towns and cities where those scraps are consumed. There, at the tail end of the commodity chain, one finds a great deal of artisanal creativity but little in the way of political power. The familiar faces of Disney characters are cut apart and reconfigured into fractured fairy tales, schizophrenic images rendered in cloth. Doris reads these stitched-up sheets and pillowcases as signifiers of Africa's many problems: its shattered economy, its asymmetrical relations with the rest of the world, its internal rivalries and prejudices. And yet he makes it clear that these cloths are extraordinarily compelling; a truer material expression of global capitalism and its effects than any squeaky-clean, mass-produced commodity could ever be.

'Destiny World: Textile Casualties in Southern Nigeria', *African Arts* 39/2 (2006).

Stay awake, don't rest your head
Don't lie down upon your bed

While the moon drifts in the skies,
Stay awake, don't close your eyes.
　　　—from *Mary Poppins*, © Disney 1964

Chaos is precariously near.
　　　—Anton Ehrenzweig (1967:31)

Walking along an Ìbàdàn roadside in July 1996, I was stopped in my tracks by the sight of a cloth displayed in a market stall—a bedsheet repeatedly printed with the recognizable face and overstuffed body of Winnie the Pooh. As I moved in for a closer look, those cartoon bears lost their legibility, competing for my attention with a clattering noise of repeating, unrelated patterns and colors, each establishing its own broken visual rhythm. Suddenly, out of the noise appeared other famous faces—Pooh's bouncing feline companion Tigger, smiling Dalmatian puppies—a hundred and one of them, all torn and scattered, submerging and rising again to the fore like an irregular heartbeat. And I thought, in a moment of gross misapprehension, "How very . . . African."[1]

But there was nothing "African" there. And the patterning of the cloth was clearly the result of industrial accident, not human agency. Yet despite that awareness, the uncanny aesthetic pleasure of my misreading was palpable, and it persisted. I soon began to collect bedsheets and pillowcases made from similarly printed fabric, hanging some of the more compelling examples on the walls of my room in a compound near Ilé-Ifè. Only later, when my displayed acquisitions drew unsolicited and intriguing appraisals from Nigerian friends and neighbors, did it become obvious that there was something here demanding further investigation. I started asking questions, and soon it became clear I had to follow the textiles to their source. The Yorùbá traders from whom I had purchased the cloths were reluctant to point the way—they thought I wanted to go into business for myself. After many assurances that such was not the case, I was on a bus headed eastward.

This is a tale about the transformative power of perception. It speaks of the strange moments of encounter with otherness, in which, without much reflection, we react to the unfamiliar, mastering it, transforming it into something we've known all along.

REJECTS

In several of southern Nigeria's larger towns, textiles featuring the printed—or more accurately, *mis*-printed—images of animated cartoon figures, super-heroes, professional sports team logos, and other icons of the contemporary American culture industry are fashioned into, and sold as, bedsheets and pillowcases. Many of these textiles are manufactured in the United States *as waste products*, never intended for sale in a legitimate consumer marketplace.

In textile industry lingo, such cloth is referred to as "leader sheeting"—a heavy, low-grade material used for gauging the accuracy of printing presses, correcting ink color and aligning design template registration, leading the way for the higher quality cloth to follow. It also is set as a spacer between cloths

receiving different designs, rolling through the presses as design templates are changed. Hundreds of yards of leader sheeting are required for the mass production of any high-quality printed cloth. To reduce waste and production costs, leader sheeting can be run through the presses several times before it is discarded. The resulting product is often a composite of several disparate designs, none coherent in its own right—a palimpsest of broken patterns, figures and colors that combine and interact in layered, random configuration.

In the U.S., leader sheeting can't be sold as second- or even third-quality goods. Instead, it is marked as trash, warehoused, and purchased in large quantities by Nigerian importers. Such entrepreneurs understand full well the topography of their country's battered economy, and know what a market of mostly poor people can bear. That's why they import these cloths—even in tough times, they know they can sell them. They're cheap, they're durable despite their defects, and they're available. And today, even after the Nigerian federal government has enacted a ban on the importation of such global refuse, they're easy enough to smuggle into the country.

For textile manufacturers in the U.S., of course, it's all just a mess, with no real market value. Likewise, the designs are not a consideration for Igbo textile importers in the town of Aba. They see the cloth not as fascinatingly random compositions, but only as a commercial opportunity. "It's very cheap," said importer Chief K.[2] "If we buy first quality, we can not market in this area. People prefer them, in fact, because it is what we can afford." With his target market in sight, K has one of his "jobbers," a commissioned agent in Atlanta, Georgia, acquire the cloths for next to nothing from factories throughout the United States. The jobber warehouses the cloths in Atlanta, and periodically dispatches them in container ships bound for Nigerian ports. Gifts of "dash," discreetly presented to customs officials, assure the contraband textiles an easy passage onto shore.[3] From there, they are trucked to Chief K's Aba warehouse in massive rolls and quarter-ton bales. The bales comprise a special opportunity for the local wholesalers who buy from Chief K. Purchased by the pound, not even the importer knows what's inside until they're cut open. When they are, what bursts out is chaos: a few pieces of luxuriously textured fabrics tossed in with scraps and strips of material often clotted with botched ink. These are the best bargains in Aba, grab-bags from the underbelly of American industry.

Obiageri, a textile retailer in Aba's Ariaria market, had no illusions regarding the cloths' value in their country of origin: "It is waste from that place," she said. "You use these *jansu* cloths to clean the engine before the nice-flowered cloth" (in Aba, "nice-flowered" means "with bright, crisply printed designs"). For Obiageri, as for many textile merchants in Aba, the significant difference between the two sorts of cloth was this: "Nice-flowered dey cost. *Jansu* dey cheap."[4]

In a shop on Aba's Msulu Street, wholesale textile merchants like Dickson Ukaegbu grade the relative quality of the cloths from "first" to "fourth," from "bright" and "best-flowered" to "dark." Most people, Ukaegbu explained, want "bright" cloths when they can afford them, and so he displays those at the front of his shop, stacked in neatly folded 10-yard bolts.[5] In the rear he piles cloths of a generally lesser grade, but Ukaegbu's distinctions are not absolute, and there's a lot of mixing in both display areas. None of the cloths in Ukaegbu's shop is perfect, but in time he sells most of them to vendors in Aba's Ariaria and New Markets, who in turn sell them to retail merchants from all over the country. Many of these merchants are Yorùbá men from the southwest, who display their inexpensive goods at roadsides, usually near the outskirts of a town, far from central markets.

And this is how Mickey Mouse comes to Nigeria.

BEAUTIFUL FLOWERS

Skilled textile workers, usually Igbo women, shape the raw cloth into finished products, cutting vast swathes down to size, stitching together strips and leftover fragments into complete bedding sets: a sheet and two pillowcases. Some of their piecework is quite artful, as we will see, but no further changes are made to the cloth—no reprinting or dyeing is involved.[6]

These specialty seamstresses are collectively known by the name of the cloth they work and sell: *jansu*, "rejects." The name is telling: *jansu* women are held in some contempt by other cloth vendors in the Aba market, who regard them as mere gleaners, scrap-collectors. Certainly, the *jansu* have the unique opportunity to collect, at no cost, the scraps of cloth that remain upon the completion of a sewing commission. They also buy cloth from wholesalers and importers' warehouses, but their choices there are limited.

As textile retailer Mrs. Grace "Madame Babyface" Okafor pointed out, the *jansu* pick through the bolts and bales of cloth that mainstream merchants like herself leave behind. "We will pick the best ones," she said, "and leave the ones we don't want for them." Similarly, Okafor suggested that the *jansu*'s economic limitations mean that they are little concerned with the design or quality of cloth. "For the *jansu* people," she explained, "it is the price they look for. They go buy anything."[7]

However, as they piece together motley collections of scraps and strips, some *jansu* (a name they do not call themselves) do indeed emphasize design in their work, and show a clear and discriminating sense of aesthetic proportion. This is most apparent in the pillowcases they produce, which are often judiciously planned and, occasionally, stunning artistic achievements.

Take, for example, a pillowcase I purchased in Ìbàdàn (25 July 1998), part of a three-piece bedding set. There is a real design sensibility at work here, a structure of aesthetic correspondences made from disparate scraps. In this work, Minnie Mouse takes center-stage, her iconic wholeness, once diminished by industrial accident, now restored by artistic intention. In a swirl of off-register color, she holds forth a blood-red blotch like a stigmata. The unknown seamstress has extended the green gingham check pattern that bisects the mouse's head—first, below, by joining to it that same pattern from another cloth, which also lengthens the red mass of Minnie's dress, and then by attaching a panel of blue and orange vertical streaks that expand the grid and rephrase the bold color of the central piece. Unplanned error is transmuted into willful design. However random the printed mishaps of the raw material, there is clearly nothing accidental in the way it is assembled into a finished product.

According to Ijeoma Chigbundu, an Aba *jansu* seamstress, other pillowcases featuring the outsized heads of Winnie the Pooh's bouncing accomplice Tigger also were notable, not only for their careful color-matching, but also for the attention paid to correspondences of texture. "These flowers," said Ijeoma, pointing to the cloud shapes that float around the cropped mass of Tigger's head, "it is like this flower here"—that is, the floral print that limns

the lower edge. "She done join am together like that," she added. "It is beautiful."[8]

THE WONDERFUL WORLD OF DISNEY

Walt Disney Company cartoon characters, benign and happy creatures all, are by far the most prevalent among the many corporate logos adorning the waste textiles sold every day in Nigerian markets. Disney is the very model of a globalizing media conglomerate, the second largest in the world, with interests in magazine and book publishing, major motion picture production, live theater, radio, internet, network television and cable broadcasting, theme parks and tourism (Wasko 2001:28–69). Unlike other such corporations, Disney makes claim to a friendly universality, with products consistently designed to appeal emotionally to the broadest possible audience. Indeed, Mickey Mouse, Disney's flagship character, may well be the most widely recognized cultural figure on the planet.

Disney spends little on product advertisement in Africa,[9] but its trademark images are everywhere, often put to use in ways not licensed, or intended, by the Company. In southwestern Nigeria, Disney characters decorate the walls of local elementary schools, and are emblazoned on hand-painted signs proffering the services of hairstylists, sign painters, mechanics, and a host of other trades-people. In the U.S., such unauthorized uses of Disney's trademarks tend to drive the company's copyright lawyers into a froth of litigation. But in Nigeria, trademarks don't stand a chance.

In one Yorùbá town in 1999, for example, the "Waltz Disney Video Club" rented pirated videos from the US, England and Nigeria, while its very name, printed in the famous Disney logo font, did a weird little dance to avoid copyright infringement—as was the good but ironic intention of the owner, an avid fan of Disney films. And just across the street, Mickey Mouse was made to serve the spiritual life of an evangelical Christian congregation, beckoning from the cement facade of the "Sanctuary of Hope" church with the smiling promise of Redemption: "YOU'RE WELCOME TO HIS WONDERFUL PRESENCE!"

The appearance in Nigeria of such dislocated images, and of those misprinted on the textiles at hand, testifies to the massive productive power of the Walt Disney Company, and to the capacity of transnational media corporations in general to replicate themselves around the world through unexpected channels. It recalls too the words of cultural critic Walter Benjamin, who, in a 1936 essay, anticipated a world at once united and dangerously lulled into unconsciousness by the universalizing creations of the burgeoning Hollywood film industry:

> The ancient truth expressed by Heraclitus, that those who are awake have a world in common while each sleeper has a world of his own, has been invalidated by film—and less by depicting the dream world itself than by creating figures of collective dream, such as the globe-encircling Mickey Mouse (2002:18).

The power of Disney animated films, and of the cartoon figures that populate them, depends on producing consistently a convincing illusion that the dream depicted on the screen is real. Evidence of manufacturer's error cannot appear anywhere in Disney's Magic Kingdom of images—such ruptures would destroy the fragile illusion, jolting viewers awake to the reality that their collective dream is in fact manufactured, a commodity to be consumed like any other. In the U.S., the epicenter of global dream production, we *never* see Mickey Mouse headless, or Minnie Mouse with three eyes and mangled arms on a printed bedsheet. Such disturbing aberrations are instead shunted to peripheral spaces such as Nigeria, where they remain invisible to everyone but Nigerians, for example, who buy them for their own reasons.[10]

QUARTER-UP: VISUAL POLYRHYTHM

In addition to the assorted scraps collected from sewing commissions, *jansu* seamstresses also purchase long, narrow strips of cloth, called "quarter-up" in Aba, directly from warehouses, where they are sold, dirt cheap, by the pound—the wider the strip, the more costly the cloth. As the name "quarter-up" suggests, the pieces have been cut from once-whole cloth. When American manufacturers discard their waste cloths, they often run them first through a shredder, assuring that the whole cloth will never be used. In Nigeria, however, where the joining of narrow strips of cloth has several long and distinguished histories, the shredded, rejected strips are revitalized, set into rhythmic motion.

Robert Farris Thompson (1983, 1996) has suggested that in many cultures throughout West and Central Africa, there is a deeply ingrained taste for disrupted, polyrhythmic patterning in a broad range of visual arts. Such a taste for visual polyrhythm, consonant with drummed and melodic idioms in music (Waterman 1990, Chernoff 1979), finds its most articulate expression in a variety of narrow-weave textiles produced across West Africa. In typical practice, horizontal weft patterns of adjacent woven strips are carefully joined together, matched in such a way that they achieve a kind of visual asymmetry. The regularly spaced pattern of one strip meets the regularly spaced pattern of another and another to create a coherent irregularity, an "offbeat phrasing" that lends movement and surprise to the surface of the cloth. In such pulsative patterning, cultured predilection, not accident, is the guiding force—though accident often provides the culturally attuned eye with delightful, welcome surprises (Ehrenzweig 1967:56–57) . . .

The resemblance of these pillowcases with Yorùbá *aso òfì* or *aso oke* textiles, or with the Ijo *popo* cloth produced south of Aba, is striking. But there's no way to know for sure if the unknown seamstresses' moves to disarticulate the boldly colored horizontal patterns was arbitrary or intentional. None of the *jansu* seamstresses with whom I spoke identified a signature style in any of these works, and none seemed particularly interested in the issue.

If we cannot access original creative intention behind such works, however, we can know the way in which others responded to them. Several Yorùbá market-women found these pillowcases particularly exciting, and were vocal in their reactions. In one exemplary and especially articulate observation, Mrs. Comfort Àdùké Títílayò, who sold spicy beancakes in Modakéké, said this about a pillowcase . . . "All the types [of designs] on this cloth are like that of *aso òfì* (*Gbogbo eya tó wà ní ara aso yìí tòfì ní*)." She continued, "We make traditional cloth with strips like this, *aso òfì*. There is no difference. It is an *òfì* pattern. But I don't think it is from here."[11]

A YORÙBÁ SOMETHING

The people with whom I spoke during my research—Igbo tailors and wholesale cloth vendors, Yorùbá retailers and consumers, a successful Igbo cloth importer, university students, and even a pair of Yorùbá divination specialists—inevitably regarded the cloths as functional: basic answers to the basic need to cover one's polyurethane mattress. Like so many aspects of life in Nigeria at the turn of the millennium, there was an air of resignation that settled heavily on the need to consume trade goods of such dubious worth. "Believe me," said Ségun Adéníji, a textile merchant at Ìbàdàn's Agodi Gate, "if the economy of this country was good, people would not be buying this kind of cloth. If I had a lot of money, I no get cloth like this."[12]

Some people, however, regarded these textiles in ways that moved beyond resignation and utility into realms of aesthetic practice and perception. For these Nigerian observers, such cloths are not only re-made but also *re-thought* as a matter of course, made subject to an aestheticizing gaze by men and women who happen now and again to consider the world in aesthetic terms. Transformation occurs in the eyes and minds of perceptive vendors and consumers, Igbo and Yorùbá, who read the surfaces of the cloths and interpret them in compelling, often conflicting ways. Significantly, this global flotsam was often considered in part as a local, and a specifically ethnic, product. In a strange and fortuitous convergence, the waste products of one society happen to correspond with the aesthetic norms, products and practices of another. They end up as something that belongs to both, and to neither.

This is particularly apparent in the bedsheets, which in most cases are simply cut to standard sizes from bolts of whole cloth. In their several strata of densely over-printed patterns, some Nigerian people do indeed see evidence of manufacturer's error. But for others, such as textile wholesaler Dickson Ukaegbu, "They are planned designs."[13] This could be a cunning sales pitch, of course, meant to allay consumer suspicion of cheap, discarded material—stuff for poor folks. But the responses of several Yorùbá men and women suggest that the cloths need no positive spin to be desirable. In Ifè, schoolteacher and part-time textile merchant Bólájí Ajíbádé offered this capsule assessment in English: "The patterns are beautiful ones. Some designs are placed on top of others, and give out good looking."[14]

How does a jumble of random and often unfamiliar patterns and figures come to be regarded as "beautiful"? Because, suggests Pierre Bourdieu, viewers are imbued with a capacity to *read through* the exotic jumble, impulsively seeking out and perceiving familiar organizations in the new and strange:

In the absence of the perception that the works are coded, and coded in another code, one unconsciously applies the code which is good for everyday perception, for the deciphering of familiar objects, to works in a foreign tradition. There is no perception which does not involve an unconscious code and it is essential to dismiss the myth of the 'fresh eye,' considered a virtue attributed to naïveté and innocence (1993:216–217).

If, in effect, individual perception is largely structured by the codes of the "cultural unconscious" that Bourdieu (1977) calls *habitus*, the spontaneous *misreading* of the unfamiliar is an inevitable first step in any process of cross-cultural understanding—a provisional and involuntary colonization of

Figure 26 a, b, c, d, e, f. Photographs of African textiles by David Doris.

Figure 26 a, b, c, d, e, f. Photographs of African textiles by David Doris.

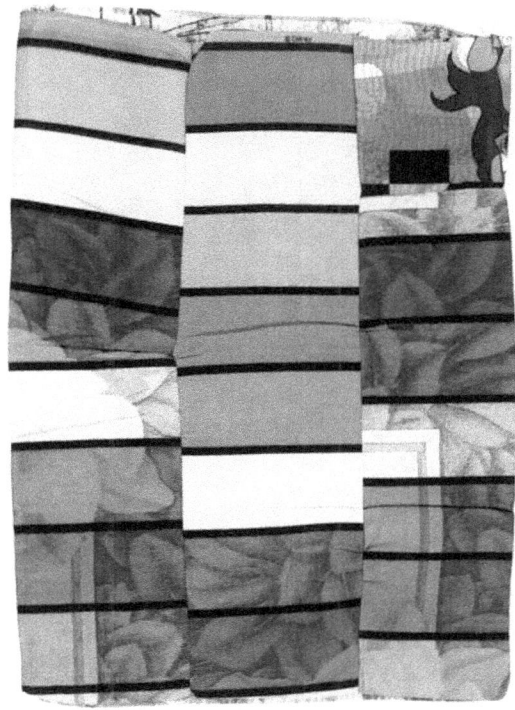

Figure 26(a) a, b, c, d, e, f. Photographs of African textiles by David Doris.

Figure 26 a, b, c, d, e, f. Photographs of African textiles by David Doris.

the unknown. That is, we perceive—*first* and *without reflection*—that which fits the templates of what we already know. Only after that do we begin to measure and translate difference. To illustrate, let's examine two bedsheets as they were described in the Yorùbá southwest.

The first is composed of two patterns: a black checkerboard, streaked and broken, superimposed over repeated Mickey Mouse icons that are ruptured by the overlaid grid, becoming secondary design elements. "Because of the small square pattern, it will be beautiful," said a marketwoman in Modakéké who continued: "This is like a traditional design, *aso gé-súgà* (sugar-cube cloth), that we have in the market now."[15]

Bólájí Ajíbádé saw this as well. "It is the sugar-cube pattern," she said, and added:

> Some call it 'block,' because one can use wood blocks in *àdìre* to make repeating patterns with wax. This one even resembles *àdìre*, but it is different because they used stencils. Maybe the wax was removed during the process of dyeing, so it caused some irregulars. But you know, they say every mistake is a design. At times mistakes will bring out other fine, beautiful patterns. It is messed up somehow, but at least it is beautiful.[16]

The cloth is beautiful, then, despite its mistakes, because it is bears a design similar to one already in the Yorùbá marketplace. But it is beautiful too *because* of its mistakes, welcome visual surprises that distinguish it from the familiar. Language provides the connective tissue, as many people described the foreign textiles in terms long familiar to students of Yorùbá culture: coolness (*ìtútù*) and

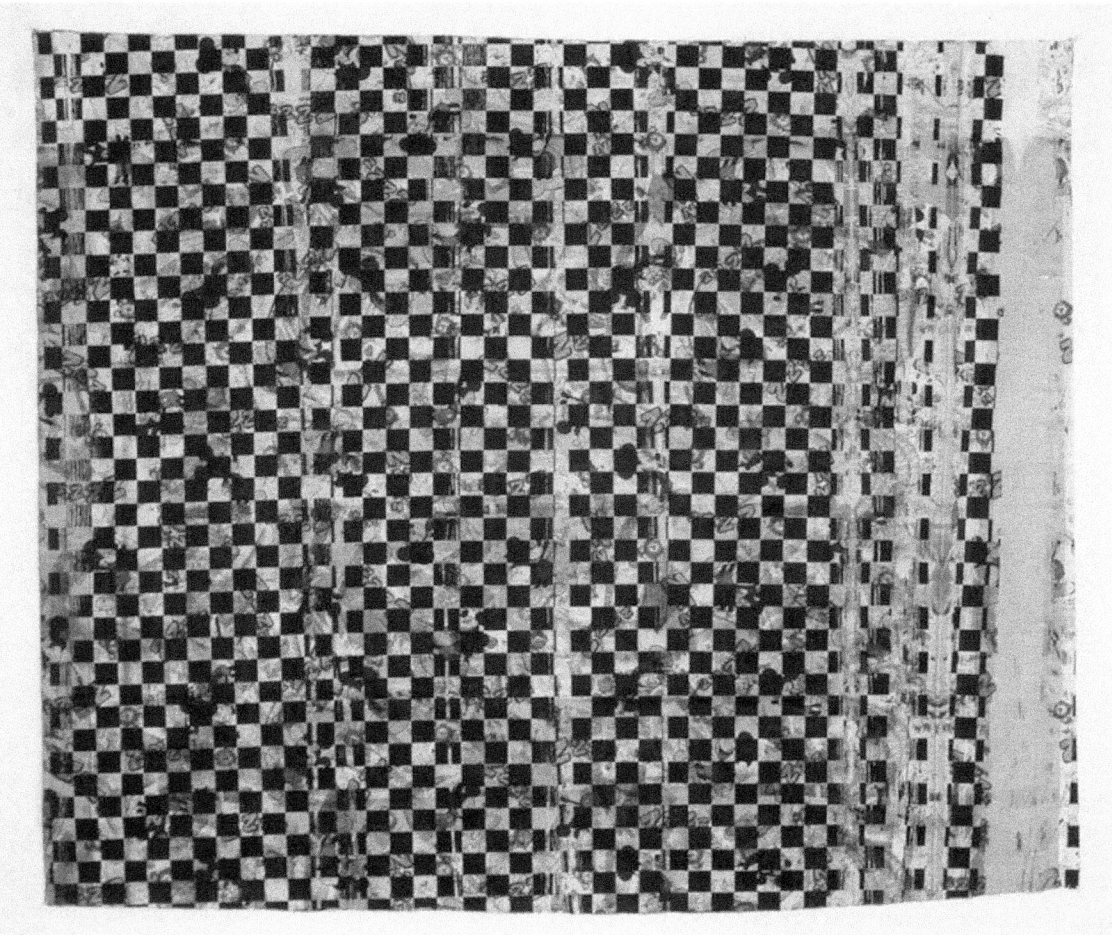

Figure 26(b)   a, b, c, d, e, f. Photographs of African textiles by David Doris.

luminosity (*titan; dídán*); that is, moderation and shining clarity as fundamental expressions of aesthetic excellence. (e.g., Abíódún 1983:23; Drewal 1980:17; Thompson 1973, 1974:37–42; Yai 1994:108)

"This one here is very, very beautiful" (*Tibi lewà gaan gaan ni*), said one Modakéké woman of a cloth imprinted with several tiers of images: Winnie the Pooh and his plush-animal entourage; Mickey Mouse and Donald Duck striking classic poses; a painterly grid in blue and green; another grid of short, thin black lines; and a plane of multi-hued rectangles. "It is cool to the eyes (*o tútù lójú*)," she said, "not too bright."[17] Similarly, Mrs. Florence Òyébamíjí of Modakéké noted, "It's very cool. Red, blue, so many colors." The profusion of colors and cartoon characters was not perceived as incoherent clutter, but rather as an attractive, balanced design, appropriate to the cloth's intended use. "Cloth that is cool" (*aso t'ó bá* cool), continued Auntie Florence, "is good for beds where we lie down in the night."[18] [. . .]

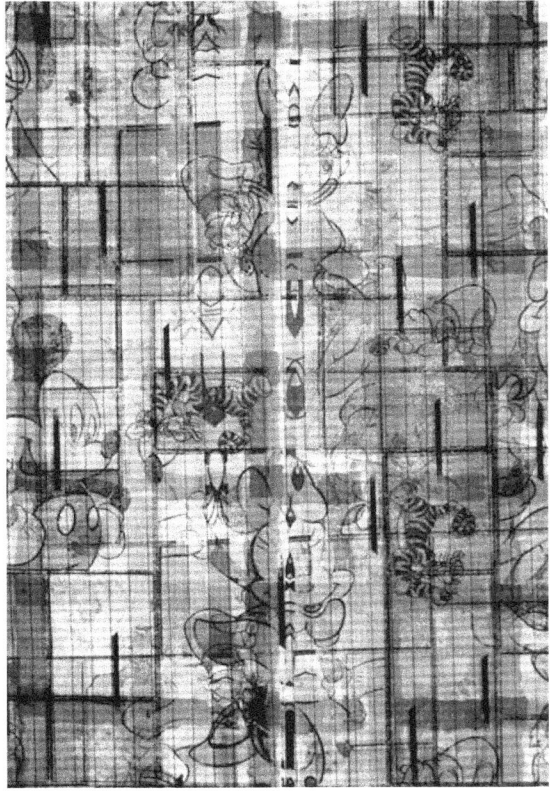

Figure 26 a, b, c, d, e, f. Photographs of African textiles by David Doris.

Clearly, among Yorùbá people I interviewed, this particular textile struck a powerful chord of recognition. Indeed, it was the cloth most often singled out for comment. Likewise, many of the other bedsheets and pillowcases I offered for display (in the form of both actual cloths and color photographs) almost inevitably elicited comments relating them to "traditional" Yorùbá textile patterns and types: àdìre, aso òfì, ankara or kampala, especially.

In these critical appraisals, Yorùbá aesthetic terms were quickly mapped out onto exotic materials, rendering them reassuringly familiar. Only afterwards was mention made of their subtle, but significant, difference from Yorùbá textiles. But difference too can be accounted for within a distinctly Yorùbá code of appreciation. Following Olabiyi Yai's (1994:113) conception of Yorùbá "tradition" (àsà) as characterized by constant departure (ìyàtò) from the given, it is not far-flung to suggest that in such creative acts of spontaneous misreading, these exotic wares become, if only for an instant, *traditional Yorùbá textiles*. "The design is traditional," said a marketwoman in Modakéké, "but they are made in the factory, and they are not from Nigeria."[19]

Surprisingly, it was one of the agents responsible for bringing such textiles into Nigeria, Chief K of Aba, an Igbo man, who perceived a certain "ethnic" flavor in the cartoon creatures or other figures of these so-called "Tom-Boy" designs (with tonal emphasis on *boy*), named after the "Tom and Jerry" cat-and-mouse-chase cartoons that appear now and again on Nigerian TV. As he leafed through my book of textile photographs, he halted abruptly at the image of [a bedsheet] and said definitively, "This one now, it is a Yorùbá something."[20]

THE MENAGERIE OF IMAGERY

"What I love most," said Bólájí Ajíbádé, regarding a bedsheet and pillowcase imprinted with Disney images of Winnie the Pooh and Piglet beneath a crumbling black checkerboard and a downpour of Navajo-style darts, "is that most of the patterns have things that are familiar and things that are not familiar."[21] At an historical moment in which American image-making corporations such as Disney have an alarming power to extend their reach to cultures throughout the globe, it might be tempting to think that their products and messages are somehow homogenizing in their effect, received intact, recognized as they were intended. But this is not the case in Nigeria, at least, where Disney character icons are subject to transpositions that detach them from their corporate source, generalize them, and replace their aesthetic and affective meanings.

[. . .]

In interviews with Yorùbá people, the Disney figures that decorated the cloths were rarely known by their trademarked character names, though they usually were recognized as animals. Among a group of market-women in Modakéké, the image of Donald Duck was regarded with some accuracy as a bird (*eye*), but the diminutive Piglet was said to be a cat (*ológbò*). Mickey Mouse was alternately discussed as a cat, an elephant (*erin*), and a dog (*ajá*), while Winnie the Pooh and Tigger were perceived as human beings (*eniyan*).[22]

In some conversations, the animal icons were glossed by the generalizing terms "toys" or "teddy-babies," acknowledging their relation to the world of childhood. "I love them because they are colorful and childish," said Déolá Ajíbóyé, a zoology student at Obafemi Awolowo University. "The teddies make me feel like a child—they are cuddly, warm and homely."[23] Tom-Boy cloths are popular with university students who, in contrast to most local market-women, cloth vendors and others, often do recognize the characters from youthful hours spent watching cartoons on television.

In many cases, however, Yorùbá market-women, especially, referred to the character images simply as "shapes" and "flowers," seeing them only as abstract pattern. It is in this de-identifying reduction—the interpretive reception of ostensibly universal icons as a complex, rhythmic interplay of repeated forms and colors—that the translation of these cloths is most astonishing, and yet most ordinary. "You have the same pattern repeating," said one Mrs. Oyinlólá in Modakéké, pointing to the grinning Mickeys, monstrous, waddling Pooh

Bears, and sleep-signifying Z's spread across a cloth surface marred by erratic streaks of black and white, and mottled bands of blue. "Very cool," she added.[24]

[. . .]

DESTINY WORLD

During one of my first visits to Nigeria, a Yorùbá man asked me, "Is it true there is a Magic Kingdom in America?" I was taken aback, obviously, because yes, there is one—two, in fact—but I didn't know how to explain to this man the function and meaning of Disney theme parks in the United States. I still don't.

It turns out it already had been explained on *Aiyé!*, a 1980 recording by the Yorùbá Fuji star, Barrister, in a hit song called "Destiny World." The title itself puns on the three-syllable Yorùbá pronunciation of Disney: *Di-si-ni*. In his Yorùbá language Barrister sings of a miraculous place he visited in Orlando, Florida, America, a Magic Kingdom where the dead speak from beyond the grave, where you can visit the moon and return again, where you travel around underwater and meet Mami Wata face-to-face, where even the architecture talks to you with a human voice. The singer is awed by all this, and praises the *oyinbos* (a term inadequately translated as "Europeans") for using their god-given wisdom to construct airplanes and useful instruments, for improving the quality of their lives through technology, for making progress in the world.

The tone of praise shifts as Barrister deals out a harsh critique to his own African people. "Let us ask ourselves, seriously," he says, "Is the God who created the whites the same God that created the blacks? We blacks are also blessed, but we are ruining ourselves . . ." He builds his argument with observations from the marketplace: "If a black man is selling lace cloth and a white man is selling lace cloth, you will buy from the white man. If an *oyinbo* is selling bad rice at a high price, and a black man is selling good rice at a low price, you will *still* buy from the *oyinbo*. This is a problem."

Despite the song's massive popularity in the Yorùbá southwest, Barrister's opinions were not shared in Nigerian political circles. Seventeen years later, Nigeria joined the World Trade Organization, opening the floodgates wider to cheap, low-quality products—especially textiles—from all around the globe. The results have been catastrophic: of the approximately 150 textile factories that existed in Nigeria in the late 1990s, only ten remain fully operational as of May 2005.[25]

"We each come into the world with our own destiny," Barrister tells us, in classic Yorùbá oratory mode, and leaves the question implicit: If technological progress and domination of a global marketplace through vehicles such as the WTO comprise the apparent destiny of *oyinbos*, what is the destiny of today's Africans? Following Barrister's logic, we might ask: what is the destiny of a people (presumably represented by their governments) who implicate themselves in their own domination by literally buying into the inequities of that marketplace, who content themselves with the discards of the Magic Kingdom, rather than responsibly building and maintaining local industry?

The redemption of broken, ruined things, of waste products subjected to ostensibly ironic re-use and revaluation in spaces far removed from the Magic Kingdom, from "Western" centers of production and consumption—over the last decade this issue has been the stuff of serious analysis in Africanist art historical discourse. At its core is a guiding metaphor: the notion of Trash-Becoming-Treasure

draws us back to the redemptive function of Art in colonized or neo-colonized spaces, and helps preserve Art as a special category of practice (see Cerny and Seriff 1996; Coote, Morton and Nicholson, et al 2000; Gundaker and McWillie 2005; Kratz 1995; Roberts 1992; and Shohat and Stam 1998). It also neatly mirrors and reifies the polarizing conception of the "First World" colonizer as monolithic, industrial oppressor and the colonized "Third World" subject as resilient, industrious bricoleur.

But in Nigeria, at least, the binaries are not so clear. As a Yorùbá proverb says, "If fire has no secret ally it cannot cross a river" (*Bi iná kò ba ni awo kì í gun òkè odò*). There is willing complicity, even a desire, among Nigerian entrepreneurs such as Chief K of Aba to import the world's discards to their country—obviously, because it's outrageously lucrative. I asked K why he imported this foreign refuse rather than support textile production in Nigeria. He responded with a scornful sneer: "The [cloths] we are producing in this country are not even up to this quality."[26]

Chief K hits on a difficult point. Among Nigerians, there has been a long-standing and pervasive lack of pride in the country's capacity to produce goods for its own consumption. Linked to a justifiable distrust of leadership in every arena of Nigerian political and economic life, this lack of pride continues to confer an additional glamour to products imported from Europe, North America and Asia. Such glamour extends even to such dubious products as the *jansu* cloths I've been discussing.

In 2002, the Nigerian government banned the import of such degraded textiles—part of a promise to revitalize the nation's ailing textile industry—but given the prevalence of corruption and the ease of smuggling in Nigerian seaports, the ban was bound from the start to fail. It is, indeed, failing miserably, despite recent government efforts to better police the ports. In April 2005, Mickey Mouse and Winnie-the-Pooh were still hanging up for display on the outskirts of Ìbàdàn, far from the Magic Kingdom, their eternally smiling faces still cracked and blasted in ways most visitors to the Magic Kingdom will never see.

But maybe there is something redemptive in all this—though, for an art historian such as myself, it's a shaky, vaporous affair. Perhaps, if the exported detritus of the Magic Kingdom does indeed make its way to Nigeria in shards and ruination, another kind of magic is spontaneously set into action to redeem those shards, for a moment, into whole, useful, and even beautiful things. This is the ordinary magic of interpretation, the transformative magic of thoughtful aesthetic practice—a thing of the mind and the hand, a thing of shared culture, a thing that somehow, despite the odds, manages to endure.

"You see?" said Dickson Ukaegbu, the Aba textile wholesaler, "These cloths no be rejects. Since they are useful here, they no be rejects like that."[27]

NOTES

1. This essay is based on a paper, "AfroDisney: Fortuitous Convergences and the Redemption of Textile Casualties in Southern Nigeria," presented at the 13th Triennial Symposium of the Arts Council of the African Studies Association, 1 April 2004. It was previously published, with more illustrations and some small textual differences, as Doris (2006). Principal research in Nigeria was enabled by a Fulbright Grant (1998–1999), administered by the Institute of International Education. Further research was made possible by a Smithsonian Institution Post-Doctoral Fellowship (2001–2002),

and by a University of Michigan Humanities Block Faculty Initiative Grant (2004). I also am profoundly grateful to the University of Michigan Department of the History of Art, and to the Center for Afroamerican and African Studies, for allowing me a semester's leave of absence so I could get it all on paper. A humble bow, too, is due to the following excellent souls for their encouragement and help along the way: Rowland Abíódún, Glenn Adamson, Lisa Aronson, Hubertus Breuer, Donald Cosentino, Navin Dadlani, Melissa A. Doris, Henry J. Drewal, Sheree Johnson, Sara Khan, Christine Mullen Kreamer, Oyinlólá Longe, Adam W. Miller, Diane Mark-Walker, Enid Schildkraut, Raymond A. Silverman and, of course, my teacher, Robert Farris Thompson. And for the generosity of each of the Nigerian men and women named on the following pages, I am forever indebted. This essay is dedicated with love to my daughter Marcella.

2. Personal interview, 23 February 1999; full name withheld.

3. Chief K would not divulge the name of the jobber, or the name of any U.S. company with which he deals. "This stops with me," he said, with a note of suspicious finality. I am grateful to Ògúnléye Taiwo, Professor of Maritime and Petroleum Law at Obafemi Awolowo University in Ilé-Ifè, for his willingness to disclose unsavory truths about endemic corruption in Nigerian seaports (personal interview, 8 April 2005).

4. Personal interview, 22 February 1999.

5. Personal interview, 28 December 1998.

6. Unlike other familiar instances of goods remade by ingenious African *bricoleurs*, such as the cans of oil and insecticide reshaped into lamps, toys and suitcases throughout the continent, the cloths do not undergo an "ironic" shift in use-value during their passage. They remain cloths throughout, and are employed for the purposes intended for their first-quality counterparts: usually

as bedclothes, but sometimes as window drapery, space dividers or even garments. The scope of this essay does not include garments made from these textiles, which I encountered only rarely during the course of my research. I have learned that in Lagos today, such garments are highly valued by fashionable women, who regard the accidental patterns on the textiles as unique (Marcia Kure, personal communication, 25 October 2002). According to Rowland Abíódún (personal communication, 22 June 2005), this also was the case in the 1970s, when the fashion for these garments "spread like wildfire" in the southwest.

7. Personal interview, 22 February 1999.
8. Personal interview, 23 February 1999.
9. A 1998 *Advertising Age* feature, "Top Global Marketers" (*Advertising Age*, 28 Sept. 1998, pp. S3-S50; cited in Wasko 2001:103), reported that Disney spent *nothing* on advertising its products on the continent. Only in 1997 did Disney establish a subsidiary, Disney Enterprises Southern Africa, to address marketing in Africa, especially in South Africa (see Burton 2001:258–259).
10. In today's global economy, indeed, there are centers and there are peripheries, a geography dutifully maintained by a ceaseless one-directional flow of corporate goods and images. For a historical critique of this construction, see Pratt (1992).
11. Personal interview, 11 March 1999.
12. Personal interview, 11 March 1999.
13. Personal interview, 22 February 1999.
14. Personal interview, 13 March 1999.
15. Personal interview, 11 March 1999.
16. Personal interview, 13 March 1999.
17. Personal interview, 11 March 1999.
18. Personal interview, 9 January 1999.
19. Personal interview, 11 March 1999. Included in a recent catalogue of Yorùbá textiles collected by Ulli Beier is an image of an "End-run of machine cotton cloth on which various colored dyes have dripped" (Abíódún, Beier and Pemberton 2004:102, plate 22). Dated "20th century," it suggests further that a Yorùbá taste for such accidental patterns extends beyond the textiles discussed in the present article.
20. Personal interview, 23 February 1999.
21. Personal interview, 13 March 1999.
22. Personal interviews, 11 March 1999.
23. Personal interview, 12 March 1999.
24. Personal interview, 11 March 1998.
25. Chris Nchakwu, "100 Textile Factories Closed, 50 in Distress—Workers," *This Day Online*, 19 May 2005, http://www.thisdayon line.com/nview.php?id=17720
26. Personal interview, 23 February 1999.
27. Personal interview, 22 February 1999.

WORKS CITED

Abíódún, Rowland. "Identity and the Artistic Process in Yorùbá Aesthetic Concept of Ìwà." *Journal of Cultures and Ideas* 1 (December): 13–30, 1983

Abíódún, Rowland, Ulli Beier & John Pemberton III (2004). *Cloth Only Wears to Shreds: Yoruba Textiles and Photographs from the Beier Collection.* Amherst: Mead Art Museum, Amherst College.

Aronson, Lisa (1982). "Popo Weaving: The Dynamics of Trade in Southeastern Nigeria." *African Arts* 15(3): 43–47, 90–91.

Barrister, Alhaji Chief Doctor Sikiru Ayinde, and his Progressive Fuji Commanders (1980). "Aiyé/Destiny World." *Aiyé*, Siky Oluyole Records, Ltd. SKOLP 010.

Benjamin, Walter (2002). "The Work of Art in the Age of Its Technical Reproducibility (Second Version)." In *Selected Writings, Volume 3: 1935–1938*, pp. 111–133. Cambridge MA & London: Harvard University Press.

Bourdieu, Pierre (1993). "Outline of a Sociological Theory of Art Perception." In *The Field of Cultural Production*, pp. 215–237. New York: Columbia University Press.

——— (1977). *Outline of a Theory of Practice.* Cambridge: Cambridge University Press.

Burton, Simon (2001). "Disney in South Africa: Towards a Common Culture in a Fragmented Society?" In *Dazzled By Disney?: The Global Disney Audiences*

Project, edited by Janet Wasko, Mark Phillips and Eileen R. Meehan, pp. 257–268. London and New York: Leicester University Press.

Cerny, Charlene and Suzanne Seriff, eds. (1996). *Recycled/Re-seen: Folk Art from the Global Scrap Heap*. New York: Harry N. Abrams.

Chernoff, John Miller (1979). *African Rhythm and African Sensibility*. Chicago: University of Chicago Press.

Coote, Jeremy, Chris Morton and Julia Nicholson, et al. (2000). *Transformations: The Art of Recycling*. Oxford: Pitt Rivers Museum, University of Oxford.

Drewal, Henry J. (1980). *African Artistry: Technique and Aesthetics in Yoruba Sculpture*. Atlanta: The High Museum of Art.

Ehrenzweig, Anton (1967). *The Hidden Order of Art*. Berkeley & Los Angeles: University of California Press.

Gundaker, Grey, and Judith McWillie (2005). *No Space Hidden: The Spirit of African American Yard Work*. Knoxville: The University of Tennessee Press.

Kratz, Corinne A. (1995). "Rethinking Recyclia." *African Arts* 28(3): 1, 7–12.

Nchakwu, Chris (2005). "100 Textile Factories Closed, 50 in Distress—Workers." *This Day Online*, 19 May (http://www.thisdayonline.com/nview.php?id=17720)

Pratt, Mary Louise (1992). *Imperial Eyes: Travel Writing and Transculturation*. London & New York: Routledge.

Roberts, Allan F. (1992). "Chance Encounters, Ironic Collage." *African Arts* 25(2): 54–63, 97–98.

Shohat, Ella, and Robert Stam (1998). "Narrativizing Visual Culture: Towards a Polycentric Aesthetics." In *The Visual Culture Reader*, edited by Nicholas Mirzoeff, pp. 27–49. London and New York: Routledge.

Thompson, Robert Farris (1996). "Impulse and Repose: the Art of Ituri Women." In Georges Meurant and Robert Farris Thompson, *Mbuti Design: Paintings by Pygmy Women of the Ituri Forest*, pp. 185–215. New York: Thames and Hudson.

——— (1983). *Flash of the Spirit: African & Afro-American Art and Philosophy*. New York: Random House.

——— (1974). "Yoruba Artistic Criticism." In *The Traditional Artist in African Societies*, edited by Warren L. d'Azevedo, pp. 18–61. Bloomington: Indiana University Press.

——— (1973). "An Aesthetic of the Cool." *African Arts* 7(1): 41–43, 64–67.

Wasko, Janet (2001). *Understanding Disney: The Manufacture of Fantasy*, Cambridge UK: Polity Press.

Waterman, Richard Alan (1990). "African Influence on the Music of the Americas." In *Mother Wit from the Laughing Barrel*, edited by Alan Dundes, pp. 81–94. Jackson: University Press of Mississippi.

Yai, Olabiyi Babalola (1994). "In praise of metonymy: the concepts of 'tradition' and 'creativity' in the transmission of Yoruba artistry over time and space," in *The Yoruba Artist: New Theoretical Perspectives on African Arts*, edited by Rowland Abíódún, Henry J. Drewal, and John Pemberton III, pp. 107–115. Washington DC: Smithsonian Institution Press.

37

ON A PARTICULAR KIND OF LOVE AND THE SPECIFICITY OF SOVIET PRODUCTION

Sergei Alasheev

In this report from the frontlines of Soviet industry, circa 1992–93, Russian anthropologist Sergei Alasheev provides a detailed, sympathetic and ultimately very moving account of what it is like to work in a ball bearing factory in Samara (a city on the Volga river, southeast of Moscow). It is packed with fascinating observations about the 'specificity' of Soviet craftsmanship. The mostly male workers have an intimacy with their machines that often outstrips any of their familial domestic relationships. Alasheev brilliantly evokes the way they coax and cajole their antiquated and recalcitrant equipment, somehow making perfectly round little spheres out of substandard raw materials. Clearly, they are enormously skilled craftsmen, despite working in a thoroughly industrial (indeed, almost uninhabitable) workplace. The essay concludes with a brilliant twist on Marxist theory. The disarray of socialist production has inadvertently produced a situation in which labor is not alienated—not because the proletariat shares the means of production, but because the system is so badly broken that they are thrown back upon their own resources.

Sergei Alasheev, 'On a Particular Kind of Love and the Specificity of Soviet Production', in Simon Clarke, ed., *Management and Industry in Russia* (Cheltenham: Edward Elgar, 1995).

One can often hear or read in the mass media that (former) Soviet people, including workers, do not know how to work, that the quality of production is low and does not in any way correspond to western standards, being at a lower level.

Scientific works also claim that that Russian production is a process of reproduction of waste and of low quality products.

In our opinion this is not quite correct. In this article we want to put forward our view of production in enterprises. In the course of carrying out our case study on the restructuring of industrial relations in Russian enterprises a thought came to us about the untechnological character of Russian production, about the absence of any well-defined technological regulation of the production process. Here we will try to provide some foundation for this point of view.

The article is based not only on interview materials but also on observation of work in shops and the activities of managers. The basic source of this article is provided by research materials from only one enterprise in Samara. This is a large ball-bearing factory. Although we will support our arguments with observations from other enterprises, nonetheless it was precisely immersion in the atmosphere of factory life in the course of the research that led us to the hypothesis proposed here.

First it is necessary to examine the aspects of the production process which affect the quality of the product and technological discipline. In our opinion the most important factors are: attitudes to work, the condition of equipment, the quality of raw materials and the technology of production in the strict sense.

ATTITUDES TO WORK

After spending eighteen months in one factory, many meetings with workers as well as with managers approximately the following picture has emerged.

Soviet (and now Russian) workers do know how to work! Yes indeed! They really know and love their job, their work.

Work is one of the most important values in people's lives (on a level with their family), according to opinion polls. According to our observations people quite often value their work above their family life. Workers get more satisfaction from carrying out their work responsibilities, and sometimes much more, than from the time they spend with their families. This basically concerns male workers.

This is all the more the case because the living conditions of the majority of workers leave something to be desired. (Quite often they live in communal flats or in rooms in hostels, but even if they live in their own flats this is not much good because the majority of workers are elderly people and the best have been passed on to their grown up children.) Five thousand of the twenty thousand people working in the factory are in the queue to receive housing . . .

Even those whose housing conditions could be considered satisfactory are not often dying to get home at the end of the working day. This is because of the low level of comfort and poor conditions for rest in our apartments, and also the lack of development of leisure facilities in the city ('after work there is nowhere to go').

Yura is a metalworker who has worked at the factory for 36 years. He has two years to go to retirement, and he continues to work in the factory, despite the attempts of the administration to cut the number of workers without a formal process of redundancy. Despite a significant fall in pay (in comparison to the growth of prices), Yura does not any under circumstances intend to leave his job, explaining his decision by the fact that, firstly, he is used to it and, secondly, he is convinced that the administration is holding down pay because it is trying to cut the numbers, but later production will return to normal and then he, as a high-grade specialist who knows the equipment thoroughly, will earn normal pay.

So Yura lives in a little room of twelve square metres in a communal flat for seven families. He lives alone, he is not married. Now and then he stays behind at work for an hour or two to finish repairing a machine (so that 'it doesn't hang on my heart'). He has a permanent and long-lasting relationship with a woman, but it is not too burdensome for him. His basic activity in his free time consists in helping his common-law wife, who has a separate one room apartment where she lives with her sick mother. He busies himself fitting out her apartment, using materials taken from the factory and tools made in the factory. He considers that there is no point in ennobling his apartment. The rest of the time he spends looking out of the window, in conversations and quarrels with his neighbours, and in drinking together with his relatives and colleagues (now and then with people he has met on the street—one cannot drink alone).

He goes to work with great pleasure. There he has many friends and acquaintances, and there are many things to talk about. At work he feels himself to be a professional, not that he is irreplaceable, but that he is needed. He talks with great enthusiasm about some unusual breakdown, which he comes across all the more rarely because the majority of them are already well known to him.

Confirmation of this loving attitude to work is provided by the fact that in one of the shops in which the case study was carried out, the workers come to the shop one and a half hours before the beginning of the shift, and spend the time chatting together on the most varied topics.

The workers find a 'safety-valve' in work, because they live in such conditions that work, if you like, is the single socially approved possibility of self-realisation . . .

Here is a quotation from an article in the factory newspaper which struck us:

'A complex multi-axis automatic machine was stopped for repair. When it was stripped down it appeared that it needed a replacement shaft, the pinions were worn out, and the ball bearings had also served their time. In another shop this would have required the machine to stand idle for repair because the repair base would only get down to making parts when they had received the drawings. But here they instructed the brigade of fitters headed by V. Barinov. The machine was repaired not only quickly, but also to a high standard. Barinov, a universal turner, can make any part without a drawing, using a sample. Take him a spindle and he will make one similar in every detail. Only a person with considerable production experience could do this. And the turner Barinov has enough experience and practical knowledge. He has worked at the factory for fifteen years, and has repaired equipment for the whole of this time. V. Barinov has another noteworthy quality: in addition to doing his

turning well, he knows grinding inside out and on these operations he over-fulfills the norms for the shift by two to two and half times.'

It was our impression that clever individualistic people who carry out not only their own narrow tasks, but who are also universal, with a wide range of skills, are respected in the factory (particularly among the veterans).

Clear evidence of the committed attitude to work is provided by the movement of worker-rationalisers. Now, as in the past, one can find many worker-rationalisers. The technologists of one shop spoke of the large number of rationalisation proposals put forward by workers, affecting both the technological potential and the efficiency of the equipment . . . On the initiative of the shop mechanic P. M. Isakov they decided to create their own design of extruding machine with a motive power of one ton and then build it directly in the shop. The task was carried out by the forces of the collective of the mechanical service of the shop, since the design of the machine had been worked out directly by P. M. Isakov. The new machine has undoubted advantages over existing production models. It is simple to adjust and repair, and provides increased speeds of extrusion—ten metres a minute, while the usual machines are only rated at six metres a minute. Now the mechanic Isakov is working on the creation of a new design of high capacity machine, intended to extrude windows in massive separators.

Workers' rationalisation proposals were encouraged by moral stimuli: the handing out of certificates, the display of photographs on the Board of Honour, the award of the title 'best rationaliser in the factory', etc., and also small monetary bonuses. It is significant that despite the insignificant material stimuli, large numbers of rationalising proposals and inventions were put forward.

In an interview I asked a fitter (former deputy chief of the shop with responsibility for technical matters), why so many people are involved in rationalisation and invention:

> Why do you have to do this? There isn't really any equipment, and there isn't—there is less to operate.
> You see it is a reflex to keep on working. I still burn with this. I still cannot exist without it. I walk around and I see that here and there I want everything to work as well as possible. Even if I do not have to do this . . .

So the love of workers for their work is on the one hand a permanent feature, and on the other it is an energetic love and not a contemplative admiration. Thus workers love their work, dedicate themselves to it completely, although in discussion they often curse it. To put it figuratively, it is a kind of 'difficult love,' and not simply sex or a fleeting passion.

Of course, in the factory there are many different kinds of people, with the most varied attitudes to their duties: there are also idlers, and dimwits and careerists etc., etc. Nevertheless the dominant quality of the majority of workers, it seems to me, is precisely this love, their commitment to their work. Even in those situations in which the real behaviour of workers is at variance with the proposition that they love their work, this attitude persists as a value of ideal behaviour; even in those situations love of one's work is considered necessary, normal and proper. 'Love' is expressed as a cultural norm, called forth by objective causes.

Those workers with whom we have met in the factory are not only good specialists: maybe they are not always high grade specialists, but specialists with specific training. They can do their work in any conditions, getting satisfaction from this. As one of the old hands at the factory said accurately of the Kadrovi workers: 'these lads are made of special stuff. They are one-offs.' They can do their work in the kind of conditions in which nobody works in the West, and even in impossible conditions.

A few years ago one of the shops being researched each month produced almost 10 million rings of 250 different types. Every day they got through about 130 tons of metal. Around one thousand people work here. According to production measurements the level of noise, and the fumes exceed the permissible level by two or three times. The uneven levels between the buildings makes it impossible to introduce mechanisation and automation. Shavings are removed on handcarts and electric trolleys. In summer it is extremely hot because the ventilation does not work properly. Twice a month the cooling system has to be cleaned of emulsion, soda. Because of the cramped conditions it is not possible to provide the workers with a place to get ready for their shift. In some operations the workers have to move around ten to twelve tons from one place to another! (From the appeal of the shop collective to the administration of the factory and deputies of the city Soviet in September 1990).

And that corresponds to our first impressions on visiting the shop.

In the shop there is a constant noise. But this is not the noise of rain or of surf, this is the noise of the ripping up of metal, the sound of blows, blending into a continuous monotonous howl. One can talk, for example, in the smoking room—this is two or three benches placed around a bucket full of cigarette ends in the corridor between departments, through which the electric trolleys pass.

In the work places themselves, in the sections, it is impossible to talk, one has to shout, and then the workers, being accustomed to the cascade of surrounding noise, may turn their attention to you, but not if they are further than twenty feet away. Then you can shout into one another's ear and can understand the words.

The workers in the shop have worked out a special way of speaking—in a very low tone, but with a kind of rich, powerful sound. This ability to suppress unnecessary sounds is a great help to the trade union activist working in the shop, when it is necessary to stop unnecessary discussion, attract attention, or at a meeting in the general din to say a necessary word.

People with such 'specific training' are becoming fewer and fewer in the factory. The director of one of the workshops in the factory said about this:

'It is no secret that it takes years to train specialists for our production. Complicated kinds of press-mould dies for the separators, moulds for consumption goods—all these are made on universal equipment, finished and polished by hand. It needs diabolical patience and the highest qualification to do it. The average age of our workers—of the basic specialists—is already more than 50. The earlier famous dynasties of Denisov, Archakov and others do not continue, and new ones have not emerged . . . Thus we lay special stress on the introduction of new equipment, on which people with lower levels of qualification can carry out their work.'

EQUIPMENT

Turning to equipment the first thing to note is that it is very specific. Our factory, like many large industrial enterprises, has its own machine construction department. More than 35% of the stock of machines were made in the factory itself. The factory has its own design department, which is concerned with the design of new equipment.

The designers of equipment receive orders from the shop the necessary technical-economic specifications for this or that planned objects, and if there is the slightest doubt they may go to the shop, department or section in which this equipment will work and introduce the necessary corrections. Thus the technologist-machine builders know well both the production and labour capacity of the production shop, and the materials which will come to be used on the given equipment. The designers of the machine building workshop work in close contact with mechanics and workers in the shop, and they adjust and finish off the equipment in the shop.

For this reason, one can say without any exaggeration that the equipment is produced at the workplace, almost for each particular worker. The machines acquired thus have their own (factory and shop) finishing touches and adaptations . . . The equipment works thanks to the fact that the workers know it inside out. It is *his* machine, it is almost his child. Kadrovi workers know how often and where it has to be lubricated, what exactly it is necessary to adjust and when, where and how it should be hit (with a sledgehammer) to eliminate a defect. The setters in the shops work on the readjustment of new types of parts, which will not happen more than once a month, and may not happen for several years, the day-to-day setting up is done by the operators themselves. We often hear talk of this or that machine having its own character, arrogance, that each one needs an individual approach . . .

The process of mastering the equipment, working conditions and relationships arising in the labour process takes three to five years, although sometimes a year is enough. To be accepted into the collective takes even longer. But then one is an important specialist who knows 1) exactly how much to tighten every nut on his machine; 2) how much wadding must be put in his ear to muffle the sound of the machine, while at the same time being able to hear the shouts of his comrades; 3) just what to say to the storewoman so that she will give him the protective mittens he needs and not be offended; 4) how it is necessary to behave with the chief and foreman so as to make sure that they don't hassle him and don't give him a bollocking if he has a hangover.

The director's idea of rotating worker's jobs, which he picked up on a visit to Japan, seems to us to be cut off from Russian reality. Workers have been immersed in this world of the shop, section, work place for many years, making it their second home. And then do it all over again? In another work place, on new equipment?

The mastering of the equipment, the finer points of the technology, this whole system of relationships allows the worker to have some time in reserve to make parts. Having mastered the finer points of the machine, the workers become practically indispensable, almost appendages of the machine . . .

The other feature of the equipment is that it is very 'Soviet.' Foreign equipment is finished off and adjusted to suit local conditions. Thus foreign machines which come into the factory are initially looted, and then parts are made in the factory by the local skilled craftsmen. As a result the new parts don't quite fit, and they have to remake the original parts too, and it turns out to be a completely different machine.

As an example one can describe the arrival of a new machine in the shop. For about a month it stood on the site while they studied the documentation, looked for a place for it, and prepared a foundation. During this time the machine was partially dismantled (looted): workers unscrewed several lamps, removed instruments, the repair kit, other parts, control buttons, various nuts were all removed bit by bit to work places or home. Even the boards from the packaging went off somewhere—for example to a dacha, where they can come in useful. When the machine was installed, it had to be finished off, completed with inadequate parts. As a result it already did not operate at the rate at which it should have done . . .

The technical rationaliser's thought of the workers does not always appear in the form of rationalisation proposals. Sometimes the realisation of their finishing touches has a personal character: the skilled craftsman does not formulate his refinement as a rationalisation proposal, but realises it independently. Moreover, they keep quiet about some of the refinements, because they lead to loss of production, but are advantageous to the worker, for example because they make it possible to save time (at the expense of quality), or because they reduce the amount of work (at the expense of the economy of raw materials).

RAW MATERIALS

The quality of the raw materials has a significant influence on the quality of the finished product. Metal arriving at the factory often does not correspond to the requirements of the production process. As a result the factory has a whole preparation workshop, which is responsible for monitoring the quality and preparing the incoming raw materials. Depending on the condition of the metal received and on which shop the metal is going to, preparation may include the following operations: repeated annealing, straightening, roughening, drawing out. There is also a smithy in the factory, where small quantities of metal can be smelted if necessary . . .

A particularly acute problem of quality of raw materials has arisen recently in connection with the breakdown of the economic links between the countries of the former Soviet Union, and correspondingly with the reorientation of the enterprise to new raw materials markets. In our factory, in place of Ukrainian metal, as its main material, they began to use metal from the Urals. The quality of the new metal was equally low, but it also had different dimensions. As a result, despite the efforts of the preparation workshop, the operating conditions for the work of the equipment which had been perfected over the years had to be changed. The shop (machines and workers) was used to working with one metal, then they had to change the operating conditions of the equipment and their skills to work with the new metal.

TECHNOLOGY

Some of the shortcomings in the quality of raw materials are revealed by checks when they arrive, but some of them are not identified at that point. Every technological inadequacy of the equipment and shortcoming in the quality of the raw materials come to light immediately in the workplace. The machine operator is faced with unexpected defects and has to decide either to: (1) remove them, or (2) to ignore them, or (3) not to carry out the task as a result of the failure of the raw material or the machine to conform to the norm. Let us say that the worker has to choose between the second and third options. If he chooses the third, i.e. not to make the part, . . . it is quite possible that after reviewing the question the chief will demand that he carry out the work with the material that he has all the same (because there

is no other, and it is not possible to remove the defect). Moreover the chief himself often knows about the low quality materials. If the worker stands on his principles, the chief will give the work to somebody else; if not, the time spent sorting it out will have been lost, which will affect his pay if he is on piece-rates.

If defects uncovered during working time are ignored, the worker loses nothing, although it is probable that these defects will have an effect on the quality of production which will show up in those parameters which are monitored. Then the defects may be exposed by the output control and as a result the part will be rejected and the workers' pay will be reduced. However you can try to prove that the failure was not your fault, but as a result of all this the part can be completed again by the worker, if the defects can be rectified.

Thus workers most often try to neutralise defects which arise in their work on their own initiative by some means, not risking the second approach, and not turning to their immediate superiors. It would be more likely that they would turn for advice to a more experienced worker (or instructor). The neutralisation of defects may be done with the aim of eliminating them completely, or of eliminating them partially, just enough to pass the output control. In this way workers correct the production technology of the parts depending on this or that inadequacy of the raw materials, equipment or components. The workers work out their own methods of removing this or that defect. Very often the foreman, senior foreman or setters told us that workers themselves know what to do and how to do it. Some of the tricks of the trade are secrets of the workers' craft.

The technology of producing one and the same part used by different workers in our factory is different. Machine operators carry out the functions of the setter. And every time the worker arrives at work for his shift he readjusts his machine.

In an interview a section foreman told us 'Every worker tunes up his machine for himself. One may set the cutting knife not in the extreme position, but a little nearer (a few millimetres), and regulate the dimensions of the cutting of the rod with the support. His replacement will arrive, set the knife in the extreme position, and his balls come out too 'hollow,' then he readjusts the machine again by controlling the support.'

Thus every worker adjusts the equipment in his own manner, and makes the products in his own particular way. The technology of production of the parts is very individual.

This technology is so individual that the foreman who took me around the shop did not know how each individual worker did it. As a result the quality of production is very varied, and not necessarily bad . . . For a new person to master a specific piece of equipment requires the development of skills, techniques, precise movements which take years to acquire. The craftsmanship of the worker and his individual methods of work is based on knowledge of the properties of production, the design of his own machine, the peculiarities of working with this or that raw material. Traditionally craftsmanship is the pride of the working person, and people share the secrets of this craftsmanship reluctantly: not because they do not feel pity or are afraid of losing something, it is simply that there is no powerful stimulus to transfer work experience other than personal sympathy.

. . . So, let us draw some conclusions from our review of the process of production in the enterprise. Equipment: 'Soviet', old and very specific. Raw materials: bad and diverse. Technology: individual. Workers: love their work.

Because to work on such equipment, with such raw materials and for such pay, and on top of that living in such conditions, is only possible if you love your work.

SPECIFICITY OF PRODUCTION

. . . The reader may raise the question: how is it possible to live in such general untechnological conditions? How is it possible to live when a person can have no confidence in the things he buys, which might be high quality or useless. Isn't it impossible to live with constant breakdowns? It is impossible to live normally without being able to have some confidence that the next thing that you take in your hands will not disintegrate. That going out into the street a balcony will not fall on your head. That the car in which you are sitting will not start to fall apart at the most inconvenient moment in the most inconvenient place. That sitting at home the chandelier won't shatter, the drains won't burst, etc.

Nevertheless it is possible to live like this.

It is possible not only to live but also to control this process.

Naturally the untechnological character of production is controlled in every workplace. That is to say, if the workers want to they can make very high quality parts (for themselves for example), using their own supply of high quality raw materials, and using the equipment in the appropriate way. All the different kinds of factors affecting the quality of production can be taken into account and made to correspond to the necessary requirements by the worker, provided that he is sufficiently motivated to get from the foreman (or through his own channels) a high quality set of parts and materials.

The manufacture of products of this or that quality can be controlled at the level of the labour process. That is to say a worker can approach other workers and ask them (or persuade them in some way) to make this part well, or even track them down.

Such regulation can also occur at the level of the foreman. The foreman in our factory has sufficient levers of pressure to make the worker work well. In such a case the foreman must through the chief of shop, or independently, obtain the necessary parts and raw materials (which is not always possible) and he must have the authority among the workers to get them to carry out this task to a high quality (particularly if the raw materials are not of the required quality).

At the level of the shop chief, he can also influence the worker to give him high quality raw materials, to organise the setters to adjust the equipment, and to follow the task through to completion. For this the chief must have sufficient influence in the shop and somehow provide an incentive to carry out the task to a high quality, now and then referring immediately to the workers.

The director of the factory can also regulate the untechnological character of production. Part of his duties are to regulate this process throughout the whole factory; in some circumstances the director can secure the production of a batch of high quality goods. And he can go personally to the workplace and by one means or another secure high quality work . . .

Thus it turns out that every individual person, with enough acquaintances, friends, relatives and personal contacts, either through having enough money or through having enough favours (or goods) to give in exchange, can get control over the process of production of this or that article, acting through the foreman, through the chief of shop, through others working in the factory, or personally.

Incidentally there is one other way of acquiring quality products—'personal' production, which is pretty widespread. A machine operator can fix any kind of broken part on a lathe, or sharpen the cutters himself, or grease the machine himself, etc. He can do other jobs than his own. One of the shop chiefs had personally repaired the roof of the building, when it began to leak again immediately after it had been repaired. Indeed it is well-known that in our daily life that it is sometimes better to make something oneself than to use industrial methods. This is particularly the case with services. Thus, it is much better to rebuild or repair your car yourself than to use a garage. The view that a man should be able to do the basic things in life himself is widespread in mass consciousness; for example, to change a broken light bulb, bang in a nail, hang a peg, repair electrical goods, repair a clock, make repairs around the home (in general to do a man's work in the home) . . .

Moreover the factory has a long tradition of heroic labour, as much after as during the war years, in the vanguard of Soviet engineering—the factory is one of the largest in its branch and was always in good repute. The traditions of self-sacrificing labour are transmitted through the existence of worker dynasties in the factory, and also through its own form of recruitment of labour, when the novice is brought (introduced to the management, invited to come) by somebody already working there.

Thus it is quite possible that the untechnological characteristics noted by us are related to this factory, in which old experienced workers and specialists dominate. And this is partly confirmed by the fact that young people, according to the old workers, do not understand the finer points of the equipment, do not have such a responsible attitude to work and to the fulfillment of their duties. And they understand their duties differently: the young person is inclined to do only his own work, but the veterans consider that their work embraces a much wider range of activities.

Taking these points into account, we have designated the specificities of production in the title of the article as *Soviet*, since it is true that the specific features of Russian production will differ from those described (in the form of a higher degree of alienation of the workers from the process of production, if one follows Marx).

FURTHER READING

Sergei Alasheev and Marina Kiblitskaya, 'How to Live on a Russian's Wage', in Simon Clarke, ed., *Labour Relations in Transition: Wages, Employment and Industrial Conflict in Russia* (Cheltenham: Edward Elgar, 1996).

Teresa Hayter and David Harvey, eds, *The Factory and the City: The Story of the Cowley Automobile Workers in Oxford* (London: Mansell, 1993).

38

ORIGINAL COPIES

Philip Tinari

The sheer scale of China's artisanal industry is mind-boggling. From the recycling of scrap metal to the hand-painting of children's toys, work done by hand is a major part of the country's rapid economic growth. This is not a sort of work that lends itself to romanticism: it tends to be repetitive, unskilled and ruthlessly exploitative. Ironically, it is this Communist economy that has most completely realized Marxist fears about debased and alienated labor. But of course, most of the country's production is driven by external demand. China sometimes defies Euro-American capitalism, particularly in the realm of intellectual property rights, but the Chinese ability to export cheap goods en masse *is what sustains the health of the world economy. In the following essay, contemporary Chinese art specialist Philip Tinari shows how one craft industry registers these economic dynamics. Tinari describes life in Dafen, where nearly every resident is a painter. The mainstay of the village is reproducing photographs in oil on canvas, often in multiple identical versions. It is a realization of a prediction made by Goethe all the way back in 1797: 'There is now to be a great painting factory, in which, they tell us, they intend to copy any painting, rapidly, cheaply, and indistinguishably from the original, by means of totally mechanical operations such as any child can be employed to perform.'¹ Rather than damning Dafen as a sweatshop center making copycat products, though, Tinari finds that the village's painters undertake subtle, skilled forms of adaptation and innovation. It is another twist in the tale of modern craft: though this business is made possible only by the globalized information economy, the craftsmen who work in it are more reminiscent of preindustrial artisans than the studio craftspeople of the West can feasibly claim to be.*

Philip Tinari, 'Original Copies', *Artforum* 46/2 (Oct. 2007).

As a city, Shenzen was almost literally painted into existence. In 1979, "Deng Xiaoping drew a circle"—or so goes the cliché immortalized in an early-'80s pop song—around a fishing village abutting Hong Kong, and proclaimed a zone of free markets for a China then beginning to awaken from its socialist reverie. Nearly thirty years later, it is a site of production on a most extraordinary scale, and the locus of a unique urban condition only possible in a place where the average resident is even younger than the fledgling city itself. Its factories turn out everything from pharmaceuticals to air conditioners; its designers invent the logos that will finally give their nation its own brands; its Window of the World theme park—where visitors amble among replicas of Angkor Wat, the Brasília parliament building, and more than a hundred other famous tourist attractions—takes the Coney Island simulacra with which Rem Koolhaas began his late-70s urban manifesto *Delirious New York* to previously unthinkable levels. And on the outskirts of this city of dreams lies the village of Dafen, a place where the notion of painting as production is pushed to its conceptual outer limits.

The Dafen Oil Painting Village—its name, in accordance with official terminology, specifies "oil painting" in order to distinguish the art practiced here from "national painting," a term denoting more traditional Chinese methods—lies just north of what is called the second line, a quasi border that once enforced a separation between the Shenzhen Special Economic Zone proper and its surrounding districts. To reach the village, one drives through a defunct checkpoint staffed by the occasional police officer standing amid the traffic islands watching the cars stream by. Having crossed this fake border, one enters Shenzhen's Longgang district, perhaps the single most productive locale in all China, with a GDP said to outstrip those of entire northwestern provinces. After a journey of a few miles, Dafen suddenly appears to the left of an eight-lane highway. It is a village only in the sense that it is a distinct pocket in an urban fabric that radiates in all directions—a "village-in-the-city," as the architectural lingo would have it, but really just a dense warren of alleyways and six- and seven-story concrete buildings containing nothing but apartments and workshops dedicated to oil painting.

Here, art links up with the market. Here, talent and fortune interchange. So pronounces a banner gracing one of Dafen's thoroughfares with a logic all too familiar to the art world. The statement could certainly be adopted as the slogan of any major art fair, and is an apt descriptor of the state of affairs in China, where the convergence of art and market, and the conversion of works into capital based on valuations of "talent," have become the *sine qua nons* of a frenzied moment. At Dafen, in less than one quarter of a square mile, some seven hundred galleries and five thousand artists convert oils and canvas into oils on canvas, realizing commissions from all around the world. Open storefront workshops are hung salon style with montages of images that mock traditional taste hierarchies with a vigor sublimated into routine. In a single stall, schlock seascapes bound for cruise ship gift shops and beach-house living rooms might vie for space with portraits of George Bush, Osama bin Laden, and Hu Jin Tao. Also typically in the mix are copies of modernist standards by artists from Vincent Van Gogh to Tamara de Lempicka; tiny icons of the "Five Friendlies," the cartoon mascots of the 2008 Beijing Olympics; and, increasingly, imitations of works by contemporary Chinese painters like Wang Guangyi and Yue Minjun culled from the pages of Hong Kong catalogues. Bulk orders for hundreds of generic landscapes, canonical images, and even made-to-order minimalism tailored to the color schemes of interior designers in South Florida are filled by the day. Here, anything can be transformed from pixels into brush-strokes with a single email. Pix2oils is the name of one of the village's most successful galleries, owned by an Australian and his Chinese wife who have developed a website of the same name where, say, a lovelorn California management consultant can have a digital photo of himself and his girlfriend transformed into art in just a few days' time—satisfaction guaranteed. Many are the young women, Chinese and Western the like, whose visages have been colored and quadrupled in the style of a Warhol silk-screen portrait, first by the ubiquitous software preinstalled on their MacBooks, then by the painters of

Dafen. These works hang briefly on display among sunflowers and waterfalls, battle scenes and poker-playing dogs, waiting for the courier service to take them away.

Dafen, perhaps not surprisingly, has proved itself highly susceptible to narration (in both the mainstream media and the art press) and to incorporation into bigger-picture discourses about both the state of art and the state of China. "Painting sweatshop," the thumbnail phrase generally first invoked, is a facile equation, one that frames the "painting workers" (as the help-wanted signs hanging in nearly every window phrase it) as drones on an assembly line of visual production. Questions of intellectual property surface quickly in most commentaries ("Own Original Chinese Copies of Real Western Art!" trumpeted one *New York Times* headline), which typically contend that Dafen is a hotbed of forgeries and knockoffs. The twin images of anonymous Chinese workers slaving away to make objects of every sort and of avaricious Chinese pirates copying the fruits of Western ingenuity loom large in the global collective unconscious at the moment, so such slippage seems almost inevitable. But even upon casual inspection, these analogies fall apart. Production in Dafen is more modular than mechanized: the system is less one of stern factory managers issuing production quotas than of commissions winding their way through the byzantine social networks that bind client to intermediary to workshop. Pigments, supports, and, indeed, painterly acumen are all meticulously classified according to quality, and single orders are often pieced together from the output of numerous small-scale studios. The shibboleth of rote, mindless copying is similarly challenged by an indigenous value chain that prizes "original creation" above all: Adapted works "in the style of" a known master command

better prices than straight replicas, and at the high end, Dafen paintings are sold, like other art works, under the name of the person who made them.

Appropriately enough, more complicated, nuanced, and, occasionally, problematic responses to Dafen might be found outside the media in the work of other artists. Chinese artist Liu Ding offered his take on the village, *Samples from the Transition-Products,* 2005, at the Second Guangzhou Triennial. (The exhibition's thematic focus was the Pearl River Delta, which includes Shenzhen.) For this project, Liu hired a group of artists from Dafen and set them to work on a three-tiered wooden stage. The thirteen participants painted furiously throughout the triennial's opening, producing copy upon copy of the same painting—a fluorescent waterfall-and-tree landscape starring two alighting cranes. The hierarchy implied by the tiers admitted that Dafen's workers are not an undifferentiated population, but individuals participating in a system that offers some hope of advancement.

It also obliquely mocked the implicit hierarchies to which artists on the biennial circuit are themselves subject. However, because the painters were displayed for an international audience like so many sideshow performers, possessing neither voice nor agency, the work could be charged with veering again toward the stereotype of the mindless minion—and if Liu was taking up the problem of exploitation here, he was doing so by brushing a bit too close to exploitation himself. The pictures produced at the triennial were later exhibited at Frankfurt's L.A. Galerie, framed in gold and hung floor to ceiling on bright red walls; visitors could contemplate them while perched on elaborately upholstered furniture. This presentation was apparently intended to critique both Chinese fantasies of opulence and European fantasies of China. But here again, things seemed too simple: Dafen "readymades," sanctioned as art through their presence in the gallery, like any number of post-Duchampian ploys.

Dafen's question for the "real art world" seems to be less about the boundaries of "real" art than about the specificities of its production—a question to which Christian Jankowski's "China Painters," shown earlier this year at Maccarone in New York, offers a more sophisticated answer. The artist's interest in Dafen was piqued by an article in a Hong Kong newspaper; having learned that a museum was under construction in the village, he traveled there to meet the architects and to photograph the site. He then showed the photos to seventeen local artists and asked each to create, in effect, a painting-within-a-painting: Each was commissioned to render a view of the museum's interior (entirely based on the selected photo) as though it were hung with an imaginary canvas of his or her own devising. The stark differences among the resultant paintings reveal the variety of mindsets and aesthetics to be found in the village. One painter chose a brightly lit wall on which to hang an image of a three-legged jade urn based on a picture he had come across in an old Christie's Hong Kong catalogue. Another placed Delacroix's *Liberty Leading the People* in a dark corner obscured by scaffolding—a not-so-subtle political critique of the regime. Perhaps the most innovative of the bunch fantasized an image that could symbolize Dafen as it is viewed by the Communist Party: "a sexy painting machine" (per the work's title), shaped like a woman's left leg and breast, spewing out a portrait of Salvador Dalí. Some of the canvases were signed by the actual painters, but all were sold as work by Jankowski. And like

most Dafen paintings, they were completed, covered with a layer of cellophane, rolled up, and sent by courier to their destination, all within a few weeks of being commissioned.

Jankowski is certainly not the first painter to outsource labor for conceptual ends, though his gesture takes on a particular political inflection in an era marked by the globalization of both the art world and industrial production. Marcel Duchamp famously used a sign painter in the execution of his final painting, the great *Tu m'*, 1918; and, perhaps more apposite, in John Baldessari's twelve "Commissioned Paintings" of 1969–70, the artist hired painters he had found at a county fair to depict photographs he had taken of a hand pointing at ordinary objects—images that made reference to Al Held's assertion that "all conceptual art is just pointing at things." Once the hands had been rendered on canvas, a sign painter added text crediting each work to the Baldessari "employee" who had created it. Dafen is essentially that county fair to the umpteenth power, as Jankowski's project suggests, and the figure of the sign painter—an anonymous technician who executes a project conceived by someone else—is perhaps the metaphor most actively at play in this village. Yet the idiosyncratic visions of the various participants in "The China Painters" effectively elaborate on the agency of the "sign painter," thereby complicating the default narrative of Dafen as an assembly line.

The individual stories of most Dafen painters do the same. Take, for example, He Liangfeng. Born in 1980, he graduated from the Shaoxing Arts and Literature College in the southeastern province of Zhejiang. Seven years ago, after a brief stint as a high school art teacher, he set up his stall in Dafen with his wife, Wu Xiaoling, also a Shaoxing graduate. Today the couple employs five painters in a facility a brief walk from the center of the village. He sees himself as an artist rather than as a craftsman, and is proud of the way his renderings of works by famous artists depart from their sources. Showing me recent paintings on his computer which runs a counterfeit version of Windows XP (as does nearly every artist's computer on which I've viewed works since 2002), he pointed to a group in the style of Wang Yuping, a Central Academy professor and member of the early '90s "New Generation" group, known for his neo-expressionist paintings of fish. "You see," he said, flipping through the images, "this is not actually a copy of a work by Wang Yuping, but an innovation on him. Among Wang Yuping's fish you will not find this fish!" But he was proudest of another painting, copied from the Chinese artist Liu Ye, a Dafen favorite whose cartoonish images of young girls and bunnies intently staring at iconic modernist paintings were on view at Sperone Westwater in New York last fall. He had edited a well-known Liu picture of a girl about to slaughter a pig, replacing her knife with a handful of vegetables. "It's much happier like this," he informed me. "Customers prefer it this way."

Amid shoptalk about how he chooses the right grade of canvas for each client (a cotton hybrid is fine for standard-issue decorative paintings, but he prefers pure linen for his original works), he aired his views on the state of art criticism; "You see, in the old system, you had painters, professors, and critics. The professors had given up their own hopes but were cultivating the next generation. The critics were where you went for approval. They don't mean anything anymore. Today, if I paint this, and someone buys it, that means it's easy to sell." This last sentence is just one degree less tautological in Chinese than it sounds in English.

Like a number of Dafen painters I met, He used to work as a museum preparatory—suggesting a degree of institutional sophistication that does not jibe with the supposition, put forward by several reviewers of "The China Painters," that Jankowski's collaborators "had never been inside a museum." Even the museum in which the painters were asked to envision their work has been repeatedly described in the Western press as a foolish totalitarian brainstorm, a ludicrous attempt to inject culture into a place beyond cultural redemption. In reality, the Dafen Art Museum, designed by Meng Yan, Liu Xiaodu, and Wang Hui, of the Shenzhen- and Beijing-based architectural collective Urbanus (best known for its research into the village-in-the-city phenomenon), is one of the more interesting products of the recent Chinese building boom. It is a well-considered intervention whose architects were clearly conscious of the absurdities of putting a museum in such a painting-ridden context. Situated on a piazza that separates it from the village proper, the museum is stratified into three levels, with a ground-floor bazaar intended to host air-fair-style booths selling paintings identical to those outside its walls. The rooftop terrace mimics the gridded layout of the village's narrow streets, with a forest of smaller square volumes intended for use as artists' studios and cafés. Pedestrian bridges connect the terrace to the school where most of the painters' children study and to a high-end condo complex. Only the middle level provides the traditional white-cube experience. And the facade, in a Jankowski-esque nod to those who work in its shadow, is punctuated by rectangular niches that the architects hope to see filled with frescoes by the winners of Dafen's annual painting competition.

Indeed, the museum is simply the grandest in a long series of state interventions aimed at championing Dafen as a "National Model Base of Cultural Industry" (as a plaque in the village proclaims). Seeing "cultural industry" as a viable economic model, the district government in 1998 set about upgrading Dafen in an attempt to control the chaos that generally characterizes villages-in-the-city. The officials lined the streets with pedestrian-friendly paving blocks, erected a giant sign shaped like an easel, and placed a bust of Leonardo da Vinci (whose Chinese moniker, Dafenqi, incidentally begins with two characters nearly identical to those that name the village) at the main intersection. Other plaques recount the village's creation myth, which turns on a Hong Kong painter named Huang Jiang who is said to have arrived in 1989. "He rented residential buildings and hired art students and artists for the creation, reproduction, collection, and export of oil paintings," we are told, "and soon, Dafen Oil Paintings had become a famous cultural brand in China and abroad." Delegations of municipal and provincial officials from around the country regularly tour the village, looking to create their own Dafens back home.

While this goal may sound risible, miniature Dafens in fact pervade the Chinese art world. Departing from the more traditional studio system now in vogue—whereby most senior painters employ teams of young assistants who are recent graduates of the art academies—a number of midcareer artists have taken advantage of the fact that paintings in China can be fabricated with ease and in bulk, and have incorporated Dafen-inspired techniques into their own conceptual practices. Take Yan Lei, an artist whose early conceptual works included a mischievous mail-art project, done in collaboration with Hong Hao, that involved sending fake invitations to Documenta 10 to a hundred Beijing artists. These days, Yan's output consists largely of paintings based on digital photographs printed onto canvas and, in a knowing nod to Baldessari, painted out by assistants, many of whom have no background in art at all. Or take Yang Yong, perhaps Shenzhen's best-known "real" artist, who made his name with photos of women posing like fashion models at construction sites, before shifting to the production of brightly colored, photo-based, thoroughly Dafen-esque realist paintings. After a long day roaming the village, I visited his studio, adjacent to that of Urbanus in the city's high-art enclave in the tree-lined Overseas Chinese Town district. There, assistants struggled to adjust a projected image of an airport so that it filled the canvas onto which it would be painted. To its left hung a purple-hued painting, completed just a few days earlier, of a foot kicking a ball: a Pix2oils transfer of a still from Douglas Gordon and Philippe Parreno's *Zidane*.

NOTE

1. Johann Wolfgang von Goethe, 'Art and Handicraft', 1797; as translated in Isabelle Frank, ed., *The Theory of Decorative Art 1750–1940* (New Haven, CT: Yale University Press/Bard Graduate Center for Studies of the Decorative Arts, 2000), pp. 151–2.

FURTHER READING

Jennifer Baichwal and Edward Burtynsky, *Manufactured Landscapes* (film, 2006).

James Elkins, ed., *Is Art History Global?* (London: Routledge, 2006).

Wu Hung, 'A Case of Being "Contemporary": Conditions, Spheres, and Narratives of Contemporary Chinese Art', in Terry Smith and Okwui Enwezor, eds, *Antinomies of Art and Culture: Modernity, Postmodernity, Contemporaneity* (Chapel Hill, NC: Duke University Press, 2008).

39

'WHAT IS CYBERNETICS?', FROM *THE HUMAN USE OF HUMANS*

Norbert Wiener

From one perspective, nothing could be more opposed to craft than artificial intelligence—the final (and as yet unrealized) stage in the process by which human skills are displaced by automation. But this is not how Norbert Wiener saw it. For him, understanding how a machine might think was a way of humanizing both the machine itself and the economy of which it was a part, and thus remediating the problems wrought by the mechanization of labor. To achieve this end, he introduced a concept that could be applied equally to an automated system or a whole social fabric: feedback. The principle is a simple one. In any system, an input creates an output. A system that incorporates feedback involves a loop, so that the output influences subsequent input. For example, when you hammer a nail you must aim the first strike consciously. But for every subsequent blow, you rely on feedback, correcting your aim based on the previous result. All craftsmanship could be said to operate on this principle of internal coherence (this is one reason it is often described as subconscious, instinctive or experiential), but the same can be said for computers, electric guitars hooked up to amplifiers, the stock market or the earth's atmosphere. Wiener's cybernetics was intended as a structural theory that would help us understand all these things and their interconnectedness in society. 'The first industrial revolution, the revolution of the "dark satanic mills", was the devaluation of the human

arm by the competition of machinery,' he wrote in his seminal 1948 book Cybernetics. *'The modern industrial revolution is similarly bound to devalue the human brain . . . The answer, of course, is to have a society based on human values other than buying and selling.'[1] Wiener was horrified when, despite this proviso, his ideas were taken (in the words of one historian) as 'a master theory for Cold War America'.[2] He therefore set out to write a corrective. This was published two years later as* The Human Use of Human Beings, *excerpted here, in which he argued that technology had to be guided by goals of social reform and integration—goals similar to those that had long motivated the craft movement.*

Norbert Wiener, 'What Is Cybernetics?', from *The Human Use of Human Beings* (Cambridge, MA: Riverside Press, 1950; rev. ed. 1954).

I have been occupied for many years with problems of communication engineering. These have led to the design and investigation of various sorts of communication machines, some of which have shown an uncanny ability to simulate human behavior, and thereby to throw light on the possible nature of human behavior. They have even shown the existence of a tremendous possibility of replacing human behavior, in many cases in which the human being is relatively slow and ineffective. We are thus in an immediate need of discussing the powers of these machines as they impinge on the human being, and the consequences of this new and fundamental revolution in technique.

To those of us who are engaged in constructive research and in invention, there is a serious moral risk of aggrandizing what we have accomplished. To the public, there is an equally serious moral risk of supposing that in stating new potentials of fact, we scientists and engineers are thereby justifying and even urging their exploitation at any costs. It will therefore be taken for granted by many that the attitude of an investigator who is aware of the great new possibilities of the machine age, when employed for the purpose of communication and control, will be to urge the prompt exploitation of this new "know-how" for the sake of the machine and for the minimization of the human element in life. This is most emphatically not the purpose of the present book.

The purpose of this book is both to explain the potentialities of the machine in fields which up to now have been taken to be purely human, and to warn against the dangers of a purely selfish exploitation of these possibilities in a world in which to human beings, human things are all-important.

That we shall have to change many details of our mode of life in the face of the new machines is certain; but these machines are secondary in all matters of value that concern us to the proper evaluation of human beings for their own sake and to their employment as human beings, and not as second-rate surrogates for possible machines of the future. The message of this book as well as its title is *the human use of human beings* [. . .]

We ordinarily think of a message as sent from human being to human being. This need not be the case at all. If, being lazy, instead of getting out of bed in the morning, I press a button which turns on the heat, closes the window, and starts an electric heating unit under the coffeepot, I am sending messages to all these pieces of apparatus. If on the other hand, the electric egg boiler starts a whistle going after a certain number of minutes, it is sending me a message. If the thermostat records that the room is too warm, and turns off the oil burner, the message may be said to be a method of control of the oil burner. Control, in other words, is nothing but the sending of messages which effectively change the behavior of the recipient.

It is this study of messages, and in particular of the effective messages of control, which constitutes the science of *Cybernetics*, which I christened in an earlier book.[3] Its name signifies the art of pilot or steersman. Let it be noted that the word "governor" in a machine is simply the latinized Greek word for steersman.

It is the thesis of this book that society can only be understood through a study of the messages and the communication facilities which belong to it; and that in the future development of these messages and communication facilities, messages between man and machines, between machine and man, and between machine and machine, are destined to play an ever-increasing part.

To indicate the role of the message in man, let us compare human activity with activity of a very different sort; namely, the activity of the little figures which dance on the top of a music box. These figures dance in accordance with a pattern, but it is a pattern which is set in advance, and in which the past activity of the figures has practically nothing to do with the pattern of their future activity. There is a message, indeed; but it goes from the machinery of the music box to the figures, and stops there. The figures themselves have not a trace of any communication with the outer world, except this one-way stage of communication with the music box. They are blind, deaf, and dumb, and cannot vary their activity in the least from the conventionalized pattern.

Contrast with them the behavior of man, or indeed of any moderately intelligent animal such as a kitten. I call to the kitten and it looks up. I have sent it a message which it has received by its sensory organs, and which it registers in action. The kitten is hungry and lets out a pitiful wail. This time it is the sender of a message. The kitten bats at a swinging spool. The spool swings to the left, and the kitten catches it with its left paw. This time messages of a very complicated nature are both sent and received. The kitten is informed of the motion of its own paw by organs called proprioceptors or kinaesthetic organs. These organs are certain nerve end-bodies to be found in its joints, in its muscles, and in its tendons; and by means of nervous messages sent by these organs, the animal is aware of the actual position and tensions of its tissues. It is only through these organs that anything like a skill is possible, not to mention the extreme dexterity of the kitten.

I have contrasted the behavior of the little figures on the music box on the one hand, and the human and animal behavior on the other. It might be supposed that the music box was an example typical of all machine behavior, in contrast to the behavior of living organisms. This is not so. The older machines, and in particular the older attempts to produce automata, did in fact work on a closed clockwork basis. On the other hand, the machines of the present day possess sense organs; that is, receptors for messages coming from the outside. These may be as simple as photo-electric cells which change electrically when a light falls on them, and which can tell light from dark. They may be as complicated as a television set. They may measure a tension by the change it produces in the conductivity of a wire exposed to it. They may measure temperature by means of a thermocouple, which is an instrument consisting of two distinct metals in contact with one another through which a current flows when one of the points of contact is heated. Every instrument in the repertory of the scientific-instrument maker is a possible sense organ, and may be made to record its reading remotely through the intervention of appropriate electrical apparatus. Thus the machine which is conditioned by its relation to the external world, and by the things happening in the external world, is with us and has been with us for some time.

The machine which acts on the external world by means of messages is also familiar. The automatic photo-electric door opener is known to every person who has passed through the Pennsylvania Station in New York, and is used in many other buildings as well. When the message constituted by the interception of a beam of light is sent to the apparatus, this message actuates the door, and opens it so that the passenger may go through.

The steps between the actuation of a machine of this type by sense organs and its performance of a task may be as simple as in the case of the electric door; or it may be in fact of any desired degree of complexity. A complex action is one in which the combination of the data introduced, which we call the *input,* to obtain an effect on the outer world, which we call the *output,* may involve a large number of combinations. These are combinations, both of the data put in at the moment and of the records taken from the past stored data which we call the *memory.* These are recorded in the machine. The most complicated machines yet

Figure 27   Martin Bodilsen Kaldahl, *Nurbs and Loop 1*, 2007.

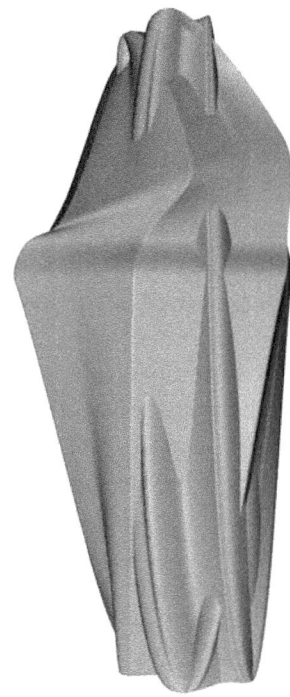

Figure 28 Martin Bodilsen Kaldahl, digital rendering, 2007.

record of the position and tensions of their muscles. For any machine subject to a varied external environment, in order to act effectively it is necessary that information concerning the results of its own action be furnished to it as part of the information on which it must continue to act. For example, if we are running an elevator, it is not enough to open the outside door because the orders we have given should make the elevator be at that door at the time we open it. It is important that the release for opening the door be dependent on the fact that the elevator is actually at the door; otherwise something might have detained it, and the passenger might step into the empty shaft. This control of a machine on the basis of its *actual* performance rather than its *expected* performance is known as *feedback,* and involves sensory members which are actuated by motor members and perform the function of *tell-tales* or *monitors*—that is, of elements which indicate a performance.

I have just mentioned the elevator as an example of feedback. There are other cases where feedback is even more essential. For example, a gun-pointer takes information from his instruments of observation, and conveys it to the gun so that the latter will point in such a direction that the missile will pass through the moving target at some time. Now, the gun itself must be used under all conditions of weather. In some of these cases the grease is warm, and the gun swings easily and rapidly. Under other conditions the grease is frozen or mixed with sand, and the gun is slow to answer the orders given to it. If these orders are reinforced by an extra push given when the gun fails to respond easily to the orders and lags behind them, then the error of the gun-pointer will be decreased. In order to obtain a performance as uniform as possible, it is

made which transform input data into output data are the high-speed electrical computing machines, of which I shall speak later in more detail. The determination of the mode of conduct of these machines is given through a special sort of input, which frequently consists of punched cards or tapes or of magnetized wires, and which determines the way in which the machine is going to act in one operation, as distinct from the way in which it might have acted in another. Because of the frequent use of punched or magnetic tape in the control, the data which are fed in, and which indicate the mode of operation of one of these machines for combining information, are called the *taping.* [. . .]

I have said that man and the animal have a kinaesthetic sense, by which they keep a customary to put into the gun a control feedback element which reads the lag of the gun behind the position it should have according to the orders given it, and which uses this difference to give the gun an extra push.

It is true that precautions must be taken so that the push is not too hard, for if it is, the gun will swing past its proper position, and will have to be pulled back in a series of oscillations, which may well become wider and wider, and lead to a disastrous instability. If the feedback is controlled and kept within limits sufficiently stringent, this will not occur, and the existence of the feedback will increase the stability of performance of the gun. In other words, the performance will become less dependent on the frictional load; or what is the same thing, on the drag created by the stiffness of the grease.

Something very similar to this occurs in human action. If I pick up my cigar, I do not will to move any specific muscles. Indeed in many cases, I do not know what those muscles are. What I do is to turn into action a certain feedback mechanism; namely, a reflex in which the amount by which I have yet failed to pick up the cigar is turned into a new and increased order to the lagging muscles, whichever they may be. In this way, a fairly uniform voluntary command will enable the same task to be performed from widely varying initial positions, and irrespective of the decrease of contraction due to fatigue of the muscles. Similarly, when I drive a car, I do not follow out a series of commands dependent simply on a mental image of the road and the task I am doing. If I find the car swerving too much to the left, that causes me to turn it to the right; and if I find it swerving too much to the right, that causes me to pull it to the left. This depends on the actual performance of the car, and not simply on the

road; and it allows me to drive with nearly equal efficiency a light Austin or a heavy truck, without having formed separate habits for the driving of the two. I shall have more to say about this in the chapter in this book on special machines, where we shall discuss the service that can be done to neuropathology by the study of machines with defects in performance similar to those occurring in the human mechanism.

It is my thesis that the operation of the living individual and the operation of some of the newer communication machines are precisely parallel. Both of them have sensory receptors as one stage in their cycle of operation: that is, in both of them there exists a special apparatus for collecting information from the outer world at low energy levels, and for making it available in the operation of the individual or of the machine. In both cases these external messages are not taken *neat,* but through the internal transforming powers of the apparatus, whether it be alive or dead. The information is then turned into a new form available for the further stages of performance. In both the animal and the machine this performance is made to be effective on the outer world. In both of them, their *performed* action on the outer world, and not merely their *intended* action, is reported back to the central regulatory apparatus. This complex of behavior is ignored by the average man, and in particular does not play the role that it should in our habitual analysis of society.

This is true whether we consider human beings alone, or in conjunction with types of automata which participate in a two-way relation with the world about them. In this, our view of society differs from the ideal of society which is held by many Fascists, Strong Men in Business, and Government. Similar men of ambition for power are not entirely unknown in scientific and educational institutions. Such people prefer an organization in which all orders come from above, and none return. The human beings under them have been reduced to the level of effectors for a supposedly higher nervous organism. I wish to devote this book to a protest against this inhuman use of human beings; for in my mind, any use of a human being in which less is demanded of him and less is attributed to him than his full status is a degradation and a waste. It is a degradation to a human being to chain him to an oar and use him as a source of power; but it is an almost equal degradation to assign him a purely repetitive task in a factory, which demands less than a millionth of his brain capacity. It is simpler to organize a factory or galley which uses individual human beings for a trivial fraction of their worth than it is to provide a world in which they can grow to their full stature. Those who suffer from a power complex find the mechanization of man a simple way to realize their ambitions. I say, that this easy path

to power is in fact not only a rejection of everything that I consider to be of moral worth in the human race, but also a rejection of our now very tenuous opportunities for a considerable period of human survival.

NOTES

1. Norbert Wiener, *Cybernetics: Control and Communication in the Animal and the Machine* (Boston: Technology Press, 1948), p. 37–8.
2. Richard Barbrook, *Imaginary Futures* (London: Pluto Press, 2007), p. 46.

FURTHER READING

Flo Conway and Jim Siegelman, *Dark Hero of the Information Age: In Search of Norbert Wiener, the Father of Cybernetics* (New York: Basic Books, 2005).

Steve J. Heims, *Constructing a Social Science for Postwar America: The Cybernetics Group, 1946–1953* (Cambridge, MA: MIT Press, 1993).

Anson Rabinach, *The Human Motor: Energy, Fatigue and the Origins of Modernity* (Berkeley: University of California Press, 1990).

40

ABSTRACTING CRAFT:
THE PRACTICED DIGITAL HAND

Malcolm McCullough

What are the borders of craft? That question has been around as long as there has been craft theory, but it has taken on a new aspect with the rise of digital culture. In 1996, digital design specialist Malcolm McCullough set the terms of the debate with an initially counterintuitive proposal: the digital restores craft to center stage. The argument was based on the rapidly increasing sophistication of design interfaces—the 'hands-on' quality of rendering and modeling software. While computers might seem to increase the mediation between the hand of the maker and the finished product, McCullough pointed to the increasing tactility of digital design processes and predicted that this would only increase in years to come. He also suggested that the increasing power of digital design and fabrication tools would make small-shop production economically competitive with large-scale manufacturers. Developments have very much borne out his ideas. Researchers today, working on the production side of the new field of 'interaction design', are developing new tools like a 'digital hammer' that simulates the process of raising metal by hand. Of course, such technologies always involve at least one remove from the physical artifact, but for McCullough that is less important than the degree of sensitivity that they support within the design process—with all the aesthetic and ethical connotations that this may entail.

Malcolm McCullough, excerpts from *Abstracting Craft: The Practiced Digital Hand* (Cambridge, MA: MIT Press, 1997).

WHAT IS CRAFT?

Tools and technologies have both assisted and opposed the hand throughout history; the relation is not necessarily adversarial. Although we find no recourse to traditional production—no more than there was in the time of Ruskin—nevertheless we must look very closely at craft. As a part of developing more engaging technology, as well as developing a more receptive attitude toward new opportunities raised by technology, we must understand what matters in traditional notions of practical, form-giving work. This will take some study of tools, some study of human-computer interaction, and some study of practicing the digital medium. But it will not require us to identify what (if anything) is truly made by hand. Nor does our praise of hands necessarily mean condemnation of technology.

As a point of departure, consider the example of a skilled computer graphics artisan—if we may use this word. His or her hands are performing a sophisticated and unprecedented set of actions. These motions are quick, small, and repetitive, as in much traditional handwork, but somehow they differ. For one thing, they are faster—in fact, their rates matter quite a bit. They do not rely on pressure so much as position, velocity, or acceleration. The artisan's eye is not on the hand but elsewhere, on a screen. The actions have a practical component, and the skill may be practiced for a livelihood and a trade identity. If we test a description of this work against Diderot's description of craft, almost every word fits.

[. . .]

Technology is—literally in the Greek—the study of skill. It is order imposed on skill, and it is also the apparatus derived from applying the results of study. Technique is a method of doing something, possibly skilled, possibly using technology. Tools, machines, computers, materials, and media all will be explored in greater detail later on; but it should be safe to say that given ubiquitous technical examples such as oil painting or motion pictures, technology can become a medium, or at least the basis of a medium.

Now there is reason to explore the possibility of craft in the emerging realm of information technology—with the computer as a medium. This hardly fits the conventional usage of the word "craft," for the usual meaning opposes high-technology processes in which the hand plays a diminished role. Thus the proposal of craft in the electronic medium is something of a paradox. But can we, here in the computer age, with fully optimistic and benevolent intent, suggest that the word needs a more inclusive definition?

This seems to be happening anyway. The word has resurfaced in popular usage—but as a verb. People "craft" everything from business memos to good stout beer. In digital production, craft refers to the condition where people apply standard technological means to unanticipated or indescribable ends. Works of computer animation, geometric modeling, and spatial databases get "crafted" when experts use limited software capacities resourcefully, imaginatively, and in compensation for the inadequacies of prepackaged, hard-coded operations. As a verb, "to craft" seemingly means to participate skillfully in some small-scale process. This implies several things. First, it affirms that the results of involved work still surpass the results of detached work. To craft is to care. Second, it suggests that partnerships with technology are better than autonomous technology. For example, personal mastery of open-ended software can take computers places that deterministic software code cannot. Third, to craft implies working at a personal scale—acting locally in reaction to anonymous, globalized industrial production—hence its appeal in describing phenomena such as microbreweries. Finally, the usage of "craft" as a verb evades the persistent stigma that has attached itself to the noun. The noun suggests class differences and amateurism. For example, craft still recalls the provincial dilution of the Arts and Crafts movement into what now consists of folk art at best, and rustic shops full of tourist trinkets at worst. Craft is seldom any longer practical trade, but it is not yet often art. It is outside of academic consideration: ever since mechanization has taken command, craft has been stranded in bourgeois territory where few self-respecting aestheticians would dare to tread. But new usage may change this situation. Based on observations of a linguistic tendency, and with a desire to explore an academically belittled area, this book is a meditation on the seeming paradox of intangible craft.

Craft remains skilled work applied toward practical ends. It is indescribable talent with describable aims. It is habitual skilled practice with particular tools, materials, or media, for the purpose of making increasingly well-executed artifacts. Craft is the application of personal knowledge to the giving of form. It is the condition in which the inherent qualities and economies of the media are encouraged to shape both process and products. It is not about standardized artifacts, however. It is not industrial design. It remains about the individually prepared artifact, which is newly practical due to digital computing. Craft is certainly an application of skill, and it may yet involve the skilled hand.

Thus the defense of skill may no longer remain a losing philosophical position. If previously it was usual to assume that computation would only worsen the hand-mind splits engendered by industrialism, now we might reconsider this problem. We might observe how software usage is restoring some respect for mastery. We might also note the invention of technologies that support the subtleties of the hand. Although most people have failed to perceive in the technology's fledgling states any capacities for new kinds of active skill, perhaps it is still early in the game, and many of these views may well shift.

The question is largely generational: younger people, for whom computing is normal, may shape the most change. As the columnist and MIT Media Lab founder Nicholas Negroponte has noted, "All that seems to count, like learning French in France, is being a child." Within computing, "The haves and have-nots are now the young and the old."[1] Anyone on the edge of this generational change—say between the ages of thirty and forty—should be keenly aware of this distinction.

DIRECT MANIPULATION

The first glimmer of digital craft, and the main breakthrough to popular computation as we know it, was the introduction of pointing. "Direct manipulation" is a term coined in 1983 by software designer Ben Shneiderman to describe a principle that we now take for granted: pointing at our work with a mouse. More specifically, the expression referred to the combination of three fundamental activities: (1) continuous visibility of the object of interest; (2) rapid, incremental, reversible, physical actions on the object; and (3) immediately visible results.[2] The slogan "What you see is what you get" popularized the essence of this technical combination, but hand-eye coordination meant more than just visual fidelity.

The Macintosh popularized the direct manipulation strategy in the mid-1980s, and MacPaint and MacDraw became the first commercially successful direct manipulation programs. Here were the first uses of tool icons, modified cursors, and realtime pixel coloration (well, black and white at least). Here the graphical objects first developed grips and intrinsic operations, such as selecting, stretching, and replicating. Here, at last, you could draw without typing in numbers on a keyboard.

This early Macintosh was commonly referred to as the first human-computer interface good enough to criticize. By now it is a familiar and storied lineage: Xerox PARC in the late 1970s, then Apple, today Microsoft. Soon the Macintosh's direct manipulation format was imitated by most of its competitors. Some expanded it into three dimensions, as with the specialized Silicon Graphics machines of the late 1980s. And when Microsoft promoted it to the mainstream millions in the form of Windows, direct manipulation based on graphical user interfaces became the unquestioned norm.

The best measure of direct manipulation as a basis for digital craft is its capacity for continuous actions. Direct manipulation's continuity depends on having enough computing speed to calculate realtime graphical feedback, so this capacity improves almost as fast as the power of the chips themselves. It may have let us start by manipulating lines and squares in MacDraw, but there is no reason why direct manipulation cannot also be applied to gestures, three-dimensional renderings, tactile textures, complex multi modal structures, or abstracted architectures of information. In research settings, and in some specialized commercial products, it already does so.

Touch technology nonetheless remains far behind other aspects of human-computer interaction. Most interaction technology has emphasized output, not input; foreground tasks, not background contexts; and visualization, not a more fully rounded sensory balance that one might call "perceptualization." So far, the much ballyhooed "look and feel" of contemporary computing is almost all look and hardly any feel.[3] For one thing, the sense of touch is relatively difficult to engineer. This is partly because it does not rely on a particular organ like the eye or ear—unless that is the hand. Pressure feedback is relatively straightforward to engineer—some arcade games do this—but temperature, texture, and wide-area contact prove more difficult. If there is something common to much research on tactile computing, that might be an emphasis on action. Researchers often use the term *haptic*, which means the exploratory and manipulative aspects of touch, as opposed to passive sensation. Some fields such as music and medicine have been advancing pressure components of touch quite rapidly, and specialized research in haptic computing is fairly easy to find within them. For example, many remote surgical operations conduct delicate touch by means of computer technology.

Haptic skill should play an equally important role in the fields of design and fabrication; but these fields have not come as far in realizing their potential. Nor have the findings of research in better-funded fields yet found their way into much merchandised software for designers. No, the two-dimensional mouse, point-and-click form of direct manipulation has prevailed for a strangely prolonged period of time. And although there is every indication that human-computer interaction is evolving toward much more satisfactory haptic engagement (among other perceptual dimensions), there is also evidence that this just might take a while.[4]

Without touch, in the meantime, perhaps we are stretching to call direct manipulation craft. There is a natural objection: What good are computers, except perhaps for mundane documentation, if you cannot even touch your work? The fact that traditional craft endures at all is because it satisfies some deep need for direct experience—and most computers are not yet providing that experience.

However, other developments are at least partially compensating for the limited role of the hands. For example, sophisticated motion tracking can incorporate gesture, and large flat-panel displays can unite the computer's metaphorical "desktop" with a real physical desktop, so as to escape the limits of screen pointing. Multimodal activities, such as coupling actions to sounds, are beginning to emerge. These many techniques first appear on the market in computer video games, for multisensory activities awaken the intuition and heighten the sense of drama, but this suggests much capacity for talent in other applications as well. As some of these interaction developments disseminate into practice, it may seem that we do not need to wait for the arrival of haptic interfaces before we raise the possibility of craft. Rather, we can begin to develop a provisional sensibility based on what we have, and wait for eventual developments in touch technology to remove our remaining reservations.

Already plenty of skills have emerged amid the application of ordinary commercial software. This is difficult to generalize because people work in so many different contexts, but obviously a lot of computer usage involves a good deal more than coded memorization of routines. Learning involves more than operational training, for practice and outlook also contribute to expertise. If you use a computer, you might observe several aspects of inarticulable skill in your everyday work. You might feel that this begins from manual dexterity. You have probably learned to find mouse positions and control key combinations by reflex. Your hands and eyes become closely coordinated despite being focused on different objects. You may recognize the importance of sequence: motions usually last only fractions of a second, but they occur in a constant stream, where their rate matters. With practice you become able to execute tightly synchronized combinations, as if you were playing an instrument.

Besides manual dexterity, you may feel some intellectual agility. You will learn to build mental models, and to switch frames of reference when necessary. You alertly monitor feedback from a variety of sources, and recognize and recover from errors before they compound themselves. You benefit from the habit of identifying patterns—and using them to work at a higher level. You learn to read system states in multiple ways, and this versatility lets you go about operations in whatever manner is currently most convenient. Your hands too, may work together in complementary modes, and each may move quickly between modes.

At times you may think that computer work mainly just tests your patience. It is incremental, like chiseling away at a piece of stone. It involves unexplained roadblocks and glitches. It is monotonous, fatiguing, and yet full of interruptions on a whole spectrum of time scales—some a couple of minutes, some a couple of seconds, some just subliminal fractions of a second. Unlike the soothing quality of continuous process in traditional work, this staccato pace is irritating. Fortunately you can compensate. For example, you may know how to slow down to match the pace of near realtime processes. You learn to cut down on unnecessary motions or state changes. You know when to put aside direct manipulations and resort to command languages or delegated agents. You *work around* problems: when one approach is blocked, you quickly find another that is open.

Meanwhile you experience new kinds of continuous actions. As computers become faster, and interfaces improve, more processes become operable by continuous strokes instead of discrete selections. This switch from discrete to continuous distinguishes digital craft from mere mechanical machine operation. When some continuous pointer motions become more precise than all but those required for the finest traditional tool applications, you might discover them quite satisfying.

Above all, you develop a contextual awareness. Like a good pianist you improve your ability to push what you have learned into a subconscious background, so that you don't have to keep so much in mind at any one time. Instead of thinking the actions, you feel the actions—and actions stir your memory, and give you a better sense of inhabiting your work. As an expert you sense what to try when; how far a medium can be pushed; when to check up on a process; which tool to use for what job. If you have used computers much you know this kind of judgment, or know that you want to learn.

Something very important is happening, and it has to do with the growing capacity of electronic symbol-processing technology for a range of skillful practices. There are three essential components to this sea change. First, the tools have become much more affordable. This reverses perhaps the greatest blow against the artisan two centuries ago, namely the establishment of means of production too large and complex for any individual to afford. As a result, industrial-age stereotypes about the complicity of technology with authoritarian or institutional agendas may soon be irrelevant. Second, human-computer interface technology has improved, and is now beginning to diversify. This means we are on the verge of much greater capacity for talent. As computers balance a greater breadth of input with their current emphasis on output (and so relieve us of too much burden of instruction), we should find it easier to work skillfully. Better gestures, more sensory combinations, and improved three-dimensional frameworks should open up many new niches of practice. Finally, there is a growing appreciation of new abstractions. Increasingly, computers let us treat abstract relations as visible, workable things. As a result, new kinds and levels of work become viable. This is partly due to better support for active skills, and partly due to better abstractions of background contextual awareness.

Histories of technology reveal the increasing abstraction of work. Successive levels of invention have freed us from hunting down our next meal, breaking our backs in the fields, sweating over the forge, and numbing our minds with accounting. Each level forms a layer over the old, rather than casting it aside, as in the stages of a natural growth. This means that even if new abstractions eventually become the most prominent methods, they do not replace existing activities so much as transform or complement them.

Because the move to electronic means of production is now in full swing, we must carefully consider potential losses and gains at a highly abstract order. Although computers are useful, are they good? But this question may be too broad and unanswerable, so let us inform it with a simpler question: Does further abstraction necessitate further decline of human skill, particularly of the hand? Let us direct our curiosities and practices in the high-tech realm toward one of the most humane of ends: craftsmanship.

If the beginnings of computing ultimately appear to future historians as the most significant outward expression of our time, they are not likely to do so on the basis of functional utility alone. Social and aesthetic concerns will matter too. The artifacts and practices that computing produces will demand—and reward—more refined interpretations. Note that traditionally it is in interpretation that we have used the word "craft" most broadly: the writer's craft, the actor's craft, and the conductor's craft join those of the cobbler and carpenter. What all such crafts share is not just technique, or hard work on form, but also a probing of their medium's capacity, a passion for practice, and moral value as an activity independent of what is produced. Is there any reason to expect these in the electronic realm? We must make them our goal.

NOTES

1. Nicholas Negroponte, *Being Digital* (New York: Knopf, 1995), p. 204.
2. Ben Shneiderman, 'Direct Manipulation: A step beyond programming languages' in

IEEE computer 16: (8), and cited in interview in Jenny Preece, Yvonne Rogers, Helen Sharp, David Benyon, Simon Holland, and Tom Cary, *Human-Computer Interaction* (Reading, MA: Addison-Wesley, 1994), p. 207.

3. These ideas have been well voiced by Bill Buxton, of Alias/Wavefront Research, among others.

4. Mark Weiser, 'The Computer for the 21st Century', *Scientific American*, special issue, 'The Computer for the 21st Century'.

FURTHER READING

Malcolm McCullough, *Digital Ground: Architecture, Pervasive Computing, and Environmental Knowing* (Cambridge, MA: MIT Press, 2004).

Malcolm McCullough and William Mitchell, *Digital Design Media* (New York: Van Nostrand Reinhold, 1991).

Bill Moggridge, *Designing Interactions* (Cambridge, MA: MIT Press, 2007).

Bruce Sterling, *Shaping Things* (Cambridge, MA: MIT Press, 2005).

41

'DIGITAL ARTISANS MANIFESTO', EUROPEAN DIGITAL ARTISANS NETWORK

Richard Barbrook and Pit Schultz

It is already difficult to recall the dot-com boom of the late 1990s—a time of vertiginous economic transformation, it seemed, when all the rules of production and consumption would change overnight. Whether this promise has been fulfilled is an open question, but the excitement and confusion of the moment come alive in the following text. It is a faux manifesto, described by one of its authors, the hypermedia theorist Richard Barbrook, as 'a sort of Situationist joke' about the fate of graphic designers in the new information age. Written with the Berlin-based artist Pit Schultz, the statement issued on behalf of an imaginary labor union of 'Digital Artisans' was written for the 1997 conference Nettime, held in Ljubljana, Slovenia. It addresses, only half-jokingly, the role that graphic designers were assuming at the time, as corporations, institutions and individuals began to establish themselves on the Web. Chagrined at the subcontracting economy that resulted and its relative lack of creative autonomy, Barbrook and Schultz borrowed liberally (even plagiaristically) from previous avant-garde manifestos, worker's rights declarations and Marxist theory. Central to their fictional organization was the principle of craft skill—reimagined for the new millennium but still founded in the sacrosanct Ruskinian principle of individual freedom in labor.

Richard Barbrook and Pit Schultz, 'The Digital Artisans Manifesto', 1997.

MAKING THE FUTURE

1. We are the digital artisans. We celebrate the Promethean power of our labour and imagination to shape the virtual world. By hacking, coding, designing and mixing, we build the wired future through our own efforts and inventiveness.

2. We are not the passive victims of uncontrollable market forces and technological changes. Without our daily work, there would be no goods or services to trade. Without our animating presence, information technologies would just be inert metal, plastic and silicon. Nothing can happen inside cyberspace without our creative labour. We are the only subjects of history.

3. The emergence of the Net signifies neither the final triumph of economic alienation nor the replacement of humanity by machines. On the contrary, the information revolution is the latest stage in the emancipatory project of modernity. History is nothing but the development of human freedom.

4. We will shape the new information technologies in our own interests. Although they were originally developed to reinforce hierarchical power, the full potential of the Net and computing can only be realised through our autonomous and creative labour. We will transform the

machines of domination into the technologies of liberation.

5. We will contribute to the process of democratic emancipation. As digital artisans, we will come together to promote the development of our trade. As citizens, we will participate within republican politics. As Europeans, we will help to break down national and ethnic barriers both inside and outside of our continent.

THE PRESENT MOMENT

6. Freedom today is now often just the choice between commodities rather the ability to determine our own lives. Over the past two hundred years, the factory system has dramatically increased our material wealth at the cost of removing all meaningful participation in work. Even poorer members of European societies can now live better than the kings and aristocrats of earlier times. However the joys of consumerism are usually constrained by the boredom of most jobs.

7. Since 1968, the desire for increased monetary rewards has increasingly been supplemented by demands for increased autonomy at work. In the European Union and elsewhere, neo-liberals have tried to recuperate these aspirations through their policies of marketisation and privatisation. If we are talented workers in the 'cutting-edge' industries like hypermedia and computing, we are promised the possibility of becoming hip and rich entrepreneurs by the Californian ideologues. They want to recruit us as members of the 'virtual class' which seeks to dominate the hypermedia and computing industries.

8. Yet these neo-liberal panaceas provide no real solutions. Free market policies don't just brutalise our societies and ignore environmental degradation. Above all, they cannot remove alienation within the workplace. Under neo-liberalism, individuals are only allowed to exercise their own autonomy in deal-making rather than through making things. We cannot express ourselves directly by constructing useful and beautiful virtual artifacts.

9. For those of us who want to be truly creative in hypermedia and computing, the only practical solution is to become digital artisans. The rapid spread of personal computing and now the Net are the technological expressions of this desire for autonomous work. Escaping from the petty controls of the shopfloor and the office, we can rediscover the individual independence enjoyed by craftspeople during proto-industrialism. We rejoice in the privilege of becoming digital artisans.

10. We create virtual artifacts for money and for fun. We work both in the money-commodity economy and in the gift economy of the Net. When we take a contract, we are happy to earn enough to pay for our necessities and luxuries through our labours as digital artisans. At the same time, we also enjoy exercising our abilities for our own amusement and for the wider community. Whether working for money or for fun, we always take pride in our craft skills. We take pleasure in pushing the cultural and technical limits as far forward as possible. We are the pioneers of the modern.

11. The revival of artisanship is not a return to a low-tech and impoverished past. Skilled workers are best able to assert their autonomy precisely within the most technologically advanced industries. The new artisans are better educated and can earn much more money. In earlier stages of modernity, factory labourers symbolised of the promise of industrialism. Today, as digital artisans, we now express the

emancipatory potential of the information age. We are the promise of history.

12. We not only admire the individualism of our artisan forebears, but also we will learn from their sociability. We are not petit-bourgeois egoists. We live within the highly collective institutions of the market and the state. For many people, autonomy over their working lives has often also involved accepting the insecurity of short-term contracts and the withdrawal of welfare provisions. We can only mitigate these problems through our own collective action. As digital artisans, we need to come together to promote our common interests.

13. We believe that digital artisans within this continent now need to form their own craft organisation. In early modernity, artisans enhanced their individual autonomy by organising themselves into trade associations. We proclaim that the collective expression of our trade will be: the European Digital Artisans Network (EDAN).

THE AIMS OF EDAN

14. We urge everyone who is working within hypermedia, computing and associated professions on this continent to join EDAN. We call on digital artisans to form branches of the network in each of the member states of the European Union and its associated countries. By forming EDAN, we will also be creating a means of forging links between European digital artisans and those from elsewhere in the world. We will strive for cooperation in work and in play with our fellow artisans in all countries.

15. We believe that the principal task of EDAN is to enhance the exercise of our craft skills. By collaborating together, we can protect ourselves against those who

wish to impose their self-interests upon us. By having a strong collective identity, we will enjoy more individual autonomy over our own working lives.

16. EDAN will celebrate our creative genius as digital artisans. The network will act as the collective memory about the achievements of digital artisans within Europe. It will publicise outstanding 'masterpieces' of craft skill made by its members among the trade and to the wider public.

17. The network will be the social meeting-place for digital artisans from across Europe. EDAN will organise festivals, conferences and congresses where we can meet to organise, discuss and party. We believe that digital artisans should express their collective identity by regularly celebrating together in private and public.

18. EDAN will collect detailed knowledge about the trade in the different regions of Europe. It will aim to provide information about best practice in contracts, copyright agreements and other business arrangements to its members. The network will also be a source of contacts in each locality for digital artisans looking for work in different areas of Europe.

19. We believe that what cannot be organised by our own autonomous efforts can only be provided through democratic political institutions. The network will lobby for changes in local, national and European legislation which can enhance our working lives as digital artisans. As concerned citizens, we will also support the fullest development of public welfare services.

20. EDAN will campaign for European governments to put more resources into the theoretical and practical education of digital artisans in schools and universities. The network will facilitate links between educational institutions teaching hypermedia and computing across the continent. EDAN also believes that

publicly-funded research is necessary for the fullest development of our industry.

21. EDAN will urge the European Union to launch a public works programme to build a broadband fibre-optic network linking all households and businesses. We believe in the principle of universal service: everyone should have Net access at the cheapest possible price. No society can call itself truly democratic until all citizens can directly exercise their right to media freedom over the Net.

22. We will campaign for the creation of 'electronic public libraries' where on-line educational and cultural resources are made accessible to everyone for free. Public investment in digital methods of delivering life-long learning is needed to create an information society. The Net should become the encyclopedia of all knowledge: the primary resource for the new Enlightenment.

23. We believe that the role of the hi-tech gift economy should be further enhanced. As the history of the Net has shown, d.i.y. culture is now an essential part of the process of social development. Without hacking, piracy, shareware and open architecture systems, the limitations of the money-commodity economy would have prevented the construction of the Net. EDAN also supports open access as means of people beginning to learn the skills of hypermedia and computing. The promotion of d.i.y. culture within the Net is now a precondition for the successful construction of cyberspace.

24. We are the digital artisans. We are building the information society of the future. We have come together to advance our collective interests and those of our fellow citizens. We are organised as the European Network of Digital Artisans. Join us.

Digital Artisans of Europe Unite!

42

CRAFT VERSUS DESIGN:
MOVING BEYOND A TIRED DICHOTOMY

Rafael Cardoso

In the following contribution to this anthology, the design historian Rafael Cardoso provides a useful review of the long history of debate about craft's relations with industry and points the way towards new approaches to that dichotomy. Cardoso, associate professor at the Pontifícia Universidade Católica, Rio de Janeiro, is a specialist on nineteenth-century design history, but he also pays close attention to new developments in technology and distribution. This combination of historical and contemporary expertise is evident in the essay, which offers a novel and important argument about craft and industry. As Cardoso points out, these two terms were—prior to the onset of modernity—synonymous with one another. Now, after two centuries of being conceptually severed, there is once again a convergence between the two terms. This is chiefly due to the fact that customized and small-batch (rather than homogenous, mass-produced) goods are becoming more and more commonplace. This trend marks a return of the bespoke, a relation between producers and consumers that has not been the norm in industrialized economies since the early nineteenth century. Hence, craft's relations with design seem to be coming full circle. Of course, this transformation owes much more to new digital technologies than to old hand techniques, but drawing on McCullough's arguments in Abstracting Craft, *Cardoso argues that the latter is implicated in the former. Theories of craft are also newly relevant in this production scenario, because they may offer models of ethical, aesthetic and authorial thinking that will be important for the future.*

Rafael Cardoso, 'Craft versus Design: Moving Beyond a
    Tired Dichotomy', 2008.

The notion of 'craft' has taken quite a beating over the past hundred years or so. Enthusiasts of craft production are routinely cast as ineffectual utopians or hopeless Romantics, vainly attempting to turn back the hands of time. Since the mid-nineteenth century, at least, 'industry' and 'progress' have been more or less equated in the public mind. The common-sense dictum that 'you can't fight progress' has most often come to the fore when the topic of discussion is mechanization or some other perceived technological imperative (and certainly not the advancement of social mores, an arena in which progress is often combated tooth and nail). Despite a revision of such ideas of 'progress' over the latter half of the twentieth century, particularly in academic circles, 'craft' continues to be viewed as a historical stage superseded by 'industry'.[1]

The mythology is familiar enough: in an industrial setting, as machines replaced workers, the role traditionally played by 'craft' would become the province of 'design'. No longer would the skilled artisan prevail, crafting his

Figure 29 Luke Limner, *Artist and Artisan*; detail from the frontispiece for *Suggestions in Design* (1853).

wares one by one, but the clever engineer who would direct machines to spew out cheap and plentiful wares, all according to a preconceived design. Like all myths, it contains a great deal of fantasy, as well as an underlying parcel of truth. Design and craft have both evolved considerably since their nineteenth-century redefinition in the immediate wake of industrialization. Changes in the paradigm of industrial production have often outpaced society's capacity to keep track of them, meaning that neither field really conforms to most people's expectations. The terms need to be unpacked—taken out of inverted commas and considered thoughtfully—if we are to advance at all in this discussion.

ORIGINS OF THE CONCEPT OF CRAFT

Craft and industry are old words, dating back several centuries in most European languages. However, their usage has changed drastically since the nineteenth century, dawn of what has come to be known as the industrial age. Prior to that time, industry usually meant skill, dexterity, diligence, assiduity. An echo of that initial meaning is still available in the application of the adjective 'industrious' to describe a personal quality. Since the nineteenth century, of course, industry has come to signify something very different, becoming inextricably bound up with the ideal of factory production. Craft once meant power, strength or skill, evolving slowly into the idea of a specific trade or calling and spawning the more persistent notion of the craftsman, probably sometime around the fifteenth century. Although the etymologies are very different, the original figure of the craftsman is equivalent in its social historical place to that of the artisan, a term finding equivalents in all Latin languages. In the usage prevalent between the sixteenth and nineteenth centuries, both craftsman and artisan were pertinent descriptions for a worker engaged in the material production of artifacts, particularly through the application of what were conceived as the mechanical arts.

Somewhat surprising, then, is the fact that craftsman, as 'one who cultivates one of the Fine Arts', is a meaning dating from as recently as 1876 (according to the OED), closely contemporary to the derivative form craftsmanship. This is no fluke of the English language. Although one of the first published uses of *artisan*, in French, is dated 1546, the derivative form *artisanat* (generally, translated as handicraft) is a nineteenth-century innovation.

The pattern repeats itself in other languages. In Portuguese, the first European language to gain global currency, *artesão* is a fifteenth-century word, but *artesanato* is twentieth-century. Clearly, a major historical divide separates a first generation of words denoting workers who manufacture artifacts from a related set of terms denoting the product of their manual labour as an abstracted concept: handicraft, in the current English usage (precisely equivalent to *artisanat* or *artesanato*), or the even more reified craftsmanship, for which there is no Latin-language equivalent. It took three to four centuries, and more than one revolution, for the current ideal of craft—i.e. the making of usable artifacts in a given material medium, done individually and by hand, preferably displaying great mastery—to develop out of the prior conception of the artisan as an ordinary worker, engaged in production that might or might not be exclusively handmade.

The altered nineteenth-century understanding of craft is the logical consequence of a prior realignment, dating back to the sixteenth century, when the original terms designating artisans came into broad usage. The turning point of both major shifts in meaning revolves around the dissociation between manual and intellectual labour.[2] The distinction between liberal arts (those pertaining to the education of a free person, *libera*) and mechanical arts is ancient. By the mediaeval period in Western history, the seven liberal arts had been classified into trivium (grammar, rhetoric, logic) and quadrivium (geometry, arithmetic, music, astronomy). Less precise was the classification of the mechanical arts, but they were always presumed to correspond to a greater degree of dexterity and lesser intellectual accomplishment. The first radical shift in this balance took place during the Renaissance period, particularly in Italy, when a consensus was built that architecture, painting and sculpture should be promoted to a status equivalent to the liberal arts. Thanks to the influence of Alberti, Da Vinci, and Vasari, among others, these arts came to be perceived as meriting special consideration, and their social status was raised to a rank befitting the exercise of elevated minds and the education of noble persons.

The first academies of art took shape at about this time, commencing a slow and painful progression away from a trade and guild model of artistic production towards the current conception of the fine artist. By the mid eighteenth century, under D'Alembert's influence, the scheme of five *Beaux-Arts* was definitively established: painting, sculpture, architecture, poetry and music. The end result of this centuries-long process was a relegation of all other skillful production of artifacts to a no-man's-land variously known as applied arts, decorative arts or even lesser arts, in which those manual trades (Fr. *métiers*, Pg. *ofícios*) perceived to be deficient in intellectual endeavour were condemned to survive as craft and industry. The extremely tenuous idea of *Beaux-Arts* capped this first great cycle of abstracting the intellectual portion of artistic labour from its manual counterpart, giving rise to the subsequent formation of an international system of academies of art in the nineteenth century.

The completion of this cycle coincides with a second major historical event: the rise of industry, in the modern sense of the term. As noted above, the word's meaning gradually changed during this same period from a primary sense of personal diligence to a more collective notion of factories and factory production. The term factory itself—distantly derived from the Portuguese *feitoria* (the root form of which is, overly literally, a place where things get done)—underwent similar transformations, moving from the idea of a trading station to a large manufactory based upon the principles of scale production, systematic division of labour and some degree of mechanization. In the factory setting—as set forth in Adam Smith's famous example of the manufacture of pins and later developed into a philosophy by the likes of Andrew Ure and Charles Babbage—manual and intellectual labour are made to achieve a maximum degree of separateness. Deskilled, workers are reduced to factory 'hands' and, in Ure's dystopian fantasy of automatic production, would eventually be replaced entirely by machines.[3] Although this never actually happened, the idea was hugely influential, particularly through its direct impact on the writings of Karl Marx.

The fact that these two pivotal points coincided in their timing did not go unnoticed. In a relatively brief historical period, 'fine art' was definitively enthroned as something higher and quite apart from mechanical work; and, in parallel, manual labour was relegated to a status beneath contempt, something to be eliminated from society through the continued perfectibility of machine technology. In the gap ripped brutally in the fabric of society by the upward movement of liberal artists and the downward spiral of working artisans, a new underclass of factory labourers was born. The ancient relationship between hand, work and art, developed over millennia of human activity and dignified by centuries of guild protection in Europe, was rather suddenly put out to pasture. The reaction was not long in coming. At this point, names like A.W.N. Pugin, John Ruskin, Richard Redgrave, Owen Jones, Henry Cole, William Morris—famously enshrined by Nikolaus Pevsner as 'pioneers of modern design' in his brilliant modernist polemic of 1936—creep inevitably into the discussion.

THE GREAT DEBATE ON ART, CRAFT AND INDUSTRY

By the middle decades of the nineteenth century, a loose consensus of opinion had formed asserting that standards of material production had declined in any number of industries, particularly in terms of the artistic value of wares traditionally manufactured by skilled artisans. It is worth pointing out that there is little hard evidence of this supposed decline. What likely did occur was a relative democratization of access to consumer goods as a result of the cheaper cost of newly industrialized areas of manufacture. Taking a cue from Eric Hobsbawm's guiding principle that industrialization 'produces in such vast quantities and at such rapidly diminishing cost as to be no longer dependent on existing demand, but to create its own market',[4] it follows that a whole new (lower) class of consumers was able to purchase, over the first half of the nineteenth century, a range of goods previously beyond their economic reach. As a phenomenon unregistered in world history, popular access to cheap wares must surely have grated upon the genteel sensibilities of the middle-class reformers who came to pioneer the great sea change of design.

Whether or not a fall in manufacturing standards actually did occur is slightly beside the point, seeing as the mere perception of declining taste proved to be such a powerful force. To speak of Pugin, Ruskin and Morris, Cole, Jones and Redgrave is to trace the development of the modern idea of design, if not of the design-related activities that sprang naturally from factory floor, printing shops and the growth of great cities.[5] However, care must be taken to distinguish between widely divergent positions in this crucial place and time that was mid-Victorian Britain. Pevsner's lasting influence on the field of design history has tended to obscure important distinctions; and there is still much confusion as to exactly who did what and when in this dubious arena of 'pioneering'. Extending to others the objection rightly made by Edward Lucie-Smith with regard to Ruskin and Morris, the very effectiveness of 'what was said and written then [has] ever since tended to conceal the real complexity of the issues'.[6]

In particular, Ruskin's thoughts on design are routinely misrepresented, as constituting an attack on industry and a defense of handicraft. The passage most often quoted is a section of *The Seven Lamps of Architecture* (1849) in which he argues that 'all cast and machine work is bad, as work' and, further on, that 'a piece of terra cotta or of plaster of Paris, which has been wrought by the human hand, is worth all the stone in Carrara, cut by machinery'.[7] Habitual readers will know that Ruskin was no stranger to inflamed rhetoric and extreme changes of opinion. Contradictions are, in fact, common in his extensive oeuvre, especially over longer lapses of time. As his political alignment matured from the Toryism of youth to a crypto-socialist outlook in the 1850s and 1860s, Ruskin duly refined his position on the relationship between machine and hand. Though he certainly never became an enthusiast of industrialism, the focus of his critique shifted from the machine as villain to the perversity of a system of manufacture that reduced the worker to an unthinking, unfeeling drone. For Ruskin, the principal error of contemporary society resided in its tacit presumption that the consuming pleasure of the few justified the dehumanization of the many. (He may still be proved right.)

By simply denying the widening gap between art and design, Ruskin hoped to stem the inevitable process of differentiation by social class that was, by then, well under way.

'Try first to manufacture a Raphael; then let Raphael direct your manufacture,' he quipped in the preface to *The Elements of Drawing* (1857), skirting the relevance of any distinction between fine and applied art.[8] This famous injunction was intended as a slap in the face of the South Kensington system and its ambitions of educating designers. His consummate statements on the relationship between art and manufactures came in lectures delivered between 1858 and 1859, and subsequently incorporated into the volume *The Two Paths* (1859). One of the central tenets of these lectures is the idea of the unity of art, subject only to distinctions of degree:

> It would be well if all students would keep clearly in their mind the real distinction between those words which we use so often, "Manufacture," "Art," and "Fine Art." Manufacture is, according to the etymology and right use of the word, "the making of anything by hands,"—directly or indirectly, with or without the help of instruments or machines. [. . .] Then, secondly, Art is the operation of the hand and the intelligence of man together: there is an art of making machinery; there is an art of building ships; an art of making carriages; and so on. [. . .] Then, Fine Art is that in which the hand, the head, and the heart of man go together.[9]

Ruskin's is a holistic vision of art, in which the practitioner must always strive for the highest result, never settling for less than the complete package, but adapting its guiding principles to the specific object at hand. In another lecture, also included in *The Two Paths*, he affirms:

> All art worthy the name is the energy—not of the human body alone, nor of the human soul alone, but of both united, one guiding the other: good craftsmanship and work of the fingers joined with good emotion and work of the heart.[10]

The fairly unusual reference to craftsmanship, as simply work of the fingers, belies any elevation of craft—or design, for that matter—to a status of autonomy.

Whereas Renaissance humanism had sought to elevate art above handwork and industrial capitalism sought further to debase handwork as merely mechanical, Ruskin discards both positions outright. It would seem like something of a losing battle to do so; but out of this essentialist refusal to accept the prevailing terms of debate, comes the paradoxical strength of the Arts and Crafts movement's reinvention of the notion of handicraft. Following Ruskin's cue, Morris conceives of craft not as a reduced form of art, but as the universal expression of human creativity, found in all places and times. His notion of popular art as 'the foundation on which all art stands', expounded in the famous 1881 lecture 'Some Hints on Pattern Designing'—delivered at the very Working Men's College, London, where Ruskin had once put into practice his ideas on art education—depicts craftwork as the essence of artistic expression, thwarted and distorted by the evils of a capitalist system in which labour is systematically debased:

> Every real work of art, even the humblest, is inimitable. I am most sure that all the heaped-up knowledge of modern science, all the energy of modern commerce, all the depth and spirituality of modern thought, cannot reproduce so much as the handicraft of an ignorant, superstitious Berkshire peasant of the fourteenth century; nay, of a wandering Kurdish shepherd, or of a skin-and-bone oppressed Indian ryot.[11]

Openly socialist and internationalist in scope, such an idea of popular art based on handicraft tradition as the true basis of all artistic expression comes very close to the craft ideal cultivated ever since by a small but dedicated band of enthusiasts. For the Arts and Crafts, craft was something of a higher calling, for it was the real art of the people.

SHIFTS IN INDUSTRIAL PARADIGM

Ruskin's contention that all machine work is bad, as work, was probably pretty nearly right in the 1840s, when it was pronounced. In those early days of industrial technology, the lure of mechanization was mainly speed and economy, certainly not quality. To this day, in fact, we have retained the usage of terms like workmanship or craftsmanship to describe a particular notion of finish, even of things made by machines. As demonstrated by David Hounshell, much of the successful machine production of the time—such as Singer sewing machines—relied on a high degree of hand finishing.[12] It should go without saying that technology has advanced considerably since the nineteenth century. However, debates on design and craft rarely take into account the implications of major shifts in industrial paradigm that have occurred over the past century and a half.

The 1850s witnessed the introduction of the so-called American system of manufactures to Britain and, subsequently, the rest of Europe. Premised on the principles of precision machine tools and interchangeable parts, the American system brought the promise of standardization, finally perfected about a half century later in the bicycle and automobile industries. It matured into the shape of Ford's full-blown assembly-line production after 1913. With the mass production of consumer durables, the possibility of design as a controlling force in industry acquired new potency. Coincidentally or not, this is precisely the historical moment when modernist architects and artists began to take an interest in industrial design.

By the early twentieth century, the technological development of machines had changed the industrial scenario considerably. Poor finish was no longer a perceived quality of mechanically made goods. Marcel Duchamp's supposed 1912 comment to Brancusi, to the effect that painting was washed up because no artist could do better than an aeroplane propeller, reveals a novel attitude to machine production.[13] Here—perhaps for the first time from the mouth of an artist—is a frank admission that industrial artifacts possess an elegance and integrity of their own, quite divorced from any considerations of the nobility of handwork. The perfection of mass-production technology signaled a new perfectibility for industrial artifacts; and designers would henceforth play the key role in ensuring that machine work was as attractive as it was efficient and cheap.

Thus, in the Modernist view, design supersedes craft as a historical stage. Through informed effort and precise methodology, design is able to guide mechanical work and render it superior even to handwork. This is, in a sense, the dream of Renaissance humanism come true: art as pure intellectual exercise. As regards the Morrisian cult of craftwork in such a context, there is little option but to accept grudgingly the technical superiority of machines and retreat to the moral high ground. Though craft is clearly unable to compete with the efficiency of machines, it purportedly retains some sort of Benjaminian aura, grounded in the uniqueness of individual manufacture. Imperfections and deviations come to be seen as legitimating characteristics. The historical roles are reversed—perfect machine work is depicted as bad and imperfect handwork as good. This is not a particularly convincing stance for a consumer society

in which more, better and cheaper artifacts are made continually available through the improvement of mass production. Craft is eventually cornered into a position of terminal nostalgia or, worse, of elitism, via a notion of consumer exclusivity.

Fortunately for those with a stake in the craft debate, the industrial paradigm has shifted once again. As Fordism played out its contradictions on a social and political scale, fundamental changes in industrial organization and manufacturing technology swept away many presumptions of mass production.[14] The most glaring example of such change is the application of information technology to production engineering. Small-batch production and even one-offs are now feasible in many industrial settings, thanks to computer-aided design and modeling, rapid prototyping (that is, solid freeform fabrication) and other related advances in digital command systems. Customization of products, at factory level, and on-demand distribution are changing the face of contemporary manufacturing. First, the consumer decides what s/he wants; then, it gets made. This is not so different from the pre-industrial relationship of buyers and artisans.

The roots of this model of flexible production go back much further than the digital technology that has made it viable in more and more industries. So-called lean production strategies, pioneered by Toyota in the 1950s and 1960s, introduced the concept of continuous flow into production engineering, making use of just-in-time strategies of inventory control and autonomation of machine regulatory systems.[15] For nontechnologists, unfamiliar with the jargon, this means that industrial manufacture is no longer obliged to resort to mass production, necessarily, as a means of achieving cost-effectiveness. As the focus shifts from quantity to quality, it becomes possible to do more with less.

We are a long way today from the 'any colour, so long as it's black' school of industrial philosophy. Design is no longer a one-way system, in which manufacturers impose products on a market without a choice. There is an unprecedented degree of reversibility to many manufacturing processes, in which consumer input is seen as a factor conditioning production. In some industries, like automobiles, consumers can do little more than choose the model or accessories they want from a prior range of possibilities. In others, like computer hardware, informed consumers can pretty well build the system they desire. In nearly all industries today, user surveys and marketing research provide feedback that will impact future design and production, sometimes almost immediately. In parallel, use is increasingly viewed as a potentially creative stage of product life cycle, via adaptation, appropriation and customization, beyond point of sale. In the fashion industry, a level of exchange of ideas between makers, distributors, sellers and users is increasingly the rule rather than the exception.

The current manufacturing ideal would seem to approximate the way software is made available online—i.e. a product the consumer acquires initially as an open package, altered and maintained by continual updates and patches. The implications of this state of affairs are multiple, profound and complex. The idea of material production as system and service possesses immense positive resonance for the deepening environmental crisis, insofar as it implies less waste. Things made in small batches, suited to specific needs and amenable to upgrading over time, are less likely to be rapidly discarded than changeless durables that are mass-produced and dumped onto the market.

DESIGN AND CRAFT IN THE CURRENT INDUSTRIAL SCENARIO

In an industrial scenario where mass production would seem to be on the wane, what happens to the dividing line between design and craft? As hinted above, flexible production can be seen as something of a return to the pre-industrial relationship between makers and users. Conceptually, what are the differences between a software patch and building an extension on a house or having the hem of a dress taken down to adapt it to a changing fashion? There are differences, of course; and they need to be teased out in order to make sense of things. Let's ask some more questions. Can a poster, designed and printed by a sole practitioner using digital technology, be construed as a craft object? Can digital manipulation of an image be considered a craft procedure, given the level of dexterity and practice it presupposes? Are automobiles customized by specialist body shops objects of design, craft or art? The answers depend on how the terms of debate are defined; and it is high time we gave serious thought to rethinking the meaning of craft in the digital era.

Malcolm McCullough's 1996 book, *Abstracting Craft*, made a convincing case that digital media might provide a new arena for craft expression. He invokes craft as a timeless ideal, hoping to enlist its values and traditions on behalf of a new medium: computing. However abhorrent the idea of a 'practiced digital hand' might appear to some, the presumption of indirect manipulation, upon which his thesis hinges, is cogent enough. If glass-blowing is a craft—despite the fact that the craftsperson never handles the material directly, but only the tools of the trade—then, clearly, an object can be manipulated indirectly and still be considered to be crafted. The fact that the artifact itself is subsequently given material existence (or not, as the case may be of immaterial objects) by a printer, plotter, rapid prototyping machine or any other digital tool is beside the point. The fact remains that its form is directly determined (i.e. informed) by the knowledge and practice of a maker. For McCullough, the essence of craft has to do with three interrelated concepts: direct experience, personal vision and mastery of a medium. His is a production-based model of craft, premised on the centrality of the craftsperson as enlightened practitioner.[16]

Many craftspeople will certainly feel uncomfortable about equating digital manipulation with handwork. The culture of craft usually presupposes a type of experience that involves bodily exertion and getting your hands dirty. These are not negligible issues, by any means; but to assume them as a theoretical bias would be akin to saying that digital surgery is not surgery because there is less blood and guts. Origami involves about as much bodily exertion as Photoshop; yet, few would question its art and craft credentials. Ruskin, for one, would not be particularly sympathetic to the claim for digital craft (or for origami, for that matter). Given his ideas on the unity of art, he might say that, yes, there is an art to manipulating digital images and models, just as there is to building ships and making carriages. But to say that such manipulation is craftsmanship is to deny the centrality of bodily presence as the defining element of craft—which, from our twenty-first-century vantage point, might seem to engender a too-close-for-comfort approximation with design. To say that it is not, on the other hand, is to reject an opening that might give craft a new breath of life in the digital era.

What are the defining features of craft? Are direct experience, personal vision and mastery of a medium enough to flesh out the specificity of this notoriously protean field? Photographers and filmmakers possess all three qualities and are rarely classified as craftspeople. How about bodily presence, physical exertion and hand skill? Musicians, athletes and dentists come to mind. Perhaps it has to do with materials and materiality? Perhaps, but so do art and industry. Most likely, craft is a term used to refer to a complex cluster of ideas that includes all of the above concepts, and more. In that case, just what is it that makes craft so different, so appealing?

The great nineteenth-century debate that redefined the field of craft set it apart both from industry, as something higher and nobler, and from art, as something broader and more democratic. The second part of this equation is often neglected. For Morris and his followers, craft is a shared practice. For it to be relevant—in the sense promoted by the Arts and Crafts ideal—it must minister to the many and engage with its popular roots. In other words, the enlightened craftsperson's potential value is realized only within a community of like-minded practitioners and users. Morris's legacy is a conception of craft as a practice shared by a network of makers, freely exchanging their wares with the elevated degree of technical appreciation that comes from a personal investment in doing it yourself, rather than merely consuming in a passive manner that which has been manufactured by others.

The digital arena, with its networks and online communities, would seem to be well suited to this dimension; so this objection, per se, can by no means be taken as dismissive of McCullough's case. Rather than stage a tardy polemic, I gladly accept his premise of abstracting craft. The past twelve years, since the publication of his book, have certainly proven that the digital manipulation of images and models, at least, is a medium of expression broad and deep enough to encompass notions of craft as well as art. I would like to go one step further here and abstract the abstraction. McCullough's argument claims a place for the digital within the realm of craft, but it stops short at the point of deriving what such a conclusion means for the club he wants to join.

The impact of digital technology on production engineering challenges the very conception we have of craft as a production-centered ideal. After all, industry is increasingly able to provide high-quality and even custom-made products—'individually crafted', in the rhetoric of much advertising. Products handmade by industrial means, though apparently paradoxical, are not unthinkable in the dawning age of pervasive computing, robotics and bionics. If craft revolves solely around the direct experience of the maker—personal, hands-on and masterful—then it has a diminished role to play in the contemporary world, increasingly impersonal, virtual and chaotic. It will continue to be the exalted pursuit of a rarefied class of practitioners and, in an indefinite future, may well become a curiosity, consigned to the same class of arcane procedures as surgery with saws and scalpels.

Direct experience is not limited to makers; it stands necessarily alongside the experience of users. After all, makers are also users, and this alone is enough to dissolve the abstract boundary separating the two experiences. The frontier—designers would call it an interface—between making and using is precisely the foreground of some of the more innovative efforts of design thinking and research today: i.e. interaction design, design for experience, emotional design. How does the use of a product over its entire life cycle, and not just up to point of manufacture, affect its meaning and feed back into planning and development? If design is seen as an ongoing process—including maintenance and redesign of previously existing products, interfaces and systems—then, it is more akin to the drawn-out and collective experience of raising a child than to a single, explosive moment of conception or birth. This idea of making as collective process possesses strong resonance with pre-industrial ideals of production.

The notion of open-ended processes, incorporating many collaborators and continual adjustments, is central to many areas today, including fine art. Why should craft be immune to this generalized breakdown of authorship and directionality? Quite on the contrary, I would argue that it has managed to survive, thus far, precisely because it appeals to a wide number of users—few of whom are master craftsmen—for its unique insights into the experience of making and as a forum for the exchange of singularities. During the very recently defunct age of Fordism—dominated by a sense of powerlessness in the shadow of mechanization—to enjoy craft was to resist the homogeneity of faceless industry. Mass production was the Goliath, and craft was a brave but insecure David. That model of manufacturing is fast being superseded; and a new industrial paradigm of individuation through consumption is taking its place. To the extent that craft is able to provide an alternative model of individuation, it may be an antidote to the worst excesses of this paradigm. For craft to survive in the face of overt consumerism, however, it must embrace the legacy of its own origin: community and shared interaction.

Craft is not primarily an individual experience, but a collective one. This is what the Arts and Crafts enthusiasts were hoping to recover when they elevated craftsmanship to the level of a reified principle, higher and broader than individual makers. This is what they were trying to restore in their quaint pursuit of mediaeval tradition and guild fellowship. Their vision was a community of producers, not consumers, somewhat like the world described in *News from Nowhere*.[17] At heart, craft aims for a type of creativity that is universal and pervasive. All people can and should be makers of some sort, even if only so that they can appreciate the mastery of truly great artists. The fact that craft has come to be seen as a mainly personal, even solitary, activity is perplexing indeed. This is perhaps to do with the attitude of resistance to change upon which the culture of craft became premised in the early twentieth century, in parallel to design's wholesale espousal of industry. At the height of Fordism's early triumph, the roles somehow got reversed. Eric Gill, as keeper of the old ways of artistic individualism, and László Moholy-Nagy, as prophet of the new faith in industrial collectivism, can perhaps be seen as the archetypes of this world turned upside down.

The field of design was, in its very earliest days, a social laboratory for the utopian ideal of production by a class of masterful makers. The faceless and anonymous designers of the mid nineteenth century were mostly artisans indeed, both in practice and by class. In a single generation—identified by names like Alfred Stevens, Godfrey Sykes, Christopher Dresser and (ironically) William Morris himself—design was introduced to the idea of authorship and liberal professional status. As more architects and artists got on board, this took the historical route with which we are familiar, resulting in the concept of the designer as brand name. Over the past twenty years or so, with increasing specialization and systemic complexity, designers have begun to revive the old lessons of collective authorship and creative commons, sharing with colleagues, intermediaries and users the responsibilities of making things work.

What, then, is the future of the craft versus design debate? With the advance of flexible models of industrial production, the designer as aesthetic autocrat would seem to be a thing of the past. Artifacts are increasingly suited to the changing experiences of a fluid community of users, and less to the predetermined designs of any one individual. Human psychology makes product differentiation desirable. Better machines and engineering make it possible. The old paradigm of mass production is on its way out; a new paradigm, the individuation of experience, arises in its place. In this scenario, the balance shifts from material to immaterial. The question is no longer what to design, but why. Craft has provided viable answers to that, historically. Designers are beginning to understand these issues and to explore them, though perhaps unwitting as to their origins. Some time way back around the sixteenth century, craft and industry were synonyms, both capable of denoting the idea of skill. Now that industry is in the process of reinventing itself, perhaps design and craft will become synonyms too: complementary aspects of the same ongoing process of shaping experience through the interaction between people and things.

NOTES

1. On the origins of craft and its multiple historical stages, see Edward Lucie-Smith, *The Story of Craft: The Craftsman's Role in Society* (New York: Van Nostrand Reinhold, 1981), pp. 11–18.
2. For further implications of this vast theme, a good start is Alfred Sohn-Rethel, *Intellectual and Manual Labour: A Critique of Epistemology* (London: Macmillan, 1978).

3. Andrew Ure, *The Philosophy of Manufactures; or, an Exposition of the Scientific, Moral, and Commercial Economy of the Factory System of Great Britain* (London: Charles Knight, 1835).

4. E. J. Hobsbawm, *The Age of Revolution, 1789–1848* (New York: Mentor, 1964), p. 50.

5. Still the most cogent argument in favour of the history of design as a largely anonymous history: Adrian Forty, *Objects of Desire: Design and Society since 1750* (London: Thames and Hudson, 1986).

6. Lucie-Smith, p. 11. A full discussion of the Victorian debate on design education is available in my *The Educated Eye and the Industrial Hand: Art and Design Education for the Working Classes in Mid-Victorian Britain* (unpublished PhD dissertation, Courtauld Institute of Art, 1995).

7. E. T. Cook and Alexander Wedderburn, eds, *The Works of John Ruskin* (London: George Allen, 1903–12), Vol. 8, pp. 81, 84.

8. *Ibid.*, Vol. 15, p. 12.

9. *Ibid.*, vol. 16, p. 294.

10. *Ibid.*, vol. 16, p. 385.

11. Asa Briggs, ed., *William Morris: Selected Writings and Designs* (Harmondsworth: Penguin, 1962), p. 106.

12. David A. Hounshell, *From the American System to Mass Production, 1800–1932* (Baltimore: Johns Hopkins University Press, 1984), esp. chapter 2.

13. Michel Sanouillet and Elmer Peterson, eds, *The Writings of Marcel Duchamp* (New York: Da Capo Press, 1973), p. 160. Duchamp's readymade *Fountain*, of 1917, is as much commentary on the triumph of industrial manufacture as it is a critique of fine art.

14. See David Harvey, *The Condition of Postmodernity: An Enquiry into the Origins of Cultural Change* (Oxford: Blackwell, 1990), esp. part II.

15. See Taiichi Ohno, *Toyota Production System: Beyond Large-Scale Production* (New York: Productivity Press, 1988); and Yasuhiro Monden, *Toyota Production System: An Integrated Approach to Just-in-Time* (Atlanta: Engineering and Management Press, 1998).

16. Malcolm McCullough, *Abstracting Craft: The Practiced Digital Hand* (Cambridge: MIT Press, 1996), pp. 21, 55, 247–50, 271–2.

17. William Morris, *News from Nowhere, or an Epoch of Rest, Being Some Chapters from a Utopian Romance* (London: Reeves & Turner, 1891), see esp. chapter 2.

FURTHER READING

Rafael Cardoso, 'The Brompton Barracks: War, Peace, and the Rise of Victorian Art and Design Education', *Journal of Design History* 8/1 (1995), p. 11–25.

Rafael Cardoso and Colin Trodd, eds, *Art and the Academy in the Nineteenth Century* (Manchester: Manchester University Press, 2000).

SECTION 5

CRAFT IN THEORY: AESTHETICS, ESSENCE, STATUS

Thus far, this book has emphasized craft's position within various frameworks: economic, social and technical. But what does craft look like when viewed in isolation, on its own terms? This might sound like an innocent question, and maybe even one that should yield easy answers. But of course nothing could be further from the truth.

It could be argued, firstly, that one should refuse the question. Craft cannot be seen 'on its own terms', because it exists only relationally. There is no autonomous, free-floating thing called craft, divorced from any particular practice. It is not a bounded sphere of activity. Craft should therefore always be seen in context and is meaningful as a concept only when it is seen in contrast to other types of production.

A second option is the mirror image of the first: craft must indeed be seen autonomously, and in fact it is imperative to see it this way, because craft captures something essential about the world around us. From this perspective, if we can just think past the manifold distractions that culture imparts to us, and see the things around us purely in terms of their materiality and the way they were made, then we will be able to see things in their true character.

Another possibility is that craft is indeed a distinct field of endeavor but best understood as a subset of a broader field—namely, the visual arts. It should enjoy a status similar to that of painting and sculpture and deserves to have a field of criticism dedicated to it, in the same way that fine arts do, as well as its own museums, galleries and publications. From this point of view, the key question about craft is its standing vis-à-vis other creative practice. It might even be something worth fighting for.

Roughly speaking, these are the areas of theory that are established in this section of the book: craft as an ideal pole within the broader field of *technique*, one way of doing things among others; craft as an intrinsic aspect of *form*, and therefore, perhaps, a way of approaching the *essence* of things and spaces; and craft's *status* in relation to other fields, particularly the field of fine art.

THEORIES OF TECHNIQUE, FORM AND ESSENCE

The first of these options is closely identified with the writings of David Pye, the British woodworker, teacher and sometime theorist. Pye's well-known distinction between the workmanship of risk and the workmanship of certainty introduced to craft discourse its own dialectic, a way of understanding activities as diverse as throwing pottery, planing a board or cutting strips of paper, all according to the same underlying dualistic principle. In any making process, Pye argued, predictability could be introduced through various self-regulating tools, with certainty established along a sliding scale. Thus paper can be cut freehand with a knife, or if a straighter line is required, with scissors (the blade acting as its own guide), or a guillotine-style paper cutter (which cuts the straightest line of all). On the basis of seemingly innocuous observations like this one, Pye was able to construct a wide-ranging account of 'workmanship', a word he used to cover all sorts of making. As he noted, all tools—from planes and scissors to injection molding machines and jacquard looms—effect some degree and type of regulation. To understand those dynamics is to understand the 'theory' of making. Pye did not use the term craft, which he regarded as unscientific, but others have associated craft directly with his phrase 'the workmanship of risk'—giving the subject a precise but relative definition, free of any romantic social ideal.

A second, much less well-known writer on technique was the French theorist of technology Gilbert Simondon. Not unlike Pye, Simondon argued that all technical production could be understood as operating between two extremes. But while Pye's pairing of risk and certainty is divorced from any claims of progression, Simondon's dialectic of the abstract and concrete was explicitly mapped on to the historical transition from craft to industry. Simondon was unapologetically modernist and progressive. He described artisanal production as a 'dead weight' that held back the project of technical development. Yet his analysis is as useful as Pye's, in that it offers a theoretical analysis of craft and industry that is equally applicable to different times, places and trades.

Going beyond the close physical observation of particular processes inevitably seems to open the floodgates to the cultural and the aesthetic. A case in point is the formalist art theory of Henri Focillon, who also wanted to look closely at the way that materials and techniques resulted in certain outcomes. Over the work of generations, he argued, each craft evolves its own characteristic forms, with the inventive artisan discovering and manipulating inherent natural principles. These ideas descended from the work of mid-nineteenth-century architectural theorist Gottfried Semper, who looked for the rationale of contemporary architecture all the way back in prehistory. Vaults, beams and posts are thick and relatively unadorned, he had argued, because they are descended from wooden support systems. Walls, by contrast, are light, thin and often patterned, because they still carry within them a kind of memory of woven textile hangings, the first means of enclosing a space for shelter.[1] On the basis of such logic, which in retrospect seems strikingly protomodernist, other writers (many also from Germany) joined in the attempt to provide a scientific basis for certain ways of making and ornamenting buildings, objects and artworks.[2]

Though he was cognizant of this German theoretical tradition, Focillon departed from it in his lack of determinism. He had an unusual reverence for the creative artisan, whose ability to devise new solutions based on the deep experience of space, materials and processes was the impetus behind humankind's aesthetic 'life of forms'. The universalist quality of this vision, which saw all art and craft as intelligible according to shared material considerations, was widespread in the 1920s and '30s. It is also exemplified in Elsie Fogerty's book *Rhythm*. As a voice trainer for theatrical and musical performers, Fogerty was not so interested in materiality per se, but rather in the cadences of creative work. Like Focillon she wanted to theorize craftsmanship, painting, dance, singing and architecture according to a single transcendental artistic principle. It is an idea that captures the tone of much modern formalist thinking, which often used narrow concepts to open up a world of aesthetic connections.

Focillon and Fogerty, who believed in a transhistorical aesthetic order, would doubtless have been pleased by one of the less expected convergences in the craft historiography: the similarity between the ideas of West African artisans and mid-century German philosophers. These two obviously very different intellectual contexts share an emphasis on what Martin Heidegger called 'the thing in itself' (*das Ding an sich*, a phrase borrowed from Kant; also called by the Greek term *noumenon*), and the way this essence can be approached through the act of making. There are certain resonances with Focillon here, as the particular qualities of certain materials (such as iron) and forms (like the jug discussed by Heidegger in his essay 'The Thing') become loaded with particular significance. But relative ideas of 'workmanship' here give way to thinking about the absolute.

A key metaphor in both the African and German writings gathered in this section is the earth, and the way that objects and people alike are bodied forth from it. This idea is applied equally to the process of forming functional ironwares in Nigeria and the act of building modernist architecture. In both cases, the craftsperson is given special priority, not (as in Focillon) as an inventive force but rather as a carrier of ancient knowledge. There is also, in both cases, a sense of loss. The presumption of many observers of material culture, whether they speak from a position outside or inside of 'traditional' cultures, is that the unified basis of life is lost as craft has been displaced by modern technologies. As Amadou Hampâté Bâ puts it, 'In traditional Africa there was no division between the sacred and the profane, as there is in our modern society.' Whether this is true can and should be debated. Yet what is clear from the writings of Hampâté Bâ, as well as Patrick McNaughton's discussion of African blacksmithing, is that craftsmanship may be imbued with 'spirit', or religious significance, in both tacit and explicit ways. Sometimes, there is a presumption that the transformative powers of the artisan are simply deserving of reverence. In other cases, the words and ritual actions of initiation, mediation or celebration might be required to make a crafted object efficacious. Perhaps the dynamic relation of craft to other aspects of ritual—the fact that its products might need to be made sacred, but that as a process it also possesses a sacredness of its own—is the nature of its power.

The idea of craft as something unspoken, but which might also become the subject of poetic or religious consciousness, was also of great importance to Walter Benjamin. He describes craft as the natural counterpart to oral tradition, pointing to the fact that traditional narratives are often recited while craft work is happening. The tacit values of one reinforce the other, in a woven fabric of knowledge. This sounds ahistorical, perhaps, but it is not necessarily antimodern. For Benjamin, the craftsperson (like the storyteller) is important because s/he embodies a particular position in relation to time—to craft is literally to embody the material qualities of inherited memory as a foil to other kinds of history. Unlike Heidegger, for whom technology was fatally out of touch with the body and the earth, Benjamin had no fear of industrial production. In fact, he argued that because mass-produced copies lacked the 'aura' of handmade originals, they actually might be preferable, as they would afford audiences and users greater cultural determination.[3] But if a copy is to be saturated with the same cultural value as, say, a traditional pot and weaving, we must have the same sense of ownership and intimacy with mass-produced objects that people of earlier times had with their own material culture. Benjamin's objective, therefore, was to update craft's traditional 'essence' for the purposes of modernity. As Terry Eagleton has written, Benjamin's idea of tradition 'is in some sense a given, yet it is always constructed from the vantage point of the present'.[4]

Theodor Adorno is often paired with Benjamin as a contrasting exemplar within Marxist cultural theory. Though he is not usually cast in the role of a craft theorist, his opinions on the matter will not surprise those familiar with his political and aesthetic writings. Like Benjamin, he saw craft in essentialist terms—not as an expressive language, but as a getting down to basics—but he was much more interested in explicit critique. A preeminent theorist of the avant-garde, he was also a formalist and, much as Semper did, saw craft as a completely logical affair: the rigorous development of material characteristics. If Adorno thought that it was important to get craft right, though, this was not for its own sake, but because only a work that was thoroughly consistent could be an effective critical instrument. 'The means have their own logic', he wrote, but it is 'a logic that points beyond them'. For Adorno, the relation between craft and art (or architecture, or design) was a simple matter: the former should simply serve the needs of the latter and draw no attention to itself in the process.[5]

ART STATUS

Adorno's view has certainly not been shared by most studio craftspeople and their supporters over the years. Instead, many members of this community have tried to establish themselves as a part of the art world—leading to a preoccupation with status that sometimes threatens to occlude other lines of debate. Focused discussion of craft's social and aesthetic position was already well developed in the late nineteenth century, when many European theorists advocated a unity of all the arts. This was sometimes motivated by egalitarianism, as in the case of the Arts and Crafts Movement. Sometimes it was an aesthetic ideal, as in the *gesamtkunstwerk* ('total work of art') of central European architects and designers. Yet in neither case was there much sentiment in favor of elevating craft objects to the status of fine art. If anything, the reverse was the case. To see painting and furniture as complementary equals, it was not necessary to see furniture as sculpture; it was, rather, necessary to dismiss fine art's claims to special status, and insist that it play its part in integrated decorative schemes.[6] The idea was to dismantle the aesthetic hierarchy, not rearrange it.

The attempt to recast ceramics, furniture, metalwork, textiles and glass as subdisciplines of fine art is more recent and quite different in its claims. The historical roots of the effort were in the post–World War II growth of craft departments within art schools and universities. In the United States, where the charge for craft-as-art has been most pronounced, there had already been several such programs in the early twentieth century. A veritable explosion occurred after the war, however, thanks to the GI Bill, which provided funding to individual veterans interested in pursuing higher education. The new craft departments tended to be located alongside painting and sculpture departments—they required similar equipment, were of a similar scale and were considered to be primarily expressive rather than practical. (It is interesting to consider what would have happened if craft departments had instead been grouped with engineering and chemistry labs.) The result, predictably, was that instructors and students in craft courses quickly began to mix with fine artists and to see themselves as equivalent to or even indistinguishable from them.

Already in the 1950s, this new generation of university-trained 'artist-craftsmen' challenged the idea that craft should be seen primarily as a preparatory stage within commercial design. This was not quite an opposition between academic and commercial interests; the 'designer-craftsman' model had, after all, emerged from the Scandinavian and German educational establishments and had been carried to American schools by emigrants from Europe in the 1930s. But there is no doubt that the new 'artist-craftsman' impulse was carried forward primarily by men and women who made their living as educators. It could hardly have been otherwise, for in the absence of a gallery system—which would not emerge until the 1970s—and without the job security and income provided by teaching posts, the idea of craft-as-art would have been purely Quixotic.

But if the 'artist-craftsmen' were ensconced in academia, they tended to be quite disconnected from contemporary art and its attendant theories. A fundamental misunderstanding emerged in the 1960s, by which time craftspeople and supportive critics and gallerists were pointedly claiming art status for their field, and accusing those who didn't see things this way of anticraft prejudice.[7] In retrospect, it is clear that this was not an accurate assessment of the situation. Theoretically speaking, the 'artist-craftsmen' and their backers were pushing against an open door, because art had by then come to be defined as a field capable of infinite absorption. In the wake of Marcel Duchamp's Readymades, which demonstrated that any object could be made into an artwork simply through an act of designation, the claims of art status for clay, fiber and the rest were not really meaningful. Even outside of this elite context, art status per se is a condition that is easily granted. We routinely describe hobbyists' paintings of rowboats and children's crayon scribbles as art, for example, and may attach great emotional and even aesthetic significance to such works, even as others may consider them to possess little intellectual and economic value. The question is not whether something counts as art, then, but which discursive constellations (which 'art worlds') it inhabits.

This leads to the obvious point that although art is an omni-inclusive category, whether a work is taken seriously as *contemporary* art is another matter. For most of the twentieth century, the status and relevance of particular artists and works have been determined mainly by an insider's game of critical practice and institutional authorization. For studio craft's supporters, it is comforting to think that this is just a matter of power relations, which secure certain fields (like painting) as art and hold others (like clay) at bay. Occasionally, one can find evidence to that effect, such as Rosalind Krauss's 1978 declaration that 'to be a ceramicist-sculptor in the 1950s and 1960s was in some essential way to be marginal to "sculpture" . . . in the semantic associations to pottery, ceramics speaks for that branch of culture which is too homey, too functional, too archaic, for the name of "sculpture" to extend to it'.[8] For the most part, though, choice of materials has not disqualified works from art status in the post–World War II period. One can enumerate many uses of so-called crafts media by fine artists—Lucio Fontana, Asger Jorn and Rosemary Trockel in clay; Robert Smithson, Kiki Smith and Fred Wilson in glass; Mark di Suvero and Doris Salcedo in wood and so on. If studio craft has not managed to make much headway within modern and contemporary art, perhaps this is not because of prejudice, but because it has been a victim of its own success in establishing a separate realm of endeavor. Once craft was defined as a separate category, it lost its purchase on the principle of infinite permission that is the theoretical baseline of contemporary art practice.[9]

The six readings about art status that are gathered in this section of the book all engage with the preceding issues, in somewhat unconventional ways. The aesthetic philosopher R. G. Collingwood and the art critic John Bentley Mays are two rare examples of authors who have explicitly argued against the idea of craft as a category within visual art. Against their texts—which have found few admirers within craft circles, unsurprisingly—are arranged the American art historian Harold Rosenberg and the British ceramicist Alison Britton, both of whom see craft's relation to art as a nuanced rather than a categorical affair. These two writers seem to recognize (as Pye did) that craft is best theorized in relational terms and therefore hold back from making unilateral claims for its status. Garth Clark, too, presents a case for thinking about craft in open-ended terms, a necessary consequence if he is right that the craft movement is effectively over. These latter arguments seem increasingly attractive. For all the talk of definitions, what seems important now is not whether craft can be defined as a sphere within art. More pressing is the more objective question of whether art institutions are willing and able to make space for craft-related thinking, as part of the subject matter for which they have been responsible all along.

NOTES

1. Gottfried Semper, *The Four Elements of Architecture and Other Writings* (Cambridge: Cambridge University Press, 1989); Gottfried Semper, *Style in the Technical and Tectonic Arts* (Santa Monica, CA: Getty, 2004).

2. Heinrich Wölfflin, *Principles of Art History* (London: Bell and Sons, 1932; orig. pub. 1915); Frederic J. Schwartz, *Blind Spots: Critical Theory and the History of Art in Twentieth-Century Germany* (New Haven, CT: Yale University Press, 2005).

3. Benjamin's best-known statement on the matter is 'The Work of Art in the Age of Mechanical Reproduction', 1936; translated in Hannah Arendt, ed., *Illuminations* (New York: Schocken, 1968).

4. Terry Eagleton, 'Capitalism, Modernism, and Postmodernism', in *Against the Grain: Essays 1975–85* (London: Verso, 1986), p. 136.

5. This is also a position held by the sociologist Thorstein Veblen—an earlier and a very different writer from Adorno, but also a critic of capitalism, best known for his codification of the concept of conspicuous consumption. Veblen also argued that craft characteristically receded into the background of human affairs and is therefore not an instrument of critique in its own right: 'The instinct of workmanship will commonly not run to passionate excess . . . it rather yields ground somewhat readily, suffers repression and falls into abeyance, only to reassert itself when the pressure of other, urgent interests is relieved.' Veblen, *The Instinct of Workmanship and the State of the Industrial Arts* (New York: Macmillan, 1914; repr. Sentry Press, 1964), p. 34.

6. For a consideration of furniture from this perspective—arguing against a purely individualist, expressive mode—see Herman Bahr, 'The Chair', 1899, as introduced and translated by Berthold Hub in *The Journal of Modern Craft* 1/2 (July 2008), pp. 279–91).

7. Rose Slivka and Lee Nordness, *Objects: USA* (New York: Viking, 1970); Mildred Constantine and Jack Lenor Larsen, *Beyond Craft: The Art Fabric* (New York: Van Nostrand Reinhold, 1972).

8. Rosalind Krauss, *John Mason: Installations from the Hudson River Series* (Yonkers, NY; Hudson River Museum, 1978), pp. 12–13.

9. For more on these issues see my own book *Thinking Through Craft* (Oxford: Berg/V&A, 2007).

43

THE NATURE AND ART OF WORKMANSHIP

David Pye

Perhaps the most widely read twentieth-century craft theorist, David Pye was from 1948 to 1974 a professor of furniture design at the Royal College of Art, London. A fiercely logical and rigorous thinker, he was a fluent writer and his writings on both craft and design continue to be widely read several decades after their publication. Pye was also a practicing craftsman, well acquainted with many materials and processes, and a specialist in making machined and fluted wooden bowls. These were made through a combination of what he describes in the following excerpt as the 'free' workmanship of risk and the 'regulated' workmanship of certainty. Looking at craft in this way is useful because it is grounded in the physical nature of the process rather than the context in which the process occurs. Though the workmanship of risk seems initially to be linked to craft and the workmanship of certainty to mass production, in fact the former is often practiced in factories and the latter in individual studios. It is typical of Pye that he resisted any attempt to project ideological values onto this distinction. Elsewhere in The Nature and Art of Workmanship, *in fact, he systematically dismantled the ideas of John Ruskin which he found to be insupportably idealistic. Pye did, however, allow one hint of aesthetics into his thought, arguing that the 'diversity' or variation that occurred as a result of free workmanship was inherently pleasing.*

David Pye, excerpts from *The Nature and Art of Workmanship* (Cambridge University Press, 1968).

THE WORKMANSHIP OF CERTAINTY AND THE WORKMANSHIP OF RISK

Workmanship of the better sort is called, in an honorific way, craftsmanship. Nobody, however, is prepared to say where craftsmanship ends and ordinary manufacture begins. It is impossible to find a generally satisfactory definition for it in face of all the strange shibboleths and prejudices about it which are acrimoniously maintained. It is a word to start an argument with.

There are people who say they would like to see the last of craftsmanship because, as they conceive of it, it is essentially backward-looking and opposed to the new technology which the world must now depend on. For these people craftsmanship is at best an affair of hobbies in garden sheds; just as for them art is an affair of things in galleries. There are many people who see craftsmanship as the source of a valuable ingredient of civilization. There are also people who tend to believe that craftsmanship has a deep spiritual value of a somewhat mystical kind.

If I must ascribe a meaning to the word craftsmanship, I shall say as a first approximation that it means simply workmanship using any kind of technique or apparatus, in which the quality of the result is not predetermined, but depends on the judgment, dexterity and care which the maker exercises as he works. The essential idea is that the quality of the result is continually at risk during the process of making; and so I shall call this kind of workmanship "The workmanship of risk": an uncouth phrase, but at least descriptive.

It may be mentioned in passing that in workmanship the care counts for more than the judgment and dexterity; though care may well become habitual and unconscious. With the workmanship of risk we may contrast the workmanship of certainty, always to be found in quantity production, and found in its pure state in full automation. In workmanship of this sort the quality of the result is exactly predetermined before a single saleable thing is made. In less developed forms of it the result of each operation done during production is predetermined.

The workmanship of certainty has been in occasional use in undeveloped and embryonic forms since the Middle Ages and I should suppose from much earlier times, but all the works of men which have been most admired since the beginning of history have been made by the workmanship of risk, the last three or four generations only excepted. The techniques to which the workmanship of certainty can be economically applied are not nearly so diverse as those used by the workmanship of risk. It is certain that when the workmanship of certainty remakes our whole environment, as it is bound now to do, it will also change the visible quality of it. In some of the following chapters I shall discuss what may be lost and gained.

The most typical and familiar example of the workmanship of risk is writing with a pen, and of the workmanship of certainty, modern printing. The first thing to be observed about printing, or any other representative example of the workmanship of certainty, is that it originally involves more of judgement, dexterity, and care than writing does, not less for the type had to be carved out of metal by hand in the first instance before any could be cast; and the compositor of all people has to work carefully and so on. But all this judgement, dexterity and care has been concentrated and stored up before the actual printing starts. Once it does start, the stored-up capital is drawn on and the newspapers come pouring out in an absolutely predetermined form with no possibility of variation between them, by virtue of the exacting work put in beforehand in making and preparing the plant which does the work: and making not only the plant but the tools, patterns, prototypes and jigs which enabled the plant to be built, and all of which had to be made by the workmanship of risk.

Typewriting represents an intermediate form of workmanship, that of limited risk. You can spoil the page in innumerable ways, but the N's will never look like U's, and, however ugly the typing, it will almost necessarily be legible. All workmen using the workmanship of risk are constantly devising ways to limit the risk by using such things as jigs and templates. If you want to draw a straight line with your pen, you do not go at it freehand, but use a ruler, that is to say, a jig. There is still a risk of blots and kinks, but less risk. You could even do your writing with a stencil, a more exacting jig, but it would be slow.

Speed in production is usually the purpose of the workmanship of certainty but it is not always. Machine tools, which, once set up, perform one operation, such for instance as cutting a slot, in an absolutely predetermined form, are often used simply for the sake of accuracy, and not at all to save time or labour. Thus in the course of doing a job by the workmanship of risk a workman will be working freehand with a hand tool at one moment and will resort to a machine tool a few minutes later.

In fact the workmanship of risk in most trades is hardly ever seen, and has hardly ever been known, in a pure form, considering the ancient use of templates, jigs, machines and other shape-determining systems, which reduce risk. Yet in principle the distinction between the two different kinds of workmanship is clear and turns on the question: 'is the result predetermined and unalterable once production begins?'

Bolts can be made by an automatic machine which when fed with blanks repeatedly performs a set sequence of operations and turns out hundreds of finished bolts without anyone even having to look at it. In full automation much the same can be said of more complex products, substituting the words 'automated factory' for 'automatic machine'. But the workmanship of certainty is still often applied in a less developed form where the product is made by a planned sequence of operations, each of which has to be started and stopped by the operative, but with the result of each one predetermined and outside his control. There are also hybrid forms of production where some of the operations have predetermined results and some are performed by the workmanship of risk. The craft-based industries, so called, work like this.

Yet it is not difficult to decide which category any given piece of work falls into. An operative, applying the workmanship of certainty, cannot spoil the job. A workman using the workmanship of risk assisted by no matter what machine-tools and jigs, can do so at almost any minute. That is the essential difference. The risk is real. But there is much more in workmanship than not spoiling the job, just as there is more in music than playing the right notes.

There is something about the workmanship of risk, or its results, or something associated with it, which has been long and widely valued. What is it, and how can it be continued? That is one of the principal questions which I hope this book may answer: and answer factually rather than with a series of emotive noises such as protagonists of craftsmanship have too often made instead of answering it.

It is obvious that the workmanship of risk is not always or necessarily valuable. In many contexts it is an utter waste of time. It can produce things of the worst imaginable quality. It is often expensive. From time to time it had doubtless been practised effectively by people of the utmost depravity.

It is equally obvious that not all of it is in jeopardy: for the whole range of modern technics is based on it. Nothing can be made in quantity unless tools, jigs, and prototypes, both of the product and the plant to produce it, have been made first and made singly.

It is fairly certain that the workmanship of risk will seldom or never again be used for producing things in quantity as distinct from making the apparatus for doing so; the apparatus which predetermines the quality of the product. But it is just as certain that a few things will continue to be specially made simply because people will continue to demand individuality in their possessions and will not be content with standardization everywhere. The danger is not that the workmanship of risk will die out altogether but rather that, from want of theory, and thence lack of standards, its possibilities will be neglected and inferior forms of it will be taken for granted and accepted.

There was once a time when the workmanship of certainty, in the form colloquially called 'mass-production', generally made things of worse quality than the best that could be done by the workmanship of risk colloquially called 'hand-made'. That is far from true now. The workmanship of a standard bolt or nut, or a glass or polythene bottle, a tobacco-tin or an electric-light bulb, is as good as it could possibly be. The workmanship of risk has no exclusive prerogative of quality. What it has exclusively is an immensely various range of qualities, without which at its command the art of design becomes arid and impoverished.

A fair measure of the aesthetic richness, delicacy and subtlety of the workmanship of risk, as against that of certainty, is given by comparing the contents of, say, the British Museum with those of a good department store. Nearly everything in the Museum has been made by the workmanship of risk, most things in the store by the workmanship of certainty. Yet if the two were compared in respect of the ingenuity and variety of the devices represented in them the Museum would seem infantile. At the present moment we are more fond of the ingenuity than the qualities. But without losing the ingenuity we could, in places, still have the qualities if we really wanted them.

THE AESTHETIC IMPORTANCE OF WORKMANSHIP, AND ITS FUTURE

In the foregoing chapter it has been suggested that the importance of good workmanship in its aesthetic aspect rests on three things:

(1) Highly regulated workmanship shows us a thing done in style: an evident intention achieved with evident success. It is anti-sordid, anti-squalid and contributes to our morale.

To do a thing in style is to set oneself standards of behaviour in the belief that the manner of doing anything has a certain aesthetic importance of its own independent of the importance of what is done. This belief is the basis of ordinary decent behaviour according to the customs of any society. It is the principle on which one keeps one's house and one's person clean and neat, and so on. Regulation which, in general, the workmanship of risk can only achieve by taking a good deal of avoidable trouble, used undoubtedly to be a part of this idea of behaviour.

With the workmanship of certainty it is becoming easier to achieve high regulation and less determination is needed to do it; but still the quality of the result is clear evidence of competence and assurance, and it is an ingredient of civilization to be continually faced with that evidence, even if it is taken for granted and goes unremarked.

(2) Free workmanship shows that, while design is a matter of imposing order on things, the intended results of design can often be achieved perfectly well without the workman being denied spontaneity and unstudied improvisation. This perhaps has special importance because our natural environment, and all naturally formed or grown things, show a similar spontaneity and individuality on a basis of order and uniformity. This characteristic aspect of nature, order permeated by individuality, was the aesthetic broth in which the human sensibility grew. Whereas in the early days of civilization highly regulated workmanship seemed admirable because it was rare, difficult, and exceptional, that situation is now completely

reversed, and we might well try to make ourselves an environment which had more concord with our natural one.

(3) Good workmanship, whether free or regulated, produces and exploits the quality I have called diversity, and by means of it makes an extension of aesthetic experience beyond the domain controlled by design, down to the smallest scale of formal elements which the eye can distinguish at the shortest range. Diversity on the small scale is particularly delightful in regulated workmanship because there it maintains a kind of pleasantly disrespectful opposition to the regulation and precision of the piece seen in the large: as when, for instance, the wild figure of the wood sets off the precision of the cabinet-work. Diversity imports into our man-made environment something which is akin to the natural environment we have abandoned; and something which begins to tell, moreover, at those short distances at which we most often see the things we use.

What changes can one foresee? Is there for instance any reason for the productive part of the workmanship of risk to continue doing highly regulated work? Why should it, when the workmanship of certainty is capable of higher regulation than ever was seen? Why, in particular, should it, considering that high regulation by the workmanship of risk is usually very expensive even where the best and most ingenious use is made of machine tools? Imagination boggles at the thought of what it might cost to build any standard family car from scratch by the workmanship of risk. How many weeks would it take to make the carburettor, for instance, or one of the head-lamps?

It should continue simply because the workmanship of risk in its highly regulated forms

can produce a range of specific aesthetic qualities which the workmanship of certainty, always ruled by price, will never achieve. The British Museum, or any other like it, gives convincing evidence of that. And one need not copy the past in order to perpetuate those qualities. People still use oil-paint, but they do not imitate Titian.

There is of course no danger that high regulation will die out in the preparatory branch of the workmanship of risk. Beyond that, the prevalence and immense capability of the workmanship of certainty will ensure that highly regulated workmanship continues and increases. Indeed there is already too much of it or, rather, there is too little diversity in it. The contemporary appetite for junk and antiques may partly be a sign of an unsatisfied hunger for diversity and spontaneity in things of everyday use. I do not think it can be quite explained either by the romantic associations of mere age or by an aversion from the ephemerality of contemporary designs. There is still comparatively so much diversity about that it is difficult to estimate how an environment quite devoid of it would strike us. The quality in design which is called 'clinical' is more or less the quality of no-diversity. A little of it, for a change, is pleasant, but a world all clinical might be fairly oppressive, and such a world of design and workmanship without diversity is decidedly a possible one, now.

Four things are going wrong:

The workmanship of certainty has not yet found out, except in certain restricted fields, how to produce diversity and exploit it.
Where highly regulated components are fitted and assembled by the workmanship of risk, in industries which are only in part 'industrialized', such as joinery for buildings, some of the workmanship is extraordinarily bad.

Some kinds of workmanship, such as the best cabinet-making, which use the workmanship of risk to produce very high regulation and the most subtle manipulations of diversity, are dying out because of the cost of what they do. But what they do has unique aesthetic qualities.

Free workmanship also is dying out, for the same reasons, and it also has unique aesthetic qualities for which there can be no substitute.

It is, I submit, quite easy to see what might be done about the last three of these things but not about the first, which is undoubtedly the most important. The workmanship of certainty can do nearly everything well except produce diversity. Its only real success in that way at present is in weaving and in making things of glass or translucent or semi-translucent plastics such as nylon or polythene which show delightful diversification because of their modulation of the transmitted light and the interplay between it and the light reflected from their surfaces. Diversity in shapes and surfaces could also, no doubt, be achieved fairly crudely by numerically controlled machine tools, and perhaps something more can be hoped for there in course of time.

Much of the diversity in highly regulated work produced by the workmanship of risk used to be achieved through the manner in which it made use of the inherent qualities of natural materials. It is very probable that, if diversity were appreciated as much as economy, synthetic or processed materials would be made with an equally rich inherent diversification.

If industrial designers and architects understood the theory and aesthetics of workmanship better, and realized the importance of it, they would surely make better use of the opportunities offered by the techniques which are now available to them. One could almost believe that some industrial designers only know of two surface qualities, shiny and 'textured'; and that to them texture means something which has to be distinguishable in all its parts three feet away! They ought to reflect that so far as the appearance of their work goes its surface qualities are not less important than its shape, for the only part of it which will ever be visible is the surface.

The want of diversity is not so much to be blamed on the technologists as on the designers, who do not think enough about it, or do not think enough of it. Perhaps I think too much of it, but it is high time somebody spoke up for it. Art is not so easy that we can afford to ignore any and every formal quality which will not go on to a drawing board. Yet, the fact remains, I can offer no better suggestion than that, if people came to love diversity, they would find out ways of producing it.

The answer to the second problem, of bad workmanship in assembly and finishing off, is much easier to see. The first thing to be grasped is that the situation now is fundamentally different from what it was in the old days of good rough workmanship. The second thing is that the force of the long traditions of the workmanship of risk is now very weak in many trades. With some honourable but rather few exceptions, it no longer concerns a joiner's self-respect and standing in the eyes of his trade, that his work shall be done properly according to those traditions, and moreover he will be paid as well as before even if it is done badly.

This situation is regrettable, but it does not necessarily mean that the joiner is a bad man. It merely means that his education in his trade has been bad (for a trade learnt according to the traditions was an education, though a circumscribed one. It taught the principles on which one should act in certain circumstances and the difference between good and bad actions). The existing situation arises from the fact that the building trade is in transition in this country from the workmanship of risk to that of certainty, to the assembly of prefabricated components so made that neither care, knowledge nor dexterity are required for their assembly; and such trades as the joiner's are in decline. There are now too few good joiners.

It is futile to hope that the process of decline can be reversed on a sufficient scale to match the size of the industry, and the action to be taken is unmistakable. We must stop designing joinery and other details of cheap buildings as though for such work we could command fully educated joiners whenever we wanted them. It is, for example, silly to design

architraves which have to be mitred round door openings. Of all joints a mitre is sure to be badly done or to go wrong in cheap work. It is necessary for the architect to understand very clearly the limitations of the workmanship which the price of the building will allow, to understand that nothing can be left to the discretion of men without education in the trade, and to design within those limitations instead of asking for highly regulated traditional joinery like mitred architraves.

As for the third and fourth problems it is again not difficult to see a line of action, but it may not be easy to arouse interest and inform opinion so that the action gets taken. It will be a great loss to the world if at least a little highly regulated work does not continue to be done by the workmanship of risk in making

Figure 30   David Pye, *Small Circular Box*, no date.

furniture, textiles, pottery, hand-tools, clothes, glass, jewellery, musical instruments and several other things. It will equally be a loss if free workmanship does not continue. Most of such work will fall within the province of what are now called 'the Crafts.' What is now required is a more realistic conception of them.

The workmanship of risk can be applied to two quite different purposes, one preparatory, the other productive. Preparatory workmanship makes, not the products of manufacture, but the plant, tools, jigs and other apparatus which make the workmanship of certainty possible. Productive workmanship actually turns out products for sale.

The preparatory branch of the workmanship of risk is, of course, already far the more important of the two, economically. Without it we should starve pretty quickly because without it the workmanship of certainty would cease, and only by way of that is mass-production possible. The productive branch on the other hand is declining, and in the course of the next two or four generations it may well have become economically negligible as a source of useful products. But, though, after that, the workmanship of risk may never again? provide our bread, it may yet provide our salt. It will no doubt provide our space-craft too, and our more enormous scientific instruments.

The term 'crafts', that sadly tarnished name, may perhaps be applied to the part of the productive workmanship of risk whose justification is aesthetic, not economic (and not space-exploratory or particle-pursuing). The crafts on that definition will still have a slight indirect economic importance, in that they will enable designers to make relatively expensive experiments which the workmanship of certainty will deny them, and also to try out materials it denies them. But economics alone will never justify their continuation.

The crafts ought to provide the salt—and the pepper—to make the visible environment more palatable when nearly all of it will have been made by the workmanship of certainty. Let us have nothing to do with the idea that the crafts, regardless of what they make, are in some way superior to the workmanship of certainty, or a means of protest against it. That is a paranoia. The crafts ought to be a complement to industry.

For the crafts, in the modem world, there can be no half measures. There can be no reason for them to continue unless they produce only the best possible workmanship, free or regulated, allied to the best possible design: in other words, unless they produce only the very best quality. That quality is never got so quickly as more ordinary qualities are. The best possible design is seldom the one which is quickest to make, or anything like it; and, even where it is, the best quality of workmanship can usually be achieved only by the workman spending an apparently inordinate amount of time on the job. There are exceptions. Pottery, some hand-loom weaving and some jewellery, for instance, can be produced relatively cheaply. Moreover, in pottery at least, industry offers no serious competition, since the aesthetic qualities of 'studio pottery' are as yet rarely attempted in industrial production. Consequently these crafts flourish—though too seldom they produce the very best quality, or the best design—and people are making a reasonable living at them. But they are exceptions. The rule is, and always was, that the very best quality is extremely expensive by comparison with things of ordinary quality.

It is very probable that most people are beginning now to associate the word 'crafts' simply with hairy cloth and gritty pots. It is not quite realized perhaps that modem equivalents of the multitude of other kinds of workmanship we see in museums could and should be made: nor how astronomically expensive many of them would be.

Now the crafts, even when they do produce the very best quality, are in direct competition with producers of ordinary quality. The crafts are in no way comparable to the fine arts, a separate domain: far from it! The crafts are a border-ground of manufacturing industry, and nearly every object they make has its counterpart and competitor in something manufactured for the same purpose. In all but a very few trades exceedingly high quality is the last remaining ground on which the crafts can now compete.

Two of the fundamental considerations which will shape the future of the crafts are the time they must take over their work and the competition they must face. The differential in price between a product of craft, of the best quality, and a product of manufacture varies, naturally, according to the trade; but it is always large and sometimes huge. It ought to be and must be. Unless it is, the craftsman has no hope of anything approaching a modest professional standard of living, and he will never be able to command a better living than that. The crafts will therefore survive as a means of livelihood only where there is a sufficient demand for the very best quality at any price.

That sort of demand still exists in some trades. Haute couture flourishes. Certain musical instruments, yachts, guns, jewellery, tailoring, and things of silver, are still in that kind of demand. But the demand is not large, by comparison, for instance, with the demand for contemporary paintings, or for antiques, at comparable prices. The situation of the craftsmen who make these things of the best quality is evidently precarious. The West End tailors and bootmakers are not finding it easy to exist any more.

In other fields that kind of demand has very nearly ceased in Britain. Cabinet- and chair-making, blacksmith's work, carving, hand-tool making, are examples. These are all cases where the differential is very large. Here the potential buyers have turned to antiques or else spend their money on things of other kinds. It is not always clear why the demand has persisted in some fields but not in others. We may suspect that where it does persist the reasons are not always very creditable ones. But we need not concern ourselves with that, for it is absolutely certain that no demand for the best quality at any price can be re-created, or stimulated where it still persists, until it becomes a fact that a fair amount of work of that quality is being done and can be had.

Now, considering the time that is needed to do it, how can such work be made? It is obvious that it must be done, at first and for a long while afterwards, for love and not for money. It will have to be done by people who are earning their living in some other way. It is sometimes hoped that a man can set up as, say, a cabinet-maker and aim at making a few pieces of the very best quality each year, so long as he keeps himself solvent by making other furniture to order, or for sale in competition with the manufacturers. This can be done and is being done. Some good furniture is being made in this way, but very, very little of the very best. The man who does it is likely to find that to make a moderate living he has to become a manager more than a maker—sales manager, works manager, despatch manager, buyer and accountant, as well as secretary, all rolled into one. Whatever he does of the very best quality will have to be done as a side line, very likely at weekends. It will not increase proportionately to the other. If it were not for being his own master he might about as well make his living working in some other office or at some other trade, and make his two or three pieces of the very best quality in his spare time.

That is the logical conclusion. With certain exceptions, some of them precarious, the crafts, like the fine arts, are not fully viable. Only a very small proportion of painters can make enough money, by painting alone, to bring up a family, and that in a time when there is a climate of educated opinion very favourable to painting, a great international trade in contemporary paintings and a whole apparatus of distribution specifically for them: and when, above all, high prices for them are paid. None of these advantages is yet available to the crafts. Moreover, they are under a disadvantage which the painters are free from: the pressure of competition just mentioned.

Nearly all craftsmen, as nearly all painters and poets already do, will have to work part-time, certainly in the opening years of their career. One of the best professional cabinet-makers in Britain, Ernest Joyce, started as an amateur and learnt his job at first from books. 'Amateur', after all, means by derivation a man who does a job for the love of it rather than for money, and that happens also to be the definition, or at least the prerequisite, of a good workman. There is only one respect in which a part-time professional need differ from a man who can spend his whole working life at the job. He who works at it part-time must be content to work more slowly in his early years. Constant practice gives a certainty quite early in life which takes much longer to attain if one is working intermittently. Until he does attain it he must make up for the want of it by taking extra care and therefore extra time.

In consequence his output will necessarily be very small; but that is unimportant. The only reason for doing this work is quality not quantity.

No one will find the patience to become a proficient workman of this sort unless he has a lively and continual longing to do it, and, given that, ways of learning the job will be found. There are books, there are examples of the work, and there are workmen. With the help of all these and with practice he will learn to do work of the highest standard. I doubt whether there is anything which a determined part-time professional could not attain to, except speed, and even that comes in time.

It is still commonly believed that a man cannot really learn a job thoroughly unless he depends on it for his living from the first and gets long experience at it. It is untrue. Two minutes experience teach an eager man more than two weeks teach an indifferent one. A man's earning hours and his creative hours can be kept separate and it may be that they are better separated. Painters and poets separate them. Are painting and poetry really so much easier than craftsmanship? Part-time seamen are making ocean voyages in small craft which any professional seaman of the days of sail would have highly respected. Is not that a parallel case? Astronomy, to take but one other example, has owed an immense debt to amateur observers and telescope makers from Newton and Sir William Herschel onwards. No one in that science would subscribe much to the idea that amateurs are apt to be amateurish. It is high time we separated the idea of the true amateur—that is to say the part-time professional—from the idea of 'do-it-yourself' (at its worse end) and all that is amateurish. The continuance of our culture is going to depend more and more on the true amateur, for he alone will be proof against amateurishness. What matters in workmanship is not long experience, but to have one's heart in the job and to insist on the extreme of professionalism.

That this kind of workmanship will be in the hands of true amateurs will be a healthy and promising state of affairs, not a *faute de mieux*, for if any artist is to do his best it is essential that his work shall not be influenced in the smallest degree by considerations of what is likely to sell profitably. What concerns us is the very best. It is that which must somehow be continued because the aesthetic quality of it is unique, and the tradition of it must be kept alive against a time when it will put out some new growth. The part-time professional will be in a position to do the very best even though he can turn out very little of it, and even though at first he will have to sell it at a price which pays him very little for his time. Why not? Whom will he be undercutting? Will there be placards saying 'Craftsmen Unfair to Automation'? That can't be helped.

Along this road there will still be pitfalls. The crafts and craftsmen have been bedevilled, ever since Ruskin wrote, by a propensity for striking attitudes. The attitude of protest I have mentioned already. Another one is the attitude of sturdy independence and solemn purpose (no truck with part time workers: they are all amateurs; social value; produce things of real use to the community); another is the attitude of holier-than-thou (no truck with machinery; no truck with industry; horny-handed sons of toil; simple life, etc.). Another is the snob attitude, learnt from the 'fine' artists (we who practise the fine crafts are not as other craftsmen are). These are ridiculous nonsense by now, but who has not felt sympathy with them, all but the last, at one time or another? For nostalgia is always in wait

for us. The workmanship of risk was in many ways better in the old days than it is now, there is no sense in pretending otherwise. Moreover, many of the trades we ought to set ourselves to continue are already taking the complexion of survivals from an older world. That should not prevent us from looking ahead. We must think of the future more than the past. Some trades which are dead economically are all alive in human terms, and still have much to show the world.

It remains to notice the most disastrous illusion which was encouraged by Ruskin's chapter, whether he meant it to be or not; and which has done the most harm: the illusion that every craftsman is a born designer. There are no born designers. People are born with or without the makings of a designer in them, but the use of those talents is only to be learnt very slowly by much practice. Any untrained but gifted man can knock up something which looks more or less passable as a design but the best design for industry is done by people who have really learnt their job; and it looks like it. The crafts are always liable to comparison with industry and they cannot afford to come off second best in design as well as in price.

Design is so difficult to learn now simply because the arts are in a state of violent flux and because there are great interests vested in constant innovation. There is no settled tradition. If there were, the profession would be far more quickly learnt. If the crafts develop as I envisage, perhaps few craftsmen will be able to go through a designer's training, but surely there will be designers who will work for them, and be glad of the chance even if they make no money by it at all. There will have to be an alliance between the craftsmen and the designers.

Some things, of course, can only be designed, or at any rate designed in detail, by the workman himself. Writing and carving are obvious examples. Other things, such as musical instruments, ought to go on being made to traditional designs (not 'reproduction' designs, which are quite a different thing. Tourte's pattern of violin bows has been in use ever since he evolved it: it is not a mere revival of something which had died out).

The whole future of the crafts turns on the question of design. If designers will only come to recognize it, the crafts can restore to them what the workmanship of certainty in quantity-production denies them: the chance to work without being tied hand and foot by a selling price: the chance to design in freedom. There is nothing more difficult or more necessary for the modem designer to attempt.

If the crafts survive, their work will be done for love more than for money, by men with more leisure to cultivate the arts than we have. Some of them will become designers, some not: that is not important: a designer is one sort of artist, a workman another. Instrumentalists do not feel any sense of inferiority because they are not composers. But the scale of what craftsmen could achieve by concerting their efforts, and the opportunity it would give designers, would be something not dreamt of. Cathedrals were built, if not with joy in the labour (*pace* Morris), quite certainly by concerted effort unaided by any plant to speak of but what the workmen made themselves. People are beginning to believe you cannot make even toothpicks without ten thousand pounds of capital. We forget the prodigies one man and a kit of tools can do if he likes the work enough. And, as for those trades by the workmanship of risk which do need plant, it is not impossible to imagine that associations of workmen will set up workshops by subscription. The great

danger is that spurious craftsmen, realizing that the workmanship of certainty can beat anyone at high regulation, will take to a sort of travesty of rough workmanship: rough for the sake of roughness instead of rough for the sake of speed, which is rough workmanship in reality. This can be seen already in some contemporary pottery.

One rather feels that painting, whatever else it does nowadays, has to take care to look as different as possible from coloured photographs. Have the crafts got to take care to look as different as possible from the workmanship of certainty? If that is the best aim they can set themselves, let them perish, and the quicker the better! If they have any sense of their purpose they will look different, right enough, without having to stop and think about it. It is infinitely to be hoped that free and rough work will continue, but-not in travesty. One works roughly in order to get a job done quickly, but all the time one is trying to regulate the work in every way that care and dexterity will allow consistent with speed.

Free workmanship is one of the main sources of diversity. To achieve diversity in all its possible manifestations is the chief reason for continuing the workmanship of risk as a productive undertaking: in other words for perpetuating craftsmanship. All other reasons are subsidiary to that one, for there is increasingly a vacuum which neither the fine arts nor industry and its designers are any longer capable of filling. The contemporary passion for anything old, for junk and antiques, is no doubt symptomatic. The crafts in their future role may yet fill the vacuum but only if craftsmen achieve some consciousness of what they are for, only if they will set themselves the very highest standards in workmanship, and only then if they attract the voluntary services of the best designers. Workmanship and design are extensions of each other.

FURTHER READING

David Pye: Wood Carver and Turner (London: Crafts Council. 1986).

Christopher Frayling and Helen Snowdon, 'Skill—A Word To Start an Argument', *Crafts* 56 (May/June 1982), pp. 19–21.

David Pye, *The Nature and Aesthetics of Design* (London: Studio Vista, 1964).

Polly Ulrich, 'The Workmanship of Risk: The Re-Emergence of Handcraft in Postmodern Art', *New Art Examiner* 25/7(April 1998), pp. 24–29.

44

'THE GENESIS OF THE TECHNICAL OBJECT: THE PROCESS OF CONCRETIZATION', FROM *DU MODE D'EXISTENCE DES OBJETS TECHNIQUES*

Gilbert Simondon

Despite his importance within theories of technology, Gilbert Simondon is little known to English-speaking audiences. A student of the phenomenologist Maurice Merleau-Ponty at the Sorbonne in Paris, he also responded to the writings of Norbert Wiener, whose theory of cybernetics plays an important role in the following translated excerpt. Simondon's central argument is that the realm of technique can be divided into the 'abstract', which he associates with craft, and the 'concrete', which he sees as the character of industry. At first, this terminology might seem confusing—a reversal of our usual way of looking at things. It only makes sense if one focuses on his idea of the 'technical object', by which Simondon means not an actual physical thing, but a process: the whole historical lineage traced by attempts to perform related tasks through related means. An abstract technical object, he argues, is composed of autonomous parts which do not act upon one another. Each of these parts is made according to its own logic. This means that the final purpose of the technical object exists only 'abstractly', that is, in the mind of the maker or operator. A concrete technical object, by contrast, is organized according to parts or subsystems that operate on one another continually, according to the principle of feedback. Working with this simple division, Simondon revisits many familiar ideas about craft—that it is inefficient and creative, and lends itself well to autonomous or small-scale
production—and recasts them in a contentious and surprising way. For example, he argues that customization can operate only within that part of a technical process that is inessential to its true purpose (the ornamentation of a car, but not its engine, for example), and suggests that it is especially vital that craft be eliminated when partial failure is disastrous (as in aeronautical engineering). Fifty years after their initial publication, his ideas seem more unexpected than ever. For example, what could be more 'concrete' than an iPod, a completely integrated electronic device that is in turn seamlessly integrated with a vast, virtual technical object? If we live in a world that is becoming more and more 'networked' but no less made up of real things, then Simondon's theories are an important means of understanding that condition.

Gilbert Simondon, 'The Genesis of the Technical Object: The Process of Concretization', from *Du Mode d'Existence des Objets Techniques* (Paris: Méot, 1958). Translated by Glenn Adamson.

ABSTRACT AND CONCRETE TECHNICAL OBJECTS

The technical object is subject to a genesis, but that genesis is difficult to define, because the technical object's individuality modifies itself in the course of the genesis. Only with difficulty can one define technical objects by their membership within a particular technique. The species are easily distinguished, for practical usage, as long as one accepts that the technical object is understood according to the practical end to which it responds; but this is an illusory specificity, as no fixed structure corresponds to any one single use. The same functional outcome can be obtained by very different means: a steam engine, a gasoline-powered engine, a turbine, a spring or weight-driven motor are all equally motors. Nevertheless, there is a greater real analogy between a spring-powered motor and a bow or crossbow than between the same motor and a steam engine; a weight-driven clock is analogous to a winch, whereas an electrically-driven clock is analogous to a doorbell or a vibrator. Usage gathers heterogeneous structures and operations together, under those genres and species that make them meaningful, thanks to the relationship between this operative system and another one, the human being in action. That which we describe by a particular name, for example "motor," can be multiple at any given moment, and can vary over time, while its individuality changes.

To try to define the laws of the technical object's genesis within this framework, rather than departing from its individuality, or its unstable specificity, it is better to turn the problem around: that is, to leave behind those criteria that one can define as the individuality and specificity of the technical object. The technical object is not such-and-such a thing at a given time and place, but that of which there is a genesis. The unity of a technical object, its individuality, its specificity, are the characteristics of consistency and convergence of its genesis. The genesis of the technical object is part of its very being. The technical object is that which is not anterior to its becoming, but presents itself at each stage of its becoming; the technical object

is a unity of becoming. The gasoline-powered engine is not any one particular motor at a particular time and place, but the fact that there is a sequence, a continuity, that goes from the first motors that we know about to those that evolved from them. As in a phylogenetic lineage, each evolutionary stage contains within it the structures and dynamic schemas that preside at the forms' evolution. Technical being advances through convergence and self-adaptation; it unifies itself through a principle of internal resonance. Today's automotive engine is not the descendant of the engine of 1910 only because the motor of 1910 was constructed by our ancestors. It is no more its descendant for being more perfected in relation to use; for some uses, in fact, a motor of 1910 is superior to that of 1956. For example, it could attain high heats without seizing up or fusing, having been constructed for sport and not from weak alloys as is the rule today. It was also more autonomous, possessing a magneto ignition. It used to be that automotive motors could be used successfully on fishing boats, well outside their normal context of usage. Only by the internal standards of one chain of causality is today's motor defined as "subsequent" to that of 1910.

In a contemporary engine, each important element is so joined to the other parts through reciprocal exchanges of energy that it could not be other than what it is. The form of the combustion chamber, the form and dimensions of the valves, the form of the piston are all part of a single system in which a multitude of reciprocal relations exist. Such forms correspond to a certain rate of compression, which itself requires a temperature determined in advance by the ignition; the form of the cylinder head, the metal from which it is made, in relation with all the other elements of cycle, produces a certain temperature in the electrodes of the ignition; in its turn, this temperature reacts with the characteristics of the starter and thus the cycle is complete. One could say that the contemporary engine is "concrete," while the old engine is "abstract." In the old motor, each part contributes at a certain moment in the cycle, then is meant to stop acting on the other parts. The parts of the motor are like people who work each in their turn, but do not know what the others are doing.

In addition, it is important to note that in the operation of older motors, this is how heat engines work. Each element is as isolated from the others as the lines that represent it on a blackboard, in a geometric space of pure externality. The old motor is a logical assemblage of parts, each defined by their complete and unique function. Each element could accomplish its own end even better if it were a perfectly finalized instrument, oriented entirely to the achievement of its function. A permanent exchange of energy between two elements would appear as an imperfection, if this exchange did not belong to its theoretical function. There is a primitive form of the technical object, *the abstract form*, in which each theoretical and material unity is treated as an absolute, achieved in an intrinsic and necessary perfection for its function, constituted within a closed system. Integration with a system presents in this case a series of technical problems to solve, which in fact are problems of compatibility between systems that already exist.

These pre-existent systems must be maintained despite their reciprocal influence. And so emerge particular structures that one can name, for each thing that constitutes itself as a unity: structures of defense. The cylinder head of an internal combustion engine bristles with fins and blades for cooling, particularly around the valves, where the temperature and pressure are highest. These cooling elements, in the earliest motors, are like exterior appendages to the theoretical, geometrical cylinder head; they perform only one function, that of cooling. In recent motors, these cooling elements play a more mechanical role, placing themselves into the pressurized engine block like veins in a deformity. In these conditions, one can no longer distinguish between the volumetric unity of the engine (the cylinder, the block) and the unity of the cooling system.

[. . .]

The technical problem is thus the convergence of functions within a structural unity, rather than a search for compromise between conflicting requirements. The conflict may be between two aspects of a single system, as in the case we have been discussing, when the most structurally sound position of fans within the engine does not necessarily provide the best airflow while the vehicle is moving. In this case the builder is perhaps obliged to retain a mixed and incomplete solution: the fan blades and tubes, if they are to be disposed optimally to cool the engine, become thicker and more rigid than they would need to be on their own. If on the contrary if they are disposed for maximum structural integrity, they will need to have a larger surface area in order to slow down the air that is lost in the thermal exchange; or finally, the fan blades might be a compromise between the two extremes, which will necessitate a larger redesign, as if this single function were the goal of the entire engine. This divergence of functional directions persists like a residue of abstraction in the technical object. It is the progressive reduction of this margin between the functions of multipurpose structures that defines progress within the technical object. It is this convergence that defines the technical object, because there is not at any given time an infinite number of possible working systems. Species of technique are of a much more restricted number than the uses to which one can put technical objects; human needs diversify themselves to infinity, while the convergences of technical types are of limited number.

The technical object thus exists as a specific type obtained through a series of convergences. This series goes from abstract to concrete modes: it tends towards a state which would make a total system, consistent with itself, entirely unified.

CONDITIONS OF TECHNICAL EVOLUTION

What are the *reasons* for this convergence which manifests itself in the evolution of technical structures?—There are probably a certain number of extrinsic causes, and particularly those which tend to produce a standardization of units and of replacement parts. However, these extrinsic causes are no more powerful than those which tend towards the multiplication of types, appropriate to an infinite variety of needs. If technical objects evolve towards a small number of specific types, that is by virtue of internal necessity and not as a result of economic influences or practical exigencies. It is not the assembly line that produces standardization, but intrinsic standardization that permits the assembly line to exist. An effort to discover, in the transition from artisanal to industrial production, the reason for the formation of the specific types of technical objects might mistake its outcome for its condition; the industrialization of production is made possible by the formation of stable types. Craft [*l'artisanat*] corresponds to the primitive state within the evolution of technical objects, that is to say, the abstract state; industry corresponds to the concrete state. The customized character

that one finds in artisans' work is inessential; it results from another, essential, character of the abstract technical object, which is based on an analytical organization, always allowing a free route to new possibilities, which are the exterior manifestations of its interior contingency. In the confrontation between the coherence of technical work and the coherence of the system of needs of use, it is the coherence of utilization that carries the day, because the technical object that is customized [*sur mesures*] is in fact without intrinsic standards [*sans mesures*]. Its norms come from outside itself; it has not yet realized its internal consistency; it is not a system driven by necessity; it is an open system with exigencies.

On the other hand, at the industrial level, the object has acquired its coherence, and it is the system of needs that is less coherent than the system of the object. Needs mold themselves on to the industrial technical object, which thus acquires the power of modeling a civilization. It is utility that is cut to fit [*taillé sur le mesures de*] the technical object. When a custom automobile is made to the whim of an individual, the builder must use a serially produced engine and chassis, and modify its exterior decorative details and accessories and join them to the automobile as an essential technical object. It is the inessential aspects that can be made to measure, because they are contingent.

The relation that exists between these inessential aspects and the true nature of the technical type is a negative one. The more the car responds to the key requirements of the user, the more its essential characteristics are dictated by an exterior servitude; the body shop groans with accessories whose forms no longer correspond to structures that allow for the best air circulation. The character of *customization* is not only inessential, it runs counter to technical being. It is like a dead weight that one imposes from without. The center of gravity of a car rises as its mass increases.

However, it is not sufficient to affirm that the evolution of the technical object passes from an analytical order to a synthetic order, from artisanal production to industrial production. Even if this evolutionary movement is necessary, it is not automatic, and one must search out its causes, which reside essentially in the imperfection of the abstract technical object. Because of its analytic character, this object uses more material and demands more work in its construction. Logically more simple, it is technically more complicated, because it is made in relation to multiple complete systems. It is more fragile than a concrete technical object, because the relative isolation of each system constitutes a subassembly that, should it fail, threatens the preservation of the other systems. Thus, in an internal combustion engine, the cooling can be realized by an entirely autonomous subassembly; if it should cease to function, the motor would be damaged. If, on the contrary, the cooling system is realized through means that are interdependent with the other functions of the assembly, the running of the engine already implies cooling. In this sense a motor cooled by air is more concrete than one cooled by water: infrared heat radiation and convection are effects which cannot but be produced; they are necessary for the engine's operation. Cooling by water is semi-concrete: if it is realized entirely by thermo-siphon, it will be almost as concrete as direct cooling by air. But the use of a water pump, receiving energy from the motor by the intermediary of a transmission belt, increases the abstraction of this type of cooling. One could say that cooling by water is concrete in its guise as a backup system (the presence of water permits cooling for a few minutes thanks to the caloric energy absorbed by vaporization, if the transmission from the motor to the pump is broken) but in its normal functioning, this system is abstract. For that matter, an element of abstraction invariably persists in the possibility of the absence of water in the cooling system's circulation. Similarly, an ignition powered by a transformer and a battery is more abstract than an ignition powered by a magneto, which itself is more abstract than an ignition powered by air compression and injection of a combustible material, as in a Diesel engine. In this sense, a magnetically-powered motor and an air cooling system are more concrete than the engine of a standard automobile; all the parts play several roles. It is not surprising that scooters developed out of the work of a specialist aviation engineer. While the automobile can be permitted to retain residues of abstraction (cooling by water, an ignition powered by a battery and a transformer), aviation is obliged to produce the most concrete possible technical objects, in order to increase the certainty of function and reduce unnecessary weight.

[. . .]

The concretization of technical objects is conditioned by the narrowing of the interval that separates the sciences from technology. The primitive artisanal phase is characterized by a weak correlation between science and technique, while the industrial phase is characterized by an increased correlation. The construction of a given technical object can become industrial when this object has become concrete, that is, when the manner in which it is understood according to constructive intention is nearly identical to the way it is understood from a scientific point of view. This explains the fact that it was possible to make some objects industrially well before others. The winch, hoist, pulley, and hydraulic press are technical objects in which the phenomena of friction, electricity, electrodynamic induction, thermal and chemical exchange could be neglected in the majority of cases without leading to the destruction of an object or to poor functionality. Classical rationalist mechanics yielded scientific understanding of the principal phenomena behind the operational system of these objects, which are called simple machines. In contrast, it was impossible to industrially manufacture a centrifugal gas pump or a heat engine in the seventeenth century. The first such engine to be made industrially, [Thomas] Newcomen's, worked only on the downstroke, because the principle of steam condensation through cooling was known only scientifically. Similarly, electrostatic machines remained artisanal nearly until the present day, because producing charges and transmitting them through insulators (and the flow of charges through the corona effect), although they had been understood qualitatively more or less since the eighteenth century, were not the object of rigorous scientific study. After the invention of [James] Wimherst's machine [in the 1880s], the Van de Graaf generator itself retained something of the artisanal, despite its huge size and power.

FURTHER READING

Gilles Deleuze, 'On Gilbert Simondon', in *Desert Islands and Other Texts, 1953–1974* (Los Angeles: Semiotext(e), 2004).

Marc DeVries, 'Gilbert Simondon and the Dual Nature of Technical Artifacts', *Techné* 12/1 (Winter 2008).

Michel Serres with Bruno Latour, *Conversations on Science, Culture and Time* (Ann Arbor: University of Michigan Press, 1995).

45

'FORMS IN THE REALMS OF MATTER', FROM *THE LIFE OF FORMS IN ART*

Henri Focillon

Why do artworks look the way they do? Posed so bluntly, that might seem an impossible question to answer, but it was essentially the one that French art historian Henri Focillon posed in his idiosyncratic modernist text La Vie des Formes. *Disputing the notion that art should be seen first and foremost as a product of its social context—a registration of political, social and climatic realities—he sought to locate the intrinsic or internal 'life of forms' themselves. This carried him close to the arguments of earlier German theorists, such as Gottfried Semper, who had defined a position that might be called 'technological determinism', in which materials and techniques were seen as leading inevitably to certain forms. A naïve version of this view would conceive the artist or artisan as a kind of servant to the object; craftsmanship, from this perspective, would not be a cause of forms but rather the effect of them. Focillon wanted to complicate this sort of formalism. He took for granted the notion that materials determined certain formal solutions, but also recognized that artisans possess form-giving power in their own right. He therefore described art as a constant exchange between the objective qualities of materiality and the subjective, problem-solving capacity of artisanal will. Formalist art theory would take a different course after World War II, most famously in the well-known work of the critic Clement Greenberg, in which the virtues of opticality were championed over the frictions imposed by materiality.[1] Subsequently, following the radical criticisms offered by Conceptualists, Feminists and Postmodernists, formalism of any kind came to seem like a dirty word. Even so, Focillon's positioning of craft at the heart of his art theory is worth revisiting.*

Henri Focillon, 'Forms in the Realm of Matter', from *The Life of Forms in Art*, translated by C. B. Hogan and George Kubler (New York: Zone Books, 1989; orig. pub. as *La Vie des Formes*, 1934), excerpted.

Unless and until it actually exists in matter, form is little better than a vista of the mind, a mere speculation on a space that has been reduced to geometrical intelligibility. Like the space of life, the space of art is neither its own schematic pattern nor its own carefully calculated abbreviation. In spite of certain illusions popularly held in regard to it, art is not simply a kind of fantastic geometry, or even a kind of particularly complex topology. Art is bound to weight, density, light and color. The most ascetic art, striving modestly and with few resources to attain to the most exalted regions of thought and feeling, not only is borne along by the very matter that it has sworn to repudiate, but is nourished and sustained by it as well. Without matter art could not exist; without matter art would be something it had never once desired to be. Whatever renunciation art makes of matter merely bears witness anew to the impossibility of its escaping from this magnificent, this unequivocal bondage. The old antitheses, spirit-matter, matter-form, obsess men today exactly as much as the dualism of form and subject matter obsessed men centuries ago. The first duty of anyone who wishes to understand anything whatsoever about the life of forms is to get rid of these contradictions in pure logic, even should they still retain some slight trace of meaning or of usefulness. Every science of observation, and in particular that which is concerned with the movements and the creations of the human mind, is, in the strictest sense of the term, essentially phenomenological. And, because of this, the opportunity is given us of grasping authentic spiritual values. A study of the surface of the earth and the genesis of topographical relief, that is, morphogeny, supplies us with admirable foundations to the poetry of landscape, but such studies do not have that object originally in view.

The physicist does not take the trouble to define the "spirit" that underlies the transformation and behavior of weight, heat, light and electricity. Then, too, nobody any longer confuses the inertia of mass with the life of matter. This is because matter, even in its most minute details, is always structure and activity, that is to say, form, and because the more we delimit the field of metamorphoses, the better do we understand both the intensity and the graph of the movements of this field. These discussions of terminology would be futile, if they did not involve methods.

In my approach to the problem of the life of forms in matter, I do not mean to separate the one concept from the other, and if I use the two terms "form" and "matter" individually, it is not to give an objective reality to a highly abstract procedure, but is, on the contrary, in order to display the constant, indissoluble, irreducible character of a true and genuine union. If we will hold this notion in mind, it will be seen that form does not behave as some superior principle modeling a passive mass, for it is plainly observable how matter imposes its own form upon form. Also, it is not a question of matter and of form in the abstract, but of many kinds of actual matters or substances—numerous, complex, visible, weighty—produced by nature, but not natural in and of themselves.

Several principles may be deduced from the preceding. The first is that all different kinds of matter are subject to a certain destiny, or at all events, to a certain formal vocation. They have consistency, color and grain. They are form, as I have already indicated, and because of that fact, they call forth, limit or develop the life of the forms of art. They are chosen not only for the ease with which they may be handled, or for the usefulness they contribute to whatever service art renders to the needs of life, but also because they accommodate themselves to specific treatments and because they secure certain effects. Thus, their form, in its raw state, evokes, suggests and propagates other forms, and, to use once again an apparently contradictory expression that is explained in the preceding chapters, this is because this form liberates other forms according to its own laws. But it must be pointed out at once that the formal vocation of matter is no blind determinism, for—and this is the second principle—all these highly individual and suggestive varieties of matter, which demand so much from form and which exert so powerful an attraction on the forms of art, are, in their own turn, profoundly modified by these forms.

Consequently, there is between the matters or substances of art and the substances of nature a divorce, even when they are bound together by the strictest formal propriety. A new order is established, within which there are two distinct realms. This is the case even if technical devices and manufactures are not introduced. The wood of the statue is no longer the wood of the tree; sculptured marble is no longer the marble of the quarry; melted and hammered gold becomes an altogether new and different metal; bricks that have been baked and then built into a wall bear no relation to the clay of the clay pit. The color, the integument, all the values that affect the sight have changed. Things without a surface, whether once hidden behind the bark, buried in the mountain, imprisoned in the nugget or swallowed in the mud, have become wholly separated from chaos. They have acquired an integument; they adhere to space; they welcome a daylight that works freely upon them. Even when the treatment to which it has been submitted has not modified the equilibrium and natural relationship of the parts, the life that seems to inhabit matter has undergone metamorphosis. Sometimes, among certain peoples, the kinship between the substances of art and the substances of nature has been the subject of many strange speculations. The Far Eastern masters, for whom space is essentially the theater of metamorphosis and migration, and who have always considered matter as the crossroads where a vast number of highways come together, have preferred among all the substances of nature those that are, as it were, the most intentional and that seem to have been elaborated only by some obscure art. And yet, these same masters, while working with the substances of art, often undertook to stamp the traits of nature upon them; they attempted, indeed, to transform them completely. And thus, by a singular reversal, nature for them is full of works of art, and art is full of natural curiosities. Their exquisite little rock gardens, for example, although composed with the utmost care, seem to have

been laid out by the mere caprice of some highly ingenious hand, and their earthenware ceramics appear to be less the work of a potter than a marvellous conglomerate created by subterranean fire or accident. In addition to this delightful emulation and to this interest in transpositions—which seeks the artificial at the heart of nature and the secret labor of nature at the heart of human invention—these men have been artisans who have worked only with the rarest of substances and who have been the most emancipated from the use of models. Nothing exists in either the vegetable or the mineral world that suggests or recalls the cold density, the glossy darkness, the burnished and shadowy light, of the lacquers made by these Eastern masters. These lacquers actually come from the resin of a certain pine, which is then long worked and polished in huts built above watercourses and perfectly protected from all dust. The raw stuff of their painting partakes both of water and of smoke, and yet is in reality neither the one nor the other, inasmuch as such painting possesses the extraordinary secret of being able to stabilize these elements and at the same time to leave them fluid and imponderable.

But this sorcery, which astonishes and delights us because it comes to us from afar, is no more captious or inventive than the labor of Western artists upon the substances of art. The precious arts, from which we might be first tempted to draw examples, do not, perhaps, offer anything at all comparable in this respect to the resources of oil painting. There, in an art seemingly dedicated to "imitation," the principle of non-imitation appears as it does nowhere else. There, in oil painting, lies the creative originality that extracts from the substances furnished by nature all the matters and the substances necessary for a new nature. This originality is, moreover, one that unceasingly renews itself. For the matter, or substance, of an art is not a fixed datum that has been acquired once and for all. From its very first appearance it is transformation and novelty, because artistic activity, like a chemical reaction, elaborates matter even as it continues the work of metamorphosis. Sometimes in oil painting we observe the spectacle of transparent continuity, of a retention of all forms, whether hard or limpid, within a delicate, golden crystallization. Or again, oil painting will nurture forms with gross abundance, and they will seem to wallow and roll in an element that is never quiescent. Sometimes oil painting can be as rough as masonry, and again it can be as vibrant as sound. Even without the introduction of color, it is obvious that the substance varies here in its composition and in the seeming relationship of its parts. But when we do call on color, it is even more obvious that the same red, for instance, takes on different properties, not only according to its use in distemper, tempera, fresco or oil, but also a different property according to the manner of its application in each one of these various processes.

This observation serves to introduce several others, but before considering them, a number of points remain still to be clarified. One might reasonably suppose that there are certain techniques in which matter is of slight importance, that drawing, for example, is a process of abstraction so extreme and so pure that matter is reduced to a mere armature of the slenderest possible sort, and is, indeed, very nearly volatilized. But matter in this volatile state is still matter, and by virtue of being controlled, compressed and divided on the paper—which it instantly brings to life—it acquires a special power. Its variety, moreover, is extreme: ink, wash, lead pencil, charcoal, red chalk, crayon, whether singly or in combination, all

constitute so many distinct traits, so many distinct languages. To be satisfied as to this, one need only imagine any such impossibility as a red chalk drawing by Watteau copied by Ingres in lead pencil or, to put it more simply (inasmuch as the names of individual masters introduce certain values that we have not yet discussed), a charcoal drawing copied in wash. The latter at once assumes totally unexpected properties; it becomes, indeed, a new work. We may at this point deduce a more general rule that invokes the principle of destiny or of formal vocation mentioned above, that is, the substances of art are not interchangeable, or in other words, form, in passing from a given substance to another substance, merely undergoes a metamorphosis.

[. . .]

That our idea of matter should, therefore, be intimately linked with our idea of technique is altogether unavoidable. They are, indeed, in no way dissociated. I myself have made this concept the very center of my own investigations, and not once has it seemed to me to restrict them in any way. On the contrary, it has been like some observatory whence both sight and study might embrace within one and the same perspective the greatest possible number of objects and their greatest possible diversity. For, technique may be interpreted in many various ways: as a vital force, as a theory of mechanics or as a mere convenience. In my own case as a historian, I never regarded technique as the automatism of a "craft," nor as the curiosities, the recipes of a "cuisine"; but instead as a whole poetry of action and (to preserve certain inexact and provisional terms used in the vocabulary of this particular essay) as the means for the achievement of metamorphoses. It has always seemed to me that in difficult studies of this sort—studies that are so repeatedly exposed both to a vagueness of judgments respecting actual worth and to extremely ambiguous interpretations—the observation of technical phenomena not only guarantees a certain controllable objectivity, but affords an entrance into the very heart of the problem, by presenting it to us in the same terms and from the same point of view as it is presented to the artist. To find ourselves in such a situation is as uncommon as it is desirable, and it is important to define wherein lies its interest. The purpose of the inquiries of a physicist or a biologist is the reconstruction of nature itself by means of a technique controlled by experiment: a method less descriptive than active, since it reconstructs an activity. But we historians, alas, cannot use experiment to check our own results, and the analytical study of this fourth "realm" which is the world of forms can amount to little more than a science of observation. But in viewing technique as a process and in trying to reconstruct it as such, we are given the opportunity of going beyond surface phenomena and of seeing the significance of deeper relationships.

Thus formulated, this methodological position appears natural and reasonable enough, and yet to understand it fully and above all to exploit its every possibility, we must still strive, within our inmost selves, to throw off the vestiges of certain old errors. The most serious and deeply rooted of these derives from that scholastic antinomy between form and subject matter, to the discussion of which there is no need to return. Next, even for the many enlightened observers who pay close attention to investigations on technique, technique remains not a fundamental element of knowledge that reiterates a creative process, but the mere instrument of form, exactly as form seems to them to be the garment and vehicle of the subject matter. This arbitrary restriction necessarily leads to two false positions, and the second

may be considered as the refuge and the excuse of the first. In regarding technique as a grammar, which unquestionably has lived and still does live, but whose rules have taken on a kind of provisional fixity—a kind of value imparted by unanimous consent—we are led to identify the rules of common speech with the technique of the writer, the practice of a craft with the technique of an artist. The second false position is to relegate every creative advance augmenting that grammar to the indeterminate world of "principles," in exactly the same way, for instance, as ancient medicine explained all biological phenomena by the action of a vital "principle." But if we no longer try to separate what is fundamentally united, and instead try simply to classify and conjoin phenomena, we see that technique is in truth the result of growth and destruction, and that, inasmuch as it is equally remote from syntax and from metaphysics, it may without exaggeration be likened to physiology.

I do not deny that I myself am using the term under discussion in two senses: techniques in particular are not technique in general, but the first meaning has exercised a restrictive influence on the second. It will be admitted that, in a work of art, these meanings represent two unequal and yet intimately related aspects of activity: that is, first, the aggregate of the trade secrets of a craft and, second, the manner in which these trade secrets bring forms in matter to life. This would amount to the reconciliation of passivity with freedom. But that fact by itself is not sufficient, for, if technique is indeed a process, we must, in examining a work of art, go beyond mere craft techniques and trace to its source an entire genealogy. This is the fundamental interest (superior to any specifically historical interest) that the "history" of a work of art has for us before it attains its ultimate form—in the analysis, that is, of the preliminary ideas, the sketches, the rough drafts that precede the finished statue or painting. These rapidly changing, impatient metamorphoses, coupled with the earnest attention given them by the artist, develop a work of art under our very eyes, exactly as the pianist's execution develops a sonata, and it is of the first importance that we should take heed of them, as they move and react within something that is still apparently static. With what do they provide us? Points of reference in time? A psychological perspective? A jumbled topography of successive states of consciousness? Far more than these: What we have here is the very technique of the life of forms itself, its own biological development. An art that yields particularly rich secrets in this respect is engraving, with its different "states" of the plates. For the amateur, these states are mere curiosities; for the student they hold a much more profound meaning. When we examine a painter's rough draft—reduced to itself alone, and irrespective of its past as a sketch or its future as a painting—we feel that it already

carries a genealogical significance and that it must be interpreted, not as an achievement in and of itself, but as an entire movement.

[. . .]

NOTE

1. Caroline A. Jones, *Eyesight Alone: Clement Greenberg's Modernism and the Bureaucratization of the Senses* (Chicago: University of Chicago Press, 2005).

FURTHER READING

George Kubler, *The Shape of Time: Remarks on the History of Things* (New Haven, CT: Yale University Press, 1962).

Robert F. Trent, 'Focillon', in *Hearts and Crowns: Folk Chairs of the Connecticut Coast 1710–1840* (New Haven, CT: New Haven Colony Historical Society, 1977).

David Summers, *Real Spaces: World Art History and the Rise of Western Modernism* (London: Phaidon, 2003).

Joanne A. Wood, 'Counter-Evolution: The Prosthetics of Early Modernist Form', *English Literary History (ELH)* 66/2 (Summer 1999), pp. 489–510.

46

RHYTHM

Elsie Fogerty

One of the paramount features of modern-ist thinking was the search for universals that could be used to bind together art made in vari-ous media and disciplines and in various parts of the world. It was a thrilling moment in art theory. Abstraction seemed to provide a com-mon language by which aesthetic experiences as diverse as a Japanese woodcut print, an African sculpture and a modern dance could be under-stood as exhibiting similar (and potentially mu-tually influential) properties. Elsie Fogerty's book Rhythm, *written just prior to the outbreak of World War II, exemplifies this optimistic appli-cation of modernist ideals. Like other writers be-fore her (such as Henri Bergson and Roger Fry), Fogerty identified* rhythm *as a transhistorical and transcultural aesthetic value and attempted to apply it as a critical tool to a diversity of art forms ranging from dance and music to painting and ceramics. Fogerty was well qualified for such an endeavor. A specialist in elocution and dra-matic speech, she had founded the Central School of Speech and Drama in 1906. (Based for its first fifty years at the Royal Albert Hall, the School is still active today in London.) A major figure in the British theatre scene, she went on to provide voice training for such luminaries as Lawrence Olivier and John Gielgud. In* Rhythm, *Fogerty attempted to link disparate subjects (poetic metre, the cadences of speech, the curvature of a pottery vessel, etc.) together under one aesthetic principle.*

The result is more evocative than persuasive, per-haps, but her ambition to link together art from many cultures and disciplines seems nothing if not contemporary.

Elsie Fogerty, excerpts from *Rhythm* (London: George Allen and Unwin, 1937). Notes deleted.

INTRODUCTION

"We all know what it means, but of course we can't express it."

"It is the profoundest mystery of the Unknowable."

The two points of view are practically alike, since both are alike pretentious. The under-lying solution may lie behind something as simple as the fall of an apple, or the swing of an altar-lamp.

A clearer understanding of Rhythm may imply a new conception of movement already dimly felt by those who have studied more nar-rowly, if more profoundly, the isolated aspects of the subject.

The simplest of all definitions of Rhythm is based on the idea of repetition. Before accept-ing this, it would be well to examine more in detail the whole range of terms associated with the word, and see how far they help us to any preliminary definition. All are of a narrower significance than Rhythm itself; "repetition" in no sense conveys the full sense of Rhythm unless we explain it as measured, significant recurrence, marked by time and force; is it not rather the means by which we become most readily *conscious* of Rhythm? We only need to picture the intolerable and irritating monotony induced by repetition of a meaningless series of beats or notes, to understand that repetition in itself conveys, and can convey, nothing rhythmic.

Periodicity refers rather to the result of rhythmic action, as in this passage from Havelock Ellis, *The Dance of Life:*

> We have but to stand on the seashore and watch the waves that beat at our feet, to observe that, at nearly regular intervals, this seemingly monotonous rhythm is accentuated for several beats that the waves are really dancing the measure of a tune.

Pattern may more properly be regarded as recurrence of design, but decorative pattern, with its strict repetition, is less rhythmic than the frieze of the Parthenon with its infinite variety. In the composition of a great plastic or pictorial design, in the intricacies of Chinese decorative art, in modern music and verse, Rhythm is present at its greatest; while repetition is implicit rather than expressed. Where pattern implies the repetition of forms in a fixed decorative or conventional order, as in the commercial "Willow Pattern" plate, or the "Egg and Dart" pattern or Victorian cornices, it easily lapses into a stencil quality to which the French appropriately apply the word *cliche.* It may be suggested that Pattern is the symmetrical result of rhythmic action on matter.

Design has a far broader meaning and covers the whole scheme of any construction before it is embodied in its appropriate material. It recalls the delightful ritual phrase in masonry, addressed to the layer of the first stone: "Will it please you to peruse the whole design?"

Here something which has a definite bearing on the significance of Rhythm may be noted. Sir Charles Barry, writing of Architectural Form, points out that "while Symmetry is architectural form (static), Rhythm is a plastic idea; symmetry implies and expresses the lasting, uniform, and inorganic. Rhythm implies change; the organic, as in Sculpture, deals with animal life."

Metre is the measure of the varied syllabic patterns obeying the significant pulse-beat of Rhythm in verse. Early verse is often strictly metric to the ear, for love of repetition is primitive; children love verbal patter, nursery rhymes, "counting out," and jingles, all based on repetition.

Time in its musical sense is constantly treated as equivalent to Rhythm; and "Tempo"—that is to say, the rate of speed at which a musical phrase is played—is also held to dominate rhythmic significance. The second is obviously inaccurate. Alteration in the pace of a phrase only affects its rhythm where that phrase is directly suggestive of a definite pace, as in a march. Time in music gives the metric or temporal beat. Imagine, however, trying to waltz to a tune played in strictly three-four time, three crotchets in each bar.

Yet it is true that Time is one of the elements of Rhythm, just as Metre is one of the elements of Poetry. Both may destroy instead of creating Rhythm.

[. . .]

So far as *Harmony* signifies the union of parts in a whole, its meaning is allied to that of Rhythm. It is in this sense we speak of the "Music of the Spheres," a phrase often used with irritating vagueness in an attempt to describe the true significance of Rhythm; but Harmony in Music signifies again the

Is it not plain that what actually relates to Rhythm in all these terms suggests one only of its manifestations rather than the thing itself? Our consciousness of Rhythm is much more profound than anyone of them, and we can say of each of them in turn that it is not Rhythm itself. But one point does become clear from such analysis and from the normal dictionary definition which includes each of these aspects in turn: that is the presence of three constant factors:

First, *The Factor of Time,* present in measured recurrence, that is to say, in the temporal interval between maximum and minimum force which is necessary to the working of a great machine, or the isochronous interval between stress and stress, which forms the basis of English prosody. Second, *The Element of Force,* without which the temporal spacing could not exist. And third, *The Element of Space* itself, without which the application of force is unthinkable, and through which the parts of the machine must travel to carry out its function.

[. . .]

RHYTHM IN CRAFTS, GAMES AND DANCING

It is inspiring to watch many occupational crafts of a rhythmic character. The swing of the scythe is still unsurpassed in grace; its automatonism is fraught with just a spice of danger to quicken attention. Endurance of fatigue depends on the breadth of spatial movement to give time for the alternate relaxation of arm and trunk muscles in time with the breathing. The problem on which the mind is fixed is the felling of the long swathes of grass, so that they lie in a static pattern of order. The exercise is perfect from the point of view of health, and so rhythmic that it needs no song to keep it steady . . .

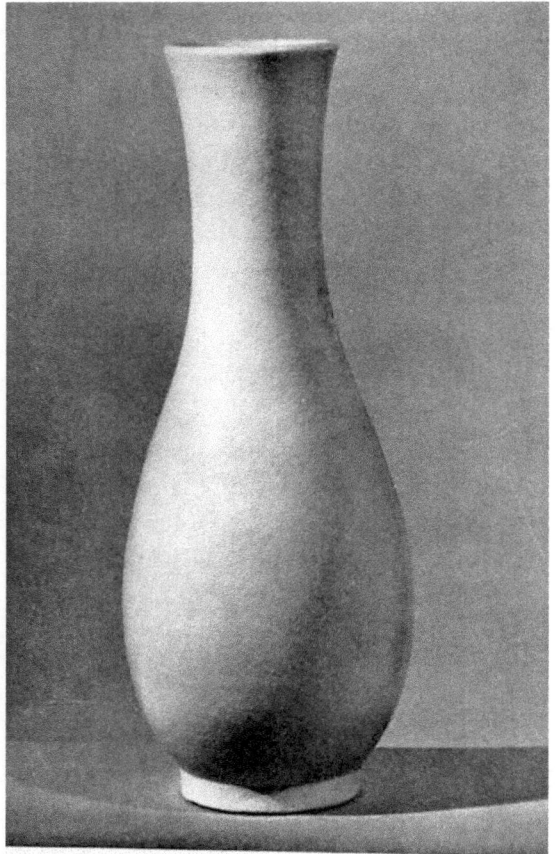

A MING VASE OF KIANG TING WARE

By permission of the Royal Academy of Arts, Chinese Exhibition, 1936

Figure 31 "A Ming Vase of Kiang Ting Ware." As pictured in Elsie Fogerty, *Rhythm*.

simultaneous accord or consonance of notes rather than the pleasant relation of certain successive tones in melody. It is Spatial, not Temporal, and gives the most massive expression of musical Rhythm. It is interesting to consider how modern Harmony departs from the older and simpler conception which first found its perfect expression in the Chorale and the Madrigal, and, instead of melody, gives us a kind of extended Harmony, in which the final note harmonizes what seemed a meaningless series of discords.

The creeping of the sickle, the rhythm of the anvil and of the woodcutter's axe, show man using the simplest apparatus in a perfection of motion. He began by willing the result: performed it with effort and fatigue, improved his mechanism, and then gradually learned to employ it easily while he supervised and determined the variety of its purposeful action. No two trees present exactly the same problem in felling, nor is the problem of the felling confined to the action of the axe. The dropping of the tree is the essential question; just as the expert bringing down an enormous factory chimney can drop it harmlessly without touching surrounding buildings and leave it lying on the chosen space in an ordered train of bricks.

All the various folk-rhythms are derived rather from instinctive action than from the mind of Man. They embody and reflect natural rhythms with a certain interchange and augmentation from their reflection, but no more. They are reflected with almost an equal degree of interest and brilliancy from every type of mind, and in every type of community. They are full of "fancy," however, and therefore unrestrained by the sense of fidelity to Nature. Theirs is the rhythm of reflection. But in the great imaginative arts we face a pattern refracted and analysed by the typical genius of the artist, as the facets of the diamond refract the light. The work of the folk-artists is "material" for genius, and as such great artists are intensely interested in it, as Shakespeare and Milton were interested in the folk-lore of their England. But once a work of genius is accomplished, it has about it a certain intactness of form, like the diamond to which it has been compared. It becomes a thing "bodied forth" by imagination. It can be an inspiration to other artists, but it can never be re-incorporated in the work of a great artist. To attempt it is to plagiarize or to deface the work of genius, as Dryden defaced Shakespeare.

At the Best it is to Produce "Derivative" Art

Transference into another medium may prove the greatness of the original inspiration, as in the illustration of a great poet by a designer. Even that cannot compare with creative interpretation, such as that of Blake. Perhaps only in great translation, where the skeleton of the original thought has been re-clothed in the subtle intellectual rhythm of another speech, can such re-handling be adequate.

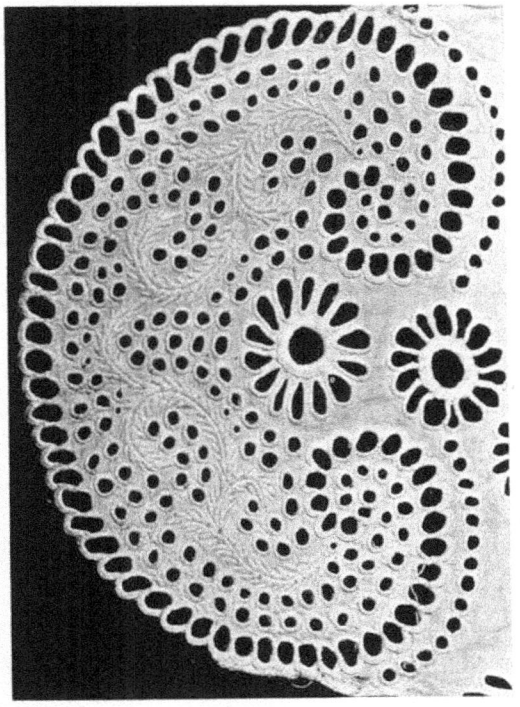

IRISH EMBROIDERY, 1863

Figure 32 "Irish Embroidery, 1863." As pictured in Elsie Fogerty, *Rhythm*.

Many occupational movements were consciously made into a dance; the girls sowed the fields to a song, and to the duller occupations song gave life . . .

In tracing the significance of Rhythm in natural law the necessity for mechanical invention became clear, but all man's earlier instruments aimed at extending the swiftness, the force, the spatial certainty of his own movements, and so he developed his rhythmic control and extended its range with every fresh invention. This is not equally true of all mechanical machine-minding. Absorption in the automatonism of movement is psychologically the most mind-shattering thing we know. It is at the back of all occupational neuroses. Often the rhythm of modern machinery is so foolproof that no one but a fool can spend his life's work in running it.

Many years ago the story was told of a group of social workers who carried out an investigation into workroom organization and the development of occupational facility in the avoidance of industrial fatigue, the object being primarily increase of output. One lady delegate had been shown how lack of hand control in feeding a delicate machine reduced the output by such and such a fraction per hour, and to what imposing figure on the other side of the dot this error could lead in a month's work. Even the interested glances thrown at the watching delegates had their unfavourable result on the sum of production, and went down on the adverse side of the balance-sheet. At last, in one corner, they noticed an absorbed figure who seemed literally part of his machine. "That," said the conductor, proudly, "is the perfect worker. That man has a perfect mechanical action, his output is higher than that of any man in the room." They gazed in admiring awe, and the lady member expressed a desire to speak

with this paragon, whereupon the group were rather hastily shepherded into another room; further enquiries brought out the apologetic explanation that unfortunately the man was definitely mentally defective!

The credibility of the story has always seemed a little doubtful, because the actions of the mentally deficient are notoriously ill-coordinated; but it does explain the need of the modern industrial demand, that ordered recreation should as nearly as possible balance occupation in time, where that occupation is monotonous. Here, in another form, we meet the closed circle of mere repetition, where the whole time of the action is employed in the reiteration of a movement without any intellectual result. No handcraftsman ever "repeats." The smith's steady hammering stroke perpetually changes its dynamic intention. In a great forging, one guiding hand taps with a little hammer to indicate the exact placing of the shattering weight of blows. Watch a blacksmith making a horseshoe, twisting and turning it with his tongs, beating together the broken pieces of the old shoe, welding it in the fire to the exact best shape and measure for the horse's hoof; then cooling it, marking it for the nail-holes, and after all his labour is over, turning farrier and soothing the nervous animal, who presently stands confidently rubbing a friendly head on the workman's shoulder while the nails are painlessly driven in, without touch to any tender point in the hoof—and the horse is shod. One pictures the great creature clumping away from the forge, after acknowledging a last friendly pat, with something of that happy mood in which we leave the dentist's room, knowing that nothing more will need to be done for the next three months! It is all rhythmic, not metric, close akin to the musical clamour of great bells, which irritates the undistinguishing ear of the Sunday city listener, but implies for the ringers a round of "changes" and variations of rhythm and tempo, bringing to life a musical disorder in the clangour of the chimes which it takes a lifetime of practice to produce and to appreciate.

There are, of course, numberless mechanical inventions which have set man's spirit free from the need of slavery and made the slave-market an economical absurdity. There are many as thrilling as a war-dance, or the riding to war of Jehu the son of Nimshi. Like the vaulting spring of the javelin thrower or the grace of the drawn bow, they leave no need for compensating corrective exercises, but set the mind of man free once more to contemplate the beauty of the aesthetic result he is creating.

This element of aesthetic delight in movement is the key to the whole matter, and it now demands special attention.

. . . "Co-ordination" is not merely an individual process or an internal activity. It is primarily the means of bringing man's being, his faculties and his powers, into relation with the rhythmic principles of the universe. It is admittedly first a coordination for existence, then for adaptation, then for enjoyment, but finally for mastery over himself and over life.

FURTHER READING

Elsie Fogerty, *Speech Craft: A Manual of Practice in English Speech* (London: Dent, 1930).

Christopher Reed, ed., *A Roger Fry Reader* (Chicago: University of Chicago Press, 1996).

47

THE MANDE BLACKSMITHS: KNOWLEDGE, POWER AND ART IN WEST AFRICA

Patrick R. McNaughton

Many cultures have a concept along the lines of the 'music of the forge'. The steady ring of the blacksmith's hammer—or, for that matter, the clack of the weaver's shuttle, or the paddling of a clay pot—are easily rendered into the beat of uninterrupted tradition. A sign of Patrick McNaughton's resistance to such sentimentality is his observation that, at least among the Mande people of west Africa, individual smiths employ distinctive rhythms with their hammers and bellows that reflect their own particular technique, and that these rhythms can be varied, in a sort of musical conversation. For those who are interested in Elsie Fogerty's ideas about rhythm but dissatisfied with her transcultural modernist bias, McNaughton's closely observed account moves from process to product, showing how metal wares such as guns, knives and even farm tools are invested with efficacy and admirableness through craft articulation.

Patrick R. McNaughton, 'The Mande Smiths as Craftsmen', in *The Mande Blacksmiths: Knowledge, Power and Art in West Africa* (Bloomington: Indiana University Press, 1988). Excerpted, and notes deleted.

Mande blacksmiths work in two domains. Their forges are in town, but they gather materials in the bush, where they used to smelt iron. In general, smiths are more mobile than other citizens; they are usually willing to resettle where business is better. A small percentage, often the younger and more adventuresome, are itinerant, but the vast majority live in towns.

Mande towns vary greatly in size. Smaller ones may only have twenty or so extended family compounds, with five to ten people living in each. The larger towns have a hundred or more family compounds and thousands of inhabitants. Even the smallest town has at least one smith family; there are just too many tools and utensils that no one else can make. Most towns have at least three to five families of smiths; many have more. Sometimes the blacksmiths congregate together, forming a special quarter called a *numusokala*. Occasionally, smiths form their own town or hamlet, called a *numu dugu*, which generally has easy access to the nearby towns of farming clans.

Small smith families will include just a husband and a wife and possibly one or two children. Large families will include a senior male smith with two or three wives and their children, his younger brothers, and their immediate families. A hamlet built for such a family by Sedu Traore's father often housed thirty smiths or more.

Generally, several male smiths in a town will work iron and wood, while at least one female member of the family will work clay. Sons and nephews of the male smiths and the daughters and nieces of the female potters are often apprentices. Sedu Traore's father trained thirteen sons, so many that most left their hometown in order not to overtax the market. Sedu, however, has trained no smiths. One of his sons became a Western-style carpenter; another decided to study the Koran.

Blacksmiths' forges are located in family compounds spread around the community, at intersections of main walking thoroughfares, or next to the clearings used as dance arenas and elders' meeting places. Often two or more smiths share a forge, or just work together by moving back and forth between the forges of each. When especially heavy labor is called for, as when an old railway tie must be reduced to hoe blades or a thick ax blade must be produced, a smith calls upon his colleagues for assistance. Every town has a "master of smiths," *numutigi,* who is generally the oldest, though not necessarily the most accomplished. He is consulted about major undertakings, and often directs joint undertakings.

Out of town, smelting furnaces used to grace the landscape with their picturesque clay forms. Now, however, they are rarely encountered, although smiths can still point out the places where they stand in ruins, or where their traces in the form of slag, the fused mineral by-product of iron smelting remain. Sometimes a single furnace stood alone. It would have been refurbished and fired once each year by one family of smiths. Sometimes several furnaces stood together, to be worked by all the smiths in a community. Morrow Campbell, a mining and metallurgy specialist working in West Africa in 1908, reported hundreds of abandoned smelting furnaces, a testimony to the past importance of iron making. Candice L. Goucher, a historian, has suggested that iron-making industries used to operate in a variety of areas across the western savanna, the smelted product being a very popular item of trade.

This chapter examines the techniques smiths employ and the products they make in their two spaces, town and bush. We proceed as a blacksmith's apprentice would, from the simplest tasks to the most complex.

TECHNIQUES

A young smith's apprenticeship begins sometime before the age of ten. By then he will have spent many of his free hours in his father's forge, working the bellows when he is needed and, in general, being exposed to *numu baara,* the "work of smiths." His father may have given him some old carving tools or made him a small set with which he could mimic the work of adults. The lad may have made himself a knife handle, with his father making a blade to fit it.

Once a youth becomes an apprentice, labor and learning begin in earnest. Over a period of some seven or eight years the apprentice moves from working the bellows to carving wood and then to forging iron, gradually becoming competent at each. It is very hard work, and some apprentices simply drop out. One of Sedu's brothers, for example, did not have the patience for it, and so switched fields dramatically, becoming something of a Muslim holy man.

In the beginning the labor is arduous indeed. Before the neophyte can master techniques and form, he first has to master pain. He begins at the bellows, where he spends many hours each day. Few tasks could be more boring—or, at the same time, more unpleasant, for the tedious repetition of the same simple moves quickly generates sore arms and shoulders. Just working the bellows is not pleasant, but Mande smiths do not learn to just work bellows; they learn to play them.

There are two basic types of bellows, clay and wood. Clay versions are large architectural constructions that rise from the floor of the smithy as a platform on which the worker sits. Before him is a broad clay area in which two clay pots are embedded. Goat skin covers the pot tops in such a way that the worker may insert his hands into slits that open when he pulls up and close when he pushes down. Iron tubes extend forward from the sides of the pots through a clay fire wall and down into the basin of the forge. Working the bellows forces large drafts of air through the pipes and into the forge.

Wood versions are portable. Y-shaped hard wood is hollowed out so that two chambers meet at the intersection. Two iron tubes are inserted there, and when the bellows is in place the tubes pass through another type of clay fire wall and down into the heart of the forge's charcoal basin. The whole skins of two small goats are attached to the open ends of the chambers, at the top of the Y. They form large bags that are split open at the back, with small pieces of wood attached to hold the slits stiff and allow the worker to insert his hands. As he pulls back on the bag he opens his hand and the skin fills with air. As he pushes he closes his hand and the air is forced through the tubes and down into the charcoal basin.

Pushing air through either type of bellows produces a wonderful sonorous blast, which can be heard all around the forge and the immediate vicinity. The skins are pushed alternately, at first in a very even fashion that results in a kind of basic two-beat rhythm. The rapidity with which air circulates through the coals determines how quickly the coals heat up and how hot they get. Both variables are significant, and the master smith lets his apprentice know how much air is needed depending on the size and state of development of the iron piece in the forge. Too much air wastes charcoal, causing it to burn too quickly.

Gradually the apprentice learns to gauge how much vigor to apply to his work, and in the process he begins to vary his basic two-beat rhythm.

The result is patterns of rhythm that resemble drum beats, except that the percussive thrust of drumming is replaced by fluid gusts of air. As a young smith grows competent, he develops a sharp, crisp precision in his rhythms, which often become astonishingly complex. Each smith has one or several favorites. The one Sedu Traore used most often is rather stately and gentle, reflecting his age . . . Several women frequently ground millet in a mortar near Sedu's forge. They alternated pounding their huge pestles into the mortar, adding hand-clapping routines and throwing the pestles up into the air when the spirit moved them. Often when they worked Sedu played his bellows rhythms in counterpoint to theirs, creating a lively atmosphere in the neighborhood.

Even though they usually play their favorites, smiths generally know many rhythms. On a trip to San in June 1978 with my research colleague Kalilou Tera, I spent a morning with a group of four smiths who shared a forge on one of the city streets. It immediately became clear that the wooden bellows worker was quite expert on the instrument, and as our enthusiasm grew, so did his repertoire and his finesse. He played for some time, drawing an appreciative audience from everyone out on the block. Then a younger smith took over, and this fellow was even better. Kalilou began calling out the names of standard drum rhythms used by the many ethnic groups that live in the region, and the smith played them all. Earlier, in 1973, I watched a very young apprentice in Bamako play rhythms on clay bellows with such speed and precision that he could have easily switched to Western trap drums and joined any band he chose.

In areas where famous blacksmiths used to live, sometimes several generations ago, young smiths of the present generation learn to play the renowned smiths' rhythms. This adds a nice historical touch to the youths' sense of profession, and it pleases their masters to no end.

In fact, learning bellows rhythms is an apprentice's first task of fundamental importance. This is even true for foreigners who become honorary apprentices. When I worked closely with Sedu in 1973, I spent many hours at his bellows. Once the town's master smith, Dramane Dunbiya, paid a visit as I was playing Sedu's standard rhythm. Dunbiya paused and then said: "Ah, now you have become a true smith."

Figure 33 *Bell*, collected in 1920. Bronze, iron.

[. . .]

Today, Mande smiths make both rifles and pistols, in a range of sizes that vary according to the length and diameter of the barrel and, therefore, the power of the charge. I know of at least six rifle sizes, each with its own name. They are used as single-charge weapons or as shotguns. In the latter instance they are loaded with a charge of small iron fragments from the blacksmith's forge; the fragments are held in place at the base of the barrel, *marafa bulu*, with small bits of rag.

The bullets, *kisew*, I saw made were not of the cartridge type. Rather, they were simply spheres forged from iron rods. Sedu made these bullets by heating a rod red-hot and cutting nearly all the way through it with a chisel at regular intervals. Then he broke each segment off by hitting it over the edge of his anvil with a hammer. Next he put these little cubes in the charcoal basin, four or five at a time. Moving back and forth between the bellows and the anvil, he removed them one at a time as they became red-hot again and worked each until they were all roughly round. Finally, with a lighter hammer and a softer touch, he worked them into neat little spheres.

Many smiths have a working knowledge of these guns, so that they can take them apart and repair them when the need arises. But good Mande guns are made by blacksmiths who specialize in the enterprise to the exclusion of every other type of work. Such specialists are peppered across the savanna. They do not live in every Mande town, but hunters know where to find them and are perfectly willing to travel fifty miles or more to buy the right gun from the smith of their choice.

In the eyes of other citizens, these specialists have attained a high level of technical expertise. Indeed, Dyula groups in northern Ivory Coast honor their rifle-making blacksmiths as the most accomplished and powerful of all the wood and iron workers, an appropriate gesture given the recent Dyula empire and the exploits of Samory. Gunmaking blacksmiths possess the typical array of tools, along with several more specialized items. Locally made files, *kaka muruw,* are commonly claimed to predate the arrival of the French and their European tool kits. Especially delicate anvils, *kulun gulimaw,* drawn to a narrow tip in front are used to model a gun's percussion pin and head. A knife-like tool called *nègènike muru* sometimes used to put threads on homemade screws. The screw is held fast in a pincer that works like a vise, then set against the wooden top of one of the smith's hammers. The blacksmith holds the hammer with his foot, leaving his hands free to make the threads.

In some instances old firing mechanisms or parts from them are incorporated into new guns. In others the smiths forge all of the parts themselves. They have no industrial aids at their disposal to ensure that the parts will fit but instead depend on the accuracy of their sight, their memories, and their very precise physical skills. At certain points in the fabrication process iron pieces must be fused to one another. To weld the firing pin to the base of the barrel, for example, smiths use an ingenious technique. They set the two pieces together, with a fine film of white powdered flux called *burasi* applied to both surfaces. Then they carefully coat and recoat the two joined pieces with a clay solution that ultimately becomes a mold much like those used across West Africa for lost wax casting. This mold is set into the charcoal basin and covered with the coals. Then a number of smiths take turns pumping furiously on the bellows. The smith in charge keeps careful watch on the charcoal, because it serves as his temperature index. After fifteen minutes or more of dynamic pumping, the master smith stops the operation and examines the surface charcoal. If it continues to glow yellow and burn even when air is no longer being forced through it, the smith judges the forge to be hot enough to have effected the weld. During the whole operation the long gun barrel has been sticking out of the forge, and shortly after the bellows work begins, smoke starts pouring out of its end. This smoking barrel and the rapid bellows rhythms make the process exciting to watch.

The stocks of these guns are hollowed out, often in the most delicate fashion, to cradle the section of the barrel that holds the firing pin and to accommodate the firing mechanism and the housing for the trigger, called *kèlè tigelan,* "the war unleasher." The stocks often run to nearly the whole length of the barrel, which is lashed firmly into place at intervals with thick, tight bands of leather. Barrels themselves are made today from the steering columns of junked cars. Formerly, they may have been constructed by hammer-forging and welding long thin sections of iron around a perfectly cylindrical hard wood core.

Some of these guns are beautiful objects. Sedu Traore has a brother named Cekòrò Traore, who lives in a small town well away from any of the main roads. With two blacksmith colleagues, Cekòrò specializes in the making and repairing of guns. When I first met Sedu he owned a large gun made by his cousin, but ultimately he found it a little too large and cumbersome for him. So he sold it and bought another from the cousin, the second smallest type, named *gwasaa.* This gun was elegant, being sensitively proportioned and delicately shaped. Cekòrò decorated the bottom and top of the stock with brass tacks and added an attractive metal inlay to the butt.

ARTICULATION

A beautiful gun is not surprising. It fits in with the products of many handcrafting cultures, and it certainly fits in with several things we have observed about Mande blacksmithing. But it is worthy of our attention a moment more, because it can be used to amplify an idea about articulation.

Mande smiths quite frequently remove their manufactures from the realm of pure utility by making them artful through addition. They do it most frequently with their door locks. Wooden locks with hollowed chambers for lock pins and sliding lock beams are not unique to this region of Africa. They are used by the Dogon nearby, and further east several Voltaic peoples use them. They are even known in North Africa, in basic versions consisting of a vertical rectangle of wood through which the lock beam passes. That basic form is elegantly but drastically transformed by smiths like Sedu Traore. Economy and abstraction make this possible. The lock is extended on top with the addition of a sculpted head. The thick vertical piece then becomes a torso or the body of an animal. It may swell at the sides to represent a turtle or extend at the bottom to suggest a lizard's tail. Through economic additions and imaginative stylizations the locks become images. The same is true of many heddle pulleys and some knives; smiths change them from objects to images that interest Western curators and collectors.

Mande smiths also embellish their manufactures to make them nicer. Here too they are economical and imaginative, although their labor does not lead to figural imagery. Sedu's gun falls into this category, but it is also something more. It is the product of great skill and great concern for form. It has been made beautiful not only by addition—the brass tacks—but also by articulation.

That gun is not unique, nor should we consider it typical. I have seen several that were not particularly attractive, and I have seen a few that appeared clumsy. Yet I have also seen several that showed as much concern for form as Sedu's and this is true of many other blacksmiths' products.

I found it most striking in knives. Nearly every Mande who goes outside of town owns a knife, and many, such as hunters and blacksmiths, carry two or three in sheaths at their sides. Even sorcerers use knives, which they have prepared for special tasks by the addition of ritual and medicine. One might not take the time to concentrate on these little objects, but when one does, one finds a surprisingly large percentage to be simply beautiful forms. They can be very nicely proportioned. Great imagination may go into developing intricately shaped handles. The blades often taper or curve with elegance. And the visual interplay between handle and blade is often lovely.

The same is true for farming tools, though less frequently, since objects like hoes are literally beaten out by the dozens. Nevertheless, the blade and handle shapes can be beautiful, with sensitive curves and flares and a wonderful play between the two-dimensionality of the business ends and the three-dimensionality of the handles. Mande hoes, like knives and guns, can be richly articulated.

Do the Mande recognize this beauty? They are not particularly keen to talk about it. There is not a large, specialized vocabulary addressed to it. The literature does not contain many references to it. Under such circumstances we might invoke the old notion that African craft and art is aimed at utility alone and that aesthetics are ancillary to our inquiries. There is, however, some evidence that we should not ignore.

First, there are the objects themselves. So many are so sensitively composed that happenstance is simply out of the question. Blacksmiths spend many years refining their skills with hammer and adze, and we must expect many of them to put these skills to work aesthetically. Even if they have no elaborate aesthetic discourse and find the questions art historians ask a little absurd, they still shape their forms deliberately and find satisfaction in the results.

People's responses to forms lend support to this position. Sedu Traore was rarely neutral about a form. He was not inclined to analyze its parts, but he always made his feelings about the whole quite clear. Like smiths, members of the farming clans would also say what they liked or disliked, in discreet but concrete fashion.

Often, too, people's behavior toward objects was very revealing. Sedu had a hunter friend we visited in a large town well south of Bamako. On one of our visits I decided to wander around to photograph architecture. A middle-aged woman of imposing disposition approached to see what I was doing.

When her suspicions were confirmed, she dragged me over to her door and demanded that I photograph it. A most elaborate door lock was attached. It had been part of her wedding trousseau, and it remained something she cherished. She stated quite plainly that if I wanted to photograph nice door locks, I would find none as beautiful as hers anywhere in town.

In 1972 I met a woman at a blacksmith's forge who had come to have her hoe blade sharpened. She owned the type called *daba muso,* and it must have been ancient because it was deeply patinated. It was also absolutely elegant in its proportions and shape, without question one of the most exquisite tools I have seen anywhere. Intending to verify its type, I asked her its name. She said *Ci wara.*

Ci wara means several things. It is the name of a farming association that used to be a secret initiation society. Zahan and Imperato discuss it at length as the name of an agricultural deity. It is also a praise-name for good farmers. As such it means "farming animal" or "farming beast," using a Mande praise formula that likens accomplishment to wildness, ferocity, and awesomeness, and often incorporates implications of great age and high levels of *nyama.* Musicians and dancers are frequently praised with similar formulas, and it is considered honorable indeed. Thus the woman applied to her hoe a form of high praise customarily used for extremely accomplished people, and in the process indicated that she found it beautiful.

Thus the concern for articulation we first encountered with apprentices at their bellows extends from the processes of making things to the products made. Good form permeates the blacksmiths' experience, and they pass it along in their tools and weapons to everyone else. Articulation, in several forms and several realms, is a central feature of the Mande smiths' identity.

FURTHER READING

Richard Anderson and Karen Field, eds, *Art in Small Scale Societies* (New York: Prentice Hall, 1993).

S. Terry Childs and David Killick, 'Indigenous African Metallurgy: Nature and Culture', *Annual Review of Anthropology* 22 (1993), pp. 317–37.

Zöe Strother, 'Invention and Reinvention in the Traditional Arts', *African Arts* 28/2 (1995), pp. 24–33.

Robert Farris Thompson, 'Yoruba Artistic Criticism', in *The Traditional Artist in African Societies* (Bloomington: Indiana University Press, 1973).

Kariamu Welsh-Asante, *The African Aesthetic: Keeper of the Traditions* (New York: Praeger, 1994).

48

AFRICAN ART: WHERE THE HAND HAS EARS

Amadou Hâmpaté Bâ

E. M. Forster's oft-cited dictum 'only connect' was a sentiment that lay at the heart of modern aesthetic theory. The twin ideals of immediacy and authenticity had, among their expressions, the seeking out of unspoiled cultures, many of which were designated as 'primitive'—a term that intertwines positive values with high-handed, superior disdain. Loving depictions of traditional culture are not only written from outsider positions or from ignorance, however, as the following text by the writer Amadou Hampâté Bâ makes clear. Born into an aristocratic Fula family in Mali, Bâ was (due to his standing as an executive officer of UNESCO from 1962 to 1970, and later as an ambassador in the Ivory Coast) among the most prominent African intellectuals of the post–World War II period. He is most often remembered for the aphorism 'when an old man in Africa dies, it is as if a library has burnt down'—a great line, and one that sums up his preoccupations. Bâ's vision was an ahistorical one, which centered on the preservation and veneration of oral traditions handed down from time immemorial. (Another of his aphorisms was, 'Writing is the photographing of knowledge, but it is not knowledge itself.') He wrote prodigiously in French, both fiction and nonfiction, addressing aspects of African culture ranging from religion and politics to aesthetics. Given its original publication context in the house organ of UNESCO, the following text might itself be considered a piece of finely tuned international cultural diplomacy. Yet, in his close analysis of the process by which objects are spiritually 'loaded', Bâ shapes a theory founded in the particularities of African craft practice.

Amadou Hampâté Bâ, 'African Art, Where the Hand Has Ears', *UNESCO Courier* (February 1976).

The meaning which we give nowadays to the words "art" and "artist" and the special place which they occupy in modern society do not entirely match the traditional African way of thinking.

"Art" was not something separate from life. It not only covered all forms of human activity, but also gave them a meaning. Ancient Africa's view of the universe was an all-embracing and religious one, and acts, particularly acts of creation, were seldom, if ever, carried out without a reason, an intention, or appropriate ritual preparations.

No one who considers traditional Africa from a strictly secular viewpoint can hope to understand it.

In traditional Africa there was no division between the sacred and the profane, as there is in our modern society. Everything was interconnected, because everything was imbued with a profound feeling of the Unity of Life, the Unity of all things within a sacred universe where everything was interrelated and mutually dependent.

Every act and every gesture were considered to bring into play the invisible forces of life. According to the tradition of the Bambara people of Mali, these forces are the multiple aspects of the Se, or Great Prime Creative Power, which is itself an aspect of the Supreme Being known as Maa Ngala. In such a context, actions, since they generated forces, were necessarily rituals, performed so as not to upset the balance of the sacred forces of the universe of which man was traditionally both the guardian and the guarantor.

The crafts of the iron-worker, carpenter, leather-worker or weaver were therefore not considered to be merely utilitarian, domestic, economic, aesthetic or recreational occupations. They were functions with religious significance and played a specific role in the community.

In the last analysis, in ancient Africa everything was considered as art, as long as knowledge of some kind was involved and also the means and methods of putting it into application.

Art was not only pottery, painting, etc. but everything at which people worked (it was called, literally, "the work of the hands") and everything which collectively could contribute to developing the individual.

These creative activities were all the more sacred since the world we live in was considered to be merely the shadow of another, higher world conceived of as a mysterious pool located neither in time nor in space. The souls and the thoughts of men were linked to this pool. In it they perceived shapes or impressions which then matured in their minds and found expression in their words or the work of their hands.

Hence the importance of the human hand, considered to be a tool which reproduced on our material plane (the "plane of shadows") what had been perceived in another dimension. The forge of the traditional ironsmith, who had been initiated into both general and secret knowledge handed down to him by his ancestors, was no ordinary workshop, but a sanctuary which one entered only after performing specific rites of purification. Every tool and instrument in the forge was the symbol of one of the active or passive life forces at work in the universe, and could be manipulated only in a certain way and to the accompaniment of ritual words.

In his workshop-sanctuary, the traditional African ironsmith was thus conscious not only of performing a task or of making an object, but of reproducing, by a mysterious analogy, the initial act of creation, thus participating in the central mystery of life.

The same was true of other crafts. In ancient traditional societies in which the concept of the "profane" was virtually non-existent, the craftsman's functions were not performed for money or to "earn a living", but corresponded to sacred functions, to paths of initiation, each of which was the medium for a body of secret knowledge patiently handed down from generation to generation.

This knowledge was always about the mystery of the primal cosmic unity, of which each trade was one particular aspect and form of expression. There were a great many craftsmen's trades, because there were also a great many possible relationships between man and the cosmos, which was the great dwelling place of God. While the art of the ironsmith is linked with the mysteries of fire and the transformation of matter, the art of the weaver is bound up with the mystery of rhythm and the creative Word acting through time and space.

In ancient times, not only was a trade or art considered to be the embodiment of a particular aspect of the cosmic forces, but it was also a means of making contact with them. To guard against an unwise mixing of powers which might prove to be incompatible, and to keep secret knowledge within the family, these various categories of craftsmen came to practise a system of marriage within their group, regulated by numerous sexual prohibitions. It is plain to see how these chains of initiation or ramifications of knowledge gradually gave rise, through marriage within the group, to the special caste system of the area formerly known as the Bafour (savanna region stretching from Mauritania to Mali). These castes enjoyed unique status within society.

Let us take a look at the middle class, which particularly concerns us here, namely the class of the craftsmen called, in Bambara, the Nyamakala. Owing to the sacred and esoteric origins of his functions, the Nyamakala could under no circumstances become a slave, and he was absolved from the obligation of war service incumbent upon noblemen. Each category of craftsmen, or Nyamakala, constituted not only a caste, but a school of initiation. The secret of their art was jealously guarded within the group and strictly handed down from generation to generation or from father to son. Craftsmen were themselves called upon to adopt a hereditary way of life, with obligations and prohibitions designed to keep alive in them the qualities and abilities required by their art.

It cannot be emphasized too strongly that ancient Africa can be understood only in the light of an occult and religious conception of the universe, where there is a living, dynamic force behind the appearances of all people and objects. Initiation taught the right way to approach these forces, which in themselves, and like electricity, were neither good nor bad, but which had to be approached in the right way so as not to cause short-circuits or destructive fires. We should remember that the first concern was not to upset in any way the balance of forces in the universe, which the First Man, Maa, had been appointed to uphold and preserve by his Creator, as were all his descendants after him.

At a time when so many dangers threaten our planet because of human folly and thoughtlessness, it seems to me that the principle thus raised by the old Bambara myth has lost none of its relevance. After the ironsmith come the traditional weavers, who also possess a high tradition of craft initiation. Initiated weavers of the Bafour work only in wool, and all the decorative patterns on their blankets or tapestries have a highly precise meaning connected with the mystery of numbers and the origin of the universe.

Woodworkers, who make ritual objects, notably masks, themselves cut the wood they need. Their initiation is thus linked to knowledge of the secrets of the African bush and of plant life. Those who make canoes must also be initiated into the secrets of water.

Then come the leatherworkers who are often reputed to be sorcerers and, finally, also belonging to the Nyamakalaw, there is the special caste of *djeliw* or "public entertainers," also known as griots.

Griots are not only musicians, singers, dancers and story-tellers. Some serve as ambassadors or emissaries, acting as intermediaries between the great families; others may be genealogists and historians. They have other roles but those I have indicated are their principal functions.

The griots as a class do not have their own initiation rites, although individually they may belong to particular societies which do have such rites. But they are nevertheless Nyamakalaw, since in fact they manipulate one of the greatest forces capable of acting on the human soul: the spoken word. While the nobles are bound by tradition to observe the utmost discretion in word and gesture, griots are completely free in this domain. As the spokesmen and intermediaries of the nobles they enjoy a special status in society.

As craftsmen in materials or in speech, transformers of natural elements, creators of objects and forms, and manipulators of forces, the Nyamakalaw occupied a place apart in traditional African society. They fulfilled a major role as mediators between the invisible worlds and everyday life.

Thanks to them, everyday or ritual objects were not simply objects but repositories of power. Such objects most often served to celebrate the glory of god and of ancestors, to open the bosom of the great sacred Mother, the Earth, or to give material form to impressions which the soul of an initiate drew from the hidden part of the cosmos and which could not be clearly expressed in language.

In the traditional religion-oriented world, fantasy did not exist. A craftsman did not make something in a spirit of fantasy, by chance or to satisfy a whim. The work had a purpose and a function, and the craftsman needed to be in a state of mind which matched the moment of its creation. Sometimes he would fall into a trance, and when he emerged from it, he would create.

In this case the object was not considered to be his handiwork. He was regarded merely

Figure 34 *Amuletic Necklace, 1880–1920.*

as an instrument or medium of transmission. People would say about his work: "God put it into you", or "God has used you to create a fine work". Art was, in fact, a religion, a form of participation in the forces of life and a way of belonging to both the visible and the invisible worlds.

The craftsman had to bring himself into a state of inner harmony before beginning his work, so that this harmony might enter the "aura" of the object and have the power to move those who saw it. He was thus obliged to perform special ablutions and recite litanies which helped to put him "into the right frame of mind". Once he had achieved this, he accomplished his task and transmitted to the work his inner "vibration". By sculpting, shaping, embroidering, drawing geometrical lines on leather or weaving symbolic patterns, the craftsman gives material form and outward expression to this inner beauty which is within him in such a way that it enters the "aura" of the object, and captures the attention of those who see it for centuries to come. This is the whole secret of his creation.

"A thing which has not kindled beauty in you", says an old adage, "cannot kindle beauty in another who looks upon it". Artistic creation was therefore the outward manifestation of an inner vision of beauty which, according to ancient tradition, was none other than a reflection of the beauty of the cosmos. Art was thus priceless because this whole creative process was something that could not be bought.

There are some statues which one cannot call "beautiful" in the aesthetic sense of the term, and yet they sometimes move us more than a lovely picture, because they are infused with a power which can attract or repel, according to the intention behind the work. Occasionally, in the bush, one stumbles unexpectedly upon a circle of statues raised by the Komo (custodians of traditional customs and beliefs among the Bambara people of Mali) which seem to have sprung out of the earth. The shock which they produce is so strong that unless their meaning has already been explained to you, your first instinctive reaction is to run away.

An object may also serve as an instrument for the transmission of knowledge by means of the symbols which it bears, such as tapestries, whose patterns may be deciphered, or carved stools whose geometrical lines have a precise meaning.

The work of art, whatever form it takes, is viewed by traditional Africans as a porthole through which one can contemplate the infinite horizon of the cosmos. One can see many things in a work of art depending on one's own degree of development. The seer can use it to contemplate the occult world. Secular art, which was certainly very rare in ancient times, differed from religious art only in the sense that the secular object was not "consecrated", and therefore not "loaded" with spiritual energy. And there can be no doubt that an object which has been consecrated and used for ritual does not make the same impression as a secular object on anyone who is at all sensitive.

Secular art was considered to be the "shadow" of religious art. It was the visible tip of the iceberg for the uninitiated. One example of the "shadow" role of secular art is the fact that copies were sometimes made of religious masks for the Kote, or traditional theatre.

It goes without saying that secular art has developed chiefly since the colonial era and that it has become very rare to discover an authentic and spiritually "loaded" object. As soon as a mask had been consecrated, in the Komo tradition, for example, or among the Dogon people, it could no longer be seen in the open. It was hidden from the eyes of the uninitiated and remained either in its hiding-place in the bush or, in the case of the Dogons, in the cave of the masks. Some Dogon masks are so meaningful and so sacred that they are taken out only once every sixty years for the great Sigui ceremony.

The conclusion to be drawn from all this is that traditional African art was not produced haphazardly, and that it served a central purpose in the human community. Most works of art, whether plastic or in the oral tradition, had several levels of meaning: a religious meaning, a meaning as entertainment and an educational meaning. So it was necessary to learn to listen to tales, teaching and legends, or to look at objects on several levels at once. This, in fact, is initiation—the profound knowledge of that which is taught through things, through appearances, and through nature itself.

Everything which is, teaches through mute speech. Form is language. Being is language. Everything is language.

But, you may say, all that was true in the past. How do things stand nowadays?

True enough, the past few decades have witnessed the destruction, or systematic disappearance, of most of the great traditional initiatory and craft centres. This has happened for several reasons: firstly, colonization policy with its usual and universally applied tendency of effacing systems of values and indigenous customs in order to replace them by its own; next, the promotion of trade by chambers of commerce. These, supported by the authorities, harassed craftsmen and drove most of the workshops out of business.

To mention only two examples, the ironsmiths were forbidden to make certain tools that competed with manufactured products imported from the colonial mother country and plant healers were prosecuted for the "illegal

practice" of medicine. Gradually, Negro-African art came to be no longer tolerated except at a "folklore" level, and, even then, only if it was remodelled and adapted to suit the tastes of the rulers. The trend became even more marked immediately after independence, with the general spread of customs and ideologies imported from abroad and the invasion of values based on money. Not only are initiation centres increasingly rare, but even where masters still exist, disciples are lacking.

Western-type studies, the attraction of large neighbouring towns and the desire to earn money draw young people like a magnet and carry them off towards other aspirations.

Traditional African custodians of the arts, sciences and ancient skill still exist. But they are few and as a rule fairly elderly. The treasure of knowledge, patiently handed down for thousands of years, can still be retrieved and rescued if we act while there is still time and are willing to listen to what the old sages have to tell us.

Since independence, the modern African artist has been struggling to assert himself. His search for authenticity and originality is both difficult and poignant, for it is not always free from outside influence. Today's African artists are on the threshold of a new era, during which they will have a vital role to play. But the importance of this role will depend on how they respond to the challenge. Ideally, no doubt, they should be able to return to the very roots of African tradition by seeking instruction from the masters who are still alive—instruction not so much in a technique as in a way of "tuning in" to the world.

The only message I have for young African artists is to draw their attention to the profound meaning of their ancestral heritage. This would lead them to take a fresh, more understanding and, above all, more receptive look at the works of art of the past, for these were not only "aesthetic" works (aestheticism had very little to do with African art) but also a means of transmitting something transcendent. Each object from the past is like a silent word. Perhaps the young artists of today, more sensitive and more receptive than most people, will be able to hear this silent word.

I can only hope that the various governments concerned, aided perhaps by international institutions, will realize the importance of this problem and at long last recognize the full educational and cultural importance of the arts.

We live in a very curious age. The amazing development of science and technology goes hand in hand contrary to all expectations, with a worsening of living conditions. Along with the conquest of space has come a sort of a shrinking of our world which has been reduced to its material and visible dimensions alone, whereas the traditional African craftsman, who had never moved from his little village had the feeling of participating in a world of infinite dimensions and being linked with the whole of the living universe.

The old African saying goes (and perhaps the artist of today can hear it): Listen! Everything speaks. Every thing is speech. Everything seeks to inform us, to give us knowledge or an indefinable, mysteriously enriching and constructive state of being.

"Learn to listen to silence", says old Africa, "and you will discover that it is music."

FURTHER READING

Ralph A. Austen, 'Amadou Hampâté Bâ: From a Colonial to a Postcolonial African Voice', *Research in African Literatures* 31/3 (2000), pp. 1–17.

Amadou Hampâté Bâ, *The Fortunes of Wangrin* (Bloomington: Indiana University Press, 2000; orig. pub. 1973).

Claude Lévi-Strauss, 'Race and History', in *The Race Question in Modern Science* (Paris: UNESCO, 1952).

49

WALTER BENJAMIN: TRACES OF CRAFT

Esther Leslie

The German cultural theorist Walter Benjamin is often associated within the Frankfurt School, a group of Jewish Marxist scholars (including Theodor Adorno, Herbert Marcuse and Hannah Arendt) who turned their attention from economic and political matters to cultural subjects such as film, music, art and literature. This is a useful way of thinking about certain aspects of Benjamin's work—such as his most famous essay, 'The Work of Art in the Age of Its Mechanical Reproducibility', which concentrates on the cultural repercussions of certain conditions of production. But Benjamin cannot be contained within the narrow concerns of Marxist cultural theory. His searching, melancholic writings range from German baroque drama, the arcaded shops of late-nineteenth century Paris, the habits of collectors, the kabbalah *and the experience of drug use. If anything tied together his interests, it was the possibility of redemption in the face of modern disenchantment. It is no surprise, then, that he returned again and again to craft. In the following indispensable guide to Benjamin's thinking about the subject, Esther Leslie traces his characteristically wide-ranging use of the figure of the artisan. Pottery and weaving make an appearance, but the most fascinating contexts are less expected ones: storytelling, Dada photomontage and (as one might expect from an author so self-reflective) the practice of writing itself.*

Esther Leslie, 'Walter Benjamin: Traces of Craft', *Journal of Design History* 11/1 (1998), excerpted.

TELLING STORIES

In 1936 Walter Benjamin completed an essay on the nineteenth-century Russian storyteller Nikolai Leskov.[1] One theme of the essay is the assertion of storytelling's interlacement in craft. Leskov, Benjamin tells us, felt bonds with craftsmanship, and faced industrial technology as a stranger. Often Leskov's stories would feature craftsmen, such as the silversmiths of Tula whose expertise exceeded the most technologically advanced nation of the time, England.[2] 'The Alexandrite' presents another craftsman, the skilful gem engraver Wenzel. Benjamin describes Wenzel as 'the perfect artisan' with 'access to the innermost chamber of the realm of created things'.[3] Craft and craftsmen do not just provide subject-matter and characters for Leskov's stories. The very act of storytelling itself he declares to be a craft.[4]

Benjamin's own braiding of craft and narration in 'The Storyteller' goes further to illuminate a historical, practical affinity between craft skills and storytelling. The ability to tell stories, Benjamin tells us, is rooted in two factors; travel to faraway places and knowledge of past local lore. Benjamin writes:

The resident master craftsman and the travelling journeymen worked together in the same rooms; and each master had been a travelling journeyman before he settled down in his home-town or elsewhere. If peasants and sailors were past masters at storytelling, the artisan rank was their master class. It combined the lore of faraway places, such as a much-travelled man brings home, with the lore of the past, as it best reveals itself to residents of a place.[5]

The habitat of the storyteller is the craft milieu, in which resident master craftsmen—who know the past, who know time—exchange experiences with travelling journeymen—who know distance, space. The wayfarer's imported ken is the key to Benjamin's ontology of experience. The German word for experience that is handed down, that is experience born of wisdom, a practical knowledge, is *Erfahrung*, and it finds its root meaning in the word for travel, *fahren*. Through travel craftsmen have experience of the world and a world of experience.

And so, Benjamin tells us, they gain audiences, lured into workshops to graft while netting experiences transmitted from mouth to ear to mouth. The best listeners, insists Benjamin, are the ones who have forgotten themselves, and while their half-conscious minds are engaged in pot-throwing, spinning and weaving, and their bodies are seized by the gentle rhythm of work, the stories they hear forego an existence on paper, imprinting themselves into the listeners' fantasy, awaiting retransmissions, after-lives.[6] Storytelling is no simple form of time-passing. It mirrors a mode of processing and reconstituting experience. It intimates how experiences pass into and out of memory. For Benjamin, to reflect on the operations of storytelling, or craft communication and experience, is to ponder the arabesque of labour, experience and selfhood.

The storyteller takes what he tells from experience, his own or others, and makes it the experience of those hearing the tale. True experience is conceived as close and practised knowledge of what is at hand. The hand touches, has practical experience of life. Recurrent in Benjamin's delineations of experience are the words tactile, tactics, the tactical, entering German, as it enters English via the Latin *tangere*, touch. To touch the world is to know the world. Pottery features here—as model and as metaphor—naturally enough as it is a form of *Handwerk*, hand work or artisan labour. Benjamin describes storytelling, the transmission of experience and wisdom, thus:

> It sinks the thing into the life of the storyteller, in order to bring it out of him again. Thus traces of the storyteller cling to the story the way the handprint [*Spur*] of the potter clings to the clay vessel.[7]

The hand marks out authentic experience, indicates Benjamin, setting himself within a tradition of humanist anatomical thought that sees the faculty of stereognosis as reliant on touch, a touch that fingers the world's textures, and hands on knowledge of those textures.[8] In 'The Storyteller', as in other essays by Benjamin, pot-throwing emerges as a figure of true experience.

That the hand, with its tactility, is central in Benjamin's comprehension of experience, or more specifically in Benjamin's vision of redemption or recovery of experience under threat, is intimated in his aphorism, 'Salvation includes the firm, apparently brutal grip'.[9] Grasping the truth, seizing the future; the hand is a political organ. But it does not work in isolation. Intrinsic to the craftsman, and the gesticulating storyteller, too, is the accord of soul, eye and hand.[10] Thinking, seeing, handling in tandem, this mesh grants a praxis. Storyteller—fashioning his material, human life—and craftsman—fashioning his—mould their raw matter, Benjamin tells us, in a solid, useful, and unique way.[11]

Stories, mirrors of true experience, and crafted objects alike are solid, useful, unique. The aesthetics of the useful and unique story or the crafted pot could not be more removed from the attributes of cheap mass-reproduction, or from those of fine art. The story and the pot are formed by a life that has something to tell. Good stories relate a practical knowledge; good potters relate a wisdom based on praxis. Here, outlining wisdom, Benjamin's metaphorical language picks up another type of craft labour, weaving. He writes: 'Counsel woven into the fabric of lived life is wisdom.'[12] It is such woven wisdom that the storyteller hands on.

In 'The Image of Proust' (1929), Benjamin correlates Proust's textual practice and weaving. Reflecting on Proust's flabelliform writings, Benjamin binds memory work, dream work and text work together in an image of handiwork; the weaving of memory. Benjamin notes that the Latin word for 'text', *textum*, means 'something woven', a web.[13] Neither plot nor personality dispatch 'strict weaving regulations', but memory, such as is activated in dreams, a tightly plaited skein tangling the linear passage of time. The individual artistic voice and the convolutions of plot are the reverse side of memory's continuum, intermittences relegated by Benjamin to the pattern on the back of the carpet.[14] In the foreground, Proust as weaver reflects on the workings of remembering, sourcing thereby social and collective structures of language and fantasy. Such *Handwerk* is a *Lebenswerk*. Weaving becomes a figure for authentic memory or the procedure of rendering the infinity of memory.

Proust verifies, for Benjamin, the textured and textual processes of memory. In dreaming we forget our conscious thoughts in order to access our memories. When we wake we remember where we left off the night before, and, Benjamin writes, the 'few fringes' of the 'carpet of lived existence that forgetting has woven in us' fall from our hands.[15] To access the crafted curlicues of dream-truth, memories, which as Proust and Benjamin recognize are infinite, utopian, curious and surreal, entails forgetting the illusion of self. Dream images and memories are the woven ornaments of self-forgetting, incubators of the story that forms itself like the pot, unconsciously, as planned, Benjamin says, 'as the lines on the palm of our hand'.[16]

That Benjamin conceives texts—and memory, too—as material, as woven, is no surprise; it is a part of his most literally understood materialism. The collector Eduard Fuchs advises Benjamin to approach history as a materialist, adumbrating the entwinement of the warp [*Textur*] of the present and the weft [*Einschuß*] of the past.[17] Materialism, historical or dialectical materialism, is alert to the fabrication of the past and the multithreaded nature of the present, shot through with that past.

WORK ON THE BODY

Benjamin's metaphors of craft, of potting and weaving, allude to a former pre-industrial mode of labouring. Of course, this mode may be romanticized, but it allows Benjamin to shade in the tendencies of an epoch, to tell a story of change, not just from past to present, but from present into future, too. This former craft mode is submerged in mass industrial society, and together with it begins to sink the mode of experience that it engendered. Technology has stormed the human body, subjecting the human sensorium to a complex training,[18] and provoking a 'crisis in perception'.[19] Soul, eye and hand are disjointed. Benjamin's anthropology of industrialized humanity submits to the discussion of experience in modernity the neurological category of shock. There are those who feel work's hard slaps on the body, while others are cushioned in the well upholstered seats of management. The techno frenzy of the First World War was made possible by nineteenth-century technological advance, and that war marks for Benjamin a re-editing of experience. From factory to battlefield the experience of shock, physical and psychic, constitutes the norm. Technology dictates a syncopated, dislocating rhythm to which workers and soldiers must permanently react. The division of labour compels a mechanical measure of labour time, the voided, homogeneous time of manufacture. The work process, especially the factory drill, de-skills operators. Industrial work processes are an 'automatic operation', wherein each act is an exact repetition of the last. Benjamin remarks:

> Marx had good reason to stress the great fluidity of the connection between segments in artisan labour [*Handwerk*]. This connection appears to the factory worker on an assembly line in a detached, reified form. Independently of the worker's volition, the object being worked upon, comes within his range of action and moves away from him just as arbitrarily. 'Every kind of capitalist production . . . ,' writes Marx, 'has this in common, that it is not the workman that employs the instrument of labour, but the instrument of labour that employs the workman. But it is only in the factory system that this reversal for the first time acquires technical and palpable reality.'[20]

Capitalist instruments of labour operate the worker, and factory machinery gives this transposition a technically concrete form. Machinery turns animate, humans become adjuncts to the machine. This is a different loss of self, an alienation, not an ingress into reverie. The modern unskilled worker, claims Benjamin, is sealed off from experience as *Erfahrung*.[21] Benjamin quotes Marx: 'In working with machines workers learn to coordinate their own "movement to the uniform and unceasing motion of an automaton".'[22] That automaton mass has liquidated its weave of memories.

The hand—so crucial to the *Handwerker* (artisan or craftsman)—is made redundant by technological advance. In 'The Storyteller' Benjamin comments that the role of the hand in production has become more modest. Again he draws the analogy with storytelling. Here the role of the hand lays waste. Benjamin continues:

> After all, storytelling, in its sensory aspect, is by no means a job for the voice alone. Rather, in genuine storytelling the hand plays a part which supports what is expressed in a hundred ways with its gestures learnt of work.[23]

Stories are lost; that is to say, textured experience, graspable experience, is lost because of the loss of the weaving and spinning activities that went on while they were heard. The web that cradled storytelling is unravelling at all its ends.[24]

Benjamin relates elsewhere the tale of the hand's redundancy for production; notably in his most famous essay 'The Work of Art in the Age of its Technical Reproducibility' (1935–8).[25] Here he tells how, until the arrival of mechanical reproduction, pictures had been made by hand, parallel to the manufacture of goods before the development of industrial machinery. Mechanical reproduction in art, beginning with wood cut technology, advances sporadically, until it attains a qualitatively new stage in lithographic reproduction. Lithographic duplication permits mass quantities and speedily changing forms. The invention of photography and film provokes a further speed-up effect, basing reproduction not on the pace of a hand that draws, but on the seeing eye in conjunction with the machinery of the lens. Culture's coordination with the body has transformed. The time of the machine, not the time of the hand, determines production. In 'The Storyteller' Benjamin quotes Paul Valéry on how once the artisan had imitated the patient processes of nature, but now no longer. Valéry writes:

> Miniatures, ivory carvings, elaborated to the point of greatest perfection, stones that are perfect in polish and engraving, lacquer work or paintings in which a series of thin, transparent layers are placed one on top of the others—all these products of sustained, sacrificing effort are vanishing, and the time is past in which time did not matter. The modern person only works at what can be abbreviated.[26]

Industrial speed-up has transformed conditions of production and standardized what is produced. An allegory from *The Trial* by Kafka evokes for Benjamin the endlessly returning bad infinity of mass reproduction. The insistent painter-dealer Titorelli impresses on Josef K. the same painting, redone again and again, modelling so capitalism's eternal return of the ever-same culture.[27] The mode of repetition of the artisan's story as it is passed on from mouth to mouth, reworked through the unique experience of listeners, degrades here into a mechanical, dead reiteration. Body accedes to machinery.

DADA-FACTURE

Do not think that this is a tragic tale of irreversibility and that Benjamin's animus is a frustrated nostalgia for the past world and past work of *Handwerk*. Benjamin recoiled from the First World War, propelled thereby to revile the economic system that he saw blasting its destructivity into being, but further he sensed a beyond that was also in some ways a restoration or a rescue of experience; and its seed-bed was the technical present. Benjamin's 'Work of Art in the Age of its Technical Reproducibility' traces the transition from *Handwerk* to *Kunstwerk*, from craft to art—from unauthored object to authentic authored valuable. Cult value is banished by authenticity—a calibre that is assured by a knowable author and translates into monetary and exhibition value. But Benjamin's essay also scents possibilities for a post-bourgeois object, a non-auratic multiple, prefigured in photography and film. This technical multiple does not squash out authentic experience but translates it into object forms and forms of experience appropriate for a modern age. These forms, like the forms that cradled craft, fan a spark of a life that is integrated harmoniously with labour. Damaged life may heal itself; through tapping recuperative energies vented in industrial culture. The trajectory continues: from *Handwerk* to *Kunstwerk* to *Kraftwerk*.

For the post-bourgeois object of the new mass art, a mass-reproduced art, the same metaphors re-surface as are encountered in 'The Storyteller'. Tactility, closeness, indexicality, at-handness mark out this new potential art for and by the masses. For Benjamin, the mass appropriation of art signals literally a manhandling of cultural products. The mass-reproduced copy can be manipulated. It is 'tactile'. Tactility, the ability to touch, are sensuous concepts that relate new art to the physical presence of the collectively receiving body. Tactility and shock—forces that act on the body—negate any ideal of artistic autonomy.

Benjamin dislodges from a bodiless idealist aesthetic based on beautiful illusion (*schöner Schein*). Idealist conceptions of culture are seen to be wound into a narcissistic ideology that argues art is born from itself. Benjamin's approach recovers the substratum of aesthetics sensuously. Locating sensuous perception as the root meaning of the Greek notion of *aisthesis*, aesthetics and art are charted along the development of the human sensorium. For Adorno, such a move is characteristic of Benjamin's behaviouristic anthropological materialism, and he labels it a positivism that takes its measure from the human body.[28]

This physico-spatial 'bringing closer' of new cultural forms allowed by mass reproduction provides a 're-modelling' of pre-industrial folkloric relations of space. Crucial to the earlier epic tradition is a reliance on the propinquity of a collective of listeners. Industrial capitalist relations corrode the oral communicability of experience, but technical reproduction reimburses that change, instituting new potential for a familiarity between receivers and producers, once more in the form of collective experience: through mediated mass-produced things. Space is recovered technically.

The artist refitted as producer is a slogan drawn from the realms of industry rather than the painter's studio. Benjamin proposes modern objects that smash through the contemplative, becoming useful, serviceable, and if not unique, then the experiences to be had with them are. The web (*Gewebe*) that Benjamin had spun as a cat's cradle of memory is evoked again as figure of reality, into which the modern culture producer penetrates.[29] Analysed, for instance, by a camera, the web of time and space is interrogated, or made knowable, but differently to the way that the travelling journeyman and the resident artisan knew it. For Benjamin, the modern work of culture finds its template in architecture, itself a penetrable space that is experienced through 'tactile reception'.[30] Hands feature again then, although it must be said that their role in cultural production is somewhat brutal, indeed invasive. Benjamin contrasts the magician who heals through the laying on of hands to the surgeon who intervenes in the body, augmented by machinery. The magician is like a painter, glossing over a surface, the surgeon is like a filmmaker who cuts in to the web of reality, and spawns thereby parameters for new ways of telling stories, new modes of reproducing experience, based in shock and mass-reproduction.

'The Work of Art in the Age of its Technical Reproducibility' professes the displacement of the authentic object under new conditions of mass reproduction. In technically reproduced art—that is, objects whose very basis is technological—there can no longer be a significant notion of an originality that is valued for its inviolate authenticity. The reproduction of an object on celluloid stands as a copy of itself, and no longer a unique representation. The essay speaks of the non-reproducible quality of authenticity—in German, *Echtheit*.[31] The presence of the original is the prerequisite to the concept of authenticity.

In 'A Small History of Photography' (1931) Benjamin also speaks of authenticity, but uses the Latin-derived word *Authentizität*. 'A Small History of Photography' underscores the indexicality of the photograph, its chemical connection to actuality that captures a moment in time and exports it into the future. The photographic object brings objects closer for inspection, providing an imprint of traces of the world.[32] It reveals traces (*Spuren*), not of the potter's handprint, but of the objective modern world.

In 'The Author as Producer' (1934) Benjamin reports that the 'revolutionary strength of dadaism' lies in its 'testing art for its authenticity' (*Authentizität*).[33] Authenticity rests on the incorporation into cultural artefacts of real-life fragments—cigarette stubs, cotton reels, bus tickets, scraps of textile such as the tatters of lace used by Hannah Höch, *Dada-monteur* by night and lace designer by day; sometimes she used her lace patterns in her photomontages. Dada frames a found segment of the world. The public, confronted by excerpted splinters from the material world, learns that 'the tiniest authentic fragment of everyday-life says more than painting'. Here again Benjamin brings in the hand, recounting a modern version of the potter's handprint. His rendition of the new authenticity of modern montage art recounts how its use of traces of the objective world is as significant, as legible, as evidencing as the bloody fingerprint (*Fingerabdruck*) of a murderer on a page of a book, a fingerprint that says more than the page's text. Fingerprints and the handprints of the potter are not signatures; such traces differ from the individuating, authenticating autographs of high art. Their virtue lies in their hinge with actuality, not their market value. Dada and Co. are modern storytellers, modern weavers and throwers of experience. Leskov, Shklovsky informs us, in 'Art as Technique' (1917), wrote in colloquialisms, not high-flown literary language.[34] Poetic truth is found in the ordinary, in the quotidian, not the sublime. And that too was the lesson of low modernism, the metro-modernism of Joyce, Duchamp or Max Ernst.

[. . .]

What emerges from all this is a sense in which Benjamin's understanding of objects—craft objects, mass-reproduced objects—includes essentially an understanding of experiences to be had with objects, and memories evoked by objects or encoded in objects—memories of objects in all possible senses. Crafted objects, specifically the pot, provide a model of authentic experience, the experience of a person imprinted on to the objects that he or she brings into being, and tapestry offers a model of authentic memory, the weave of past and present experience and utopian possibility. In the case of the modern mass reproduced object, however, despite new conditions of production, such intimacy and imaginative investment in objects may still be possible. Craft as mode of activity translates into craft as a power, an obscure power, nestling in the imaginatively conceived object.

BROKEN POTS

And to end then, back to the beginning and thoughts on pots and telling stories. In an early essay, titled 'The Task of the Translator', Benjamin alludes to pottery. This is in the course of contending the impossibility of literal translation, of transmitting a story unaltered from one language to another. He speaks here of translation as the gluing together of fragments of a vessel. These fragments must match one another in the smallest details, but they need not be like one another.[35] This image draws on Issac Luria's cabbalistic concept of *tikkun*. According to the doctrine of *tikkun*, vessels of God's attributes were broken and this breaking of the vessels scattered divine sparks in fragments throughout the material world. These fragments must be brought together, the pots remade, a task both secular and divine. Much like the meshing of shards of montage, or the restorative practice of Benjamin's Angelus Novus, the angel of history, the world is to be put back together—but it is a montage praxis, using debris and rubbish, the broken pots and torn scraps, not the high, sublime reordering of harmony in a bloodless, hands-off aestheticism.

NOTES

1. See Walter Benjamin, 'The storyteller', in *Illuminations*, Fontana, 1992 or 'Der Erzähler', in *Gesammelte Schriften* (hereafter G.S.), II: 2, Suhrkamp, 1991. [...]
2. See Nikolai Leskov, 'The left-handed artificer', in *The Enchanted Pilgrim* (translated by David Margarshack), Hutchinson International Authors, 1946.
3. 'The storyteller', p. 106 or 'Der Erzähler', p. 463.
4. 'The storyteller', p. 91 or 'Der Erzähler', p. 447?
5. 'The storyteller', p. 85 or 'Der Erzähler', p. 440. Note that, on occasion, for example here, I have modified Harry Zohn's translation of Benjamin.
6. 'The storyteller', pp. 90–1 or 'Der Erzähler', pp. 446–7.
7. 'The storyteller', p. 91 or 'Der Erzähler', p. 447.
8. See footnote 9 in J. H. Prynne's 'A Discourse on Willem de Kooning's *Rosy-fingered Dawn at Louse Point*', in *Act 2*, edited by Juliet Steyn, Pluto Press, 1996, p. 53.
9. See *Passagenwerk*, G.S., V: 1, p. 592.
10. The storyteller', p. 107 or 'Der Erzähler', p. 464.
11. See 'The storyteller', p. 107 or 'Der Erzähler', p. 464.
12. 'The storyteller', p. 86 or 'Der Erzähler', p. 442.
13. 'The image of Proust', p. 198 or 'Zum Bilde Prousts', pp. 311–12.
14. Barthes uses a similar metaphorical language in discussing Proust. In 'From work to text' (1971) he writes of the 'textual' novelist: 'If he is a novelist, he is inscribed in the novel like one of his characters, figured in the carpet, no longer privileged, paternal, aletheological, his inscription is ludic. He becomes, as it were, a paper-author: his life is no longer the origin of his fictions but a fiction contributing to his work; there is a reversion of the work on to the life (and no longer the contrary); it is the work of Proust, of Genet which allows their lives to be read as a text.' See Barthes' text reprinted in *Art In Theory 1900–1990: An Anthology of Changing Ideas*, edited by Paul Wood and Charles Harrison, Blackwell, 1992, pp. 944–5.
15. 'The image of Proust', p. 198 or 'Zum Bilde Prousts', p. 311.
16. 'The image of Proust', p. 208 or 'Zum Bilde Prousts', p. 322.
17. 'Eduard Fuchs, collector and historian', in *One-Way Street*, New Left Books, 1979, p. 362 or 'Eduard Fuchs, der Sammler und der Historiker', in G.S., II: 2, p. 479.
18. See *Charles Baudelaire: A Lyric Poet in the Era of High Capitalism*, New Left Books, 1973, p. 132 or 'Über einige Motive bei Baudelaire' (1939), in G.S., I: 2, p. 630. See also Benjamin's 1939 review of the *Encyclopédie Française*, in G.S., III, p. 583n.
19. *Charles Baudelaire*, p. 147 or 'Über einige Motive bei Baudelaire', p. 645.
20. *Charles Baudelaire*, pp. 132–3 or 'Über einige Motive bei Baudelaire', p. 631.
21. See *Charles Baudelaire*, p. 133 or 'Über einige Motive bei Baudelaire', p. 632.
22. *Charles Baudelaire*, pp. 132–3 or 'Über einige Motive bei Baudelaire', p. 631.
23. 'The storyteller', p. 107 or 'Der Erzähler', p. 464.
24. See 'The storyteller', p. 91 or 'Der Erzähler', p. 447.
25. 'The work of art in the age of its technical reproducibility' was written in the same period as 'The storyteller'. Both were begun in the mid-1930s. They are frequently analysed in tandem and shown to provide quite contradictory, indeed irreconcilable, accounts of Benjamin's stance toward modernity.
26. The storyteller', p. 92 or 'Der Erzähler', p. 448.

27. See *Passagenwerk*, in G.S., V: 2, p. 686 and pp. 675–6.

28. See Adorno's letter to Benjamin of 6 September 1936 in G.S., VII: 2, p. 864.

29. See 'The work of art in the age of mechanical reproduction', in *Illuminations*, p. 227 or 'Das Kunstwerk im Zeitalter seiner technischen Reproduzierbarkeit', in G.S., I: 2, pp. 459 and 496, and G.S., VII: 1, p. 374.

30. See 'The work of art in the age of mechanical reproduction', p. 233 or 'Das Kunstwerk im Zeitalter seiner technischen Reproduzierbarkeit', in G.S., I: 2, pp. 466 and 505, and G.S., VII: 1, p. 381.

31. For example, see 'The work of art in the age of mechanical reproduction', p. 214 or 'Das Kunstwerk im Zeitalter seiner technischen Reproduzierbarkeit', in G.S., I: 2, pp. 437 and 476, and G.S., VlI: 1, p. 352.

32. 'A small history of photography', in *One-Way Street*, p. 256 or 'Kleine Geschichte der Photographie', in G.S., II: 1, p. 385.

33. 'The author as producer', in *Reflections*, Schocken Books, 1986, p. 229 or 'Der Autor als Produzent', in G.S., II: 2, p. 692.

34. See Shklovsky's text reprinted in *Art In Theory 1900–1990: An Anthology of Changing Ideas*, edited by Paul Wood & Charles Harrison, Blackwell, 1992, p. 278.

35. See 'The task of the translator', in *Illuminations*, p. 79 or 'Die Aufgabe des Übersetzers' (1921), in G.S., IV: 1, p. 18.

FURTHER READING

Walter Benjamin, *Illuminations*, edited and introduced by Hannah Arendt (New York: Schocken Books, 1968).

Susan Buck-Morss, *The Dialectics of Seeing: Walter Benjamin and the Arcades Project* (Cambridge, MA: MIT Press, 1991).

David Frisby, *Fragments of Modernity: Theories of Modernity in the Work of Simmel, Kracauer and Benjamin* (Cambridge: Polity Press, 1985).

Ulrich Lehmann, *Tigersprung: Fashion in Modernity* (Cambridge, MA: MIT Press, 2000).

Michael P. Steinberg, *Walter Benjamin and the Demands of History* (Ithaca, NY: Cornell University Press, 1996).

50

FUNCTIONALISM TODAY

Theodor Adorno

The Marxist cultural theorist Theodor Adorno considered himself, first and foremost, a musicologist. This helps to explain the uncompromising character of his thought, because during his formative years, the opposition within contemporary music must have seemed stark. One the one hand there was the austere atonality of modernist composers like Arnold Schoenberg and Alban Berg; on the other, the explosion of commercial recorded music due to the rise of phonographs and radio.[1] Adorno saw this opposition as absolute, a battle between the forces of intellectual rigor and demeaning, manipulative kitsch. In his most characteristic mode, whatever the subject, he tended to see a similar Manichean struggle at work. For him the fundamental struggle of modernity pitted the self-awareness of the avant-garde against the debased anti-intellectualism of a pervasive 'culture industry', channeled through mass media. This expressly elitist worldview would seem incompatible with a high regard for craft. Yet when Adorno was asked toward the end of his life to deliver an address to the German Werkbund, he spoke passionately on the subject, picking up where Adolf Loos left off in his identification of craft as a key consideration for the modernist avant-garde. The issue for him, as always, was the capacity for self-understanding (as opposed to 'imagination', which he dismissed as an excuse for lack of discipline). Just as Adorno wanted to see composers delve into the foundations of their art form, he also exhorted architects, artists and designers to engage truthfully with the conditions of their own making, whether that was considered from physical, social or political points of view. This was by no means the same as creating a kind of expressive theater of production or, worse, fetishizing certain 'authentic' ways of making. Thus Adorno was no partisan; he was not 'for' craft. He saw all means of production as potentially susceptible to self-examination or false consciousness. But Adorno did see craft skill as a requisite for making consistent and self-aware work: nothing more than that, it is true, but also nothing less.

Theodor Adorno, 'Functionalismus Heute', delivered to the German Werkbund on 23 October 1965. Originally published in *Neue Rundschau* 77/4 (1966); translated by Jane O. Newman and John H. Smith, and reprinted as 'Functionalism Today' in *Oppositions* 17 (Summer 1979). Excerpted.

I would first like to express my gratitude for the confidence shown me by Adolf Arndt in his invitation to speak here today. At the same time, I must also express my serious doubts as to whether I really have the right to speak before you. *Métier*, expertise in both matters of handicraft and of technique, counts in your circle for a great deal. And rightly so. If there is one idea of lasting influence which has developed out of the Werkbund movement, it is precisely this emphasis on concrete competence as opposed to an aesthetics removed and isolated from material questions. I am familiar with this dictum from my own *métier*, music. There it became a fundamental theorem, thanks to a school which cultivated close personal relationships with both Adolf Loos and the Bauhaus, and which was therefore fully aware of its intellectual ties to objectivity (*Sachlichkeit*) in the arts. Nevertheless, I can make no claim to competence in matters of architecture. And yet, I do not resist the temptation, and knowingly face the danger that you may briefly tolerate me as a dilettante and then cast me aside. I do this firstly because of my pleasure in presenting some of my reflections in public, and to you in particular; and secondly, because of Adolf Loos's comment that while an artwork need not appeal to anyone, a house is responsible to each and everyone. I am not yet sure whether this statement is in fact valid, but in the meantime, I need not be holier than the pope.

I find that the style of German reconstruction fills me with a disturbing discontent, one which many of you may certainly share. Since I no less than the specialists must constantly face this feeling, I feel justified in examining its foundations. Common elements between music and architecture have been discussed repeatedly, almost to the point of ennui. In uniting that which I see in architecture with that which I understand about the difficulties in music, I may not be transgressing the law of the division of labor as much as it may seem. But to accomplish this union, I must stand at a greater distance from these subjects than you may justifiably expect. It seems to me, however, not unrealistic that at times—in latent crisis situations—it may help to remove oneself farther from phenomena than the spirit of technical competence would usually allow. The principle of "fittingness to the material" (*Materialgerechtigkeit*) rests on the foundation of the division of labor. Nevertheless, it is advisable even for experts to occasionally take into account the extent to which their expertise may suffer from just that division of labor, as the artistic naiveté underlying it can impose its own limitations.

Let me begin with the fact that the anti-ornamental movement has affected the "purpose-free" arts (*zweckfreie Künste*) as well. It lies in the nature of artworks to inquire after the essential and necessary in them and to react against all superfluous elements. After the critical tradition declined to offer the arts a canon of right and wrong, the responsibility to take such considerations into account was placed on each individual work; each had to test itself against its own immanent logic, regardless of whether or not it was motivated by some external purpose. This was by no means a new position. Mozart, though clearly still standard-bearer and critical representative of the great tradition, responded in the following way to the minor objection of a member of the royal family—"But so many notes, my dear Mozart"—after the premier of his "Abduction" with "Not one note more, Your Majesty, than was necessary." In his *Critique of Judgment*, Kant grounded this norm philosophically in the formula of "purposiveness without a purpose" (*Zweckmässigkeit ohne Zweck*). The formula reflects an essential impulse in the judgment of taste. And yet it does not account for the historical dynamic. Based on a language stemming from the realm of materials, what this language defines as necessary can later become superfluous, even terribly ornamental, as soon as it can no longer be legitimated in a second kind of language, which is commonly called style. What was functional yesterday can therefore become the opposite tomorrow. Loos was thoroughly aware of this historical dynamic contained in the concept of ornament. Even representative, luxurious, pompous, and in a certain sense, burlesque elements may appear in certain forms of art as necessary, and not at all burlesque. To criticize the Baroque for this reason would be philistine. Criticism of ornament means no more than criticism of that which has lost its functional and symbolic signification. Ornament becomes then a mere decaying and poisonous organic vestige. The new art is opposed to this, for it represents the fictitiousness of a depraved romanticism, an ornamentation embarrassingly trapped in its own impotence. Modern music and architecture, by concentrating strictly on expression and construction, both strive together with equal rigor to efface all such ornament. Schönberg's compositional innovations, Karl Kraus's literary struggle against journalistic clichés, and Loos's denunciation of ornament are not vague analogies in intellectual history; they reflect precisely the same intention. This insight necessitates a correction of Loos's thesis, which he, in his open-mindedness, would probably not have rejected: the question of functionalism does not coincide with the question of practical function. The purpose-free (*zweckfrei*) and the purposeful (*zweckgebunden*) arts do not form the radical opposition which he imputed. The difference between the necessary and the superfluous is inherent in a work, and is not defined by the work's relationship—or the lack of it—to something outside itself.

In Loos's thought and in the early period of functionalism, purposeful and aesthetically autonomous products were separated from one another by absolute fact. This separation, which is in fact the object of our reflection, arose from the contemporary polemic against the applied arts and crafts (*Kunstgewerbe*). Although they determined the period of Loos's

development, he soon escaped from them. Loos was thus situated historically between Peter Altenberg [a Viennese writer, and close associate of Loos] and Le Corbusier. The movement of applied art had its beginnings in Ruskin and Morris. Revolting against the shapelessness of mass-produced, pseudo-individualized forms, it rallied around such new concepts as "will to style," "stylization," and "shaping," around the idea that one should apply art, reintroduce it into life in order to restore life to it. Their slogans were numerous and had a powerful effect. Nevertheless, Loos noticed quite early the implausibility of such endeavors: articles for use lose meaning as soon as they are displaced or disengaged in such a way that their use is no longer required. Art with its definitive protest against the dominance of purpose over human life suffers once it is reduced to that practical level to which it objects, in Hölderlin's words: "For never from now on/Shall the sacred serve mere use." Loos found the artificial art of practical objects repulsive. Similarly, he felt that the practical reorientation of purpose-free art would eventually subordinate it to the destructive autocracy of profit, which even arts and crafts, at least in their beginnings, had once opposed. Contrary to these efforts, Loos preached for the return to an honest handicraft which would place itself in the service of technical innovations without having to borrow forms from art. His claims suffer from too simple an antithesis. Their restorative element, not unlike that of the individualization of crafts, has since become equally clear. To this day, they are still bound to discussions of objectivity.

In any given product, freedom from purpose and purposefulness can never be absolutely separated from one another. The two notions are historically interconnected. The ornaments, after all, which Loos expulsed with a vehemence quite out of character are often actually vestiges of outmoded means of production. And conversely, numerous purposes, like sociability, dance, and entertainment, have filtered into purpose-free art; they have been generally incorporated into its formal and generic laws. Purposefulness without purpose is thus really the sublimation of purpose. Nothing exists as an aesthetic object in itself, but only within the field of tension of such sublimation. Therefore there is no chemically pure purposefulness set up as the opposite of the purpose-free aesthetic. Even the most pure forms of purpose are nourished by ideas—like formal transparency and graspability—which in fact are derived from artistic excellence. No form can be said to be determined exhaustively by its purpose . . .

The belief that a substance bears within itself its own adequate form presumes that it is already invested with meaning. Such a doctrine made the symbolist aesthetic possible. The resistance to the excesses of the applied arts pertained not just to hidden forms, but also to the cult of materials. It created an aura of essentiality about them. Loos expressed precisely this notion in his critique of batik. Meanwhile, the invention of artificial products—materials originating in industry—no longer permitted the archaic faith in an innate beauty, the foundation of a magic connected with precious elements. Furthermore, the crisis arising from the latest developments of autonomous art demonstrated how little meaningful organization could depend on the material itself. Whenever organizational principles rely too heavily on material, the result approaches mere patchwork. The idea of fittingness to the materials in purposeful art cannot remain indifferent to such criticisms. Indeed, the illusion of purposefulness as its own purpose cannot stand

up to the simplest social reality. Something would be purposeful here and now only if it were so in terms of the present society. Yet, certain irrationalities—Marx's term for them was *faux frais*—are essential to society; the social process always proceeds, in spite of all particular planning, by its own inner nature, aimlessly and irrationally. Such irrationality leaves its mark on all ends and purposes, and thereby also on the rationality of the means devised to achieve those ends. Thus, a self-mocking contradiction emerges in the omnipresence of advertisements: they are intended to be purposeful for profit. And yet all purposefulness is technically defined by its measure of material appropriateness. If an advertisement were strictly functional, without ornamental surplus, it would no longer fulfill its purpose as advertisement. Of course, the fear of technology is largely stuffy and old-fashioned, even reactionary. And yet it does have its validity, for it reflects the anxiety felt in the face of the violence which an irrational society can impose on its members, indeed on everything which is forced to exist within its confines. This anxiety reflects a common childhood experience, with which Loos seems unfamiliar, even though he is otherwise strongly influenced by the circumstances of his youth: the longing for castles with long chambers and silk tapestries, the utopia of escapism. Something of this utopia lives on in the modern aversion to the escalator, to Loos's celebrated kitchen, to the factory smokestack, to the shabby side of an antagonistic society. It is heightened by outward appearances. Deconstruction of these appearances, however, has little power over the completely denigrated sphere, where praxis continues as always. One might attack the pinnacles of the bogus castles of the moderns (which Thorstein Veblen despised), the ornaments, for example, pasted onto shoes; but where this is possible, it merely aggravates an already horrifying situation.

[. . .]

A general demystification, which began in the commercial realm, has encroached upon art. With it, the absolute difference between inflexible purposefulness and autonomous freedom has been reduced as well. But here we face another contradiction. On the one hand, the purely purpose-oriented forms have been revealed as insufficient, monotonous, deficient, and narrow-mindedly practical. At times, of course, individual masterpieces do stand out; but then, one tends to attribute the success to the creator's "genius," and not to something objective within the achievement itself. On the other hand, the attempt to bring into the work the external element of imagination as a corrective, to help the matter out with this element which sterns from outside of it, is equally pointless; it serves only to mistakenly resurrect decoration, which has been justifiably criticized by modern architecture. The results are extremely disheartening. A critical analysis of the mediocre modernity of the style of German reconstruction by a true expert would be extremely relevant. My suspicion in the *Minima Moralia* that the world is no longer habitable has already been confirmed; the heavy shadow of instability bears upon built form, the shadow of mass migrations, which had their preludes in the years of Hitler and his war. This contradiction must be consciously grasped in all its necessity. But we cannot stop there. If we do, we give into a continually threatening catastrophe. The most recent catastrophe, the air raids, have already led architecture into a condition from which it cannot escape.

The poles of the contradiction are revealed in two concepts, which seem mutually exclusive: handicraft and imagination. Loos expressly rejected the latter in the context of the world of use: "Pure and clean construction has had to replace the imaginative forms of past centuries and the flourishing ornamentation of past ages. Straight lines; sharp, straight edges: the craftsman works only with these. He has nothing but a purpose in mind and nothing but materials and tools in front of him." Le Corbusier, however, sanctioned imagination in his theoretical writings, at least in a somewhat general sense: "The task of the architect: knowledge of men, creative imagination, beauty. Freedom of choice (spiritual man)." We may safely assume that in general the more advanced architects tend to prefer handicraft, while more backward and unimaginative architects all too gladly praise imagination. We must be wary, however, of simply accepting the concepts of handicraft and imagination in the loose sense in which they have been tossed back and forth in the ongoing polemic. Only then can we hope to reach an alternative. The word "handicraft," which immediately gains consent, covers something qualitatively different. Only unreasonable dilettantism and blatant idealism would attempt to deny that each authentic and, in the broadest sense, artistic activity requires a precise understanding of the materials and techniques at the artist's disposal, and to be sure, at the most advanced level.

Only the artist who has never subjected himself to the discipline of creating a picture, who believes in the intuitive origins of painting, fears that closeness to materials and technical understanding will destroy his originality. He has never learned what is historically available, and can never make use of it. And so he conjures up out of the supposed depths of his own interiority that which is merely the residue of outmoded forms. The word "handicraft" appeals to such a simple truth. But quite different chords resonate unavoidably along with it. The syllable "hand" exposes a past means of production; it recalls a simple economy of wares. These means of production have since disappeared. Ever since the proposals of the English precursors of "modern style" they have been reduced to a masquerade. One associates the notion of handicraft with the apron of a Hans Sachs, or possibly the great world chronicle. At times, I cannot suppress the suspicion that such an archaic "shirt sleeves" ethos survives even among the younger proponents of "handcraftiness;" they are despisers of art. If some feel themselves superior to art, then it is only because they have never experienced it as Loos did. For Loos, appreciation of both art and its applied form led to a bitter emotional conflict. In the area of music, I know of one advocate of handicraft who spoke with plainly romantic anti-romanticism of the "hut mentality." I once caught him thinking of handicrafts as stereotypical formulas, practices as he called them, which were supposed to spare the energies of the composer; it never dawned on him that nowadays the uniqueness of each concrete task excludes such formalization. Thanks to attitudes such as his, handicraft is transformed into that which it wants to repudiate: the same lifeless, reified repetition which ornament had propagated. I dare not judge whether a similar kind of perversity is at work in the concept of form-making when viewed as a detached operation, independent from the immanent demands and laws of the object to be formed. In any case, I would imagine that the retrospective infatuation with the aura of the socially doomed craftsman is quite compatible with the disdainfully trumped-up attitude of his successor, the expert. Proud of his expertise and as unpolished as his tables and chairs, the expert disregards those reflections needed in this age which no longer possesses anything to grasp onto. It is impossible to do without the expert; it is impossible

in this age of commercial means of production to recreate that state before the division of labor which society has irretrievably obliterated. But likewise, it is impossible to raise the expert to the measure of all things. His disillusioned modernity, which claims to have shed all ideologies, is easily appropriated into the mask of the petty bourgeois routine. Handicraft becomes hand craftiness. Good handicraft means the fittingness of means to an end. The ends are certainly not independent of the means. The means have their own logic, a logic which points beyond them. If the fittingness of the means becomes an end in itself, it becomes fetishized. The handworker mentality begins to produce the opposite effect from its original intention, when it was used to fight the silk smoking jacket and the beret. It hinders the objective reason behind productive forces instead of allowing it to unfold. Whenever handicraft is established as a norm today, one must closely examine the intention. The concept of handicraft stands in close relationship to function. Its functions, however, are by no means necessarily enlightened or advanced.

The concept of imagination, like that of handicraft, must not be adopted without critical analysis. Psychological triviality—imagination as nothing but the image of something not yet present—is clearly insufficient. As an interpretation, it explains merely what is determined by imagination in artistic processes, and, I presume, also in the purposeful arts. Walter Benjamin once defined imagination as the ability to interpolate in minutest detail. Undeniably, such a definition accomplishes much more than current views which tend either to elevate the concept into an immaterial heaven or to condemn it on objective grounds. Imagination in the production of a work of representational art is not pleasure in free invention, in creation *ex nihilo.* There is no such thing in any art, even in autonomous art, the realm to which Loos restricted imagination. Any penetrating analysis of the autonomous work of art concludes that the additions invented by the artist above and beyond the given state of materials and forms are *miniscule* and of limited value. On the other hand, the reduction of imagination to an anticipatory adaptation to material ends is equally inadequate; it transforms imagination into an eternal sameness. It is impossible to ascribe Le Corbusier's powerful imaginative feats completely to the relationship between architecture and the human body, as he does in his own writings. Clearly there exists, perhaps imperceptible in the materials and forms which the artist acquires and develops something more than material and forms. Imagination means to *innervate* this something. This is not as absurd a notion as it may sound. For the forms, even the materials, are by no means merely given by nature, as an unreflective artist might easily presume. History has accumulated in them, and spirit permeates them. What they contain is not a positive law; and yet, their content emerges as a sharply outlined figure of the problem. Artistic imagination awakens these accumulated elements by becoming aware of the innate problematic of the material. The minimal progress of imagination responds to the wordless question posed to it by the materials and forms in their quiet and elemental language. Separate impulses, even purpose and immanent formal laws, are thereby fused together. An interaction takes place between purpose, space, and material. None of these facets makes up any one Ur-phenomenon to which all the others can be reduced. It is here that the insight furnished by philosophy that no thought can lead to an absolute beginning—that such absolutes are

the products of abstraction—exerts its influence on aesthetics.

[. . .]

The concern of functionalism is a subordination to usefulness. What is not useful is assailed without question because developments in the arts have brought its inherent aesthetic insufficiency into the open. The merely useful, however, is interwoven with relationships of guilt, the means to the devastation of the world, a hopelessness which denies all but deceptive consolations to mankind. But even if this contradiction can never be ultimately eliminated, one must take a first step in trying to grasp it; in bourgeois society, usefulness has its own dialectic. The useful object would be the highest achievement, an anthropomorphized "thing," the reconciliation with objects which are no longer closed off from humanity and which no longer suffer humiliation at the hands of men. Childhood perception of technical things promises such a state; they appear as images of a near and helpful spirit, cleansed of profit motivation. Such a conception was not unfamiliar to the theorists of social utopias. It provides a pleasant refuge from true development, and allows a vision of useful things which have lost their coldness. Mankind would no longer suffer from the "thingly" character of the world, and likewise "things" would come into their own. Once redeemed from their own "thingliness," "things" would find their purpose. But in present society all usefulness is displaced, bewitched. Society deceives us when it says that it allows things to appear as if they are there by mankind's will. In fact, they are produced for profit's sake; they satisfy human needs only incidentally. They call forth new needs and maintain them according to the profit motive. Since what is useful and beneficial to man, cleansed of human domination and exploitation, would be correct, nothing is more aesthetically unbearable than the present shape of things, subjugated and internally deformed into their opposite. The raison d'être of all autonomous art since the dawning of the bourgeois era is that only useless objects testify to that which may have at one point been useful; it represents correct and fortunate use, a contact with things beyond the antithesis between use and uselessness. This conception implies that men who desire betterment must rise up against practicability. If they overvalue it and react to it, they join the camp of the enemy. It is said that work does not defile. Like most proverbial expressions, this covers up the converse truth; exchange defiles useful work. The curse of exchange has overtaken autonomous art as well. In autonomous art, the useless is contained within its limited and particular form; it is thus helplessly exposed to the criticism waged by its opposite, the useful. Conversely in the useful, that which is now the case is closed off to its possibilities. The obscure secret of art is the fetishistic character of goods and wares. Functionalism would like to break out of this entanglement; and yet, it can only rattle its chains in vain as long as it remains trapped in an entangled society.

I have tried to make you aware of certain contradictions whose solution cannot be delineated by a non-expert. It is indeed doubtful whether they can be solved today at all. To this extent, I could expect you to criticize me for the uselessness of my argumentation. My defense is implicit in my thesis that the concepts of useful and useless cannot be accepted without due consideration. The time is over when we can isolate ourselves in our respective tasks. The object at hand demands the kind of reflection which objectivity (Sachlichkeit) generally rebuked in a clearly non-objective manner. By demanding immediate legitimation of a thought, by demanding to know what good

that thought is now, the thought is usually brought to a standstill at a point where it can offer insights which one day might even improve praxis in an unpredictable way. Thought has its own coercive impulse, like the one you are familiar with in your work with your material. The work of an artist, whether or not it is directed toward a particular purpose, can no longer proceed naively on a prescribed path. It manifests a crisis which demands that the expert—regardless of his prideful craftsmanship—go beyond his craft in order to satisfy it. He must do this in two ways. First, with regard to social things: he must account for the position of his work in society and for the social limits which he encounters on all sides. This consideration becomes crucial in problems concerning city planning, even beyond the tasks of reconstruction, where architectonic questions collide with social questions such as the existence or non-existence of a collective social subject. It hardly needs mentioning that city planning is insufficient so long as it centers on particular instead of collective social ends. The merely immediate, practical principles of city planning do not coincide with those of a truly rational conception free from social irrationalities; they lack that collective social subject which must be the prime concern of city planning. Herein lies one reason why city planning threatens either to degenerate into chaos or to hinder the productive architectonic achievement of individuals.

Secondly, and I would like to emphasize this aspect to you, architecture, indeed every purposeful art, demands constant *aesthetic* reflection. I know how suspect the word "aesthetic" must sound to you. You think perhaps of professors who, with their eyes raised to heaven, spew forth formalistic laws of eternal and everlasting beauty, which are no more than recipes for the production of ephemeral, classicist kitsch. In fact, the opposite must be the case in true aesthetics. It must absorb precisely those objections which it once raised in principle against all artists. Aesthetics would condemn itself if it continued unreflectively, speculatively, without relentless self-criticism. Aesthetics as an integral facet of philosophy awaits a new impulse which must come from reflective efforts. Hence recent artistic praxis has turned to aesthetics. Aesthetics becomes a practical necessity once it becomes clear that concepts like usefulness and uselessness in art, like the separation of autonomous and purpose-oriented art, imagination, and ornament, must once again be discussed before the artist can act positively or negatively according to such categories. Whether you like it or not, you are being pushed daily to considerations, aesthetic considerations, which transcend your immediate tasks. Your experience calls Molière's Monsieur Jourdain to mind, who discovers to his amazement in studying rhetoric that he has been speaking prose for his entire life. Once your activity compels you to aesthetic considerations, you deliver yourself up to its power. You can no longer break off and conjure up ideas arbitrarily in the name of pure and thorough expertise. The artist who does not pursue aesthetic thought energetically tends to lapse into dilettantish hypothesis and groping justifications for the sake of defending his own intellectual construct. In music, Pierre Boulez, one of the most technically competent contemporary composers, extended constructivism to its extreme in some of his compositions; subsequently, however, he emphatically announced the necessity of aesthetics. Such as aesthetics would not presume to herald principles which establish the key to beauty or ugliness itself. This discretion alone would place the problem of ornament in a new light. Beauty today can have no other measure except the depth to which a work resolves contradictions. A work must cut through the contradictions and overcome them, not by covering them up, but by pursuing them. Mere formal beauty, whatever that might be, is empty and meaningless; the beauty of its content is lost in the pre-artistic sensual pleasure of the observer. Beauty is either the resultant of force vectors or it is nothing at all. A modified aesthetics would outline its own object with increasing clarity as it would begin to feel more intensely the need to investigate it. Unlike traditional aesthetics, it would not necessarily view the concept of art as its given correlate. Aesthetic thought today must surpass art by thinking art. It would thereby surpass the current opposition of purposeful and purpose-free, under which the producer must suffer as much as the observer.

NOTE

1. See Robert W. Witkin, 'Why did Adorno 'Hate' Jazz?', *Sociological Theory* 18/1 (March 2000), pp. 145–70.

FURTHER READING

Theodor Adorno, *Aesthetic Theory*, trans. Robert Hullot-Kentor (London: Athlone Press, 1997; orig. pub. 1970).

Theodor Adorno, *The Culture Industry: Selected Writings on Mass Culture* (London: Routledge, 1991).

Martin Jay, *The Dialectical Imagination: A History of The Frankfurt School and the Institute for Social Research* (Boston: Little, Brown, and Company, 1973).

51

THE THING

Martin Heidegger

One necessarily approaches the work of Martin Heidegger with mixed feelings. He was complicit with the Nazi government's rise to power in Germany in the early 1930s (though there is debate about the extent of his fascist sympathies), and his philosophy has often been dismissed as essentialist (albeit highly complex) mysticism. And yet his writings have made him among the most influential philosophers of the twentieth century. Heidegger is best classified as a phenomenologist—that is, a philosopher of existence as it is perceived from a subjective position. The key questions within this field have to do with the relation between what is perceived and what is 'real'. How do we know what an object truly is, when we experience it only within certain frameworks (such as visuality, tactility and temporality)? In the following lecture, delivered shortly after the end of World War II, Heidegger offers one of the most concise statements of his thinking on this matter, using the example of a handmade ceramic jug. What he wants to know is what this thing is—not how it appears to us, but what it actually is. The jug as an 'object' (that is, as we regard it or use it) has a function and a form, and it is made of a particular material (fired earth). But these qualities are all derived from a fundamental 'thingness' which must precede any understanding of it by a subject. How are we to think about this 'thingness'? Heidegger's answer to this conundrum is a poetic one; he sees the jug as constituted fundamentally by the void inside it. The jug shapes the void and is in turn shaped by it. (As Michael Taussig puts it, 'Heidegger thought you could get a handle on Being by sneaking up on it backward, so to speak, by approaching it through the Nothing.')[1] That 'thingly' meeting of absence and presence conditions, and is prior to, all our relations to it as an object. In this sense, our subjective experience of things could be compared to our subjective experience of time, which is a continuous unfolding of presence framed the absences of past and present. This is rather abstract, but in this style of thinking Heidegger clearly commits himself to a concern for essences; and it is no coincidence that he chooses a handcrafted pot as his example, rather than (say) a plastic bowl. In another celebrated essay, 'The Question Concerning Technology', Heidegger argued that there are ways of making that connect us to existence, to the world we inhabit, and others that separate us. The former of these 'technologies', of course, are the crafts.

Martin Heidegger, 'The Thing', originally delivered as a lecture to the Bayerischen Akademie der Schonen Kunste, 1950. Translated by Albert Hofstadter in *Poetry Language Thought* (New York: Harper and Row, 1971).

All distances in time and space are shrinking. Man now reaches overnight, by plane, places which formerly took weeks and months of travel. He now receives instant information, by radio, of events which he formerly learned about only years later, if at all. The germination and growth of plants, which remained hidden throughout the seasons, is now exhibited publicly in a minute, on film. Distant sites of the most ancient cultures are shown on film as if they stood this very moment amidst today's street traffic. Moreover, the film attests to what it shows by presenting also the camera and its operators at work. The peak of this abolition of every possibility of remoteness is reached by television, which will soon pervade and dominate the whole machinery of communication.

Man puts the longest distances behind him in the shortest time. He puts the greatest distances behind himself and thus puts everything before himself at the shortest range.

Yet the frantic abolition of all distances brings no nearness; for the nearness does not consist in shortness of distance. What is least remote from us in point of distance, by virtue of its picture on film or its sound on the radio, can remain far from us. What is incalculably far from us in point of distance can be near to us. Short distance is not in itself nearness. Nor is great distance remoteness.

What is nearness if it fails to come about despite the reduction of the longest distances to the shortest intervals? What is nearness if it is even repelled by the restless abolition of distances? What is nearness if, along with its failure to appear, remoteness also remains absent?

What is happening here when, as a result of the abolition of great distances, everything is equally far and equally near? What is this uniformity in which everything is neither far nor near—is, as it were, without distance?

Everything gets lumped together into uniform distancelessness. How? Is not this merging of everything into the distanceless more unearthly than everything bursting apart?

Man stares at what the explosion of the atom bomb could bring with it. He does not see that the atom bomb and its explosion are the mere final emission of what has long since taken place, has already happened. Not to mention the single hydrogen bomb, whose triggering, thought through to its utmost potential, might be enough to snuff out all life on earth. What is this helpless anxiety still waiting for, if the terrible has already happened?

The terrifying is unsettling; it places everything outside its own nature. What is it that unsettles and thus terrifies? It shows itself and hides itself in the way in which everything presences, namely, in the fact that despite all conquest of distances the nearness of things remains absent.

What about nearness? How can we come to know its nature? Nearness, it seems, cannot be encountered directly. We succeed in reaching it rather by attending to what is near. Near to us are what we usually call things. But what is a thing? Man has so far given no more thought to the thing as a thing than he has to nearness. The jug is a thing. What is the jug? We say: a vessel, something of the kind that holds something else within it. The jug's holding is done by its base and sides. This container itself can again be held by the handle. As a vessel the jug is something self-sustained, something that stands on its own. This standing on its own characterizes the jug as something that is self-supporting, or independent. As the self-supporting independence of something independent, the jug differs from an object. An independent, self-supporting thing may become an object if we place it before us, whether in immediate perception or by bringing it to mind in a recollective representation. However, the thingly character of the thing does not consist in its being a represented object, nor can it be defined in any way in terms of the objectness, the over-againstness, of the object.

The jug remains a vessel whether we represent it in our minds or not. As a vessel the jug stands on its own as self-supporting. But what does it mean to say that the container stands on its own? Does the vessel's self-support alone define the jug as a thing? Clearly the jug stands as a vessel only because it has been brought to stand. This happened during, and happens by means of, a process of setting, of setting forth, namely, by producing the jug. The potter makes the earthen jug out of earth that he has specially chosen and prepared for it. The jug consists of that earth. By virtue of what the jug consists of, it too can stand on the earth, either immediately or through the mediation of table and bench. What exists by such producing is what stands on its own, is self-supporting. When we take the jug as a made vessel, then surely we are apprehending it—so it seems—as a thing and never as a mere object.

Or do we even now still take the jug as an object? Indeed. It is, to be sure, no longer considered only as object of a mere act of representation, but in return it is an object which a process of making has set up before and against us. Its self-support seems to mark the jug as a thing. But in truth we are thinking of this self-support in terms of the making process. Self-support is what the making aims at. But even so, the self-support is all thought of in terms of objectness, even though the over-againstness of what has been put forth is no longer grounded in mere representation, in the mere putting it before our minds. From the objectness of the object, and from the product's self-support, there is no way that leads to the thingness of the thing.

What in the thing is thingly? What is the thing in itself? We shall not reach the thing in itself until our thinking has first reached the thing as a thing.

The jug is a thing as a vessel—it can hold something. To be sure, this container has to be made. But its being made by the potter in no way constitutes what is peculiar and proper to the jug insofar as it is *qua* jug. The jug is not a vessel because it was made; rather, the jug had to be made because it is this holding vessel.

The making, it is true, lets the jug come into its own. *But* that which in the jug's nature is its own is never brought about by its making. Now released from the making process, the self-supporting jug has to gather itself for the task of containing. In the process of its making, of course, the jug must first show its outward appearance to the maker. But what shows itself here, the aspect (the *eidos*, the *idea)*, characterizes the jug solely in the respect in which the vessel stands over against the maker as something to be made.

But what the vessel of this aspect has this jug, what and how the jug *is* as this jug-thing, *is* something we can never learn—let alone think properly—by looking at the outward appearance, the *idea*. That is why Plato, who conceives of the presence of what is present in terms of the outward appearance, had no more understanding of the nature of the thing than did Aristotle and all subsequent thinkers. Rather, Plato experienced (decisively, indeed, for the sequel) everything present as an object of making. Instead of "object"—as that which stands before, over against, opposite us—we use the more precise expression "what stands forth." In the full nature of what stands forth, a twofold standing prevails. First, standing forth has the sense of stemming from somewhere, whether this be a process of self-making or of

Figure 35 *Jug*, ca. 1560–75.

being made by another. Secondly, standing forth has the sense of the made thing's standing forth into the unconcealedness of what is already present.

Nevertheless, no representation of what is present, in the sense of what stands forth and of what stands over against as an object, ever reaches to the thing *qua* thing. The jug's thingness resides in its being *qua* vessel. We become aware of the vessel's holding nature when we fill the jug. The jug's bottom and sides obviously take on the task of holding. But not so fast! When we fill the jug with wine, do we pour the wine into the sides and bottom? At most, we pour the wine between the sides and over the bottom. Sides and bottom are, to be sure, what is impermeable in the vessel. But what is impermeable is not yet what does the holding. When we fill the jug, the pouring that fills it flows into the empty jug. The emptiness, the void, is what does the vessel's holding. The empty space, this nothing of the jug, is what the jug is as the holding vessel.

But the jug does consist of sides and bottom. By that of which the jug consists, it stands. What would a jug be that did not stand? At least a jug *manqué*, hence a jug still—namely, one that would indeed hold but that, constantly falling over, would empty itself of what it holds. Only a vessel, however, can empty itself.

Sides and bottom, of which the jug consists and by which it stands, are not really what does the holding. But if the holding is done by the jug's void, then the potter who forms sides and bottom on his wheel does not, strictly speaking, make the jug. He only shapes the clay. No—he shapes the void. For it, in it, and out of it, he forms the clay into the form. From start to finish the potter takes hold of the impalpable void and brings it forth as the container in the shape of a containing vessel. The vessel's thingness does not lie at all in the material of which it consists, but in the void that holds.

NOTE

1. Michael Taussig, *My Cocaine Museum* (Chicago: University of Chicago Press, 2004), p. 61.

FURTHER READING

Jacques Derrida, *The Truth in Painting*, trans. Geoff Bennington and Ian McLeod (Chicago: University of Chicago Press, 1987; orig. pub. 1978).

Martin Heidegger, 'Building Dwelling Thinking' (1951), translated in Neil Leach, *Rethinking Architecture: A Reader in Cultural Theory* (London: Routledge, 1997).

Martin Heidegger, 'The Question Concerning Technology' (1954), translated in Heidegger, *Basic Writings* (San Francisco: Harpers, 1977).

Adam Sharr, *Heidegger's Hut* (Cambridge, MA: MIT Press, 2006).

52

RAPPEL A L'ORDRE: THE CASE FOR THE TECTONIC

Kenneth Frampton

Of the countless perceptive observations that the historian, critic and theorist Kenneth Frampton has made about architecture over the course of his career, perhaps the most important is a simple matter of etymology. Architecture, he points out, derives from the Greek term for 'master craftsman' (arch meaning 'master,' technê meaning 'art, craft or technique'). As with so many other figures whose writings appear in this book, Frampton's strategy has been to move forward by going back, in this case to the very origins of building. Drawing inspiration from the writings of the nineteenth-century design theorist Gottfried Semper and the existentialist philosopher Martin Heidegger, he has argued for an architecture that constantly reacquaints itself with constructional logic, the aspect of the discipline that he calls 'tectonics'. The following essay expounds on this idea, using some favored figures from the history of modern architecture as examples. Elsewhere in Frampton's voluminous writings, especially on the subject of 'critical regionalism', he has argued that the tectonic involves inscription into the particularities of a given site. Following Heidegger, he opposes any use of technology to create an architecture of placelessness, as with a prefabricated house or standardized chain store that can be dropped down anywhere on the surface of the earth. In this sense, Frampton holds fast to the notion that craft always involves a connection to the local. What at first appears to be a statement of aesthetics on his part—a preference for well-made, organic and expressive structure—proves to be a deeply political position. For Frampton, craft is the only way forward to a viable built environment.

Kenneth Frampton, 'Rappel a l'Ordre: The Case for the Tectonic', *Architectural Design* 60/3–4 (1990), excerpted. Reprinted in Frampton, *Labour, Work and Architecture* (London: Phaidon, 2002).

I have elected to address the issue of tectonic form for a number of reasons, not least of which is the current tendency to reduce architecture to scenography. This reaction arises in response to the universal triumph of Robert Venturi's decorated shed; that all too prevalent syndrome in which shelter is packaged like a giant commodity. Among the advantages of the scenographic approach is the fact that the results are eminently amortizable, with all the consequences that this entails for the future of the environment. We have in mind, of course, not the pleasing decay of nineteenth-century Romanticism but the total destitution of commodity culture. Along with this sobering prospect goes the general dissolution of stable references in the late-modern world; the fact that the precepts governing almost every discourse, save for the seemingly autonomous realm of techno-science, have now become extremely tenuous. Much of this was already foreseen half a century ago by Hans Sedlmayr, when he wrote, in 1941:

The shift of man's spiritual centre of gravity towards the inorganic, his feeling of his way into the inorganic world, may indeed legitimately be called a cosmic disturbance in the microcosm of man, who now begins to show a one-sided development of his faculties. At the other extreme there is a disturbance of macrocosmic relationships, a result of the especial favour and protection which the inorganic now enjoys—almost always at the expense, not to say ruin, of the organic. The raping and destruction of the earth, the nourisher of man, is an obvious example and one which in its turn reflects the distortion of the human microcosm from the spiritual.[1]

Against this prospect of cultural degeneration, we may turn to certain rear-guard positions, in order to recover a basis from which to resist. Today we find ourselves in a similar position to that of the critic Clement Greenberg who, in his 1965 essay 'Modernist Painting', attempted to reformulate a ground for painting in the following terms:

Having been denied by the Enlightenment of all tasks they could take seriously, they [the arts] looked as though they were going to be assimilated to entertainment pure and simple, and entertainment itself looked as though it was going to be assimilated, like religion, to therapy. The arts could save themselves from this leveling down only by demonstrating that the kind of experience they provided was valuable in its own right, and not to be obtained from any other kind of activity.[2]

If one poses the question as to what might be a comparable ground for architecture, then one must turn to a similar material base, namely that architecture must of necessity be embodied in structural and constructional form. My present stress on the latter rather than the prerequisite of spatial enclosure, stems from an attempt to evaluate twentieth-century architecture in terms of continuity and inflection rather than in terms of originality as an end in itself.

In his 1980 essay 'Avant-Garde and Continuity', the Italian architect Giorgio Grassi had the following comment to make about the impact of avant-gardist art on architecture:

. . . as far as the vanguards of the Modern Movement are concerned, they invariably follow in the wake of the figurative arts . . . Cubism, Suprematism, Neoplasticism, etc., are all forms of investigation born and developed in the realm of the figurative arts, and only as a second thought carried over into architecture as well. It is actually pathetic to see the architects of that 'heroic' period and the best among them, trying with difficulty to accommodate themselves to these 'isms'; experimenting in a perplexed manner because of their fascination with the new doctrines, measuring them, only later to realize their ineffectuality.[3]

While it is disconcerting to have to recognize that there may well be a fundamental break between the figurative origins of abstract art and the constructional basis of tectonic form, it is, at the same time, liberating to the extent that it affords a point from which to challenge spatial invention as an end in itself: a pressure to which modern architecture has been unduly subject. Rather than join in a recapitulation of avant-gardist tropes or enter into historicist pastiche or into the superfluous proliferation of sculptural gestures—all of which have an arbitrary dimension to the degree that they are based in neither structure nor in construction—we may return instead to the structural unit as the irreducible essence of architectural form.

Needless to say, we are not alluding here to mechanical revelation of construction but rather to a potentially poetic manifestation of structure in the original Greek sense of *poesis* as an act of making and revealing. While I am well aware of the conservative connotations that may be ascribed to Grassi's polemic, his critical perceptions none the less cause us to question the very idea of the new, in a moment that oscillates between the cultivation of a resistant culture and a descent into value-free aestheticism. Perhaps the most balanced assessment of Grassi has been made by the Catalan critic Ignasi Sola Morales, when he wrote:

Architecture is posited as a craft, that is to say, as the practical application of established knowledge through rules of the different levels of intervention. Thus, no notion of architecture as problem-solving, as innovation, or as invention ex novo, is present in Grassi's thinking, since he is interested in showing the permanent, the evident, and the given character of knowledge in the making of architecture.

. . . The work of Grassi is born of a reflection upon the essential resources of discipline, and it focuses upon specific media which determine not only aesthetic choices but also the ethical content of its cultural contribution. Through these channels of ethical and political will, the concern of the Enlightenment . . . becomes enriched in its most critical tone. It is not solely the superiority of reason and the analysis of form which are indicated, but rather, the critical role (in the Kantian sense of the term) that is, the judgement of values, the very lack of which is felt in society today . . . In the sense that his architecture is a meta-language, a reflection on the contradictions of its own practice, his work acquires the appeal of something that is both frustrating and noble . . .[4]

The dictionary definition of the term 'tectonic' to mean 'pertaining to building or construction in general; constructional, constructive used especially in reference to architecture and the kindred arts', is a little reductive to the extent that we intend not only the structural component *in se* but also the formal amplification of its presence in relation to the assembly of which it is a part. From its conscious emergence in the middle of the nineteenth century with the writings of Karl Bötticher and Gottfried Semper, the term not only indicates a structural and material probity but also a poetics of construction, as this may be practised in architecture and the related arts.

The beginnings of the Modern, dating back at least two centuries, and the much more recent advent of the Post-modern, are inextricably bound up with the ambiguities introduced into Western architecture by the primacy given to the scenographic in the evolution of the bourgeois world. However, building remains essentially tectonic rather than scenographic in character and it may be argued that it is first and foremost an act of construction rather than a discourse predicated on the surface, volume and plan, to cite Le Corbusier's 'Three Reminders to Architects', Thus one may assert that building is ontological rather than representational in character and that built form is a presence rather than something standing for an absence. In Martin Heidegger's terminology we may think of it as a 'thing' rather than a 'sign'.

I have chosen to engage with this theme because I believe it is necessary for architects to reposition themselves given that the predominant tendency today is to reduce all architectural expression to the status of commodity culture. In as much as such resistance has little chance of being widely accepted, a 'rear-guard' posture would seem to be an appropriate stance to adopt rather than the dubious assumption that it is possible to continue with the perpetuation of avant-gardism. Despite its concern for structure, an emphasis on tectonic form does not necessarily favour either Constructivism or Deconstructivism. In this sense it is astylistic. Moreover it does not seek its legitimacy in science, literature or art.

Greek in origin, the term tectonic derives from the term *tekton*, signifying carpenter or builder. This in turn stems from the Sanskrit *taksan*, referring to the craft of carpentry and to the use of the axe. Remnants of a similar term can also be found in Vedic, where it refers to carpentry. In Greek it appears in Homer, where it again alludes to carpentry and to the art of construction in general. The poetic connotation of the term first appears in Sappho where the *tekton*, the carpenter, assumes the role of the poet. This meaning undergoes further evolution as the term passes from being something specific and physical, such as carpentry, to the more generic notion of construction and later to becoming an aspect of poetry. In Aristophanes we even find the idea that it is associated with machination and the creation of false things. This etymological evolution would suggest a gradual passage from the ontological to the representational. Finally, the Latin term *architectus* derives from the Greek *archi* (a person of authority) and *tekton* (a craftsman or builder).

The earliest appearance of the term 'tectonic' in English dates from 1656 where it appears in a glossary meaning 'belonging to building', and this is almost a century after the first English use of the term architect in 1563. In 1850 the German oriental scholar K. O. Muller was to define the term rather rudely, as 'A series of arts which form and perfect vessels, implements, dwellings and places of assembly'. The term is first elaborated in a modern sense with Karl Bötticher's *The Tectonic of the Hellenes* of 1843–52 and with Gottfried Semper's essay 'The Four Elements of Architecture' of the same year. It is further developed in Semper's unfinished study, *Style in the Technical and Tectonic Arts or Practical Aesthetic*, published between 1863 and 1868.

The term 'tectonic' cannot be divorced from the technological, and it is this that gives it a certain ambivalence. In this regard it is possible to identify three distinct conditions: 1) the *technological object*, which arises directly out of meeting an instrumental need; 2) the *scenographic object*, which may be used equally to allude to an absent or hidden element; and 3) the *tectonic object*, which appears in two modes. We may refer to these modes as the *ontological and representational* tectonic. The first involves a constructional element that is shaped so as to emphasize its static role and cultural status. This is the tectonic as it appears in Bötticher's interpretation of the Doric column. The second mode involves the representation of a constructional element which is present, but hidden. These two modes can be seen as paralleling the distinction that Semper made between the *structural-technical* and the *structural-symbolic*.

Aside from these distinctions, Semper was to divide built form into two separate material procedures: into the *tectonics* of the frame, in which members of varying lengths are conjoined to encompass a spatial field; and the *stereotomics* of compressive mass that, while it may embody space, is constructed through the piling up of identical units (the term *stereotomics* deriving from the Greek term for solid, *stereos* and cutting, *-tomia*). In the first case, the most common material throughout history has been wood or its textual equivalents such as bamboo, wattle and basketwork. In the second case, one of the most common materials has been brick, or the compressive equivalent of brick such as rock, stone or rammed earth and later, reinforced concrete. There have been significant exceptions to this division, particularly where, in the interest of permanence, stone has been cut, dressed and erected in such a way as to assume the form and function of a frame.

While these facts are so familiar as to hardly need repetition, we tend to be unaware of the ontological consequences of these differences; that is to say, of the way in which framework tends towards the aerial and the dematerialization of mass, whereas the mass form is telluric, embedding itself ever deeper into the earth. The one tends towards light and the other towards dark. These gravitational opposites, the immateriality of the frame and the materiality of the mass, may be said to symbolize the two cosmological opposites to which they aspire: the sky and the earth. Despite our highly secularized techno-scientific age, these polarities still largely constitute the experiential limits of our lives. It is arguable that the practice of architecture is impoverished to the extent that we fail to recognize these transcultural values and the way in which they are latent in all structural form. Indeed, these forms may serve to remind us, after Heidegger, that inanimate objects may also evoke 'being', and that through this analogy to our own corpus, the body of a building may be perceived as though it were literally a physique. This brings us back to Semper's privileging of the joint as the primordial tectonic element, as the fundamental nexus around which building comes into being, that is to say, comes to be articulated as a presence in itself.

Semper's emphasis on the joint implies that fundamental syntactical transition may be expressed as one passes from the *stereotomic* base to the *tectonic* frame, and that such transitions constitute the very essence of architecture.

They are the dominant constituents whereby one culture of building differentiates itself from the next.

There is a spiritual value residing in the 'thingness' of the constructed object, so much so that the generic joint becomes a point of ontological condensation rather than a mere connection. The work of Carlo Scarpa would seem to exemplify this attribute.

[. . .]

Semper's 'Four Elements of Architecture' brings the discussion full circle in as much as Semper added a specific anthropological dimension to the idea of tectonic form. Semper's theoretical schema constitutes a fundamental break with the 400-year-oId humanist formula of *utilitas, firmitas, venustas* [usefulness, strength and beauty] that first served as the intentional triad of Roman architecture and then as the underpinning of post-Vitruvian architectural theory. Semper's radical reformulation stemmed from his seeing a model of a Caribbean hut in the Great Exhibition of 1851. The empirical reality of this simple shelter caused Semper to reject Laugier's primitive hut, adduced in 1753 as the primordial form of shelter with which to substantiate the pedimented paradigm of Neoclassical architecture. Semper's 'four elements' countermanded this hypothetical assumption and asserted instead an anthropological construct comprising: 1) a hearth, 2) an earthwork, 3) a framework and a roof, and 4) an enclosing membrane.

While Semper's elemental model repudiated Neoclassical authority it none the less gave primacy to the frame over the load-bearing mass. At the same time, Semper's four-part thesis recognized the primary importance of the earthwork, that is to say, of a telluric mass that serves in one way or another to anchor the frame or the wall, or *Mauer*, into the site.

This marking, shaping and preparing of ground by means of an earthwork had a number of theoretical ramifications. On the one hand, it isolated the enclosing membrane as a differentiating act, so that the *textual* could be literally identified with the proto-linguistic nature of textile production that Semper regarded as the basis of all civilization. On the other hand, as Rosemary Bletter has pointed out, by stressing the earthwork as the fundamental basic form, Semper gave symbolic import to a non-spatial element, namely, the hearth, which was invariably an inseparable part of the earthwork. The term 'breaking ground' and the metaphorical use of the word 'foundation' are both obviously related to the primacy of the earthwork and the hearth.

In more ways than one Semper grounded his theory of architecture in a phenomenal element having strong social and spiritual connotations. For Semper the hearth's origin was linked to that of the altar, and as such it was the spiritual nexus of architectural form. The hearth bears within itself connotations in this regard. It derives from the Latin verb *aedisficare* which in its turn is the origin of the English word *edifice*, meaning literally 'to make a hearth'. The latent institutional connotations of both hearth and edifice are further suggested by the verb to *edify*, which means to educate, strengthen and instruct.

Influenced by the linguistic and anthropological insights of his age, Semper was concerned with the etymology of building. Thus he distinguished the massivity of a fortified stone wall, as indicated by the term *Mauer*, from the light frame and in-fill—wattle and daub, say—of medieval domestic building, for which the term *Wand* is used. This fundamental distinction has been nowhere more graphically expressed than in Karl Gruber's reconstruction of a medieval German town.

Both *Mauer* and *Wand* reduce to the word 'wall' in English, but the latter in German is related to the word for dress, *Gewand*, and to the term *Winden*, which means to embroider. In accordance with the primacy that he gave to textiles, Semper maintained that the earliest basic structural artefact was the knot, which predominates in nomadic building form especially in the Bedouin tent and its textile interior. There are etymological connotations residing here of which Semper was fully aware, above all, the connection between knot and joint, the former being in German *die Knoten* and the latter *die Verbindung*, which may be literally translated as 'the binding'. All this evidence tends to support Semper's contention that the ultimate constituent of the art of building is the joint.

[. . .]

As we have already indicated, the tectonic lies suspended between a series of opposites, above all between the *ontological* and the *representational*. However, other dialogical conditions are involved in the articulation of tectonic form, particularly the contrast between the culture of the heavy—*stereotomics*, and the culture of the light—*tectonics*. The first implies load-bearing masonry and tends towards the earth and opacity. The second implies the dematerialized A-frame and tends towards the sky and translucence. At one end of this scale we have Semper's earthwork reduced in primordial times, as Gregotti reminds us, to the marking of ground. At the other end we have the ethereal, dematerialized aspirations of Joseph Paxton's Crystal Palace, that which Le Corbusier once described as the victory of light over gravity. Since few works are absolutely the one thing or the other, it can be claimed that the poetics of construction arise, in part, out of the inflection and positionings of the tectonic object. Thus the earthwork extends itself upwards to become an arch or a vault, or alternatively withdraws first to become the cross-wall support for a simple light-weight span and then to become a podium, elevated from the earth, on which an entire framework takes its anchorage. Other contrasts serve to articulate this dialogical movement further—such as smooth versus rough at the level of material (see Adrian Stokes's study *Smooth and Rough*, 1951), or *dark* versus *light* at the level of illumination.

Finally, something has to be said about the signification of the 'break' or the 'dis-joint' as opposed to the signification of the joint. I am alluding to that point at which things break against each other rather than connect: that significant fulcrum at which one system, surface or material abruptly ends to give way to another. Meaning may be thus encoded through the interplay between 'joint' and 'break', and in this regard rupture may have just as much meaning as connection. Such considerations sensitize the architecture to the semantic risks that attend all forms of articulation, ranging from the over-articulation of joints to the under-articulation of form.

POSTSCRIPT: TECTONIC FORM AND CRITICAL CULTURE

As Sigfried Giedion was to remark in the introduction to his two-volume study *The Eternal Present* (1962), among the deeper impulses of modern culture in the first half of this century was a 'transavantgardist' desire to return to the timelessness of a pre-historic past; to recover in a literal sense some dimension of an eternal present, lying outside the nightmare of history and beyond the processal compulsions of instrumental progress. This drive insinuates itself again today as a potential ground from which to resist the commodification of culture.

Within architecture the tectonic suggests itself as a mythical category with which to acquire entry to an anti-processal world wherein the 'presencing' of things will once again facilitate the appearance and experience of men. Beyond the aporias of history and progress and outside the reactionary closures of historicism and the neo-avant-garde lies the potential for a marginal counter-history. This is the primeval history of the logos to which Vico addressed himself, in his *Nuova Scienza*, in an attempt to adduce the poetic logic of the institution.[5] It is a mark of the radical nature of Vico's thought that he insisted that knowledge is not just the province of objective fact but also a consequence of the subjective, 'collective' elaboration of archetypal myth, that is to say, an assembly of those existential symbolic truths residing in the human experience. The critical myth of the tectonic joint points to just this timeless, time-bound moment, excised from the continuity of time.

NOTES

1. Hans Sedlmayr, *Art in Crisis: The Lost Centre* (New York and London: Hollis and Carter Spottiswoode, Ballantune & Co., Ltd., 1957), p. 164.
2. Clement Greenberg, 'Modernist Painting' (1965), republished in Gregory Battcock, ed., *The New Art*, New York: Dutton, 1966, pp. 101–2.
3. Giorgio Grassi, 'Avant-Garde and Continuity,' in *Oppositions*, no. 21, summer 1980, pp. 26–7.
4. Ignasi Solà Morales, 'Critical Discipline,' in *Oppositions*, no. 23, winter 1981, pp. 148–50.
5. See Joseph Mali, 'Mythology and Counter-History: The New Critical Art of Vico and Joyce,' in *Vico and Joyce*, Donald P Verene, ed., Albany: State University of New York Press, 1987.

FURTHER READING

Kenneth Frampton, 'Towards a Critical Regionalism: Six Points for an Architecture of Resistance', in Hal Foster, ed., *Postmodern Culture* (London: Pluto Press, 1983).

Kenneth Frampton, *Studies in Tectonic Culture* (Cambridge, MA: MIT Press, 1995).

Eduard Sekler, 'Structure, Construction, and Tectonics', in Georgy Kepes, ed., *Structure in Art and in Science* (New York: Braziller, 1965).

Mark Swenarton, *Artisans and Architects: The Ruskinian Tradition in Architectural Thought* (Basingtone, Hampshire: Macmillan Press, 1989).

Katie Lloyd Thomas, ed., *Material Matters: Architecture and Material Practice* (London: Routledge, 2007).

53

'ART AND CRAFT', FROM *THE PRINCIPLES OF ART*

R. G. Collingwood

The philosopher Robin George Collingwood was in the business of making careful distinctions. In his Essay on Philosophical Method *of 1933, he drew boundaries around the discipline of philosophy itself, distinguishing it from other realms of intellectual labor. In his writings as an ambitious amateur archaeologist, he subjected Roman inscriptions and excavation evidence to inspection. And, in his closely argued book on aesthetic theory, he expresses the opinion that art and craft are quite separate things. Disputing what he calls the 'technical theory' of aesthetics (the idea that art consists of nothing more than the techniques employed in its realization), Collingwood tries to isolate the seemingly ineffable component that renders something artistic. He focuses not on issues like functionality or social status, which he considers to be incidental, but rather the nature between means and ends. In one characteristic passage, he distinguishes poetry from blacksmithing, noting that only the former can justifiably be said to trade in emotion and expression: 'If the two kinds of conversion were the same, a blacksmith could make horseshoes out of his desire to pay the rent.' But one should not be tempted into reading Collingwood as dismissive of craft. Not only did he have impeccable credentials to write on the subject, his father having been a personal secretary to John Ruskin, but he was also an avid amateur craftsman in his own right, interested mainly in woodworking and nautical crafts. It*

may not be coincidental that his style of argumentation strongly resembles that of David Pye, another theorist who developed his ideas out of actual artisanal experience. With the benefit of hindsight, Collingwood's theory of art seems naïve. He was not able to foresee the institutional theories of art that would emerge in the work of theorists like George Dickey, who persuasively argued that art is a socially constructed category—it is simply that which is nominated as art by an authority, or included within the institutionalized artistic framework or discourse.[1] For those who think of craft as a necessary (but not sufficient) aspect of all art-making, however, Collingwood's ideas still have much to recommend them.

R. G. Collingwood, 'Art and Craft', from *The Principles of Art* (Oxford: Clarendon Press, 1938), excerpted, notes deleted.

The first sense of the word 'art' to be distinguished from art proper is the obsolete sense in which it means what in this book I shall call craft. This is what *ars* means in ancient Latin, and what *techne* means in Greek: the power to produce a preconceived result by means of consciously controlled and directed action. In order to take the first step towards a sound aesthetic, it is necessary to disentangle the notion of craft from that of art proper. In order to do this, again, we must first enumerate the chief characteristics of craft.

(1) Craft always involves a distinction between means and end, each clearly conceived as something distinct from the other but related to it. The term 'means' is loosely applied to things that are used in order to reach the end, such as tools, machines, or fuel. Strictly, it applies not to the things but to the actions concerned with them: manipulating the tools, tending the machines, or burning the fuel. These actions (as implied by the literal sense of the word means) are passed through or traversed in order to reach the end, and are left behind when the end is reached. This may serve to distinguish the idea of means from two other ideas with which it is sometimes confused: that of part, and that of material. The relation of part to whole is like that of means to end, in that the part is indispensable to the whole, is what it is because of its relation to the whole, and may exist by itself before the whole comes into existence; but when the whole exists the part exists too, whereas, when the end exists, the means have ceased to exist. As for the idea of material, we shall return to that in (4) below.

(2) It involves a distinction between planning and execution. The result to be obtained is preconceived or thought out before being arrived at. The craftsman knows what he wants to make before he makes it. This foreknowledge is absolutely indispensable to craft: if something, for example stainless steel, is made without such foreknowledge, the making of it is not a case of craft but an accident. Moreover, this foreknowledge is not vague but precise. If a person sets out to make a table, but conceives the table only vaguely, as somewhere between two by four feet and three by six, and between two and three feet high, and so forth, he is no craftsman.

(3) Means and end are related in one way in the process of planning; in the opposite way in the process of execution. In planning the end is prior to the means. The end is thought out first, and afterwards the means are thought out. In execution the means come first, and the end is reached through them.

(4) There is a distinction between raw material and finished product or artifact. A craft is always exercised upon something, and aims at the transformation of this into something different. That upon which it works begins as raw material and ends as finished product. The raw material is found ready made before the special work of the craft begins.

(5) There is a distinction between form and matter. The matter is what is identical in the raw material and the finished product; the form is what is different, what the exercise of the craft changes. To describe the raw material as raw is not to imply that it is formless, but only that it has not yet the form which it is to acquire through 'transformation' into finished product.

(6) There is a hierarchical relation between various crafts, one supplying what another needs, one using what another provides. There are three kinds of hierarchy: of materials, of means, and of parts. (a) The raw material of one craft is the finished product of another. Thus the silviculturist propagates trees and looks after them as they grow, in order to provide raw material for the felling-men who transform them into logs; these are raw material for the sawmill which transforms them into planks; and these, after a further process of selection and seasoning, become raw material for a joiner. (b) In the hierarchy of means, one craft supplies another with tools. Thus the timber-merchant supplies pit-props to the miner; the miner supplies coal to the blacksmith; the blacksmith supplies horseshoes to the farmer; and so on. (c) In the hierarchy of parts, a complex operation like the

manufacture of a motor-car is parcelled out among a number of trades: one firm makes the engine, another the gears, another the chassis, another the tyres, another the electrical equipment, and so on; the final assembling is not strictly the manufacture of the car but only the bringing together of these parts. In one or more of these ways every craft has a hierarchical character; either as hierarchically related to other crafts, or as itself consisting of various heterogeneous operations hierarchically related among themselves.

Without claiming that these features together exhaust the notion of craft, or that each of them separately is peculiar to it, we may claim with tolerable confidence that where most of them are absent from a certain activity that activity is not a craft, and, if it is called by that name, is so called either by mistake or in a vague and inaccurate way.

[. . .]

BREAKDOWN OF THE THEORY

(1) The first characteristic of craft is the distinction between means and end.

Is this present in works of art? According to the technical theory, yes. A poem is means to the production of a certain state of mind in the audience, as a horseshoe is means to the production of a certain state of mind in the man whose horse is shod. And the poem in its turn will be an end to which other things are means. In the case of the horseshoe, this stage of the analysis is easy: we can enumerate lighting the forge, cutting a piece of iron off a bar, heating it, and so on. What is there analogous to these processes in the case of a poem? The poet may get paper and pen, fill the pen, sit down and square his elbows; but these actions are preparatory not to composition (which may go on in the poet's head) but to writing. Suppose the poem is a short one, and composed without the use of any writing materials; what are the means by which the poet composes it? I can think of no answer, unless comic answers are wanted, such as "using a rhyming dictionary," "pounding his foot on the floor or wagging his head or hand to mark the metre," or "getting drunk." If one looks at the matter seriously, one sees that the only factors in the situation are the poet, the poetic labour of his mind, and the poem. And if any supporter of the technical theory says "Right: then the poetic labour is the means, the poem the end," we shall ask him to find a blacksmith who can make a horseshoe by sheer labour, without forge, anvil, hammer, or tongs. It is because nothing corresponding to these exists in the case of the poem that the poem is not an end to which there are means. Conversely, is a poem means to the production of a certain state of mind in an audience? Suppose a poet had read his verses to an audience, hoping that they would produce a certain result; and suppose the result were different; would that in itself prove the poem a bad one? It is a difficult question; some would say yes, others no. But if poetry were obviously a craft, the answer would be a prompt and unhesitating yes. The advocate of the technical theory must do a good deal of toe-chopping before he can get his facts to fit his theory at this point.

So far, the prospects of the technical theory are not too bright. Let us proceed.

(2) The distinction between planning and executing certainly exists in some works of art, namely those which are also works of craft or artifacts; for there is, of course, an overlap between these two things, as may be seen by the example of a building or a jar, which is made to order for the satisfaction

of a specific demand, to serve a useful purpose, but may none the less be a work of art. But suppose a poet were making up verses as he walked; suddenly finding a line in his head, and then another, and then dissatisfied with them and altering them until he had got them to his liking: what is the plan which he is executing? He may have had a vague idea that if he went for a walk he would be able to compose poetry; but what were, so to speak, the measurements and specifications of the poem he planned to compose? He may, no doubt, have been hoping to compose a sonnet on a particular subject specified by the editor of a review; but the point is that he may not, and that he is none the less a poet for composing without having any definite plan in his head. Or suppose a sculptor were not making a Madonna and child, three feet high, in Hoptonwood stone, guaranteed to placate the chancellor of the diocese and obtain a faculty for placing it in the vacant niche over a certain church door; but were simply playing about with clay, and found the clay under his fingers turning into a little dancing man: is this not a work of art because it was done without being planned in advance?

All this is very familiar. There would be no need to insist upon it, but that the technical theory of art relies on our forgetting it. While we are thinking of it, let us note the importance of not over-emphasizing it. Art as such does not imply the distinction between planning and execution. But (a) this is a merely negative characteristic, not a positive one. We must not erect the absence of plan into a positive force and call it inspiration, or the unconscious, or the like. (b) It a permissible characteristic of art, not a compulsory one. If unplanned works of art are possible, it does not follow that no planned work is a work

of art. That is the logical fallacy that underlies one, or some, of the various things called romanticism. It may very well be true that the only works of art which can be made altogether without a plan are trifling ones, and that the greatest and most serious ones always contain an element of planning and therefore an element of craft. But that would not justify the technical theory of art.

(3) If neither means and end nor planning and execution can be distinguished in art proper, there obviously can be no reversal of order as between means and end, in planning and execution respectively.

(4) We next come to the distinction between raw material and finished product. Does this exist in art proper? If so, a poem is made out of certain raw material. What is the raw material out of which Ben Jonson made *Queene and Huntresse, chaste, and faire*? Words, perhaps. Well, what words? A smith makes a horseshoe not out of all the iron there is, but out of a certain piece of iron, cut off a certain bar that he keeps in the corner of the smithy. If Ben Jonson did anything at all like that, he said: "I want to make a nice little hymn to open Act v, Scene vi of *Cynthia's Revels.* Here is the English language, or as much of it as I know; I will use *thy* five times, *to* four times, *and, bright, excellently,* and *goddesse* three times each, and so on." He did nothing like this. The words which occur in the poem were never before his mind as a whole in an order different from that of the poem, out of which he shuffled them till the poem, as we have it, appeared. I do not deny that by sorting out the words, or the vowel sounds, or the consonant sounds, in a poem like this, we can make interesting and (I believe) important discoveries about the way in which Ben Jonson's mind worked when he made

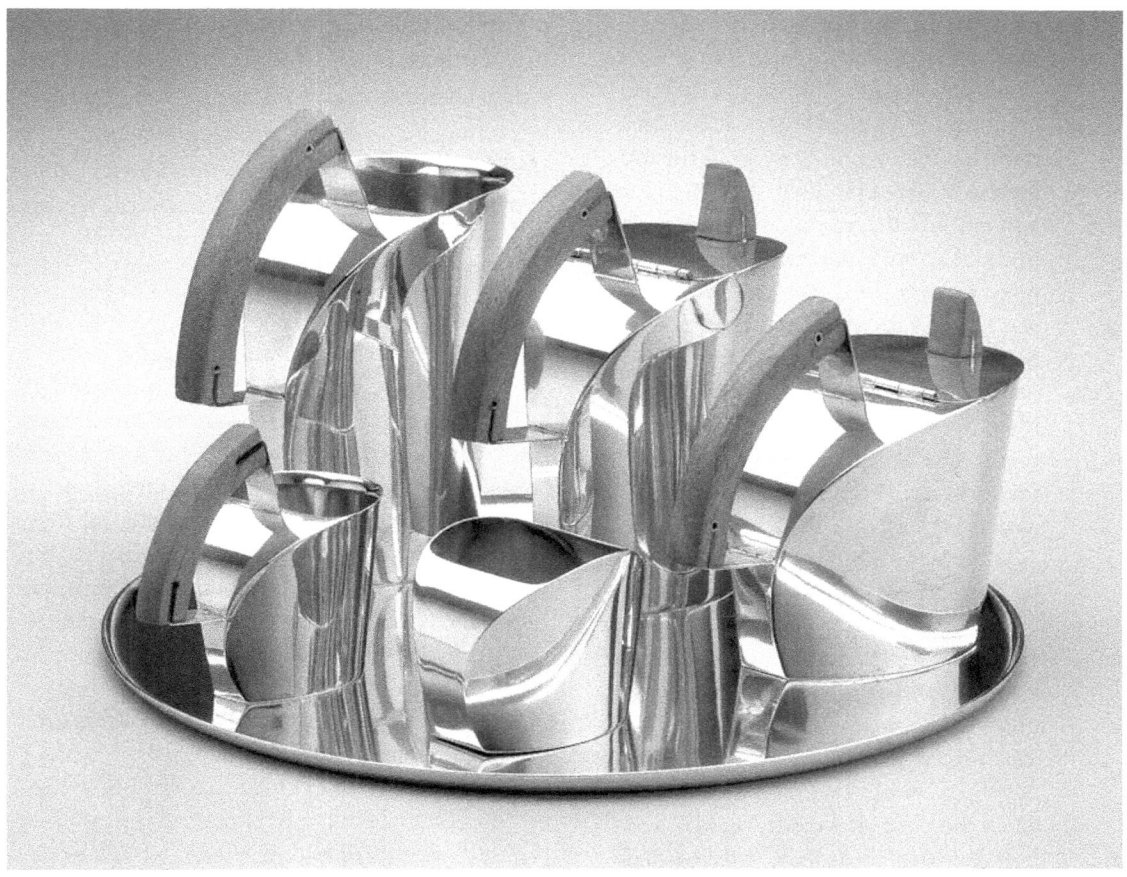

Figure 36   Harry Bertoia, *Tea and Coffee Service*, 1940.

the poem; and I am willing to allow that the technical theory of art is doing good service if it leads people to explore these matters; but if it can only express what it is trying to do by calling these words or sounds the materials out of which the poem is made, it is talking nonsense.

But perhaps there is a raw material of another kind: a feeling or emotion, for example, which is present to the poet's mind at the commencement of his labour, and which that labour converts into the poem. "Aus meinem grossen Schmerzen mach' ich die kleinen Lieder," ["from my great anguish I make little

songs"] said Heine; and he was doubtless right; the poet's labour can be justly described as converting emotions into poems. But this conversion is a very different kind of thing from the conversion of iron into horseshoes. If the two kinds of conversion were the same, a blacksmith could make horseshoes out of his desire to pay the rent. The something more, over and above that desire, which he must have in order to make horseshoes out of it, is the iron which is their raw material. In the poet's case that something more does not exist.

(5) In every work of art there is something which, in some sense of the word, may

be called form. There is, to be rather more precise, something in the nature of rhythm, pattern, organization, design, or structure. But it does not follow that there is a distinction between form and matter. Where that distinction does exist, namely, in artifacts, the matter was there in the shape of raw material before the form was imposed upon it, and the form was there in the shape of a preconceived plan before being imposed upon the matter; and as the two coexist in the finished product we can see how the matter might have accepted a different form, or the form have been imposed upon a different matter. None of these statements applies to a work of art. Something was no doubt there before a poem came into being; there was, for example, a confused excitement in the poet's mind; but, as we have seen, this was not the raw material of the poem. There was also, no doubt, the impulse to write; but this impulse was not the form of the unwritten poem. And when the poem is written, there is nothing in it of which we can say, "this is a matter which might have taken on a different form," or "this is a form which might have been realized in a different matter."

When people have spoken of matter and form in connexion with art, or of that strange hybrid distinction, form and content, they have in fact been doing one of two things, or both confusedly at once. Either they have been assimilating a work of art to an artifact, and the artist's work to the craftsman's; or else they have been using these terms in a vaguely metaphorical way as means of referring to distinctions which really do exist in art, but are of a different kind. There is always in art a distinction between what is expressed and that which expresses it; there is a distinction between the initial impulse to write or paint or compose and the finished poem or picture or music; there is a distinction between an emotional element in the artist's experience and what may be called an intellectual element. All these deserve investigation; but none of them is a case of the distinction between form and matter.

(6) Finally, there is in art nothing which resembles the hierarchy of crafts, each dictating ends to the one below it, and providing either means or raw materials or parts to the one above. When a poet writes verses for a musician to set, these verses are not means to the musician's end, for they are incorporated in the song which is the musician's finished product, and it is characteristic of means, as we saw, to be left behind. But neither are they raw materials. The musician does not transform them into music; he sets them to music; and if the music which he writes for them had a raw material (which it has not), that raw material could not consist of verses. What happens is rather that the poet and musician collaborate to produce a work of art which owes something to each of them; and this is true even if in the poet's case there was no intention of collaborating.

Aristotle extracted from the notion of a hierarchy of crafts the notion of a supreme craft, upon which all hierarchical series converged, so that the various "goods" which all crafts produce played their part, in one way or another, in preparing for the work of this supreme craft, whose product could, therefore, be called the "supreme good." [*Nicomachean Ethics*] At first sight, one might fancy an echo of this in Wagner's theory of opera as the supreme art, supreme because it combines the beauties of music and poetry and drama, the arts of time and the arts of space, into a single whole. But, quite apart from the question whether Wagner's opinion of opera as the greatest of the arts is justified, this opinion does not really rest on the idea of a hierarchy of arts. Words, gestures, music, scenery are not means to opera, nor yet raw materials of it, but parts of it; the hierarchies of means and materials may therefore be ruled out, and only that of parts remains. But even this does not apply. Wagner thought himself a supremely great artist because he wrote not only his music but his words, designed his scenery, and acted as his own producer. This is the exact opposite of a system like that by which motorcars are made, which owes its hierarchical character to the fact that the various parts are all made by different firms, each specializing in work of one kind.

NOTE

1. George Dickey, *Art and the Aesthetic: An Institutional Analysis* (Ithaca, NY: Cornell University Press, 1974). See also Thierry de Duve, *Kant After Duchamp* (Cambridge, MA: MIT Press, 1996); and Michael Grenfell and Cheryl Hardy, *Art Rules: Pierre Bourdieu and the Visual Arts* (Oxford: Berg, 2007).

FURTHER READING

R. G. Collingwood, *An Autobiography* (Oxford: Oxford University Press, 1939).

John Dillworth, 'Is Ridley Charitable to Collingwood?', *The Journal of Aesthetics and Art Criticism* 56/4 (Autumn 1998), pp. 393–6.

Aaron Ridley, *R. G. Collingwood: A Philosophy of Art* (London: Phoenix, 1998).

54

ART AND WORK

Harold Rosenberg

In 1964, the leading American art critic Harold Rosenberg was invited to address the First World Congress of Craftsmen in New York. His talk, transcribed in Craft Horizons *the following year, is one of the earliest statements about craft by a knowledgeable supporter of the postwar avant-garde. Like later fine art specialists who were drafted in to speak to the subject (such as Clement Greenberg, Arthur Danto and Donald Kuspit), Rosenberg's direct knowledge of craft history and theory was probably quite limited.[1] He nonetheless offered a thoughtful and nonprejudicial consideration of the relationship between art and craft. His words are very much of their time, shadowed by memories of World War II, and looking back at the New York school of 'Action Painters' (a term that Rosenberg himself had coined) and askance at the current emergence of Pop Art.[2] At this moment of uncertainty, Rosenberg developed an extremely individualistic theory of contemporary art—that it should be done not for art's sake, but rather for the artist's sake, as a form of self-development. Given the potential erasure of the line between art and life (an idea much in the air at the time), Rosenberg was bold enough to offer this as a general principle of late industrial culture. With this rather utopian end in view, he connected craft to the ideal of nonalienated labor: "Work done because the worker wants to do it, when he wants to do it, how he wants to do it." It is an ideal that William Morris would have had no difficulty recognizing.*

Harold Rosenberg, 'Art and Work', in *Craft Horizons* (May/June 1965), pp. 26, 54–5.

The prospect for the arts is bound up, of course, with the direction of the culture as a whole. There are many people who have a rather rosy picture of America's artistic future. In contrast to such an outlook, there is an opinion, apparently growing in strength, that the way things are going the arts have no future. We live, we are being told constantly, in a scientific and technological civilization which is systematically reducing man to the most primitive appetites and functions. Cultural decline used to be the theme of European romanticists and Americans visiting the great Renaissance cathedrals. Since the war, novelists, playwrights, thinkers, both here and abroad, have been competing with one another in finding the absolute symbol with which to represent the nonentity of the modern person. He has been dramatized as rubbish in a trash can—and as a volunteer rhinoceros. Philosophers have written volumes on loss of the self, and when committed by modern man even the utmost unspeakable crimes have been held to be "banal."

An element in this mood has no doubt been the memory of the Nazi death camps and the threat of nuclear war. But the dark prophecy of the fall into subhumanity antecedes Auschwitz and Hiroshima. Indeed it is customary to see these atrocities as effects of the current human condition rather than as its initiators. The true source of the man-made disasters of the twentieth century lies, philosophers of the fall inform us, is what the advance of technology has done to man himself. This is another way of saying that the so-called decline of man has to do with the subject of this Congress—that is to say, with the crafts. To many of the critics of contemporary civilization the practice of the crafts is the activity by which the human creature is defined. Man is a maker, *homo faber*, an artist. Put this proposition in reverse–when man ceases to be a maker he is no longer man—and our present crisis is explained. The fall began not in Eden, when man was condemned to labor, but in the nineteenth century when the machine first threatened him with leisure. With "soulless manufacture," as Ruskin called it, turning out endless quantities of copies of objects to which the human touch was alien, man-the-maker commenced to lose his skills and with them his dignity and independence. He was converted into an atom of mass society, a unit of energy susceptible of being put to use for any purpose—and to being replaced by other energy units generated from nature.

The dissolution of the human essence based on man's handling of materials now reaches its climax in automation, by which even the most rudimentary operations are eliminated. In regard to man as a tool-using fabricator the outlook is thus one of absolute blackness. He may preserve his skills out of sentimentality, and even revive abandoned ones, but these exercises no longer have a role in the serious realm of necessity and can no longer hold the human being to an ultimate definition.

If the arts are identified with the crafts—and they can never be altogether detached from them—their role too is in doubt. This doubt pervades the art of the past hundred years and is the essential content of the advanced writings and paintings of the half century since the First World War. Our time has given birth to the concept of the last artist. His work takes the form of reducing to zero the tradition of skillfully contrived objects. A splash or a ruled rectangle asserts the terminal proposition reached by the logic of art history. Then art and the artist are no more.

This doubting of the kind of making called art has turned up in all the arts. Under the name of anti-art it has exerted a constantly revolutionizing influence. It has resulted in the endeavor to carry art beyond fabrication into the realms of action and revelation. In painting and the drama it has caused the psychology of the gesture and the metaphysics of the void to emerge as keys to form. The symbol of art in which making has reached its last gasp is the work of the New York painter who literally conceives art as an all-black picture, which he repeats in a uniform size, shape, and surface: five feet wide, five feet high, five thousand dollars. This artist could announce if he chose, "I have met the machine and it is I."

I should like to be able to prove that the prophecy of mechanized man and the death of the arts is baseless. As against aristocrats who lost everything when their craftsmen went into the factories and the ideologists who became disillusioned when the workingmen ceased to be an object of commiseration, my interests put me on the side of equality and of plenty of free time for all, such as cultures based on the crafts could not allow. As to art, the masterpieces of modern music, painting, literature are not to my mind mere shadow images of a megapolitan spiritual desert.

Still, some kind of deterioration in the quality of people, their behavior and their products does seem to be taking place. This may be an effect of transition. The bureaucracy is growing, and the arts are by no means immune to it. The man who handles and shapes the materials, be they leather or steel, or words, paint, or sounds, has less and less control over the use to which his product is put, including its intellectual use. The artist is isolated from his public by the very processes and institutions through which his work is brought before it. The larger the influence exerted by his work the less that influence communicates the idea or sense of things embodied in the work itself.

All we have on the positive side is the individual's capacity for resistance. Resistance and criticism. Most modern masterpieces are critical masterpieces. Joyce's writing is a criticism of literature, Pound's poetry a criticism of poetry, Picasso's painting a criticism of painting.

This art also criticizes the existing culture.

Because it lives through criticism, modern art cannot be used as an argument to prove the case in favor of modern times and modern man. Like the art of all periods, it has the characteristics, including the negative characteristics, of the culture in which it was created. One who hates the modern world will find its most odious qualities mirrored in Joyce or Picasso, and will see only disintegration, distortion, and the absence of form and nobility.

In sum, vis-à-vis the past, the future and its creations are, to say the least, in question. When labor is no longer needed, or when its character as machine-tending has reached its ultimate—for example, when work consists of watching lights flash on and off and pressing a button when something goes wrong—when the use of skill in production is no longer even a rarity, changes in mankind of the profoundest magnitude may be postulated. All relations between man and nature, as well as between the individual and himself, will be transformed in unimaginable ways. Whether one considers this to be the bottom most point of human history or its height, one thing is certain: that the values of the future, including its aesthetic values, are all in the making. We may be headed toward a society of dehumanized robots or toward a community of intellectual supermen. But whatever be the outcome, the arts cannot be expected to carry on as they did before the industrial age. And even works produced in earlier epochs are bound to appear in a state of alteration and serve different functions from those they did in their own time.

A situation so drastic will, naturally, tend to generate extreme ideas. Given the forecast of Everyman as an utterly passive consumer inhabiting a vast supermarket that fills up automatically each morning with synthetic goods, it is not absurd to retort with the prophecy of Everyman as creative artist. The effect of universal automation must be to make geniuses of us all—as an alternative to converting us into amoeba-like digestive apparatuses. There is evidence, of course, that both conditions are in the process of being realized-which means, of course, that neither can be.

For the purpose of this discussion, I shall therefore assume that history will continue to behave like history; that it will bring forth everything except what is logically expected of it. With history as history, the vista consists of a mixup of the new and the old, of the outworn, the revived and the original. Thus the development of art in the decades to come will tend to parallel the three major phases of production in general: the crafts, scientific technology, free creation (including improvisation, games, totems). I shall comment on these briefly in relation to art today.

In much of contemporary painting and sculpture, art retains its ancient tie with the crafts, Reacting against machine-produced copies of things, art functions as a workshop for fashioning handmade ornaments and pushing forward possibilities in design. In this approach the fine artist and the inventive craftsman are indistinguishable from each other. It is regrettable that an inherited hierarchy makes it more desirable to be an artist than an artisan—for instance, much of the fuss about Pop art has been due to the ignorance among critics and curators of what is being done in advertising and in the display industry. For an art historian to justify his admiration for Lichtenstein by praising the latter's draftsmanship is laughable—as if the art departments of Madison Avenue and Hollywood were not full of the prize students of America's art academies. (To appreciate Lichtenstein's contribution to the art gag is something else.) That the painter or sculptor creates a single object rather than a model for machine production is not in itself sufficient to distinguish his work from that of the designer-craftsman in the automobile or space industry or in "the communications." The enormous improvement in techniques of reproduction in art further decreases the significance of the handmade as such in determining what is and what is not art. In their originals both artist and craftsman preserve the quality of the human touch and a control of their materials more flexible than that of the machine.

What defines art as craft is placing the emphasis on the object and its qualities to the exclusion of the personality of the artist, his unique consciousness, his dilemmas. For instance, since Leonardo painters have consciously made use of accident to arouse suggestions that would either help them to begin a painting or to redirect it during its creations. Certain effects of accident in modern art—splashes, runs, etc.—are now used by potters to enhance their surfaces. The accident is induced for the sake of what it does to the appearance of the object, rather than as an element in the artist's thinking and feeling: it becomes a category of decoration and a technical device. But painters in the last few years have been using accident in the same way, that is, not as part of a searching or imagining effort but in order to obtain a certain look. With the intellectual-emotional motive eliminated, so is the difference between the painter and the potter.

There is nothing in art that cannot be reduced to inconsequence if understood in how-to-do-it terms. Like accident, "painterliness" ceases to be a virtue in the hands of a recipe painter. But by the same token so does non-painterliness. Feeble painting can be one in pigment an inch thick or in a wash as thin as the reasoning of certain art critics.

Today, in connection with Pop and Gag art and various kinds of pattern-making abstraction in painting, there is much talk of an "anonymous" approach to art. This "new" idea reasserts the aboriginal relation of the craftsman to his produce, a relation of skill in making an object for us. Under present-day conditions anonymous art can be nothing else than a euphemism for commercial or industrial art brought into cultural areas formerly occupied by serious work. Art of the anonymous type, and the attitudes that produce it, may be expected to appear under any social conditions that provide a market for visual novelty and ornament.

The link of art with the crafts will persist no doubt as a force in art criticism and education.

Regardless of its mode of creation, any work of art, even the most expressive or exploratory, can be interpreted as a fabrication and judged in terms of the recipes by which it achieves its effects. Criticism seems to be much slower than art itself in casting off the spell of Greek and Latin terms that identify the poetry and the artist with making and the maker.

The merger of the studio and the workshop goes back to the beginnings of Roman history. Since the Renaissance the studio has also been linked with the laboratory. The use of art to investigate nature led almost from the start to investigation of the means of conducting that investigation—for example, study of perspective as a system for apprehending the physical world. The research of art into its own means has constituted a powerful motive in modern painting, poetry, music, the novel, the theater. In our century artists have often emphasized these means as part of the content of the work itself—as in the exaggeration of brushstrokes in painting, or in making the composition and acting of the play part of the plot of the play itself.

Research into the means falls roughly under two heads: 1) experimenting with the formal elements of the art in question—space, scale, and color relations in painting, non-metrical rhythms in poetry; 2) free play with the characteristic materials employed by each art—for example, pigment, sound, words; 3) the introduction of new raw materials, such as found objects in sculpture or streets sounds in music.

There can be little doubt that art will continue to experiment with the means of art and with the independent capacities of these means to simulate nature and evoke aesthetic response.

Perhaps the latest among the conscious interests of art is the formative effect of its creative processes upon the artist himself and upon his audience as individuals. Free work, whether in the studio, the workshop, the laboratory, or the industrial plant, is work done because the worker wants to do it, when he wants to do it, how he wants to do it. It is done not in obedience to external need but as a necessity of the worker's personality. It is work for the sake of the worker, his means of appropriating nature, and the heritage of other men's ideas and skills—this his means of developing himself.

Faced with the dissipation of local cultures and the mass recruitments of modern industry, art has found in its own practices the discipline for a continuous formation of individuality, as well as a direct means of communion with artists of other times, places, and cultures. Art aimed at self-creation has thus been inseparable from awareness of man's changed relation to production. Conversely, this historical consciousness has led art in our time toward an increased subjectivity. As against art conceived as the making of attractive objects, or as the vision of things as they are or are believed to be, history-conscious art has isolated in painting or in poetry the psychic experience of creation. It is art for making artists. In that it seeks to change the quality of living it is art that is political in the deepest sense—as contrasted with propaganda art which delivers preconceived messages through craftsman-like presentations. At the same time, it reawakens the primitive motive of art as magic and celebration. In work inspired by this new-old motive, art goes against its past as a making of things and takes on the characteristics of action. At the same time it preserves the essence of artisanship through its respect for the traditional media and its need to keep them alive as sources of suggestion. Thus art today continues to supply models for the crafts at the same time the latter often take on the gratuitous quality of art.

Whether or not art for artist's sake will remain an important strain in the future depends on the strength of our will to individual independence and social freedom. Art as craft and, to a lesser degree, art as experiment can function under any social system. Art as action, however, is the offspring of this revolutionary epoch and can flourish only so long as individuals are determined to be responsible for their own development and to interpret the past in relation to this aim. The ideal vista for the future is clear: it is that self-development shall be the motive of all work. If that ideal prevails, the distinction between the arts and other human enterprises will become meaningless.

NOTES

1. Clement Greenberg, "Status of Clay," 1979, in Garth Clark, ed., *Ceramic Millennium: Critical Writings on Ceramic History, Theory and Art* (Halifax: Press of the Nova Scotia College of Art & Design, 2005); Arthur C. Danto, "Philosophizing with a Hammer," in *Made in Oakland: The Furniture of Garry Knox Bennett* (New York: American Craft Museum, 2001); Donald Kuspit, "Craft as Art, Art as Craft," in Kuspit, *Redeeming Art: Critical Reveries* (New York: Allworth Press, 2000).

2. Harold Rosenberg, "The American Action Painters," *Art News* 51/8 (Dec. 1952), pp. 22–23, 48–50.

55

'COMMENT' AND RESPONSES

John Bentley Mays

Did the Canadian art critic John Bentley Mays know what he was getting into when he wrote the following short comment for American Craft *magazine? In his short, sharp, piece of writing, much despised in craft circles then and since, he tried to dismantle studio craft's pretensions to contemporary art status. Contemporary art, he argued, is expected to engage theoretically with its own condition: its claims upon the audience, the way it takes form within space (both physical and institutional), its status as a commodity for sale and so forth. Contemporary craft, in his view, did none of this. He condemned it as exuberant but unreflective. This was not so much an argument from theory—a definition of what craft was intrinsically. The assessment was, rather, based on his observations of what was actually happening in the craft marketplace. Mays's case is weakened by sloppy history and reductive logic, but twenty years later, it is hard to deny that he had a point. The 1980s were a period of rapid expansion in the North American craft economy (as in the Reagan-era economy generally), and criticality was not the watchword of the day. In the hotly collected field of glass especially, and to a lesser degree in other crafts media, Mays's harsh judgment of '80s craft as 'arty sculptures destined for the trash heap of history' has been more or less borne out. Certainly Mays touched a nerve, as is evident from the furious letters sent to American Craft in response. In retrospect, neither he nor his critics was taking the full measure of craft; there were elements of caricature on both sides. But the intensity of the debate cannot be doubted.*

John Bentley Mays, 'Comment', in *American Craft* 45/6 (December 1985/January 1986); and selected readers' responses in subsequent issues.

A spectre haunts the craft world of America, the writing about craft, the gatherings of artisans, the studios in schools and universities: the spectre of art.

Season by season, articles are written, conference goers put their heads together; and artisans give and hear speeches about art and the art world, and the pros and cons of the validation of craft by the art world's apparatus of critics, dealers, curators and so on. For the most part, it appears, artisans are inclined to think such blessing is a good thing; however, from time to time a knifemaker or weaver turns up who is willing to argue pluckily that a laying on of hands by the art museums is nothing craftspeople ought to be worrying their heads about.

I hear about all this because in the course of my work as art critic of *The Globe and Mail*, a daily published in Toronto, I occasionally run into artisans and representatives of professional craft organizations who are worrying their heads about such matters. These energetic, often urgent people have identified press coverage of their exhibitions as an important step toward validation, exposure and recognition, and they want it regularly. They are not getting it from the *Globe,* from other Toronto newspapers, from the local art magazines or, for that matter, from art magazines or art critics across North America. And they want to know why.

In this informal ramble through the mine field of the art-craft controversy, I am going to explain why I don't regularly cover crafts. And, later on, I will suggest reasons artisans should do their work, ignore the art world and forget about craft-as-art. Indeed, everything I have to say here is addressed to the potter who is tired of his wheel and the weaver weary of his loom. I have nothing to say to or about the contented, devoted craftsman-who will, in turn, be interested in nothing I have to say.

But before getting on to my theme, a bit of history to put my own views and decision in perspective. The early 1980s, when I joined the *Globe,* were exciting years to be on the art beat in Toronto. Ambitious commercial galleries were promoting a new generation of sophisticated artists in all media. The city's network of artist-run centers—a far more considerable phenomenon in Canada than in the United States—was the scene of many important experimental projects and programs. Artists were moving to Toronto from all over the country to take part in the market, the discussion, the vivid activity.

I soon concluded that, given the sheer weight of this new work and energy, the public's interest would hardly be served by limiting my columns to exhibitions of painting and sculpture in the venerable museums and well-established private galleries, The first task I set for myself was the extension of our regular local coverage to video and performance art, environmental installation, experimental film, photography and other forms of innovative work, wherever in the city they might be taking place.

My decision to expand in the direction of newer forms of visual art, incidentally, had nothing to do with materials. The stuff something is made from was (and is) a matter of complete professional indifference to me. The ideas, strategies and consciousness of history in a work of art is, finally, the only thing that matters—along with its place in the great tradition of Modern art making.

The art of the tiny screen from the mid-1960s onward, for example, was a completely recognizable innovation within the Modern art movement, even though the technology hailed from outside the salons of the more deluxe, luxurious arts such as painting and sculpture. Though video didn't look like conventional art to many gallery-goers, I certainly did not have to invent an ingenious new art history to justify covering it. In their anarchic attitudes toward medium, inquisitive interest in the body and the commonplace, and their Modernist insistence on properties of seeing, video artists declared their kinship in the great tribe of Modern art, whose patriarchs include Courbet and Manet, the Impressionists and Duchamp.

And that brings us back to craft, or the absence thereof, in my columns. It is worth remembering that modem art criticism, as we journalists practice it, began in popular reports on the news of this tribe—who was who and what was what at salons and salons des refuses in Paris during the mid-19th century. Whatever else we do, art critics will probably always be reporters on the complex, rambunctious tribe of Modernist artistry and sensibility, as it has descended and proliferated from the time of the great

French Moderns down to the present. It's a big and wonderful job. But it does mean that art critics will never be paying as much attention to crafts as craftspeople (and even some artists) think they should.

This is so not because craft or craft-as-art (as I have experienced it) are inferior to art, but because they are not art. Ceramic and fiber artists, like novelists and composers and physicists, belong to other tribes of creative discourse, with peculiar languages, technical strategies, codes and histories. These may be engaging to the critical imagination. But beyond my clear mandate as a critic, there is another, more compelling reason for writing about art, and not about crafts, science, theater, cooking or something else. The reason is the deep, abiding interest that art has been nourishing ever since I belatedly discovered it in my late 20s, about 15 years ago. Like many another art critic of my generation, I was lured into this career by the endless variety and invention, and the contentious spirit of liberty, displayed in the work of Pop artists, Minimalists Conceptualists, and in the immensely engaging art of the neo-Dadaists of the 1960s and early 1970s. At first I merely loved the joyous havoc of it all. But what held me (and has never let me go) was not the antic newness of contemporary art, but its deeply humane, critical spirit.

That spirit (like art criticism itself) was born in mid-19th-century Paris, in an atmosphere of skepticism and inquiry, curiosity about the new and impatience with the conventional, and deep negativity toward established authority in matters artistic, political, intellectual and ethical. Artists such as Courbet, Manet and Monet were not merely brilliant reinventors of painting. They were also apostles of art's new freedom from the tyrannies of academy, church, nobility, conventional ways of seeing and, alas! the established

art critics of the day. Whenever artists have wanted to renew their art, they have trekked back up country to the source, and reclaimed the radicalism, honesty and disruptive critical spirit of the beginning.

It should be noted here that Modern art's peculiar dislike for the hand—along with Modern criticism's avoidance of crafts—is a heritage of this beginning. The pioneers of Modernism were enchanted by the breathtaking changes in economic and civil society wrought by industrialization, The machine, not the hand, would become the emblem of these transformations, as the artists sought to create an art equal in power to the machine, and to the tonic, exciting dislocations of industrial culture; they were moved by skepticism and science, not by old-fashioned pieties. It is easy to see why the haughty eye (with its inevitable quizzical arched eyebrow) was given precedence over the strong, humble, preindustrial hand. Hands cannot contemplate; and the creation of works for disinterested, hands-off contemplation has traditionally been a central concern of all Modern art production.

Nothing that has been discovered since Courbet invalidates the central wisdom of this strategy of disinterestedness as a way to achieve an understanding of what is real. And Modern art itself, in all its variety, is proof that the historically anti-hand, anti-craft strategy continues to be radical and greatly rewarding. All this is a matter of passionate conviction among many art critics and curators, alike, if it is not usually put as bluntly as I've done here. The well-lit, empty, free and austere space in which painting and sculpture conventionally exist is cherished because it is there (and not in the familiar clutter of life) that Modern art yields up its complex, ironic truth about the word—not in being handled and known intimately, but in being contemplated by the educated eye.

The distinction I am drawing between hand and eye is a philosophical one, but not *merely* that. Like many high-minded notions bandied about by critics, this one also has a peculiarly American history which no craftsman, and no artist, can afford to ignore. Even if you don't make up your mind about things based on where you live or what historians say, it helps to keep this history in mind when looking at the literature of the art-craft discussion.

Writing in The *New Yorker* five years ago, for example, Calvin Tomkins noted that "the California clay artists and their descendants are widely known, as artists, on the west coast and in many areas of the United States and Europe. In fact about the only place their fine-art status is still uncertain is New York. Why this should be so is anybody's guess."

Is New York's "uncertainty" about crafts as fine art really so incomprehensible? At the immediate level, the phenomenon Tomkins observed reflects the well-known division in American critical culture itself, between New York's intellectually rigorous, theoretical understanding of art—one I share—and California's breezy, populist notions about art activity. This geographical distinction is sometimes disguised by language. Postmodernism, for instance, may be just a highfalutin' way to say California.

But however it is paraphrased and sublimated, this division is part of the more elaborate conflict precipitated by the recent general shift of power, money and population from the eastern seaboard to the Sun Belt. This reorientation (or dislocation) of American cultural focus, which has been going on in fits and starts for the last 25 years, has been immensely exciting, whatever your allegiances. It has occasioned the rise of original, star-spangled American developments of artistic Modernism in performance. video, experimental film and, of course, in clay, fiber and other craft media. It has challenged orthodoxies, artistic and other. Peter Voulkos's abstract ceramic sculpture, Ant Farm's famous Los Angeles performance involving a pyramid of blazing television sets rammed by a Cadillac, and the froggie clay universe of David Gilhooly, pop Buddhism and acid—all these things came crashing in, for better or worse, on the wave of Pacific Rim consciousness from the mid-1950s onward, changing the way we think about American art and culture forever.

In recent years, however, the exuberant culture of that hot strange America, and especially California, that once bedazzled us has come under cooler scrutiny by North American critics of art and culture, for several reasons. The current President of the United States and the swing to the right he has helped set in motion are among them. It is simply not possible to look with unqualified delight at the American culture of populism, nostalgia, sexiness and superficial charm which somehow managed to produce Ronald Reagan. Another (and related reason) is the recent surge of big, but avid, ignorant and hedonistic money into the contemporary art market—California cruising gone greedy. The resistance of some art critics at *October*, *Art in America* and elsewhere to graffiti art, Julian Schnabel's variety of thuggish, decorative painting and other philistine, crowd-pleasing American art seemed perhaps too stringent and puritanical only a few years ago. Today, with the paradox of America's squalor and spectacle more extreme than ever, the reservations of such critics seem more apt than ever.

But all is not Spenglerian doom. Many critics of contemporary culture find themselves looking at new art that embodies the qualities which have always made the difference between the best Modern art and all its counterfeits and alternatives: the will to human liberty, and the determined resistance to authority (including art-world authority), convention and complacent accommodation to human unfreedom. That we are finding such embodiments—especially in art at the edges of America's empire, in Europe, Canada, South America, Australia—makes the work of critical inquiry especially interesting these days.

But is it any wonder that, by the same token, many of us are less vulnerable than ever to the calls of craftspeople to extend our interest in their direction? If the North American craft press presents a true picture, ceramic and fiber artists are principally concerned nowadays with issues of style, surface, technique, and are largely unconcerned (at least in their work) about the cultural issues very much on the minds of art critics these days. They have come by their nonchalance honestly; to my knowledge, American craft-as-art has never undergone critical pitched battles comparable to the ones painting and sculpture have endured during the last 100 years. The emergence of the new ceramic and fiber arts in California, for example, may have been greeted by grumbling from some potters and weavers, but the reception accorded these developments in the national craft press, as well as in the art magazines, appears to have been generally enthusiastic. Right from the postwar days, when it decided to hanker after art's prestige and language and high profile, craft-as-art has been smiled on by sunny days,

Perhaps for that reason, its practitioners have never developed a self-critical attitude capable of pushing it into the total reversals and radical renewals which has heralded each new dawn in the history of Modern art. There

appears to be no force in the craft-as-art movement comparable to the urgency which, again and again, has pulled Modem art back from complicity with aristocratic privilege, self-satisfied Biedermeier comfortableness, and the suffocating pieties of ruralism—the three principal enemies of art, and of human liberty, in the liberal democratic countries of the west.

It is surely not my intention to hammer artisans for their complicity with these anti-Modern forces. But in a crafts community apparently bewitched by the prospect of certification as art, what power is protecting crafts from becoming merely the fiefdom of these forces, or of any art critic or curator, however reactionary, who will confer the validation artisans appear to want? What defenses do craftspeople have against exploitation by art-world opportunists? For example, the desirability of recognition by the high-art machinery of museums, market and the art press is taken for granted by most writers about craft I've read—as though the motives of art-world institutions were somehow above suspicion and reproach, or even some sober analysis.

And who among the craftspeople is counting the cost of "validation" by the museums and art press? The artisans who take the dictates of art critics seriously are bound to suffer, simply because the worst art in the world is made by those trying to please or second-guess art critics. Whatever the outcome on that score, the lust for recognition has already had the effect of lulling talented young artisans away from their wheels and looms and condemning them to obscurity as producers of arty sculptures destined for the trash heap of history. It is dismaying to visit craft studios and find energetic young men and women busily turning out dull imitations of Voulkos or John Mason or Gilhooly, when all America is crying out for a terrific five-dollar cookie jar.

Also, the quest for certification has undammed a sea of incredibly vulgar imitative "clay art" and "fiber art"—a flow that continues to the present day, unchecked by a craft press too cozy with the people it should be criticizing, and far too enchanted by the goal of validation itself to say much about emperors and new clothes.

Or, for that matter to deal with emperors at all. The successful careers of Voulkos, Robert Arneson, Gilhooly and other senior California ceramic sculptors are routinely presented as foretastes of the good things waiting for artisans just over Validation Mountain. The message of the craft press has been clear for 30 years: these are the masters. Follow them, oh ye potters of Dubuque and Pasadena!

What is lacking in the celebrations I have read, however, is serious consideration of the received pieties about these artists. Who is questioning the final creative importance, for example, of Gilhooly's facetious frog statues? Or of the originality of Voulkos's appropriation of Abstract Expressionism?

It has always seemed to me that Voulkos merely borrowed the swagger and hot-licks stylistics of action painting without much understanding of their precise, inalienable relations to the history of painting, then deployed these technical gestures in a kind of popular, stylish pastiche—a comfortable version of Modern art for people who feel intimidated by it, but who still wish to appear chic and knowledgeable. Am I missing something? Or is Voulkos really (as I suspect) the Mantovani of Ab-Ex?

Clearly the craft press has its work cut out for it. Because I have my work cut out for me, I will not be involved in the rethinking of craft priorities that is so urgently needed. But I do hope that one outcome of this rethinking is a fresh appreciation for the work of potter and weaver and jeweler, who must be exempted from everything negative I have said about the practitioners of craft-as-art. The quality of mercy in great pottery and weaving is much needed in a visual culture which, under the steady bombardment of television and advertising, has become hugely wordy, demanding and obsessive, and saturated with insatiable desires. The artisan's commitment to the physical stuff of his craft is his only hope for salvation from the brushfires of fashion and the art world's endless poodle parade. It remains an exemplary commitment, with the power to inspire all creative people with its high seriousness, and its intelligent detachment from the astonishments and empty pageantry of contemporary mass culture.

LETTERS TO THE EDITOR

In a way I welcome John Bentley Mays's hard-bitten critique of 'craft-as-art' under "Comment," in the December 1985/January 1986 issue of *American Craft*. I appreciated his openness, which will help us to recognize who the enemies are and what ideas of theirs are potentially destructive to us. My own resolve as an artist working in wood remains unshaken, but I see the lure in his brand of elitism which has fatefully attracted craftspeople to cross over into the arena where the activity called art is practiced in search of the promised mysteries.

Let me give a little background on myself. I attended the Boston Museum School of Fine Arts, and studied painting. The position from which Mays expresses his artistic point of view is one that I held once. For the last 30 years I have been a wood craftsman working almost exclusively in furniture and doing it for a living most of the time. In the last few years I have arrived at a clearer understanding of what is important for me in my work and I have adjusted my work priorities in order to accomplish some artistic goals.

In his book *Art as Experience* John Dewey wrote, "Craftsmanship to be artistic in the final sense must be 'loving'; it must care deeply for the subject-matter upon which skill is exercised." I quote this because it contrasts so sharply with Mays's prerequisites for artistic production—"empty, free and austere space," and "not being handled and known intimately." Mays's prerequisites recommend dehumanizing one's self for the sake of art; he makes it sound as if art wasn't, after all, for people but done for a concept of an abstract society. I think art can be more fun than that.

But I believe Mays is right when he writes: "American craft-as-art has never undergone critical pitched battles comparable to the ones painting and sculpture have endured during the last 100 years." Somehow, and I see it in furniture more clearly, crafts have not gotten out from under the shroud surrounding them from the past, not crossed bravely into the 20th century with fresh forms and structures supporting new ideas. When furnituremakers get abstract or painterly, they borrow heavily from the more revolutionary painters and sculptors, which results in the "pastiche" Mays refers to.

We furnituremakers might feel smug today that we no longer kowtow to elitist demands, but in fact a narrow and powerful clique is still with us seeking to influence productions (and being very successful at it) in the crafts to reflect their image of power and money. In order to progress, furnituremakers need to resist the powerful urges to conform totally to conventional methods and materials. The products that materialize from the axis of client and tradition become boring and numb our sense to everything except pretty wood and the small enticements of nostalgia.

Anyone having difficulty seeing today's hold-overs from the past should examine ancient Egyptian furniture from this point of view and he will find every form, every technique, virtually as they are practiced today. In this particular craft, on another continent, 4,000 years later,

things have changed very little. These remnants from the past and their manifestations are part of what Mays calls 'anti-modern forces.'

Mays is not a new phenomenon either. He is following in the tradition of the aristocracies, keeping up the elitism and the exclusionary practices of the past. Art critics, at least since Clive Bell, the super elitist, have been bent on establishing themselves and their media clients on rarified high ground so as to weaken other potential contenders with volleys of missives sent down in regular bursts. The most effective tool they have is to keep strict control of who appears in their art columns.

There is one particular advantage for the press and the critics to continue to exclude certain groups from coverage in the art sections of their newspapers. This sets them up as the arbiters of art, as it were. They become the experts, repositories of the unknown. If you wish to crack the mystery, come look in your daily newspaper. In this next quote from Mays he presents himself as the expert communicating with the mystery: "Nothing that has been discovered since Courbet invalidates the central wisdom of this strategy of disinterestedness as a way to achieve an understanding of what is real. And Modern art itself, in all its variety, is proof that the historically anti-hand, anti-craft strategy continues to be radical and greatly rewarding." This esoteric innuendo embodies the flimsy framework that Mays holds up as the reason that "crafts-as-art" should be excluded from his art columns. There is no clear evidence at all, to me, that craftsmanship in art retards or hinders the creative process.

Let me quote Dewey again: "Art involves the molding of clay, shaping of marble, casting of bronze, laying on of pigments, construction of buildings, singing of songs, playing of instruments, enacting roles on the stage, going through rhythmic movements in dance."

Is Mays aware that the other arts are thriving? In glass, for instance, what is going on now is a phenomenal revival. Artists are energetically responding with new forms, new techniques, using modern idioms in ways never before experienced. We will be counting the aesthetic contributions from these high-energy sophisticated experiments for years.

I am not impressed by Mays when he pits New York's "intellectual rigorous theoretical understanding of art" against California's "breezy populist notions about art activity." It makes him sound as if he too is afflicted by one of the enemies of art, the "suffocating piety of ruralism." In this case the East Coast variety. His attack on Peter Voulkos seems partly to spring from his own severely circumscribed view of art. His trouble with the medium actually prevents him from seeing Voulkos as a visual artist. Voulkos has a national reputation; I'm sure he will have his defenders. I believe that Mays and his public will miss one of the most important and exciting movements in art in this century: the emerging artists liberating the traditional crafts through artistic experiments with mediums. There is just about enough time left, 15 years; it will happen fast.

Let me quote Dewey once more, in a statement particularly apropos for addressing Mays's rigid definition of art: "Rigid classifications are inept (if they are taken seriously) because they distract attention from that which is esthetically basic—the qualitatively unique and integral character of experience of an art product. But for a student of esthetic theory they are also misleading. There are two important points of intellectual understanding in which they are confusing. They inevitably neglect transitional and connecting links; and in consequence they put insuperable obstacles in the way of an intelligent following of the historical development of any art."

If you are going to be an artist that is what you are. The idea of an artist not being an artist because he or she is a craftsperson is a historical idea with no place in our contemporary art world which has effectively liberated mediums from categorical identification with value judgments. That Mays doesn't understand this is evident in his use of the word "artisan," which, aside from being insensitive, chiefly implies skilled labor—not really what we are about. We as a group that appear in this magazine are striving for a set of mind that will enable us to make original contributions of substance through our work, will take and use whatever we need to do this, including the word *artist* for its valuable psychological assets for all craftspersons.

—John Marcoux, Providence, RI

I read about a page and a half of "Comment" by John Bentley Mays. What a chore. So pretentious, so verbose and grandiose and sooo boring.

I have a degree in art history; therefore, I am not unfamiliar with this kind of statement. So self-serving, in this case almost a mea culpa. Who really cares whether Mays reviews or doesn't review anything?

—Theo Portnoy, New York, NY

I felt as though I were being patted on the head by [Mays]. However, I do not feel any particular indignation, because I felt his commentary was purposely limited in perception. His views of the work being produced by artists-craftspeople was by his own definition very narrow, and seemed to be an attempt to relegate the art produced within the craft world into a very small frame, in order to dismiss it more easily. Mays implies that craft artists ought to stick to pure craft because those who have tried to bridge the gap between "art" and "craft" thus far have failed. This is unduly hopeful on his part; this apparent failure is only temporary.

Mays bewails the lack of revolutionary thought, yet suggests that we peasants not concern ourselves with such a rigorous occupation. That patriarchal and patronizing stance is familiar to women and other minorities; now artists in craft media must learn to recognize this same approach applied to themselves. Presented in a kindly and quasi-encouraging guise, it is a homily intended to keep us in our place. I, for one, never occupied that place.

Five-dollar cookie jars? Shall we stay in the kitchen, on the farm and in the ghetto to make them? Being dismissed as presumptuous and uppity is familiar to minorities, yet these qualities are what often drive minorities to become revolutionaries.

Nonetheless, Mays makes some necessary points about the present craft vision. The concern with surface and technique can be boring and repetitive, and I have seen more of the fried-rice spot-and-dash school of decoration than I ever needed to. But the same forces are visibly rampant in *ARTnews* and *Art in America*; the mad dog and skeletal visage imagery of many recent paintings, for example, with jagged lines and teeth everywhere.

The worst failing that I think Mays is accusing the craft world of is provincialism, illustrated by his comparison of California art with the "legitimate" New York art scene. In this I think he is correct. However, his implication is that artists in craft media can never transcend their materials-focused roots. I think this can be done, and needs to be done now: I perceive craft and fine art made in traditional craft as being presently on a plateau, possibly in a stage of assimilation and synthesis of technical competence with the lessons learned from art. Possibly the craft world is also in a state of smugness, at present, taking few risks, addressing minor issues. Just as there are inferior watercolorists and aspiring but pedestrian painters, there are ranks of craftspeople whose work, despite their intent, will never attain the status of good art. Similarly, an art school background or a concern with contemporary issues in art can produce merely academic or trendy statements with no long-term or critical validity. We accept that, and are as aware of the pitfalls as any art school graduate who

finds him/herself a loft in Soho from which to launch an art career. But to dismiss the validity of the statements by artists in somewhat unexpected media merely because of the media is in itself provincial.

I think Mays is miscalculating in his determined shortsightedness and his refusal to concern himself with the art produced in craft media today. While he assiduously turns his attention to the passing art parade, and patronizingly congratulates craftspeople on their luck to be out of it, there is good art being created. It stands on its own terms, not begging a glance from art critics, nor plaintively protesting that it is not craft any more. Art is being produced, transcending the funky and the trendy, coming from a vision that reached for the nearest materials to make a statement. Watch for it, Mr. Mays. Revolutions come from unexpected places, and good art does as well.

—Janice Anthony, Brooks, ME

I was delighted to see John Bentley Mays's piece on crafts and modern art. The essay was a remarkable mixture of perspicacity, energy, passion, smugness, naiveté, wit, joy, erudition, understanding and misunderstanding. He hit so many nails on the head he can be forgiven for once or twice striking his thumb. Although, after his clarity of vision about lust and validation, and important hints at the real differences between the fields of modern art and craft he does do some damage to our confidence by suggesting that the alternative to imitative mediocrity is a bargain in cookie jars. An amusing conceit or not, he patronizes, and leaves an appreciable segment of the crafts community mournfully regarding their wheels. I say those who are less than supremely masterful need not be banished to the crafts activity room, there to make clay pot holders.

But if he is a little morally and practically wanting, the weight of theoretical perception is on his side. He is so happily right about the importance—or "salvation," as he likes to put it—of the "artisan's commitment to the physical stuff of his craft." But if that is catechism, it is only the first question.

Yes, Mr. Mays, the crafts press does have "its work cut out for it." And so, when it understands how and sees its way, shall that larger press that pretends to any concern with aesthetically critical events.

—Lisa Hammel, New York, NY

Is it in the best interest of the visual arts community to promote the "informal ramblings" of a visual bigot? If the editors of *American Craft* are attempting to create a dialogue to preface the upcoming American Craft Council conference in Oakland, divisive comments, such as those presented in the December/January issue, merely play into the hands of those who wish to see everything segregated from paintings to pots to the color of one's skin.

If John Bentley Mays wishes to criticize individual artists in the craft world or the art world I wish him all the success affordable to those of his profession. But to lump together everyone working in traditional craft media as non-artists is patently absurd. His notions on how art is created are a mystery. As if with some "hands-off" magic, paintings and sculptures materialize, for the fascination of waiting critics and historians to be intellectually dissected, ismed, schismed, theorized and classified. Art is conceived in the mind, but brought to fruition by hands and materials. It was a pair of hands that made Brancusi's "birds" take flight, that painted Picasso's *Demoiselles d'Avignon*, that assembled Duchamp's *Large Glass* and on and on. To list the contradictions and misinformation in the commentary of Mays would be like going over a speech by Ronald Reagan—with whom Mays seems to be a kindred spirit—in his delight to eliminate all but the narrowest vision of culture.

Fortunately, like the artists that brought us such critically debased movements as Impressionism, Cubism, Abstract Expressionism to name a few, "craft artists" will not be intimidated by the verbal abuse so easily bandied about by those with little creative ability of their own. For those of us who have chosen to express ourselves through the use of a visual language, all too often we expect others to understand us immediately. But history has pointed out time and time again that it is long after the moment of conception that a work of art is understood by those who are not open-minded. The ability to create is in our hands and minds. It is our responsibility to resist becoming complacent in view of recent achievements in our field and to continue the natural evolution which is set in motion.

—Hap Sakwa, Baywood Park, CA

John Bentley Mays's "Comment" makes me absolutely joyous that I am an authentic person—a crafts artist. I am not a superficial one who plugs in tattered intellectual formulas—an art critic. I visualize your creative energies, Mr. Mays, pulsing down to your fingertips and instead of being released in the beautiful toolmaking and decorating tradition of crafts, they short-circuit and circulate back up to make carbon cholesterol in your little black heart.

It is unfortunate that you didn't spare time in your 20s (before your life was cast in stone as an "art critic") learning a little anthropology. The development of technology, i.e., toolmaking, is one of the most fascinating and creative and art-full of human endeavors, and it is the tradition that is carried on by crafts artists. Today's post-technology crafts express some very human ideas about tools and their makers. Nonfunctional "art" crafts can express the absurdity of the uses we make of our technological/scientific information. It is craft that has made the bomb, not some dilettante who dabbles around with a brush to express himself or paint barns. Unfortunately, I see the bomb as significant.

If I had been invited to write two pages for a magazine of the caliber of *American Craft* I hope I would have prepared myself with a little research into the subject before I presented myself in print. I would wager that all the ideas you presented have already been in print before in your columns. Tsk.

Working with materials keeps us craftspeople honest, human and quite discerning about quality. When you want to really understand art, make something. For now, to the trash heap with your words and ideas.

—Mary Byington, Santa Barbara, CA

FURTHER READING

Edward S. Cooke, Jr., 'Wood in the 1980s: Expansion or Commodification?', in Davira Taragin et al., Contemporary Crafts and the Saxe Collection (New York: Hudson Hills, 1993).

Bruce Metcalf, "Replacing the Myth of Modernism," *American Craft* (February/March 1993), pp. 40–7.

56

THE MAKER'S EYE

Alison Britton

Studio craftspeople have long been frustrated by the lack of critical response to their activities. A few have taken matters into their own hands, writing about their own work and that of their peers. British ceramic artists have been particularly avid in this regard, both as historians— Edmund De Waal and Emmanuel Cooper are important examples—and as critics. Alison Britton is the most thoughtful of the latter. A formidable potter in her own right, Britton trained at the Royal College of Art, London (where she has also taught for many years), in the late 1970s, a moment of generative ferment for British ceramics. Poststructuralist semiotic theory, the Feminist-derived Pattern and Decoration movement, and affinities with historical ceramics all combined to produce a generation of Postmodernist potters. Britton's handbuilt vessels, with their walls akimbo and features such as handles and spouts rendered into abstract compositional anchors, and festooned with patterns of every description, offered a uniquely satisfying combination of charismatic form and eclectic appropriation. That dialectic has remained at the core of her work ever since. The following short comment is taken from the catalogue for The Maker's Eye, *an exhibition held by the British Crafts Council in which makers were asked to choose objects according to their own aesthetic predilections. (David Pye was among the other participating artist-curators.) Britton, characteristically,*

presented herself as torn between competing impulses. It would be too simplistic to say that she was drawn towards both craft-as-art and craft for its own sake. Rather, she wrote of her attraction towards different, irreconcilable models of integrity and communication. In retrospect, this declaration of internal conflict—like her pots of the same period—makes her seem simultaneously honest, indeterminate and, given the aesthetics of ambivalence common in ceramics today, weirdly prescient.

Alison Britton, curatorial comment from *The Maker's Eye* (London: Crafts Council, 1982).

Many of the objects that have had a powerful effect on me and my work have been ancient, foreign, or outside the definition of 'craft', and so it has been something of a shock to have to choose mainly from amongst what has been made by craftspeople in Britain during this century. I find, to my surprise, that within these confines what seems to me to be most important reveals a narrow and specific interest: almost all vessels, many of these in my own material, ceramics, and mainly produced in the last few years.

My work may in the future be seen to have belonged to a 'group', and I think that the objects I have chosen reinforce this idea of a hypothetical group of artist craftspeople, having certain trains of thought in common, whether or not these have been articulated or brought to the surface. I would say that this group is concerned with the outer limits of function; where function, or an idea of a possible function, is crucial, but is just one ingredient in the final presence of the object, and is not its only motivation. I think that this preoccupation, which can be perceived in various fields and materials, stands out as a distinct contribution of the last ten years: something has happened that is only gradually being described and recognised. Some people will certainly feel that it represents the last decadent throes of an artistic crafts movement of dwindling relevance, where over self-conscious makers turned in on themselves for want of a real sense of necessity. But perhaps to others it will be seen as something closely in line with 'modernism' in the other arts, in painting or literature for example. A 'modern' novel (one following such writers as Proust and Joyce) is both made of, and about, language. Some of the objects I have chosen are similarly self-referential, that is, they perform a function, and at the same time are drawing attention to what their own rules are about. (As Michael Rowe, in the notes for his 1978 exhibition at the Crafts Council, says: 'The boxes are purely about their own space and the characteristics of the sheet metal that they are made from.') In some ways such objects stand back and describe, or represent, themselves as well as being. In the analogy with the novel 'function' stands for 'story' as the central content.

However, I am not only concerned with this rather elusive category in my selection. I have chosen vessels or containers that are 'ordinary' too, and to me supremely and powerfully so. I would like to make a comparison evident between 'prose' objects and 'poetic' objects; those that are mainly active and those that are mainly contemplative. To me the most moving things are the ones where I experience in looking at them a frisson from both these aspects at once, from both prose and poetry, purpose and commentary. These have what I call a 'double presence'.

I would like now to focus on my own subject, ceramics. Clay is a material that has been prone to metaphor for centuries. It is such a malleable, versatile, simulatory substance, that practical objects have been formed in some disguise or other for probably almost as long as clay has been used. The first vessels to use clay are thought to have been woven reed containers plastered with a layer of sticky clay to make them waterproof. Even in this there is an ambiguity between 'pot' and 'basket'. Being a vessel is not very demanding. Once the functional requirements of holding are fulfilled, there is still plenty of room for interpretation and variety of outer form. Chronologically, my selection begins with a Martin Brothers jar; a striking example of an object that is not what it seems to be, where decoration is carried to dominating extremes. A jar becomes a bird (and a Carol McNicoll plate becomes a piece of origami), function unimpaired. I am suggesting that what I have been trying to define is a descendant of this tradition. A jar can be also a representation of a jar, function unimpaired.

Many of these objects have been made, it seems, in the light of modern painting. Still life painting is such a strong tradition, at its most seductive perhaps with Matisse and Morandi. Why have painters cared so much about depicting pots and pans? What do they stand for—slices of life? Symbols of domestic or intimate selves? I find myself aware of a sequence here: from ordinary everyday objects, to painting, and through re-depiction in three dimensions of not-so-ordinary objects. Andrew

Figure 37 Alison Britton, *Pair with Black Lines*, 1981.

Lord, working in ceramics, reconstructs the objects that might have been the subject of a painting. They are presented in prescribed groups. He concerns himself with the way light falls on objects, and builds in clay with a painter's eye on the transmutations of light and tone. Physically, his objects retain some of the imprecision and loose, impressionistic quality of a two-dimensional representation. Function is hardly a characteristic at all—though they are still hollow and waterproof, and more than that, eloquent about the relation between inside and outside. Life has been translated through still life and beyond into some ghostlier, more ambivalent form of object. Andrew Lord's work stands at the far edge of my spectrum from ordinary to magical. The objects in the middle, my main concern, are about life and still life at once. They can be used, but their function is partly frozen in reflection about themselves. Technically, I think this effect is usually achieved through some kind of physical distortion, some thwarting of our expectations of form, and is perhaps particularly to do with flattening of form. A Steven Newell jug, for instance, simply by being flattened (and helped by being transparent) moves towards being a representation of a jug as well as being a jug in fact. An Erik de Graaff chair (could a chair be described as a container?) gives a similar jolt to the expectations, and is at once business-like and questioning. Objects such as these fill the gap between prose and poetry, between ordinary and breathtaking, combining both. These are the things that matter to me most.

It is hard to explain my own inability to stop making vessels. It could be somehow inherent in the training of a potter, something one is lumbered with as part of the equipment. Or it could be that the inclusion of a function is a crutch for one lacking the courage to make a piece of work that is entirely aesthetic; I may be clinging to the residue of use as a justification. Or I may have an irresistible (and fairly abstract) preoccupation with something very deep-rooted. Vessels are basic, archetypal, timeless. A container is a fundamental prop (and symbol) of civilization. A container is an object made with a specific relation to people in mind. Two-faced objects such as I have described are giving more than was demanded of them. That seems to be worth doing.

FURTHER READING

Garth Clark, *The Potter's Art: A Complete History of Pottery in Britain* (London: Phaidon, 1995).

Edmund DeWaal, *20th Century Ceramics* (London: Thames and Hudson, 2003).

Peter Dormer, *The New Ceramics: Trends and Traditions* (London: Thames and Hudson, 1988).

The Raw and the Cooked: New Work in Clay in Britain (Oxford: Museum of Modern Art, Oxford, 1993).

57

HOW ENVY KILLED THE CRAFTS

Garth Clark

The historian, art dealer and critic Garth Clark has spent his career championing the cause of ceramics as an art form. Born and raised in South Africa, Clark opened his New York gallery together with partner Mark Del Vecchio in 1981. There had been other significant dealers of ceramics (Helen Drutt of Philadelphia, for example) but the Garth Clark Gallery introduced a new sophistication and ambition to the presentation of the medium, much as the Peter Joseph Gallery would do for furniture a decade later. Clark has also produced a large and important body of scholarly writing, charting the history of modern ceramics and helping to construct a canon of key artists. Working through the nonprofit Ceramic Art Foundation, he has organized conferences on the history of the medium. He has consulted widely with museums and collectors and wrote short, sharp critical assessments of individual artists. Not all of Clark's enterprises have met with success—a foray into jewelry, for example, was short-lived—and he has certainly had his detractors over the years. But it is impossible to imagine contemporary ceramics without him. This background adds considerable interest to the following text, which Clark first composed as a lecture to be delivered at the Museum of Contemporary Craft in Portland, Oregon, directly following his retirement from gallery work. It is writing borne of long experience and deep frustration. Clark identifies a fundamental contradiction in terms at the heart of the craft movement. The longstanding ambition of craftspeople to be accepted as artists was worse than Quixotic, he argues: it led to self-loathing and ethical rudderlessness. Clark doesn't deny that the craft movement has had its heroes, or that it has produced objects and ideas of tremendous power. (Notably and somewhat controversially, he also is at pains to preserve a discrete category of genuine artists who just happen to work in ceramics, and are not to be confused with crafters.) But he argues that the days of those achievements are over. Even if the institutions that the craft movement produced—museums, magazines, medium-based organizations—manage to find new roles, the movement itself is over. Only time will tell whether this obituary will prove to be premature; the recent explosion of DIY activity, for example, is clearly a crafts movement of sorts, and might be considered a variation on studio craft rather than a complete departure. But it is certainly worth paying attention when someone in Clark's position is willing to say 'goodbye to all that'.

Garth Clark, 'How Envy Killed the Crafts', 2008.

For most of the modern craft movement's one hundred and fifty year life it has wrestled with a debilitating condition, an unhappy, contentious relationship with the fine arts. In 1939 *Fortune* magazine ran a survey of ceramics in America and titled it "The art with an inferiority complex". That was true then as now, not just of ceramics but of craft across the board as it dealt with the status of being "less than" art. Craft has moved constantly between resentment and envy with the relationship growing increasingly acrimonious as art moved away from craft-based values in the mid-century and closer to post-1950 conceptualism and the dematerialization of the art object.

The roots of craft's art envy are long and complex and begin at birth when it was named the Arts and Crafts Movement, a title determined essentially by class. Craftsmen in early Victorian England were mostly rural and lower working class. The members of the new Movement were middle and upper class. So they needed something that said "better than just craft" hence the role of the term "arts". Right there and then the strange and unhappy dance-macabre between art and craft begins. Over the decades it grew from an annoying neurosis to a full-blown pathological obsession that ultimately, in the late twentieth century, killed the movement.

[. . .]

By 1970 craft's marketplace began to explode. Over the next decade a strong three-tier structure emerged: at the lower populist end was the craft fair, the most populist level; in the middle there was a hybrid—the craft shop and gallery—a bridge to the top end; and at the top of the pyramid, true galleries that were modeled on the fine arts. Prices quickly soared after 1980 from hundred of dollars to tens of thousands and eventually even hundreds of thousands. The collector base, once tiny, grew in leaps and bounds and the new collectors were affluent.

The field was represented by a powerful, effective New York–based organization, the American Crafts Council. It published a respected magazine, *Craft Horizons*, ran the Museum of Contemporary Craft in New York and a thriving craft shop, America House, organized national and regional conferences, touring exhibitions, arranged fairs and generally promoted the field.

This appeared to be the perfect success story. But below the surface a damaging disorder was festering. While craft was doing well, fine art was doing better. It was much more glamorous, had better museums and institutions and, of course, a better rewards program. Ego-driven envy was fueled by resentment that craft, while successful, was not as respected or as valued as the fine arts.

This resentment was justifiable with a sub-genre, those who were not crafters, but because of their material choices were confined to the crafts by a material apartheid instituted by the Modernist regime: Robert Arneson, Ron Nagle and others. Certain activities, such as specializing in ceramics, were considered de facto "craft" no matter whether the maker was producing art or not. For these few artists the insistence that that they be taken seriously as fine artists was just.

For the rest of the crafters who climbed aboard this bandwagon, it was wishful thinking, a Quixotic journey that ended badly. They too attempted to cross over, as had Ken Price and others, loading their craft with footnotes from Janson's *History of Art* and festooning it with quotes from Michel Foucault. Craft was now more self-consciously influenced by art than ever before, underlining this connection at every turn.

But it did not become fine art. And this was difficult for some to understand. If the dialogue they had engaged was similar, why were craft and art not equals? The key term is "influenced" by art. For example a sculptor can spend his entire career being influenced by architecture but he never becomes an architect. The same is true of craft and its input from the fine arts.

But the unending desire to escape craft had, by the beginning of the 21st century, left the movement in tatters. Craft lost its flagship museum, the ACC had been moribund for over a decade, the market had fallen apart, education was shrinking and failing to produce young crafters, graduating multi-media sculptors. Craft today is completely overshadowed by design and is a less influential element of the visual arts than ever before.

While the "movement" died, craft itself lives on today and obviously the estimated 500,000 professional crafters in America have also survived. What is gone is certain idealism, a mission begun over a century ago to produce high craft that was the peer of high art.

To blame this on a single problem, art envy, may seem an exaggeration. But the evidence is compelling. More than any other single factor it poisoned the movement and brought it to its knees.

I had a unique vantage point from which to view this battle. My partner Mark Del Vecchio and I ran a gallery in Los Angeles and New York for twenty-seven years. We dealt in ceramics. Some of what we exhibited was unquestionably craft and we identified it unashamedly as such as such. But we also handled ceramics by Fontana, Caro, Noguchi and others that was fine art. This meant that we worked both sides of the art-craft divide. We showed art at the SOFA craft fairs as well as the blue chip art fair, The Art Show, organized by the exclusive American Art Dealers Association of America, of which we were members. Hence we were privy to the backroom arguments and gossip of both.

From 1980 onwards, the argument that craft was really art became fevered and relentless. It was blind to any logic, to a rational view of art history and to the opinions of the fine arts itself. Resistance to this notion was blamed on fine art's elitism but rarely did one hear the argument and simple truth that it was so because was craft was finally, and beneficially, different.

Two relatively small groups within the crafts, some leading artists and their collectors, drove the argument. Even though they were a minority within the community, they were a large percentage of its leadership; vocal, influential and driven. It would be nice to say that they were being selfless, motivated by a desire to upgrade the entire movement. But that is not true.

Hubris in both cases was the motivation. The makers wanted higher prices and more prestige and the collectors wanted their increasingly costly collections to be taken seriously by the fine arts. Far from wanting to improve the crafts, the real goal was to escape the field and let those left behind in low-rent craftsland survive as best they could. What it finally did was to push craft into to a bloody civil war against itself.

The ACC's Museum of Contemporary Craft in New York was the major battlefield and what happened there is the most instructive case study. But at the same time, on a smaller level, a hundred similar battles were taking place across America.

It began in earnest in the late 1970's when the Council was forced to look for patrons. Until then the New England bluestocking Mrs. Aileen Osborn Vanderbilt Webb had generously funded the organization, writing a check at the end of each year to cover the inevitable shortfall. When she ran out of funds the Council turned to collectors, the only pool of affluence in the craft world. These were the new collectors (the earlier variant was from a more modest economic class): captains of industry, retail mavens, property developers, financial wizards and hugely wealthy. But alas they were also often cheap, in part because some saw craft much as an undervalued property that could be gentrified, upgraded into art with a resultant increase in value and prestige.

This would give the collectors heightened stature as cultural czars, something they nakedly craved, without paying the same high entrance fees as in the fine arts. This may sound cynical but I was there and had hundreds of conversations with frustrated anguished collectors who were angry at being rejected by the fine arts establishment. There are many exceptions, collectors who viewed craft more realistically, but they did not prevail.

Once on the board the collectors took command. Firstly they imposed a corporate style makeover. Under the reign of board president, Ted Nieremburg (founder of Dansk Design), *Craft Horizons* was renamed *American Craft* and the Museum of Contemporary Craft became American Craft Museum. It all looked logical and sleekly organized on paper but at the price of creating a bland institutional character for an organization, which for all of its faults, had a lively, funky identity.

It was soon apparent that the trustee/collectors were less interested in the Council than its museum. This was the prize, right in the middle of New York City and opposite the Museum of Modern Art. They decided to build a new museum building, exchanging the two townhouses on 53rd street that had housed their museum and headquarters for a condominium museum in the new Deutsche bank building on the site of one of these brownstones.

It opened in 1986 and was impressive if one remained outside, admiring its two-story glass façade. The interior was dominated by a vast entrance, more than a third of the space, three stories deep, with a large arc of curving stairs, nicknamed the "staircase to nowhere" because it led to tiny, claustrophobic galleries. The staff was housed in a dank windowless basement. There was no coat check (essential in any New York public space) and bizarrely, they decided there would be no gift shop even though the Council, burdened by construction costs, was in a perilous financial state. (As a result in later years they had to create an ugly ad-hoc craft bazaar in the entrance foyer.) It was a dazzling minuet of missteps.

But the new home had a certain superficial glamour and now with a temple for their collections, the collectors had no need of the Council. They were disinterested in bedrock craft, the kind that was made by the bulk of the Council's membership. Indeed, this artisanal world was an embarrassment, a reminder of craft's peasant roots. Their investment was in craft that looked like art, and in their minds, given a rather primitive understanding of cultural politics, was art. So the museum sued for divorce.

In 1990, after a particularly vicious and acrimonious separation, they parted company. The Council moved to a lonely canyon on the fourth floor of an old industrial building in Soho and never recovered from this bruising and demoralizing fight. Council's spirit had been severely wounded and ACC lapsed into dormancy except for its two for profit enterprises, the magazine and the fairs.

With the Council out of the way the museum had only one stumbling block to get rid of, the name above their door. In 2002 it finally became the Museum of Arts and Design, which gave them their curiously beloved acronym MAD. The palace had finally defeated the cottage and craft officially became the art that dare not speak its name.

None of this turned craft into art. The more vigorously the Museum argued that craft was art, delivered with Palinesque bluster (I understand fine art because I can see a collection in an apartment across the street from mine), the more they exposed their ignorance. The art world saw (and talked of them) as philistines at the gate, which they kept tightly locked so the crafts could not enter. Philippe De Montebello the former director of the Metropolitan Museum of Art gave them a name, "the homeless ones," unable to live in the crafts, not allowed to reside in the arts. Behind the scenes the museum was becoming the laughing stock of New York's arts.

At the end of this war against itself (there was no attack from without) the field has ended up in a shambles. While the leadership was focused on its upward migration policy, the marketplace was slipping, education was failing, criticism had become a sham (it was almost impossible to write honestly about the field while it pretended to be something else) and crafts identity was fatally compromised. The field lost respect, credibility, direction, purpose and what had always been one of its sterling qualities, authenticity.

Apologists will tell you that as a result of these efforts, no matter how misplaced, craft is now more accepted in the fine arts. This is not true. Crafters are no more accepted today than decades before. What has changed is that craft materials, processes and contexts can now be used in art. An artist working in these materials will not be automatically ejected as in the past. But this was the result of the liberalizing impact of postmodernism and is promiscuous approach to means and matter, not the victory of crafts.

The fight resulted in distorted values. Craft became the only community outside the penitentiary to give its greatest respect to escapees.

Ken Price, Ron Nagle, Betty Woodman, Jun Kaneko, Richard DeVore are the field's most admired players. What they all have in common is that they "got away" and joined the fine arts. This became the ultimate goal leading craft to be viewed as purgatory, where weavers, potters, metalsmiths, glass blowers and jewelers waited impatiently to be allowed into nirvana.

Eventually, wearied by this artificial, self-loathing and thankless crusade, the craft movement weakened and died. I would place the time of death at or about 1995. Of course craft itself continues but this venerable old movement, with its desire to establish high craft as a peer of high art was failing the breath-on-the-mirror test. All its vital organs (its institutions) have failed or are failing and it had been reduced to the outer margins of American culture. Compared to art and design, craft is so marginalized that it is practically irrelevant.

[. . .]

Craft's demise has benefits. One can do things to a dead movement that one cannot do to a living one, such as taking a knife and saw and opening it to examine the viscera and organs. And when one does this, probing around looking for damage, the forensic examination produces some surprising results. Yes, craft did die from the toxicity of art envy, but other findings were more unexpected. It turns out that corpus craft also suffered from the aesthetic equivalent of advanced diabetes and dangerously hardened arteries.

Craft has been overdosing on nostalgia, the equivalent of sugar in art. This is craft's Achilles heel. It was born as a revivalist movement and these activities (like a historic house museum) are powered by nostalgia. Some degree of this "ye olde craftsman" romance is unavoidable in craft. Used with restraint it can add charm and a rich connection to the past.

But when it is overdone it turns into syrupy restoration village sentimentality. This is the reason why craft is so afflicted with cloying whimsy and saccharine cuteness.

Being hooked on nostalgia also seems to have stunted craft's ability to engage in a contemporary aesthetic, which is a more astringent approach. We have witnessed craft aesthetics becoming more regressive and anachronistic in this new century. Its own audiences are now complaining about this retreat into the depths of a romantic yesterday craft aesthetic. It is the natural impulse of the conservative, and craft is fundamentally conservative, to fall back on the past when challenged by the future. Art fairs report that buyers are looking for a fresher, younger vision that speaks of a new century and to a new generation of buyers. Craft seems to be incapable of delivering this reasonable expectation. The hardened arteries are caused by a different excess: academic influence. Craft, and this is the big surprise, may be the most academically dependent activity in the arts . . .

And lastly, the autopsy revealed clear indications of incest and resultant signs of severe brain damage. For decades the tradition in craft was to have a close friend, and fellow crafters write one's reviews. Crafters have written most of the books, curated the bulk of the exhibitions, organized the conferences. Little light was shone on craft from without, much to its detriment. Indeed, as arts movements go, craft is so inbred that it is just one cousin away from becoming a cyclops.

It could well be that all of this talk about art, envy, nostalgia, academicism and incest is moot. The American Craft Movement is roughly 129 years old, ancient by art movement standards where the average lifespan from inception to peak and decline is a mere seven years. Could one not in all good conscience write "natural causes" on the death certificate and let go?

There is a problem in taking this position. Craft has a twin from whom it was separated at birth and that twin, the same age, involved in the same issues of function and decoration, has never been healthier, more potent or more relevant. Why then did one thrive and the other fail?

Both came out of the Reform Movement's incubator, The Great Exhibition in London in 1851. This paean to industry raised considerable concern about the poor standard of machine produced design. While the craft movement made the decision to fight industry, its twin took a more prosaic view of the situation, realized that industry was going to triumph no matter what, and chose to fight from within. This produced the first generation of industrial design or as the field was known then, the applied arts. The means were different but the end had an identical purpose, devising gracious, intelligent objects for the home.

In the early 20th century Applied Art took on the name Modern Design and forged a close working relationship with both art and architecture without compromising its own identity. From the outset it had a good relationship with museums beginning in the 1930's when design became an active department within the Museum of Modern Art. Many progressive museums followed suit and opened design departments. Design's scholarship was, by and large excellent.

Design suffered less from hardened arteries because it was not as intimately connected to, and dependent on, the university system. Design had to live or die in the capitalist world where the tolerance for academic posturing was slight. Design was driven by a desire to be new, inventive and flexible, constantly adjusting to the desires of its audience and to changes in lifestyle. This is not just an intellectual position. Design had to connect with its market's needs or fail.

Nor is art envy a problem for design although there are signs that it might be in the early stages of this malaise. Design, as long as it kept to its own identity and purpose, was a welcome part of the art club so it had nothing to prove. And the relationship has grown cozier. Today major art galleries are increasingly including design in their exhibition programs. Gagosian Gallery in New York recently held an exhibition of furniture by Britain's design star, Marc Newson. Aside from the fact that the four week exhibition grossed $50 million, what was impressive was that Larry Gagosian, when interviewed about the event and asked "is design the new art," bluntly said that design did not need art to give it importance. If only craft had the same confidence.

Selling mass-produced furniture, ceramics, or Newson's distinctive sneakers did not produce Gagosian's sales. He offered Newson's limited edition furniture, tables, chairs and bookcases carved from solid blocks of marble. These sold from $350,00 to $750,000 each. Recently a limited edition of his now classic metal chaise sold for $1.4 million. These special editions are practically handmade and intrude on a market for higher priced handmade craft furniture that was once the sole purview of crafters like Wendell Castle.

Design is undermining the craft market at every level. It can deliver handsome ceramics, fabric and jewelry at low cost. It can produce work that to the average eye seems to be handcrafted and can program machines to produce objects that are to some extent, unique.

On the other hand, fine artists are working more frequently with craft materials. If one wants a sculpture in glass, wood, ceramics or fiber, one can get from an art gallery with the added advantage that it carries the imprimatur of an internationally known artist.

This pincer moment from both fine art and design, while not intended to kill craft, is doing a good job of making it redundant by crushing its market. Also, one cannot blame craft's slowing market on general economic conditions in the past ten pre-recession years. During that period more Americans spent unprecedented amounts on distinctive contemporary home furnishings, decoration and art than ever before. While this market waxed the interest in craft waned. This is sobering.

Now that the autopsy is complete what about [the future]?

First, do not try and bring this movement back to life. It will be about as much fun, and as pretty, as the craft version of *Night of the Living Dead*. Death in this case is a blessing, a mitzvah. The rot of death is the food for new life. Its demise presents an opportunity to rethink craft from the ground up.

Andrew Glasgow, the new director of the ACC, craft-smart and clear-eyed, has a great opportunity on his hands, to reinvent the Council. How can he give craft and his Council a new life? It all depends upon what he decides to jettison from its past and what he decides to keep.

Here is short list of possibilities. Let go of New York. Move the ACC to a small city where craft can have a higher profile and not have to scrape by like a struggling shade plant beneath Manhattan's interlinked design, fashion and fine art monoliths . . . Deal with two issues; one is develop together with working crafters, a viable new business model for the craft studio. The other, encourage craft into the 21st century aesthetically speaking. Then post a definition of craft that is accurate and unambiguous. State that craft is in the same fields as decorative arts and design. Sculptors (unless they make decorative sculptural objects) should not be welcome. They should live or die in the sculpture community, which they have for so long claimed to be their home while sheltering in the craft cottage. If you fail and if you learn from this experience and become a devout born-again crafter, you can always return.

Only accept members that are self-identifying. Make the new entity an unwelcoming place for failed sculptors to live. Create a place for traditional, or the term I prefer, classical craft, mainly rustic wares, which is not contemporary aesthetically yet deserves respect and a home.

Forge an alliance with design. This is the winning marriage, not the unhappy, fruitless stalking of the fine arts. Craft is too small and its institutions are too diminished to survive alone. It needs access to design's market clout and its highly developed infrastructure.

The Dutch noted the compatibility between craft and design some time ago. One of their best crafters, the jeweler Gijs Bakker, was a founder of the immensely influential Droog design movement, and he keeps a foot firmly planted in both worlds. The Dutch also came up with a great new name for craft, "Free Design" meaning that the crafter is released from the demands of industrial production. A free designer can make one-of-a-kind pieces or work in series without the pressure to sell millions of units or having to please the marketing department. They can also design for industry as well. This marriage will take craft into a more sophisticated and urbane world, removing its "little house on the prairie" blinkers . . .

For those who feel alienated from craft and yet prefer to work outside the fine arts, there is a healthier option than changing craft.

Recently a loosely defined Applied Art movement has begun in Europe. It's a mix of art and design and has many recovering ex-crafters as members. The aesthetic is mainly industrial and includes artists Marek Cecula in Poland and Barnaby Barford in England who deal with the transformation of domestic objects, taking their familiarity and placing them in a different critical context. Some of these "mutant housewives," my playful name for them, make applied art, fine art and design.

The Applied artists are confident, non-hierarchal and are mining a seam that is also being explored in the fine arts by artists like Timothy Horn with his jewelry on steroids, Cornelia Parker's steamrolled silver tea services and by Ai Weiwei with his demolition derby, destroying and mutating treasures such Sung furniture and seven thousand year old pots.

When I rejoice that the old craft's movement is dead ridding us of its mountains of heavy baggage, I am not being disrespectful. Craft's remains should not simply be tossed on the scrapheap of cultural detritus from the past century. There is much to salvage that is underrated and of immense value.

The period from 1945 up until 1980 is particularly golden. Think of the best work of Peter Voulkos, Wendell Castle, Albert Paley, Dale Chihuly and many others. (And I vouch only for their pre-1980 work, after that art envy brought about some ghastly art-wannabe objects into their oeuvre.) It was a period of extraordinary inventiveness, deep conviction and material magic that will only become more revered if we do our scholarly duty. And the best of this movement is unquestionably art, but it is craft art.

The continued legacy of these artists rests on how successfully and intelligently the craft community completes the scholarship surrounding this period. The fine arts will not do it for us. Nor will design. This is the field's sacred trust . . .

So there is life after death but only once craft becomes proud, confident and easy in its own skin. This will enable craft to get its horse in front of the cart. In 1979 the great art critic Clement Greenberg pointed out a failing in the crafts during his keynote address for the Ceramic Art Foundation's first international conference in 1979. He told the assembled delegates "you strike me as a group that is more concerned with opinion than achievement." If craft can finally reverse that imbalance, it will do just fine.

Craft is dead, long live craft.

FURTHER READING

Garth Clark, ed., *Ceramic Millennium: Critical Writings on Ceramic History, Theory and Art* (Halifax: Press of the Nova Scotia College of Art, 2006).

Garth Clark and Margie Hughto, *A Century of Ceramics in the United States 1878–1978* (New York: E. P. Dutton, 1975).

John Pagliaro, ed., *Shards: Garth Clark on Ceramic Art* (New York: D.A.P./Ceramic Arts Foundation, 2004).

SECTION 6

CRAFT IN ACTION: LIFE, ART, DESIGN

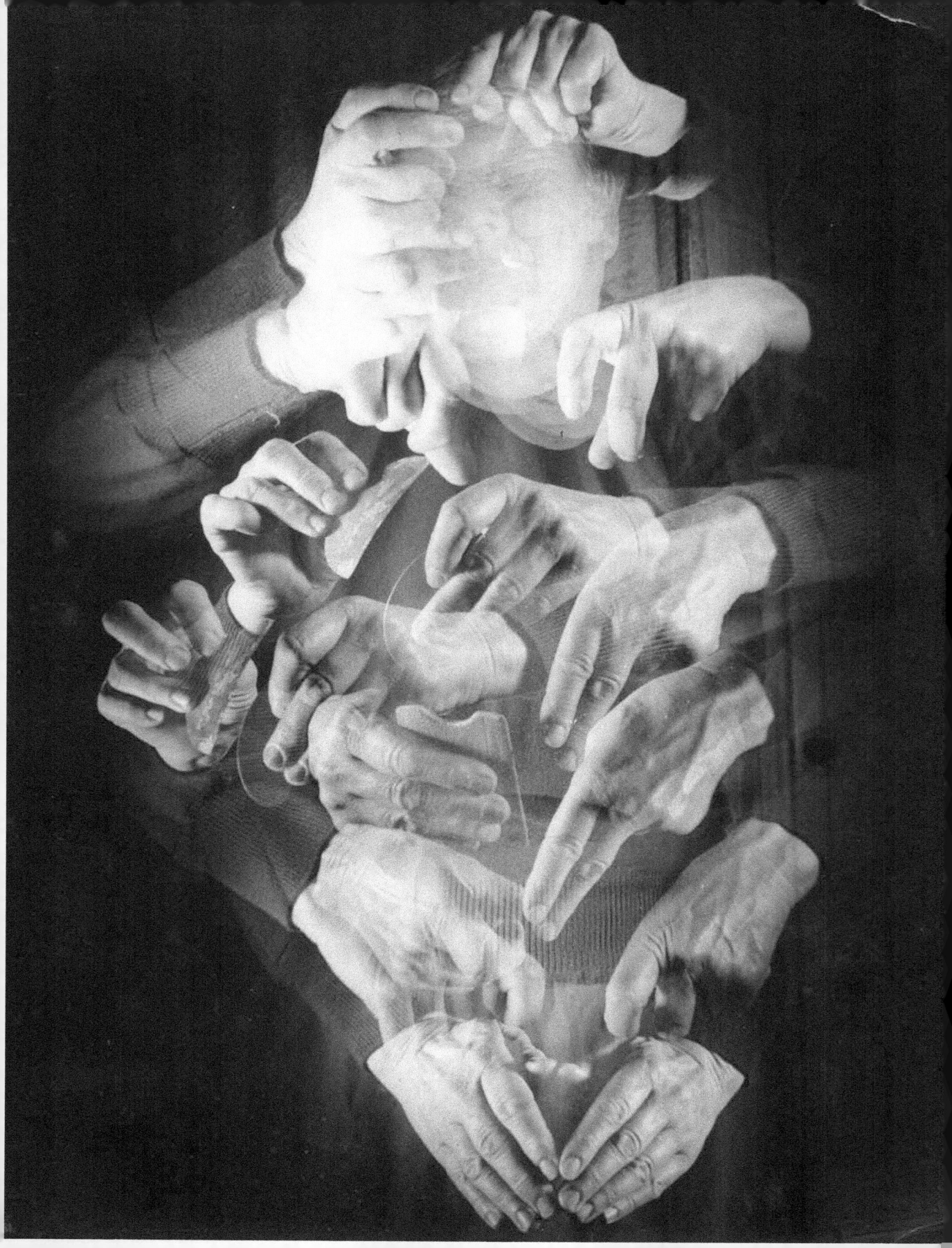

SECTION INTRODUCTION

Moving from theory to practice, this section of *The Craft Reader* looks at craft's importance to creative fields outside the 'craft movement' itself. The definitional questions that we encountered in the last section fall away here, as we turn to writers who value materials, process and skill only insofar as they can be applied in a given situation. The selections fall roughly into three groups, which deal with the role of craft in everyday life, contemporary art and industrial design.

One way to approach this section of the book is as a prehistory of our contemporary moment, which is frequently described as 'postdisciplinary' (see Section 7). From this perspective, terms like art, craft and design would have to be seen as historical points of reference rather than as ongoing categories of practice. This section traces a history of this open-endedness through an examination of three interrelated areas of practice, in which craft can be seen in action.

CRAFT AND THE EVERYDAY

For many contemporary observers, the main appeal of craft is its connection to the rhythms and realities of what has been called the 'everyday'. We might first associate this term with anthropology, which (unlike most types of history) studies not the exceptional and the historically significant, but rather the tacit, typical and quotidian. Understanding everyday experience—especially that in an unfamiliar culture—is a supremely difficult challenge. Linguistic difference and a distrust of outsiders present certain obstacles, but the key problem is that many core cultural assumptions are unspoken, taken for granted. Anthropologists have often looked to artisanal products as a way of getting around these problems, because they seem to make cultural beliefs concrete, but in a seemingly unselfconscious way.[1] Structuralist anthropologists such as Claude Lévi-Strauss could sometimes make this seem a mechanistic affair: the craft object carried religious belief and cultural ritual much in the manner of a diagram.[2] More recently, this view has been challenged, preeminently by the British anthropologist Alfred Gell, who sought to introduce 'agency' into the picture. The mentalities of the makers and users of a craft, he argued, had to be seen as dynamically related and mediated through the object. His writing revisits the ancient sense of craft as a means of achieving potency—it's worth recalling that our English word derives from the German *Kraft*, meaning power—and ascribes that power to the object itself, which he sees as an active agent in the process of social formation. The implication of Gell's work is that artisanal skill has no meaning, no value, outside of a specific context; craft is therefore as relative and variable a phenomenon as culture itself.

Today, when 'ethnographic' crafts are more likely to be encountered in a museum or a tourist shop than in a context that could be described as traditional, retaining a sense of that specificity is more crucial than ever.[3] This brings us to the question of what capitalism has done to everyday life, and thus to the more politically charged concept of the 'everyday' that was introduced by the philosopher Henri LeFebvre in his book *Critique de la Vie Quotidenne* (1958).[4] LeFebvre was writing from a Marxist perspective, and his investigations were intended as a way of grappling with the transformative processes of modernity. For him the everyday was itself a nineteenth-century creation, a product of the alienation wrought by capitalism, in the sense that it was only under these conditions that mundane experience became subject to unceasing manipulation. Everyday life was thoroughly, if subtly, reconditioned by the techniques of capital—preeminently the domination of standardized commodities within the private realm, and new ways of producing and reproducing power within public space. While LeFebvre wrote little about craft, he initiated an important trend in postwar Marxist thought by arguing that the seemingly trivial, unconscious actions of daily life such as sleeping, eating, walking, working, dressing and undressing—even if they were deadeningly repetitive and demeaning for many ordinary people—retained a possibility for radical change. Their intrinsic bodily character lent them an authenticity that the spectacular products and representations of the culture industry do not. If that authenticity could be reclaimed, it might be a source of revolutionary agency. Although he wrote little about craft per se, LeFebvre clearly regarded it as sharing in this potential. Like the idealist theorists of the nineteenth century he saw medieval guilds and peasant life as having had an integrated character sadly lacking in modern life. He was no revivalist and would have detested the top-down model of design reform as practiced by Morris or Ashbee. But there is a commonality between his ideas and the more political strands of thought within the Arts and Crafts Movement.

French Marxists writing in LeFebvre's wake, notably Michel DeCerteau and Guy DeBord, extended his thought by trying to imagine how everyday culture could be repurposed in the direction of revolutionary action.[5] Political transformation lay at the heart of their ideas, from DeCerteau's notion of 'tactics' (small-scale actions in which he saw the potential to subvert the 'strategies' used by governments and corporations) to the seemingly irrational proposals of DeBord's Situationist art movement (for example, to drift aimlessly through the city, getting lost on purpose as a way of detaching oneself from established social configurations). The Situationist slogan *sous les pavés, la plage* ('under the paving stones, the beach') is a good example of this thinking during the heady days of 1968, when revolution seemed imminent.[6] These ideas became widely influential within cultural studies of the 1980s. Theorists like Stuart Hall, Dick Hebdige and André Gorz also followed LeFebvre in conceding that 'everyday life has splintered into isolated pockets of time and space, a succession of excessive, aggressive demands, dead periods and periods of routine activity', and continued to see the quotidian as the critical sphere in which power relations could be expressed and contested. The workplace might be driven by the imperatives of capitalism, but dress, music and hobbies could be reimagined in subcultural terms as 'free self-activity with goals of its own'.[7]

The Marxist interest in everyday life as a seat of radicalism was a key precursor to Feminist thought as it has developed since the 1970s. LeFebvre pointed out that the 'generalized passivity' produced by modern consumer culture 'weighs more heavily on women, who are sentenced to everyday life'.[8] Many Feminists have come to the same conclusion, but also saw in that very repetitive mundanity a possible source of techniques for the disruption of patriarchy. Rita Felski, for example, has expanded upon LeFebvre's ideas by arguing that 'change is often imposed on individuals against their will; conversely, everyday rituals may help to safeguard a sense of personal autonomy and dignity . . . Repetition can signal resistance as well as enslavement'.[9] But Feminist writers have also departed in important ways from Marxist thinking about the everyday. While Marxist theory tends to imply that shared experiences of production will necessarily result in a shared politics, Feminism offers a theoretical matrix premised on the politics of difference. Within this 'oppositional consciousness', as Donna Haraway explains, there is no presumption that a woman and a man in the same job will have the same interests, or even that one woman's experience will necessarily map on to another's. (Even the category 'women', with its implication of a unity across space, time and ethnicity, is suspect from this perspective.) The Feminist challenge to Marxist analyses of production therefore involves thinking in terms of 'contradictory locations and heterochronic calendars, not about relativisms and pluralisms'.[10]

For authors such as the critic Lucy Lippard and the historian Rozsika Parker, craft history offered a route into such experiential particularities. Perhaps there were few great women artists in the canon, these authors pointed out, but there were countless great craftswomen and designers.[11] The fact that their quilts, samplers and handwoven baskets have traditionally not been accorded the same respect as the fine arts began to seem more a matter of sexism than aesthetics. The Feminist reevaluation of domestic crafts—which had traditionally been consigned, like women themselves, to the margins of history—has made them a vital subject of scholarship. Specificity is crucial here. Often, as in the article by design historian Carole Tulloch included here, it is achieved through oral histories, including interviews of the author's own family. This is best seen, perhaps, as a scholarly application of that other great counterculture slogan: 'the personal is the political'.

CRAFT IN CONTEMPORARY ART

The historian Ingrid Rowland has recently written that among the important developments from the early Feminist emphasis on craft is the 'setting the work of women painters and sculptors within a broader range of women's—or human—handiwork'.[12] And indeed, one of the many effects of the Feminist movement (the impact of which is only now beginning to be recognized) is the increasingly common usage within contemporary art of craft techniques, imagery and materials.[13] This is a vexed topic, because it inevitably raises the old hobbyhorses of craft versus art all over again. But things have definitely changed. In the 1960s, when Rose Slivka, the pioneering editor of *Craft Horizons*, began to advance the idea that craft materials could be used unprejudicially in a contemporary art context, she seemed to most of the magazine's readers to be either visionary or deluded. Now this viewpoint is almost taken for granted, at least in the context of art schools where tomorrow's leadings practitioners are training.

This momentous development has a long history. As John Roberts has recently written, one way of looking at twentieth-century art would be to see it as a triangulation between three modes of art-making: traditional artisanal skill (the sculptor with a chisel); deskilling (exemplified by Marcel Duchamp's Readymades, industrial products which were named as artworks rather than made by the artist); and reskilling (by which the artist might become an orchestrator of other hands, something like a film producer).[14] The importance of Roberts's argument is that he insists on the continuing relevance of all three of these modes, and the intricacies of their interrelation. The Readymade does not make skill obsolete but on the contrary opens up new configurations. As Roberts points out, Duchamp himself often combined the mechanically reproduced and the artisanally made within a single artwork. In the postwar period, artists as diverse as Jasper Johns, Carl Andre, Andy Warhol, Mary Kelly, Janine Antoni, Takeshi Murakami and Ai Weiwei have moved freely between skilled, deskilled and reskilled modes of production.[15] The message is that craft-based making has been a vital aspect of postwar and contemporary art, albeit always in a dynamic relation to other possibilities.

This multiplicity is certainly borne out by the readings in this section, which situate craft in relation to art in myriad ways but never with the directness that one customarily finds among craft's institutional supporters, who are often content to argue, simply, that 'craft is art'. The complexity of the situation is suggested by Philip Leider's essay 'What I Did On My Summer Vacation'. Just a few years after Slivka used the avant-garde, particularly Abstract Expressionism, as a stick to beat the conventionality of America ceramics, Leider—another pioneering magazine editor, this time at *Artforum*—pronounced himself antagonized by the conventionality of the avant-garde itself. He turned to everyday craft, as it was practiced in a Bay Area commune, as a possible alternative. Inspired by the 'woodbutcher' houses he saw in the hippie town of Canyon, he eloquently spoke of craft's appeal for an art world in search of political efficacy. Robert Morris, meanwhile, was often featured in *Artforum*'s pages and shared Leider's concern for art as a political instrument. His solution, however, was to make art as directly as possible, in a craftsmanly way. He therefore constructed his works entirely via the logic of process and materials. Finally, Lee Ufan, the leader of the *Mono-ha* art movement in Japan, writes of the contemporary artwork as something that mediates between a person and his surroundings—a notion that is reminiscent of Alfred Gell's anthropological view of craft as enchantment.

CRAFT IN INDUSTRIAL DESIGN

The 'designer-craftsman' movement of the immediate postwar period was rooted in the arguments of prewar Modernism. Despite the anticraft reputation that Modernists have somehow gained, most of them regarded handwork as existing on a continuum with Modernist industrial design, and sometimes as the beating heart at its center. The activities of Walter Gropius, László Moholy-Nagy, and their colleagues at the Bauhaus pointed to a working model by which craft served as a preparatory stage for mass production, a kind of laboratory where forms could be developed by people who had a real understanding of process and materials. Whether this role for craft was ever really viable is an open question, but it was certainly the subject of a great deal of optimistic discussion through the 1950s, as the readings selected for this section attest. Places such as such Germany, Italy, Japan and Scandinavia, all of which were incompletely industrialized at the end of World War II, all saw concerted attempts to deploy craft materials and techniques within a design context.[16] Sometimes these efforts were backed by institutions devoted to the purpose, such as the revived German Werkbund, the Slöjdföreningen in Sweden and many regional 'designer-craftsman' organizations in America.

In the 1960s, whatever promise the 'designer-craftsman' model held seemed to gradually slip away. On the one hand craftspeople became increasingly concerned with advancing claims to fine art status, and on the other industrial design was becoming an increasingly formalized profession. Companies relied less and less on external consultancies and developed their own in-house branding departments, squeezing out independent industrial designers and craftspeople alike. Furthermore, the ideology of the craft movement, with its respect for the traditional and the local, its generally left-wing politics and its suspicion of abstraction and artificiality, made working for industry seem increasingly distasteful. Even in the past few years, authors sympathetic to the crafts have argued that product design inevitably tends to lose touch with the humane and immediate work of the artisan. Howard Risatti, for example, argues that 'in the process of designing, the designer does not encounter directly the physical world of matter; no dialogical/dialectical process occurs, no give and take, as it were, between idea, form and matter through which they eventually come together as a fully fledged design object'.[17] The suggestion here is that craft and design, the cultures of the workshop and the corporation, are fundamentally opposed.[18]

The readings collected here, however, suggest otherwise. The process of designing, while it certainly involves semiosis (the production of meaning through signs), nonetheless always has a basis in material interaction. Drawing with mechanical pencils on tracing paper, modeling a prototype in clay or plaster, or (as we saw in Section 4 of this reader) even making a rendering using computer software could all be said to be crafts in their own right. Such material engagement may well be quite distinct from the material of the finished product, but the designerly crafts still have their own rhythms, skills and effects on the final product. This may be largely implicit when it comes to the latest redesign of a corporate logo, but that is really a matter of style and choice, not inevitability. Indeed, contemporary trends in graphic design draw on the intentional disjunctions of Postmodern design but employ hand-lettered text and other craft techniques to achieve a distinctive and personalized style.[19]

In furniture and product design too, there has been a steady stream of work since the early 1990s that achieves the look, and sometimes the productive reality, of the handmade. Tord Boontje, Ron Arad, Marc Newson, Fernando and Humberto Campana and especially the designers in the Dutch collective Droog could all be taken as examples of this trend.[20] They use various craftsy materials and techniques ranging from stitched-together rags and riveted metal plates to basketweaving, enamel and cut paper. Objects are typically sold in limited editions, with restricted access helping to maintain high prices and a sense of exclusiveness. Companies focusing on the manufacture and sale of this so-called 'Design Art', such as Established & Sons and Meta, function much in the same way that art fabricators do—marshalling highly trained (but typically anonymous) craftspeople to realize the vision of someone who has few if any of the skills and equipment necessary to make their own designs.[21] The marketplace success of such objects may be the result of changes in the economics of distribution, which allow a return to small-batch production; it may reflect a shift in the market for high-style design, which has now begun to approach the heady heights of contemporary art; or it may simply be a matter of diversification within the market, with the handmade carving out its niche through its stylistic opposition to overly refined, 'slick' computer-based work. It's hard to believe any of this is what Gropius and his colleagues at the Bauhaus had in mind, but the new alliance between craft and design seems to be here to stay—at least for a little while.

NOTES

1. In a well-known essay on the study of material culture, Jules Prown likened this process of reading an object to the analysis of a dream, which carries content without being consciously shaped by the dreamer. Jules Prown, 'Style as Evidence', *Winterthur Portfolio* 15/3 (1980), pp. 197–210.

2. On Levi-Strauss and crafted form see Thomas Crow, *The Intelligence of Art* (Chapel Hill: University of North Carolina Press, 1999).

3. For discussions of tourism, souvenirs and craft see Susan Stewart, *On Longing: Narratives of the Miniature, the Gigantic, the Souvenir, the Collection* (Baltimore: Johns Hopkins University Press, 1984); Dean MacCannell, *The Tourist: A New Theory of the Leisure Class* (Berkeley: University of California Press, 1976); Stephen Williams, *Tourism: Critical Concepts in the Social Sciences* (London: Taylor and Francis, 2004); Daniel J. Crowley, 'The Traditional Art Market in Southeast Asia', *African Arts* 16/4 (August 1983), pp. 65–8, 70; Emily Moore, 'The Silver Hand: Authenticating the Native Alaska Art, Craft and Body', *Journal of Modern Craft* 1/2 (July 2008), pp. 197–220; Kristina Dziedzic Wright, 'Cleverest of the Clever: Coconut Craftsmen in Lamu, Kenya', *Journal of Modern Craft* 1/3 (November 2008), p. xx.

4. Henri Lefebvre, *Critique of Everyday Life*, trans. John Moore (London: Verso, 1991; orig. pub. 1958). For an overview of this historiography, see Ben Highmore, *Everyday Life and Cultural Theory: An Introduction* (London: Routledge, 2002); for a general anthology see Ben Highmore, ed., *The Everyday Life Reader* (London: Routledge, 2001); for applications in the German context see Alf Lüdtke, *The History of Everyday Life: Reconstructing Historical Experiences and Ways of Life*, trans. William Templer (Princeton, NJ: Princeton University Press, 1985).

5. Michael de Certeau, *The Practice of Everyday Life*, trans. Steven Rendall (Berkeley: University of California Press, 1984; orig. pub. 1980); Guy Debord, *The Society of the Spectacle* (New York: Zone Books, 1995; orig. pub. 1967).

6. The slogan, first seen in the form of anonymous graffiti, also referred to the use of paving stones as projectiles in riots—when they were picked up to be thrown, they revealed a bed of sand beneath.

7. André Gorz, *Critique of Economic Reason* (London: Verso, 1988), pp. 91–3. See also Dick Hebdige, *Subculture: The Meaning of Style* (London: Routledge, 1979); Stuart Hall and Tony Jefferson, eds, *Resistance through Rituals: Youth Subcultures in Post-War Britain* (London: Routledge, 1993); Lawrence Grossberg, *Dancing in Spite of Myself: Essays on Popular Culture* (Durham, NC: Duke University Press, 1997).

8. Henri Lefebvre, 'The Everyday and Everydayness', trans. Christine Levich, *Yale French Studies* 73 (1987), pp. 7–11: 10.

9. Rita Felski, *Doing Time: Feminist Theory and Postmodern Culture* (New York: New York University Press, 2000), p. 84.

10. Donna Haraway, 'A Manifesto for Cyborgs: Science, Technology and Socialist Feminism in the 1980s', in Linda J. Nicholson, *Feminism/Postmodernism* (London: Routledge, 1990), p. 197. See also Chela Sandoval, *Methodology of the Oppressed* (Minneapolis: University of Minnesota Press, 2000).

11. Pat Kirkham, ed., *Women Designers in the USA, 1900–2000. Diversity and Difference* (New Haven, CT: Yale University Press in association with Bard Graduate Center, 2001).

12. Ingrid Rowland, 'Women Artists Win!', *New York Review of Books* (19 May 2008), pp. 26–29: 27.

13. Cornelia Butler and Lisa G. Mark, eds, *Wack! Art and the Feminist Revolution* (Los Angeles: Museum of Contemporary Art, 2007).

14. John Roberts, *The Intangibilities of Form: Skill and Deskilling in Art after the Readymade* (London: Verso, 2007). See also Molly Nesbit, 'Ready-Made Originals: The Duchamp Model', *October* 37 (Summer 1986), pp. 53–64.

15. For discussions of skill in the work of these and other artists, see Leo Steinberg, 'Jasper Johns: The First Seven Years of His Art', in Steinberg, *Other Criteria: Confrontations with Twentieth-Century Art* (New York: Oxford University Press, 1972); Caroline Jones, *Machine in the Studio: Constructing the*

Postwar American Artist (Chicago: University of Chicago Press, 1996); Sabine Breitweiser, ed., *Mary Kelly: Re-reading Post-Partum Document* (Vienna: Generali Foundation, 1999); Martha Buskirk, *The Contingent Object of Contemporary Art* (Cambridge, MA: MIT Press, 2003); Pamela M. Lee, 'Economies of Scale: Takashi Murkami's Technics', *Artforum* 46/2 (October 2007), pp. 336–43; *Ai Weiwei: Works, Beijing 1993–2003* (Beijing: Timezone 8 Ltd., 2003). For general overviews of 'reskilled' labor in postwar art see Helen Molesworth et al., *Work Ethic* (Baltimore: Baltimore Museum of Art/Penn State Press, 2003).

16. Penny Sparke, 'The Straw Donkey: Tourist Kitsch of Proto-Design? Craft and Design in Italy, 1945–1960', *Journal of Design History* 11/1 (1998), pp. 59–69; Yuko Kikuchi, 'Russel Wright and Japan: Bridging Japonisme and Good Design through Craft', *Journal of Modern Craft* 1/3 (November 2008), p. xx; Marianne Aav and Nina Stritzler-Levine, eds., *Finnish Modern Design: Utopian Ideals and Everyday Realities, 1930–1997* (New Haven, CT: The Bard Graduate Center for Studies in the Decorative Arts/Yale University Press, 1998).

17. Howard Risatti, *A Theory of Craft: Function and Aesthetic Expression* (Chapel Hill: University of North Carolina Press), p. 171.

18. This is also the position taken by Richard Sennett in *The Craftsman* (London: Allen Lane, 2008).

19. Robert Klanten, Sven Ehmann and Matthias Hubner, eds, *Tactile: High Touch Visuals* (London: Die Gestalten Verlag, 2007); Michael Perry, *Hand Job: A Catalogue of Type* (Princeton, NJ: Princeton Architectural Press, 2007).

20. Gareth Williams, *The Furniture Machine: Furniture Design Since 1990* (London: V&A Publications, 2006); Martina Margetts, *Tord Boontje* (New York: Rizzoli, 2007); *Home Made Holland: How Craft and Design Mix* (London: Crafts Council, 2002); Gijs Bakker and Renny Ramakers, eds, *Droog Design: Spirit of the Nineties* (Rotterdam: 010 Publishers, 1998).

21. Alex Coles, *Design and Art* (Cambridge, MA: MIT Press, 2007); Michelle Kuo, 'Industrial Revolution: The History of Fabrication', *Artforum* 46/2 (October 2007), pp. 306–15; Patsy Craig, ed., *The Mike Smith Studio: Making Art Work* (London: Trolley, 2003).

58

THE ENCHANTMENT OF TECHNOLOGY AND THE TECHNOLOGY OF ENCHANTMENT

Alfred Gell

The anthropologist Alfred Gell had many talents. His fieldwork was highly regarded by his peers; he was a witty and influential lecturer, spending most of his professorial life at the London School of Economics; and he was a brilliant stylist, so that even his most complex theoretical writings are a pleasure to read. But most of all, he was able to think from the ground up. His characteristic method was to infuse the simplest of observations—truisms such as 'art is a means of communication'—with unexpected depth and nuance. He died, tragically, just prior to the publication of his most influential book, Art and Agency, *but while working towards that manuscript, he wrote the following hugely imaginative essay, which anticipates some of the book's arguments in condensed form. Gell begins from the premise that if one wants to approach art anthropologically, one must not fall into the trap of believing in its culturally specific claims (just as if one cannot approach a religion anthropologically while believing in its teachings). One must therefore seek a culturally universal premise on which to build a theory. Gell concludes that this can only mean thinking about art in terms of its making. What gives an artwork its power within culture—within the milieu of everyday life—is, he argues, technique: a particular way of making that sets the work apart from the rest of daily experience. Gell applies this idea widely, moving from wood carvings in New Guinea to sculptures by Picasso, without losing any sense of their cultural specificity. While it would be a mistake to simply equate his idea of technique with craft, it is clear that the latter plays an important role in Gell's discussion. In many cultures, the skill of an artisan is literally seen as enchanted. One of the many lessons of his essay is that members of the modern and contemporary 'art cult' may not be as far removed from that mindset as we think.*

Alfred Gell, 'The Enchantment of Technology and the Technology of Enchantment', in J. Coote, ed., *Anthropology, Art, and Aesthetics* (Oxford: Oxford University Press, 1992).

INTRODUCTION: METHODOLOGICAL PHILISTINISM

The complaint is commonly heard that art is a neglected topic in present-day social anthropology, especially in Britain. The marginalization of studies of primitive art, by contrast to the immense volume of studies of politics, ritual, exchange, and so forth, is too obvious a phenomenon to miss, especially if one draws a contrast with the situation prevailing before the advent of [Bronislaw] Malinowski and [A. R.] Radcliffe-Brown. But why should this be so? I believe that it is more than a matter of changing fashions in the matter of selecting topics for study; as if, by some collective whim, anthropologists had decided to devote more time to cross-cousin marriage and less to mats, pots, and carvings. On the contrary, the neglect of art in modern social anthropology is necessary and intentional, arising from the fact that social anthropology is essentially, constitutionally, anti-art. This must seem a shocking assertion: how can anthropology, by universal consent a Good Thing, be opposed to art, also universally considered an equally Good Thing, even a Better Thing? But I am afraid that this is really so, because these two Good Things are Good according to fundamentally different and conflicting criteria.

When I say that social anthropology is anti-art, I do not mean, of course, that anthropological wisdom favours knocking down the National Gallery and turning the site into a car park. What I mean is only that the attitude of the art-loving public towards the contents of the National Gallery, the Museum of Mankind, and so on (aesthetic awe bordering on the religious) is an unredeemably ethnocentric attitude, however laudable in all other respects.

Our value-system dictates that, unless we are philistines; we should attribute value to a culturally recognized category of art objects. This attitude of aestheticism is culture-bound even though the objects in question derive from many different cultures, as when we pass effortlessly from the contemplation of a Tahitian sculpture to one by Brancusi, and back again. But this willingness to place ourselves under the spell of all manner of works of art, though it contributes very much to the richness of our cultural experience, is paradoxically the major stumbling-block in the path of the anthropology of art, the ultimate aim of which must be the dissolution of art, in the same way that the dissolution of religion, politics, economics, kinship, and all other forms under which human experience is presented to the socialized mind, must be the ultimate aim of anthropology in general.

Perhaps I can clarify to some degree the consequences of the attitude of universal aestheticism for the study of primitive[1] art by drawing a series of analogies between the anthropological study of art and the anthropological study of religion. With the rise of structural functionalism, art largely disappeared from the anthropological bill of fare in this country, but the same thing did not happen to the study of ritual and religious belief. Why did things happen this way? The answer appears to me to lie in an essential difference between the attitudes towards religion characteristic of the intelligentsia of the period, and their attitudes towards art.

It seems to me incontrovertible that the anthropological theory of religion depends on what has been called by Peter Berger 'methodological atheism' (Berger, 1967: 107). This is the methodological principle that, whatever the analyst's own religious convictions, or lack of them, theistic and mystical beliefs are subjected to sociological scrutiny on the assumption that they are not literally true. Only once this assumption is made do the intellectual maneuvers characteristic of anthropological analyses of religious systems become possible, that is, the demonstration of linkages between religious ideas and the structure of corporate groups, social hierarchies, and so on. Religion becomes an emergent property of the relations between the various elements in the social system, derivable, not from the condition that genuine religious truths exist, but solely from the condition that societies exist.

The consequences of the possibility that there are genuine religious truths lie outside the frame of reference of the sociology of religion.

These consequences—philosophical, moral, political, and so on—are the province of the much longer-established intellectual discipline of theology, whose relative decline in the modern era derives from exactly the same changes in the intellectual climate as have produced the current efflorescence of sociology generally and of the sociology of religion in particular. It is widely agreed that ethics and aesthetics belong in the same category. I would suggest that the study of aesthetics is to the domain of art as the study of theology is to the domain of religion. That is to say, aesthetics is a branch of moral discourse which depends on the acceptance of the initial articles of faith: that in the aesthetically valued object there resides the principle of the True and the Good, and that the study of aesthetically valued objects constitutes a path toward transcendence. In so far as such modern souls possess a religion, that religion is the religion of art, the religion whose shrines consist of theatres, libraries, and art galleries, whose priests and bishops are painters and poets, whose theologians are critics, and whose dogma is the dogma of universal aestheticism.

Unless I am very much mistaken, I am writing for a readership which is composed in the main of devotees of the art cult, and, moreover, for one which shares an assumption (by no means an incorrect one) that I too belong to the faith, just as, if we were a religious congregation and I were delivering a sermon, you would assume that I was no atheist.

If I were about to discuss some exotic religious belief-system, from the standpoint of methodological atheism, that would present no problem even to non-atheists, simply because nobody expects a sociologist of religion to adopt the premises of the religion he discusses; indeed, he is obliged not to do so. But the equivalent attitude to the one we take towards religious beliefs in sociological discourse is much harder to attain in the context of discussions of aesthetic values. The equivalent of methodological atheism in the religious domain would, in the domain of art, be *methodological philistinism* and that is a bitter pill very few would be willing to swallow. Methodological philistinism consists of taking an attitude of resolute indifference towards the aesthetic value of works of art—the aesthetic value that they have, either indigenously, or from the standpoint of universal aestheticism. Because to admit this kind of value is equivalent to admitting, so to speak, that religion is true, and just as this admission makes the sociology of religion impossible, the introduction of aesthetics (the theology of art) into the sociology or anthropology of art immediately turns the enterprise into something else. But we are most unwilling to make a break with aestheticism—much more so than we are to make a break with theology—simply because, as I have been suggesting, we have sacralized art: art is really our religion.

We can not enter this domain, and make it fully our own, without experiencing a profound dissonance, which stems from the fact that our method, were it to be applied to art with the degree of rigour and objectivity which we are perfectly prepared to contemplate when it comes to religion and politics, obliges us to deal with the phenomena of art in a philistine spirit contrary to our most cherished sentiments. I continue to believe, nonetheless, that the first step which has to be taken in devising an anthropology of art is to make a complete break with aesthetics. Just as the anthropology of religion commences with the explicit or implicit denial of the claims religions make on believers, so the anthropology of art has to begin with a denial of the claims which objects of art make on the people who live under their spell, and also on ourselves, in so far as we are all self-confessed devotees of the Art Cult.

But because I favour a break with the aesthetic preoccupations of much of the existing anthropology of art, I do not think that methodological philistinism is adequately represented by the other possible approaches: for instance, the sociologism of Bourdieu (e.g. 1968), which never actually looks at the art object itself, as a concrete product of human ingenuity, but only at its power to mark social distinctions, or the iconographic approach (e.g. Panofsky, 1962) which treats art as a species of writing, and which fails, equally, to take into consideration the presented object, rather than the represented symbolic meanings. I do not deny for an instant the discoveries of which these alternative approaches are capable; what I deny is only that they constitute the sought-for alternative to the aesthetic approach to the art object. We have, somehow, to retain the capacity of the aesthetic approach to illuminate the specific objective characteristics of the art object as an object, rather than as a vehicle for extraneous social and symbolic messages, without succumbing to the fascination which all well-made art objects exert on the mind attuned to their aesthetic properties.

ART AS A TECHNICAL SYSTEM

In this essay, I propose that the anthropology of art can do this by considering art as a component of technology. We recognize works of art, as a category, because they are the outcome of technical process, the sorts of technical process in which artists are skilled. A major deficiency of the aesthetic approach is that art objects are not the only aesthetically valued objects around: there are beautiful horses, beautiful people, beautiful sunsets, and so on; but art objects are the only objects around

which are *beautifully made*, or *made beautiful*. There seems every justification, therefore, for considering art objects initially as those objects which demonstrate a certain technically achieved level of excellence, 'excellence' being a function, not of their characteristics simply as objects, but of their characteristics as *made* objects, as products of techniques.

I consider the various arts—painting, sculpture, music, poetry, fiction, and so on—as components of a vast and often unrecognized technical system, essential to the reproduction of human societies, which I will be calling the technology of enchantment.

In speaking of 'enchantment' I am making use of a cover-term to express the general premise that human societies depend on the acquiescence of duly socialized individuals in a network of intentionalities whereby, although each individual pursues (what each individual takes to be) his or her own self-interest, they all contrive in the final analysis to serve necessities which cannot be comprehended at the level of the individual human being, but only at the level of collectivities and their dynamics. As a first approximation, we can suppose that the art-system contributes to securing the acquiescence of individuals in the network of intentionalities in which they are enmeshed. This view of art, that it is propaganda on behalf of the status quo, is the one taken by Maurice Bloch in his 'Symbols, Song, Dance, and Features of Articulation' (1974). In calling art the technology of enchantment I am first of all singling out this point of view, which, however one refines it, remains an essential component of an anthropological theory of art from the standpoint of methodological philistinism. However, the theoretical insight that art provides one of the technical means whereby individuals are persuaded of the necessity and desirability of the social order which encompasses them brings us no closer to the art object as such. As a technical system, art is orientated towards the production of the social consequences which ensue from the production of these objects. The power of art objects stems from the technical processes they objectively embody: the *technology of enchantment* is founded on the *enchantment of technology*. The enchantment of technology is the power that technical processes have of casting a spell over us so that we see the real world in an enchanted form. Art, as a separate kind of technical activity, only carries further, through a kind of involution, the enchantment which is immanent in all kinds of technical activity. The aim of my essay is to elucidate this admittedly rather cryptic statement.

PSYCHOLOGICAL WARFARE AND MAGICAL EFFICACY

Let me begin, however, by saying a little more about art as the technology of enchantment, rather than art as the enchantment of technology. There is an obvious prima-facie case for regarding a great deal of the art of the world as a means of thought-control. Sometimes art objects are explicitly intended to function as weapons in psychological warfare; as in the case of the canoe prow-board from the Trobriand Islands—surely a prototypical example of primitive art from the prototypical anthropological stamping-ground. The intention behind the placing of these prow-boards on Kula[2] canoes is to cause the overseas Kula partners of the Trobrianders, watching the arrival of the Kula flotilla from the shore, to take leave of their senses and offer more valuable shells or necklaces to the members of the expedition than they would otherwise be inclined to do. The boards are supposed to dazzle the beholder and weaken his grip on himself. And they really are

very dazzling, especially if one considers them against the background of the visual surroundings to which the average Melanesian is accustomed, which are much more uniform and drab than our own. But if the demoralization of an opponent in a contest of will-power is really the intention behind the canoe-board, one is entitled to ask how the trick is supposed to work. Why should the sight of certain colours and shapes exercise a demoralizing effect on anybody?

The first place one might seek an answer to such a question is in the domain of ethnology, that is, in innate, species-wide dispositions to respond to particular perceptual stimuli in predetermined ways. Moreover, were one to show such a board to an ethnologist, they would, without a doubt, mutter 'eye-spots!' and immediately start pulling out photographs of butterflies' wings, likewise marked with bold, symmetrical circles, and designed to have much the same effect on predatory birds as the boards are supposed to have on the Trobrianders' Kula partners, that is, to put them off their stroke at a critical moment. I think there is every reason to believe that human beings are innately sensitive to eye-spot patterns, as they are to bold tonal contrasts and bright colours, especially red, all of them features of the canoe-board design. These sensitivities can be demonstrated experimentally in the infant, and in the behavioural repertoire of apes and other mammals.

But one does not have to accept the idea of deep-rooted phylogenetic sensitivity to eyespot patterns and the like to find merit in the idea that the Trobriand canoe-board is a technically appropriate pattern for its intended purpose of dazzling and upsetting the spectator. The same conclusion can follow from an analysis of the *Gestalt* properties of the canoe-board design. If one makes the experiment of attempting to fixate the pattern for a few moments by staring at it, one begins to experience peculiar optical sensations due to the intrinsic instability of the design with its opposed volutes, both of which tend to lead the eye off in opposite directions.

In the canons of primitive art there are innumerable instances of designs which can be interpreted as exploiting the characteristic biases of human visual perception so as to ensnare us into unwitting reactions, some of which might be behaviourally significant. Should we, therefore, take the view that the significance of art, as a component of the technology of enchantment, derives from the power of certain stimulus arrays to disturb normal cognitive functioning? I recall that Ripley's *Believe It Or Not* (at one time my favourite book) printed a design which was claimed to hypnotize sheep: should this be considered the archetypal work of art? Does art exercise its influence via a species of hypnosis? I think not. Not because these disturbances are not real psychological phenomena; they are, as I have said, easily demonstrable experimentally. But there is no empirical support for the idea that canoe-boards, or similar kinds of art objects, actually achieve their effects by producing visual or cognitive disturbances. The canoe-board does not interfere seriously, if at all, with the intended victim's perceptual processes, but achieves its purpose in a much more roundabout way.

The canoe-board is a potent psychological weapon, but not as a direct consequence of the visual effects it produces. Its efficacy is to be attributed to the fact that these disturbances, mild in themselves, are interpreted as evidence of the magical power emanating from the board. It is this magical power which may deprive the spectator of his reason. If, in fact, he behaves with unexpected generosity,

it is interpreted as having done so. Without the associated magical ideas, the dazzlingness of the board is neither here nor there. It is the fact that an impressive canoe-board is a physical token of magical prowess on the part of the owner of the canoe which is important, as is the fact that he has access to the services of a carver whose artistic prowess is also the result of his access to superior carving magic.

THE HALO-EFFECT OF TECHNICAL 'DIFFICULTY'

And this leads on to the main point that I want to make. It seems to me that the efficacy of art objects as components of the technology of enchantment—a role which is particularly clearly displayed in the case of the Kula canoe—is itself the result of the enchantment of technology, the fact that technical processes, such as carving canoe-boards, are construed magically so that, by enchanting us, they make the products of these technical processes seem enchanted vessels of magical power. That is to say, the canoe-board is not dazzling as a physical object, but as a display of artistry explicable only in magical terms, something which has been produced by magical means. It is the way an art object is construed as having come into the world which is the source of the power such objects have over us—their becoming rather than their being.

Let me turn to another example of an art object which may make this point clearer. When I was about eleven, I was taken to visit Salisbury Cathedral. The building itself made no great impression on me, and I do not remember it at all. What I do remember, though, very vividly, is a display which the cathedral authorities had placed in some dingy side-chapel, which consisted of a remarkable model of Salisbury Cathedral, about two feet high and apparently complete in every detail, made entirely out of matchsticks glued together; certainly a virtuoso example of the matchstick modeller's art, if no great masterpiece according to the criteria of the salon, and calculated to strike a profound chord in the heart of any eleven-year-old. Matchsticks and glue are very important constituents of the world of every self-respecting boy of that age, and the idea of assembling these materials into such an impressive construction provoked feelings of the deepest awe. Most willingly I deposited my penny into the collecting-box which the authorities had, with a true appreciation of the real function of works of art, placed in front of the model, in aid of the Fabric Fund.

Wholly indifferent as I then was to the problems of cathedral upkeep, I could not but pay tribute to so much painstaking dexterity in objectified form. At one level, I had perfect insight into the technical problems faced by the genius who had made the model, having myself often handled matches and glue, separately and in various combinations, while remaining utterly at a loss to imagine the degree of manipulative skill and sheer patience needed to complete the final work. From a small boy's point of view this was the ultimate work of art, much more entrancing in fact than the cathedral itself, and so too, I suspect, for a significant proportion of the adult visitors as well.

Here the technology of enchantment and the enchantment of technology come together. The matchstick model, functioning essentially as an advertisement, is part of a technology of enchantment, but it achieves its effect via the enchantment cast by its technical means, the manner of its coming into being, or, rather, the idea which one forms of its coming into being, since making a matchstick model of

Salisbury Cathedral may not be as difficult, or as easy, as one imagines.

Simmel, in his treatise on the *Philosophy of Money* (1979: 62 ff.), advances a concept of value which can help us to form a more general idea of the kind of hold which art objects have over us. Roughly, Simmel suggests that the value of an object is in proportion to the difficulty which we think we will encounter in obtaining that particular thing rather than something else. We do not want what we do not think we will ever get under any set of circumstances deemed realizable. Simmel (ibid. 66) goes on to say:

> We desire objects only if they are not immediately given to us for our use and enjoyment, that is, to the extent to which they resist our desire. The content of our desire becomes an object as soon as it is opposed to us, not only in the sense of being impervious to us, but also in terms of its distance as something not yet enjoyed, the subject aspect of this condition being desire. As Kant has said: the possibility of experience is the possibility of objects of experience—because to have experiences means that our consciousness creates objects from sense-impressions. In the same way, the possibility of desire is the possibility of objects of desire. The object thus formed, which is characterised by its separation from the subject, who at the same time establishes it and seeks to overcome it by his desire, is for us a value.

He goes on to argue that exchange is the primary means employed in order to overcome the resistance offered by desired objects, which makes them desirable, and that money is the pure form of the means of engaging in exchange and realizing desire.

I am not here concerned with Simmel's ideas about exchange value and money; what I want to focus on is the idea that valued objects present themselves to us surrounded by a kind of halo-effect of resistance, and that it is this resistance to us which is the source of their value. Simmel's theory, as it stands, implies that it is difficulty of access to an object which makes it valuable, an argument which obviously applies, for example, to Kula valuables. But if we suppose that the value which we attribute to works of art, the bewitching effect they have on us, is a function, at least to some extent, of their characteristics as objects, not just of the difficulties we may expect to encounter in obtaining them, then the argument cannot be accepted in unmodified form. For instance, if we take up once again the instance of the matchstick model of Salisbury Cathedral, we may observe that the spell cast over me by this object was independent of any wish on my part to gain possession of it as personal property. In that sense, I did not value or desire it, since the possibility of possessing could not arise: no more am I conscious today of any wish to remove from the walls and carry away the pictures in the National Gallery. Of course, we do desire works of art, the ones in our price bracket, as personal property, and works of art have enormous significance as items of exchange. But I think that the peculiar power of works of art does not reside in the objects *as such*, and it is the objects as such which are bought and sold. Their power resides in the *symbolic* processes they provoke in the beholder, and these have *sui generis* characteristics which are independent of the objects themselves and the fact that they are owned and exchanged. The value of a work of art, as Simmel suggests, is a function of the way in which it resists us, but this 'resistance' occurs on two planes. If I am looking at an old master painting, which, I happen to know, has a saleroom value of two million pounds, then that certainly colours my reaction to it, and makes it more impressive than would be

the case if I knew that it was an inauthentic reproduction or forgery of much lesser value. But the sheer incommensurability between my purchasing power and the purchase price of an authentic old master means that I cannot regard such works as significant exchange items: they belong to a sphere of exchange from which I am excluded. But nonetheless such paintings are objects of desire—the desire to possess them in a certain sense, but not actually to own them. The resistance which they offer, and which creates and sustains this desire, is to being possessed in an intellectual rather than a material sense, the difficulty I have in mentally encompassing their coming-into-being as objects in the world accessible to me by a technical process which, since it transcends my understanding, I am forced to construe as magical.

THE ARTIST AS OCCULT TECHNICIAN

Let us consider, as a step up from the matchstick model of Salisbury Cathedral, J. F. Peto's *Old Time Letter Rack,* sometimes known as *Old Scraps,* the notoriously popular *trompe-l'oeil* painting, complete with artfully rendered drawing-pins and faded criss-cross ribbons, letters with still-legible, addressed envelopes to which lifelike postage stamps adhere, newspaper cuttings, books, a quill, a piece of string, and so on. This picture is usually discussed in the context of denunciations of the excesses of illusionism in nineteenth-century painting; but of course it is as beloved now as it ever was, and has actually gained prestige, not lost it, with the advent of photography, for it is now possible to see just how photographically real it is, and all the more remarkable for that. If it was, in fact, a colour photograph of a letter rack, nobody would give tuppence for it.

But just because it is a painting, one which looks as real as a photograph, it is a famous work, which, if popular votes counted in assigning value to paintings, would be worth a warehouse full of Picassos and Matisses.

The popular esteem in which this painting is held derives, not from its aesthetic merit, if any, since nobody would give what it represents (that is, a letter rack) a second glance. The painting's power to fascinate stems entirely from the fact that people have great difficulty in working out how coloured pigments (substances with which everybody is broadly familiar) can be applied to a surface so as to become an apparently different set of substances, namely, the ones which enter into the composition of letters, ribbons, drawing-pins, stamps, bits of string, and so on. The magic exerted over the beholder by this picture is a reflection of the magic which is exerted inside the picture, the technical miracle which achieves the transubstantiation of oily pigments into cloth, metal, paper, and feather. This technical miracle must be distinguished from a merely mysterious process: it is miraculous because it is achieved both by human agency but at the same time by an agency which transcends the normal sense of self-possession of the spectator.

Thus, the letter rack picture would not have the prestige it does have if it were a photograph, visually identical in colour and texture, could that be managed. Its prestige depends on the fact that it is a painting; and, in general, photography never achieves the popular prestige that painting has in societies which have routinely adopted photography as a technique for producing images. This is because the technical processes involved in photography are articulated to our notion of human agency in a way which is quite distinct from that in which we conceptualize the technical

processes of painting, carving, and so on. The alchemy involved in photography (in which packets of film are inserted into cameras, buttons are pressed, and pictures of Aunt Edna emerge in due course) are regarded as uncanny, but as uncanny processes of a natural rather than a human order, like the metamorphosis of caterpillars into butterflies. The photographer, a lowly button-presser, has no prestige, or not until the nature of his photographs is such as to make one start to have difficulties conceptualizing the processes which made them achievable with the familiar apparatus of photography.

In societies which are not over-familiar with the camera as a technical means, the situation is, of course, quite different. As many anthropologists who have worked under such conditions will have occasion to know, the ability to take photographs is often taken to be a special, occult faculty of the photographer, which extends to having power over the souls of the photographed, via the resulting pictures. We think this a naive attitude, when it comes to photography, but the same attitude is persistent, and acceptable, when it is expressed in the context of painting or drawing. The ability to capture someone's likeness is an occult power of the portraitist in paint or bronze, and when we wish to install an icon which will stand for a person—for example, a retiring director of the London School of Economics—we insist on a painted portrait, because only in this form will the captured essence of the no-longer-present Professor Dahrendorf continue to exercise a benign influence over the collectivity which wishes to eternalize him and, in so doing, derive continuing benefit from his *mana*.

Let me summarize my point about Peto's *Old Scraps* and its paradoxical prestige. The population at large both admire this picture and think that it emanates a kind of moral virtue, in the sense that it epitomizes what painters 'ought' to be able to do (that is, produce exact representations, or rather, occult transubstantiations of artists' materials into other things). It is thus a symbol of general moral significance, connoting, among other things, the fulfilment of the painter's calling in the Protestant-ethic sense, and inspiring people at large to fulfil their callings equally well. It stands for true artistry as a power both in the world and beyond it, and it promotes the true artist in a symbolic role as occult technician. Joined to this popular stereotype of the true artist is the negative stereotype of the false ('modern') artist of cartoon humour, who is supposed not to know how to draw, whose messy canvases are no better than the work of a child, and whose lax morality is proverbial.

Two objections can be made to the suggestion that the value and moral significance of works of art are functions of their technical excellence, or, more generally, to the importance of the fact that the spectator looks at them and thinks, 'For the life of me, I couldn't do that, not in a million years.' The first objection would be that *Old Scraps,* whatever its prestige among *hoi polloi,* cuts no ice with the critics, or with art-cultists generally. The second objection which might be raised is that, as an example of illusionism in art, the letter rack represents not only a particular artistic tradition (our own) but also only a brief interlude in that tradition, and hence can have little general significance. In particular, it cannot provide us with any insight into primitive art, since primitive art is strikingly devoid of illusionistic trickery.

The point I wish to establish is that the attitude of the spectator towards a work of art is fundamentally conditioned by his notion of the technical processes which gave rise to it, and

the fact that it was created by the agency of another person, the artist. The moral significance of the work of art arises from the mismatch between the spectator's internal awareness of his own powers as an agent and the conception he forms of the powers possessed by the artist. In reconstructing the processes which brought the work of art into existence, he is obliged to posit a creative agency which transcends his own and, hovering in the background, the power of the collectivity on whose behalf the artist exercised his technical mastery.

The work of art is inherently social in a way in which the merely beautiful or mysterious object is not: it is a physical entity which mediates between two beings, and therefore creates a social relation between them, which in turn provides a channel for further social relations and influences. This is so when, for instance, the court sculptor, by means of his magical power over marble, provides a physical analogue for the less easily realized power wielded by the king, and thereby enhances the king's authority. What Bernini can do to marble (and one does not know quite what or how) Louis XIV can do to you (by means which are equally outside your mental grasp). The man who controls such a power as is embodied in the technical mastery of Bernini's bust of Louis XIV is powerful indeed. Sometimes the actual artist or craftsman is quite effaced in the process, and the moral authority which works of art generate accrues entirely to the individual or institution responsible for commissioning the work, as with the anonymous sculptors and stained-glass artists who contributed to the glorification of the medieval church. Sometimes the artists are actually regarded with particular disdain by the power elite, and have to live separate and secluded lives, in order to provide ideological camouflage for the fact that theirs is the technical mastery which mediates the relation between the rulers and the ruled.

I maintain, therefore, that technical virtuosity is intrinsic to the efficacy of works of art in their social context, and tends always towards the creation of asymmetries in the relations between people by placing them in an essentially asymmetrical relation to things. But this technical virtuosity needs to be more carefully specified; it is by no means identical with the simple power to represent real objects illusionistically: this is a form of virtuosity which belongs, almost exclusively, to our art tradition (though its role in securing the prestige of old masters, such as Rembrandt, should not be underestimated). An example of virtuosity in non-illusionistic modern Western art is afforded by Picasso's well-known *Baboon and Young*, in which an ape's face is created by taking a direct cast from the body-shell of a child's toy car. One would not be much impressed by the toy car itself, nor by the verisimilitude of Picasso's ape just as a model of an ape, unless one were able to recognize the technical procedure Picasso used to make it, that is, commandeering one of his children's toys. But the witty transubstantiation of toy car into ape's face is not a fundamentally different operation from the transubstantiation of artists' materials into the components of a letter rack, which is considered quite boring because that is what artists' materials are for, generically. No matter what avant-garde school of art one considers, it is always the case that materials, and the ideas associated with those materials, are taken up and transformed into something else, even if it is only, as in the case of Duchamp's notorious urinal, by putting them in an art exhibition and providing them with a title *(Fountain)* and an author ('R. Mutt', alias M. Duchamp, 1917). Amikam Toren, one of the most ingenious contemporary artists, takes objects like chairs and teapots, grinds them up, and uses the resulting substances to create images of chairs and teapots. This is a less radical procedure than Duchamp's, which can be used effectively only once, but it is an equally apt means of directing our attention to the essential alchemy of art, which is to make what is not out of what is, and to make what is out of what is not.

THE FUNDAMENTAL SCHEME TRANSFER BETWEEN ART PRODUCTION AND SOCIAL PROCESS

But let us focus our attention on art production in societies without traditions and institutions of 'fine art' of the kind which nurtured Picasso and Duchamp. In such societies art arises particularly in two domains. The first of these is ritual, especially political ritual. Art objects are produced in order to be displayed on those occasions when political power is being legitimized by association with various supernatural forces. Secondly, art objects are produced in the context of ceremonial or commercial exchange. Artistry is lavished on objects which are to be transacted in the most prestigious spheres of exchange, or which are intended to realize high prices at market. The kind of technical sophistication involved is not the technology of illusionism but the technology of the radical transformation of materials, in the sense that the value of works of art is conditioned by the fact that it is difficult to get from the materials of which they are composed to the finished product. If we take up the example of the Trobriand canoe-board once more, it is clear that it is very difficult to acquire the art of transforming the root-buttress of an ironwood tree, using the rather limited tools which the Trobrianders have at their disposal, into such a smooth and refined finished product. If these boards could be simply cast in some plastic material, they would not have the same potency, even though they might be visually identical. But it is also clear that in the definition of technical virtuosity must be included considerations which might be thought to belong to aesthetics.

Let us consider the position of a Trobriand carver, commissioned to add one more to the existing corpus of canoe-boards. The carver does not only have the problem of physically shaping rather recalcitrant material with inadequate tools: the problem is also one of visualizing the design which he mentally follows in carving, a design which must reflect the aesthetic criteria appropriate to this art genre. He must exercise a faculty of aesthetic judgment, one might suppose, but this is not actually how it appears to the artist in the Trobriands who carves within a cultural context in which originality is not valued for its own sake, and who is supposed by his audience, and himself, to follow an ideal template for a canoe-board, the most magically efficacious one, the one belonging to his school of carving and its associated magical spells and rites. The Trobriand carver does not set out to create a new type of canoe-board, but a new token of an existing type; so he is not seeking to be original, but, on the other hand, he does not approach the task of carving as merely a challenge to his skill with the materials, seeing it, instead, primarily as a challenge to his mental powers. Perhaps the closest analogy would be with a musician in our culture getting technically prepared to give a perfect performance of an already existing composition, such as the 'Moonlight' Sonata.

Carvers undergo magical procedures which open up the channels of their minds so that the forms to be inscribed on the canoe-board will flow freely both in and out. Campbell, in an unpublished study of Trobriand (Vakuta) carving (1984), records that the final rite of carving initiation is the ingestion of the blood of a snake famed for its slipperiness. Throughout the initiation the emphasis is placed on ensuring free flow (of magical knowledge, forms, lines, and so on) by means of the metaphoric use of water and other liquids, especially blood and bespelled betel-juice. It is, of course, true that the Melanesian curvilinear carving style is dominated by an aesthetic of sinuous lines, well-represented in the canoe-board itself; but what for us is an aesthetic principle, one which we appreciate in the finished work, is from the carver's point of view a series of technical difficulties (or blockages of the flow) which he must overcome in order to carve well. In fact, one of the carver's initiatory rites represents just this: the master carver makes a little dam, behind which sea-water is trapped. After some magical to-do, the dam is broken and the water races back to the sea. After this, the initiate's mind will become quick and clear, and carving ideas will flow in similarly unimpeded fashion into his head, down his arms, out through his fingers, and into the wood.

We see here that the ability to internalize the carving style, to think up the appropriate forms, is regarded as a matter of the acquisition of a kind of technical facility, inseparable from the kind of technical facility which has to be mastered in order for these imagined forms to be realized in wood. Trobriand carving magic is technical-facility magic. The imaginative aspect of the art and the tool-wielding aspect of the art are one and the same. But there is a more important point to be made here about the magical significance of the art and the close relationship between this magical significance and its technical characteristics.

It will be recalled that these boards are placed on Kula canoes, their purpose being to induce the Kula partners of the Trobrianders to disgorge their best valuables, without holding any back, in the most expeditious fashion. Moreover, these and the other carved components of the Kula canoe (the prow-board, and the wash-board along the side) have the additional purpose of causing the canoe to travel swiftly through the water, as far as possible like the original flying canoe of Kula mythology.

Campbell, in her iconographic analysis of the motifs found on the carved components of canoes, is able to show convincingly that slipperiness, swift movement, and a quality glossed as 'wisdom' are the characteristics of the real and imaginary animals represented, often by a single feature, in the canoe art. A 'wise' animal, for instance, is the osprey, an omnipresent motif: the osprey is wise because it knows when to strike for fish, and captures them with unerring precision. It is the smooth, precise efficiency of the osprey's fish-getting technique which qualifies it to be considered wise, not the fact that it is knowledgeable. The same smooth and efficacious quality is desired for the Kula expedition. Other animals, such as butterflies and horseshoe bats, evoke swift movement, lightness, and similar ideas. Also represented are waves, water, and so on.

The success of the Kula, like the success of the carving, depends on unimpeded flow. A complex series of homologies, of what Bourdieu (1977) has called 'scheme transfers', exists between the process of overcoming the technical obstacles which stand in the way of the achievement of a perfect 'performance' of the canoe-board carving and the overcoming of the technical obstacles, as much psychic as physical, which stand in the way of the achievement of a successful Kula expedition. Just as

carving ideas must be made to flow smoothly into the carver's mind and out through his fingers, so the Kula valuables have to be made to flow smoothly through the channels of exchange, without encountering obstructions. And the metaphoric imagery of flowing water, slippery snakes, and fluttering butterflies applies in both domains, as we have seen.

We saw earlier that it would be incorrect to interpret the canoe-board ethologically as an eye-spot design or, from the standpoint of the psychology of visual perception, as a visually unstable figure, not because it is not either of these things (it is both) but because to do so would be to lose sight of its most essential characteristic, namely, that it is an object which has been made in a particular way. That is, it is not the eye-spots or the visual instabilities which fascinate, but the fact that it lies within the artist's power to make things which produce these striking effects. We can now see that the technical activity which goes into the production of a canoe-board is not only the source of its prestige as an object, but also the source of its efficacy in the domain of social relations; that is to say, there is a fundamental scheme transfer, applicable, I suggest, in all domains of art production, between technical processes involved in the creation of a work of art and the production of social relations via art. In other words, there exists a homology between the technical processes involved in art, and technical processes generally, each being seen in the light of the other, as, in this instance, the technical process of creating a canoe-board is homologous to the technical processes involved in successful Kula operations. We are inclined to deny this only because we are inclined to play down the significance of the technical domain in our culture, despite being utterly dependent on technology in every department of life.

Technique is supposed to be dull and mechanical, actually opposed to true creativity and authentic values of the kind art is supposed to represent. But this distorted vision is a by-product of the quasi-religious status of art in our culture, and the fact that the art cult, like all other cults, is under a stringent requirement to conceal its real origins, as far as possible.

THE ENCHANTMENT OF TECHNOLOGY: MAGIC AND TECHNICAL EFFICACY

But just pointing to the homology between the technical aspect of art production and the production of social relations is insufficient in itself, unless we can arrive at a better understanding of the relation between art and magic, which in the case of Trobriand canoe art is explicit and fundamental. It is on the nature of magical thought, and its relation to technical activity, including the technical activity involved in the production of works of art, that I want to focus in the last part of this essay.

Art production and the production of social relations are linked by a fundamental homology: but what are social relations? Social relations are the relations which are generated by the technical processes of which society at large can be said to consist, that is, broadly, the technical processes of the production of subsistence and other goods, and the production (reproduction) of human beings by domesticating them and breeding them. Therefore, in identifying a homology between the technical processes of art production and the production of social relations, I am not trying to say that the technology of art is homologous to a domain which is not, itself, technological, for social relations are themselves emergent characteristics of the technical base on which society rests. But it would be misleading to suggest that, because societies rest on a technical base, technology is a cut-and-dried affair which everybody concerned understands perfectly.

Let us take the relatively uncontentious kind of technical activity involved in gardening—uncontentious in that everybody would admit this is technical activity, an admission they might not make if we were talking about the processes involved in setting up a marriage. Three things stand out when one considers the technical activity of gardening: firstly, that it involves knowledge and skill, secondly, that it involves work, and thirdly, that it is attended by an uncertain outcome, and moreover depends on ill-understood processes of nature. Conventional wisdom would suggest that what makes gardening count as a technical activity is the aspect of gardening which is demanding of knowledge, skill, and work, and that the aspect of gardening which causes it to be attended with magical rites, in pre-scientific societies, is the third one, that is, its uncertain outcome and ill-understood scientific basis.

But I do not think things are as simple as that. The idea of magic as an accompaniment to uncertainty does not mean that it is opposed to knowledge, i.e. that where there is knowledge there is no uncertainty, and hence no magic. On the contrary, what is uncertain is not the world but the knowledge we have about it. One way or another, the garden is going to turn out as it turns out; our problem is that we don't yet know how that will be. All we have are certain more-or-less hedged beliefs about a spectrum of possible outcomes, the more desirable of which we will try to bring about by following procedures in which we have a certain degree of belief, but which could equally well be wrong, or inappropriate in the circumstances. The problem of uncertainty is, therefore, not opposed to the notion of knowledge and the pursuit of rational technical solutions to technical problems, but is inherently a part of it. If we consider that the magical attitude is a by-product of uncertainty, we are thereby committed also to the proposition that the magical attitude is a by-product of the rational pursuit of technical objectives using technical means.

MAGIC AS THE IDEAL TECHNOLOGY

But the relationship between technical processes and magic does not only come about because the outcome of technical endeavours is doubtful and results from the action of forces in nature of which we are partially or wholly ignorant. Work itself, mere labour, calls into being a magical attitude, because labour is the subjective cost incurred by us in the process of putting techniques into action. If we return to Simmel's ideas that 'value' is a function of the resistance which has to be overcome in order to gain access to an object, then we can see that this 'resistance' or difficulty of access can take two forms: (i) the object in question can be difficult to obtain, because it has a high price at market or because it belongs to an exalted sphere of exchange; or (ii) the object can be difficult to obtain because it is hard to produce, requiring a complex and chancy technical process, and/or a technical procedure which has high subjective opportunity costs, i.e. the producer is obliged to spend a great deal of time and energy producing that particular product, at the expense of other things he might produce or the employment of his time and resources in more subjectively agreeable leisure activities. The notion of 'work' is the standard we use to measure the opportunity

cost of activities such as gardening, which are engaged in, not for their own sake, but to secure something else, such as an eventual harvest. In one sense, gardening for a Trobriander has no opportunity cost, because there is little else that a Trobriander could conceivably be doing. But gardening is still subjectively burdensome, and the harvest is still valuable because it is difficult to obtain. Gardening has an opportunity cost in the sense that gardening might be less laborious and more certain in its outcome than it actually is. The standard for computing the value of a harvest is the opportunity cost of obtaining the resulting harvest, not by the technical, work-demanding means that are actually employed, but effortlessly, by magic. All productive activities are measured against the magic-standard, the possibility that the same product might be produced effortlessly, and the relative efficacy of techniques is a function of the extent to which they converge towards the magic-standard of zero work for the same product, just as the value to us of objects in the market is a function of the relation between the desirability of obtaining those objects at zero opportunity cost (alternative purchases forgone) and the opportunity costs we will actually incur by purchasing at the market price.

If there is any truth in this idea, then we can see that the notion of magic, as a means of securing a product without the work-cost that it actually entails, using the prevailing technical means, is actually built into the standard evaluation which is applied to the efficacy of techniques, and to the computation of the value of the product. Magic is the baseline against which the concept of work as a cost takes shape. Actual Kula canoes (which have to be sailed, hazardously, laboriously, and slowly, between islands in the Kula ring) are evaluated against the standard set by the mythical flying canoe, which achieves the same results instantly, effortlessly, and without any of the normal hazards. In the same way, Trobriand gardening takes place against the background provided by the litanies of the garden magician, in which all the normal obstacles to successful gardening are made absent by the magical power of words. Magic haunts technical activity like a shadow; or, rather, magic is the negative contour of work, just as, in Saussurean linguistics, the value of a concept (say, 'dog') is a function of the negative contour of the surrounding concepts ('cat', 'wolf', 'master').

Just as money is the ideal means of exchange, magic is the ideal means of technical production. And just as money values pervade the world of commodities, so that it is impossible to think of an object without thinking at the same time of its market price, so magic, as the ideal technology, pervades the technical domain in pre-scientific societies.[3]

It may not be very apparent what all this has got to do with the subject of primitive art. What I want to suggest is that magical technology is the reverse side of productive technology, and that this magical technology consists of representing the technical domain in enchanted form. If we return to the idea, expressed earlier, that what really characterizes art objects is the way in which they tend to transcend the technical schemas of the spectator, his normal sense of self-possession, then we can see that there is a convergence between the characteristics of objects produced through the enchanted technology of art and objects produced via the enchanted technology of magic, and that, in fact, these categories tend to coincide. It is often the case that art objects are regarded as transcending the technical schemas of their creators, as well as those of mere spectators, as when the art object is

considered to arise, not from the activities of the individual physically responsible for it, but from the divine inspiration or ancestral spirit with which he is filled. We can see signs of this in the fact that artists are not paid for 'working' for us, in the sense in which we pay plumbers for doing so. The artists' remuneration is not remuneration for his sweat, any more than the coins placed in the offertory plate at church are payments to the vicar for his praying on behalf of our souls. If artists are paid at all, which is infrequently, it is as a tribute to their moral ascendancy over the lay public, and such payments mostly come from public bodies or individuals acting out the public role of patrons of the arts, not from selfishly motivated individual consumers. The artist's ambiguous position, half-technician and half-mystagogue, places him at a disadvantage in societies such as ours, which are dominated by impersonal market values. But these disadvantages do not arise in societies such as the Trobriands, where all activities are simultaneously technical procedures and bound up with magic, and there is an insensible transition between the mundane activity which is necessitated by the requirements of subsistence production and the most overtly magico-religious performances.

THE TROBRIAND GARDEN AS A COLLECTIVE WORK OF ART

The interpenetration of technical productive activity, magic, and art, is wonderfully documented in Malinowski's *Coral Gardens and Their Magic* (1935). Malinowski describes the extraordinary precision with which Trobriand gardens, having been cleared of scrub, and not only scrub, but the least blade of grass, are meticulously laid out in squares, with special structures called 'magical prisms' at each corner, according to a symmetrical pattern which has nothing to do with technical efficiency, and everything to do with achieving the transcendence of technical production and a convergence towards magical production. Only if the garden looks right will it grow well, and the garden is, in fact, an enormous collective work of art. Indeed, if we thought of the quadrangular Trobriand garden as an artist's canvas on which forms mysteriously grow, through an occult process which lies partly beyond our intuition, that would not be a bad analogy, because that is what happens as the yams proliferate and grow, their vines and tendrils carefully trained up poles according to principles which are no less 'aesthetic' than those of the topiarist in the formal gardens of Europe.[4]

The Trobriand garden is, therefore, both the outcome of a certain system of technical knowledge and at the same time a collective work of art, which produces yams by magic. The mundane responsibility for this collective work of art is shared by all the gardeners, but on the garden magician and his associates more onerous duties are imposed. We would not normally think of the garden magician as an artist, but from the point of view of the categories operated by the Trobrianders, his position is exactly the same, with regard to the production of the harvest, as the carver's position is with regard to the canoe-board, i.e. he is the person magically responsible, via his ancestrally inherited *sopi* or magical essence.

The garden magician's means are not physical ones, like the carver's skill with wood and tools, except that it is he who lays out the garden originally and constructs (with a good deal of effort, we are told) the magic prisms at the corners. His art is exercised through his speech. He is master of the verbal poetic art, just as the carver is master of the use of visual metaphoric forms (ospreys, butterflies,

waves, and so on). It would take too long, and introduce too many fresh difficulties, to deal adequately with the tripartite relationship between language (the most fundamental of all technologies), art, and magic. But I think it is necessary, even so, to point out the elementary fact that Trobriand spells are poems, using all the usual devices of prosody and metaphor, about ideal gardens and ideally efficacious gardening techniques. Malinowski (1935: i. 169) gives the following ('Formula 27'):

I

Dolphin here now, dolphin here ever!
Dolphin here now, dolphin here ever!
Dolphin of the south-east, dolphin of the
  north-west.
Play on the south-east, play on the north-west,
  the dolphin plays!
The dolphin plays!

II

The dolphin plays!
About my *kaysalu*, my branching support, the
  dolphin plays.
About my *kaybudi*, my training stick that
  leans, the dolphin plays.
About my *kamtuya*, my stem saved from the
  cutting, the dolphin plays. About my *tala*, my
  partition stick, the dolphin plays.
About my *yeye'i*, my small slender support, the
  dolphin plays. About my *tamkwaluma*, my
  light yam pole, the dolphin plays. About my
  kavatam, my strong yam pole, the dolphin
  plays.
About my *kayvaliluwa*, my great yam pole, the
  dolphin plays. About my *tukulumwala*, my
  boundary line, the dolphin plays. About my
  karivisi, my boundary triangle, the dolphin
  plays.
About my *kamkokola*, my magical prism, the
  dolphin plays.
About my *kaynutatala*, my unchanned prisms,
  the dolphin plays.

III

The belly of my garden leavens, The belly of
my garden rises,

The belly of my garden reclines,

The belly of my garden grows to the size of a
bush hen's nest, The belly of my garden grows
like an ant-hill,

The belly of my garden rises and is bowed
down,

The belly of my garden rises like the iron-wood
palm, The belly of my garden lies down,

The belly of my garden swells,

The belly of my garden swells as with a child.

and comments (1935: ii. 310–11):

> the invocation of the dolphin . . . transforms, by
> a daring simile, the Trobriand garden, with its
> foliage swaying and waving in the wind, into a
> seascape . . . Bagido'u [the magician] explained
> to me . . . that as among the waves the dolphin
> goes in and out, up and down, so throughout
> the garden the rich garlands at harvest will wind
> over and under, in and out, of the supports.

It is clear that not only is this hymn to super-
abundant foliage animated by the poetic devices
of metaphor, antithesis, arcane words, and so
on, all meticulously analysed by Malinowski,
but that it is also tightly integrated with the
catalogue of sticks and poles made use of in the
garden, and the ritually important constructions, the magic prisms and boundary triangles
which are also found there. The garden magician's technology of enchantment is the reflex
of the enchantment of technology. Technology
is enchanted because the ordinary technical
means employed in the garden point inexorably towards magic, and also towards art, in that
art is the idealized form of production. Just as
when, confronted with some masterpiece, we
are fascinated because we are essentially at a loss
to explain how such an object comes to exist in
the world, the litanies of the garden magician

express the fascination of the Trobrianders with
the efficacy of their actual technology which,
converging towards the magical ideal, adum-
brates this ideal in the real world.

NOTES

1. 'Non-Western' has been suggested to me as
   a preferable alternative to 'primitive' in this
   context. But this substitution can hardly be
   made, if only because the fine-art traditions of
   Oriental civilizations have precisely the characteristics which 'primitive' is here intended
   to exclude, but cannot possibly be called
   'Western'. I hope the reader will accept the
   use of 'primitive' in a neutral, non-derogatory
   sense in the context of this essay. It is worth
   pointing out that the Trobriand carvers who
   produce the primitive art discussed in this
   essay are not themselves at all primitive; they
   are educated, literate in various languages,
   and familiar with much contemporary technology. They continue to fabricate primitive
   art because it is a feature of an ethnically
   exclusive prestige economy which they have
   rational motives for wishing to preserve.

2. The Kula is a system of ceremonial exchanges
   of valuables linking together the island communities of the Massim district, to the east of
   the mainland of Papua New Guinea (see Malinowski, 1922; Leach and Leach, 1983). Kula
   participants (all male) engage in Kula expedi-
   tions by canoe to neighbouring islands, for the
   purpose of exchanging two types of traditional
   valuable, necklaces and arm-shells, which may
   only be exchanged for one another. The Kula
   system assumes the form of a ring of linked
   island communities, around which necklaces
   circulate in a clockwise direction. Kula men
   compete with other men from their own community to secure profitable Kula partnerships
   with opposite numbers in overseas communities in either direction, the object being to
   maximize the volume of transactions passing
   through one's own hands. Kula valuables are

not hoarded; it is sufficient that it should become public knowledge that a famous valuable has, at some stage, been in one's possession. A man who has succeeded in 'attracting' many coveted valuables becomes famous all around the Kula ring (see Munn, 1986).

3. In technologically advanced societies where different technical strategies exist, rather than societies like the Trobriands where only one kind of technology is known or practicable, the situation is different, because different technical strategies are opposed to one another, rather than being opposed to the magic-standard. But the technological dilemmas of modern societies can, in fact, be traced to the pursuit of a chimera which is actually the equivalent of the magic-standard: ideal 'costless' production. This is actually not costless at all, but the minimization of costs to the corporation by the maximization of social costs which do not appear on the balance sheet, leading to technically generated unemployment, depletion of unrenewable resources, degradation of the environment, etc.

4. In the Sepik, likewise, that growing of long yams is an art-form, and not just metaphorically, because the long yam can be induced to grow in particular directions by careful manipulation of the surrounding soil: it is actually a form of vegetable sculpture (see Forge, 1966).

REFERENCES

Berger, Peter (1967). *The Social Reality of Religion.* Harmondsworth, Middlesex: Penguin.

Bloch, Maurice (1974). 'Symbols, Song, Dance, and Features of Articulation: Is Religion an Extreme Form of Traditional Authority?', *Archives Européennes de Sociologie*, 15/1: 55–81.

Bourdieu, Pierre (1968). 'Outline of a Sociological Theory of Art Perception', *International Social Science Journal*, 20/4: 589–612.

—— (1977). *Outline of a Theory of Practice.* Cambridge: Cambridge Univ. Press.

Campbell, Shirley (1984). 'The Art of the Kula'. Ph.D. thesis, Australian National Univ., Canberra.

Forge, Anthony (1966). 'Art and Environment in the Sepik', *Proceedings of the Royal Anthropological Institute for 1965.* London: Royal Anthropological Institute, 23–31.

Leach, Jerry W., and Leach, Edmund (1983). *The Kula: New Perspectives on Massim Exchange.* Cambridge: Cambridge Univ. Press.

Malinowski, Bronislaw (1922). *Argonauts of the Western Pacific: An Account of Native Enterprise and Adventure in the Archipelagos of Melanesian New Guinea.* London: Routledge.

—— (1935). *Coral Gardens and their Magic. A Study of the Methods of Tilling the Soil and of Agricultural Rites in the Trobriand Islands.* 2 vols. London: Allen & Unwin.

Munn, Nancy (1986). *The Fame of Gawa: A Symbolic Study of Value Transformation in a Massim (Papua New Guinea) Society.* Cambridge: Cambridge Univ. Press.

Panofsky, Erwin (1962). *Studies in Iconology: Humanistic Themes in the Art of the Renaissance.* New York: Harper & Row.

Simmel, Georg (1979). *The Philosophy of Money.* Boston: Routledge & Kegan Paul.

FURTHER READING

Mary Butcher, 'Eel Traps Without Eels', *Journal of Design History* 10/4 (1997), pp. 417–29.

Alfred Gell, *Art and Agency: An Anthropological Theory* (Oxford: Clarendon Press, 1998).

Alfred Gell, 'Vogel's Net: Traps as Artworks and Artworks as Traps', *Journal of Material Culture* 1 (1996), pp. 15–38.

Francesco Pellizzi, ed., *Res: Anthropology and Aesthetics* 36. Thematic issue on 'Factura' (Autumn 1999).

59

MAKING SOMETHING FROM NOTHING (TOWARD A DEFINITION OF WOMEN'S 'HOBBY ART')

Lucy Lippard

The critic Lucy Lippard has navigated the twists and turns of art for over four decades without losing her open-mindedness, writerly skill and unflagging sense of political engagement. She first gained prominence as an observer of the New York City avant-garde, publishing short pieces in the art press and curating groundbreaking exhibitions such as 'Eccentric Abstraction' (1966), a prescient gathering of artists like Eva Hesse and Barry LeVa who were later grouped as Postminimalists. Her Six Years: The Dematerialization of the Art Object *(1973), one of the first books on conceptual art, remains an invaluable survey of that movement. In ensuing years, Lippard emerged as the key critical voice of the Feminist movement. She was among the first to propose that there were characteristic qualities in women's art—'a uniform density, or overall texture, often sensuously tactile or detailed to the point of obsession'.[1] Still later, she would distance herself from this 'essentialist' position, championing a multicultural Postmodernism in art. As major art fairs have sprouted up in Asia, Africa and Latin America, her early calls for a decentered art world have come to seem prescient. The following essay, originally published in the Feminist journal* Heresies *(which she had helped to found), sees Lippard tipping over a sacred cow, as usual. She reframes amateurism not as an embarrassing condition which women artists need to transcend, but as a measure by which to judge the extent of gender and class prejudice. By turns funny, outraged and perceptive, the article is a great example of early Feminist writing that still feels urgent today.*

Lucy Lippard, '"Making Something from Nothing (toward a Definition of Women's "Hobby Art")', *Heresies* 4 (1978).

In 1968 Rubye Mae Griffith and Frank B. Griffith published a "hobby" book called *How to Make Something from Nothing*. On the cover (where it would sell books) his name was listed before hers, while on the title page (where it could do no harm), hers appeared before his. It is tempting to think that it was she who wrote the crypto-feminist dedication: "To the nothings—with the courage to turn into somethings." The hook itself is concerned with transformation—of tin cans, beef knuckle bones, old razor blades, breadbaskets and bottlecaps into more and less useful and decorative items. As "A Word in Parting," the authors state their modest credo: "This book . . . is simply a collection of ideas intended to encourage your ideas . . . We want you to do things your way . . . Making nothings into somethings is a highly inventive sport but because it is inventive and spontaneous and original it releases tensions, unties knots of frustration, gives you a wonderful sense of pleasure and accomplishment. So experiment, dare, improvise—enjoy every minute—and maybe you'll discover, as we did, that once you start making something from nothing, you find you can't stop, and what's more you don't want to stop!"

Despite the tone and the emphasis on enjoyment—unpopular in serious circles—this "sport" sounds very much like fine or "high" art. Why then are its products not art? "Lack of quality" will be the first answer offered, and "derivative" the second, even though both would equally apply to most of the more sophisticated works seen in galleries and museums. If art is popularly defined as a unique and provocative object of beauty and imagination, the work of many of the best contemporary "fine" artists must be disqualified along with that of many "craftspeople," end in the eyes of the broad audience, many of the talented hobbyist's works *would* qualify. Yet many of these, in turn, would not even be called "crafts" by the purists in that field. Although it is true that all this name calling is a red herring, it makes me wonder whether high art by another name might be less intimidating and more appealing. On the other hand, would high art by any other name look so impressive, be so respected and so commercially valued? I won't try to answer these weighted queries here, but simply offer them as other ways of thinking about some of the less obvious aspects of the art of making.

Much has been made of the need to erase false distinctions between art and *craft,* "fine" art and the "minor" arts, "high" art and "low" art—distinctions that particularly affect women's art. But there are also "high" crafts and "low" ones, and although women wield more power in the crafts world than in the fine art world, the same problems plague both. The crafts need only one more step up the aesthetic and financial respectability ladder and they

will be headed for the craft museums rather than for people's homes.

Perhaps until the character of the museums changes, anything ending up in one will remain a display of upper class taste in expensive and doubtfully "useful" objects. For most of this century, the prevailing relationship between art and "the masses" has been one of paternalistic *noblesse oblige* along the lines of "we who are educated to *know* what's correct must pass our knowledge and good taste *down* to those who haven't the taste, the time, or the money to know what is Good." Artists and craftspeople, from William Morris to de Stijl and the Russian Constructivists, have dreamed of socialist Utopias where everyone's life is improved by cheap and beautiful objects and environments. Yet the path of the Museum of Modern Art's design department, also paved with good intentions, indicates the destination of such dreams in a capitalist consumer society. A pioneer in bringing to the public the best available in commercial design, the Museum's admirable display of such ready-mades as a handsome and durable 39¢ paring knife or a 69¢ coffee mug has mostly given way to installations more typical of Bonnier's, DR, or some chic Italian furniture showroom.

It is, as it so often is, a question of audience as well as a question of categorization. (One always follows the other). Who sees these objects at MOMA? Mostly people who buy $3.00 paring knives and $8.00 coffee mugs which are often merely "elevated" examples of the cheaper versions with unnecessary refinements or simplifications. Good Taste is once again an economic captive of the classes who rule the culture and govern its institutions. Bad Taste is preferred by those ingrates who are uneducated enough to ignore or independent enough to reject the impositions from above. Their lack of enthusiasm provides an excuse for the aesthetic philanthropists, their hands bitten, to stop feeding the masses. Class-determined good/bad taste patterns revert to type.

Such is the process by which both "design" objects and the "high" crafts have become precisely the consumer commodity that the rare socially conscious "fine" artist is struggling to avoid. Historically, craftspeople, whose work still exists on a less exalted equilibrium between function and commerce, have been most aware of the contradictions inherent in the distinction between art and crafts. The distinction between design and "high" crafts is a modern one. Both have their origins in the "low" crafts of earlier periods, sometimes elevated to the level of "folk art" because of their usefulness as sources for "fine" art. A "designer" is simply the craftsperson of the technological age, no longer forced to do her/his own *making*. The Bauhaus *became* the cradle of industrial design, but the tapestries, furniture, textiles and tea sets made there were still primarily works of art. Today, the most popular housewares all through the taste gamut of the American lower-middle to upper-middle class owe as much stylistically to the "primitive" or "low" crafts—Mexican, Asian, American Colonial—as to the streamlining of the international style. In fact, popular design tends to combine the two, which meet at a point of (often spurious) "simplicity," and to become "kitsch"—diluted examples of the Good Taste that is hidden away in museums, expensive stores, and the homes of the wealthy, accessible to everyone else.

The hobby books reflect the manner in which Good Taste is still unarguably set forth by the class system. Different books are clearly aimed at different tastes, aspirations, educational levels. For instance, Dot Aldrich's *Creating with Cattails, Cones and*

Pods is not aimed at the inner city working-class housewife or welfare mother (who couldn't afford the time or the materials) or at the farmer's wife (who sees enough weeds in her daily work) but at the suburban upper middle-class woman who thinks in terms of "creating," has time on her hands and access to the materials. Aldrich is described on the dust jacket as a garden club member, a naturalist, and an artist; the book is illustrated by her daughter. She very thoroughly details the construction of dollhouse furniture, corsages and "arrangements" from dried plants and an occasional orange peel. Her taste is firmly placed as "good" within her class, although it might be seen as gauche "homemade art" by the upper class and ugly and undecorative by the working class.

Hazel Pearson Williams' *Feather Flowers and Arrangements,* on the other hand, has the sleazy look of a mail-order catalogue; it is one of a craft course series and its fans, birdcages, butterflies and candles are all made from garishly colored, rather than natural materials. The book is clearly aimed at a totally different audience, one that is presumed to respond to such colors and to have no aesthetic appreciation of the "intrinsic" superiority of natural materials over artificial ones, not to mention an inability to afford them.

The objects illustrated in books like the Griffiths' are neither high art nor high craft nor design. Yet such books are myriad, and they are clearly aimed at women—the natural *bricoleurs,* as Deena Metzger has pointed out. The books are usually written by a woman, and if a man is co-author he always seems to be a husband, which adds a certain familial coziness and gives him an excuse for being involved in such blatantly female fripperies (as well as dignifying the frippery by his participation). Necessity is the mother, not the father of invention. The home *maker's* sense of care and touch focuses on sewing, cooking, interior decoration as often through conditioning as through necessity, providing a certain bond between middle-class and working-class housewives *and* career women. (I am talking about the making of the home, not just the keeping of it; "good housekeeping" is not a prerogative for creativity in the home. It might even be the opposite, since the "houseproud" woman is often prouder of her house, her container, than she is of herself.) Even these days women still tend to be brought up with an exaggerated sense of detail and a need to be "busy," often engendered by isolation within a particular space, and by the emphasis on cleaning and service. A visually sensitive woman who spends day after day in the same rooms develops a compulsion to change, adorn, *expand* them, an impetus encouraged by the "hobby" books.

The "overdecoration" of the home and the fondness for bric-a-brac often attributed to female fussiness or plain Bad Taste can just as well be attributed to creative restlessness. Since most homemade hobby objects are geared toward home improvement, they inspire less fear in their makers of being "selfish" or "self-indulgent," there is no confusion about pretentions to Art, and the woman is freed to make anything she can imagine. (At the same time it is true that the imagination is often stimulated by exposure to other such work, just as "real" artists are similarly dependent on the art world and the works of their colleagues.) Making "conversation pieces" like deer antler salad tongs or a madonna in an abalone shell grotto, or a mailbox from an old breadbox, or vice versa, can be a prelude to breaking with the "functional" excuse and the making of wholly "useless" objects.

Now that the homebound woman has a little more leisure, thanks to so-called labor-saving devices, her pastimes are more likely to be cultural in character. The less privileged she is, the more likely she is to keep her interests *inside* the home with the focus of her art remaining the same as that of her work. The better off and better educated she is, the more likely she is to go *outside* of the home for influence or stimulus, to spend her time reading, going to concerts, theatre, dance, staying "well informed." If she is upwardly mobile, venturing from her own confirmed tastes into foreign realms where she must be cautious about opinions and actions, her insecurity is likely to lead to the classic docility of the middle-class audience, so receptive to what "experts" tell them to think about the arts. The term "culture vulture" is understood to apply mainly to upwardly mobile *women*. And culture, in the evangelical spirit of the work ethic, is often also inseparable from "good works."

Middle- and upper-class women, always stronger in their support of "culture" than any other group, seem to *need* aesthetic experience in the broadest sense more than men—perhaps because the vital business of running the world, for which educated women, at least to some extent, have been prepared, has been denied them, and because they have the time and the background to think—but not the means to act. Despite the fact that middle-class women have frequently been strong (and anonymous) forces for social justice, the earnestness and amateur status of such activities have been consistently ridiculed, from the Marx Brothers' films to the cartoons of Helen Hokinson.

Nevertheless, the League of Women Voters, the volunteer work for underfunded cultural organizations, the garden clubs, literary circles and discussion groups of the comfortable classes have been valid and sometimes coura-geous attempts to move out into the world while remaining sufficiently on the fringes of the system so as not to challenge its male core. The working-class counterparts are, for obvi-ous reasons, aimed less at improving the lot of others than at improving one's own, and, like hobby art, are more locally and domestically focused in unions, day care, paid rather than volunteer social work, Tupperware parties—and the PTA, where all classes meet. In any case, the housewife learns to take derision in her stride whether she intends to be socially effective or merely wants to escape from the home now and then (families are jealous of time spent elsewhere).

Women's liberation has at least begun to erode the notion that woman's role is that of the applauding spectator for men's creativity. Yet as makers of (rather than housekeepers for) art, they still trespass on male ground. No won-der, then, that all over the world, women priv-ileged and/or desperate and/or daring enough to consider creation outside traditional limits are finding an outlet for these drives in an art that is not considered "art," an art that there is some excuse for making, an art that costs little or nothing and performs an ostensibly useful function in the bargain—the art of making something out of nothing.

If one's only known outlets are follow-the-number painting or the ready-made "kit art" offered by the supermarket magazines, books like the Griffiths' open up new ter-ritory. Suggestions in "ladies'" and handi-work magazines should not be undervalued either. After all, quilt patterns were pub-lished and passed along in the 19th century (just as fashionable art styles are in today's art world). The innovative quilt maker or group of makers would come up with a new idea that broke or enriched the rules, just as

the Navajo rug maker might vary brilliantly within set patterns (and modern abstrac-tionists innovate by sticking to the rules of innovation).

The shared or published pattern forms the same kind of armature for painstaking hand-work and for freedom of expression within a framework as the underlying grid does in contemporary painting. Most modern women lack the skills, the motive and the discipline to do the kind of handwork their foremoth-ers did by necessity, but the stitch-like "mark" Harmony Hammond has noted in so much recent abstract art by women often emerges from a feminist adoption of the positive aspects of women's history. It relates to the ancient, sensuously repetitive, Penelopean rhythms of seeding, hoeing, gathering, weaving, spinning, as well as to modern domestic routines.

In addition, crocheting, needlework, embroi-dery, rug-hooking and quilting are coming back into middle- and upper-class fashion on the apron strings of feminism and fad. Ironi-cally, these arts are now practiced by the well-off out of boredom and social pressure as often as out of emotional necessity to make connec-tions with women in the past. What was once *work* has now become art or "high" craft—museum-worthy as well as commercially valid. In fact, when Navajo rugs and old quilts were first exhibited in New York fine arts museums in the early 1970s, they were eulogized as neu-tral, ungendered sources for the big bold geo-metric abstractions by male artists like Frank Stella and Kenneth Noland. Had they been presented as exhibitions of women's art, they would have been seen quite differently and probably not have been seen at all in fine art context at that time.

When feminists pointed out that these much-admired and "strong" works were in fact women's "crafts", one might have expected traditional women's art to be taken more seriously yet such borrowings from "below" must still be validated from "above". William C. Seitz's "Assemblage" shown at the Museum of Modern Art in 1961 had acknowledged the generative role of popular objects for Cubism, Dada and Surrealism, and predicted Pop Art, but he never considered women's work as the classic *bricolage*. It took a man, Claes Oldenburg, to make fabric sculpture acceptable, though his wife Patty did the actual sewing. Sometimes men even dabble in women's spheres in the lowest of low arts—hobby art made from throwaways by amateurs at home. But when a man makes, say a macaroni figure or a hand-tooled *Last Supper*, it tends to raise the sphere rather than lower the man, and he is likely to be written up in the local newspaper. Women dabbling in men's spheres, on the other hand, are still either inferior or just freakishly amazing.

It is supposed to be men who are "handy around the house," men who "fix" things while women "make" the home. This is a myth, of course, and a popular one. There are certainly as many women who do domestic repairs as men, but perhaps the myth was devised by women to force men to invest *some* energy, to touch and to care about *some* aspect of the home. The fact remains that when a woman comes to make something, it more often than not has a particular character—whether this originates from role-playing, the division of labor, or some deeper consciousness. The difference can often be defined as a kind of "positive fragmentation" or as the collage aesthetic—the mixing and matching of fragments to provide a new whole. Thus the bootcleaner made of bottlecaps suggested by one hobby book might also be a Surrealist object.

But it is not. And this is not entirely a disadvantage. Not only does the amateur status of

hobby art dispel the need for costly art lessons, but it subverts the intimidation process that takes place when the male domain of "high" art is approached. As it stands, women—and especially women—can make hobby art in a relaxed manner, isolated from the "real" world of commerce and the pressures of professional aestheticism. During the actual creative process, this is an advantage, but when the creative ego's attendant need for an audience emerges, the next step is not the gallery but to become a "cottage industry". The gifte shoppe, the county or crafts fair and outdoor art show circuit is open to women where the high art world is not, or was not until it was pried open to some extent by the feminist art movement. For this reason, many professional women artists in the past made both "public art" (canvases and sculptures acceptable to galleries and museums, conforming to a combination of the two current art world tastes) and "private" or "closet" art made for "personal reasons" or "just myself"—as if most art were not.

With the advent of the new feminism, private has either replaced or merged with the public in much women's art and the delicate, the intimate, the obsessive, even the "cute" and the "fussy" in certain guises have become more acceptable, especially in feminist circles. A striking amount of the newly discovered "closet" art by amateur and professional women artists resembles the *chotchkas* so universally scorned as women's playthings and especially despised in recent decades during the heyday of neo-Bauhaus functionalism. The objects illustrated in *Feather Flowers and Arrangements* bear marked resemblance to what is now called Women's Art, including a certainly unconscious bias toward the forms that have been called female imagery.

Today we are resurrecting our mothers', aunts' and grandmothers' activities—not only in the well-publicized areas of quilts and textiles, but also in the more random and freer area of transformational *rehabilitation*. On an emotional as well as on a practical level, rehabilitation has always been women's work. Patching, turning collars and cuffs, remaking old clothes, changing buttons, refinishing or recovering old furniture are all the traditional private resorts of the economically deprived woman to give her family public dignity. This continues today, even though in affluent Western societies cheap clothes fall apart before they can be rehabilitated and inventive patching is more acceptable (to the point where expensive new clothes are made to look rehabilitated and thrift shops are combed by the well-off). Thus "making something from nothing" is a brilliant title for a hobby book, appealing as it does both to housewifely thrift and to the American spirit of free enterprise—a potential means of making a fast buck.

Finally, certain questions arise in regard to women's recent "traditionally oriented" fine art. Are the sources direct—from quilts and county fair handiwork displays—or indirect—via Dada, Surrealism, West Coast funk, or from feminist art itself? Is the resemblance of women's art-world art to hobby art a result of coincidence? Of influence, conditioning, or some inherent female sensibility? Or is it simply another instance of camp, or fashionable downward mobility? The problem extends from source to audience. Feminist artists have become far more conscious of women's traditional arts than most artists, and feminist artists are also politically aware of the need to broaden their audience, or of the need to broaden the kind of social experience fine art reflects. Yet the means by which to fill these needs have barely been explored. The greatest lack in the feminist art movement may be for contact and dialogue with those "amateurs" whose work sometimes appears to be imitated by the professionals. Judy Chicago and her co-workers on *The Dinner Party* and their collaboration with china painters and needleworkers, Miriam Schapiro's handkerchief exchanges and the credit given the women who embroider for her, the "Mother Art" group in Los Angeles which performs in laundromats and similar public/domestic situations, the British "Postal Art Event," and a few other examples are exceptions rather than the rule. It seems all too likely that only in a feminist art world will there be a chance for the "fine" arts, the "minor" arts, "crafts," and hobby circuits to meet and to develop an *art of making* with a new and revitalized communicative function. It won't happen if the feminist art world continues to be absorbed by the patriarchal art world.

And if it does happen, the next question will be to what extent can this work be reconciled with all the varying criteria that determine aesthetic "quality" in the different spheres, groups, and cultures? Visual consciousness raising, concerned as it is now with female imagery and, increasingly, with female process, still has a long way to go before our visions are sufficiently cleared to see all the arts of making as equal products of a creative impulse which is as socially determined as it is personally necessary; before the idea is no longer to make nothings into some things, but to transform and give meaning to all things. In this utopian realm, Good Taste will not be standardized in museums, but will vary from place to place, from home to home.

NOTE

1. Lucy Lippard, 'Why Separate Women's Art?', in *Pink Glass Swan* (New York: The New Press, 1995; essay orig. pub. 1973), p. 57. For other connections between craft and Feminism, see

Grace Glueck, 'Women Artists 80', *Art News* 79/2 (October 1980), pp. 59–60. As recently as 1993, fiber work has been held up as exemplary of 'women's efforts to appropriate space—a (sewing) room of their own—for their quiltmaking activities seem very much linked to a desire for an experience of time that is uninterrupted, and an experience of self that is sensuous, desiring and whole'. Jane Przybysz, 'Quilts and Women's Bodies: Diseased and Desiring', in Katherine Young, ed., *Body Lore* (Knoxville: University of Tennessee Press, 1993), p. 172.

REFERENCES

The Griffiths' book was published by Castle Books, NYC; the Aldrich book by Hearthside Press, Great Neck, N.Y.; the Williams book by Craft Course publishers, Temple City, California. I found all of them in the tiny local library in a Maine town with a population of circa three hundred.

The Hammond article appeared in *Heresies* No. 1; her "Class Notes" in No. 3 are also relevant. The Metzger appeared in *Heresies* No. 2 and is an important contribution to the feminist dialogue on "high" and "low" art. The British Postal Event, or "Portrait of the Artist as a Housewife," is a "visual conversation" between amateur and professional women artists isolated in different cities. They send each other art objects derived from "non-prestigious folk traditions," art that is "cooked and eaten, washed and worn" in an attempt to "sew a cloth of identity that other women may recognize." It is documented in *MAMA!,* a booklet published by a Birmingham collective and available from PDC, 27 Clerkenwell Close, London EC 1.

I have also been indebted in this series of articles (which includes "The Pink Glass Swan" in *Heresies* No. 1) to Don Celender's fascinating *Opinions of Working People Concerning the Arts,* 1975, available from Printed Matter, 7–9 Lispenard St., NYC 10013.

FURTHER READING

Stephanie Cash, 'The Art Criticism and Politics of Lucy Lippard', *Art Criticism* 9/1 (1994).

Lucy Lippard, *Changing: Essays in Art Criticism* (New York: E. P. Dutton, 1971).

Lucy Lippard, *Pink Glass Swan* (New York: The New Press, 1995).

Adrienne Rich, 'Conditions for Work: The Common World of Women', *Heresies* 3 (1977).

60

'THE CREATION OF FEMININITY', FROM *THE SUBVERSIVE STITCH: EMBROIDERY AND THE MAKING OF THE FEMININE*

Rozsika Parker

In the hopes of building a less sexist future, the Feminist movements of the 1970s and '80s engaged with inequalities both past and present. Linda Nochlin's essay 'Why Have There Been No Great Women Artists?' is the most famous example of an attempt to redress the historical quashing of female talent, with an obvious lesson for the art world of her own day.[1] In Britain, Rozsika Parker and Griselda Pollock's Old Mistresses: Women, Art and Ideology *showed how the writing of art history itself had led to the exclusion of women. Similarly, Feminists undertook a revaluation of amateur (usually domestic and unpaid) crafts historically practiced by women, such as spinning, quilting, embroidery, fancywork, lace making and china painting. All of these were featured in publications like the* Feminist Art Journal *and* Heresies, *alongside accounts of women's contemporary art. Feminists approached these crafts with an acute ambivalence, seeing them as 'trivialized and degraded categories of "women's work" outside of the fine arts', but also as an arena for self-expression in the face of oppression.[2] Historical female creativity was not simply celebrated, but was also seen as the result of social constraint. Though Parker's* The Subversive Stitch *was a fairly late entry into this literature, it was the most thoroughly researched and remains the best known today. It identified the historical hierarchical division of fine art and craft as a major*

force in the marginalization of women's work. Taking embroidery as a case history, she demonstrated the cultural downgrading of a craft as it became increasingly associated with women. Parker was among the first British Feminists in the arts—she joined the magazine Spare Rib *at its inception in 1972, and the following year was a co-founder of the Women's Art History Collective. For her, too, craft was a double-sided thing—'both a source of pleasurable creativity and [an] oppression'—and hence a way into the complex reality of past women's lives. The book would later inspire an exhibition of both historical and contemporary women's textile art, held in Manchester.*

Rozsika Parker, excerpts from *The Subversive Stitch: Embroidery and the Making of the Feminine* (London: Women's Press, 1984).

'Has the pen or pencil dipped so deep in the blood of the human race as the needle?' asked the writer Olive Schreiner. The answer is, quite simply, no. The art of embroidery has been the means of educating women into the feminine ideal, and of proving that they have attained it, but it has also provided a weapon of resistance to the constraints of femininity.

In this book I examine the historical processes by which embroidery became identified with a particular set of characteristics, and consigned to women's hands. By mapping the relationship between the history of embroidery and changing notions of what constituted feminine behaviour from the Middle Ages to the twentieth century, we can see how the art became implicated in the creation of femininity across classes, and that the development of ideals of feminine behaviour determined the style and iconography of needlework. To know the history of embroidery is to know the history of women.

THE CREATION OF FEMININITY

"Needlework is the favourite hobby of two percent of British males, about equal to the number who go to church regularly. Nearly one man in three fills in football coupons, in an average month, or has a bet."[3]

The Guardian was no doubt confident that its coverage of a Government survey of changing trends in leisure activities was eye-catching, and that this opening sentence was guaranteed to amuse by its incongruity. The unspoken assumption implied by the juxtaposition of male needleworkers and church-goers is that these men are pious, prim and conformist. Real men gamble and fill in football coupons; only sissies and women sew and swell congregations.

The sexual division that assigns women to sewing is inscribed in our social institutions, fostered by school curricula which still direct boys to carpentry and girls to needlework. Even in today's progressive schools the assumptions and divisions remain intact. An enthusiastic report on a large suburban primary school praised the diligent, pioneering teaching practised by the staff. Two photographs illustrated science teaching methods: in one, a small group of boys were shown unselfconsciously engrossed in a 'wave power

machine'; in the other, two smiling girls displayed copper atoms embroidered in silk.[4]

The role of embroidery in advertising and commercial design also endorses the notion that a man who practises embroidery is imperiling his sexual identity. Embroidery is invariably employed to evoke the home. The cover of a brochure produced by a British home removal firm illustrates an embroidery of a house, the stock motif of so many samplers, and bears the embroidered words 'Home Moving Guide'. Embroidery connotes not only home but a socially advantaged home, securely placed in the upper reaches of the class structure. An advertisement for embroidery patterns promises that 'the tapestries are a pleasure to make and once completed will elegantly grace any home and become much valued family heirlooms'.

It is not only home and family that embroidery signifies but, specifically, mothers and daughters. Heinz based an advertising campaign for tomato ketchup on a picture of a sampler stitched with the words, 'If other ketchups were as rich, then I'd say so stitch by stitch. Ann and Lucy James (but mostly Lucy).' The sampler associates tomato ketchup with the ideal of childhood as sincere, innocent and pure.

Embroidery also evokes the stereotype of the virgin in opposition to the whore, an infantilising representation of women's sexuality. Thus Lil-lets the menstrual tampons were recently packed in a box masquerading as fabric, embroidered with pastel flowers to represent menstruation as natural and entirely non-threatening. The conflation of embroidery and female sexuality, both innately virginal and available for consumption, is blatantly expressed in the title bestowed on a porn magazine, the *Rustler Sampler*, which offered 'nearly two hundred, yes, two hundred juicy, picture-packed pages'. The word 'Sampler' evokes an image of innumerable passive, powerless women just waiting to be selected and roped in by the 'Rustler'. Embroidery has become indelibly associated with stereotypes of femininity.

I shall define briefly what I mean by femininity. In *The Second Sex*, 1949, Simone de Beauvoir wrote: 'It is evident that woman's "character"—her convictions, her values, her wisdom, her morality, her tastes, her behaviour—are to be explained by her situation.[5] In other words, femininity, the behaviour expected and encouraged in women, though obviously related to the biological sex of the individual, is shaped by society. The changes in ideas about femininity that can be seen reflected in the history of embroidery are striking confirmation that femininity is a social and psychosocial product.

Nevertheless, the conviction that femininity is natural to women (and unnatural in men) is tenacious. It is a crucial aspect of patriarchal ideology, sanctioning a rigid and oppressive division of labour. Thus women active in the upsurge of feminism which began in the 1960s set out to challenge accepted definitions of the innate differences between the sexes, and to provide a new understanding of the creation of femininity. In consciousness-raising groups and campaigns we compared our experiences at work, at school, at home, in relationships, as mothers, as daughters and sisters. The workings of sexism were scrutinised in the division of labour in and out of the home, in sexuality, the family, healthcare, child care, language, the law, education, the arts, the media and government policy. How race, class and sex intersect to shape women's lives became clearer.

Institutional discrimination co-exists and interacts with the mechanisms and effects of psychic subordination, though obviously rigid divisions cannot be drawn between internal and external oppression. The complex of emotional attitudes of passivity, submission and masochism which guarantee the subordination of women cannot simply be shrugged off or discounted. Juliet Mitchell, in *Psychoanalysis and Feminism*, 1974, observed that:

> . . . the status of woman is held in the heart and the head as well as in the home: oppression has not been trivial or historically transitory—to maintain itself so efficiently it courses through the mental and emotional bloodstream. To think that this should not be so does not necessitate pretending it is already not so.[6]

Many feminists have looked to psychoanalysis and Marxist theory to provide an account of how masculinity and femininity are constructed and reproduced historically. The family was identified as the place where the 'inferiorised psychology'[7] of women was reproduced and the social and economic exploitation of women as wives and mothers legitimised. Writing of the construction of femininity in the family, anthropologist Gayle Rubin in an essay in *Towards an Anthropology of Women*, 1975, commented: 'One can read Freud's essay on femininity as a description of how a group is prepared to live with oppression', and she makes clear how painful the process is. 'It is certainly plausible to argue that the creation of "femininity" in a woman in the course of socialisation is an act of psychic brutality.'[8]

It is, however, important to distinguish between the construction of femininity, lived femininity, the feminine ideal and the feminine stereotype. The construction of femininity refers to the psychoanalytic and social account of sexual differentiation. Femininity is a lived identity for women either embraced or resisted. The feminine ideal is an historically changing concept of what women should be, while the feminine stereotype is a collection of attributes which is imputed to women and against which their every concern is measured. Millicent Fawcett, the nineteenth-century British feminist, declared, 'We talk about "women and women's suffrage", we do not talk about Woman with a capital W. That we leave to our enemies.'[9]

In other words, there is a significant difference between acknowledging the construction of femininity in the family and its maintenance in social institutions, and accepting the cultural representation of women imposed upon us. The feminine stereotype categorises everything women are and everything we do as entirely, essentially and eternally feminine, denying differences between women according to our economic and social position, or our geographical and historical place. In fact, what Gayle Rubin termed 'the act of psychic brutality' meets with resistance at all levels, in different ways at different historical moments.

What, then, is the purpose of the feminine stereotype? In *Old Mistresses: Women, Art and Ideology*, 1981, Griselda Pollock and I looked at the role of the feminine stereotype in the writing of art history. We asked why painting by women has been set apart from painting by men and why women's art, in all its diversity, has been described as homogeneous. We revealed the feminine stereotype to be one of the major elements in the construction of the current view of the history of art.[10] The particular way women's work is presented—the constant assertion of the feminine weakness of women's art—sustains the dominance of masculinity and male art.

The situation of embroidery is more elusive. When women paint, their work is categorised as homogeneously feminine—but it is acknowledged to be art. When women embroider, it is seen not as art, but entirely as the expression of femininity. And, crucially, it is categorised as craft. The division of art forms into a hierarchical classification of arts and crafts is usually ascribed to factors of class within the economic and social system, separating artist from artisan. The fine arts—painting and sculpture—are considered the proper sphere of the privileged classes while craft—or the applied arts—like furniture-making or silver-smithery—are associated with the working class. However there is an important connection between the hierarchy of the arts and the sexual categories male/female. The development of an ideology of femininity coincided historically with the emergence of a clearly defined separation of art and craft. This division emerged in the Renaissance at the time when embroidery was increasingly becoming the province of women amateurs, working for the home without pay. Still later the split between art and craft was reflected in the changes in art education from craft-based workshops to academies at precisely the time—the eighteenth century—when an ideology of femininity as natural to women was evolving.

The art/craft hierarchy suggests that art made with thread and art made with paint are intrinsically unequal: that the former is artistically less significant. But the real differences between the two are in terms of where they are made and who makes them. Embroidery, by the time of the art/craft divide, was made in the domestic sphere, usually by women, for 'love'. Painting was produced predominantly, though not only, by men, in the public sphere, for money. The professional branch of embroidery, like that of painting, was, from the end of the seventeenth century to the end of the nineteenth century, largely in the hands of working-class women, or disadvantaged middle-class women. Clearly there are huge differences between painting and embroidery; different conditions of production and different conditions of reception. But rather than acknowledging that needlework and painting are different but equal arts, embroidery and crafts associated with 'the second sex' or the working class are accorded lesser artistic value.

The classification of embroidery is a difficult task. To term it 'art' raises special problems. Moving embroidery several rungs up the ladder of art forms could be interpreted as simply affirming the hierarchical categorisations, rather than deconstructing them. Moreover, to describe embroidery as 'art' is to fail to distinguish it from painting, concealing the profound differences that have developed historically between the two media. However, to call it 'craft' is no solution. Embroidery fails to comply with the utilitarian imperative that defines craft—because much of it is purely pictorial. Traditionally, women have called embroidery 'work'. Although to some extent an appropriate term, it tends to confirm the stereotypical notion that patience and perseverance go into embroidery—but little else. Moreover, the term was engendered by an ideology of femininity as service and selflessness and the insistence that women work for others, not for themselves. I have decided to call embroidery art because it is, undoubtedly, a cultural practice involving iconography, style and a social function.

That embroiderers do transform materials to produce sense—whole ranges of meanings—is invariably entirely overlooked. Instead embroidery and a stereotype of femininity have become collapsed into one another, characterised as mindless, decorative and delicate; like the icing on the cake, good to look at, adding taste and status, but devoid of significant content.

The association between women and embroidery, craft and femininity, has meant that writers concerned with the status of women have often turned their attention towards this tangled, puzzling relationship. Feminists who have scorned embroidery tend to blame it for whatever constraint on women's lives they are committed to combat. Thus, for example, eighteenth-century critical commentators held embroidery responsible for the ill health which was claimed as evidence of women's natural weakness and inferiority. In the nineteenth century, women wanting to be taken seriously in supposedly 'male' spheres deliberately declared their rejection of embroidery to distance themselves from the feminine ideal. In Helen Black's late nineteenth-century publication, *Notable Authors of the Day*, 1893, consisting of interviews with novelists, Adeline Sargeant stated, 'I have done some elaborate embroidery in my time but now I never use the needle for amusement, only for necessity.'[11] She asserts her seriousness and her disdain for feminine frivolity. The majority of interviewees, however, stress their needlework. For although writing novels was, by then, an acceptable activity for women, professionalism was frowned upon. Women therefore covered themselves with their amateur 'Work'. Mrs L. B. Walford, for example, is described as 'wearing a pretty blue tea gown richly embroidered in silk by her own hand'[12] and Helen Mather is offered as 'a great needlewoman, not only are the long satin curtains by her own hand but the pillows, cushions, and dainty lampshade.'[13]

To reject embroidery, as Adeline Sargeant did, was to run the risk of appearing to disparage other women, or to endorse the typical view of the art propounded by a

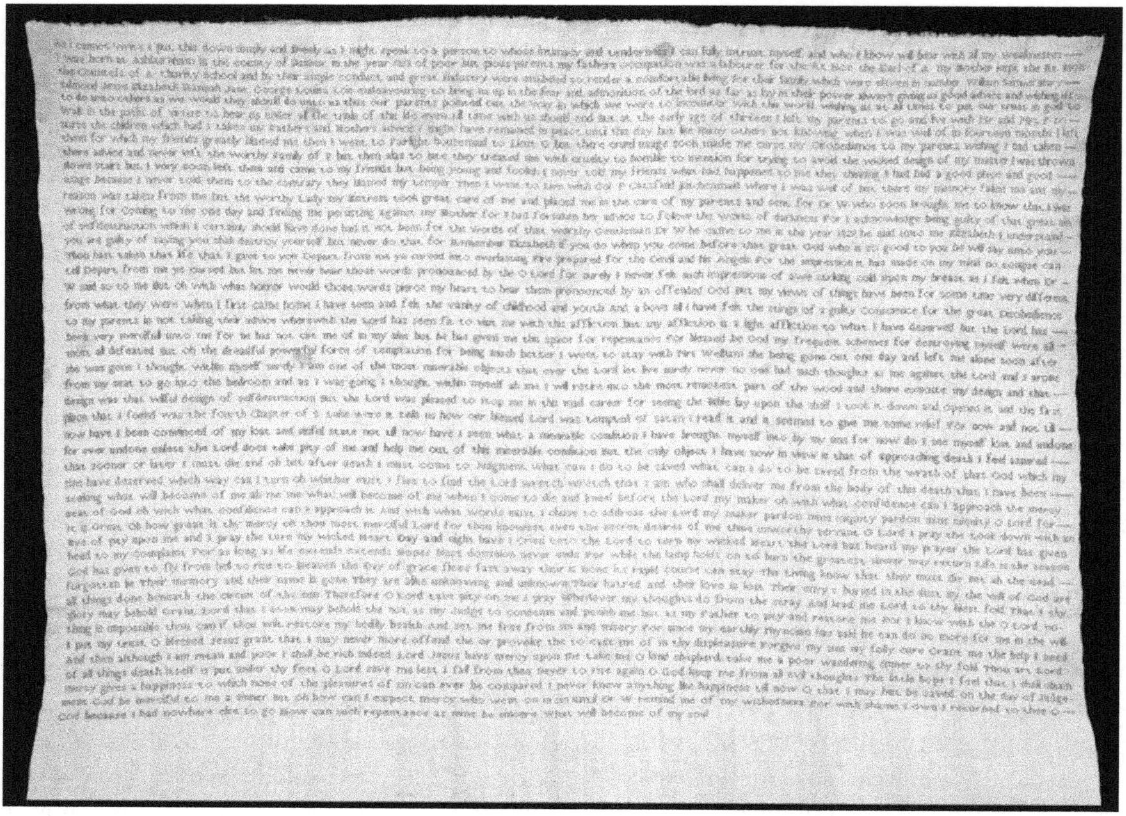

Figure 38 Elizabeth Parker, *Sampler*, 1830.

male-dominated society. For purely tactical reasons therefore, women who might have been critical of embroidery praised it. Thus the more enlightened seventeenth-century women educationalists had included needlework in their curriculum largely to provide an acceptable face for women's education. Nineteenth-century writers defended embroidery, claiming it as an unappreciated art form. Some believed in raising the status of women, not by dismissing women's traditional creative activity, but by demanding that its true worth be recognised. In her novel *The Beth Book*, 1897, Sarah Grand offers embroidery as evidence of women's superiority. Beth embroiders, selling her work secretly through the

discreet commercial outlets provided by the Arts and Crafts Movement to market 'ladies' work'. For Sarah Grand embroidery represents the beauty of the female imagination, its spiritual clarity in contrast to male pedestrian rationalism. But this attempt to validate women's work ultimately reinforces the rigid sexual categorisation and justifies the separate spheres.

[. . .]

The manner in which embroidery signifies both self-containment and submission is the key to understanding women's relation to the art. Embroidery has provided a source of pleasure and power for women, while being indissolubly linked to their powerlessness. Paradoxically, while embroidery was employed to inculcate femininity in women, it also enabled them to negotiate the constraints of femininity. Observing the covert ways embroidery has provided a source of support and satisfaction for women leads us out of the impasse created by outright condemnation or uncritical celebration of the art. Nevertheless, it would be a mistake to underestimate the importance of the role played by embroidery in the maintenance and creation of the feminine ideal. During the seventeenth century the art was used to inculcate femininity from such an early age that the girl's ensuing behaviour appeared innate. By the eighteenth century embroidery was beginning to signify a leisured, aristocratic life style—not working was becoming the hallmark of femininity. Embroidery with its royal and noble associations was perfect proof of gentility, providing concrete evidence that a man was able to support a leisured woman. Moreover, because embroidery was supposed to signify femininity—docility, obedience, love of home, and a life without work—it showed the embroiderer to be a deserving, worthy wife and mother. Thus the art played a crucial part in maintaining the class position of the household, displaying the value of a man's wife and the condition of his economic circumstances. Finally, in the nineteenth century, embroidery and femininity were [entirely] fused, and the connection was deemed to be natural. Women embroidered because they were naturally feminine and were feminine because they naturally embroidered. Then embroidery was blamed for the conflicts provoked in women by the femininity the art fostered. By the end of the century, Freud was to decide that constant needlework was one of the factors that 'rendered women particularly prone to hysteria'

because day-dreaming over embroidery induced 'dispositional hypnoid states'.[14]

The subject matter of a woman's embroidery during the eighteenth and nineteenth centuries was as important as its execution in affirming her femininity (and thus her worth and worthlessness in the world's eyes). It was expected to reflect the current feminine ideal, which was held to be the highest, yet paradoxically most natural, achievement of women. If the content conformed to the ideal it supposedly won the needlewoman love, admiration and support.

[. . .]

The iconography of women's work is rarely given the serious consideration it deserves. Embroidery is all too often treated only in terms of technical developments. One reason why the subject matter of embroidery is summarily dismissed is that embroiderers employ patterns. The interpretation, adaptation and variation of pattern is an integral aspect of the activity and it is therefore assumed that stylistic and technical properties are all that concern the embroiderer. However, needlewomen chose particular patterns, selecting those images which had meaning for them. The enormous popularity of certain images at different moments indicates that they had specific importance and powerful resonance for the women who chose to stitch them. Where embroiderers have actually employed contemporary paintings as patterns, we can perceive what could or could not be stitched by women, and how they were able to make meanings of their own, by observing which they selected and where they departed from their models. Nevertheless, the meanings of any embroidered picture have to be carefully considered within their historical, artistic and class context. What a picture conveys often relates to the needs of a woman's class as much as to her experience as a woman at that time, as well as to the dominant concerns of contemporary paintings and to the history of embroidery.

Sometimes embroiderers reinforced the feminine ideal in their work; comfortingly concealing the disjunctures between the 'ideal' and the 'real' by the words and images they stitched—'Home Sweet Home'. At other times they resisted or questioned the emerging ideology of feminine obedience and subjugation, as in the following seventeenth-century sampler verse:

> When I was young I little thought
> That wit must be so dearly bought
> But now experience tells me how
> If I would thrive then I must bow
> And bend unto another's will
> That I might learn both care and skill
> To Get My Living with My Hands
> That So I Might Be Free From Band
> And My Own Dame that I may be
> And free from all such slavery.
> Avoid vaine pastime fle youthful pleasure
> Let moderation allways be thy measure
> And so prosed unto the heavenly treasure.

The verse is a curious mixture of piety and rebellion, resentment and acquiescence. Because samplers were becoming the place where moral sentiments were impressed upon young girls, they were sometimes also the place where conflicts underlying the ideology were expressed.

Such overt recognition of the clash between individual ambition and the ideology of femininity is rare indeed. More often the embroiderers' desire was to achieve exactly what was expected of them, developing satisfying and praiseworthy levels of skill. From our vantage point, it is all too easy to sneer at the Victorian embroiderer completing yet another pair of slippers stitched with a fox

head, or at the eighteenth-century embroiderer reproducing in thread the moralising, sentimental domestic genre paintings of her time. But rather than ridiculing them, or turning embarrassed from our history, we should ask why they selected such subjects, what secondary gains they accrued from absolute conformity to the feminine ideal, and how they were able to make meanings of their own while overtly living up to the oppressive stereotype . . .

While recognising the varied ways in which women have conformed to and resisted the dictates of femininity in their work, it is important to remember that embroidery has been and is a source of artistic pleasure to many women. Olive Schreiner, in her novel *From Man to Man*, 1927, evoked the satisfaction of needlework, particularly the narcissistic pleasures it provided:

> All her life she had dreamed of having a dress made of thick black silk, with large blue daisies with white centres embroidered in raised silk work all over it at intervals. Her mother had had such a bit of silk in a patchwork quilt she had brought from England with her.[15]

Embroidery summons up both 'advanced' civilisation and very early childhood when a primal, unproblematic unity with the mother still existed. However, the work provides narcissistic pleasure not only because it evokes the love and unity of early childhood; but also because women were taught to embroider as an extension of themselves; and quite crudely because embroidery is used on clothing where it provokes admiration. Urged to embroider clothing and furniture, encouraged to see it as the natural expression of their nature, women were still accused of vanity when they embroidered for themselves. The stereotype of embroidery as a vain and frivolous occupation, like the stereotype of the silent, seductive needlewoman, controls and undermines the power and pleasure women have found in embroidery, representing it to us negatively. Nevertheless, women have found gratification in the activity. Olive Schreiner conveys the immense creative satisfaction it provides:

> Slowly the Scores of little tucks and fine embroidery shaped themselves. At the end of the week there were two tiny armholes. At the end of a fortnight the long white rope with its delicate invisible stitching was also complete . . .[16]

She also perceived the bond that embroidery forged between women; sewing allowed women to sit together without feeling they were neglecting their families, wasting time or betraying their husbands by maintaining independent social bonds:

> They were unalike physically and mentally but they had tastes which harmonised. While Veronica sat upright on a high-backed chair knitting heavy squares for a bed quilt, Mrs Drummond, on a low settee, with her head a little on one side, chose carefully the shades of silk for an altarcloth which she was making.[17]

The women's choice of work indicates the different personalities; that they both engaged in domestic art reveals what they share as women in society.

After placing embroidery at the centre of women's lives, Olive Schreiner makes a plea that it be recognised as art, as a creative expressive activity, but nevertheless betrays her kinship to the attitude towards needlework manifested by the contemporary novelists Sarah Grand and May Sinclair:

> The poet, when his heart is weighted, writes a sonnet, and the painter paints a picture, and the thinker throws himself into the world of action; but the woman who is only a woman, what has she but her needle? In that torn bit of brown leather brace worked through and through with yellow silk, in that bit of white rag with the invisible stitching, lying among fallen leaves and rubbish that the wind has blown into the gutter or street corner, lies all the passion of some woman's soul finding voiceless expression. Has the pen or pencil dipped so deep in the blood of the human race as the needle?[18]

While placing embroidery as an art like poetry and painting, Olive Schreiner reasserts its association with femininity. It is the bearer of women's soul. The images of the white rag with the invisible stitching, the yellow silk besmirched and trodden underfoot silently suggest a comparison with women's fate in the streets. Olive Schreiner maintains the link between embroidery and feminine purity, thus presenting it as a sexual characteristic, and failing to establish it as an art form equal to painting and poetry. In part this reflects Olive Schreiner's own ambivalence towards the domestic labour she describes: 'The worst of this book of mine is that it's so womanly. I think it's the most womanly book that ever was written, and God knows I've willed it otherwise.'[19] But the effect of the passage is largely determined by the hierarchical categorisation of art forms in our culture. By claiming that embroidery should be valued because of its intimate associations with women's lives and domestic tradition, Olive Schreiner inevitably though unwittingly discounted it as art.

The extraordinary intractability of embroidery, its resistance to re-definition, is the result of its role in the creation of femininity during the past five hundred years.

NOTES

1. Linda Nochlin, "Why Have There Been No Great Women Artists?," *Art News* 69/9

(Jan. 1971), reprinted in Thomas Hess and Elizabeth Baker, eds, *Art and Sexual Politics* (New York: Macmillan/Collier Books, 1973).

2. Arlene Raven, "Blood Sisters: Feminist Art and Criticism," in Lydia Yee, et al., *Division of Labor: "Women's Work" in Contemporary Art* (New York: Bronx Museum of Art, 1995), p. 47.

3. John Ezard, 'Victorian Touch to Credit Cold Britain', *The Guardian,* 6 December 1979.

4. Adrian Hopkins, 'Firm but Not Fixed in Their Ways', *The Guardian,* 30 March 1979.

5. Simone de Beauvoir, *The Second Sex.* London: Penguin Books, 1972, p. 635.

6. Juliet Mitchell, *Psychoanalysis and Feminism.* London: Penguin Books, 1974, p. 363.

7. Ibid.

8. Gayle Rubin, 'The Traffic in Women: Notes on "The Political Economy of Sex"', in Rayna R. Reiter, editor, *Toward an Anthropology of Women.* New York: Monthly Review Press, 1975, p. 196.

9. Millicent Fawcett, cited in Theodore Stanton, *The Woman Question in Europe.* London: 1884, p. 6.

10. Rozsika Parker and Griselda Pollock, *Old Mistresses: Women, Art and Ideology.* London: Routledge & Keegan Paul, 1981.

11. Helen Black, *Notable Authors of the Day.* London: 1893, p. 169.

12. Ibid. p. 27.

13. Ibid. p. 78.

14. Joseph Breuer and Sigmund Freud, *Studies on Hysteria,* in James Strachey, editor, *The Complete Works of Sigmund Freud.* London: The Hogarth Press, Vol 2, p. 12.

15. Olive Schreiner, *From Man to Man.* London: Virago, 1982, Ch XI.

16. Ibid. Ch IX.

17. Ibid. Ch VI.

18. Ibid. Ch IX.

19. Ruth First and Ann Scott, *Olive Schreiner.* London: André Deutsch, 1980, p. 175.

FURTHER READING

Pennina Barnett, *The Subversive Stitch* (Manchester: Whitworth Art Gallery/Cornerhouse, 1988), and 'Afterthoughts on Curating "The Subversive Stitch"', in Katy Deepwell, ed., *New Feminist Art Criticism* (Manchester: Manchester University Press, 1995).

Mary C. Beaudry, *Findings: The Material Culture of Needlework and Sewing* (New Haven, CT: Yale University Press, 2006).

Heresies 4: Women's Traditional Arts: The Politics of Aesthetics (1978).

Rozsika Parker and Griselda Pollock, *Old Mistresses: Women, Art and Ideology* (London: Rivers Oram Press, 1981).

61

THERE'S NO PLACE LIKE HOME: HOME DRESSMAKING AND CREATIVITY IN THE JAMAICAN COMMUNITY OF THE 1940s TO THE 1960s

Carole Tulloch

The Culture of Sewing, *edited by Barbara Burman, was a landmark publication in the study of domestic crafts. The book includes a series of essays on home needlecrafts, with case studies drawn from Britain, the United States and Jamaica and ranging from the early nineteenth century onwards. Many of the authors employ a feminist approach, featuring some of the strategies pioneered by writers in the 1970s such as first-person narratives; a focus on and validation of amateur craft activity; and a consideration of the institutional frames that structure domestic crafts, for example by researching the sale of sewing machines and dress patterns. In this selection from the volume, London-based design historian Carole Tulloch draws on her own family history to sketch a compelling picture of the place of one home craft—dressmaking—in the lives of West Indian immigrants to Britain. As Tulloch points out, what might at first seem like a matter of economic necessity was, for many women, an important way to assert their identity during the transition to a new life.*

Carole Tulloch, 'There's No Place Like Home: Home Dressmaking and Creativity in the Jamaican Community of the 1940s to the 1960s', in Barbara Burman, ed., *The Culture of Sewing: Gender, Consumption and Home Dressmaking* (Oxford: Berg, 1999).

INTRODUCTION: THE FAMISHED CREATIVE SPIRIT

The African-American author Alice Walker asks how and when did her mother 'feed her creative spirit'. (Walker 1984: 239) As co-provider, mother and wife, Walker's mother worked all day in the fields, made all her children's clothes, the sheets and the quilts for the beds, in addition to the 'traditional' duties expected of her. It was in what Walker terms, the 'ambitious gardens' which her mother cultivated around and in their 'shabby house', working on them before she left for her field work and in the spare time available on her return. Her 'hobby' was to draw admirers from miles around. Walker maintains that her mother's ingenuity in finding time and space to cultivate her creativity, to champion the creative spirit, saved her mother's sanity:

I notice that it is only when my mother is working in her flowers that she is radiant, almost to the point of being invisible—except as Creator: hand and eye. She is involved in work her soul must have [. . .] being an artist has still been a daily part of her life. This ability to hold on, even in very simple ways, is work black women have done for a very long time. (Walker 1984: 241–2)

Walker's memorialization of her mother's need and ability to feed the creative spirit offers a personal anecdote of an individual as an example to consider other cases and, in turn, a group. Walker contextualizes this inquisition into the thirst of the creative spirit amongst African-American women and acknowledges Virginia Woolf's earlier discussion of the feminist creative discourse in her seminal work of 1929, *A Room of One's Own.* Both authors lament the loss to history of the material epitaphs borne out of the creative practices conducted by women. This has been attributed to either a lack of public recognition, or women's inability to fulfil their creative potential due to patriarchal control or social deprivation. (Walker 1984: 235–7) Debate on this issue is extensive. (Hooks 1982; Nunn 1987; Darwent 1998) Walker recommends that 'we must fearlessly pull out of ourselves and look at and identify with our lives the living creativity some of our great-grandmothers were not allowed to know.' (Walker 1984: 237)

The focus of this essay is a consideration of how home dressmaking can feed the creative spirit of working-class Jamaican women who have lived in Jamaica and Britain.[1] If an Oxford English Dictionary definition of the word spirit is 'the vital animating essence of a person', then the creativity which emanates from an individual is not simply concerned with creating and making objects, but is simultaneously about maintaining and representing the individual, the self. This compound of creativity, the aesthetic-self and its impact on a collective identity, provides the structure for this work. Essentially, it is located in contextualizing the practice of home dressmaking and the finished garment within wider cultural issues. The limitation of this work's length has necessitated such a reductive outline and does not afford an extensive study into the practice of home dressmaking by a large number of Jamaican women.

By the early twentieth century, home dressmaking was a characteristic cultural and social feature of Jamaica.[2] It was practised by most racial and ethnic groups as a means of socialization amongst the middle class and a necessity for the working class. It was a compulsory subject within the education system for boys and girls up to the age of ten. Independent professional dressmakers were to be found in their thousands—the 1891 census lists 3,656 professional dressmakers based in Jamaica's capital, Kingston, and 18,966 throughout the island.[3] The paper will concentrate on the oral testimony and dressmaking history of Mrs. Anella James[4], an exponent of this cultural legacy. Anella was the second youngest of eleven children of a working-class family. Her story spans from 1936 to 1965, from the rural town of Slygoville in the parish of St Catherine, Jamaica, to Ilford, Essex, where Anella emigrated in 1961.

Although the case study of an individual 'cannot speak for the collective', (Chamberlain 1995: 95) it can, I argue, contribute to the overall identity of a culture and to a specific recreational practice such as home dressmaking.

The subjective study of Anella's interaction with home dressmaking allows concentration on the relativity of the practice to an individual's sense of aesthetic-self. A supplementary consideration is the consequential effect on, or interpretation of, the collective identity. To this end I want to expand on Mary Chamberlain's thesis that:

Memory and the individual are indivisible [. . .] If we recognise that memory rather than being confined to, and by the individual, manifests elements of a shared consciousness and is part of the process of social production, then oral sources offer the potential for entering into a wider cultural milieu. In that sense, the individual voice may be representative of the collective voice and provide evidence of broader attitudes, values and patterns of behaviour.

Several such voices may confirm cultural practices [. . .] What may appear to be an individual and fragmented account, is representative of the totality and it is the totality which provides, through affirmation or denial, meaning. (1995: 96)

Memory, then, is shaped by language and images, priorities and expectations which are in turn influenced by the collective as 'culturally and socially determined', therefore the individual and the collective memory affect one another. (Chamberlain 1995: 95–6)

I acknowledge the problems associated with the empirical resources of oral history, such as whether statements are true or false. Inexact dates and a propensity to expound other areas are among the 'peculiarities of oral history'. Nonetheless, 'oral sources tell us less about events as such than about their meaning, and their value lies in the areas of language, narrative and subjectivity.' (Chamberlain 1995: 95)[5] The photographs of Anella's home-produced designs have enabled me to arrive at different historical evidence and the associated meanings of her work. One of the advantages of oral history and memory is the insight they give into groups generally not catered for in history books, such as black women. Additionally, this intimate method, whether between interviewer and interviewee or from autobiography can extract not only facts but subliminal emotions and meaning, as Walker has demonstrated.

MAKING AND MEANING

The act and energy of making and the associated meanings of the garment produced permeate this introduction to the home dressmaking culture of Jamaican women. In the early 1960s Paul Holbourne, a nineteen-year-old British Mod, experienced a cultural revelation and personal awakening, following the observation of the dress styles worn by West Indian immigrants.

> We have to get all our clothes made because as soon as anything is in the shops it becomes too common. I once went to a West Indian club where everyone made their own clothes. It was fantastic, everyone was individual, everyone was showing themselves as they really wanted to be [. . .] They were just expressing themselves as everyone should be entitled to do, be it in homes or private clubs or in the streets. (Hamblett and Deverson 1964)

Within the context of this essay the making of clothes by an individual, as opposed to a ready-made garment, has a double-edged quality. The process applied to the making of clothes as an individualized and private action and that of the finished garment in the use of an individual's image, as a personalized, forceful agency, is imperative to the establishment of a collective identity. Add to this equation the maker producing their own designs and the formula becomes potently expressive. The questions, 'What went into the making?' and 'What does it mean?' add to our understanding of the subjectivity of home dressmaking and the creation of new clothes.

Andrew Harrison argues that making things is part of a constructive cultural activity and is a part of the fabric of a society. In this sense it is valid to consider the home-made garments worn by Jamaican women as 'objects of communication' and 'a vehicle for thought . . . such objects demand our understanding and interpretation, and in doing so demand at the same time . . . an understanding of the maker of them'. (Harrison 1978:1) Anella calls the process of home dressmaking she practices, 'freehand dressmaking'. The very term 'freehand', in association with creativity and making, possesses an innate sense of liberation for the practitioner. The method rejects the use of paper dress patterns, thereby requiring the dressmaker to possess competence, skill and confidence in order to draw onto, or even cut directly into, the fabric. Anella does not lay much store by patterns as she feels they can lead one astray.[6] Her philosophy is, 'if you aim for what you want, you just get it . . . whatever you're sewing, it's inside of you.'

This is the nub of the system. Freehand dressmaking is the creation of individualized designs which may be inspired by a variety of sources, not predetermined by a bought paper pattern. These are produced as quick sketches or, according to Anella, 'from your knowledge'. In bypassing the use of paper patterns, this practice relates to certain definitions of so called 'high culture'. Harrison has applied the term 'free design' to the act of designing while making in fine art. He explains that 'free design' is possible through the build-up of extensive knowledge and skill in all areas of the subject; but only through practical exploration can this knowledge be developed and amplified. In order for it to be successful, free design relies on the competence of the maker, their extensive knowledge and a range of procedures and ideas. (Harrison 1978)

The freehand method of home dressmaking and the manipulation of designs to create something new challenges established rules of copyright. From Anella's experience, and that of other Jamaican women, the design and styling of clothes associated with this method consists of observing designs from a variety of sources—in a shop window, magazines or from films—as the basis for inspiration and adapting the design details to personal taste and body shape. The oral history of Mrs. Gloria Bennett substantiates this. Gloria emigrated to Britain from Jamaica in 1959 and settled in Doncaster, South Yorkshire, where she has remained a well patronized dressmaker to the black female community.[7] She learnt the freehand method as a teenager by watching others at their work and making clothes for herself:

> You used to get your styles out of a magazine out in Jamaica and you used to see nice little styles in there and then I would just sit down and cut out a dress without using a pattern then, but now I use patterns because I think it is much easier than messing about, laying out the fabric, measuring from here to here, how far your darts should go. (Bennett 1991)

In Britain Anella had access to a wealth of inspiration—shop windows, magazines, television, mail-order catalogues—but in rural Slygoville of the 1940s and 1950s, she visited the cinema only once, and unlike Gloria, did not have access to magazines. Her main source of information on the changes in cultural tastes and values was the radio. Therefore Anella relied very much on her own ideas and observations of the city and its people when she visited Spanish Town and the capital of Jamaica, Kingston:

> If a person asks you to make them a nice dress, they would normally give you no idea what style, I would look at that person and think, this material would be nice in a lovely square neck, or this material would be nice in a lovely V-neck or off the shoulder style. (James 1996)

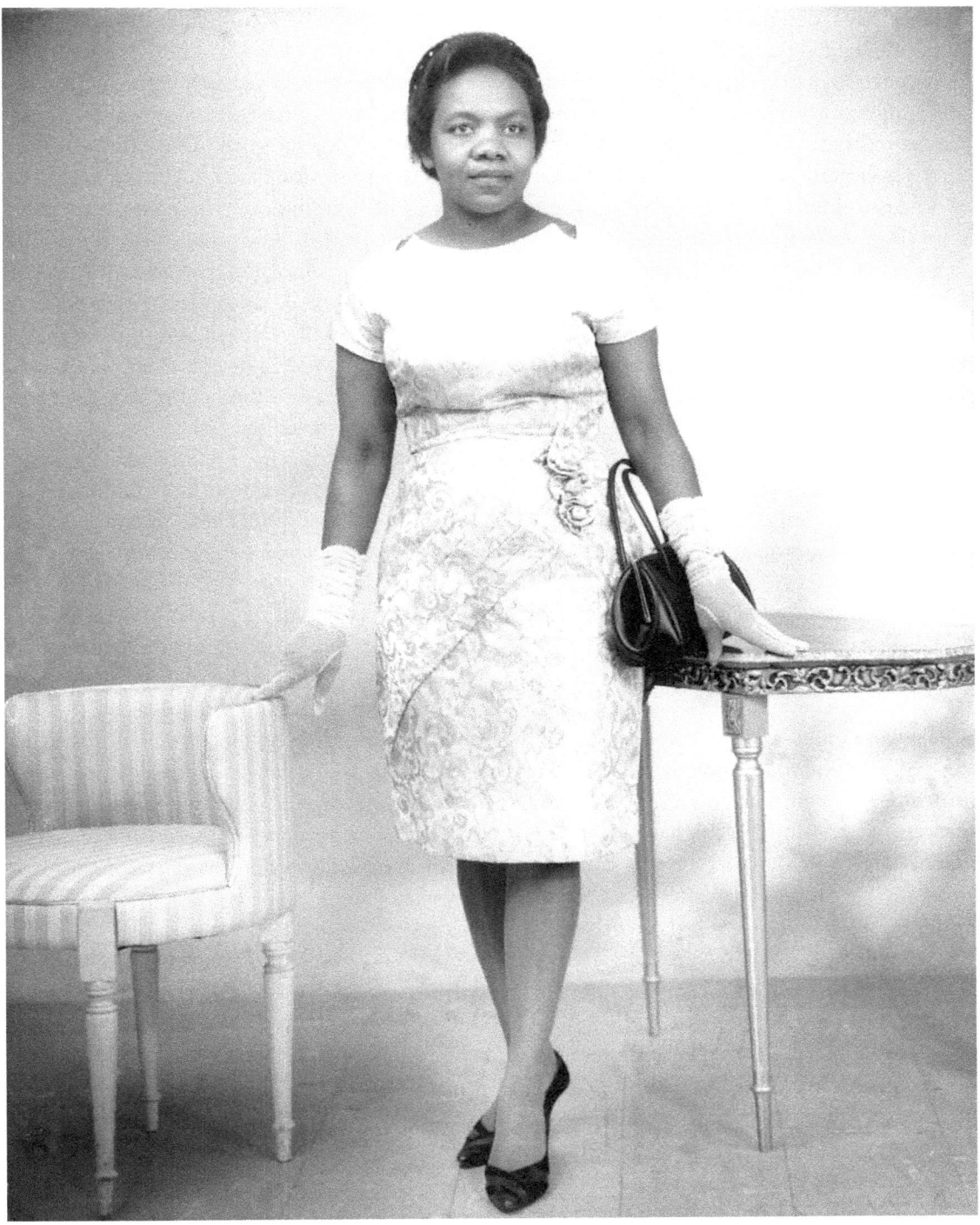

Figure 39 Anella James in her handmade dress.

When Anella designs and makes clothes to match the character of a client, the process is not interrupted or corrupted. Because of the liberating nature of freehand dressmaking the procedure is one of synchronization between creative thought and creative action. Once the measurements have been taken there is little time delay between the desired design and the creative action of cutting it out in the actual fabric. There is no lengthy paper pattern manipulation to detain the process—it is nigh on automatic.[8] If a definition of design is 'the practice of organising various elements to produce a desired result' and design 'deals exclusively with organisation and arrangement of form'[9], in this instance then the process of making is not separated 'from decisions concerning the form being made'. (Lambert 1993: 45) Therefore I propose that Anella is best defined as a designer-maker.

THE WONDER YEARS

Anella was taught sewing skills at school from the age of eight.[10] The process was slow and impractical, beginning with making bookmarkers. Once in the 'higher class' Anella made a doll's petticoat, followed by the accompanying dress. Her home was to be the more serious training ground. Anella's mother was also a dressmaker, whom she describes as 'a clever person' who just measured, cut, basted and sewed. Along with her sisters, they would practice on their mother's hand Singer sewing machine, 'I would use it to sew little skirts and things for myself.' In the late 1940s, due to her increased output of homemade and designed garments for herself and clients, Anella purchased her own Pinnock foot-operated sewing machine for about £30. As a tool and enhancement in the creative act of dressmaking, the sewing machine also signifies values invested in it by the individual. For Anella, her sewing machine is supreme.[11] Her recollections of her first sewing machine, the Pinnock purchased in Spanish Town, are full of detailed minutiae. The terms available for buying a sewing machine during the 1940s and 1950s were hire purchase or a percentage would be paid in advance and the remainder within thirty days and the machine would remain at the cash price. The Pinnock came with fourteen discs. Each one represented a particular stitch style—zig zag, straight stitch, etc. To use a stitch, a disc would be placed into the top of the machine.

Conversations with Anella and Gloria, my mother and my grandmother,[12] indicate that during the period under consideration clothes made by a dressmaker had far more cachet than ready-made garments. I argue the attraction lay in the fact that the whole process alluded to glamour and social status: the personal attention and the creation of an individualized garment which fits the individual well—both of which avoided the social stigma of wearing something too common and ill-fitting. This distinction between the two clothing types available for working-class Jamaican women surfaces in the derisory nomenclature given to a range of ready-made clothes:

> In the market place they used to sell ready-made or 'Wretch-e-dung' dresses hanging up . . . The reason why they were called 'Wretch-e-dung' you would find these things hung up, dresses overlapping one another . . . when somebody saw a dress they liked they asked the market holder to '"Wretch-e-dung", fetch that one down.' Those dresses were cheaper than one from the dressmakers. If you went to the store to buy a piece of cloth to make a dress it would cost you more money to go and buy it and make it, but at the same time it was made-to-fit and to look stylish in. (James 1996)

Anella's clients in Jamaica and Britain included men and women of all age groups, who generally depended on her to produce the ideas. Anella was particularly proud of the fact that she was adept at catering for the older female age group:

> I used to sew for them, I knew exactly what they wanted. In those days old ladies didn't wear short sleeves they preferred three-quarter-length sleeves.
>
> They liked pockets because most of the old ladies used to smoke a pipe (white chalk pipes) . . . These were patch pockets, or sometimes I would give them a little style, something special when they were going to a wedding, a little inserted pocket, or flap pocket because they would take their pipes with them, women in their 60s and 70s, people even older. They liked their skirts full and nice. Plain and print [fabrics], no dark colours, pinks and blues. (James 1996)

'Print' during the 1940s and 1950s was the generic term for all floral designs. Generally customers would ask Anella to choose the fabric for them and the request for a print was a clear indication to Anella that a floral design was required. Another colloquial term was 'the Blue Dress'. As Anella explains, this was the equivalent of what is now described as casual wear: 'Some of the old ladies would ask for a "Blue Dress". Not that it was actually a "Blue Dress", but they asked for a dress to wear to market, everyday wear. It might be in gingham, chambray or "old iron blue", [that is] denim.' When thinking of a design for the older ladies, Anella had to also bear in mind that they generally did not like their necklines too high or too low, and the opening should be at the front, not the back, for ease of accessibility. The garments Anella designed and produced for her elderly clients were a concerted attempt to achieve an 'expression of function'

and 'self-justifying aesthetic'. (Lambert 1993: 23–4) In essence she had pursued the solution of the combination of 'functional form with beauty'. (Lambert 1993: 4) This is partly attributable to Anella's sensitivity to the needs of the individual client.

By the 1950s Anella was a wife and the mother of six children. In order to make extra cash, she gave sewing lessons to girls in her home, in addition to producing and selling her own range of men's and children's clothing:

> I would buy a bale of cloth with all kinds of fabric [in it]. The amount of garments you could get out of it . . . could make a lot of money. From the bits and pieces I would make children's clothes, and use the good pieces to make things for men. I would sometimes have a length of khaki which could make a pair of trousers and a shirt . . . I would cut out about a dozen, say, pink knickers . . . I would just stitch them on the machine in a long line. I would buy a little decorative edging, elastic for the legs and waist. I would sell these to people at the weekends, for 6d and 4d, going to people's homes because I was looking after the children in the week. (James 1996)

This unusual relationship of men engaging a seamstress to make their clothes cannot be dealt with fully within this brief chapter. Her ability to make a good pair of trousers won over the confidence of the notoriously fashion-conscious black man in Jamaica and in Britain. (Tulloch 1992) Anella's skills were considered more than adequate to suppress the fear and stigma generally associated with 'home-made' clothes produced by a woman.[13] The matriarchal subtext of the social character of Jamaica may have some bearing on this.

HOME-MADE RESPECTABILITY

After 1945, Britain experienced the disconcerting status of full employment. All who wanted work had access to jobs, with thousands to spare. Britain's solution to this was to look to its colonies and invite British subjects to fill the plethora of blue-collar vacancies. The West Indians responded slowly at first. On 21 June 1948 only 547 Jamaican men arrived on the SS *Empire Windrush*. In 1959 approximately 29,397 West Indians, including some 12,573 Jamaican men, women and children, had migrated. (Glass 1960: 3–7) The resounding effect of this phenomenon was recorded widely throughout the media at the time and up until the present day.[14] Photographs of smiling yet bewildered men, women and children arriving at Southampton docks or Waterloo station, all of whom were immaculately dressed in tailored suits or crisp dresses, appropriately accessorized, became documentary evidence of the 'Colour Problem'.[15] To a working-class woman from rural Jamaica such as Anella, who had previously only travelled a few miles to Kingston, this was an incredible adventure and opportunity, weighted by the ambivalence of fear and excitement and managed with courage and trepidation. For although Britain was a foreign land, to Anella and thousands of other West Indians, Britain was also viewed as 'Home' and the 'Mother Country'.

My reference to home stretches beyond the four walls of a house and a home, to incorporate geography and a sense of one's place in society. In the context of Anella's story, and consequently that of the other women who emigrated to Britain from the West Indies, 'home' is an ambiguous, contested site—torn between the physical and geographical definitions in Jamaica and Britain. The construct of 'home' was a duplicitous layer of ambiguities based on their dual identity as Jamaicans and British subjects, who emigrated to Britain from their native homes, leaving their physical

homes behind to establish new ones. The geographical and material home of Britain was both strange and familiar, it held positive and negative qualities, it offered protection and aggression, home was connection and disconnection. West Indians preferred the term 'migrant' to 'immigrant'; because, as British citizens, they were only flitting between one part of the Empire and another. (Glass 1960: 1) All of which, I argue, affected and reflected Anella's sense of identity and place as a black woman in Jamaica and Britain which was to find expression in part through the practice of freehand home dressmaking. Home, then, is not purely considered as the physical building which stands as a symbolic and symptomatic expression of the good mother and/or wife, rather what is primary here is 'the identity of the home dweller'. (Pile 1996: 55)

Dress and self-image acted as an accessible conduit to all of this. They enciphered their desires and values, to be *seen* by the British public as respectable, and also cultural, social and economic values—on a more prosaic level, simply coping. (Hall 1984: 2–9) 'When I was coming here I made a tight hobble dress, straight dress, there was a bolero attached to the front of the dress with hand embroidery on the front of the dress and around the hem, [hand embroidered by Anella]. It was mauve.' (James 1996) This kind of testimony is supported by Gloria who remembered vividly the dress style she composed for her entrance into Britain:

> It was aquamarine trimmed with black. The dress was narrowly gathered at the waist, with black dropped in the front, a sweetheart neck. Because, you know, you were travelling, you came dressed up, black gloves, black shoes and I think I even came in a bloody hat. Coming so posh you know, I came in a hat. I made that dress. (Bennett 1991)

Such evocative descriptions of sheath-like dresses shamelessly accentuating the female form, and their co-ordinated accessories, the *frisson* of decorative highlights and the universal hat, were amongst the material trappings of respectable femininity demanded by the global conservative social etiquette of the period. This was supported by high fashion and such taste-making publications as *Vogue* and *Queen*. In this instance, Anella's engagement with fashion was founded on three things: the ethnic values of her black community, a limited budget and, lastly, on the signification of 'a distance from the world of necessity and an ability to indulge in a level of luxury [and] the idea of fashion as display was integral to the pursuit of femininity.' (Sparke 1995: 44)

Anella emigrated to Britain, along with her husband, out of necessity, because of the crippling state of underemployment in Jamaica (Glass 1960; Pryce 1979: 10–13) and the desire to gain a better standard of living for themselves and their children. The negative reaction to the presence of the 'Coloured Guests' (Glass 1960:1) brought the feasibility and complexities of this move into sharp relief. The issues of acceptance and survival of black people since 1948 and ensuing racial debates have been widely documented. (Johnson 1985; Gilroy 1987) Freehand home dressmaking, with its associated subjective aesthetics, was a way of negotiating these realities. In this creative act Anella could attain the pleasures of display and sensuality. I suggest that what Anella expressed in the presentation of herself as well-dressed was an ingrained belief in the power of glamour. The Hollywood movie, more than any other medium, projects the power of glamour and throughout Anella's lifetime Hollywood has maintained that to be well-dressed, and to project glamour, is to wield a sexual power, albeit subversive.

'Movies clearly state that fashion and glamour are fundamental to a woman's definition—in her own eyes, [. . .] and in the eyes of society.' (Basinger 1993: 114) Anella's wholehearted engagement with the aesthetics of glamour, I believe, was to achieve a positive definition of her own self and as a representative of her community.

A photograph of Anella taken in the mid-1960s is an image entirely composed by her. The fully lined, gold lame dress was styled with a 'boat neck to make it a little different'. Anella dyed the white gloves a mustard colour to match her dress. Her handbag and shoes were black, the latter suede. To crown her head, Anella wears a bandeau that she covered with velvet to match the lavish texture of the suede shoes. For me, the most pertinent design motif of the dress is the asymmetrical panel detail across the front of the dress, and positioned by a self-fabric corsage on the left of the waist. This styling detail certifies the 'difference' most strongly—but what is this difference? To return to Chamberlain:

> Appearances represent identity; they signal femininity. On a broader level, clothes are part of the iconography of womanhood. But they also indulge the imagination and the senses. Clothes represent a definition or statement of difference, independence and autonomy. They may also signal defiance and deception. 'I ent show poor' . . . such definitions may be illusory, but dressing well places women in the centre, as creators of the illusion . . . The signal may be subtle, but then the best deceptions are. (Chamberlain 1995: 106)

The difference being indicated by this generation of Jamaican women, encoded by the time and place of mid-1960s Britain, was not a conscious subversive action to flout the accepted values of British dress codes and fashions. It was, I propose, a means to integrate these codes with their own idiosyncratic inflections that advocated *their* cultural values, *their* 'colouredness',[16] *their* 'Jamaicanness'. What freehand home dressmaking had facilitated for Anella and her fellow Jamaican dressmakers was the subliminal emotions and meanings in being a Jamaican woman in Britain and the assertion of her own aesthetic-self and by extension a collective identity. Here in the oral history of Anella, freehand home dressmaking extends beyond Anella and her family, to produce goods for sale to her local community and to reach and to support a kinship based on a particular identity constructed and inspired by the peculiarities of a given place. The parameter of the home and home dressmaking is extended into wider significance and the aesthetic self.

NOTES

1. I concentrate on Jamaican women, as my research and the interviews conducted over the past ten years have been associated with the Jamaican community in Britain and Jamaica. There is of course evidence of issues raised in this essay applicable to women from other Caribbean Islands. See Bryan (1991) and Shepherd et al. (1995).

2. The dressmaking skills of Jamaican women were highly recommended in travel guides. British travel writer Bessie Pullen-Berry assured potential female visitors to the island that Jamaican dressmakers were 'excellent copyists and clever machinists who, provided they were given a good pattern, would turn out a well-made skirt for about 6s and a blouse for a little less, in 2 or 3 days'. (1903: 47)

3. *Jamaica Gazette*, Kingston, 18 June 1891. The first store to offer a costumes and dresses made-to-order department was Alfred Pawsey & Co. in 1905. See Robertson (1987–88).

4. The paper will also make reference to other Jamaican women to counterbalance the oral history of Anella James.

5. Allessandro Portelli, 'The Peculiarities of Oral History', *History Workshop Journal*, No. 12, Autumn 1981. Quoted by Mary Chamberlain. (1996: 95)

6. In the mid-1970s Anella attended an evening class at Waltham Forest College, as she felt it was time to learn what she termed, 'the British Method of dressmaking'. It was also a way of interacting with other black and white women. She discovered that, fundamentally, the basic concepts and procedures were the same as the freehand method, apart from the ubiquitous paper pattern. The classes did not convert Anella to the use of paper dress patterns. Only for an extremely complex style will she draft her own pattern.

7. Mrs. Gloria Bennett practiced dressmaking as a supplementary income to her full-time work as a bus conductor; Her client base has decreased since the 1980s. Reasons for this are purely speculative. I suspect easier access to more reasonably-priced clothing, the pervasive culture of 'designer clothing', desired by a younger Black British market, though this group continue to turn to Gloria for bridal wear, which, in the 1990s, is her main area of production.

8. It was a tradition in my family, and not unusual amongst other West Indian and some of the white working-class families I knew as a child, to have a new outfit for Christmas Day. For the special day of 1972 I purchased a new top, but had no 'bottom' to go with it. The actual chain of events is difficult to recall, but I do remember one minute, on Christmas Eve, my mum preparing the food for Christmas lunch, and the next she was laying out some cream fabric on the floor and cutting out a circular skirt for me. My mother used no pattern, neither did she draw an outline of the desired skirt onto the fabric. She just checked my waist measurement, and the

length and the circumference of the hem that I wanted. I do remember a deal being struck that my mother would complete the body of the skin if I sewed the hem. What is vivid in my memory is the apparent ease with which my mother dexterously executed the skirt amid the commotion of Christmas preparations: she cut out the skirt as if following an imaginary outline, then assembled the skirt, sewing the centre back seam, securing the zip and then waistband, passed the skirt to me to complete whilst she returned to the kitchen to complete the festive food preparations.

9. Norman Bel Geddes, *Horizons in Industrial Design,* 1934. Quoted by Susan Lambert. (1993: 45)

10. Anella's experimentation with the restyling of her clothes began at the age of six. She individualized the standard dress styles reserved for children, to the chagrin of her mother. Anella reduced the tent-like dress into a waisted silhouette by cutting and 'bad' hand sewing, so that she could look more like her older sisters.

11. My interview with Anella in January 1996 was conducted beside her industrial sewing machine, without any prompting from me.

12. Mrs. Roslyn Agatha Simpson, my grandmother, emigrated to Britain from Jamaica in 1953. My mother emigrated to Britain from Jamaica in 1954.

13. Anella taught herself to make men's trousers by unpicking a pair in order to study how they were cut. In Britain this knowledge proved indispensable to West Indian men. She could reduce the voluminous 'Wind Breaker' trousers into the slimmer and cropped cigarette pant styles of the 1960s. This Jamaican nickname derived from the fact that due to the voluminous style of the trouser, the trousers filled up like a balloon when the wind blew.

14. The fiftieth anniversary, in 1998, of the arrival of the SS *Empire Windrush* attracted significant coverage in the popular media, which examined its continuing historical and cultural legacy.

15. *Picture Post,* June 1956.

16. I have used this term as indicative of the political thinking amongst the average West Indian at this time and for consideration of its relevance to the ethnicity and identity of this community. Many of this generation still use the term 'coloured' over 'black' in reference to themselves, because of the political connotations of the latter.

FURTHER READING

Fiona Hackney, '"Use Your Hands for Happiness": Home Craft and Make-do-and-Mend in British Women's Magazines in the 1920s and 1930s', *Journal of Design History* 19/1 (2006), pp. 23–38.

62

HOUSE-TRAINED OBJECTS: NOTES TOWARDS WRITING AN ALTERNATIVE HISTORY OF MODERN ART

Tanya Harrod

Tanya Harrod's survey The Crafts in Britain in the Twentieth Century, *published in 1999, set a new standard for craft history. Wide-ranging in its coverage and contextual in its approach, the book provides a subtle and unillusioned narrative of debate, disappointment, possibility and radicalism. It is a must-read for any serious student of craft history, theory or practice. Because of its reach and complexity, however, it is also impossible to excerpt. Here instead is presented a more condensed example of Harrod's lateral thinking. Written for a volume on curatorial approaches to domesticity, the essay examines the unacknowledged role that craft objects played in the formation of canonical modernism. This 'alternative history' includes such subjects as the architecture of Le Corbusier, the paintings of Kirchner and Matisse and the wallpaper designs of the British Pop artist Eduardo Paolozzi. Through such examples, Harrod demonstrates that modern art's theoretical separation from craft, its aesthetic predilection for the machine, was by no means observed in practice. The argument is a crucial one for understanding the present moment, which is often described as 'postdisciplinary' as a means of distinguishing it from earlier periods. Such a sense of newness is justifiable, given the tremendous shifts that have occurred in art's media, educational institutions and markets. As other selections in this volume argue, one result of these changes has been a new prominence for craft*

within art, curatorial practice and subculture. But, as Harrod shows us, this productive instability has deep roots: as with all the best history, her writing provides not only an account of the past but also a lens through which to view the present.

Tanya Harrod, 'House-Trained Objects: Notes Towards Writing an Alternative History of Modern Art', in Colin Painter, ed., *Contemporary Art and the Home* (Oxford: Berg, 2002), excerpted.

To start with a truism: homes in Britain contain objects valued by their owners. Some of these objects will have been purchased, some may be gifts. Some may be defined as luxuries, others as necessities, but it is the interplay between these goods which is important and through which individuals display discrimination, responsibility and agency in their role as consumer-collectors.[1] A proportion of this *stuff* will have been diverted from an original use and re-presented, in effect taking the form of a souvenir. Such objects are framed by what has been called the "aesthetics of decontextualisation"[2] and might include horse brasses and Toby jugs (mementoes of an indigenous past) as well as totemic objects like blown glass gondolas signifying Venice or tribal rugs from Iran suggesting travel, embodying geographical, as opposed to temporal distance.[3] These might be modestly priced—a glass knick-knack from Venice—or have great rarity value—a kelim rug with a romantically narrative provenance.

Especially valued contemporary objects in any one British home may include paintings, prints, kitchen equipment, sound systems and so on. In Britain especially there are likely to be "craft" or applied art objects in many homes, made and designed by the same person and related to a family of objects made in the same material. As tracking the fortunes of the crafts and applied arts is one sub-text of this discussion, perhaps I should elaborate. There may be striking differentials in the material value of this category of handmade objects. Take for instance, objects seen as part of the British studio pottery movement. This encompasses a broad band of creative activity and could include a humble hand-thrown mug costing as little as 5 pounds or a Hans Coper pot which at auction might reach sums as high as 60,000 pounds. But, as is argued elsewhere in this volume, few British homes contain objects from the world of contemporary art, even if, paradoxically, many contemporary artists characteristically adapt or replicate "homely" objects in their work.

Another truism: elite one-off works of contemporary art are not widely available and are extremely expensive. But there are imaginative schemes like The Multiple Store set up in 1998 to produce limited editions of three-dimensional objects designed by leading sculptors. Although these cost between 950 and 90 pounds, in the view of Sally Townsend, the project's organiser, they appeal primarily to a highly informed audience, only really making sense if the consumer has some knowledge of the contributing artists' wider careers.[4] We may regret this state of affairs and seek to redress it through projects that offer artworks on loan to members of the public or to institutions such as schools or hospitals. We might even argue that the absence of art in a broad sense from British homes is disputable. Mediated forms of art enter our homes indirectly in the form of cheap reproductions and posters and, rather differently, in the form of television programmes, CD covers, designer objects, clothes and graphics in magazines, all of which reflect the visual culture of our time. But these broad definitions do not obtain in what the American philosopher of art Arthur Danto has identified as the "art-world"—a network of major collectors, dealers, museums, critics whose consensus defines what are accepted as appropriate art genres.[5]

This matter of appropriateness, a kind of modern version of mannerist decorum, is important because fine art is highly and effectively commoditised and to that end its boundaries are stringently policed by the art-world's gatekeepers. On the whole artists themselves do not operate as gatekeepers. During the past one hundred years many of them have crossed boundaries frequently, either deliberately or innocently. More recently, however, they have taken on many of the activities of curators and critics and have tended to endorse subtle, equally exclusive, variants of this taxonomic process. Take, for instance Richard Wentworth's remarkable 1998 exhibition *Thinking Aloud* which was an extended meditation on the nature of objects. It was inclusive but also exclusive, including architectural plans, fine art, maps, plans, flags, signs, toys and traps but, for example, no objects of contemporary applied art.[6] This may signal that currently the applied arts or crafts occupy a difficult cultural position—not quite art objects but not innocent workaday objects either, ripe for discovery by an artist curator.

SOME BOUNDARIES

Most of us, for whatever reason, do not purchase works of contemporary art for our homes (although we may buy reproductions, craft and applied art). Our major museums and galleries have their own systems of separation. The premier site for viewing contemporary art in Britain, London's Tate Modern,[7] does not collect (and rarely displays) applied art, craft or design, even if made or designed by acknowledged fine artists. In fact the demanding experience of visiting Tate Modern, a former power station reconfigured as a gallery of art, vividly underscores these divisions. Unlike the former Tate Gallery at Millbank (now Tate Britain), Tate Modern is the antithesis of a domestic space. Perhaps that is why its galleries are difficult places in which to view most art made before 1945.

Boundaries have to be drawn and, of course, London's Victoria and Albert Museum is where we expect to find twentieth and twenty-first century applied art and design intelligently presented. This is not particularly surprising or dismaying, but at times the exclusion distorts our perception of relatively recent historical moments. Curatorial taxonomies demonstrate that the museum of modern art necessarily has an ambivalent attitude to cultural history. Its purpose is partly to ring-fence and protect an activity called fine art—not to explain visual culture as a whole. It is a protectionist attitude that was also, and arguably remains, characteristic of the academic discipline called art history—particularly amongst the first scholars and curators who contributed to the construction of histories of the modern period.

Let us look at some examples. In 1997 the exhibition *Modern Art in Britain 1910–1914* (Barbican Art Gallery, London) attempted to recreate the crucial series of exhibitions of modern European art in London just before the First World War. Roger Fry's assemblage of continental art staged at the Grafton Galleries in 1910–11 under the title *Manet and the Post-Impressionists* is generally thought to be of crucial importance in this context and, of course, included ceramics as well as paintings and sculpture. But the curator of the Barbican show did not retrieve the sizeable number of pots by figures like Henri Matisse, Andre Derain and Maurice de Vlaminck shown in 1910 nor were they mentioned in the catalogue essays. In effect, the curator failed to cultivate a period eye. Fry's decision to include ceramics in 1910 was important, signalling that in the first decade of the twentieth century making and decorating pots was one way of avoiding the potential academicism of easel painting.[8]

When it comes to monographic exhibitions the complexity of an artist's interests is almost invariably censored. The 1993 exhibition on the work of *Ben Nicholson* (Tate Gallery, London)[9] was something of a landmark in its inclusion of Nicholson's applied art in the form of textiles and painted boxes. But they really only appeared in an archival context and Barbara Hepworth and Nicholson's shared intense involvement in the interiors which they created was therefore marginalised. Similarly the 1985 exhibition *St Ives 1939–64* (Tate Gallery, London),[10] which looked at the postwar artistic community in St Ives, included ceramics essentially as a footnote giving little idea of the fruitful interrelation of a range of applied arts and fine art in West Cornwall at that time.

Even when an artist's forays into the applied arts are allowed to become the focus of an exhibition, fine art curators have a way of missing the point. For instance, the 1998 exhibition *Picasso: Painter and Sculptor in*

Clay (Royal Academy of Arts, London) was limited to that artist's "unique" work in fired clay. But Picasso's one-off pieces were only a small part of the story. He also produced ceramics editions and series in collaboration with the Madoura Pottery for reasons that were utopian—linked to his membership of the Communist Party—and specific to a time and a place—the south of France just after the Second World War.[11]

We accept the kind of curatorial decisions I have outlined almost without a second thought. Clearly though, the relationship between craft, design, the fine arts and architecture needs to be addressed when writing the history of modern visual culture. Why do certain activities get ignored? Why have attitudes towards objects that appear intended for a domestic environment fluctuated so markedly in the past hundred and fifty years? After all, in the late nineteenth century and the early part of the twentieth century craft and applied art activity was seen as a way of expressing ambivalence towards bourgeois industrial society.[12]

CRAFT, ART AND EARLY MODERNISM

In 1905 the Arts and Crafts painter, designer and socialist Walter Crane published a collection of essays entitled *Ideals in Art* (1905). Trained as an illustrator, Crane sought a more democratic union of the arts and worked most fruitfully as a book illustrator and a designer of textiles, wallpapers, interiors and stained glass. He also created vivid graphic designs for the socialist cause.[13] We do not think of Crane as a radical artist in the context of early modernism—he belongs to an earlier period—but he inspired intelligent younger artists all over Europe by associating fine art

with decline and decadence. In Crane's view what he called "the art of portable picture painting" lacked serious contemporary links to architecture and a wider world.[14] There was an ambivalence about Crane's slighting references to art 'enclosed in gilt frames or supported on pedestals"[15] because Crane worked hard as a painter and by the 1890s has a reputation as a fine artist in Germany.[16] And to an extent he was borrowing and updating ideas expressed by William Morris as early as the mid-1870s.

Morris's linking of art and politics and morality and his concern with a context for art was taken up by avant-garde thinkers all over Europe and North America and in the Far East, particularly Japan. Morris's political writing led serious young artists to think how their work was consumed and by whom. Morris's imaginative valorisation of design encouraged a new interest in the quotidian, the everyday, in *things* at the expense of easel painting.

[. . .]

These experiments in the applied arts were also inspired by another early modern trope—that familiar anti-modern nostalgia for archaic and non-European cultures. Anti-industrial yearnings were commonplace amongst European novelists, critics and artists from Rainer Maria Rilke[17] to Walter Benjamin[18] to Le Corbusier. On his journey through eastern Europe in 1911—an important journey of self-education—Edouard Jeanneret (he was to adopt the pseudonym Le Corbusier in 1920) was deeply moved by vernacular architecture and craft as he travelled eastwards towards Turkey. He bought quantities of peasant pots as he travelled through the Balkans, as well as traditional rugs and embroideries, all of which were shipped back to Switzerland at great expense.[19] His *Voyage d'Orient* shaped his subsequent thinking.

Le Corbusier's frequently quoted identification of the home as a "machine to live in" and his negative attitude towards female domestic taste obscures the complexity and poetry of his attitude to housing and towards interiors. The crucial model was vernacular and peasant domestic architecture. Thus art in the home would chiefly take the form of craft, in the form of vernacular pots and Romanian and Berber rugs. Then there were objects that he deemed beyond style like the archaic and non-European artefacts and sculptures shown in his studio in 1935 in the informal exhibition *Les arts dit primitives dans la Maison d'aujourd'hui*. Other admissible objects included mural paintings by himself and a handful of artists he respected, tapestries (christened *le mural du nomade*) and found objects *(objets a reaction poetique).*[20]

Le Corbusier's discovery of vernacular ceramics in 1911 had been prefigured in 1908 by Adolf Loos who saw in the act of making a pot "chance, passion, dreams and the mystery of creation."[21] Their interest was shared by other artists and was part of a wider exploration of processes such as direct carving in stone and the creation of crude wood blocks to print paper or textiles. This was art for the home, both actual and potential. Intimacy was important as was bohemianism. The two were braided in photographs taken in 1910–12 of the Expressionist Ernst Ludwig Kirchner's studio homes in Dresden and Berlin—cave-like interiors filled with African carvings and hung with Kirchner's hand block printed textiles.[22]

CERAMICS: A SPECIAL CASE?

In the case of ceramics made by (rather than collected by) artists in the early modern period, there was a disjunction between a desire to experiment and the capacity of the art world to take in craft genres. Ceramics could synthesise painting and sculpture but this very hybridity proved problematic. Gauguin's ceramics in particular confused the boundaries between fine art and applied art. Mostly made between 1886 and 1895, they were not particularly well received and even today his ceramics still seem "difficult" and adventurous with their odd, inelegant conjoining of abstract vessels and realistic figures. Gauguin clearly hoped for an audience and was bitter about the way in which the public appeared to prefer safer kinds of experimentation in the pure forms of Auguste Delaherche's handsome neo-oriental pots.

The ceramics decorated by the painters known as Les Fauves around 1907 similarly tended to be little discussed either when they were exhibited or subsequently. They were made in most instances with the encouragement of the dealer Ambrose Vollard who suggested that his artists work with the self-taught tin-glaze potter Andre Metthey at his studio north of Paris in 1907.[23] On one level, involvement in making or decorating ceramics alerted artists to deficiencies in industrial design. Gauguin had been particularly critical of the historicist production at Sevres for instance—he called it "the death of ceramics"[24]. But an involvement with ceramics worked in other ways too. In the case of Matisse, in his early paintings he depicted little worlds of objects which deserve further investigation.

Ceramics and small sculptures made by his hand and the textiles, furniture and carpets that he collected animated his interior scenes, portraits and still lives. For instance, *The Red Studio* of 1911, *Girl with Green Eyes* of 1908 and his *Still life on an Oriental Rug* of 1907 all include ceramics of the kind that he decorated in collaboration with Andre Metthey. At that early date Matisse was also depicting archetypal figures against empty flattened backgrounds like *La Danse* of 1909–10. In this context his ceramic experiments were important in a different way, for his painting on pots suggested a new kind of dancing figure inhabiting a new kind of space. Matisse was not the only artist to learn from his applied art experiments during this Fauve period. In a reverse process, Andre Derain combined the decoration of pots like *The Dancers* and *Three Seated Nudes* of 1907 with the making of woodcuts and direct carving. All these works in non–fine art media allowed him to develop the kind of heightened colour and flattened space that he was pursuing in paintings like *The Dance* of 1906.

Apart from Gauguin none of these early modern painters left accounts of what ceramics meant to them and for each artist it seems likely that ceramics played a different role, ranging from spatial to colouristic experimentation. The best account of the possibilities of ceramics is provided by Pablo Picasso in a letter to the sculptor Henri Laurens. Picasso explained to Laurens in 1948 that while painting should create a sense of space he had found that by painting a ceramic form he was able to create the multiplicity of flattened view points which he demanded from sculpture.[25] As Picasso's dealer Kahnweiler astutely observed, some of the paradoxes that Picasso had first explored in sculptures like the 1914 *Glass of Absinthe* series were further investigated in his post-war ceramics. But the possibilities were many and complex. With Picasso, as Kenneth Silver has pointed out, there was a romantic political dimension to his decision to make editions of ceramics. He liked the idea that "anyone might buy them,

use them, and maybe hang them on the wall like a souvenir."[26]

WRITING MODERNISM

The desire of early modern artists to experiment and escape the stranglehold of easel painting and expensive sculptural processes like bronze casting by using a range of craft media deserves further exploration. We associate a move into the applied arts with the British Arts and Crafts Movement and with *art nouveau* but, as I have suggested, it had a more extended history than that. While we have histories of the role of the "ready-made" as an avant-garde challenge to accepted art practice, the more complex, messier and less conceptually transparent world of the "hand-made" remains under-documented.[27] In effect we can trace two early twentieth century avant-garde histories: an institutionalised fine art version, fully constructed by the 1950s, which continues to be policed predominantly by non-practitioners influenced by the art market. Then there is the less manageable actuality of painters and sculptors with uneasy links with the crafts, design and architecture.

Wandering through London's Tate Modern or Tate Britain, we would not expect to find the sculptor Eduardo Paolozzi's wallpapers of the early 1950s nor his wallpaper, textile, tile and ceramic work undertaken in collaboration with the remarkable photocollagist Nigel Henderson. We would be surprised to see the jewellery and silver-smithing work of the painter Alan Davie, likewise Henry Moore's cast concrete wall lights made speculatively in 1932 and his magnificent silk-screen wall hangings created in collaboration with Zika and Lida Ascher in 1948. These exclusionary policies often have gender implications as with the absence of Margaret Traherne's remarkable small stained glass panels and Frances Richard's spare embroidered pictures. These were just two amongst many women who trained as painters and who in the 1950s went into the applied arts because the area offered more opportunities for women.[28] All these objects of applied art have manifest links to concurrent developments in the fine art world of the 1950s but they were made from inappropriate and highly specific materials—cloth, fired clay, glass. In some instances they were made by inappropriate people, by which I mean women. For instance Richards, the wife of the painter Ceri Richards, was operating in a male dominated art world at that date. These objects were also, with few exceptions, manifestly made for the domestic environment.

We do not know much about this kind of work because, by contrast with the fine arts, the crafts in the twentieth century, in the form of a movement with roots in the nineteenth century Arts and Crafts Movement, have had an uncertain, complex identity and are under or ineffectively commoditised. Craft practice requires an ethnography rather than an evolutionary history. Thus in the first part of the twentieth century, up to say 1939, craft could include blind ex-servicemen making nets just after the First World War at the philanthropic workshops set up by the charity St Dunstans as well as a hand-block printed textile called *Log* of 1915 by Phyllis Barron inspired by Vorticist painting. Any definition of craft could also take in lots of handwork in industry and surviving vernacular craft such as hurdle making or basketry or the manufacture of turned and carved spoons—all examples of good design arrived at by non-design, by a tradition of making. After the Second World War craft could take in an anarchist counter-culture in which the workshop confers freedom and autonomy from the system as well as ceramics which conflated painterly mark-making and sculptural presence, sharing many of the tropes of fine art while operating on a domestic scale.

Only comparatively recently has it become apparent that twentieth century modernism was far from monolithic and that, indeed, we need to recover some lost modernisms. Figures like Phyllis Barron, Margaret Traherne and Frances Richards suggest that craft made by women and based on hand processes constitute one such lost modernism.[29] The situation is complicated by the way that all the visual arts tended to be in constant interplay, making it inadvisable to write about painting without considering design or to discuss sculpture without reference to ceramics.

Thus the critic Herbert Read moved from admiring the studio pots of William Staite Murray in the late 1920s to dismissing handmade objects in favour of the perfection of cast ceramics by 1934. Similarly in the late 1920s and early 1930s the painter Paul Nash took an interest in a whole range of craftsmen and women—in potters like Bernard Leach and hand-block printers like Barron, Dorothy Larcher and Elspeth Little and saw them as the one group in Britain who could effectively furnish a modern interior. But in 1934 he founded Unit One. Its membership was limited to artists and architects and it was to be "a practical unit in an industrial system."[30] In effect in about 1934 both fine art and the crafts were marginalised in favour of design and architecture by progressive thinkers faced with mass unemployment and slump and a flood of positive propaganda about the command economies of totalitarian regimes.

But the home, the domestic, still continued to be of interest to artists like Paul Nash, Barbara Hepworth and Ben Nicholson and commentators like Herbert Read. Problematically for the crafts however the home was to be furnished with a restricted mixture of advanced fine art, mass produced goods and mostly non-European art and craft. These were the components of the classic progressive interior, good examples being Herbert Read's flat in Hampstead in about 1933 and another Hampstead dwelling, the architect Erno Goldfinger's house at 2, Willow Road.[31] Their agenda would have found favour with figures abroad like Le Corbusier but, as we have seen, he carried on a far fuller Arts and Crafts tradition of integration, particularly in the fine detailing of his exteriors and interiors, his fondness for murals and tapestries and through his continuing interest in the vernacular, particularly in relation to building types and techniques.

If the situation was relatively fluid between the two world wars, after the Second World War figures in the art world began fully to articulate the belief that fine art should be viewed apart from other visual disciplines and that a discipline like painting was inherently alienated from domestic space. Clement Greenberg, the paramount fine art critic of the 1950s and 1960s, beautifully encapsulates the mood in his 1948 essay "The Situation at the Moment." He dismisses the Paris art world—the talk, the cosy literary and art magazines, and goes on to say "what is much more real at the moment is the shabby studio on the first floor of a cold-water, walk-up tenement on Hudson Street; the frantic scrabbling for money; the two or three fellow painters who admire your work; the neurosis of alienation that makes you such a difficult person to get along with." The art produced in these difficult conditions would not however be for private ownership; "abstract pictures rarely go with the furniture".

[. . .]

AWKWARD CUSTOMERS

Greenberg's strictures were slow to take root in Europe. I have suggested that in the 1950s artists in Britain experimented with a range of craft media for domestic interiors. Sometimes the aim was to gain a steady income. In the case of women the world of applied art could offer more creative opportunities. For an artist like the Danish Asger Jorn, Greenberg's exclusivity had little significance. In the early 1950s Jorn was briefly involved with a new design school modelled on Bauhaus lines, the Hochschule fur Gestaltung at Ulm in West Germany. But he soon fell out with its director Max Bill. Jorn's response was to set up his own counter cultural *Mouvement International pour un Bauhaus Imaginiste*. From 1953 he spent time in Albisola near Genoa where he and former COBRA group members made wild, expressive ceramics that were subsequently shown at the Milan Triennale of 1954. In Milan he made his speech *Contre le Fonctionalisme*—a heart-felt plea for the inclusion of what he called the "free artist" in the shaping of the post-war world. Jorn's ceramics were emblematic of that longing and their intense chaotic quality stood for everything which was lacking in the encroaching technocracy and warrior politics of the Cold War.[32] The way in which Jorn exhibited ceramics in Albisola is of interest. They were shown informally in garden and house settings. In 1955 he exhibited 500 plates decorated by school children.[33]

Jorn's activities had a good deal in common with a body of later work that mounted similar socio-political attacks against the art market and the primacy of painting and sculpture. But the joyful democratisation characteristic of Jorn's work was absent—as was its domesticity and craftedness. One question that might be asked of the conceptual and experimental art of the 1960s and 1970s is "where was it meant to go?" Jorn's work done in Albisola is still housed in his home—now a museum. But that kind of setting hardly seems appropriate for the conceptual art of the 1970s. The answer seems to be an archive, or at least a kind of site that is neither gallery nor home. Yet the facture of 1970s conceptual art employed a domesticity of process. As John Stezaker points out: "Conceptual art opened up the use of non-specialist processes of everyday life in the production of work (for me photography, collecting, captioning etc.) . . . The work which I most value from this time (of mine and others) is the work which remains closest to the everyday procedures employed and to the confrontation with the everyday which these allow."[34] Stezaker's investigations could in theory be done without a studio. The kitchen table would do. But even if everyday processes were employed it would be absurd to see domesticity in the art or to think much about the work being displayed in a domestic environment.

Perhaps the anti-domesticity of conceptualism explains why there was a craft revival from 1973, under the patronage of the government funded Crafts Advisory Committee (later Crafts Council). Certainly unbuyable fine art and the uniformity of product design created a space for complex, referential objects. Most of the crafts of the 1970s and 1980s essentially drew on early modernism in painting and sculpture for inspiration. All the genres of craft were enlivened by this symbiotic relationship with twentieth century fine art. Ceramics, in particular, functioned as affordable abstract sculpture for the home. But surprisingly, many of the tropes that we associate with contemporary art—a fondness for creating ghostly doppelgangers of existing objects for instance or for using textiles to carry confessional communications—were already to be found in the more radical crafts of the 1970s.[35] "Radical craft"—the phrase seems designed to raise a smile. But its existence deserves serious documentation.

THE SITUATION AT THE MOMENT

If we look over the past century and a half, it is possible to discern an alternative history of the visual arts, which is inclusive rather than exclusive and which honours the variousness of artists' approaches and which includes crafts and applied arts in the story. One way of noting this catholicity is to be attentive to the home and objects of domestic scale. As we have seen, this approach results in tales of the unexpected in which Le Corbusier emerges as a pottery enthusiast and Eduardo Paolozzi writes busily to the Council of Industrial Design with details of his wallpaper designs.[36]

If we pursue this alternative history up to the present, activities that seemed overlooked and marginal now appear to have taken centre stage. It would be hard to ignore the recent development of interest among fine artists in domestic objects, particularly in their oddness and instability. It is a tendency inspired by, as much as anything, the cultural and theoretical studies taught in art schools that introduce students to sources as varied as Marx's haunting opening lines on *The Fetishism of the Commodity and the Secret Thereof*, Freud's 1919 essay on the uncanny and a powerful body of writing by figures as diverse as Mary Douglas, Alfred Gell and Susan Stewart. At the heart of much of this writing is an emphasis on consumption as a process through which individuals construct identities and in which objects, like people, have unstable and unpredictable careers. Artists reading these texts find oblique confirmation of their rejection of avant-garde notions of originality in favour of replicating, reconstituting and remaking everyday "things". This kind of activity, however, does not herald a situation in which the home becomes a site of experimental endeavour. Quite the reverse.

[. . .]

It would be innocent to suggest that many artists would want their work to be exhibited in an actual home. Indeed the everydayness (with a twist) of most of the work suggests that it would be lost in a cluttered "real" home. This was made abundantly clear in the 2001 exhibition and project *Close Encounters of the Art Kind*.[37] Its curator, the artist Colin Painter, recruited six sculptors and six North London households and over six months a work by each sculptor was rotated round each home. This was a worthy project that appears to have brought pleasure to both householders and artists. But the photographic record of the project is revealing; most of the sculptures are rendered virtually invisible by the home environment and are only properly recuperated in the exhibition at the Victoria and Albert Museum where they are displayed apart from the everyday objects that clutter most homes. The V&A exhibition also included three or four much loved objects from each household. These were shown in museum vitrines and ended up looking as odd and as challenging as the sculptors' contributions—a distinctly "uncanny" outcome!

This is where we return to the applied arts and in particular to ceramics. I have suggested that ceramics are a 'special case', outlining instances when fine artists were particularly drawn to the genre. But now I want to focus on ceramics made by ceramicists—artists who devote themselves pretty exclusively to the medium of fired clay. I have already suggested that ceramics have a role as affordable abstract sculpture for the home. But ceramics are historically part of a rich tradition of ornament. In particular, the extraordinary plasticity of clay means that ceramics have re-represented all kinds of artifacts—from bronze statuettes to silverware. As a result ceramics have, over the centuries, carried all kinds of high and low art references into the domestic space. Although modernism in design narrowed the ornamental range of industrial ceramics, from the 1920s onwards studio pottery was able to carry on this process of re-representation.

Currently, a few ceramicists, like their cousins in the fine art world, are commenting on the world of things and on consumption itself. But the fact that ceramics have always been part of the domestic environment as part of a culture of display and collecting largely orchestrated by women raises an important question. Does an object's status as an artwork depend upon context? In the exhibition "Art/Artifact" in New York in 1988 the curator Susan Vogel put a Zande hunting net, tightly rolled for transport, into a clean white art gallery space. A practical African object immediately began to look like a conceptual sculpture.[38] Very few "homes', indeed only those configured to resemble galleries, could have worked that kind of tranformation on the Zande net. In most domestic settings the bundled up net might have stood out, but a viewer would have been unlikely to freely associate it with a similar looking artwork. As we have established, few homes contain artworks. The net in a home would be more likely to spark off thoughts of practical objects—a tent for instance, or even a net!

Which brings me to work of the ceramicist Richard Slee. Like the Zande net, his work looks well in art galleries. Slee's ceramics are sensitive to the whole range of ornamental objects that do 'memory work' for us—the horse brasses, Toby jugs and glass gondolas that I mentioned at the beginning of this paper. To run through our original truisms: homes are full of practical things like tents, mixers, and washing machines. They frequently lack artworks as defined by the art-world. We expect instead to find a range of fairly assertive ornaments. Slee, by working from ornament to create ornament, is able to enter the home with ease. And without anyone much noticing Slee manages to investigate big themes like national identity, landscape, 1940s animated film, cheap British industrial pottery and icons of power and governance. So what have we got here? An object made of glazed fired clay, which, by virtue of its facture, material and its maker's background, generically does not quite fit into the art-world. Or is this a house-trained art object? But here I rest my case because relations between art, craft and the home fluctuate and will continue to do so. Inscribed in this instability is an alternative history of modern art.

NOTES

1. On the responsible consumer see D. Miller, *Consumption and its Consequences* in (ed) Hugh Mackay, *Consumption and Everyday Life*, Sage 1997; on "consumers-collectors" see Guido Guerzoni and Gabriele Troilo, *Silk Purses Out Of Sows' Ears: Mass rarefaction of consumption and the emerging consumer-collector* in (ed) Marina Bianchi, *The Active Consumer: Novelty and Surprise in Consumer Choice*, Routledge, 1998.

2. Arjun Appadurai, *Introduction: commodities and the politics of value* in (ed), Arjun Appadurai, *The Social Life of Things: commodities in cultural perspective*, CUP 1986, p. 28.

3. On souvenirs see Dean MacCannell, *The Tourist: A new theory of the leisure class*, (1976) University of California Press, 1999, pp. 145–160; for the value and distance see Pennina Barnett, *Rugs R Us(And Them): The Oriental Carpet as Sign and Text*, Third Text, 30, Spring 1995, p. 17.

4. For more on The Multiple Store see www.multiplestore.org.

5. Arthur Danto, *The Art-World*, Journal of Philosophy, 61, 1964, pp. 571–84: also Alfred Gell, *Vogel's Net: Traps as Artworks and Artworks as Traps*, Journal of Material Culture, vol. 1, no. 1, March 1996.

6. Richard Wentworth, *Thinking Aloud*, Hayward Gallery Publishing, London 1998.

7. I am focusing on London here; major galleries outside the metropolis include art, craft and design under one roof, but invariably in different areas of the museum.

8. See Anna Gruetzner Robins, *Modern Art in Britain 1910–1914*, Merrell Hoberton/Barbican Art Gallery 1997; for other reasons for the inclusion of ceramics see S. K. Tillyard, *The Impact of Modernism: Visual Arts in Edwardian England*, Routledge 1988, pp. 127–128.

9. Jeremy Lewison, *Ben Nicholson*, Tate Gallery 1993.

10. *St Ives 1936–64: Twenty Five Years of Paintings, Sculpture and Pottery*, Tate Gallery, 1985.

11. See the important review of the exhibition by Kenneth E. Silver, *Pots, Politics, Paradise*, Art

in America, March 2000, pp. 78–141. For negativity towards Picasso's ceramic practice see Tanya Harrod, *Picasso's Ceramics*, Apollo, May 1989.

12. For an elegant analysis of this phenomena see Amy Ogata, *Artisans and Art Nouveau in Fin-de-Siecle Belgium* in (ed) Lynda Jessup, *Antimodernism and Artistic Experience: Policing the Boundaries of Modernism*, University of Toronto Press, 2001.

13. (ed) Greg Smith and Sarah Hyde, *Walter Crane 1845–1915: Artist, Designer and Socialist*, Lund Humphries/Whitworth Art Gallery, Manchester, 1989, p. 108.

14. See Tillyard, op. cit., p. 109.

15. Walter Crane in *Arts and Crafts Exhibition Society*, 1888.

16. My thanks to Alan Crawford for clarifying my thoughts on Crane.

17. See Idris Parry, *Rilke and Things* in (ed) Tanya Harrod, *Obscure Objects of Desire: reviewing the crafts in the twentieth century*, Crafts Council 1999.

18. See Esther Leslie, *Walter Benjamin: Traces of Craft*, Journal of Design History, vol. 11, no. 1, 1998.

19. See H. Allen Brooks, *Le Corbusier's Formative Years*, University of Chicago Press 1997, pp. 263–300, 326.

20. On Le Corbusier and the applied arts see Pierre Saddy/Claude Malecot, *Le Corbusier: le passé a reactions poetique*, Ministere de la culture et de la communication, Paris 1988.

21. Adolf Loos, *Pottery* (1908) in (ed) Adolf Loos, *Ornament and Crime: Selected Essays* (ed) Adolf Opel, Ariadne Press, California 1998.

22. Colin Rhodes, *Through the Looking-Glass Darkly; Gendering the Primitive and the Significance of Constructed Space in the Practice of the Brucke* in Louise Durning and Richard Wrigley, *Gender and Architecture*, John Wiley & Sons, Chichester 1996.

23. Musee Matisse, Nice, *La Ceramique Fauve: Andre Methey et les Peintres,* Reunions des Musees Nationaux, 1996.

24. Ibid., p. 15

25. Daniel-Henry Kahnweiler, *Picasso: Keramik*, check 1957.

26. Quoted in Silver, op. cit., p. 141.

27. But see Joan Key, *Readymade, or Handmade?* in Richard Salmon/Kettle's Yard, *Craft*, Edwardes Square Studios, 1997.

28. See Tanya Harrod, *The Crafts in Britain in the Twentieth Century*, Yale University Press, 1999, pp. 245, 295, 324–5.

29. Harrod, 1999, *Ibid.*, pp. 11, 116–118.

30. Herbert Read(ed), *Unit One: The Modern Movement in English Architecture, Painting and Sculpture, Cassell, 1934.*

31. See Gillian Naylor, *Modernism and Memory: Leaving Traces* in (eds) Marius Kwint, Christopher Breward, Jeremy Aynsley, *Material Memories: Design and Evocation*, Berg 1999.

32. See Tanya Harrod, *British Ceramics: a discussion document* in (ed) Lise Seisboll, *Britisk Keramik.2000.dk*, Rhodos, 2000.

33. On Jorn see Janet Koplos, Max Borka, *The Unexpected: Artists' Ceramics in the 20th Century*, Museum het Kruithuis, s'Hertogenbosch, 1992; Guy Atkins, *Asger Jorn: The Crucial years 1954–1964*, Lund Humphries 1977; Asger Jorn, *Pour la Forme: Ebauche d'une methodologie des arts*, Edite par l'Internationale Situationniste, n.d.

34. Clive Philpot, Andrea Tarsia, *Live in Your Head: Concept and Experiment in Britain 1965–75*, Whitechapel Art Gallery, 2000, p. 153.

35. Harrod, 1999, *Ibid.*, Chapter 10. See also Rosemary Betterton, *Undutiful Daughters: Avant-Gardism and Gendered Consumption in Recent British Art*, Visual Culture in Britain, vol. 1, no. 1, 2000.

36. Paolozzi to Peter Hatch, Council of Industrial Design, August 22, 1955, List 1, no. 2, Nigel Henderson papers, Tate Gallery Archive.

37. Victoria & Albert Museum, *Close Encounters of the Art Kind*, 2002.

38. For a brilliant discussion of ART/ARTIFACT see Alfred Gell, *Vogel's Net: Traps as Artworks and Artworks as Traps*, Journal of Material Culture, vol. 1, no. 1, March 1996.

FURTHER READING

Tanya Harrod, *The Crafts in Britain in the Twentieth Century* (New Haven, CT: Yale University Press, 1999).

Tanya Harrod, ed., *Obscure Objects of Desire: Reviewing the Crafts in the Twentieth Century* (London: Crafts Council, 1997).

Janet Koplos et al., *The Unexpected: Artists Ceramics of the 20th Century* (s'-Hertogenbosch: Museum Voor Hedenaagse, 1999).

63

THE NEW CERAMIC PRESENCE

Rose Slivka

Though assertions as to the art status of craft are not in short supply, convincing critical writing that makes the case is rare indeed. Perhaps the best example is 'The New Ceramic Presence', published in Craft Horizons *(the magazine of the American Craftsmen's Council) by its editor, Rose Slivka. A good candidate for the single most famous piece of writing on American studio craft, Slivka's impassioned essay concerns the new ceramics of the moment—mostly emanating from the circle of Peter Voulkos through his teaching at the Otis Art Institute in Los Angeles. She presents this work as a sort of painting-in-the-round, an explicit departure from all precedent in ceramics and an implicit assault on the qualities of the medium itself. For many readers, the article was a shocking affront, but given her background it should have been no surprise: prior to joining* Craft Horizons *in 1955, and taking over the helm in 1959, she had been a freelance journalist and editor for several New York fine art publications. A cynic might argue that her support of what would come to be called 'Abstract Expressionist Ceramics' betrayed a degree of self-interest, a search for a niche among the many writers who made their names through criticism of the New York School. But even if this did play into her initial motivations, Slivka would go on to be a committed, intelligent and sometimes even visionary editor. Until 1979, when* Craft Horizons *was replaced by the more commercially oriented* American Craft *and she departed (briefly editing another quarterly,* Craft International*), she ran one of the most interesting and multivalent journals published on any aspect of the visual arts.*

Rose Slivka, 'The New Ceramic Presence', *Craft Horizons* 21/4 (July/August 1961).

American ceramics—exuberant, bold, irreverent—has excited admiration and controversy among craftsmen in every field both here and abroad. The most populated, aggressively experimental, and mutable area of craft expression, it is symptomatic of the vitality of United States crafts with its serious, personal, evocative purposes.

As in the other arts, ceramics, also, has broken new ground and challenged past traditions, suggested new meanings and possibilities to old functions and habits of seeing, and has won the startled attention of a world unprepared for the unexpected. (At the second International Ceramic Exhibitions at Ostend, Belgium, in 1959, the United States exhibit, circulated abroad for the last two years, became the focus of the show.) To attempt some insight into what is happening—for it is a happening, peculiar to our time and to American art as a whole—to probe the complex sources of our ceramics and its vigorous new forms is the aim of this investigation.

What is there in the historical and philosophic fabric of America that engendered the unique mood of our expression?

America, the only nation in the history of the modern world to be formed out of an idea rather than geographic circumstance or racial motivations—the country compelled by the electrifying and still new idea of personal freedom that cut through geographic, racial, and economic lines to impel people everywhere in unparalleled scope, rate, and number—was a philosophic product of the Age of Reason and the economic spawn of the Industrial Revolution. In the two hundred years of our short history our expanding frontier kept us absorbed in the problems of practical function and pressured us to solve them in a hurry. We have, as a result, become the most developed national intelligence in satisfying functional needs for the mass (in a massive country), with availability an ideal. The rapidity, the scale, and the intense involvement in mechanization have been unprecedented. If there is, in fact, any one pervasive element in the American climate, it is that of the machine—its power, its speed, its strength, its force, its energy, its productivity, its violence.

Not unified by blood or national origin (everyone is from someplace else), or a sense of place (with many generations of a family history identified with one place, as in Europe), we are a restless people. A nation of immigrants with a continuing history of migration, we are obsessed by the need for arrival—a pursuit that eludes us. And so, we are always on the go. (Our writers—Walt Whitman, Herman Melville, Thomas Wolfe, and, most recently, Jack Kerouac—have struggled for a literary art form to express this.) Having solved our need for mobility by mechanical means, we love engineering and performance and the materials and tools by which we have achieved them.

In our involvement with practical matters, we were too busy really to cultivate the idea of beauty. Beauty as such—the classical precepts of harmonious completion, of perfection, of balance—is still a Western European idea, and it is entirely possible that it is not the aesthetic urgency of an artist functioning in an American climate—a climate that not only has been infused with the dynamics of machine technology, but with the action of men—ruggedly individual and vernacular men (the pioneer, the cowboy) with a genius for improvisation. Our environment, our temperament, our creative tensions do not seem to encourage the making of beauty as such, but rather the act of beauty as creative adventure—energy at work—tools and materials finding each other—machines in movement—power and speed—always incomplete, always in process.

As far back as 1870, a Shaker spokesman declared that Shaker architecture ignored "architectural effect and beauty of design" because what people called "beautiful" was "absurd and abnormal."[1] It had been stated by others before and was restated many times since, including the declaration in the 1920s by famous architect Raymond M. Hood: "This beauty stuff is all the bunk."[2] A typical American attitude, it may well have expressed the beginning of a new American aesthetic rather than gross lack of appreciation for the old one.

This is the ebullient, unprecedented environment of the art that, particularly in the fifteen years since World War II, has asserted itself on every level.

First manifested in painting—the freest of the arts from the disciplines of material or function—it projected such a presence of energy, new ideas, and methods that it released a chain reaction all over the world, and for the first time we saw the influence of American painting abroad. But nowhere has the impact of contemporary American painting been greater than here at home. Feeding on itself, it has multiplied and grown in vitality and daring to penetrate every field of creative activity.

Pottery, of course, has always served as a vehicle for painting, so this in itself is nothing new. The painted pottery of Greece strictly followed the precepts of the painting of the time in style and quality, while that of Japan was often freer and in advance of its other media of painting, even anticipating abstract modern approaches. Contemporary painting, however, has expanded the vocabulary of abstract decoration and given fresh meaning to the accidental effects of dipped, dripped, poured, and brushed glazes and slips on the pot in the round.

But its greatest and most far-reaching effect in ceramics has been the new emphasis it gave to the excitement of surface qualities—texture, color, form—and to the artistic validity of spontaneous creative events during the actual working process—to everything that happens to the clay while the pot is being made.[3] Clay, perhaps more than any other material, undergoes a fabulous creative transformation—from a palpable substance to a stonelike, self-supporting structure—the self-recorded history of which is burned and frozen into itself by fire.

More than in any other form of art, there is a tradition of the "accident" in ceramics—the unpremeditated, fortuitous event that may take place out of the potter's control, in the interaction between the living forces of clay and fire that may exercise mysterious wills of their own. The fact that the validity of the "accident" is a conscious precept in modern painting and sculpture is a vital link between the practice of pottery and the fine arts today. By giving the inherent nature of the material greater freedom to assert its possibilities—possibilities generated by the individual, personal quality of the artist's specific handling—the artist underscores the multiplicity of life (the life of materials and his own), the events and changes that take place during his creative act.

Painting shares with ceramics the joys and the need for spontaneity in which the will to create and the idea culminate and find simultaneous expression in the physical process of the act. Working with a sense of immediacy is natural and necessary to the process of working with clay. It is plastic only when it is wet and it must be worked quickly or it dries, hardens, and changes into a rigid material.

The painter, moreover, having expanded the vistas of his material, physically treats paint as if it were clay—a soft, wet, viscous substance responsive to the direction and force of the hand and to the touch, directly or with tool; it can be dripped, poured, brushed, squeezed, thrown, pinched, scratched, scraped, modeled—treated as both fluid and solid. Like the potter, he even incorporates foreign materials—such as sand, glass, coffee grounds, crushed stone, etc.—with paint as the binder, to emphasize texture and surface quality beyond color. (We are aware that the application of paint as color, with its inherent qualities and dependency for a supporting structure by adhesion to a plane in another material, makes a fundamental difference between the two arts—between it and all other practices of the plastic arts. We are not trying to simplify or equate. We are pointing to those common denominators that have profoundly affected and influenced the new movement in pottery.) It is corollary that the potter today treats clay as if it were paint. A fusion of the act and attitudes of contemporary painting with the material of clay and the techniques of pottery (the potter's hand if not always his wheel is there), it has resulted in a new formal gesture that imposes on sculpture.

In the past, pottery form, limited and predetermined by function, with a few outstanding exceptions, has served the freer expressive interests of surface. Today, the classical form has been subjected and even discarded in the interests of surface—an energetic, baroque clay surface with itself the formal "canvas." The paint, the "canvas," and the structure of the "canvas" are a unity of clay.

There are three extensions of clay as paint in contemporary pottery:

1. the pot form is used as a "canvas";
2. the clay itself is used as paint three-dimensionally—with tactility, color, and actual form;
3. form and surface are used to oppose each other rather than complement each other in their traditional harmonious relationship—with color breaking into and defining, creating, destroying form.

This has led the potter into pushing the limits of paintings on the pot into new areas of plastic expression: sculptured painting, with the painted surface in control of the form. The potter manipulates the clay itself as if it were paint—he slashes, drips, scrubs down, or builds up for expressive forms and textures. Or around the basic hollow core he creates a continuum of surface planes on which to paint. In so doing, he creates a sculptural entity whose form he then obliterates with the painting. This, in turn, sets up new tensions between forms and paint. It is a reversal of the three-dimensional form painted in two. Now the two-dimensional is expressed in three—on a multiplaned, sculptured "canvas." As a result, modern ceramic expression ranges in variety from painted pottery to potted painting to sculptured painting to painted sculpture to potted sculpture to sculptured pottery. And often the distinctions are very thin or nonexistent.

The current pull of potters into sculpture—in every material and method, including welded metals, cast bronze, plaster, wood, plastics, etc.—is a phenomenon of the last five years. So great a catalyst has been American painting that the odyssey from surface to form has been made through its power. Manipulating form as far as it could go to project the excitement of surface values, the potters found even the slightest concession to function too limiting. From painter-potters, they were impelled to become painter-sculptors. Instead of form serving function, it now serves to develop the possibilities of the new painting. However, while this painting generates the creation of forms for itself—often massive in scale—it tolerates the dominance of no presence other than—itself. In his new idea of a formal synthesis, the potter is inevitably pushing into space—into the direction of sculpture.

As a fusion between the two dimensional and the three dimensional, American pottery is realizing itself as a distinct art form. In developing its own hybrid expression, it is like a barometer of our aesthetic situation.

Involvement in the new handling of surface with form, however, cannot rest on traditional categorizing. The lines cross back and forth continuously. While the painter, in building and modeling his surface has reached toward the direction of sculpture, so, too, the sculptor has been independently reaching away from the conventional bounds of sculptural form toward an energy of space and the formal possibilities of an activated surface (with or without color). The hybrid nature of this expression, however, has always been within

Figure 40 Peter Voulkos, *Soleares II*, 1958.

the realm of sculpture, only to be released as an entity in our time.

Sculpture, as every area of the plastic arts, is reevaluating the very idea that gave it birth—monumentality. A traditional sculptural aspiration, its values, too, have changed. The sculptor today places greater emphasis on event rather than occasion, in the force of movement and the stance of dance rather than in the power of permanence and the weight of immobility, in the metamorphosis of meanings rather than in the eternity of symbols.

Specific to the kinship between potter and sculptor is the fact that clay is a primary material for both (for the potter, the sole material; for the sculptor, one of several). Its tools and methods impose many of the same technical skills and attitudes on both. In general, potter and sculptor share a creative involvement in the actuality of material as such—its body and dimension—an experience of the physicality of an object that in scale and shape relates to the physicality of the artist's own body in a particular space.

The developments in abstract sculpture have decidedly affected the formal environment of ceramists everywhere. The decision of the sculptor to reinterpret the figure as well as all organic form through abstraction and even to project intellectually devised forms with no objective reference inevitably enlarged the formal vistas of every craftsman and designer working in three dimensions.

To pottery, sculpture has communicated its own sense of release from the tyranny of traditional tools and materials, a search for new ways of treating materials and for new forms to express new images and new ideas.

In addition to painting and sculpture, other influences that contributed decisively to the new expression in American pottery were: The bold ceramic thrust of Picasso, and Miró with Artigas, gave encouragement and stimulation to the movement that had already begun here. The Zen pottery of Japan, furthermore, with its precepts of asymmetry, imperfection (crude material simplicity), incompleteness (process), found profound sympathy in the sensibilities of American potters.

The freedom of the American potter to experiment, to risk, to make mistakes freely on a creative and quantitative level that is proportionally unequaled anywhere else has been facilitated, to a large extent, by this country's wealth and availability of tools and materials. It gave further impetus to the potter's involvement in total process—in the mastery of technology and the actual making of the object from beginning to end—in marked contrast to the artist-potters of European countries who leave the technology and execution to the peasant potter and do only the designing and finishing. Aside from the fact that we have no anonymous peasant potters in this country to do only the technical or preparatory work, the American loves his tools too much to leave that part of the fun to someone else. For him, the entire process contains creative possibilities. Intimacy with the tools and materials of his craft is a source of the artist's power.

Spontaneity, as the creative manifestation of this intimate knowledge of tools and their use on materials in pursuit of an art, has been dramatically articulated as an American identity in the art of jazz—the one medium that was born here. Always seeking to break through expected patterns, the jazzman makes it while he is playing it. With superb mastery of his instrument and intimate identification with it, the instrumentalist creates at the same time he performs; the entire process is there for the listener to hear—he witnesses the acts of creation at the time they are happening and shares with the performer the elation of a creative act.

Crafts that functioned in the communal or regional culture of an agrarian society do not have the same meaning in the internationalized culture of an industrial society. Thus, all over the modern world, the creative potter has been reevaluating his relation to function. Certainly, the potter in the United States is no longer obliged to produce for conventional function, since the machine has given us so many containers for every conceivable variety of purposes and in every possible material—plastic, paper, glass, ceramic, fiber, metal—with such quick obsolescence and replacement rates that they make almost no demands on our sensibilities, leaving us free—easy come, easy go—from being possessed by the profusion and procession of objects that fill our lives today. We are accustomed to our functional problems being solved efficiently and economically by mechanical means; yet we are acutely aware of our particular need for the handcrafts today to satisfy aesthetic and psychological urgencies. The painter-potter, therefore, engages in a challenge of function as a formal and objective determinant; he subjects design to the plastic dynamics of interacting form and color and even avoids immediate functional association—the value by which machine-made products are defined—a value that can impede free sensory discovery of the object just as its limitations can impede his creative act. And so, the value of use becomes a secondary or even arbitrary attribute.

Then comes the inevitable question: Is it craft? In the view of this writer, as long as it is the intent of the craftsman to produce an object of craft (the execution of which he performs with the recognized tools, materials, and methods of craftsmanship), and he incorporates acknowledgment, however implied, of functional possibilities or commitments (including the function of decoration)—as long as he maintains personal control over the execution of the final product, and he assumes personal responsibility for its aesthetic and material quality—it is craft. At the point that all links with the idea of function have been severed, it leaves the field of the crafts.

Ceramics, perhaps more than any other craft, throughout its long history has produced useful objects that are considered fine art. Time has a way of overwhelming the functional values of an object that outlives the men who made and used it, with the power of its own objective presence—that life invested quality of being that transcends and energizes. When this happens, such objects are forever honored for their own sakes.

We are now groping for a new aesthetic to meet the needs of our time, or perhaps it is a new anti-aesthetic to break visual patterns that no longer suffice. The most powerful forces of our environment—electronic and atomic, inner and outer space, speed—are invisible to the naked eye. Our aesthetic tradition, involved as it has been with visual experience, does not satisfy the extension and growth of reality in our time. Our greatest sensory barrier to a new aesthetic is visual enslavement in a subvisual world. The aspect of man is no longer the center of things, and his eyes are only accessories of his own growing sense of displacement.

Throughout the arts in America we are in the presence of a quest for a deeper feeling of presence.

The American potter, isolated from the mass market, which makes no demands on his product as a material necessity, is motivated by a personal aesthetic and a personal philosophy. Lacking an American pottery tradition, he has looked to the world heritage and made it his

Figure 41   Ken Shores, *Little Red i*, 1962.

own. For this, he has had to study and travel. Today, with his knowledge about himself, his craft, and his art—historically, contemporaneously, and geographically—cumulatively greater than ever before, the United States craftsman, a lonely, ambitious eclectic, is the most eager in search of his own identity.

All this, then, has made him most susceptible and responsive to the startling achievements of contemporary American painting and sculpture. For better or worse, he has allied himself with a plastic expression that comes from his own culture and his own time, and from an attitude toward work and its processes with which he can identify. The American potter gets inspiration from the top—from the most developed artistic, intuitive consciousness in his society. As always, the artist is led—not by the patron, not by populace, certainly not by the critic—artist is led by artist. The artist is his own culture.

Briefly, the characteristic directions of the new American pottery are: the search for a new ceramic presence, the concern with the energy and excitement of surface, and the attack on the classical, formal rendering.

Pottery, with a continuity that reaches back to the very beginnings of man, has always had a tradition for variety. If there is any one traditional characteristic of American pottery, it is this enormous variety. And if there is anything that distinguishes American plastic expression, it is the forthrightness, the fearlessness, the individuality, the aloneness of each man's search.

NOTES

1. John Kouwenhoven's documented study of American aesthetics, *Made in America* (Newton Centre, Mass.: Charles T. Branford Co., 1948).
2. Ibid.
3. The writer does not wish this article to be interpreted as a statement of special partisanship for those potters working with the new forms and motivations. It is an attempt to treat a direction of work which, with its provocative attitudes, has evoked strong response—for it as well as against it. Our partisanship is for creative work in all its variety. We recognize that pottery has as many faces as the people who make it.

FURTHER READING

Garth Clark, 'Otis and Berkeley: Crucibles of the American Clay Revolution', in Jo Lauria, ed., *Color and Fire: Defining Moments in Studio Ceramics 1950–2000* (Los Angeles: LACMA/Rizzoli, 2000).

Andrew J. Perchuk, 'Time in Mid-Twentieth Century Ceramics', *American Art* 21/1 (Spring 2007), pp. 10–13.

Rose Slivka, *Peter Voulkos: A Dialogue with Clay* (New York: New York Graphic Society, 1978).

Cheryl White, 'Towards an Alternative History: Otis Clay Revisited', *American Craft* 53/4 (August/September 1993).

64

HOW I SPENT MY SUMMER VACATION OR, ART AND POLITICS IN NEVADA, BERKELEY, SAN FRANCISCO AND UTAH

Philip Leider

Philip Leider's essay 'How I Spent My Summer Vacation' was a goodbye letter of sorts, published in the leading American art magazine, Artforum, *just prior to his acrimonious resignation as chief editor. It is a deceptively informal, deeply complex piece of writing. Though it seems simple enough—a picaresque retelling of a trip that Leider took across the West Coast—it is in fact a bittersweet summation of Leider's own passion for, and disillusionment with, the political potential of advanced art. Craft functions as a foil within the essay. It is at once a means toward self-sufficient integrity, but also a sign of naïve disengagement. In the manner of a medieval pilgrimage, every stop in Leider's tour is bestowed with allegorical qualities. A sublimely natural, craftsy, hippie town called Canyon functions as the heart of the narrative. Framing it are Leider's visits to two 'land art' works,* Double Negative *by Michael Heizer and* Spiral Jetty *by Robert Smithson. Created at the scale of industry rather than craft, these enormous sculptures are treated as exemplars of the avant-garde's confrontational encounter with the environment. The people in Canyon are conspicuously humble in comparison. Dropouts from capitalism, they are constantly hounded by the authorities for the supposed inadequacies of their artisanally made homes. For Leider, their countercultural village raises serious issues: is it the truest form of political expression, or just a Utopian fantasy? Is Canyon a valuable rebuke to the conceits of the art world or a tragic withdrawal from intellectual discourse? Leider's essay offers no easy answers to these questions—his path begins with dueling aphorisms and ends in an inconclusive meditation about the 'ever-deepening spiral of politics'. What he does show, however, is that craft and the avant-garde have much to gain from their mutual encounter.*

Philip Leider, 'How I Spent My Summer Vacation or, Art and Politics in Nevada, Berkeley, San Francisco and Utah (Read About It in Artforum!)', *Artforum* 11/1 (September 1970).

Art has never been a question of life and death . . .

—Barbara Rose

Art is the only thing worth dying for.

—Abbie Hoffman

We took this really nice house in Berkeley that some friends were vacating for the summer. Lots of rooms, a few pieces of old furniture, dark wood paneling, and the basic item of Bay Area life, a round oak table around which there always seems to be a lot of people. Shortly after our arrival I was supposed to meet Richard Serra and Joan Jonas to drive down to Nevada to see Heizer's *Double Negative*.

I had been talking to Serra on and off for about two years. He has a gargantuan appetite for art and its problems. Ideas explode in his head with the regularity of Dexedrine spansules popping. He has a fine sense of art world theatrics and times his art world (life) actions with the precision of an Abbie Hoffman. As a matter of fact, Hoffman's name came up in the ride to Nevada pretty frequently. Serra had gone to school at Santa Barbara and, after Isla Vista, was having serious doubts about whether he was the most revolutionary thing that ever came out of that campus.

What, we argued, was the most revolutionary thing to do?[1] Serra was wondering whether the times were not forcing us to a completely new set of ideas about what an artist was and what an artist did. I argued for Michael Fried's idea that the conventional nature of art was its very essence, that the great danger was the delusion that one was making art when in fact you were doing something else, something of certain value but not the value of art. That's where Hoffman came in:

> I'm more interested in art than in politics, but, well, see, we are all caught in a word box. I find it difficult to make these kinds of divisions. Northrup, in Meeting of East and West, said, "Life is an undifferentiated esthetic continuum." Let me say that the Vietcong attacking the U.S. Embassy in Saigon is a work of art. I guess I like revolutionary art.

Serra wasn't quite ready to absorb even elegant military actions into art, but neither was he ready to dismiss the idea that there are certain moments when what artists do is suddenly thrown up for grabs. Was it possible that Hoffman had seen where a whole lot of art, from the Happenings on, had been leading?

> Throwing money onto the floor of the Stock Exchange is pure information. It needs no explanation. It says more than thousands of anti-capitalist tracts and essays.

The car broke down about fifteen miles outside of Bakersfield, and we had to spend the night. As we walked across the parking lot of a truck stop toward the diner, Serra said, "Jesus Christ, look at that—bombs!" A huge truck, parked in the lot, was stacked full with open-slatted crates containing, sure enough, bombs. We walked over to it and continued our political discussion:

"B-O-M-B-S," spelled Serra, reading the stencils on the crates. He looked at me. "They're bombs."

"Look, they pack the nose cones separately," I said, meaning the warheads, or the lips, or whatever they were.

"A whole truckload of bombs," said Serra.

"Maybe they only travel at night," I said.

We ate in the diner. When we came out, the bombs had left, off to Cambodia. Would they have gotten past Abbie Hoffman that easily?

Heizer's piece was on a giant mesa high behind the town of Overton, Nevada. We were all expecting something strong, but none of us were quite prepared for it, as it turned out. We were all yipping and yowling as if Matisse had just called us over to look at something he was thinking of calling *Joy of Life*. The sun was down; we wound up slipping and sliding inside the piece in the dark. The piece was huge, but its scale was not. It took its place in nature in the most modest and unassuming manner, the quiet participation of a man-made shape in a particular configuration of valley, ravine, mesa and sky. From it, one oriented oneself to the rest in a special way, not in the way one might from the top of the mesa or the bottom of the ravine, but not in a way competing with, or at odds with them either. The piece was a new place in nature. That seemed to me a risky kind of art; there was a range of consequences in doing it wrong that one wasn't used to contemplating in relation to art. But *Double Negative* was not doing it wrong.

[. . .]

The first rule for a good community is bad roads.
—Canyon saying

I ran into David Lynn in a community called Canyon, about fifteen miles north of Berkeley. I remembered his sculpture from *Artforum*. He had taught at the University, and his sculpture had been more abstract than a lot of the work being done in the Bay Area at that time. In Canyon, he was working on a house, with one helper and a broken-down crane. The house was four stories high. The frame was well on the way, and was being made of the hugest beams I'd ever seen. The corner beams kept the original shape of the trees intact: they were not even planed. Lynn got his beams from piers that were being demolished and a number of other inconvenient sources: he isn't into cutting down trees,

(When he runs out of money, Lynn contracts to design and build a house for someone else, and stops or slows down work on his own house. For one of these jobs, in Pleasanton, Lynn used Canyon labor, thus giving Canyon people some good carpentry training and also bringing a little money into the community. Local construction bosses heard about it and demanded to see union cards, so all the workers went and joined the I.W.W., including Lynn. He showed me his Wobbly card ("Master Builder"). "It has a Preamble," he said, "and a slogan, and there's all my dues stamps." I looked at it with incredible curiosity.)

Lynn hadn't been doing sculpture for a while, but a lot of his friends were Bay Area artists. I had been to an opening in San Francisco and saw some people that Lynn knew, and we more or less brought each other up to date. The opening had been of a collection of ceramic work by Dave Gilhooley. They were very whimsical pieces, with titles like "A Thousand Frogs Dance on the Head of a Nail." I told Lynn that I'd met Gilhooley, who'd told me he was living in Saskatchewan. He didn't like living in the States very much any more, and Saskatchewan seemed just fine. Arlo Acton and Mel Moss were building a house way off in the country about 100 miles away from San Francisco. We'd heard that Win Ng was living in a pottery commune on a Canadian island. Goodness, I thought, lots of artists seem to be disappearing.

Canyon is a peculiar community, almost all of it being illegal. People like David Lynn build mostly without permits, because no permits are issued. Others have their houses condemned out of hand because their houses bear no relation to anything described in the California Building Codes as a house. The nearest community is a non-site collection of real-estate developments called Moraga, which has Muzak piped onto the sidewalks of its shopping center. Canyon can't come close to Moraga for safe and sound housing. Moraga levels off hilltops like a barber. In Canyon it is worth your life to cut down a tree. Moraga homes are neighborly, near one another, laid out in "courts." In Canyon you have to climb a mountain and wander around in the underbrush for hours to find out where your best friend lives. In Moraga the paved highway is laid down even before the houses are built. Canyon's roads are cemeteries of automobiles that tried to negotiate them. Moraga passes its sewage right into San Francisco Bay. Canyon has offered the county a fully worked out plan for the recycling of its sewage. Moraga has discreet bathrooms. In Canyon the open-air bathtub is all the rage.

Canyon people don't like Moraga very much, and try to have to go there as little as possible. They know there's a lot of prejudice in their attitudes, but many of them nevertheless seem to feel that their neighbors to the south are sexually desperate, physically ugly, unavailable to reason, and capable at any moment of instant, murderous violence. But Canyon kids have to go to high school in Moraga, and that's the trouble. The Canyon point of view seems to have been taking uncommon hold among the Moraga kids, and it looks strictly like it's going to be a one-generation town for sure unless something is done about it.

Other neighbors also feel that something has to be done about Canyon. Abbie Hoffman had said, "Always create Art and destroy Property," and while the art world may not be sure whether Canyon is doing either, the real estate people are pretty sure they're doing both. To them, Canyon itself, with its dense brush and uncountable trees, has that long-haired look to which real estate people so itch to give that old subdivided crew cut. Brush shaved off to reveal that smooth concrete beneath, trees trimmed drastically from the sides and the back of the neck and there it is, all ready for the ranch house. So there are lots of reasons why concerned authorities should move against Canyon, and they have, repeatedly and consistently, beginning, of course, with the condemning of most of the houses they know about, and the self-evident illegality of those they didn't. Canyon people spend a lot of time in court, patiently explaining that the housing code is financially repressive and ecologically disastrous, that more concrete means less grass and more automobiles mean less air and therefore they have not felt honor-bound to provide off-street parking,

there being, in any event, little auto traffic in Canyon, and fewer streets. They try to suggest that the houses they live in are beautiful, strong, economical and designed to fit the needs of the persons occupying them in a way that no house in Moraga or all of California for that matter could even approximate; that they effect no change in the natural ecology of the region; that, not unmindful of their duty to their neighbours in the outside world, they must therefore attempt to use this courtroom to indict the building codes, the real estate interests, the water departments, the sewer departments and all the other interests and departments that don't seem to realize that there's a war going on. They can't seem to get it across, and lose all their cases.

The thing is, as soon as court lets out the Canyon people rush home and start building, not as if there was no tomorrow, but as if there were an infinite number of tomorrows. Those houses just don't look as if the people that live in them plan to give them up, and, in a state that kills a boy and gasses its population just because people made a park where there should have been a parking lot, that's a grim thought.

The last time I saw Dave Lynn, he didn't look grim at all, chortling over his Wobbly book and swinging another monstrous beam into place. We didn't talk about sculpture at all; it seemed pretty clear that as far as Lynn was concerned, every sculptural idea he had ever had was in his building. The revolution in Lynn's art, if there was one, was dictated by the terrain: with Moraga just three miles down the road, and coming closer all the time, what serious artist could do otherwise? Whether this meant that Lynn wasn't an artist anymore or whether he had undergone that complete redefinition of what an artist is and does that Serra worried about was my problem, not his.

I'm interested in the politics of the Triassic period.

—Robert Smithson

John Coplans and I met Robert and Nancy Smithson in Salt Lake; we were going to drive from there to see *Spiral Jetty*, a piece Smithson had made on the north shore. Smithson told us that Serra had called from Missouri, where he was tearing his hair out trying to make a piece. Smithson has a very slow and very evil grin, which he breaks out at gleeful moments. "I told him," he said, grinning, "it was going to be tough." Every time you thought you found your place in a site the site kicked you out of it. Makes you feel like a fool. That's what Serra was going through. (I *think* that's what Smithson was saying.) Smithson had had site trouble too. He had been looking around the vicinity of the Great Salt Lake for two months without a hit. "Then this guy told me he knew a place where there was a red lake. I said 'Where?'"

On the way we talked about the ecological groups, which Smithson finds confused. There had been a lot of ecological language used in the furor that had preceded the Canadians' decision to cancel the island of broken glass, an ecologically harmless piece. And ecology-minded people had grumbled against Serra's Pasadena piece, for wasting trees. Smithson felt that in both cases the community had made of the art scapegoats for their own failure to come to grips with what they knew was killing them. It was true, I thought. The ecological conscience of Moraga would be outraged by both pieces; people from Canyon, on the other hand, would simply have taken them as ecology pieces, pretty good ones. (To which Serra might say, "*Ecology* pieces! Where's *that* at?")

The handwriting was on the wall for ecology, Smithson felt. "All those sins. And here's 2000 coming so near. Sin everywhere. The dead river, with its black oil slime. The crucified river instead of the crucified man. When do you think they'll start burning polluters at the stake?" Such talk makes me nervous, so I said something about *Spiral Jetty*. Smithson had been making fun of something I'd written about the "ever-deepening" political crisis. He thought there was a phony moral urgency in the use of terms like that. "Yeah," he said balefully, "the ever-deepening spiral of politics."

The red lake is on Rozel Point, described in one of Smithson's geology books as ". . . a small, blunt peninsula . . . extending southward on the north shore of the Great Salt Lake." The Great Salt Lake, Smithson told me happily, had successfully resisted any and all attempts by man to put it to any constructive use whatsoever, from the day men first laid eyes on it up to now. I had also discovered that for a long time it had been an oasis of chills and thrills in the humdrum desert of geology: "The notion," says the *Guidebook to the Geology of Utah* (a must for art critics), "that the lake must be connected to the Pacific by a subterranean channel, at the head of which a huge whirlpool threatened the safety of lake craft was not dispelled until the 1870s." A bad decade for geology, the 1870s, worse for the useless and now not even interesting Great Salt Lake. Is art supposed to give back what science takes away? Smithson, I remembered, gets into conversations with Mexican gods.

Art is nature, re-arranged. Like everyone else, Smithson learned it in high school. In a free society, artists get to re-arrange nature just like everyone else, lumber kings, mining czars, oil barons; nature, a kind of huge, placid Schmoo, just lays there, aching with pleasure. Smithson, reaching for his artistic birthright, kept turning up another kind of nature: "The non-sites let you know about the entropy of the urban." Planting a tree upside down, a relatively elementary rearrangement, turned out to change not only the object, but the subject: "art for the flies." Holding the mirror up to nature in the Yucatan, an even more rudimentary re-arrangement, reflected a vast conspiracy of pre-scientific forces moving over the face of the earth. The system obviously wasn't working right: you were supposed to re-arrange nature, not join it.

Art is also art re-arranged, and *Spiral Jetty* does what it can. There was Andre's *Lever*, and Brancusi's *Endless Column* before that. You don't get a piece like *Lever* to turn in on itself by fooling around with a length of rubber hose, as Smithson had undoubtedly discovered by looking at New York art for the last few years.

It took a long time walking out onto *Spiral Jetty*. Smithson kept being amazed at all the changes the piece had gone through since he'd last seen it. Thick deposits of salt had outlined the piece in white. A completely unexpected yellow mineral had appeared, mixing with the rosy water and the white salt crystals along many edges of the piece. Best of all, an electric storm was coming up across the lake, lightning and all. The piece was a fantasy. In the middle of Utah. Well, isn't that what artists do? Make fantasies?

> The truth makes you "hip."
> —Charles Manson

Back in Berkeley the weather was nice, and a nice time began, Berkeley was coming up, slowly and cautiously, from two serious downers: Altamont and Manson. In both cases, a subterranean, criminal class, inextricably involved in the Movement, had made its move. It would no longer settle for a position behind the photographers. This was one revolution

that was not going to betray it. But by midsummer, acute anxiety had become mild uneasiness, and people were telling cheerful stories. A Berkeley psychiatrist had been kicked out of the army when they discovered he'd been discharging the troops at the rate of one every five minutes. (The part that Berkeley liked was that the shrink was suing for an honorable discharge.) The Bay Area was leading the nation in draft resistance—34%, someone said. But the issue kept coming up, like a toothache. The Tribe printed a letter from a girl named HN drumming Manson out of the hip community, and devoted its center page to Manson's answer. Tom Hayden called Weatherman the Id of its generation, because it supported Manson (there is violence and there is Violence). The ever-deepening spiral of politics. Just before I left, Smithson had given me a Xerox of his lease on Rozel Point, for a souvenir.

NOTE

1. "Revolution" was the most often-used word I ran into this summer. Nobody used it to mean the transfer of political power from one class to another. Most of the time it seemed to refer to those activities which would most expeditiously bring America to her senses and force her to stop the war, and racism and begin to take the lead among nations in rescuing the planet from the certain destruction toward which it is headed.

FURTHER READING

Thomas Crow, *The Rise of the Sixties: American and European Art in the Era of Dissent* (New York: Weidenfeld and Nicolson, 1996).

Amy Newman, *Challenging Art: Artforum 1962–1974* (New York: Soho Press, 2000).

Amy Baker Sandback, ed., *Looking Critically: 21 Years of Artforum Magazine* (Ann Arbor: UMI Research Press, 1984).

65

SOME NOTES ON THE PHENOMENOLOGY OF MAKING: THE SEARCH FOR THE MOTIVATED

Robert Morris

'When I sliced into the plywood with my Skilsaw, I could hear, beneath the ear-damaging whine, a stark and refreshing "no" reverberate off the four walls: no to transcendence and spiritual values, heroic scale, anguished decisions, historicizing narrative, valuable artifact, intelligent structure, interesting visual experience.'[1] This is how Robert Morris recalled his early experiments with plywood sculpture, which today rank as defining statements of Minimalist art. For the ensuing decade, he would help to define key tendencies of avant-garde sculpture in America: Process Art, Earthworks, and Conceptualism. As the following essay makes clear, Morris saw the act of making as an uncharted terrain of artistic experimentation. Interviewed in 1968, he had noted his growing interest in 'a working process which did not in any way equate with the image'.[2] This was because image, and its implied corollaries of form and content, always involved a degree of imposition—an external set of ideas and associations that a viewer brings to the work. What if one were to imagine an artwork that could speak entirely for itself? The answer, Morris suggested, was that the work would have to be developed entirely from and through the means of its own making. If nothing extraneous to that process were allowed to intrude, then the resulting work would be completely integrated, nonrepresentational and self-reliant. This was a typical piece of Morris thinking—derived from the logical consistency of modern art theory, but also incipiently postmodern in its arrival at an 'antiform' art situation in which even the artist himself could not determine the outcome. Morris put his theories into practice in sculptures using cut, torn, folded and hung industrial felt and a related group of 'scatter' works in which heterogeneous materials were repetitively manipulated and then dropped on the gallery floor. Like much Process Art, these works looked both backwards and forwards—back to the 'chance operations' of Dada and to the serial rigors of Minimalism, but also ahead to the expressive, often installation-based work of the 1970s and 1980s.

Robert Morris, 'Some Notes on the Phenomenology of Making: The Search for the Motivated', *Artforum* 8/8 (April 1970), excerpted.

Art tells us nothing about the world that we cannot find elsewhere and more reliably.
—Morse Peckham

Between the two extremes—a minimum of organization and a minimum of arbitrariness—we find all possible varieties.
—Ferdinand de Saussure

I

A variety of structural fixes have been imposed on art—stylistic, historical, social, economic, psychological. Whatever else art is, at a very simple level it is a way of making. So are a lot of other things. Oil painting and tool making are no different on this level, and both could be subsumed under the general investigation of technological processes. But it is not possible to look at both in quite the same light because their end functions are different, the former being a relation to the environment, oneself, society, established by the work itself, while a tool functions as intermediary in these relations. Perhaps partly because the end function of art is different from the intermediary function of practical products in the society, a close look at the nature of art making remains to be undertaken. Authors such as Morse Peckham[3] have looked at art as behavior, but from the point of view of discovering its possible social function. He and others divide the enterprise into two basic categories: the artist's role playing on the one hand and speculations on the general semiotic function of the art on the other. My particular focus lies partly within the former category and not at all within the latter. Psychological and social structuring of the artist's role I will merely assume as the contextual ground upon which this investigation is built. The interest here is to focus on the nature of art making of a certain kind as it exists within its social and historical framing. I think that previously, probably beginning with Vasari, such efforts have been thought of as a systemless collection of technical, anecdotal, or biographical facts that were fairly incidental to the real "work," which existed as a frozen, timeless deposit on the flypaper of culture.

Much attention has been focused on the analysis of the content of art making—its end images—but there has been little attention focused on the significance of the means. George Kubler in his examination of Machu Picchu[4] is startlingly alone among art historians in his claim that the significant meanings of this monument are to be sought in reconstructing the particular building activity—and not in a formal analysis of the architecture. I believe there are 'forms' to be found within the activity of making as much as within the end products. These are forms of behavior, aimed at testing the limits and possibilities involved in that particular interaction between one's actions and the materials of the environment. This amounts to the submerged side of the art iceberg. The reasons for this submersion are probably varied and run from the deep-seated tendency to separate ends and means within this culture to the simple fact that those who discuss art know almost nothing about how it gets made. For this and perhaps other reasons the issue of art making, in its allowance for interaction with the environment and oneself, has not been discussed as a distinct structural mode of behavior organized and separate enough to be recognized as a form in itself.

The body's activity as it engages in manipulating various materials according to different processes has open to it different possibilities for behavior. What the hand and arm motion can do in relation to flat surfaces is different from what hand, arms, and body movement can do in relation to objects in three dimensions. Such differences of engagement (and their extensions with technological means) amount to different forms of behavior. In this light the artificiality of media-based distinctions (painting, sculpture, dance, etc.) falls away. There are instead some activities that interact with surfaces, some with objects and a temporal dimension, etc. To focus on the production end of art and to lift up the entire continuum of the process of making and find in it "forms" may result in anthropomorphic designations rather than art categories. Yet the observation seems justified by a certain thread of significant art which for about half a century has been continually mining and unearthing its means, and these have become progressively more visible in the finished work.

Ends and means have come progressively closer in a variety of different types of work in the twentieth century. This resolution re-establishes a bond between the artist and the environment. This reduction in alienation is an important achievement and accompanies the final secularization that is going on in art now.[5] However, what I wish to point out here is that the entire enterprise of art making provides the ground for finding the limits and possibilities of certain kinds of behavior, and that this behavior of production itself is distinct and has become so expanded and visible that it has extended the entire profile of art. This extended profile is composed of a complex of interactions involving factors of bodily possibility, the nature of materials and physical laws, the temporal dimensions of process and perception, as well as resultant static images.

Certain art since World War II has edged toward the recovery of its means by virtue of grasping a systematic method of production that was in one way or another implied in the finished product. Another way of putting it is that artists have increasingly sought to remove the arbitrary from working by finding a system according to which they could work. One of the first to do this was John Cage, who systematized the arbitrary itself by devising structures according to deliberate chance methods for ordering relationships. Cage's deliberate chance methods are both prior to and not perceptible within the physical manifestation of the work. The kind of duality at work here in splitting off the structural organization from the physically perceived still has strains of European Idealism about it.

However, for Cage such Idealism was forged into a dual moral principle: on the one hand, he democratized the art by not supplying his ordering of relationships; on the other, by his insertion of chance at the point of decision about relationships, he turned away the engagement with "quality"—at least at the point of structural relationships where it is usually located. It is not possible to mention Cage without bringing in Duchamp, who was the first to see that the problem was to base art making on something other than arrangements of forms according to taste. It is not surprising that the first efforts in such an enterprise would be to embrace what would seemingly deny certain aspects of preferred relationships—chance ordering. The entire stance of a priori systems according to which subsequent physical making followed or was made manifest are Idealist-oriented systems that run from Duchamp down through the logical systems of Johns and Stella to the totally physically paralyzed conclusions of Conceptual art. This has been one thread of how the systematic has been enlisted to remove the arbitrary from art activity.

Another thread of system-seeking art making, distinct enough to be called a form of making, has been built on a more phenomenological basis where order is not sought in a priori systems of mental logic, but in the "tendencies" inherent in a materials/process interaction. Pollock was the first to make a full and deliberate confrontation with what was systematic in such an interaction. Until Pollock, art making oriented toward two-dimensional surfaces had been a fairly limited act so far as the body was concerned. At most it involved the hand, wrist, and arm. Pollock's work directly involved the use of the entire body. Coupled to this was his direct investigation of the properties of the materials in terms of how paint behaves under the conditions of gravity. In seeing such work as "human behavior," several coordinates are involved: nature of materials, the restraints of gravity, the limited mobility of the body interacting with both. The work turned back toward the natural world through accident and gravity and moved the activity of making into a direct engagement with certain natural conditions. Of any artist working in two dimensions it could be said that he, more than any others, acknowledged the conditions of both accident and necessity open to that interaction of body and materials as they exist in a three-dimensional world. And all this and more is visible in the work.

II

[. . .]

Any process implies a system, but not all systems imply process. What is systematic about art that reduces the arbitrary comes out as information, revealing an ends-means hookup. That is, there is about the work a particular kind of systematizing that process can imply. Common to the art in question is that it searches for a definite sort of system that is made part of the work. Insofar as the system is revealed, it is revealed as information rather than esthetics, Here is the issue stated so long ago by Duchamp: art making has to be based on terms other than those of the arbitrary, formalistic, tasteful arrangements of static forms. This was a plea as well to break the hermeticism of "fine art" and to let in the world on terms other than image depiction.

III

The two modes of systematizing employed by American art over the last half century have been briefly sketched. The materials/process approach tends to predominate now. American art, unlike American thought, has occasionally had a strong idealist bias, but the a priori has so far proved unnerving and uncomfortable tools for the American artist. To pursue a more material route was, in the late 1940s, to be up against the formalism of Cubism. Pollock was the first to beat his way out of this. But all art degenerates into formalism, as Pollock himself found out. The crisis of the formalistic is periodic and perpetual, and for art to renew itself, it must go outside itself, stop playing with the given forms and methods, and find a new way of making.

Certain artists are involved in the structures I am discussing here. They form no group. The nature of the shared concerns does not mold a movement nor preclude the validity of other approaches and concerns. The term "mainstream" is political. Several present-day critics can be observed wading down one, hoping one day to float on the back of the oarsmen they have in tow. In fact, the current art swamp is awash with trickling mainstreams. Art that has or is participating in the structures articulated here is, to me, either interesting or strong or both.[6] Of the many concerns in art, the ones dealt with here have given powerful leverage in opening up possibilities, whether as mere tendencies in past work or self-conscious methods in present work. Other kinds of art making focus on other concerns—the nature of color in art making would, it seems, be totally outside these investigations.

The issues here stretch back into art history, but in a particularly linear way. The concerns are partly about innovative moves that hold in common a commitment to the means of production. Duchamp, Cage, Pollock, Johns, and Stella have all been involved, in different ways, in acknowledging process. Quite a few younger artists are continuing to manifest the making process in the end image. But the tendencies to give high priority to the behavior end of making can be found in much earlier artists. Rather than modeling parts of the costume in Judith and Holofernes, Donatello dipped cloth in hot wax and draped it over the Judith figure. This meant that in casting, the molten bronze had to burn out the cloth as well as the wax. In the process some of the cloth separated from the wax and the bronze replaced part of the cloth, revealing its texture. This was a highly finished work, and corrections could have been made in the chasing phase had the artist wanted to cover it up. It has also been claimed that the legs of the Holofernes figure were simply cast from a model rather than worked up in the usual way.[7] Evidence of process in this work is not very apparent and could have been noticed only by the initiated. But here is an early example of a systematic, structurally different process of making being employed to replace taste and labor, and it shows up in the final work. Draping and life casting replace modeling. Michelangelo's "unfinished" marbles give far more evidence of process, but with the important difference that no structurally new method of making is implied.

What is particular to Donatello and shared by many twentieth-century artists is that some part of the systematic making process has been automated. The employment of gravity and a kind of "controlled chance" has been shared by many since Donatello in the materials/process interaction. However it is employed, the automation serves to remove taste and the personal touch by co-opting forces, images, and processes to replace a step formerly taken in a directing or deciding way by the artist. Such moves are innovative and are located in prior means, but are revealed in the a posteriori images as information. Whether this is draping wax-soaked cloth to replace modeling, identifying prior "found" flat images with the totality of a painting, employing chance in an endless number of ways to structure relationships, constructing rather than arranging, allowing gravity to shape or complete some phase of the work—all such diverse methods involve what can only be called automation and imply the process of making back from the finished work.

Automating some stage of the making gives greater coherence to the activity itself. Working picks up some internal necessity at those points where the work makes itself, so to speak. At those points where automation is substituted for a previous "all made by hand" homologous set of steps, the artist has stepped aside for more of the world to enter into the art. This is a kind of regress into a controlled lack of control. Inserting the discontinuity of automated steps would not seem, on the face of it, to reduce the arbitrary in art making. Such controlled stepping aside actually reduces the making involvement or decisions in the production. It would seem that the artist is here turned away from the making, alienated even more from the product. But art making cannot be equated with craft time. Making art is much more about going through with something. Automating processes of the kind described open the work and the artist's interacting behavior to completing forces beyond his total personal control.

The automated process has taken a variety of forms in various artists' work. Jasper Johns focused very clearly on two possible ways for painting. One was to identify a prior flat image of a target or flag with the total physical limit of the painting. Another sequentially systematic mode that implied process was the number and alphabet works. These, and Stella's subsequent notched striped paintings, present total systems, internally coherent. Both imply a set of necessary sequential steps which, when taken, complete the work. Less painterly and far more deliberate, Stella's work of the early 1960s was some of the first to fold into a static, "constructed" object its own means of production. I have discussed elsewhere[8] how the work of both artists, with its deliberateness of execution according to an a priori plan, implies a mode of making, or form of behavior, that can be more fully realized in the making of three dimensional objects.

So-called Minimal art of the early and mid-1960s was based on the method of construction. The structure necessary to rectilinear forming precludes any "arranging" of parts. The "how" of making was automated by accepting the method of forming necessary to rectilinear things. What is different about making objects, as opposed to applying a surface, is that it involves the body, or technological extensions of it, moving in depth in three dimensions. Not only the production of objects, but the perception of them as well involves bodily participation in movement in three dimensions. It might be said that the construction of rectilinear objects involves a split between mental and physical activity and a simultaneous underlining of the contrast: on one hand, the obviousness of the prior plan, and on the other, the extreme reasonableness of the materials used to manifest the structure. A certain strain of constructed art of the 1960s continued an emphasis on refined or colored surfaces and optical properties—essentially an art of surfaces moved into three dimensions.

Other constructed art opted for the emphasis on more traditionally sculptural values—volume, mass, density, scale, weight. The latter work tended to be placed on the floor in one's own space. This is a condition for sculptural values in materials to register most fully, as it is under this condition that we make certain kinesthetic, haptic, and reflexive identifications with things. I have discussed the nature of this perceptual bond to things in our own body space before.[9] For the argument here it is only necessary to reiterate a few points. The body is in the world, gravity operates on it as we sense it operating on objects. The kinds of identification between the body and things initiated by certain art of the 1960s and continued today was not so much one of images as of possibilities for behavior. With the sense of weight, for example, goes the implicit sense of being able to lift. With those estimations about reasonableness of construction went, in some cases, estimations of the possibility for handling, stability or lack of it, most probable positions, etc. Objects project possibilities for action as much as they project that they themselves were acted upon. The former allows for certain subtle identifications and orientations; the latter, if emphasized, is a recovery of the time that welds together ends and means. Perception itself is highly structured and presupposes a meaningful relation to the world. The roots of such meanings are beyond consciousness and lie bound between the culture's shaping forces and the maturation of the sense organs that occurs at a preverbal stage. In any event, time for us has a direction, space a near and far, our own bodies an intimate awareness of weight and balance, up and down, motion and rest, and a general sense of the bodily limits of behavior in relation to these awarenesses. A certain strain of modern art has been involved in uncovering a more direct experience of these basic perceptual meanings, and it has not achieved this through static images, but through the experience of an interaction between the perceiving body and the world that fully admits that the terms of this interaction are temporal as well as spatial, that existence is process, that the art itself is a form of behavior that can imply a lot about what was possible and what was necessary in engaging with the world while still playing that insular game of art.

Recent three-dimensional work with its emphasis on a wide range of actual materials and locating the making of possibilities of behaving or acting on the material in relation to (rather than in control of) its existential properties brings very clearly into focus that art making is a distinct form of behavior. This is underlined even more now that the premises of making shift from forming toward stating. Around the beginning of the 1960s the problem presented itself as to what alternatives could be found to the Abstract Expressionist mode of arranging. The Minimal presented a powerful solution: construct instead of arrange. Just as that solution can be framed in terms of an opposition (arrange/build), so can the present shift be framed dialectically: don't build . . . but what? Drop, hang, lean-in short, act. If the static noun of "form" is substituted for the dynamic verb "to act" in the priority of making, a dialectical formulation has been made. What has been underlined by recent work in the unconstructed mode is that since no two materials have the same existential properties, there is no single type of act that can easily structure one's approach to various materials. Of course the number of possibilities for the basic kinds of interactions with the materials are limited, and processes do tend to become forms that can be extrapolated from one material interaction to another. But what is clear in some recent work is that materials are not so much being brought into alignment with static a priori forms as that the material is being probed for openings that allow the artist a behavioristic access. What ties a lot of work together is its sharing of the "automated" step in the making process, which has been enlisted as a powerful ally in the recovery of means or time and in increasing the coherence of the making phase itself.

Not only in plastic art but in art that specifically exists in time there have been recent moves made to reduce that existential gap between the studio preparation and the formal presentation. Some theater and dance work now brings rehearsal and literal learning sessions for the performers into the public presentation. One could cite other instances in film and music where the making process is not behind the scenes but is the very substance of the work.

Peckham speaks of the necessity of preserving a "psychic" insulation within which the strain of disorienting art moves can be made.[10] Studios, galleries, museums, and concert halls all function as insulated settings for such experience. Much recent art that is being discussed does not require a studio, and some recent plastic art does not even fit inside museums. In contrast to the indoor urban art of the 1960s, much present work gets more and more beyond studios and even factories. As ends and means are more unified, as process becomes part of the work instead of prior to it, one is enabled to engage more directly with the world in art making because forming is moved further into the presentation. The necessary "psychic insulation" is within one's head.

NOTES

1. Quoted in Hal Foster, Rosalind Krauss, Yve-Alain Bois, and Benjamin H. D. Buchloh, *Art Since 1900: Modernism, Antimodernism, Postmodernism* (New York: Thames and Hudson, 2004), p. 492.
2. Robert Morris interviewed by Paul Cummings, March 10, 1968; *Archives of American Art,* Smithsonian Institution.

3. Morse Peckham, *Man's Rage for Chaos* (New York: Schocken, 1967).

4. George Kubler, "Machu Picchu," *Perspecta* 6, 1960.

5. Annette Michelson, "Robert Morris," *The Corcoran Gallery of Art* (Baltimore, Md.: Garamond/Pridemark Press, 1969), p. 23.

6. Marcel Duchamp, "The Creative Act," Robert Lebel, *Marcel Duchamp* (Paris: Trianon Press. 1959), p. 77.

7. Bruno Bearzi, *Donatello, San Ludovico* (New York: Wildenstein, n.d. [1948]), p. 27.

8. Robert Morris, "Beyond Objects," *Artforum,* Vol. VII, No. 7 (April 1969).

9. Robert Morris, "Notes on Sculpture, Part II," *Artforum,* Vol. V, No. 2 (October 1966).

10. Peckham, p. 82.

FURTHER READING

Julia Bryan-Wilson, 'Hard Hats and Art Strikes: Robert Morris in 1970', *The Art Bulletin* 89/2 (June 2007), pp. 333–59.

Anna C. Chave, "Minimalism and the Rhetoric of Power", *Arts* 64 (January 1990), pp. 44–63.

James Meyer, *Minimalism: Art and Polemics in the Sixties* (New Haven, CT: Yale University Press, 2001).

Robert Morris, *Continuous Project Altered Daily* (Cambridge, MA: MIT Press, 1993).

66

THE ART OF ENCOUNTER

Lee Ufan

The sculptor and painter Lee Ufan was born in Korea, came to Japan at the age of twenty and pursued a degree in philosophy at Nihon University in Tokyo before embarking upon a career in the avant-garde. He became, alongside Sekine Nobuo (b. 1942), a leader of the group of artists later designated by the term Mono-ha, or 'the school of things', an appellation that (like earlier labels such as the Impressionists and the Fauves) was originally intended to mock the group rather than praise it. Lee's intellectual grounding in the thought of such modern philosophers as Martin Heidegger, Maurice Merleau-Ponty and Nishida Kitarô comes through clearly in his theoretical writings, and also in his art—sculptures composed of unworked industrial steel and found stones, and large blank canvases marked with only a single, wide brushstroke. The formal restraint of such works has often led commentators to see a typically Asian austerity in Lee's work, but this is a stereotype that he categorically rejects: 'When something is placed in the category of a foreign culture in the west, this may well be a way of rejecting it with praise. In any case, the existence of an artist named Lee is submerged in the sea of "orientalism" and the issues I raise are ignored when people choose to describe me as "oriental"'. Lee's profundity on the subject of 'the hand' may come as more of a surprise, as he studiously avoids craftsmanship of any kind. However, it is clear that like Robert Morris, Lee has made a study of the process of art making from the ground up. He proposes an 'art of encounter' in which things that are 'made' (like big plates of steel) and things that are 'unmade' (like stones found in a riverbed) are brought into juxtaposition with the body and the surrounding space, with the goal of opening a space that 'eliminates everydayness and arouses fresh perceptions'. In this respect Lee's work should be seen in relation to Minimalist and Postminimalist artists like Morris, Carl Andre, Richard Serra, all of whom are his contemporaries—the difference being, perhaps, that for Lee the encounter has little to do with artistic ego or persona and everything to do with, as he puts it, 'things in themselves'.

Lee Ufan, selections from *The Art of Encounter* (London: Lisson Gallery, 2004), translated by Stanley N. Anderson.

"ON THE HAND," 1973 (REVISED 1987)

The hand is a friend of the brain. The hand and brain work together to paint a picture or make a sculpture. The hand is extremely important to the brain, but at times betrays it. That is because the hand is part of the body. Like the eyes, mouth, feet, ears, buttocks, the brain itself, and internal organs, it is an organ of the body. Because the hand is an organ connected to the other parts of the body, it can see and feel and think.

The body is an ambivalent entity that belongs to the world as well as to 'me.' In this body, a boundary area connecting the inside and the outside, the hand serves at the most forward outpost.

There is a way of thinking that sees the body, and the hand, as subordinate to my self. This view derives from the attitude that everything belongs to me as a conscious subject. The soul, an extreme form of the concept of the self, or God, is everything and the existence of the world as externality is not recognized. The world is created endlessly through self-realization. In Christianity, the tendency to respect the spirit and denigrate the flesh is an expression of the view of the non-existence of the outside world. Here, the outside exists only as an extension of the spirit. In philosophies where the outside world is not accepted, like those of Descartes or Marx, everything is material for self-realization and has no intrinsic value. That is, only what is made is the world and takes on value. This making is performed by the hand, which executes my commands. The hand is an extension of myself, my tool. In reality, tools are a point of contact between the world and myself, but this fact is twisted in the common view that tools are a means of carrying out my ideas.

The paintings of artists who see the hand as a tool are uninteresting. Paintings that ignore physicality are ultimately limited to an all-over reproduction of the self.

Everyone, however, has encounters with the outside world. Encounters are entailed in the fact that other things than oneself exist and it is possible to enter into a dialogue with them. An encounter is an interaction with externality/otherness. Through such encounters, I take on otherness, that is, I accept or reject the outside world. When I encounter what is external to myself, various powers come together and a world that transcends me opens up. *Tekhnê* (Greek for art or skill) refers to methods of concentrating the power of otherness. *Tekhnê* originally meant the wisdom connected to otherness, and it requires training and discipline of the self. My own power can be multiplied 20 or 30 times by the *tekhnê* that entails otherness. By utilizing the power of otherness, the work becomes non-transparent and much greater than myself. This is a matter of profound passivity, the true meaning of *hariki hongan* (salvation through dependence upon a greater power). It is still possible to develop an open computer that is not devoted only to the reproduction of the self.

For the artist, the hand is connected to many parts of the outside world, including the brush, the paint, the canvas, the air, time, and space. It is an intermediary that provides experience of the world, produces thought, and leads me into an unknown otherness. Also, the hand of the artist performs the role of creating physical intermediaries (artworks) in the interval between the world and the self.

I can paint with a 'hand' that is not my true hand. However, it does not produce the same sense of an exciting encounter between myself and the world, so I do not want to use this 'hand.' I want to join forces with my true hand to demonstrate the ambiguity of the physical body.

"ART AND THE BODY," 1978 (REVISED 1999)

I think with my head, paint with my hand, and walk on my own feet in order to find what I am looking for. This means that I train my body and refine my consciousness, bringing

Figure 42   Lee Ufan, *Relatum*, 1979–1996.

them together in a tense relationship as I engage in the venture of making art. This stance and method of working may seem crude nowadays. It might even imply poverty or seem overly troublesome or bothersome to some people. However, I do not think there is a better way to create art than with my own body.

In today's industrial urban society, brain-centered thinking is the norm and the body is often ignored. It is typical of the age for artists to have their art made by other people in a factory or with a machine. It is hard to maintain precision when one uses one's own body to create art, and however it is done, many artists consider it essential to make something that is identical with what is planned. Emotion and accident are eliminated from

works of art, just as if they were industrial products, and it is thought to be important to fabricate them with the same perfection as the products of an automated manufacturing system. There is also an emphasis on concept, precision, complexity, uniformity, scale, and the physical presence and volume of materials. More and more, thoughts and materials are turned into programs and data. They are made to expand and proliferate automatically, and it is necessary to reproduce them accurately. This approach to art clearly conveys the characteristics of brain-centered thinking.

To me, this sort of art is boring. Sometimes it even frightens me.

It seems that many artists turn an idea into a work of art with no modification. That is, they think everything is alright if the idea can be recognized in the work. Afterwards, even if the viewer feels overwhelmed by various forms, colors, techniques, or scale, it is not considered proper to feel the art with the body or see it with the eyes. Such art is a thoroughly controlled world, like a transparent language. It rejects foreign objects, otherness, and the outside world. It is constructed so as to create identity between the rational mind and the world and is composed only of things that can be made into data, strictly distinguished, and given material form. If there is anything of interest in such art, it could only be the effective method of political rule by a powerful dictator.

Here, seeing means expressing agreement rather than making discoveries or being watched. Insular collective thinking is perpetuated in the name of universalism. Seeing has no meaning except as an operation of confirmation or acknowledgement. The motto is complete replication of data, so there are no discrepancies or non-transparent elements. Everything is clear.

This clarity is what I find uninteresting. It is frightening and makes me feel that I am suffocating. This concern with clarity has its source in brain-centered thinking that cuts off all relationships with the outside world and despises and alienates the body.

The body, as Merleau-Ponty proclaimed, is an ambiguous thing that belongs not only to me but is also connected to the outside world. This is self-evident. It is because of the existence of their physical bodies that human beings exist in the world. Therefore, strictly speaking, it is a mistake to speak of my hand or my eye. It should be remembered that this is just a convenient way of speaking.

The body mediates between the inside and the outside, and can arouse us to the possibility of a more open situation. There is mutual cooperation between consciousness and the body, but they are not identical. The body is more strongly connected to the greater world than consciousness. The body is part of the outside world.

Therefore, human beings can know the outside world and experience transcendence by taking advantage of the existence of the body and the role it plays. This is the reason that I am so concerned about the body.

In order to join the body and consciousness at a high level, I highlight this ambiguity and always do my best to act physically. I make art by performing repeated actions with the body, thus bringing about transformations and changes in my ideas and increasing the depth and expansiveness of the work. My personal capacity may be only 10, but it is magnified to 20 or 30 by the operation of the body in relation to the outside world. It is only through the existence of the body that I can make a work of art that transcends my self. As more externality is incorporated, transparency fades, the unknown appears, and the work becomes richer in content and allusion.

The body and consciousness are sometimes in conflict and sometimes in accord. This leads to divergence in the process of making. The work takes on otherness as externality penetrates it via the body. An even more important aspect of employing the body is the fact that art-making is a site of encounter with externality. Experience is contact with the outside world, through which one can sense infinity. Making art is truly living, and touching infinity, through physical action.

"THE PROCESS OF MAKING ART," 1979

Ordinarily, one carries out the process of making art after planning. However, there are times when an artist begins working as he is inspired by things or a place. Both methods are ways of starting and do not lead immediately to making a work. If one starts by planning, one will, in reality, gradually move away from the plan. If one starts out from place, a plan will gradually emerge.

This process is not mechanistic but depends on a sensibility of organic generation. It is only when there are varied encounters that the work matures and develops.

What is most important is that the process of making evolve through negotiations between consciousness and the outside world and the work come into being while still containing the unknown.

FURTHER READING

Lee Ufan: The Art of Margins (Yokohama: Yokohama Museum of Art, 2005).

Simon Groom, *Mono-ha* (Cambridge: Kettle's Yard, 2001).

Janet Koplos, *Contemporary Japanese Sculpture* (New York: Abbeville Press, 1991).

Alexandra Munroe, ed., *Japanese Art After 1945: Scream Against the Sky* (New York: Harry N. Abrams, 1994).

67

LET THE ARTISANS CRAFT OUR FUTURE

Grayson Perry

For those who like their craft taken straight, the British artist Grayson Perry may be a tough sell. But he has no shortage of admirers. He won the prestigious Turner Prize in 2003 (on which occasion he commented, 'Well, it's about time a transvestite potter won the Turner Prize. I think the art world had more trouble coming to terms with me being a potter than my choice of frocks') and has been a frequent commentator in the London newspapers; his sardonic and perceptive voice has been influential in recent years. This is no less true of his art, which has been well received precisely because it is amateurishly made and disturbing in its content. Perhaps because Perry's authority is premised on a public airing of his own private insecurities, he has been unusually effective in dissecting the crafts' marginal position within the contemporary art world, and craftspeople's unacknowledged anxieties about themselves. He has used the resonant metaphor of the crafts as a lagoon, a secure space in which those unable to cut it in the bruisingly competitive art world can enjoy a measure of security. But Perry doesn't see this as a permanent state of affairs. For all his critical acidity, this is one transvestite potter who is genuinely committed to the idea that craft (if only it had a little more 'bolshie passion') could assume, as he has, a viable role within the contemporary avant-garde.

Grayson Perry, 'Let the Artisans Craft our Future', *London Times* (5 April 2006).

Last week I was invited to a press lunch at Somerset House by the Crafts Council. I wore a jacket and skirt as I expected it to be a ladylike do, and I was not disappointed. Out of twenty or so present, there were only four men, if you count me. Most things to do with the craft world have this gender imbalance. For me as a machophobe this was always one of its chief attractions.

Yet I have always felt a little uneasy about the Crafts Council. Despite my playing at ladies who lunch, I sense that the version of craft that it promotes needs to be more muscular. I'd like to give it an injection of testosterone, some bluff confidence, some bolshie passion. It promotes quality in the applied arts and has put on some excellent shows, but I think that it has failed to pay attention to the wider picture, and the meaning of "craft" in our society has suffered. The council has somehow ended up standing for a kind of watered-down art and design rather than championing relevant craft skills. I love and admire craftsmanship but "craft" has become a concept that I do not always want to be identified with. I fear it has become the domain of ladies in dangly earrings.

Craft, I feel, to people outside the exquisitely constructed ring fence of the Craft Council has become a hobby. It is a leisure activity practised by exhibitors at craft fairs who fashion novelty covers for vacuum cleaners and children who are bought bead jewellery kits by well-meaning aunts who think they watch too much television. Craft has become an overblown *Blue Peter* project. Craft is becoming a zoo-bred animal that could not survive in the wild of the market place.

Craftsmanship is no longer central in the bodged-together service economy. Craft used to be something integral to many walks of life. Guilds of apprentices would parade through the streets with special examples of their work held proudly aloft. Quilts were prized family heirlooms hand-stitched by cabals of women. Even as recently as the 1950s Teddy Boys got kitted out for their subculture by their local bespoke tailors. To be a craftsman was to be working-class nobility.

Maybe I am being sentimental and nostalgic about horny-handed men in leather aprons wielding a spokeshave, but I think that there is a place for the commissioned one-off handmade object in our future because, as we know, the future has to be green.

Handmade is often a byword for pricey, and local means unadventurous or a lack of choice. But what about when the oil runs out and the forests are all cut down, when we can't just drive to Furniture Barn and buy a table, designed in Scandinavia, made in China with wood from South America, for the price of a round of drinks?

Maybe in the future the distinctiveness that consumerism promises will be concentrated not on choice but on customisation. Maybe ethical concerns will force us to turn to local craftsmen to give added value to our status-defining objects. Maybe it will be OK again for local labour to be the major cost factor instead of importers, distributors, fuel and marketing. Instead of being sucked along by instant gratification and fashions driven by

manufacturers, maybe people will collaborate with a skilful designer/maker. Together they can work on producing the perfect artefact that fulfils their functional needs and reflects their values.

And don't tell me that this will be affordable only by the super rich. The spendaholic middle classes could use their ever-increasing discretionary buying power to commission three bespoke items a year instead of 30 throwaway fads. Maybe we would think about how things would look when they get worn and old and factor that into the design. Maybe we will buy a suit for the decade and a table for life.

We are becoming increasingly visually sophisticated, so let's put it to good use. Instead of browsing through a catalogue, let us sketch our ideal. Instead of having houses bulging with stuff most of which has gone out of fashion through designer obsolescence, we could invest in our cherished individualism. Each unique object could be an expression of who lives there, flaws and all. Maybe we will then do something more useful with our weekends than shopping for insubstantial tat.

Ever since I had a set of motorcycle leathers made for me in 1989, one of my greatest pleasures is commissioning craftsmen and women to make me things. I find this immensely satisfying. In the course of the past quarter century I have worked with dressmakers, printers, bicycle-frame builders, kinky cobblers, rubber couturiers, carpenters and foundry men, and I'm about to take possession of my first architectural commission.

Few of these artefacts have the computerised sheen of the mass-produced. They have their faults; they bear the mark of the human hand; they are the evidence of a relationship; and I love them all. None of them had anything to do with the Crafts Council, which is sad.

68

MANIFESTO OF THE BAUHAUS AND EDUCATION AND THE BAUHAUS

Walter Gropius and László Moholy-Nagy

Just at the moment when one might expect the 'machine aesthetic' to have been at its height—in the founding of the quintessential Modernist design school—we read this: 'Architects, painters, sculptors, we must all return to the crafts!' This was the memorable proclamation of Walter Gropius's first Bauhaus manifesto. With their orientation to rationalism and automated manufacture, Modernist designers are sometimes caricatured as anticraft; but on the whole, nothing could be further from the truth. Even if Gropius sometimes tried to downplay the Bauhaus's largely artisanal production, in order to emphasize its few successes in working directly with industry, the school was nevertheless the logical culmination of a tradition in which craft was viewed as a preparatory stage within mass manufacture. This idea was installed in the school's curriculum, which carried students from material-based craft exercises in a 'foundation course' and gradually to more formal design and fine art coursework. The same pedagogical system was used at the New Bauhaus in Chicago (now the Institute of Industrial Design) by the Hungarian émigré designer and artist László Moholy-Nagy, who had decamped there in 1937. These principles became widely influential in the English-speaking world, not only in design education but also within the craft movement, partly via the work of Moholy-Nagy's American students, such as the wood-turner James Prestini and the jeweller Margaret DePatta.

Walter Gropius, *Manifesto of the Bauhaus* (Weimar, 1919).

The ultimate aim of all creative activity is the building! The decoration of buildings was once the noblest function of fine arts, and fine arts were indispensable to great architecture. Today they exist in complacent isolation, and can only be rescued by the conscious co-operation and collaboration of all craftsmen. Architects, painters, and sculptors must once again come to know and comprehend the composite character of a building, both as an entity and in terms of its various parts. Then their work will be filled with that true architectonic spirit which, as "salon art", it has lost.

The old art schools were unable to produce this unity; and how, indeed, should they have done so, since art cannot be taught? Schools must return to the workshop. The world of the pattern-designer and applied artist, consisting only of drawing and painting must become once again a world in which things are built. If the young person who rejoices in creative activity now begins his career as in the older days by learning a craft, then the unproductive "artist" will no longer be condemned to inadequate artistry, for his skills will be preserved for the crafts in which he can achieve great things.

Architects, painters, sculptors, we must all return to the crafts! For there is no such thing as "professional art". There is no essential difference between the artist and the craftsman. The artist is an exalted craftsman. By the grace of Heaven and in rare moments of inspiration which transcend the will, art may unconsciously blossom from the labour of his hand, but a base in craft is essential to every artist. It is there that the original source of creativity lies.

Let us therefore create a new guild of craftsmen without the class-distinctions that raise an arrogant barrier between craftsmen and artists! Let us desire, conceive, and create the new building of the future together. It will combine architecture, sculpture, and painting in a single form, and will one day rise towards the heavens from the hands of a million workers as the crystalline symbol of a new and coming faith.

László Moholy-Nagy, 'Education and the Bauhaus', *Focus* No. 2. (1938), as reprinted in Krisztina Passuth, ed., *Moholy-Nagy* (London: Thames and Hudson, 1985). Excerpted.

SECTORS OF HUMAN DEVELOPMENT

A human being is developed only by the crystallization of the sum total of his experiences. Our present system of education contradicts this axiom by stressing preponderantly a single field of application.

Instead of extending our milieu, as the primitive man was forced to do, combining as he did in one person hunter, craftsman, builder, physician, etc., we concern ourselves only with one definite occupation—leaving unused other faculties.

Tradition and the voice of authority intimidate man today. He no longer dares to venture into certain fields of experience.

Figure 43   Theodor Bogler, *Combination Teapot and Sugar Bowl*, 1923.

He becomes a man of one calling; he no longer has first-hand experience elsewhere. In constant struggle with his instincts, he is overpowered by outside knowledge. His self-assurance is lost. He no longer dares to be his own physician, not even his own eye. The specialists—like members of a powerful secret society—obscure the road to all-sided individual experiences, the possibility for which exists in his normal functions, and the need for which arises from the center of his being.

Today, the accent lies on the sharpest possible definition of the single vocation, on the building up of specialized faculties; the 'market demand' is the guide. Thus a man becomes a locksmith or a lawyer or an architect or the like (working inside a closed sector of his faculties) and is at best a happy exception if, after he has finished his studies, he stretches to widen the field of his calling, if he aspires to expand his special sector.

At this point our whole system of education has hitherto been found wanting—notwithstanding all our vocational guidance, psychological testing, measurement of intelligence. Everything functions—and functions alone—on the basis of the present system of production, which recognizes only motives of material gain.

A 'calling' means today something quite different from following one's own bent, quite different from solidarity with the aims and requirements of a community. One's personal life goes along outside the 'calling', which is often a matter of compulsion and is regarded with aversion.

THE FUTURE NEEDS THE WHOLE MAN

Our specialized training cannot yet be abandoned at this time when all production is being put on a scientific basis. However, it should not start too soon, and it should not be carried so far that the individual becomes stunted—in spite of all his highly prized professional knowledge. A specialized education becomes full of meaning only if a man of integration is developed along the lines of his biological functions, so that he will achieve a natural balance of his intellectual and emotional power, and not along those of an outmoded educational aim of learning unrelated details. Without this aim the richest differentiations of specialized study—the 'privilege' of the adult—are mere quantitative acquisitions, bringing no intensification of life, no widening of its scope. Only a man equipped with the clarity of feeling and the sobriety of knowledge will be able to adjust himself to complicated requirements. and to master the whole of life. Working only from this basis can one find a plan of life which places the individual rightly within his community.

THE PRESENT SYSTEM OF PRODUCTION

All educational systems are the result of economic structure. In the frenzied march of the industrial revolution, the industrialists set up specialized schools to produce quickly the badly-needed specialists.

These schools favored the development of men's powers only in very few instances, and offered no opportunity to penetrate to the essential kernel of things and the individual himself. But—to tell the truth—no one concerned himself with this because no one could foresee its destructive results. Thus today neither education nor production springs from an inner urge, nor from an urge to make products which satisfy the requirements of one's self and those of society in a mutually complementary way.

Figure 44   Gunta Stölzl, *Wall Hanging*, 1926–7.

Our modern system of production is imposed labor, mostly a mad pursuit, without plan in its social aspects. Its motive is merely to squeeze out profits to their limit, in most cases a complete reversal of its original purpose. Not only the working class finds itself in this position today; all those caught within the workings of the present economic system are basically just as badly off. At most there are slight degrees of difference. The chase after rewards in money and power influences the whole form of life today, even to the basic feelings of the individual. He thinks only of outward security, instead of concerning himself with his inner satisfaction. On top of all this there is the penning up of city dwellers in treeless barracks, the extreme contraction of living space. This cramping of living space is not only physical; city life has brought with it herding into barren buildings, without adequate open space.

BUT HOW ABOUT TECHNICAL PROGRESS?

It might easily be judged from the foregoing remarks that present-day industrial production, and especially our technical progress, is to be condemned. In fact there are numerous writers and politicians who suggest this. They mix the effect with the cause. In the nineteenth century some people tried to make a right diagnosis but suggested a wrong therapy. Gottfried Semper declared in the 1850's, for example, that if iron ever was to be used in building it would have to be used (because of the static nature of iron) in a fashion of transparent spider web. But, he continued, architecture must be 'monumental', thus 'we never shall have an iron-architecture.' (!) A similar mistake was made by the Ruskin-Morris circle in the 1880's. They found that industrial mass production killed quality in craftsmanship. Their remedy was to kill the machine, go back to the handwork exclusively. They opposed machines so strongly that to deliver their handmade products to London, they ran a horse coach parallel with the hated railway. In spite of this rebellion against the machine, technical progress is a factor of life which develops organically. It stands in reciprocal relation to the increase in the number of human beings. That is its real justification.

Notwithstanding its manifold distortion by profit interests, the struggle for mere accumulation and the like, we can no longer think of life without such progress. It is an indispensable factor in raising the standard of life.

The possibilities of the machine—with its abundant production, its ingenious complexity on the one hand, its simplification on the other, has necessarily led to a mass production which has its own significance.

The task of the machine—satisfaction of mass requirements—will in the future be held more and more clearly in mind. The true source of conflict between life and technical progress lies at this point. Not only the present economic system, but the process of production as well, calls for improvement from the ground up. Invention and systematization, planning, and social responsibility must be applied in increased measure to this end. [. . .] The solution lies accordingly not in working against technical advance, but in exploiting it for the benefit of all. Through technique man can be freed, if he finally realizes the purpose: a balanced life through free use of his liberated creative energies.

[. . .]

THE BAUHAUS

The first Bauhaus, founded by Walter Gropius in 1919, attempted to meet this shortcoming, not placing 'subjects' at the head of its curriculum, but man, in his readiness to grasp the whole of life.

Although for reasons of convenience a division into terms was retained, the old concept and content of 'school' was discarded, and a community of work established. The powers latent in each individual were to be welded into a free collective body. Also the pattern of a community of students who learn 'not for school, but for life' had to be worked out and converted into a cross-section of full, organic, and adaptable living. Such a society implies practice in actual living. Its individual members have to learn to master not only themselves and their own powers, but also the living and working conditions of the environment.

The foundation of the educational program of the Bauhaus, or, more appropriately, its working program, rested upon the recognition of this fact.

The first year was directed toward the development and enrichment of feeling, sensation and thought—especially for those young people who, in consequence of the usual childhood education, brought with them a sterile hoard of textbook information. Only after this first year of development and enrichment did the period of occupational training begin, based on free selection within the Bauhaus shops. During the period of occupational training the ultimate end still was: man was a whole. Man, who, if he but works from his biological center, when faced with all the material things of life, can again take his position with instinctive sureness, who does not allow himself to be intimidated by industry, the rush tempo, external evidences of an often misunderstood 'machine-culture.'

[. . .]

FURTHER READING

Walter Gropius, *The New Architecture and the Bauhaus* (Cambridge, MA: MIT Press, 1965).

Kathleen James-Chakraborty, *Bauhaus Culture: From Weimar to the Cold War* (Minneapolis: University of Minnesota Press, 2006).

Rainer K. Wick, *Teaching at the Bauhaus* (Ostfildern-Ruit: Hatje Cantz, 2000).

Hans Wingler, *Bauhaus* (Cambridge, MA: MIT Press, 1969).

SHAPING AMERICA'S PRODUCTS

Don Wallance

The industrial designer Don Wallance was best known for his work with flatware and furniture, including an early tubular steel and plastic chair for the Museum of Modern Art (MoMA) and the seating at Lincoln Center. But his 1956 book Shaping America's Products *looked far beyond his own specialisms, covering many different fields and scales of production. The book perfectly captured the 'designer-craftsman' ideal in America, which held that craft should be integrated into manufacturing as a way of improving quality and functionality. Wallance offered a series of profiles, from George Nakashima and Charles and Ray Eames to manufacturers like Heath Ceramics, Corning Glass Works, Jantzen (a knitwear firm specializing in swimming garments), and his own client, H. E. Lauffer. Each of these was presented as an exemplification of the integration of design and craft. The book itself was a product of such interdisciplinarity; it was co-sponsored by the Walker Art Center, which had a design program closely based on MoMA's, and the American Craftsmen's Educational Council. Wallance's ideas were not necessarily unusual. They were clearly derived from Bauhaus theory, which was sweeping American design schools at the time. But his book is unique in providing concrete instances of the 'designer-craftsman' ideal in action.*

Don Wallance, *Shaping America's Products* (New York: Reinhold, 1956).

"Art is not a special sauce applied to cooking; it is the cooking itself, if it is good. Most simply and generally, art may be thought of as The Well-Doing Of What Needs Doing."
—William R. Lethaby, *Art and Workmanship*

DESIGN AND CRAFTSMANSHIP IN AN INDUSTRIAL SOCIETY

Everyday life in America is carried on with a complement of equipment whose quantity, variety and technical complexity seem to be growing at an accelerating pace. Mechanization of numerous household functions such as house cleaning, laundering, cooking and food preservation has introduced many new objects into the home that were unheard of a short time ago. New methods of communication and entertainment have added the telephone, radio, television and phonograph to the possessions considered essential to a decent standard of living. The forms of these objects have no counterparts in the past. They are as unprecedented as their functions and have helped to transform our visual environment as well as our way of life.

On the other hand many of the things commonly used today have their roots in tradition and their forms retain a continuity with the past. Such common things as dishes, pots, knives, forks, chairs, tables, fabrics and wearing apparel are still used pretty much as they were several centuries ago. But the social changes brought about by mechanization and the introduction of new materials and production techniques are creating new forms for old things. Copper, cast iron and pottery still have a place in the modern kitchen, but plastics, aluminum and stainless steel are transforming the character of the commonest kitchen utensils, whose use has remained unchanged for centuries. Chairs can be made by assembling bits and pieces of wood as our forefathers did, or by spraying plastic on metal mesh. Fabrics may be woven of plant or animal fibers such as cotton and wool, or of man-made plastic, glass and metallic fibers in bewildering variety and endless combinations.

More significant perhaps than the number, variety and novelty of the things available for everyday use is the scale on which these objects are made and distributed to nearly all levels of the population. The most significant contribution to the growth of democracy made by 20th century America has been not in politics or government but in the widespread distribution of material goods. The Sears Roebuck catalogue might be called the Magna Charta of our civilization, and some cynics might add—its Bible too.

The forms of everyday objects are a key to the character of a democratic industrial culture. If the invention, production, distribution and acquisition of things are among the principal preoccupations of modern life, the industrial arts which give form to these things are a characteristic artistic expression of our time. For most people the ordinary objects around them provide the main source of esthetic experience. Toward the end of the 19th century William Morris enunciated the doctrine that the most important role of the arts is in the creation of the everyday things used by multitudes of people. This association of art and utility, of artist and technician, was not a new idea but the revival of a very old one. Before the Renaissance art and craft, artist and artisan, were identical concepts. The word "artist," as distinguished from "artisan," apparently did not come into use until the 16th or 17th century. The expression "fine arts," as distinguished from "useful arts," did not come into use until the 18th century. Although these distinctions still prevail today, one finds many areas in which art and technics seem to be moving toward each other even within the framework of specialization imposed by a highly industrialized society.

Our preoccupation with sheer quantity of production and distribution has been tempered by a growing interest in the design and quality of ordinary objects. This trend has been stimulated from several sources. It is largely motivated by business considerations—the desire to make products more salable in a competitive market sensitive to an increasingly selective and discriminating public. Rising standards of taste have received substantial stimulus and direction from a variety of educational institutions, consumer publications, art museums and proselytizing individuals who see a need for greater order, beauty and convenience in daily living. Perhaps this movement is merely one aspect of a growing awareness of the need to apply our technological resources toward human ends. Whatever the case may be, rising standards of taste and increasing selectivity on the part of the general public are making product design and quality not only a matter of cultural interest, but also a matter of increasing concern to business management—a decisive factor in the competition for markets. Product creation has always been a prime function of business management. Under present conditions it is taking on heightened significance for business and for the community at large.

How then do the myriads of products turned out by our factories and workshops get to be the way they are? What are the forces that shape their forms? The production of these things has been increasingly mechanized and automatized. But their creation still remains a function of the human mind, eye and hand despite mechanization and specialization. Nevertheless, the revolution in production which has taken place during the last 150 years has been accompanied by marked changes in the ways in which the things produced are created. The preindustrial master craftsman was designer, maker and seller of his handiwork. Integration of these functions was simple and direct. But changing patterns of production, distribution and business organization have broken up the hitherto united functions of designing, making and marketing into specialized functions which must be integrated in new and often more complex ways. In the more advanced industrial countries the hand craftsman was supplanted as a productive force of any consequence quite early in the industrial era. But the passing of the preindustrial craftsman who was both designer and maker left a creative void in the industrial process which is still being filled.

This book is about the patterns of product creation which are taking shape today. Insofar as product creation is inevitably bound up with technical skills and with the human "instinct of workmanship" this book is also an inquiry into the nature of craftsmanship and how it finds its expression in some areas of our industrial society. It is based on a study of the origins and development of a diversified group of consumer products that are typical of the best in contemporary design and workmanship.

[. . .]

"Design" in the sense that the term is used here always involves both the technical and esthetic aspects of an object. But all design for human use, whether the conscious emphasis is primarily technical or primarily esthetic, is actually a combination of both. Even the engineer who determines the form of an object on the basis of what appear to be purely technical considerations is probably influenced by esthetic bias and visual memory, though this may be unconscious. The esthetic elements of order, balance and rhythmic variation are also found in the structure of all matter—in physical, biological and technological forms. Subjective response to these elements in one area of experience is undoubtedly carried over into the other. As Herbert Read points out in the preface to a recent book of essays by scientists on form in nature: "The increasing significance given to form or pattern in various branches of science has suggested the possibility of a certain parallelism, if not identity, in the structures of natural phenomena and of authentic works of art . . . Aesthetics is no longer an isolated science of beauty; science can no longer neglect aesthetic factors." [Lancelot L. Whyte, ed., *Aspects of Form*, 1951] During the last war the Naval Department of Special Devices, which was concerned with the development of highly complex technical equipment involving advanced scientific concepts, had on its staff a number of artists with no specialized scientific training. These artists worked on the development of fresh approaches to the solution of technical problems that baffled the purely technical mind. The author knew one painter on the staff who acquired quite a number of patents in radar. Incidentally, he went right back to painting at the end of the war. This does not mean that artists should supplant scientists and engineers, or vice versa.

But esthetic design and technical design, as a creative process, are not as unrelated as they may seem. Product creation in modern large-scale industry usually requires that artist and engineer be separate persons. A question arises at this point. If esthetic design and technical design are the functions of specialists, can art and technics be integrated in modern industry as they were in the work of the preindustrial master-craftsman? This brings us to that aspect of product creation which is concerned with execution—craftsmanship.

In its broadest sense the term "craftsmanship" describes the common human drive toward perfection in the doing of any task. Whether this human trait is a biological or instinctual endowment of the species or whether it is an acquired characteristic which each generation in turn absorbs from its cultural environment might be argued at some length. Whatever the genetics of the matter may be, there is much to show that the sense of workmanship or craftsmanship has always been an essential component of the human spirit. The urge to do things well for the sake of doing them well, has found its expression in one way or another with each succeeding generation. Craftsmanship in this sense may find its characteristic expression in one way during an era of handcraft production, and in quite another way during the present era of machine production. It is always an essential ingredient of well-done things.

The term "craftsmanship" is commonly associated with skilled hand work. But we do not use the term here with reference to any particular method of doing things or to the nature of the tools used. What we mean by craftsmanship might be briefly summarized as will and skill—the will to take pains and the skill and resourcefulness to take pains creatively and effectively.

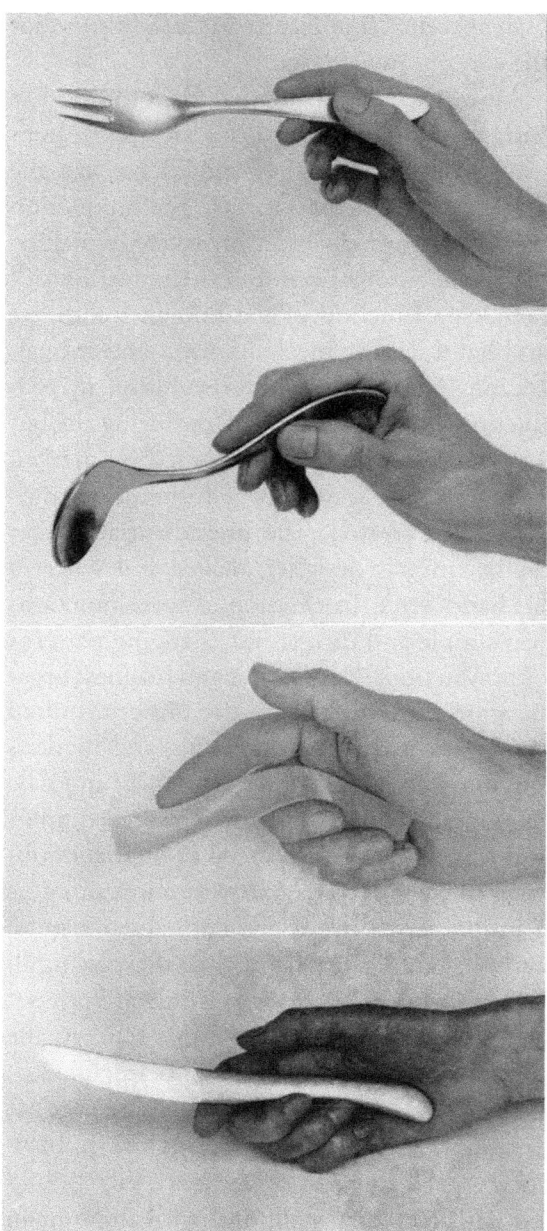

Figure 45 Don Wallance, studies for *Design One* Cutlery, 1956, as depicted in *Shaping America's Products*.

Design is in its essence a conceptual process. But design is interwoven with craftsmanship throughout the creative process. This is just as true in the case of an object to be produced by the machine as it is in the creation of hand-made things. Before a design concept finds its ultimate expression in the concrete form of an actual object, even a machine-made object, models must be made, tools and dies designed and made, standards of quality established, and production procedures worked out. These are not merely routine steps in the mechanical translation of a design concept into the production of the end object itself; they are a creative aspect of the development of form, whether in the creation of a painting or of a chair, without which only sterile forms can emerge.

The foregoing distinctions between design and craftsmanship as aspects of product creation are made for purposes of discussion and analysis. In actual practice design cannot readily be considered apart from craftsmanship. Design, the conception of forms that fulfill given ends, also requires conception of the means necessary to achieve these ends, as well as standards of excellence to be met. The very act of designing itself involves procedures and techniques that can be executed with a greater or lesser degree of craftsmanship. Product creation, in the sense we are concerned with, is the interaction of design and craftsmanship.

Unity of design and craftsmanship, of form and technique, is perhaps most readily perceived in the work of primitive craftsmen. Franz Boas, in his introduction to *Primitive Art*, writes: "So far as our knowledge of the works of art of primitive people extends, the feeling for form is inextricably bound up with technical experience . . . The manufactures of man the world over prove that the ideal forms are based essentially on standards developed by expert technicians." Gene Weltfish, in her more recent book *The Origins of Art*, concludes from a wealth of anthropological research material that "the historical evidence repeatedly shows the close relationship between art and industrial production. Art emerges not as an abstract vision but as a celebration of the skill of work well done and enjoyed."

A simple, direct relationship between art and technics continues so long as the tools required to produce an entire object can be acquired and manipulated by a single individual who is both artist and technician. The preindustrial craftsman produced excellent and beautiful things as a matter of course and without much conscious concern with "design." This was so because he worked within a well-established tradition, using forms which had been perfected through continuous adaptation by successive generations of craftsmen. Time, measured in decades and centuries, was an integral aspect of the "design" process.

[. . .]

Patterns of Product Creation

American conditions have been peculiarly favorable to the growth of large-scale industry. The size of the American population and its high standard of living have provided a large and ever expanding internal market which has stimulated highly mechanized production methods and has supported the large capital investment required. The wealth of self-contained natural resources essential to industrial growth has provided an ample material basis for the development of a mass production technology. But American industry also continues to be surprisingly diverse in the size and organizational patterns of its production units. Small-scale industry (employing less than 100 workers) and intermediate size industry (employing less than 500 workers) together account for about half the gross national product. For purposes of discussion in this book the term "small-scale industry" is therefore used flexibly but, in general, applies to firms employing less than 500 workers.

Apart from their role in the economy, small-scale industry and handcrafts, as we shall see later, have exerted an important influence on the nature of mass-produced products in many fields by exploring and pioneering new directions in design. Nevertheless, the fact remains that there has been an increasing concentration of production in large units. With the automatic production line already a matter of current planning, the indications are that this trend will be accelerated. Let us take a look at some salient aspects of large-scale industry as they affect present-day patterns and standards of product creation.

No individual in large-scale industry creates a product. A product is created by activating and synchronizing the thoughts and operations of a great many people. From the initial conception of the product to its use by the ultimate consumer, a complex of interrelated activities is brought into play involving market appraisal or analysis; product analysis, design and development; technical research; cost analysis; materials specification and procurement; tool and die making; organization of production facilities; training of workers; choice and preparation of distribution channels; planning of promotional activities. Each of these phases may in itself be a complex operation involving planning and coordination by many people. The planning of the whole process or any part of it can be a highly creative activity. Those who participate in this process must be capable of subordinating their own individuality to the discipline of the team. The "rugged individualist" may have played a key role in the original accumulation of power and capital from which today's giant enterprises have arisen. But the functioning of these organizations necessarily imposes patterns of

work in which the individual is subordinated to the group. This trend toward group activity rather than individual exploit is not limited to production planning and the assembly line but extends to the more recondite aspects of industrial activity. Even in the research laboratory, whose contributions to industrial development are commonly associated with individual geniuses such as Edison and Bell, we find individual brilliance and initiative absorbed into the group. In his book *The Genius of Industrial Research*, D. H. Killeffer writes: "Actually, the pattern of successful industrial research as practiced in leading laboratories has assumed quite definite form. While no one can deny that genius is extremely valuable to a research worker, yet most modern developments are achieved with far less of this invaluable attribute than is commonly supposed." In an article for *Chemical Engineering News*, Sir Ivan Morris Heilbron writes: "Gone are the days of Priestley and Faraday . . . Almost all of our major chemical problems now require the coordinated efforts of research teams in which each individual forges an important link."

Walter Gropius, pioneer architect and design educator, has long been an advocate of the group approach in architecture and design. In an article on group architectural practice for the British *Architectural Review*, Gropius writes:

"Synchronizing all individual effort by a continuous give and take of its members, a team can raise its integral work to higher potentials than the sum of the work of so many individuals." This point of view is strikingly similar to that expressed to the writer by W. H. Martin, vice president of the Bell Telephone Laboratories, largest industrial research laboratory in the world. In tracing the origins of the forms of many of today's mass-produced products, we find numerous instances in which the contributions of various individuals is so interdependent that it is impossible to ascribe the form of the object to anyone person. Some of the product case studies which appear later in the book illustrate this point. Many other instances could be cited. The patterns of product development may vary considerably, and in some instances a highly creative and forceful individual may have a dominating role, but the further we probe the various factors that shaped the product the more apparent their multiplicity usually becomes. This does not mean that individual creativity and genius are less important today than they were in the past. They are being absorbed into working relationships which are increasingly collective in nature.

The current tendency to focus on personalities in nearly all of the arts and to play up designers' names in product publicity seems to contradict the foregoing observations. To some extent this may reflect a survival of the Renaissance outlook which glorified the godlike individual hero and genius as the independent source of all great creative achievement. It may also reflect the survival of 19th century attitudes when individual exploit was still a dominating factor in progress during the early stages of the industrial era, as shown by the key role of the brilliant inventors and millwrights who created the technological basis for the industrial revolution; and by the industrialists who forged the economic power from which today's giant enterprises have arisen. This outlook received articulate expression early in the industrial era in the works of Thomas Carlyle. It survives as current dogma even though today's industrial complex imposes patterns of work in which individual exploit is increasingly merged with that of the group.

The current practice of featuring designers' names can be attributed in part to the

Figure 46   Marianne Strengel, *Taj Mahal* automobile upholstery fabric, 1959.

undoubted sales promotional value of associating an otherwise anonymous and impersonal product with a colorful personality. It is natural and human to react to personalities, and this human trait has been further stimulated by movies, newspapers and advertising. But a study of product origins in large-scale industry shows that it is usually a great oversimplification to say that a given product was designed by this or that designer. This is not to deny the creative contribution of many designers of great talent and forceful personality who have left their mark on hundreds of today's products. But it is a far cry from the outlook of the Renaissance artist, responsible to a single patron, or today's independent artist or craftsman, responsible to a small following, to that of the designer for large-scale industry, who must subordinate his own personality to the exigencies of a complex industrial process and his personal taste to the requirements of a mass market.

A trend away from emphasis on the individual designer is already evident. It is interesting in this connection to note that Walter Dorwin Teague Associates, one of the best-known industrial design offices, has increasingly tended to deemphasize identification of its name with the products it has helped to create. This trend toward anonymity on the part of even the largest and most publicized offices seems to be increasing.

[. . .]

The designer for large-scale industry necessarily accepts the fact that sales considerations take precedence over intrinsic values. But the? designer is a human being and an artist. His inner convictions, pride in work and sensitivity to the opinions of others cannot be completely dismissed. The designer's inner motivations may impel him to try to reconcile the conflict between the realities of the market place and the standards of his own craft. Some designers fully accept their responsibility to assist the development of salable products in an idiom which has mass acceptance but they also attempt to find forms within this idiom which are sound in themselves. A number of industrial designers feel that by staying close to the most advanced limits permitted by public acceptance levels—Raymond Loewy calls this the MAYA (Most Advanced Yet Acceptable) stage—they are discharging their responsibility to industry while slowly helping to raise the level of public taste. Henry Dreyfuss, while committed to this general approach, goes even further and has on occasion refused to accept a design job when he felt that the conditions imposed on him would compromise minimum standards as he sees them.

Whether a consciously calculated approach to product creation on a higher or lower level, however well motivated, can result in vital forms is another question. Can forms which have vitality and lasting quality be created without inner conviction? In the long run it may be that this conflict can be resolved only when designer, producer and consumer achieve a common outlook and response to visual form.

One of the earliest criticisms leveled against mechanized production was its debasement of craftsmanship and the shoddiness of its products. This bias against the machine is still widespread. "The common association of quality with handwork is recognized by the advertising profession which frequently tries to give machine-made products the aura of hand craftsmanship or to emphasize the personal skill and craftsmanlike attitude of the client company's production workers, as in the case of the Studebaker Company's father-and-son advertisements. But craftsmanship both as a motivational force and in its more concrete sense as skilled hand work is a commonplace aspect of mechanization from its earliest beginnings. The early European clock makers, forerunners of the later machine designers and tool makers, represent the highest level of skill, ingenuity and meticulous perfection of detail. Craftsmanship as an expression of mechanization has flourished increasingly until the present day. The English millwrights and tool makers of the late 18th and early 19th centuries, who were the pioneers of mechanization in production and the forerunners of the modern engineer, were highly creative craftsmen both on a mechanical and manual level. In the words of one of them, Sir William Fairbairn, reminiscing in 1861: "A good millwright was a man of large resources; he was generally well educated, and could draw out his own designs and work at the lathe; he had a knowledge of mill machinery, pumps and cranes, and could turn his hand to the bench or the forge with equal adroitness and facility." A contemporary of Henry Maudslay, wishing to illustrate the superb craftsmanship? of this brilliant inventor and maker of machine tools, described him as "quite splendid with an 18" file!"

As modern technology has become less empirical and more mathematical and theoretical in its procedures, mechanical design and execution have tended to become separate functions of the engineer or designer on the one hand and the tool maker on the other. Craftsmanship as an expression of personal skill in the manipulation of tools and materials is to be found primarily in the tool and die maker. It is clear, of course, that modern machine tools, and the dies, instruments and other precision equipment essential to the production process, represent an exceedingly high degree of craftsmanship as well as technical knowledge. The relationship of craftsmanship to mechanization becomes less clear when we turn from the means of production to the end products themselves. One could walk into almost any department store, not to mention five-and-ten-cent stores, and find a proliferation of cheap, shoddily made products to support the contention that mechanization has debased craftsmanship. The plastics industry is currently trying to overcome strong public prejudice against plastics materials because of the irresponsibility of some molders who have used materials unsuited to their purpose or who have skimped on thickness of wall sections to save

a few pennies. But these examples of debased quality in mass produced products result neither from limitations of the machine process nor the inferiority of synthetic materials, but from economic pressures in a competitive market. Anyone familiar with the properties of plastics knows that, properly used, they are not inferior substitutes for other materials but have unique qualities of their own that can provide superior quality at low cost.

It is true, of course, that the pressures to undersell one's competition, to widen profit margins by lowering quality or to create built-in obsolescence have often tended to lower the quality of mass-produced goods as compared to their handmade equivalents. But there are counteracting forces as well. Perhaps the most important of these is technological progress. As the technical virtuosity and precision of the machine improves, quality becomes less dependent on manipulatory skill. A machine-made chair mass-produced at a low price from bits and pieces of wood glued together like its hand-made prototype is obviously inferior in durability and finish to its higher priced hand-made equivalent. But a chair molded or stamped as a jointless stressed skin shell supported by a welded steel undercarriage will probably outlast both the hand-made chair and its machine-made imitation and need be no poorer in the perfection of its form and surface finish.

Despite mechanization, human factors are still crucial to the attainment of quality in the modern factory. There are few industries so completely mechanized that achievement of quality can be removed completely from the province of the production worker. In the manufacture of most objects there are innumerable operations, however simple or repetitive, as well as others requiring some skill, whose proper execution depends to some extent on worker attitudes and application. It is a matter of common observation, however, that the deterioration of workmanlike attitudes is one of the most vexing problems encountered in modern industry. Numerous studies have been undertaken by industrial psychologists and sociologists to understand the motivational forces behind worker attitudes and to devise plans which will stimulate improved workmanship as well as productivity. Industrial psychologists have found that work attitudes are poorer in operations of a semi-automatic nature than in the case of those which are either fully automatic or fully non-automatic. Whatever the case, the fact remains that under prevailing work relationships in the factory the production worker feels little interest stake, or personal identification with the end product to which he contributes some minor detail.

In designing for mass production the designer must take into account not only the machines and processes available in the plant but the capabilities of the production workers themselves. Design details which require a degree of precision and pains beyond the capacity of habitual practice of the workers, and which are not controlled completely by the machine process, will fail to come off, however beautiful they may have appeared on the drawing board or hand-made model. Furthermore, if the workers on the production line come to feel that a product imposes unusual or unfair difficulties it will create a morale problem for the plant as a whole as well as an animus toward the product and the designer. These attitudes are soon reflected in higher production costs and failure to meet the original intent of the designer.

The writer does not know of any instances in which the production workers are represented directly on the team responsible for product creation, along with management, engineering, sales and design. A few designers, however, do recognize the value of understanding the human limitations of the production line and even cultivate the acquaintance of individual workers. W. Archibald Welden, in recounting the development of the Revere stainless kitchen utensils says that he made it a practice to know the production workers individually and found their suggestions of great value. Attainment of quality on the production line devolves mainly on the engineers who design the production tools and formulate technical specifications and methods of quality control, and on the skilled tool makers who actually make the tools, dies, jigs, and fixtures which will give final shape, down to the minutest detail and subtlest contour, to thousands of identical objects. The tool and die maker, mold maker and pattern maker represent the apex of craftsmanship as manipulative skill in the plant. It is on their skill, sensitivity and co-operation that the realization of the designer's intent ultimately rests.

FURTHER READING

Everyday Art Quarterly (Minneapolis: Walker Art Center, 1946–1954; replaced by *Design Quarterly*, 1954–1996).

Victor Papanek, *Design for the Real World* (New York: Pantheon, 1972).

Herbert Read, *Art and Industry: The Principles of Industrial Design* (London: Faber and Faber, 1934).

Terence Riley and Edward Eigen, 'Between the Museum and the Marketplace: Selling Good Design', in *The Museum of Modern Art at Home and Abroad* (New York: Museum of Modern Art, 1994).

ASILOMAR CONFERENCE PROCEEDINGS 1957

Marguerite Wildenhain and Charles Eames

In the heady early days of the American studio craft movement, anything seemed possible. So much is clear from the impressive lineup gathered at Asilomar, an idyllic conference center in Monterey, California, for the American Craftsmen's Council in 1957. Among the attendees were leading representatives from every craft medium—Anni Albers, Sam Maloof, Peter Voulkos, Michael Higgins, John Paul Miller—as well as leading design figures such as Jay Doblin, Arthur Pulos and Asger Fischer, director of the Danish firm Den Permanente. Like most conferences, the event was as important for the connections it made as the discussions that were had. Indeed, attendees sometimes contested the notion that there even was a substantive difference between the two fields. In a panel on woodwork, Maloof commented, 'If I make fifty chairs, that may not be mass production but it is production in quantity,' prompting wood turner Jake May to observe, 'It's amazing the amount each worker in the furniture factory knows. These men are still craftsmen.' Two voices from either side of the nominal craft/design divide are excerpted here: Marguerite Wildenhain, the Bauhaus-trained potter, who had come to the United States in 1940 as part of the Jewish flight from Germany; and Charles Eames, who through partnership with his wife, Ray (1912–1988), had become the best-known industrial designer in America. The consonance of their ideas is striking. To find a dissenting voice at the conference, one would have had to pay close attention—it was only evident in the response of a few mavericks like Voulkos, who was clearly unimpressed by what he was hearing: 'This brings to mind why I try to avoid most organizations. You start getting all kinds of rules and regulations. The only reason I do what I do is that I like to do it.' At Asilomar, few could have predicted that within little more than a decade, it was this minority view that would come out on top.*

Asilomar conference proceedings, 1957. Unpublished transcript, courtesy of the American Craft Council.

MARGUERITE WILDENHAIN, "A CERAMIST SPEAKS ON DESIGN"

The very fact that we craftsmen are coming here to-day to discuss the relation of design to techniques should amply prove our low standards of crafts. That those two things which are the characteristics of any good object, and which should be insolubly connected with each other, got separated in our times would, taken by itself, prove this to be true. And since this division became possible, many have slipped into the habit of thinking of design as a thing in itself, that one adds on top of whatever is made, rather than as an inherent quality of all well-made things; a not self-conscious characteristic that comes from long and laborious trial and error, and from the entire and daily concentration and devotion of the craftsman to his work.

Ever since man has made objects with his hand, there was always the problem of design in relation to the techniques used; more or less primitive as they may have been. As the ages changed, so did the techniques, and the forms of the things that man made. The machine in modern times was only one more new tool. It need not have changed the attitude towards workmanship, design and technique, but man got confused. With the power of industrial production in his hand, he lost faith in himself and in what should be the main issue of life and put up instead an adoration of machines and mass-production for their own sakes. The craftsman let himself be pushed into a competition with the machine and was almost ground up in the process. In that painful transaction one thing that had been his pride and pleasure through centuries, namely the unity of design and technique in a well-made piece of his hands, was the first thing to disappear. Design went flying out the window, for time had become money, and there was none left for those well-designed objects of former centuries.

Obviously something should be done to get back the unity of design and technique, both for the craftsmen themselves and also for industry. There is no real reason to look at the crafts and industry as enemies. They are not, and if it should seem to be so in many cases today, it is due to misunderstanding of the basic problem. For good design, both in a unique hand-made object or in the millions produced by machine will depend on

Figure 47   The fiber panel at the American Craftsmen's Council, Asilomar, 1957.

the same few basic qualities. I name them quickly:

1. The fusion of material and object: that means extensive knowledge of what materials are available and suited, how they can be treated so as to get the most characteristic and alive expression and quality.
2. The competent use of techniques, processes or methods, of tools or machines for the one special purpose.
3. The solving of all problems of use and function adequately and masterfully, but not only on the surface and roughly so, but in all details (knobs included!).
4. Creative use of lines, colors, volumes, tensions, form and decor. And last but not least, artistic integrity.

Why then is it so difficult to get really good products, both in the crafts and in industry. Apparently there must be somewhere along the line of our arts-and-crafts-education something missing, something wrong in the approach, so that neither good craftsmen nor good designers for the industry are developed. What is it?

Obviously there is nothing anomalous for a well-trained craftsman to make models for industry; or for him to use machines and mass-production methods if he wants to go in for larger production in his own shop. But it should be clear to all of us that to have machines, and to be able to use them efficiently, does not in itself mean that we are partaking of a higher civilization or of a way of life that is more creative, nor that makes for more enlightened, better and wiser men. If we agree that the means of production cannot as such be the aim of a man's life, it should not be difficult to see what sort of place the machine should hold in our lives.

It is a tool, not more. So it is necessary to use the machine creatively, and this should be the craftsman's aim.

We will have to educate craftsmen in such a way that they do not go into mass-production before they have thoroughly explored the unlimited possibilities that handwork gives, before they have explored every alley and every depth hand-processes allow (far more diversified than any machine can make), before they have learned what form means, what lines can convey, what a rim or a foot on a pot can express, what tension and volumes are, what a tighter or looser handling of a detail can convey, before they know what black and white make possible, or horizontal or vertical divisions produce, before they have explored the near unlimited field of materials, of textures, of color and decor, and before they have been able to develop into personal and expressive objects more or less basic and common ones.

A craftsman who knows his crafts has an unlimited store of forms in his mind and hands that have evolved out of long years of intimate contact with his materials and all the problems of his craft. He will look at a new machine, new materials and new production-methods with the same creative thoughtfulness and with the deep alert interest that is the essence of his life. His designs for industry will have a real relation to materials and techniques, to function, to effective use of methods of production; and also an understanding of the cost and the time involved. All that has been the sum total of all his efforts all his life, and he will thus not have a dilettante approach. Nor will he fall for modish non-conformity in itself, for his designs will be solidly based and developed from sound craftsmanship and artistic integrity. He will also not become a superficial contortionist nor exhibitionist for the sole sake of being different, for he has a deep humility towards the things he is making and their live quality becomes the essence of his own life too.

This live quality required more than a course in "design" or in "industrial design" and it takes more than a pencil and some "experimenting or exploring into the problems of design for reproduction." Designs cannot be taught; only the principles can. Designs that are really good are the human and artistic reaction of a gifted person towards the total of materials, forms, use, methods, tools, objects. They do not exist in themselves, detached from all these essential requirements in some abstract pattern nor rule. All this the able craftsman knows. But if we go into mass-production and into industrial design before we are that full-fledged, alert, skillful and able craftsman, we will just design bad objects. Every time has its own way of making corn. Ours is relatively easy, and could be called "unconformity and cheap streamlining." So, let us beware of what can be a very tempting pitfall, namely this "design for industry"; for many who have not got the knowledge of the craftsman nor the talent of a designer of quality.

Yes, when we are really ready to make models for mass-production, let us humbly see if we have anything to give that is good enough for all of us and that is worth being repeated by the millions, but do not let us think that every little student can make "designs for industry" before he knows anything about his own field. The negative reaction of most students against a "plaster-wheel and a jigger, against exploring new clay forms for architectural use" is absolutely sound. They had never got the basic facts of their craft, not even barely touched the "forms and techniques recorded in history"—how could they invent new ones? They felt that instinctively, and rejected the superficiality of the project.

And let us not underestimate industry. If so much bad stuff is still being made, it is not because industry cannot produce better things, but because there are so many people who like and buy the badly-designed things. Don't let us be adding to their numbers by educating students to design those bad objects, may they be ever so elegantly "streamlined" or "modern". Let us rather educate students to know what a good piece of craftsmanship means, and little by little the standards of products will rise accordingly. As I see it, I know of no better way to educate those future producers or buyers of mass-produced things than to take them through the always excellent, basic and instructive discipline of handwork. But this must be done in the most basic and thorough way, that will require years of intense training and not only courses counted in hours and credits.

"Creative approach" does not necessarily mean that you have to use new materials, nor new machines, it is a mental attitude not depending on things only. Let us also beware of the new cliches, all the more so because they are fashionable and bring success. Not everything that has stood the test of centuries is obsolete, not everything that man makes today is of value. Far from it. Inside of the one basic pattern of mankind, there is ample scope for the saint and the criminal, the genius and the idiot, the tall and the small, for all races, for all forms and for all grades of human development. In the same way, man's work and the objects he forms will change and evolve slowly, in the same way as he matures and develops as a man.

It seems to me thus that we have to start at this very point: change the need of man for badly designed objects to his request for well-designed ones, and he will see to it that these well-designed objects are being produced.

Figure 48 Otto Hagel, *Marguerite Wildenhain throwing at the wheel*, ca. 1945.

Education thus is the ultimate and main problem.

For let us beware of trying to change the outer-form of man's life, the objects that surround him, without doing anything to develop and raise his deepest human essence.

On the contrary let us resolve and pledge ourselves to actually do what is needed to take the crafts out of their usual mediocrity and their dilettante approach, and build them up anew on a fundamental and competent basis.

Sound craftsmanship, honestly modern artistic expression and excellent design for industry will be the inevitable result.

CHARLES EAMES, "THE MAKING OF A CRAFTSMAN"

I do not think there is any single group who use the words "craft" and "craftsman" with deeper respect than architects do. And if there is a name which I would like to copyright, it is the name—"craftsman." It is a name which places a tremendous responsibility on those who claim it. When I say that the papers we have just heard were prepared in a craftsman-like way I am saying what I feel—that a tremendous amount of thought and work went into their preparation, and that they merit careful reading and thought by all of us who are concerned in any way with crafts and craftsmen.

Yesterday afternoon I dropped into the Wood Panel in time to hear some of the discussion. Something happened there that I believe must go on in all the discussions that come up at this conference. [James] Prestini was saying, "All right for so and so, and for this and that but really, what do you feel is the place of a craftsman in our society today?"

There were a number of statements made in reply to this, and some perfectly good ones, most of them ending with man's relationship to the thing he was producing, the amount of satisfaction he got from it, and the sincerity with which he did it, as well as the value of the opportunity for individual expression. And so on. But the real question that Prestini had put remained unanswered.

In the course of the discussion Prestini was questioned about his bowls—what they mean to him, what they give him, and so on. Yet when I think of Prestini as a craftsman (and I do), his bowls are not what I think of first. Nor even second. I have to sort of work my way around to them because, fundamentally, and first and second, I think of Prestini as a craftsmanlike guy with a terrifically humble attitude to the materials that he works with and to the problems that surround him.

I use Prestini as an example because the question he brought up is closely related to the way I think of the building of the values that go to make a craftsman. And believe me, in our world and in our time, we are deeply in need of the values which come under the head of "craftsmanship". I would venture to say that society today is more in need of these values than of any other thing.

The richness that a craftsman working for himself in a simple and direct way gradually brings to our society cannot be overestimated. Society needs this. The only contention that I would have with the individual craftsman, or with a group of them working for themselves and for the satisfactions they get from their work, is in those areas in which the individual craftsman is uncraftsmanlike. And this is the most damning thing that you can say about the majority of craftsmen.

The same thing is true of architects. Architects are just as bound as craftsmen. So are

Figure 49   Charles Eames and Eero Saarinen, *Side chair*, 1940.

educators. So when I point the finger of un-craftsmanship to some of the craftsmen, I am pointing it all around the circle, to all other areas. This is the problem of our time. It is the problem that confronts us as a nation and it is the problem of the human race. If we are going to survive we have to become craftsman-like people; and in that word's deepest, fullest sense.

It is urgent therefore that we start thinking and talking about craftsmen and craftsmanship in a much broader and unlimited way.

The field of design is undergoing changes. It would be interesting to see how a designer of great value to a company functions today. A great number of them, functioning as designers in the best sense of the term, for various companies producing material are not even drawing a line. It's not necessary. Nor do they consider a form. Nor do they dictate necessarily any special relationship. Very often they work within areas, and there are so many, that no individual would possibly presume or want to be responsible for their form and structure himself. Such a presumption would be ridiculous.

Because of this, the responsibility for a decision—aesthetic, practical and connected with craft—has to be set back in the industry itself. So we have the designer in industry feeding back basic responsibility to management, to the purchasing agent, to the engineer, to the analyst. There is an area in design that has to do with not taking on individual responsibility, but making a sense of responsibility run through industry and management, making those who have previously thought of themselves as responsible only in those areas—management and industry—more and more responsible in the peripheral areas.

This is the sort of thing that must happen to the whole of society. There is an area in large industry which has this and that in abundance, but which lacks craft. And there is a similar lack among craftsmen in regard to what industry has. Craftsmen need to examine the mechanics of production and the results and relationships in industry. Mutual understanding will do much for both. It will help build up a feeling of craft in industry, to industry's and the craftsman's and society's benefit.

I began by speaking of architects, and now, at the close, I want to go back to architecture again just to mention Mies van der Rohe. I want to mention Mies because here is a man who is essentially a superb craftsman. It has been said of Mies that the ambition of his life would be to take the most perfect brick in the world and lay it in the most perfect bed of mortar in the most perfect relationship to the next brick.

Mies himself has something to say which bears on this, and which bears much thinking about. He said, "I don't want to be interesting, I just want to be good."

This, I think, a craftsman should have tattooed across his chest.

FURTHER READING

Pat Kirkham, *Charles and Ray Eames: Designers of the Twentieth Century* (Cambridge, MA: MIT Press, 1998).

Joyce Lovelace, 'Recalling Asilomar'. *American Craft* 57/3 (June/July 1997), pp. 66–9.

Marguerite Wildenhain, *Pottery: Form and Expression* (New York: American Craftsmen's Council, 1959).

71

'THE NEW HANDICRAFTS', FROM *HOT HOUSE*

Andrea Branzi

The emergence of Italian Postmodern design in the late 1970s has a twofold importance for craft history. First and most obviously, the creations of Alessandro Mendini, Ettore Sottsass, Nathalie du Pasquier and their peers for the design groups Studio Alchymia and Memphis were deeply influential on the crafts. American studio furniture, British ceramics and Central European jewelry were just a few of the areas where the style was decisive in encouraging new directions. Less obvious, perhaps, is the importance of craft for the Italian radical designers themselves. Though few of them worked with their hands, all had a deep interest in the logic of materials, which is often forgotten in the rush to present the new Italian design as concerned principally with semiotics and signification. (The materials were unfamiliar ones—neon tubes and plastic laminates—but they were still subject to evaluation on their own terms.) In the following passage from The Hot House, *the most widely read and intellectually rigorous book to emerge from the Memphis moment, architect and designer Andrea Branzi provides an insider's account of Postmodernist attitudes towards craft production. As he makes clear, Sottsass and his colleagues took a 'small batch' approach to their work and readily acknowledged the fact that the difference between their objects and crafts was principally a matter of rhetoric, achieved in the realms of distribution and discourse rather than in the workshop.*

Andrea Branzi, 'The New Handicrafts', from *The Hot House: Italian New Wave Design* (London: Thames and Hudson, 1984).

Interest in handicrafts is a phenomenon that crops up regularly in the history of design and modern architecture, as a sort of confrontation with a production system that is seen as "lost innocence." Throughout the twentieth century, the fact that industrial work has been continually spurned has led to a perception of often non-existent values in handicrafts. People nostalgic for pre-industrial society, for example, are now looking to the values of craftsmanship for a new pattern of organization based on the rediscovery of the individual and on the myths of humanism.

The urban models of Leon Krier and the academic school of Luxembourg, for instance, resurrect the idea of an artisan city in which eclectic monuments are used to reconstitute an urban landscape abounding in symbols, where architecture may be saved by the sacrifice of social and technical progress and where it is given back the space to get across its own message; making a clean sweep of all the encumbrances of modernism.

In this way handicrafts resume their place at the heart of a very broad concern with culture, as a touchstone for different hypotheses of historical development. But it turns out to be extremely difficult to isolate handicrafts as a category of production today; none of their supposed "purity" stands up to close examination. Complex and ambiguous relationships with manufacturing and the market have led on the one hand to a decline in the expressiveness of traditional products and on the other to an astonishing growth in the capacity for technological experimentation, unknown to industrial mass-production.

In his book *Dov'è l'artigiano*, Enzo Mari has attempted the first thorough examination of what goes by the name of handicrafts today, trying to retrace and rehabilitate the common thread linking those areas where "diffuse craftsmanship" is to be found in the industrial world. In the educational exhibition that he mounted for the Tuscan Region on the occasion of the 54th International Handicrafts Exhibition, Mari brought together the two extremes between which modern handicrafts are defined today, combining in a mathematical fashion the various factors that underlie differing patterns of production. He starts out with an example of the optimal state in which ownership of the means of production, design and execution coincide in the figure of a single operator: "An artist in modern society; a builder of arches in a primitive society." The other extreme is represented by total dissociation of all the different phases of the manufacturing process, in which designer, manager, executor and owner of the means of production are all different people, with separate responsibilities and fields of knowledge. As Mari puts it: "A manufacturer of domestic appliances in modern society, the constructors of pyramids in an ancient society." At this pole the involvement of the artisan has fallen to zero; but if just one of these conditions changes, one can fairly speak of an "ingredient of craftsmanship" in industrial production. The assumption of a clear-cut technical and ideological rift between handicrafts and industry does not stand up to the evidence of a vast program of functional collaboration between the two systems of manufacturing.

This is the setting in which the old controversy over the use of the machine that so exercised the greatest theoreticians of the last century, from Pugin to Morris, and then from Muthesius to Gropius, has been superseded by the technological flexibility of modern craftsmanship, which permits experimental processes that are beyond the grasp of the rigid structures of industry. Being an artisan does not mean not using machines in the process of manufacturing; on the contrary it means using all the machines in the workshop in rotation, maintaining direct control over all phases of production by passing—unlike the assembly line worker who is confined to a single stage of construction—from one machine to another, just as the carpenter moves directly from the planer to the electric saw and the drill as the stages and methods of his own scheme of assembly require.

Making "by hand" today, if it does not coincide with the possession of the means of design and construction, means using the hand just like any other machine; it is no surprise that under such working conditions the category of handicrafts is often a cover for the most alienated section of the black economy.

The product of "diffuse craftsmanship" in a modern industrial civilization should not therefore be confused with the naïve or poor product. It is better represented by the experimental prototype with a high technological content: from the injection mould for the manufacture of plastic goods to the prototype of a Formula 1 racing car. Such craftsmanship is indispensable to and indeed the basic premise of industrial manufacturing.

When on the other hand the artisan is employed in making handicrafts, i.e. producing goods aimed at a separate and alternative market to the industrial one, he is making use of cultural models far less susceptible to analysis, and the products that he offers have highly distinctive structural qualities, though they are often more symbolic than real. In fact the so-called arts and crafts are of considerable commercial importance today and are linked to a stable market with its own specific range of merchandise. People demand handmade goods, or goods that look handmade, precisely because they suppose it involves old and consequently better techniques of construction than those used in mass-production and also because they attribute a continuity with traditional culture and its methods to the design of this kind of product. The question of the technology and cultural models used in arts and crafts involves such a tangled and complex series of equivocations that it is difficult to unravel; perhaps the only feasible approach to making a critical examination of this question is one that involves going back to the period in which the market for handicrafts as a separate category from that of industrial products was formed.

One way to do this is to adopt the model of historical analysis recently proposed by Argan in connection with "peasant culture," which has also been the victim of widespread critical misunderstanding of an ideological origin. Argan holds that the origin of peasant culture lies in the courtly tradition of Byzantium, i.e. in a theocratic culture? that, displaced by the emerging urban mercantile civilization, fled into agricultural territory, where it clung to stable values of living and working. Far from being spontaneous and primitive, it stands as a tradition of high cultural import with its own concepts of work and ethics that are in conflict with those of a mercantile civilization where the products of culture are viewed as goods of exchange with their own market value. So the peasant culture's system of symbols and techniques derives from Byzantine ideology, in which a system of immutable values was determined to exclude new technical and social experiences.

The same method of analysis can be applied to the process by which an artisan technology of marketable goods was formed over the last century, in an attempt to create an alternative to industrial mass-production and above all to the program of cultural adaptation and impoverishment that this involved.

Modern artistic handicrafts have been vaunted from their birth as an area where

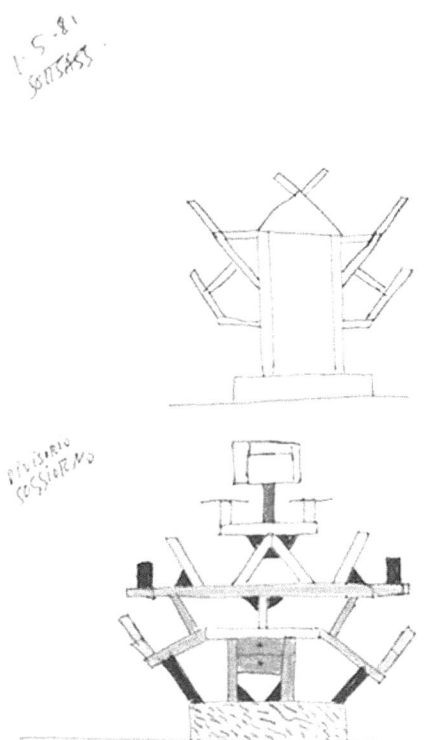

Figure 50 Ettore Sottsass, *drawing for the Carlton bookcase*, 1981.

traditional values are conserved; dissociating themselves from any culture of design the arts and crafts have ground out a range of products that are represented as lying outside the consumer market, responding to the real exigencies of age-old human needs rather than to the fashions of the market of induced needs. The supposed creativity of the artisan, or at least his "ancient wisdom," is contrasted with today's consumption-oriented technology. In reality the arts and crafts, shifting the responsibility for research into new forms of merchandise and consumption onto industry, stick to the pure and simple "reproduction" of existing models, i.e. those already shaped by tradition. The myth of artisan creativity crumbles before an analysis of the ways in which it functions on the market; all innovative processes are alien to it, and it only enters the market with formal models that are already stabilized.

Paradoxically, then, industry absorbs the experimental flexibility of a technological craftsmanship, utilizing it as a research phase within the industrial cycle, and at the same time eviscerates the creativity of artistic craftsmanship by transforming it into an alienated system for the reproduction of historical models for the market. The result is a total loss of political and cultural autonomy on the part of modern handicrafts, which are obliged to aid the development of markets and techniques in industry and at the same time make room for the latter on the market, renouncing any innovative role for its own artistic product.

In this way artisan production recycles even the modern style, though it is treated as a historical style, interchangeable with other formal models drawn from tradition; in any case any ingredient of research or design is extraneous, and construction becomes synonymous with imitation.

Curiously, we are seeing today a repetition of this phenomenon that occurred a hundred years ago: within an advanced industrial society a theoretical debate of great significance is once again developing over handicrafts, seen as the area of confrontation between different theories about the development of architecture and the city. And this is taking place, again, at a moment when handicrafts are in the throes of an extreme crisis.

The crisis of the crafts as an independent field of research and design, together with the collapse of the myth of the artisan workshop, seriously complicates any research in this direction. We can approach handicrafts only as a privileged area within industrial production, where a particular organization of work permits the realization of products in which experimentation plays a greater part or the use of industrial materials that do not fall within the requirements of mass-production.

Handicrafts can be viewed, then, as a major experimental instrument, as an industrial area open to new patterns and new designs that mass-production would be unable to create owing to the rigidity of its technical and manufacturing structure. Possession of this instrument became, towards the end of the seventies, indispensable to the renewal of the culture of design. In 1979 Studio Alchymia of Milan presented the first "Bauhaus" collection, made up of prototypes and models by Sottsass, Mendini and De Lucchi, as well as myself: that moment saw the birth of a new formula of production and distribution to which we have given the name "new handicrafts."

The "new handicrafts," which were first put on show by Alessandro Guerriero at his Alchymia, possess certain very precise characteristics: the craftsmanship employed, given that production is made up of small runs or unique pieces, does not depend on the use of particular techniques, but rather on the speed with which the models whose design makes no concessions to the possibility of future mass production are constructed by craftsmen using the most advanced techniques of modern joinery.

The explicitly cultural aspect of the models does not derive from "artisan culture," but rather from the way the latter is used as an area for experimentation. In fact the prototype and the limited run make no pretence of being an alternative to mass-production, but treat it as a possible subsequent phase to the experiments in design permitted by the new handicrafts. The "original," the model that can be reproduced only as a limited repetition of a prototype, is a consequence of the experimental nature of the design and not a theoretical premise. In this sense, the correct place for the new handicrafts is alongside, or before, mass-production and not in opposition to it, since by nature they involve an experiment not in technique or production but in expression.

The new handicrafts are born out of a realistic appraisal of the present industrial production of furniture and in particular the essentially artisan nature of a large part of it. Often, in fact, the adjective "industrial" is intended as an indication more of a style than of a genuine mass-production of models. This is the result of a large number of factors, not least of which is the necessity of testing the commercial viability of a design before going into automated production, with the enormous amount of time and expense that this entails. What this means in practice is the production of handcrafted prototypes of potential mass runs that will require substantial modification if they are ever to go into mass-production. Simulated mass-production runs, i.e. those envisaged but not realized, make up a large part of the present-day furniture industry. The new handicrafts accept the positive side of this somewhat ambiguous situation, at least from the stylistic point of view, and turn it to advantage in a production that is free from the problems of mass scale and involves a high degree of experimentation and research. This makes clear the commercial rather than productive difference between the new handicrafts and industry, where the one aims to sell a few prototypes of high cultural quality and the other aims to sell many prototypes of mediocre cultural value. There is no conflict here, only a difference in marketing strategies.

FURTHER READING

Volker Fischer et al., *Design Now: Industry or Art?* (London: Prestel, 1989).

Fredric Jameson, *Postmodernism, or the Cultural Logic of Late Capitalism* (London: Verso, 1991).

Ezio Manzini, *The Material of Invention* (London: Design Council, 1989; orig. pub. in Italian, 1986).

Barbara Radice, *Memphis: Research, Experiences, Results, Failures and Successes of New Design* (New York: Rizzoli, 1984).

Penny Sparke, *Italian Design* (London: Thames and Hudson, 1988).

SECTION 7

CONTEMPORARY APPROACHES

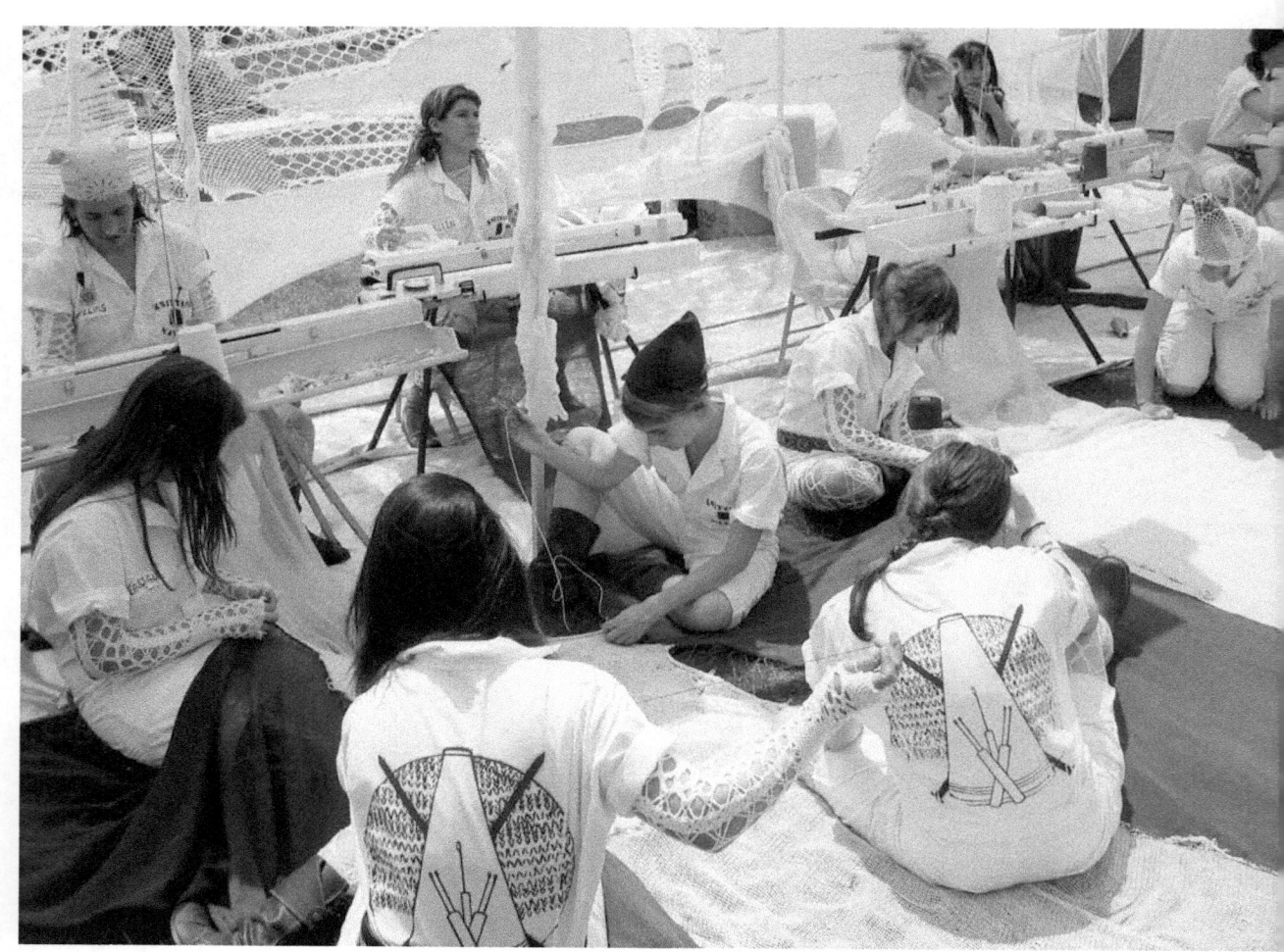

SECTION INTRODUCTION

If the history charted in this reader is sometimes confusing, with ideas leading in contradictory directions, then the year 2009 certainly offers no less clarity. Craft today is still a subject marked by infinite diversity and internal conflict. This is partly because the twenty-first-century picture of craft is an accumulative one, a palimpsest. It is still possible to find 'traditional' crafts across the globe, and people and institutions devoted to their preservation and/or exploitation. Several of the Arts and Crafts organizations founded in the late nineteenth century—the Art Workers Guild in London and the Boston Society of Arts and Crafts among them—still exist, and in some respects preserve the intentions of the men and women who founded them. The Modernist dream of designing through craft practice is still alive, too, in forms both new ('DesignArt') and old (Bauhaus designs are more available than ever at a home furnishing store near you). And the studio craft movement, though it now looks long in the tooth and possibly out of ideas, still makes for the livelihoods of countless craftspeople, gallerists, curators and administrators. In this sense, twenty-first-century craft contains within it all the models of craft that have gone before.

As ever, there is also a sense of crisis. Recently I attended a meeting of the heads of several London-area art and design schools, to discuss what could be done about the steady disappearance of craft programs in UK secondary education. The concerns expressed at the meeting were the old Ruskinian ones: new generations ever more divorced from material experience, as new technologies (now iPhones, not replicating lathes) arise to mediate the world. And the solution, too, was Ruskinian: craft, and its evergreen promise to connect us to all that modernity seems to wrench away. I raise this anecdote not to imply that the antiquity, and seeming inevitability, of these anxieties should induce us to ignore them. Indeed, their persistence is one of the most compelling indications that we need to study modern craft's rich literature, for the problems that faced the nineteenth century are in many ways still with us today.

THE POLITICAL AND THE POSTDISCIPLINARY

But of course there are new developments as well, of which two in particular stand out: politicization and postdisciplinarity. The first of these is still a trend in search of a definitive label: neo-craft, craftivism (or craftism), DIY culture, microrevolt and subcultural craft have all been proffered. More unanimous, and more telling perhaps, is the preferred term of self-identification for those involved: they describe themselves as 'crafters'. In its active, verblike quality, the word seems designed to achieve differentia-tion from the older term 'craftsperson', with its connotation of fixed and permanent identity. If studio craft strove to achieve professionalization, the new crafters happily embrace vocationalism. Nor are all (or even most) of the crafters explicitly political, as a breeze around the Web to sites such as Etsy.com will attest. But even in its commercial variants, the crafter wave wears its small-is-beautiful ideology on its hand-embroidered sleeves. It may be no coincidence that this development has emerged in a time of war and economic instability, but ironically, what is more important is the sustaining environment of the virtual. The new crafter wave is fueled by an intriguing alliance of the oldest and newest of social technologies, the sewing circle and the blog. In a sense, the twenty-first century of craft is beginning the way that the twentieth century did: by finding in tradition the possibility for social change.

A second trend, towards postdisciplinarity, is no less wide-ranging and even more difficult to comprehend. The term is best taken to mean not a border-crossing between discrete fields, but rather a situation in which such borders no longer really exist. Postdisciplinary practitioners are free to call themselves whatever they wish, or indeed not call themselves anything at all. Effectively, this puts an end to the old 'craft versus art' debate because in an undifferentiated field of practice, no one activity has any more right to be called art than another. (The same would hold true for design.) As we have seen, this idea of the 'unity of the arts' has precedents stretching back to the nineteenth century, but this latest version arguably took shape in the early 1990s. On the one hand, there was a theoretical drive within the academy in the ultrapermissive message of Postmodernism, which taught that there was no barrier that could not be deconstructed. On the other hand, there was a shift in practice; a new generation of art and design school instructors began to distance themselves from the last vestiges of Modernist thinking. Truth to materials and fitness to function were long gone. The next logical step was to leave behind prejudgment of material selection and utility altogether. Students began to learn skills as and when they were needed, rather than as a standard foundational repertoire that could be turned to various ends.

Whether we have really reached a state of total flux is debatable. Proclamations of postdisciplinary culture may well be premature, in that distinctions of status based on genre clearly persist. The marketplace is still divided into discrete sectors, each with their own power structures. Craft itself still acts as an important impediment; wanting to make something and being able to do so are definitely two different things. Indeed, it may be that considerations of production are destined to be the last real brake on postdisciplinarity. Forms of making exert a friction that cannot be overcome by wishful thinking—a fact that explains why so much contemporary art today is made in the absence of skill on the part of its author.[1] Yet it seems certain that particular modes of making will always constitute a source of value, precisely because of their difficulty. It is hard to say how craft's position within the arts will be conditioned by these inescapable facts, but they certainly cannot be ignored.

GETTING IT TOGETHER: CURATORIAL APPROACHES TO CRAFT TODAY

There are many possible ways to approach the previously described themes. As a sometime curator myself, I have elected to stress that way of looking at the subject, and so most of the selections in this section deal specifically with strategies of museum and gallery display. This is also a natural way to test the idea of postdisciplinarity, for if artists can freely define and redefine themselves at will, most institutions are not so lucky. With long histories of division by genre, mainstream art and design museums have responded to the present ecumenical environment sluggishly, and in some cases with suspicion and recalcitrance. On the other hand, many institutions have gone out of their way to capture the excitement of the present moment in craft, among them the Museum of Arts and Design (formerly the American Craft Museum) with 'Radical Lace and Subversive Knitting', the Design Museum, London, with 'My World', and the Victoria and Albert Museum with 'Out of the

Ordinary: Spectacular Craft' (all of these exhibitions 2007). Less predictable venues have included the Art Gallery at Haifa University, which picked up on the trend early in 2003 with the exhibition 'OverCraft'. Contemporary art spaces have also approached the subject of postdisciplinary craft, though usually emphasizing issues of labor and obsession rather than skill. Notable examples include 'Work Ethic' at the Wexner Art Center, Columbus (2004), 'Over and Over: Passion for Process' at the Krannert Art Museum, Illinois (2005), and 'Poetics of the Handmade', focusing on Latino artists, at the Los Angeles Museum of Contemporary Art (2007).[2]

Perhaps unsurprisingly, the curating of contemporary craft tends to be done most interestingly where it is least expected. Scandinavia, which last took the lead in formulating craft's identity at mid century as the model for the 'designer-craftsman' ethos, and which retains a reputation for being a sort of nature preserve for Modernism, has become the improbable site of the world's most convincing craft avant-garde. This is largely thanks to structural issues: an unusual combination of healthy government funding, top-flight art schools, relatively numerous exhibition venues, and a complete lack of private collectors. The result is that craft-based artists often work collaboratively and must compete for the attention of museums as patrons, rather than private clients. Predictably, their work is more adventurous than that of their peers in the United States and Britain. Imaginative curators have taken full advantage of the situation, conjuring one show after another that convincingly chart a future for avant-garde craft. A similar situation prevails in New Zealand, which also has a reputation for conservative craft abroad but supports the innovative venue Objectspace, an experimental noncommercial space in Auckland, and Te Papa, a museum in Wellington that intermixes craft, art and design in an attempt to draw together English and Maori cultures.

Again, it is difficult to say where all this activity might lead. Who could have predicted the mature writings of William Morris in 1850, the modernist weavings of Anni Albers in 1900 or the vigorously incoherent pots of Peter Voulkos in 1950—or, for that matter, the present fashionability of craft in 2000? And so, who can say what historians in 2050 will make of today's many versions of craft—and what they might therefore conclude about our own times?

While *The Craft Reader* has pretended to a sort of comprehensiveness, and to a detached, Archimedean point of view, it is itself a product of its moment. At some point in the not too distant future, as craft studies expand and diversify still further, this book will no doubt come to seem trapped inside the classic debates. Its contents will seem too narrow in their range, its geography too weighted toward Europe and America, its plotlines too few and all too predictable. For me, as editor, this is a weirdly comforting thought. Any anthology is intended first and foremost to spur new thinking—and doubly so given that this is an anthology about craft, that apparently simple but actually protean thing.

NOTES

1. On this point see my own 'Sloppy Craft', *Crafts* 211 (March/April 2008).
2. Helen Molesworth et al., *Work Ethic* (College Park: Pennsylvania State University Press, 2003); Alma Ruiz, *Poetics of the Handmade* (Los Angeles: Los Angeles Museum of Contemporary Art, 2007). See also Shu Hung and Joseph Magliaro, eds, *By Hand: The Use of Craft in Contemporary Art* (New York: Princeton Architectural Press, 2007).

72

AFFECTIVITY AND ENTROPY: PRODUCTION AESTHETICS IN CONTEMPORARY SCULPTURE

Johanna Drucker

'I mean, how can an artist make anything as amazing, in sheer production terms, as a pink plastic laundry basket from K-mart?' This question might be considered the slogan of the following text by Johanna Drucker. A contemporary art historian, critic, book artist, poet and theorist, Drucker is now the head of the Media Studies Program at the University of Virginia. If a single theme could be said to link her activities, it is the materiality of art—even art that is billed as Conceptual. By focusing on questions of production rather than ideation, Drucker makes the byzantine complexity of contemporary practice at least a bit more legible. If one wants to make sense of the bewildering visual cacophony of a major contemporary art fair, for example, it's sometimes helpful to ignore how things look, and what they are supposedly about, and instead focus on how they are made. Until recently, contemporary art required little in the way of skill to produce: found objects, outsourced fabrication, bad painting, big photos printed on expensive equipment, 'relational' situations requiring only rudimentary props, and outsized sloppy sculpture were the dominant forms of the 1990s. Drucker was among the first to spot the more recent fashion for more highly crafted sculpture. She argues that the sheer challenge of fabricating things that can capture public attention, in an environment of massively capitalized production, has driven sculptors into certain characteristic patterns. This allows her to reconceptualize skill, or the lack thereof, as a matter of strategy rather than absolute value.

Johanna Drucker, 'Affectivity and Entropy: Production Aesthetics in Contemporary Sculpture', in Anna Fariello and Paula Owen, eds, *Objects and Meaning: New Perspectives on Art and Craft* (Lanham, MD: Scarecrow Press, 2004).

A tension exists in contemporary sculpture between work that possesses seductive production values (looks good, is obviously well made, and has the degree of fine finish worthy of a showroom floor) and work that aggressively disregards these values (so that it looks like cast-off debris beneath the level of roadside junk). In both cases, I would argue, sculptors are struggling to make their attitudes towards production so conspicuous that they differentiate their work from other consumable objects of mainstream material culture in formal as well as conceptual terms. Whether contemporary sculptors make a show of attending not at all to formal, material values or attending too much to these values, they seem united in their efforts to use this self-consciousness to make clear that theirs is an art product. This suggests that no matter what a sculptor's work is about thematically, it is always trying to balance the values of production and the production of value(s). Paraphrased, this suggests a struggle to keep standards of making (fabrication, finish, the intrinsic worth of the stuff, and the finished quality of the reworked object) from becoming values that simply reinforce and reiterate the prevalent standards of the standard market for consumables in mainstream culture. At the same time, the object has to demonstrate enough value not to be dismissed from all aesthetic or critical notice.

Overall, this work pushes aside the shadow of negation that has been so central to contemporary art and its critical reception as a legacy of the classic avant-garde. The aesthetic differentiation this work embodies presents an alternative to the values of mainstream consumer culture. A feature that distinguished work in the last decade of the 20th century from now-classic postmodern work—with its allegiances to a cool, critical, differentiation from the mass-culture objects it was at pains to critique—is an attitude that can only be described as affirmative. The idea that affirmation is a viable premise on which to continue the romantic project of art as that which allows us to imagine otherwise is so unfamiliar that there is almost no critical scaffolding on which to begin its discussion. Much of the work I will discuss here enacts its transformations of the materials and objects used for its production through an enthusiastic engagement with studio practice, the flirtatious appreciation of artists for mass-culture artifacts and iconography, and a very positive playfulness. All are elements of an attitude that can't be adequately contained within the critical parameters of negativity. Nor can it be subsumed under a claim to be "political" in the old-fashioned sense that still floats through much of the rhetoric of contemporary art and critical practice. That rather simpleminded elision constrains the concept of political to a notion of instrumental transformation of institutions, sites, and relations of power

and as an intervention in the symbolic order (with its considerably more distant, muted efficacy). Almost all contemporary art practice that claims to be political falls into the latter category, though the category struggles to pass itself off as "instrumental" in a way that falls apart at the first serious scrutiny. Clearly, an alternative theoretical foundation is needed that construes transformations and alternatives within the symbolic order as significant without requiring that they conform to a negative criticality.

This introduction is meant to frame the discussion of production values and to show that they embody a shift in attitude within artistic practice that calls for a radically different critical premise. This is an attempt to imagine otherwise, to think ourselves out of the cul-de-sac of received tradition and its adherence to a set of critical principles that simply don't apply (and haven't for a long time) to current work. Recent sculpture is no longer framed within the critical precepts of modernism and its supposed negativities.[1] The challenge is to think about recent work in a way that moves beyond this moribund critical legacy premised on refusals and rejections towards another, equally insightful but differently premised, critical point of view.

Writing about the late-1990s exhibition of sculpture she titled *Blunt Object,* curator Courtenay Smith used the phrase "latent entropy" to describe the condition presented by several of the pieces. She was referring to an evident material contradiction in the works between their appearance of solidity and the fact that in many cases they were highly fragile. Alternatively, they gave the appearance of being tremendously ephemeral and delicate objects-though in fact many were virtually indestructible pieces made by new technological production processes whose longevity can be measured against the half-life of current land-fills. Smith's emphasis was not on temporality or the physical life expectancy of these works, but on the interrelation between production values and the production of values in new sculpture. The challenge she identified is the one most fundamental to contemporary artists: how to make an image or object that has any clear identity as a work of art in the context of a culture of maxed-out consumption in its material and visual domains. How do you call attention to production as something that matters? In particular, how can production claim attention for itself as a defining element of a work of art when the high end of the spectrum of production values is populated by mass-produced objects? The idea of "latent entropy" suggests that though it may not be easy to maintain this stance of differentiation, a crucial aspect of making the attempt is to enact a strategy of displacement and transformation at the level of material production of the object. This strategy allows the aesthetic domain of art activity to effectively exercise a measure of critical transformation with respect to the broader cultural field of the production of objects.

Such strategies of differentiation, I suggest, gravitate towards two poles—that of "entropy," invoked by Smith, and the counterpole of "affectivity" invoked in my title. To understand the way these are defined, it is useful to invoke the classical Aristotelian distinction between form (as organization and structure) and matter (as that which is possessed of qualities, even without having form). In my discussion of sculpture, form is meant to carry the resonance of that philosophical legacy, particularly in the role it plays in making matter into something meaningful.[2] Transposed into contemporary parlance, these traditional concepts have become enriched with other critical attributes: the definition and understanding of matter is also imbued with the characteristics of materiality, with all the social and cultural implications the latter term invokes. Like form, matter can be read through semiotic and cultural codes in which its specific attributes are given some kind of value. Therefore, matter is never outside a cultural (economic and historical) system of production, but always inscribed within it, as well as providing the substantive foundation of that system. That is, the value of any material comes through its place in a structural system, but the system has no means of articulating such values without the presence of material through which to do so. Form is the signifying configuration in my discussion, and matter the inherently valuable stuff. Form configures and thus pulls material into a system of meaning, but material provides a basis with intrinsic properties that can be worked upon. By linking this discussion back to the investigation of contemporary art and Smith's curatorial project, the dualism of form and matter as two variables that are always, in every work, in some specific (but distinct) relation to each other becomes apparent.

The distinction between affectivity and entropy embodies a contrast in approaches to production. In the first, affectivity, form is pushed to the fore to add extra value through organization, while in an entropic approach, matter insists upon itself in an attempt to undermine normative ideas of meaning. In current sculpture, this production aesthetic plays a crucial part in putting fine art on a viable footing with contemporary material culture. This is, of course, a longstanding issue. Far from cropping up as a symptom of late-1990s work, it has its roots within the foundation of modern art—itself always in dialogue with industrial production and commercial and popular culture. As I've written elsewhere, by the 1960s, with pop, minimalist, and, more importantly, conceptual art, a serious gesture of capitulation was made by fine art to the overwhelming superiority of the production values of material culture. At that historical moment, fine art retreated to the high ground of conceptual premises to distinguish fine art objects from material culture objects. What is interesting in this recent work is the shift back to an engagement with production as a viable means for articulating artistic identity for a sculptural object. This move builds on the solutions arrived at in postmodern sculpture, pushing them into new territory in an explicit enthusiasm for material culture.

I should also point out that I've deliberately identified the two poles of affectivity and entropy to echo the title of Wilhelm Worringer's famous essay of 1907, *Abstraction and Empathy.*[3] This is more than mere parody since the structural underpinnings of Worringer's approach were organized around the interpretation of form as a specific cultural response to a natural environment. I argue that the works in *Blunt Object* are specific responses to a cultural environment, each working symbolically to describe a relationship to its natural-seeming systems of consumption and production.

Worringer sketched out two opposing poles of form, the geometric and the organic, and suggested that the first is created out of a response of fear of the natural world, the second as an expression of harmony with it. The geometric enacts an aesthetic of distance and control, the organic a synchronous ease of relations. Worringer had elicited his analysis from the study of motifs and patterns of decoration and device. His insights, reductive though they may be, belong to the phase of art historical scholarship intent upon finding meaning in style. In defining affectivity and entropy, my interest is not in providing a key to the meaning of forms, but rather, a way to understand how it is that a use of material can effectively show up—that is, register as significant within the overwhelming field of other things. At base my argument is simply an attempt to understand how objects can answer the challenge to art posed by material culture. Art cannot possibly compete on its own terms, so it must arrive at some mode that belongs entirely to it. Affectivity and entropy are axes of critical insight along which art gestures slide the habits of thought into a condition of surprise to disturb the epistemological conventions. These are radical gestures, but not too radical, and the affirmative aspect of this new criticality marks its distance from the old-fashioned radicality of resistant avant-gardism, with its shrilly pitched defense of difficulty as the essence of resistant work.

The sculpture of mainstream 1980s postmodernism made clear that the aesthetic austerity that minimalism and conceptualism employed no longer served to provide an engaging formal basis for contemporary artists. The capitulations to consumer culture that came to the fore in the 1980s were overt in their flirtation with the material mainstream, even if they had a repressive quality in relation to the culture's explicit attitude towards pleasure. Modes of appropriation and display (Haim Steinbach) or fabrication (Jeff Koons) differentiated art works from commercial works through a self-consciousness about the framing effect of the gallery and the formal means of restructuring meaning through rearrangement and reordering. Other sculptors of the 1980s created a trash aesthetic as a comment on or alternative to the glitzy superficiality of material culture (Cady Noland and Mike Kelley). But to some extent, all of these artists kept a cool distance from the objects and world they commented upon. If Noland loved the aluminum cans in her installations or the rubbish in her pilfered supermarket cart or the cheap American flags that festooned a work, she didn't show it. A sense of critical objectification pervaded her production, as it did Koons's supercilious self-indulgence in kitsch and Steinbach's designer-perfect eye. Even Kelly's fetishization of worn material wasn't about adoring the material culture origins of store-bought Steiff toys, but about the display of emotional history that had worn the fuzz off the felted eyes of the favorite crochet bear in his collection. Tellingly, none of these artists was involved in any serious tradition of studio practice. None was engaged with the transformation of matter into form, with any artisanal skill base, or with any addition of value through a process of facture in the largest sense of the term.

Thus, for all its apparent engagement with mass material culture, the sculptural mainstream of 1980s postmodernism maintained a degree of distance from it. This isn't surprising. This attitude continues the self-identified intellectual's disdain for television so familiar from the academy or art circle. Such postures of distance don't admit to the seductive intensity of contemporary material forms or objects. I mean, how can an artist make anything as amazing, in sheer production terms, as a pink plastic laundry basket from K-mart? This isn't a historical moment in which artisanal superiority resides within the artistic community so that its products can be readily distinguished from those of "low" culture. Artisanal skill is positioned in a very different spectrum of production values than those that characterized early modernism and the avant-garde. The cool critical stance belies, or at least qualifies, any charge of genuine attraction between the artist and the artifacts of mass culture. I am

not suggesting that this is a false stance or that it enacts a critical bad faith—but I do want to suggest that the critical paradigm that emerges from the attitudes can't explain works that proceed from an affirmative or even qualifiedly positive relation to mass culture and its artifacts.

In the late 1990s, the urge to play with mass material culture on its own terms eclipsed the alienated critiques of earlier generations' resistant gestures. This is not to say that contemporary sculpture (or image production) is absolutely one with or indistinguishable from the mass culture with which it flirts so openly. Strategies of displacement and transformation remain crucial features of the art move, the distinguishing feature of a process that creates a space for reflection and criticism between perceived phenomena and the critical act of understanding. Social and cultural networks of production may overwhelm the casual consumer. But the artist's capability, what I am terming the affirmative, positive capability of the art move, is in the still potent capacity to jar the familiar senses and cognitive channels long enough to produce a moment of dissonant sensation and insight. Not quite the avant-gardist's dream of social revolution or cultural transformation, but the continuing blip on the radar screen of otherwise complacent, complicit, or confused (un)consciousness in the dazed consumer-overload state of current culture.

Art practice is still a significant, institutionalized, cultural space in which such critical positions are articulated and rewarded. These attitudes serve as a major justification for the existence of an entire sphere of cultural production. Here again the terms of an affirmative criticality have not been clearly articulated, so we still suffer from the persistent model of negation within the aesthetic field—suffer because the model is inadequate descriptively and critically. It's easy to assert that artists should do things through materials, forms, and objects that succeed in performing and communicating some kind of displacement from the habitual modes of use in which they are encountered. It is another proposition altogether to construe the means of doing so in ways sufficiently effective to actually achieve such displacements.

The work in the *Blunt Object* exhibition is at once consumable and critical. The dreadful inflexible arbiters of avant-gardism, who dictated that to be important and political art must be difficult, are shown up as cheerless schoolmen of a new academicism by this work, since it has humor, lightness, and beauty and makes use of the qualities that make material culture attractive, rather than eschewing them in favor of dreary esotericism. This work dispenses with making a virtue out of difficulty, for its own sake, as if the reward were in the very resistance it offers, a too-Adorno-esque extreme that has come to be the hallmark of academic writing and academically acclaimed artistic practice.

But, if sculpture were to become too consumable, would it run the risk of simply passing for any other object, being absorbed and consumed without causing the least ripple in the surface of consciousness? How can art status and art identity be embodied in a piece, not merely by its place in the gallery setting or museum hall, that indicates the distinction between the art object and the thing of the world? This was the challenge originally answered by minimalism in the embrace of the least-gesture of distinction and difference (arrangement, configuration, displacement). With the upped ante of high-stakes materialism, a mere stack of bricks can't compete with a rack of barrettes or new running shoes for our attention.

Affectivity and entropy are intensified gestures of differentiation. Each defines a relation to the contemporary world of production and consumption that allows visual art to distinguish itself from mainstream consumer and material culture while still engaging with its very means. Rather than defining art as an entirely separate domain, affectivity and entropy suggest that fine art is a use, a way of working, a gesture of distinction, within the realm of material culture and of its objects, things, and stuff. Each describes a distinctly different mode of transformation.

The affective gesture puts material objects (or stuff, that is, cloth, string, things that are often shaped by use after they are purchased in bulk) into an organized construction. In so doing, the affective gesture brings the inert to life, it rehumanizes material, not in the romantic sense, but in a production sense. Affectivity gives material a sense of intention and form, of sentience and action; it shifts it out of the mere material while engaging with it, tweaking the stuff, making it active. Affectivity takes what looked like matter already formed and uses it as simple matter to give rise to another level of organization and structure. Thus, a laundry basket or a soap dish becomes the marble of the new sculpture and becomes organized into another level of form. Meanwhile, the associations invoked by the functional identity of these objects-as-material are also part of the final communicative whole of the piece.

Entropy, on the other hand, is a deconstruction of normative identity through material means. It demonstrates the effect of removing things from the system of production and consumption in which they normally circulate. By rendering objects nonuseful, the entropic gesture forces attention back onto its "mere" materiality as an object, as a thing, so that it can't be pulled back into the form of the usual "commodified" (and readily consumable) object. The affective gesture is active. It is the positive dynamic of doing; it is action on the processes and materials of consumption. The entropic gesture is passive. It is constituted by the negative dynamic of undoing and the taking apart of the processes and materials of consumption. Many look-alike precedents for this exist within the tradition of the modern readymade, the assemblage mode, the postmodern acts of appropriation and display. But the new modes of affectivity and entropy can be distinguished from these by the imaginative reuse to which they subject the materials and forms of these works. The postmodernism of the 1980s had no place for the massage of form and extra-emotional seductive charge that affectivity adds to an object, any more than it would have permitted the wanton deconstructive negativity of an entropic act. Neither affectivity nor entropy is about pulling materials of real life into art as a transgression of boundaries, as if to shock the fine art world by an act of material bad behavior. Those gestures belong to the classical period of modern art, in which the break with academic tradition was marked so clearly by a slumming-it superficial flirtation with mass culture, not a relation on terms of equality, or one in which the advantage is assumed to lie with the mass rather than the elite term.

This new work continues the exploration of non-orthodox materials that has been part of modern art and, in particular, 20th-century sculpture. Again, the significant point is that the underlying anxiety in the current climate of art production is how to make art count, how to make it show up on the culture screen. How, in fact, can any act or art making compete in a manner that is sufficiently interesting

to be distinct from non-art products? In answering this question, the artists of *Blunt Object* have gone straight to the source of their anxiety and pleasure: they shop for their materials at Waldbaums, Stop and Shop, Home Depot, and Toys R Us, selecting the stuff of their production from the aisles of bright, vividly colored plastics, paints, and other already made things. No fine cherry wood, no Carrara marble, no bronze, no gold. The point of departure for production is mass-produced and processed stuff, generally plastic, synthetic, indestructible-seeming artifacts.

By the mid-1990s, sculptor Jessica Stockholder was using plastic laundry baskets, vivid latex paint, hardware, and general merchandise as the basis of a formal approach to installation-scale sculpture. A recognizable and distinctive feature of this work was its distance from the merely appropriative and display-oriented sculptures of the late 1980s, of which Haim Steinbach is exemplary. Steinbach's arrangements eschew the presence of hand or the transformative gestures of making or remaking. In short, Steinbach's work is separated from Stockholder's by his refusal to engage with studio practice or to let any of its trademarks show through in the final form of the work. No trace of hand or of artistic facture is present. They are systematically and conspicuously absent.

All the work in *Blunt Object* manifests some trace of studio practice as a part of its production, though often through materials or processes that would have been taboo as elements of studio work the last time production values were considered important to artistic objects. This would correspond to the hand painted phase of pop art, of late abstract expressionist canvases and sculptures, and of assemblage and funk work. Feminist aesthetics and work invoking ethnic traditions also tended towards the evident display of studio practice, which usually meant handwork (though it would be a mistake to characterize all feminist or ethnically rooted art in this way). A stark division exists between work that foregrounds studio practice and work that aggressively represses or denies it. The latter category belongs to the definitive phase of New York postmodernism, from about 1979 through the early 1980s, in which "the photographic impulse" and implications of mechanical and distancing production held sway. Such work displays a conspicuous absence of affectivity, a deliberate absence of feeling, emotion, and even individual subjectivity.

A veritable catalogue of possibilities for reinventing traditions of art making and of shifting the relation of critical opposition to mass media into a different key can be enunciated, one in which the pleasures of consumption are an acknowledged part of aesthetic production rather than a repressed one. Both affectivity and entropy are part of a taking out of service of material, a displacement from the use and activity in which the material or object originally was found or for which it was intended within the cultural context of its production. As Wilhelm Worringer indicated in his description of the empathic and abstract modes, these ways of working were both means of making a significant relation to the natural world through artistic form. Form had meaning in its specific qualities as well as in the mere fact of being made as art at the intersection of individual and collective expression. In the current phase of contemporary life, when all of nature is culture, when the line between these two once-conceived-to-be-separate domains is all but erased, then

affectivity and entropy act out the fundamental gestures that produce aesthetic value and render aesthetics significant through transformations and alternatives to mainstream value and production.

NOTES

1. There is an enormous problem of exclusion from critical consideration within mainstream art history of works produced within the historical period of modern art—particularly the early 20th century—because they do not conform to these supposed negativities. See my article "Who's Afraid of Visual Culture?" *Art Journal* Vol. 58, No.4 (Winter 1999): 36–47.

2. Bertrand Russell's summary of Aristotle's positions is succinctly cited: "We may start with a marble statue; here marble is the matter, while the shape conferred by the sculptor is the form." He goes on to emphasize, however, that "'form' does not mean 'shape'" in a reductive or literal sense, but in the sense of defining border and delimited identity. Bertrand Russell, *A History of Western Philosophy* (New York: Simon and Schuster, 1945), 165–66, where Russell summarizes Aristotle thusly: "Form is 'more real than matter'—and a form can exist, as per Plato's notion of idea, outside of matter" (p. 166).

3. Wilhelm Worringer, *Abstraction and Empathy* (New York: International Universities Press, 1953).

FURTHER READING

Johanna Drucker, *Theorizing Modernism: Visual Art and the Critical Tradition* (New York: Columbia University Press, 2004).

Johanna Drucker, *Sweet Dreams: Contemporary Art and Complicity* (Chicago: University of Chicago Press, 2005).

73

'CRAFTSMEN IN THE FACTORY OF IMAGES', FROM *BOYSCRAFT*

Tami Katz-Frieberg

Tami Katz-Freiman is a curator who pays attention. So much is clear from the catalogue essay for her recent exhibition in Haifa, Israel. BoysCraft was an international round-up of male artists who use craft techniques more often associated with women. Katz-Freiman mentions a wide array of touchstones which informed her thinking about the exhibition, from Arts and Crafts theory to outsider art, but it is clear that the Feminist art movement of the 1970s takes pride of place. What are we to make of the prevalence of men adopting not just the particular craft techniques, but also the aesthetic and even the political strategies associated with that earlier moment? She offers no easy answer to this question (among the possibilities, she notes, are gay and 'metrosexual' identities), but it is clear that, for the purposes of her project at any rate, craft still operates as a gendered term, albeit one that is open for use by anyone and towards a diversity of ends. BoysCraft was the follow-up to Katz-Freiman's 2003 OverCraft, also shown in Haifa, which was among the first exhibitions to take note of the recent international trend towards craft-based contemporary art.

Tami Katz-Freiman, 'Craftsmen in the Factory of Images', in *BoysCraft* (Haifa: Art Gallery of the University of Haifa, 2007).

Twice a week, on Mondays and Thursdays, the classes at the PIKA elementary school in Petach Tikva were divided in two during the sixth period: the girls walked up to the second floor for two hours of "crafts for girls," while the boys went down to the basement for "crafts for boys." We learned how to sew, embroider, bead and knit; it never crossed our minds to object to this gendered division, and to march down the two flights of stairs leading to the basement. The sound of sawing and the smells of sawdust and sweat that wafted up the stairs filled us with endless curiosity about the masculine enigma. Once every few months, while we were hard at work deciphering the secret of the hidden stitch, the boys completed one of their (entirely impractical) "projects" and came back upstairs proudly carrying some carved camel or donkey that made us green with envy.

BoysCraft, the title of this exhibition, alludes to the implications of this gendered division, which characterized the Zionist education system of the 1950s and 1960s. Back then "the girls" were taught skills that were gendered as female, and which were related to housekeeping (sewing, weaving, knitting, home economics and cooking). "The boys," meanwhile, were taught skills gendered as male—such as carving, using a screwdriver, cutting, sawing and building. The "crafts for girls" were designed to prepare the female pupils for life, and to supply them with the knowledge necessary to become good wives who could skillfully thread a needle and darn socks. The "crafts for boys," by contrast, were dedicated to the acquisition of "male" skills such as working with wood and metal, so that they could eventually become accomplished workers both at home and in the outside world. What this exhibition focuses upon, however, is not the skills acquired in those decades-old craft classes, but rather the ability of male artists to excel at handicrafts traditionally associated with women. The 41 artists participating in this exhibition create their works using techniques of embroidery, weaving, knitting, spinning yarn, beading, sewing, quilting, cutting and pasting. Until recently, such handicrafts were still associated with strictly feminine practices, with "folk art" and with functional art. *Boys-Craft*, by contrast, brings together works by contemporary male artists who share a preoccupation with labor-intensive work processes and with the sensual experience of excess, materiality and multiple details. This exhibition thus reflects some of the complex processes that have taken place in the artworld in the wake of the feminist revolution—and presents a new generation of Israeli and international artists whose works are informed by feminist, gender and postcolonial theories. These artists all make unconventional use of various materials in order to transform traditional craft techniques into key artistic strategies. In this case, for a change, it is the men who engage in social criticism—and raise gender-related questions from a male point of view.

The range of voices that are given expression in this exhibition creates a rich tapestry, a patchwork that slowly comes together to form a sensual and complex panoply of different cultures, styles and skills. The works of each of these artists are characterized by time-consuming, labor-intensive processes that involve repetitive and monotonous actions, based on age-old traditions of manual craft. Most of the works are centered upon a world of images based on "male" or "macho" stereotypes, yet their creation involves techniques that are culturally associated with "female" or "childlike" forms of expression. This disruption undermines stereotypical gender divisions and dismantles old-fashioned dichotomies, charging them with new meanings.

So, for instance, qualities and behavioral patterns that are identified as quintessentially "masculine"—such as violence and the abuse of power, physical strength, competitiveness and the consumption of pornography—are related in a number of works to traditional weaving, beading or paper cutting techniques (Assaf Rahat, Guy Goldstein, Ashraf Fawakhry, Ben Ben Ron and Tom Gallant); in addition, the exhibition includes quintessentially male symbols such as an American car and a Harley Davidson motorcycle, which are represented by means of laborious manual craft processes (Ramazan Bayrakoglu and Guy Zagursky); several works include allusions to patriotic or military subjects whose meaning is subverted or ironically examined through the use of handicrafts such as sewing and quilting (Lior Shvil, Haim Maor and Dave Cole); a number of other works transform hegemonic male domains identified with the shaping of the cultural sphere—such as textual, scientific and architectural bodies of knowledge—through delicate handicrafts such as embroidery, carving and cutting (Shaul Tzemach, Jonathan Gold, Jonathan Callan and Tomás Rivas). These creative strategies—together with a wide range of expressive means and related themes that shall be explored below—stem from a cultural nostalgia for the predigital age; they are related to the sweeping social and gender-related changes that have taken place in recent decades, and especially over the past ten years.

THE RETURN TO MANUAL CRAFTS: A LONGING FOR A PRE-INDUSTRIAL PAST

Manual crafts and folk art are related in the Western worldview to the pre-industrialized, premodern world. The contemporary preoccupation with, and reevaluation of, these traditions are part of a global trend that reflects a longing for a simpler life that stands out in contrast to today's global, commercialized artworld—and which involves a great deal of nostalgia and ecological idealization.

The search for a unique, "authentic" source that could serve as the basis for a community or family-based visual genealogy was one of the factors that has led to the resuscitation of craft traditions. This trend is also characterized by a deep yearning for a unique artistic signature—for artworks that bespeak a personal and direct form of individual expression and a significant investment on the part of the artist. Indeed, over the last decade handicrafts have assuredly entered the canonical field of contemporary art, and leading artists are using traditional crafts in order to create socially engaged works.

The works assembled in this exhibition respond to the longing for manually created works produced with the artist's own hands. Such works are perceived to be imbued with a sensual quality that is provided by the artist's unique touch, and which adds a tactile dimension to the clinical, alienated world represented in the images that surround us—many of which originate in the sphere of digital technologies. In an age in which almost every boundary has been transgressed, and in which the uniform and anonymous colors

of the global village dominate everything we see, it seems at times that art itself strives to conform to a uniform model shaped by market demands. These works, by contrast, point to a clear emphasis on uniqueness and individuality. The historical parallel to this contemporary trend is the social agenda of the late 19th century Arts and Crafts Movement, which propagated anti-modern and conservative values and saw labor-intensive, manual crafts as a therapeutic catalyst for inner renewal. This English-born movement was led by John Ruskin and William Morris, who supported the revival of medieval arts and crafts—and of related techniques and skills that were seen as a miracle cure for the ills of mass production and of industrial capitalism. In this context, one must recall that prior to the industrial revolution, the range of crafts that later came to be associated with women's work were dominated by male guilds (such, for instance, as the first knitting guild founded in Paris in 1527).

These efforts, however, failed to endow manual crafts with significant prestige; to this day, hierarchical distinctions between crafts regarded as "folk art," "outsider art" or "women's art" and between "high," elitist art continue to shape the field of contemporary art. Above all, the perpetuation of these hierarchies has to do with the fact that manual crafts came to be seen as functional, decorative and resolutely old fashioned by the modernist movement, which zealously guarded against them. Their multiplicity of mesmerizing detail; their labor-intensive production process, which is associated with physical exertion and with materiality; their visually accessible quality; their seductive and pleasure-inducing aspects; and their lack of intellectual sophistication—all these led manual crafts to be marginalized outside of the modernist canon.

Figure 51 Kristian Kozul, *Wheelchair I*, 2003.

THE FEMINIST REVOLUTION: SUBVERSION AND ACTIVISM AS TOOLS FOR SOCIAL HEALING

The painstakingly slow process of composing works based on small details and "ignoble" materials, which transforms each artwork into a labor of love, embodies the exact opposite of the modernist (male) approach to artmaking as a process based on large, assertive gestures and on a combination of sublime emotion and analytical thinking. Modern art, and especially conceptual art, disregarded the time-consuming dimension of artmaking—attributing the work's value to its conceptual makeup and devaluating the process of its execution. In recent years, it seems that this dimension of creative processes, which is related to a penchant for details, is once again in vogue; indeed, one is frequently

tempted to evaluate the final product based on the amount of work that was invested in its creation. Yet as the feminist critic Naomi Schor has noted, details and embellishments were long viewed by society as an excessive, decadent and tiresome form of expression, as: "women's matters." Art similarly reflected the male view that a penchant for small details amounted to a subversion of ideal, sublime or classical forms. Details were seen as threatening to undermine the internal hierarchical order of the artwork, and to blur the relations between center and margins, between major and subordinate elements and between foreground and background.[1]

Indeed, many of the skills required to manually produce richly detailed works were traditionally perceived as quintessentially female; they supposedly required—in addition to ample leisure time—developed motor skills, a high level of concentration, meditative qualities, patience and a good eye. The products of these craft processes were looked down upon as decorative, mindless and devoid of content. Embroidery, for instance, was considered lowlier than other handicrafts, because it was historically identified as a quintessential form of women's work. In her important 1984 book *The Subversive Stitch*, Rozsika Parker paralleled the construction of conceptions of femininity with the separation between "fine art" and "craft" that occurred with the advent of the Renaissance. During the 18th century, these constructions were reflected in the changes that took place in art education—with the transition from apprenticing in artist's workshops to the academic study of art, and the regendering of craft traditions. A substantial change in this conception occurred with the first feminist wave of the late 1960s and early 1970s, when radical women artists attempted to restore the culture of traditional female crafts as part of their effort to define what is often referred to as female "essentialism." Artists such as Harmony Hammond, Faith Wilding, Judy Chicago, Miriam Shapiro and others turned to manual crafts as a political act that challenged the modernist hierarchy.

This feminist contribution was essential to the launching of a wide-ranging postmodernist strategy, which enabled handicrafts to be used in a manner no longer considered to be inferior. Beginning in the 1980s, the use of craft techniques became an increasingly legitimate part of the artistic canon. Women artists such as Rosemarie Trockel, and later Ann Hamilton, gave new meaning to the term "labor intensive." During the 1990s, male artists such as Mike Kelley, Lucas Samaras and Jim Hodges adopted similar strategies, and began using craft techniques in order to destabilize the modernist canon. This trend may also be related to the development of queer theories—which followed in the wake of the feminist discourse that undermined preexisting gender categories and offered alternative, flexible and liberating ways of thinking about gender. The emergence of queer theories in the late 1980s and early 1990s is also related to the AIDS crisis, which played a significant role in postmodern developments. This crisis led to a substantial change in social attitudes towards homosexuality, which paradoxically enhanced the visibility of this form of otherness. The culture of drag and camp, and its relation to queer and alternative practices, gradually filtered into art. The transmutation of kitsch into high art, and the charging of mass imagery with subversive and critical meanings that allowed it to penetrate into an elitist discourse, both characterize the art of recent years. In the context of *BoysCraft*, this homosexual discourse is given expression in the installation of embroidered laundry by Gil & Moti, which relates homoeroticism to multicultural political contexts; in Gil Yefman's bizarre knit dolls, whose touching character manages to ridicule our ideas of normalcy; in the photographs by Uri Gershuni, who compassionately and devotedly knit a woolen cape for the photographed figure; in the work of Stephan Goldrajch, who created meticulously knit masks of grotesque creatures; and in the wild camp movies by Francesco Vezzoli, in which he embroiders nonchalantly beside a well known Italian diva.

BoysCraft thus reflects the fact that more and more male artists today are adopting traditional handicrafts and using them in a new, refreshing and thought-provoking manner. Indeed, it is difficult to imagine someone today relating to this preoccupation as marginal to the artistic discourse. The very existence of an exhibition based on "gender-related discrimination," moreover, could only take place from the point of view of a postfeminist "achievement" or "victory." Several of the works included in the exhibition (such as those by Roee Rosen, Daniel Silver, Izhar Patkin, Jonathan Shilo, Servet Koçyigit and Eliahou Eric Bokobza) involve (or represent in painting) a range of understated "feminine" practices such as weaving, sewing, knitting and quilting. These craft practices undermine familiar hierarchies between "high" and "low," and undo hegemonic relationships in the field of artmaking. The prominent influence of feminist strategies may also be detected in the widespread use of handicrafts in community-related contexts, where they appear as an expression of loss and healing. During the 1980s, the AIDS crisis shaped a social and cultural sphere in which a growing number of artists experienced profound solitude and pain.

Collective knitting, quilting or embroidery projects came to constitute rituals of mourning and expressions of grief, as well as tools for commemoration, for protest and for raising public awareness. In this context, the engagement with labor-intensive handicrafts was related to their process-oriented, time-consuming and meditative qualities, which were seen as therapeutic. Indeed, in recent decades a growing number of social and community-related projects have centered upon craft works created in a chain process that cuts across national, religious and geographical boundaries, and which emblematizes the ideals of human solidarity and social awareness. In the aftermath of the events of September 11, this trend became even more widespread. It continued to evolve in the context of a growing need for community-based action and for various kinds of support groups, and of a desire to reconnect to the past. Dave Cole's knitting machine installation—whose documentation is included in this exhibition—was created for the first anniversary commemorating the attack on the Twin Towers, and touches directly upon these events. In other works, the labor-intensive actions of knitting and beading are given expression in the context of loss and healing connected to the AIDS epidemic (Oliver Herring); in relation to nursing, paralysis, violence and pain (Kristian Kozul); and in relation to Israeli symbols of morning (Erez Israeli). This preoccupation with emotions and with a collective human vulnerability—as well as with some themes that are clearly identified with a masculine reality—is based, to a large degree, on changing perceptions of masculinity influenced by the feminist revolution.

LABORIOUSNESS + OBSESSION = "AUTHENTICITY"

The most striking characteristic of handicrafts is related to the term "laboriousness," which is often associated, in this context, with the term "obsession." The demanding work process, focus on details and compulsive repetition involved in some of the works included in the exhibition may indeed be termed "obsessive." "Obsession" is defined as a "compulsive preoccupation with a fixed idea or an unwanted feeling or emotion, often accompanied by symptoms of anxiety; in clinical, psychiatric terms, it is described as a form of neurosis whose main characteristic is clinging to a disturbing thought, impulse or image that persists and imposes itself on one's consciousness. The compulsive actions are meant to diminish the anxiety caused by the obsession, and express a desperate effort to achieve a semblance of control over an uncontrollable world. This clinical definition relates obsessive expressions in the field of folk art to the work of outsider artists—untrained artists who are unaware of the contemporary art discourse. Many of them create during states of psychosis, which activate their creative imagination in an extraordinary manner. In "Hotel Utopia-Dystopia," published in a special issue of *Studio* edited by writer and artist Meir Agassi, the world of outsider artists was defined as "a world experienced and perceived as if through an autistic sheet of glass—a convoluted, crowded, labyrinthine world whose intensity immediately causes the viewer a sense of discomfort, a sudden loss of equilibrium. Formal and narrative labyrinths lure the eye into a complex trap, a deluge of images that floods the paper and creates a tension-filled fusion between imagination and reality."[2]

Historically, Western culture treated various expressions of otherness—ranging from the work of outsider artists to that of non-Western artists—with a colonialist, exoticizing approach rife with contradictions: enthusiastic consumption to the point of overwhelming acceptance went hand in hand with a dismissive attitude, and with a lack of understanding concerning the cultural context of these works. At the basis of this sweeping interest in outsider and non-Western art was the value of "authenticity," which postmodernism has denied to the point of making it nearly extinct.[3] The preoccupation with "otherness," and the postcolonial approaches that have developed in the course of recent decades, finally led to an assimilation of this subject into the iconography of contemporary art. Third-world art has become increasingly appealing to a satiated Western world, and the contemporary art market has been flooded with African, Asian and Latin-American artists. A number of the artists participating in the exhibition treat this exoticized perception of authenticity with a great degree of irony: El Anatsui, for instance, does so by means of a rich tapestry of corks and labels collected from alcohol bottles. His work relates to the historical, ritual dimensions these materials have had in an African context, as well as to their modern, commercial aspects; Ohad Meromi uses weaving in order to reexamine the "authentic" Zionist-Israeli identity forged in the context of the "Maskit" arts and crafts project, which fused authentic Yemenite crafts with modern chic; Tim Curtis' work is a homage to the inventive imagination and talent for improvisation that characterize the works of third-world artisan-vendors; Nick Cave enhances a ritual costume in order to create a magical effect, which relates it to conflicts concerning identity, gender and race; and Nicholas Hlobo's installation and sculptures combine queer themes with postcolonial cultural criticism, and with a focus on social rituals and norms related to his South African roots. In all of these works, the use of "authentic" materials and obsessive, labor-intensive processes sheds light on the problematic and on the disruptions that characterize contemporary cultures and identities.

EXCESS AND THE PLEASURES OF ORNAMENT

The manual, labor-intensive investment evident in a large number of the works included in this exhibition naturally results in a wide-ranging emphasis on the work's material qualities and texture. Most of the artists make use of cheap, recycled and unglamorous materials—which are sometimes surprisingly simple—in order to create the illusion of rich, glamorous and luxurious surfaces and thus to redefine their meaning. Iron threads, concrete blocks, plastic sewage tubes, fabrics, tire rubber, various sewing notions, wallpaper, galvanized netting, aluminum plates, stones, thread, fabric, paper cutouts, plastic sheeting, nylon, wood, glass beads and mirrors—all serve as the basis or support for time-consuming, repetitive actions that enhance the work's tactile qualities. The commitment to a long and exhausting work process, and the ability to surprisingly transform materials in unfamiliar ways, characterize the work of many of the participating artists. The sculptural works in the exhibition fit the definition of "soft sculpture," and undermine the traditional definition of sculpture as a solid, heavy mass. Their presence in space bespeaks a fragile, tentative and ephemeral existence. These qualities are given expression, for instance, in the works of Goran Tomcic, Gal Weinstein, Ron Aloni, Haimi Fenichel and Lionel Estève, in which a rigid material mass (such as barbed wire, stone, etc.) is destabilized or metaphorically undermined so that it comes to radiate fragility and softness.

Alongside these works, the exhibition also includes new media works that document the practice of particular handicraft processes, or which create digital simulations charged with symbolic meaning: digital "lace" as a metaphor for cyberspace (Leon David); the knitting of the American flag as a symbol of patriotic allegiance (Dave Cole); or embroidery as a metaphor for obsession (Francesco Vezzoli).

One of the striking aspects of this use of traditional handicrafts is the excessive, decorative quality of the works, whose ornamental complexity affords an experience of pure pleasure. The excess that characterizes some of the works exhibited in *BoysCraft* does indeed imbue them with a pleasureful sensual quality, and re-evokes concepts related to beauty that were excluded from the modernist discourse. At the same time, these works provoke thoughts about the relations between ornament, eroticism and fetishism—and between decorativeness, disintegration and sickness. Research shows that the visual examination of a richly colored and textured ornament provokes a pleasureful stimulus in the brain; the beauty embedded in a crowded weave of different colors causes the viewer sensual excitement that cannot be verbally described. Such is the case with the wallpaper works of avaf [Assume Vivid Astro Focus], which resemble psychedelic and kaleidoscopic collages; with the breathtaking assemblages by El Anatsui and Kristian Kozul; with Nick Cave's work, which combines a range of exotic materials into a densely decorated ritual costume; and with Gean Moreno's work, which creates an effect of excess based on the visual cacophony of flea markets and on a street aesthetic.

THE RELATIONS BETWEEN "HIGH" AND "LOW": THE COLLAPSE OF HIERARCHICAL CATEGORIES

Over the past two decades, numerous exhibitions have been concerned with the blurring of distinctions between "high" (elitist) and "low" (popular) art, and have revealed the complex relations that exist between these different categories. The most important of these exhibitions was the 1992, which was exhibited at MoMA in New York, and which summed up the dialogue between these two categories. Another exhibition that took place two years later, and which focused on the influence of outsider art on modern art, was *Parallel Visions: Modernist Artists and Outsider*

Figure 52 Haimi Fenichel, *Passive Aggressive*, 2007.

Art, which was exhibited at the Los Angeles County Museum in 1992. The exhibition closest in spirit to *BoysCraft*, however, was *A Labor of Love*, which was exhibited at the New Museum of Contemporary Art in New York in 1996. This exhibition, which focused on the adoption of labor-intensive crafts and folk art traditions by contemporary artists, examined the complex reciprocal relations between these different categories.

The most recent exhibition that touched upon related issues was *Radical Lace, Subversive Knitting*, which opened at the Museum of Arts & Design in New York in February 2007. Focusing on the domains of craft and design, this exhibition drew attention to the pervasive use of handicrafts, and further blurred the validity of norms distinguishing art from craft. Curator David McFadden attempted to show that fiber-based handicrafts such as lace and knitting could be charged with radical content. This exhibition essentially celebrated the collapse of categorical boundaries between various arts and between art and design, which has taken place in recent years. In this context, traditional handicraft techniques were engaged for the purpose of social criticism, and the intimate craft of lace making, which was once a private, domestic activity, was translated into monumental architectural installations. The exhibition *The Height of the Popular*, which was shown at the Tel Aviv Museum of Art in 2001 (curator: Ellen Ginton), was concerned with similar themes. One may also mention the 2001 solo exhibition of Elaine Reichek at the Tel Aviv Museum of Art, which included embroidery works with feminist and anti-racist messages (curator: Edna Mosenson). Yinka Shonibare's 2002 solo exhibition at the Israel Museum also preceded *BoysCraft* in terms of its concern with the use of textile to examine themes related to postcolonialism, identity and otherness (curator: Suzanne Landau).

Notwithstanding these affinities, however, *BoysCraft* is a direct and complementary sequel to the 2003 exhibition *OverCraft: Obsession, Decoration and Biting Beauty*, which was shown at the Art Gallery of the University of Haifa and at the Tel Aviv Artists House.[4] *OverCraft* was concerned with labor intensive processes in a feminist context. *BoysCraft*, by contrast, focuses on the resonances of the sweeping change that has recently taken place in Western and in Israeli perceptions of masculinity. No longer a gruff, macho "sabra" whose life experience revolves around his military service, the Israeli man has been transformed into a self-aware metrosexual who is not afraid to express emotions or to groom his body. This exhibition may thus also reflect the gains of the feminist revolution, and the ways in which they have been integrated into the conflicted psyche of the new male with the gradual eclipse of the machoist age.

Documenta XII in Kassel (2007) raised questions concerning the essence of "modernity," and attempted to demonstrate that the avant-garde does not necessarily constitute the opposite of tradition. Bringing together creative ideas from different historical periods, the curators combined centuries-old manual crafts (kilims, carpets and embroidery) with contemporary art. Without judging this unusual curatorial act, it is possible to state that one of the outstanding experiences at Documenta was the overwhelming presence—and reevaluation of—manual crafts and skills. One of the artists, Danica Dakic, for instance, filmed her work in the city's wallpaper museum, which was established in 1923 and was obviously never frequented by contemporary art connoisseurs. She also created a sound work that called attention to the ultimate decorative and labor-intensive craft tradition—wallpaper manufacturing. The templates used to produce the wallpaper; the multitude of colors; the precise and repetitive production process; the covering of large surface expanses; and the obsessive and decorative quality of this endeavor perfectly melded with the sounds and texts emanating form the loudspeakers, creating a unique and surprising experience.

The exhibition *BoysCraft* thus reflects the manner in which handicrafts have been integrated into the language of the artistic canon. Once associated with folk, functional and "outsider" art, and with women's leisure activities and hobbies, such labor-intensive work processes have been transformed into fully accepted and highly valued contemporary art practices. Strategies that in previous decades were identified with women artists attempting to liberate themselves of the male hegemony have been integrated into contemporary artmaking as a legitimate form of self expression, a celebration of manual production in a world that has wildly over-computerized itself.

NOTES

1. Naomi Schor, *Reading in Detail: Aesthetics and the Feminine* (New York: Routledge, 1989), pp. 4, 15.
2. Meir Agassi, "Hotel Utopia-Dystopia," *Studio* 89 (Jan. 1998), p. 6. [in Hebrew].
3. Two key exhibitions that have examined the notion of otherness were the modernist *Primitivism in 20th Century Art* (New York: Museum of Modern Art, 1984), and the postmodernist *Les Magiciens de la Terre* (Paris; Centre Pompidou, 1989).
4. In addition to its dialogue with *OverCraft*, *BoysCraft* also relates to a number of previous exhibitions I curated in Israel, Including *Antipathos* (The Israel Museum, 1993) and *Metasex* (Ein Harod Museum of Art, 1994), which similarly examined the non-canonical margins of local art.

FURTHER READING

Tami Katz-Freiman, *OverCraft: Obsession, Decoration and Biting Beauty* (Haifa: Art Gallery of the University of Haifa, 2003).

Marcia Tucker, *A Labor of Love* (New York: New Museum of Contemporary Art, 1996).

74

AND WHAT IS YOUR TITLE?

Zandra Ahl

Some of the most memorable exhibitions of the twentieth century were carried off not by curators, but by artists: the total art environments (gesamtkunstwerke) of the turn-of-the-century Viennese Secessionist movement; Marcel Duchamp's installation Sixteen Miles of String *for the First Papers of Surrealism exhibition in 1942; or the Feminist project* Womanhouse, *a collective project by Judy Chicago, Miriam Schapiro and their students staged in a suburban house slated for demolition. All three of these examples involved vivid encounters between the languages of craft and fine art that prevailed at the time. But only in recent years have avant-garde artist-curators who focus specifically on craft emerged. One of the first was Zandra Ahl, who trained at Konstfack, the art and design school in Stockholm, in the mid 1990s. Even as a student, she began organizing exhibitions as well as producing her own work and wrote her first short publication,* Fult & Snyggt (Ugly and Cute), *in 1998. But it was her next book,* Svensk Smak: Myter Om den Moderna Formen *(Swedish Taste: Myths about Modern Design), co-authored with journalist Emma Olsson in 2001, that established her reputation as the enfant terrible of the Swedish art and craft scene. An attack on the country's ongoing adherence to Modernist design principles, the text had a bracing effect within the art school–based community. Ahl's own work, composed mainly of found glass and ceramic objects in loose sculptural arrangements decorated with ribbons and bows, draws on a language of kitsch and gender instability. Her curatorial activities have been numerous: one-off exhibitions in mainstream museums; collaborative work with the avant-garde glass and ceramics group* We Work in a Fragile Material; *long-term projects in fringe spaces; and the editing of a fanzine titled* Slicker *(Swedish for slip, the liquid clay used in ceramics). The tone of her writing and projects also varies widely. She continues to launch satires of the design establishment and critiques of 'business as usual' museum practice but also celebrates unconventionality in collecting and artistic style. In the following short text, written for this reader, Ahl reflects on her own multivalent, postdisciplinary practice.*

Zandra Ahl, 'And What Is Your Title?', 2008.

AND WHAT IS YOUR TITLE?

A well-known curator asked me that question at a seminar. Is it ice-skating-princess-and-craft-maker? Obviously, this was a joke. But I get the question all the time. Depending on the event, my profession changes names, but the core questions of my work do not. In fact my profession slides between different genres that provide a perfect platform for my investigation of taste, power, hierarchy, class

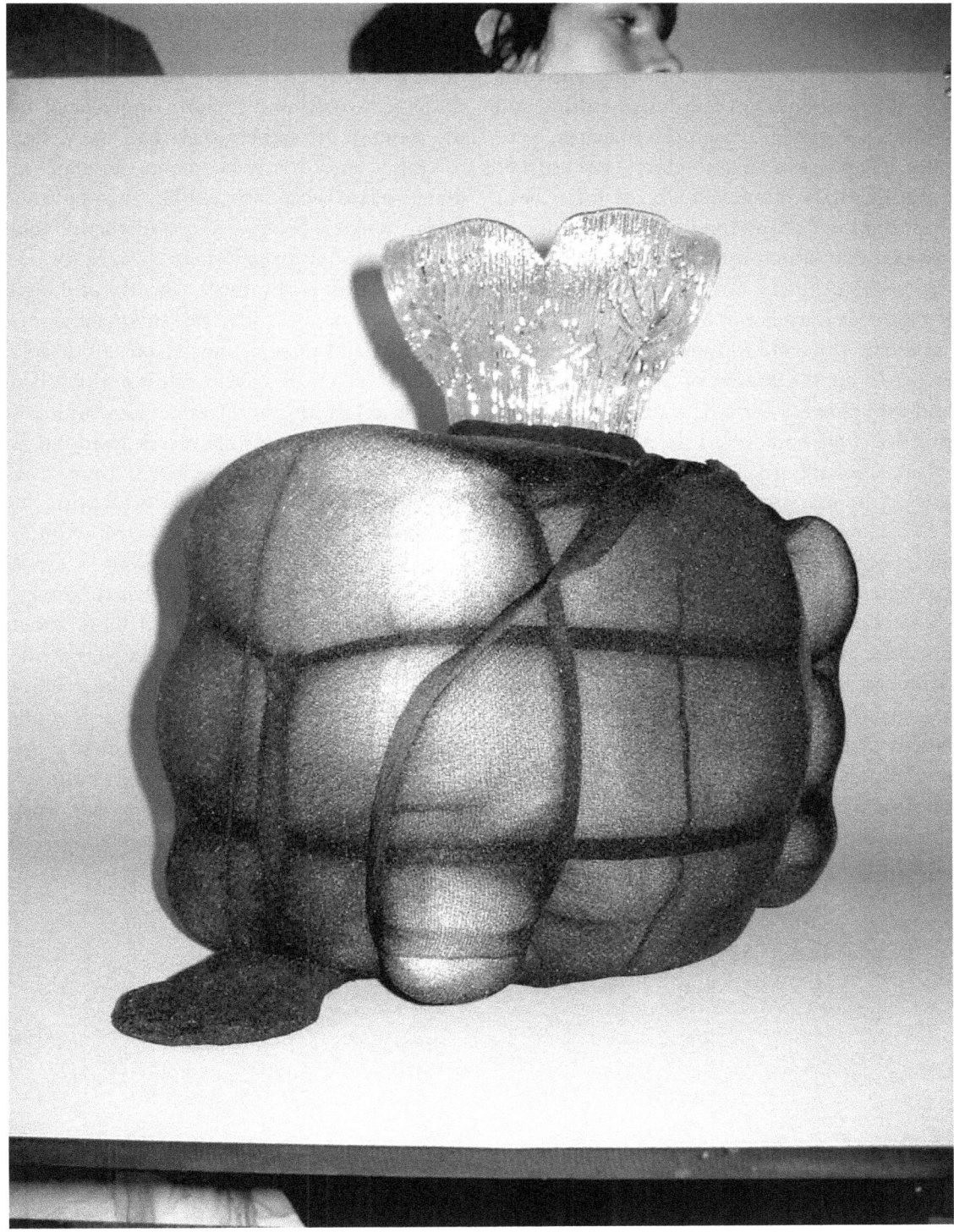

Figure 53 Zandra Ahl with her work *Autoshape*, 2008.

and gender linked to material culture. I call it craft.

In my practice I can be anything. I investigate different layers of the design culture and politics in Sweden, as a project manager, a curator, a writer; as a fanzine editor, craft artist or a documentary filmmaker. To me it has been important not to interpret or look at different aspects of production from the outside, but instead to do them myself. I'm a maker, and if I want to learn about institutions I go work for them. The same with manifestos or exhibitions. To me the process of forming questions and thoughts is a form of craft; it is linked to practice. One could say I do not belong anywhere. Probably that is true. But to me that kind of homelessness has been a key disguise. I have done my homework the past ten years and soon I think I will have the picture done.

Design in Sweden has always been in close connections to those in political power. I will not bore you by talking about the old social democrat model, which used design as a tool for making a dream society, and how this failed. But that history plays an important role in understanding why some things, traditions or even people are regarded as good design, and others as just bad DIY—misfits and outcasts.

Another question I often get is whether art and design and craft are the same thing, since my work sometimes takes forms other than the object-based and is more appreciated by the Swedish art scene rather than the design or craft scenes. Many people in the craft- or design-related scenes would like it to be true. Today design is being used as a marketing tool for those who want to kidnap the area for political reasons, or by those who think design should only be about lifestyle. This makes some questions and themes unpalatable. And sadly very few makers disagree. Craft is not polished by political dreams in the same way, but it has other problems. And this means freedom, as long as you are willing to claim your work belongs to the craft scene and deals with the traditions and questions linked to it. And as long you don't ask disturbing questions. Once craft is attached to art it will always be controlled by systems that makes it more about *looking* at craft instead of being a part of the material culture and everyday life, the practice of making. To me being a craft maker is about the obligation to ask the questions no one else asks, and use my insight to ask even more inconvenient and essential questions. My work is like political mismatched ice skating—an elusive DIY Holiday on Ice.

75

CRAFT HARD, DIE FREE: RADICAL CURATORIAL STRATEGIES FOR CRAFTIVISM IN UNRULY CONTEXTS

Anthea Black and Nicole Burisch

The politicization of craft in the present moment raises serious questions: of precedent, of potency and (as always in matters to do with craft) of identity. In this essay, originally a paper developed for the conference 'New Craft: Future Voices' held in Dundee, Scotland, in 2007, two Canadian curators and 'craftivists' take on all three issues. They also provide a brief international survey of activities which seek to deploy craft for the purposes of protest. Knitting, and other textile arts traditionally associated with communal crafting, plays the leading role. The concept of the 'revolutionary knitting circle' recalls the 1970s feminist use of a similar group exchange as a form of consciousness raising. Black and Burisch also cite the AIDS Quilt project of the 1980s as an important precursor for the present moment. So much for precedents; what about the future? Clearly, efficacy and identity are interwoven in this essay, which takes for granted another '70s concept—that the personal is the political—and offers real-world strategies for increasing the efficacy of symbolic craft. It is too early to say whether craftivism will have staying power in the cultural imagination, like the Arts and Crafts, studio and countercultural craft movements before it. But there is little doubt that Black, Burisch and their peers have breathed new life into this old set of ideas.

Anthea Black and Nicole Burisch, 'Craft Hard, Die Free: Radical Curatorial Strategies for Craftivism'. A version of this paper will be published in Maria Elena Buszek, *Extra/Ordinary: Craft Culture and Contemporary Art* (Durham, NC: Duke University Press, forthcoming).

While craft historians, feminist historians and fine craft practitioners argue for the recognition of craft within art and academic dialogues, craft supplies are simultaneously mass produced and packaged as hobby-commodities for affluent consumers, and craft practices are appropriated into the mainstream marketing of alternative and DIY lifestyles. In addition, the accessibility of global communication networks have contributed to the increased sharing of craft knowledge and skills, and created an overall democratization of crafting practices. This current academic and popular interest in "craftivism" calls for a discussion of productive strategies to maintain its radical potential.

This research is focused in particular on those practices that defy tidy classification, or that straddle traditionally constructed distinctions between art, craft, curatorial, and activist practices. Craftivists and artists who use politically engaged crafting methods continue to hybridize their practices by joining craft, technology, the politicization of digital space, public spaces and traditional arts venues. Many of the following examples and perspectives emerge from our experience within Canadian artist-run culture.[1]

Politically engaged crafting practices and many contemporary curatorial approaches share a common ability and imperative to challenge the dominant economies in which they are situated. While these approaches generally do not seek legitimacy within mainstream economies or spaces of display, this radical activity continues to be commodified into corporate and institutional cultures. Radical curatorial strategies that deliberately blur the boundaries between artist, crafter and curator are of particular interest here, as are their locations within the network of Canadian artist-run centres, where artists, activists and crafters are often employed as administrators, programmers and curators. In such sites, we argue that craftivism can be riotously, ethically and effectively included in—and used to redefine—contemporary curatorial strategies and contribute to a politicization of these spaces and cultures of display.

As the essays in this volume make clear, the perception of the crafted object as old-fashioned or traditional has now been eschewed in favor of crafting as a strategy to examine and challenge contemporary issues. Rather than viewing craft as pre-industrial, current craftivist practices are situated within the challenges of urbanity, globalization and capitalism in a post-industrial, technology-saturated world. The proliferation of craft on the Internet and in the gallery is further evidence of shifting views on appropriate technologies and spaces for the exchange and contextualization of craft. Diverse spaces, approaches, and materials now exist alongside each other in the maker's political tool kit as effective strategies for sharing skills, techniques, and information.

SURVEY OF PRACTICES

The rise of craftivism and other politically engaged crafting practices—which value the radical potential of a particular craft activity rather than its finished end-product—shift traditional emphasis away from polished, professionally-made craft objects themselves and toward a political and conceptual focus, positioning, and deployment of the work involved in making them. This emphasis has made room for reconsiderations of crafts(wo)manship, performativity, mindfulness, tacit knowledge, skill sharing, DIY, anti-capitalism and activism.

In an expanding and diverse field of craftivism, the extraordinarily transformative NAMES Project AIDS Quilt can be seen as an early craftivist precedent. In "NAMING NAMES: The art of memory and the NAMES Project AIDS Quilt," Peter Hawkins' celebration and critical discussion of the AIDS quilt offers a comprehensive survey of the contradictory and sometimes problematic ways in which the project is deployed, read and understood as a politicized craft object.[2] It should be noted that the concerns around the project[3]—particularly around commodification of the project, exclusion of people of color and the working class—reoccur in contemporary forms of craftivism. Despite the inability of the project to represent all the aspects of AIDS and the overwhelming growth and cost of displaying the project, the political impact of the quilt continues to occupy a prominent space within mainstream America's awareness of AIDS.[4]

For the founder of the project, Cleve Jones, the tradition and formal qualities of quilting were important links to "comfort, humanity and warmth,"[5] and despite the use of traditional media, the quilt is assembled as a collaged, non-hierarchical mix of quilting styles, skill levels and aesthetics.[6] The NAMES Project Quilt employs what has come to be a common craftivist method of group making, where each participant contributes one panel to be assembled together into a diverse, collaborative whole.[7]

THE REVOLUTIONARY KNITTING CIRCLE

Calgary's well-known Revolutionary Knitting Circle and other radical crafting groups effectively include the sharing of tacit knowledge in their emphasis on teaching knitting, quilting or crafting skills in a collaborative or workshop-based environment. The interactions and discussions that take place during group knitting also act as an accessible forum for teaching, sharing, and promoting activist strategies and politics. The Revolutionary Knitting Circle manifesto advocates knitting (and other crafts) as constructive and non-violent tools for opposing the dominant corporate models of production.[8]

The group also participates in marches, rallies, and protests by conducting group "knit-ins" or by carrying the large, co-operatively knitted *Peace Knits* banner. The public knit-ins and the banner both serve as peaceful and accessible rallying points for action, discussion, and awareness. The banner is an overtly collaborative project that emphasizes a multiplicity of voices and the democratization of its making. The *Peace Knits* banner has something of an intentional anti-aesthetic: the individually knitted panels of thrifted and donated yarn speak to its cooperative construction and its multiple creators.

It is the act of knitting peacefully in public groups or of carrying the knitted banner where the true revolutionary strength of this group's activity lies. In the text for the exhibition *SUPERSTRING*, curator Anthea Black notes that "it is the simultaneous unruliness and gentleness of public knitting—when a large group of knitters occupies a public space or a place of power with a non-violent action— that creates a constructive dialogue."[9]

MARIANNE JØRGENSEN AND THE CAST OFF KNITTERS

Similarly, Danish artist Marianne Jørgensen's *Pink Tank* project directly situated the collaborative work of crafters and activists within a public space. The artist collaborated with the Cast Off knitters group and several knitters from around the world to knit and assemble over 4000 squares into a covering for a World War II combat tank as a protest against the current war in Iraq.[10] For this project, the Internet functioned as an important tool for spreading the word, recruiting new knitters and sharing techniques and specifications. The tank was covered from the cannon to the caterpillar tracks with knitted and crocheted squares made with pink yarn.[11]

As Jørgensen has put it, "The main impression of the knitted tank is that it consists of hundreds of patches knitted by many different people in different ways: single colored, stripes with bows or hearts, loosely knitted, closely knitted, various knitted patterns." As such, it represents "a common acknowledgement of a resistance to the war in Iraq."[12] The project uses a democratic process where each contributor is able to knit their dissent mindfully. If the blanket is read as a petition, each individual panel of the blanket acts as a stand-in for a signature, but instead of a petition to be delivered pleadingly to a government elite, this gesture defiantly occupies public space.

The deployment of such a large blanket on such a threatening object in a public space may seem like a disarmingly absurd gesture, but the dramatic use of the crafted object to call attention to what is underneath creates a rupture in the ways in which the public interacts with the tank as a public war monument. The blanket gives the tank a physical presence, rather than a purely symbolic one. While addressing the pink tank cover, Jørgensen relies on the contrast between the symbols of tank and blanket: "For me, the tank is a symbol of stepping over other people's borders. When it is covered in pink, it becomes completely unarmed and it loses its authority."[13] It links remembrance of war with our collective ability to reinterpret and affect it through public action, dissent, and dialogue.

BARB HUNT

Barb Hunt's series of knitted and stuffed landmines conjures similar associations, using the tension between the formal character of the crafted object and the political context in which it is considered. As part of her project to knit soft pink replicas of the more than 350 different anti-personnel landmines that exist, Hunt has researched the technical aspects of landmine construction, their production, and the devastating number of casualties that they claim each year. During the construction of each she will "sit and knit for a few hours and enjoy it a lot, then suddenly realize that during that time about half a dozen people were injured or killed by a land mine somewhere in the world."[14] Many politically engaged crafters share this awareness of the paradox of linking sedentary leisure activity with political action, and like Jørgensen's pink tank, the knitted landmines are a stark contrast to the original object, in both form and intent. Hunt's project "refocuses attention on the value of small personal gestures that can accumulate into a declaration of caring and hope," and juxtaposes the mindfulness and time dedicated to a knitting project with the contemplation of "knowledge that is otherwise too difficult to bear."[15] In this way, Hunt highlights the usefulness of this labour in bringing politicized content into the gallery.

Unlike many craftivist works, Hunt's *antipersonnel* series is designed for the formal space of the gallery exhibition, where she reappropriates the conventions of museum display to enhance the possibilities for political engagement with the work. While the objects may at first appear playfully soft and seductively crafted as part of a pristine institutional display, the viewer recoils in horror[16] as they discover the global context of the project through a booklet that is distributed as part of the work. The booklet includes detailed information about landmines and their use worldwide (specifically implicating the countries who continue to manufacture landmines or who have not signed the Mine Ban Treaty).

In a museum or public gallery space, the conventional text would be written by an exhibition curator and interpret the works on display in an accessible language. When *antipersonnel* was exhibited as part of the *Museopathy* project at the Royal Military College of Canada Museum, the work took on an additional layer of meaning. By including information about the political context of *antipersonnel*, the artist was able use the institutional space—both the museum, and the specific context of the military museum—as a site for productive resistance. *antipersonnel* directly contradicts the ways that we tend to view works of art and craft and historical objects in gallery and museum spaces as timeless and apolitical.

WEDNESDAY LUPYPCIW'S HANDICRAFTS FOR HANDICATS WORKSHOP

Craftivist practices share common ground with the DIY aesthetic and its political roots.

Ideas around the use of craft practices as political tools are again circulating within the growing indie craft movement, where handcrafted (and often locally produced) items are championed as alternatives to mass-produced, globally distributed goods. The indie craft scene includes numerous approaches to making and selling handcrafted goods, as well as varying degrees of political engagement.[17]

Calgary artist Wednesday Lupypciw uses a variety of craft-related materials and processes in her performance and video work to critique craft traditions as well as recent DIY and indie craft trends. In September of 2006 she conducted a craft-based workshop for TRUCK, a Calgary artist-run gallery. Lupypciw's workshop took place in CAMPER (Contemporary Art Mobile Public Exhibition Rig), a converted 1979 motorhome stationed in downtown Calgary. Offering hands-on demonstrations in various craft techniques, Lupypciw's workshop exposed visitors to an array of crochet, hot-loops, pipe cleaners, plastic lacing, knitting and craft projects, encouraging them to (re)consider the esthetic, social, tactile, and visual potential of these materials. Lupypciw uses materials that are often considered more appropriate for hobby craft or kitsch, and through their use she calls into question the conventionally acceptable materials and outcomes for "fine craft" work.

As DIY and indie craft approaches continue to grow in popular appeal, the ways in which these objects and markets relate to contemporary cultures of craft consumption, display, and understanding are often overlooked. By

Figure 54 Marianne Jørgenson, *Pink M.24 Chaffee*, 2006.

focusing on the processes of crafting and the use of easily-learned techniques, Lupypciw's workshop shifted emphasis away from the production of polished, completed, or functional craft objects (and the positioning of craft objects as saleable commodities). Lupypciw states: "I am interested in dissecting the contemporary DIY ethic with this project instead of being outright cynical about it. The 'IT' in 'Do It Yourself' literature is portrayed far too many times as a gorgeously photographed object, which creates an underlying competitive/consumer mentality."[18]

CONSUMING CRAFT

Throughout the survey of practices presented above, we see that the key features of craftivism include: participatory projects that value democratic processes, the use of various cross-disciplinary media, and an ongoing commitment to politicized practices, issues, and actions. Sustainable community-based activity and relationships are emphasized in the creation of politically engaged craft projects. The values and methods of craftivism are located outside of, and used to critique both corporate and institutional cultures. However, as the handcrafted aesthetic gains popularity, it is increasingly being used within corporate marketing cultures as a way to affiliate their products with alternative lifestyles without any connection to the political/activist aims of these subcultures.

This may be true also of appropriation of craft into museum and institutional contexts. Politically engaged crafting is now being presented in formal exhibition spaces such as major museums, national galleries and educational institutions, and is in turn being contextualized within the respective curatorial frameworks and marketing strategies of these institutions. Artist, writer, and performer Andrea Fraser's essay "A museum is not a business. It is run in a businesslike fashion," is a prescient analysis of the ways in which the corporate model is applied to arts institutions. She suggests that the ways in which institutions are administered and governed have drastic impacts on curatorial autonomy. She cautions that "the political arguments against global corporate expansion apply to the art world as well—that despite the rhetoric associated with niche marketing, such expansion is producing an institutional monoculture that's destroying diversity."[19] This level of corporate involvement in culture, branding, and audience development initiatives is now often considered as a necessary part of adapting to changing funding structures for non-profit and public arts organizations. When the business model is applied to program development, emphasis is placed on defined outcomes and quantitative measures of the success of a project or exhibition—approaches that are a direct contrast to the values of politically engaged crafting projects. Alanna Heiss points out that "Museums are, to a greater extent than alternative spaces, in the audience business, a business that often includes subsuming a work of art to the composition of a room or a theme."[20]

The broad appeal and accessibility of crafting practices is central in achieving the political aims of the craftivist project and presents interesting avenues from which to critique the institutionalization of public cultural organizations. However, Fraser charges that "Critical discourse and the politics of democracy no longer appear to provide ready arguments against the corporatization of museums."[21] For public institutions with mandates to involve increasingly broader publics, curating new craft is often positioned as a way to tap into something of popular significance or interest and make the institution seem visionary and fresh. Exhibition press releases for new craft "blockbusters" presented at larger museums and public institutions would suggest that democratized and politically engaged craft works are indeed being recognized for their inclusive appeal.[22] When these works are used to engage and develop new audiences, markets, and communities that might have previously been marginalized within the imagined publics of an arts institution, it shifts the value of the work away from the original practices.[23] This presents the potential for institutional appropriation, in which the institution or persons working there stand to gain something (economic, social, cultural or intellectual) by adapting the original context of the works for their own aims. The danger is that this shift in context can also result in the erasure of community identities or of the activist issues that these practices seek to address.

CURATORIAL STRATEGIES

Craftivism and contemporary curatorial practices both critique the mainstream economies that govern their respective disciplines. With the inclusion of craftivist works in these spaces, exhibitions and their venues must be considerate of the fact that craftivist practice is often deliberately situated outside of traditional approaches and models of presentation. Curators must work to preserve and communicate the original context in which craftivist objects and artworks are made to function. Maintaining curatorial dialogue with artists and activists is a key strategy for avoiding the ethical problems of institutional appropriation. The emphasis on dialogue and participation should be a prominent component of any exhibition that includes craftivist or politically engaged

craft work, and applies to each of the following strategies:

1. Provide unmediated opportunities for craftivists to speak about their practices and to disseminate information about the political context of their works.
2. Build opportunities for teaching/learning crafting skills, sharing knowledge and participatory making into the exhibition.
3. Situate craftivist works within an organization that is truly committed to community-driven, structural changes.
4. Use printed matter, documentation, archiving and diverse distribution networks as a means to preserve and distribute information.

Among the works in *SUPERSTRING*, an exhibition at the Stride Gallery in Calgary, the Revolutionary Knitting Circle's *Peace Knits* banner became the flashpoint for a debate about the legibility of craftivist projects or actions within a gallery space. As with many craftivist projects, the *Peace Knits* banner is primarily intended for use by the group in particular contexts of protest and public action, and does not "require, or ask for, legitimacy within the art gallery system to be considered productive and politically charged."[24] However, in the context of the exhibition space, the banner was read and assessed as an art object, particularly when considered alongside the other works in the gallery.

Yet including this work in the exhibition alongside those of other artists and crafters undermined the exclusivity commonly associated with fine art spaces, including artist-run centres, which admittedly—despite their history and mandates—have become institutionalized in many ways. The members of the Revolutionary Knitting Circle were also invited to host a series of knitting workshops within the gallery, opening up a space for discussion and action that might not have been possible with the display of the banner alone. By providing space for communicating directly with the public, the workshop format enabled members of the group to speak about their work. These workshops also expanded the network of members of the Revolutionary Knitting Circle, and allowed the group to mobilize new participants around issues of independent production, labour and community building.

Both the Revolutionary Knitting Circle's workshops and Wednesday Lupypciw's practice place participatory making and process-driven approaches at the centre of their activities. Their use of skill-sharing and hands-on demonstrations in publicly accessible spaces keeps the focus on the interactions and relational potential of craft making rather than on a passive relationship between viewer and object. Furthermore, the interactive format challenges the notions of solitary artistic work and the emphasis on a highly skilled individual creator. These approaches teach and encourage lifelong and sustainable involvement with both crafting and activist practices, and distribute the means for empowerment and creation to a broader circle of participants.

Situating craftivist works within an organization that is committed to community-driven, structural changes that address activist concerns is another strategy that can preserve the intent and spirit of the work and instigate political dialogue at an organizational level. Exhibiting craftivism in arts spaces also poses important questions about how these spaces can be used by activists as sites for affecting political and social change.[25] Collaborations between activist groups and arts organizations have the potential to counter the increased corporatization of public spaces and arts spaces alike by providing alternate methods of generating community support.

The artist-run working atmosphere functions simultaneously to critique and to exist as a parallel system to other cultural institutions such as museums, commercial and public galleries. This network of centres is well positioned to facilitate the presentation of craftivist works, while maintaining their unruly manifestations and politicized contexts. Furthermore, the programming focus for many artist-run centres often emphasizes relational, performative, collaborative and community-based practices as well as the skill sharing and workshopping of ideas, processes and projects that are common to many craftivist methods. The staff and committees in artist-run centres are particularly well placed to respond to activist concerns: through their hands-on participation in programming, or curatorial directive of the centre, and interaction with the centre's constituency of members on a day-to-day basis. Among the methods by which true institutional change is achieved, "activists identify concerns important to them to which a particular arts institution may respond."[26] Pressing dialogues that are seeking representation and voice within the centre often are reflected through programming and operations that are responsive to community needs.

Exhibitions which extend their reach through the dissemination of printed matter, media and web based resources ensure that politically engaged projects connect with a broad network and remain accessible outside the confines of institutional space. As part of the Pink Tank project, Marianne Jørgensen recruited collaborators using calls for submission, contact and dialogue with Internet knitting groups, and produced a DVD of the process. She documented the process of covering the tank with the completed pink blanket on video and displayed this documentation as an integral part of the exhibition. The video now functions as an archive of the intervention, and can be exhibited, archived or distributed separately to raise awareness about the project. All these methods extended the reach of the action and connected crafters, artists, and activists who share common goals to contribute their work to one unified project. Barb Hunt's installation of *antipersonnel* also used printed matter to engage viewers beyond the scope of the exhibition, providing further information, reading and resources with a call to action that encouraged the viewer to take action directly.[27] Maintaining an archived history of craftivist works is important, not only to document the process, objects and exhibitions, but also to serve and a resource for future projects and to preserve the history of this political action.

CONCLUSIONS

Radical approaches to curating politically engaged craft have the potential to suggest new ways of discussing, critiquing, making, exhibiting, deploying, and resituating these practices while preserving the vigour of the maker's political intent. The organizational methods of craftivism offer new and productive spaces where practitioners create situations for their work outside of dominant institutional or corporate models: protests and marches, websites or web-based exhibitions, zines collectives, workshops and off-site events held outside of traditional gallery spaces. These new spaces provide open and accessible channels for the creation of alternative economies, new communities, and creative investigation. Even as crafting practices and methods are commodified, these examples offer productive strategies for the exhibition of craft that challenge institutional systems. In turn, continued cross-pollination between the practices of craftivists and politically engaged crafters and the curatorial strategies of galleries, independent artists, crafters, and curators continues to build the pool of appropriate methods for the creation, display, and understanding of craftivist works. By maintaining and allowing for rigorous self-directed strategies within these sites where the critique of dominant arts economies and craftivist practices might cohabit, politically engaged practices will maintain their radical potential.

NOTES

1. Over its 35-year history, the Canadian artist-run culture network has supported the development of new artistic practices, such as video, performance and new media, and fostered the development of hybrid curatorial practices.
2. Peter S. Hawkins, "NAMING NAMES: The art of memory and the NAMES Project AIDS Quilt," in *Thinking About Exhibitions*, ed. Reesa Greenberg et al. (London: Routledge, 1996), 133–156.
3. Additional criticisms of the AIDS Quilt have included: the historical association of quilting with repressed sexuality, the self-censorship and sanitization of queer identities, the fact that the majority of quilting work is carried out by women, and yet they aren't well represented within the memorialization of AIDS dead, the failure of the quilt to represent both the racial and socio-economic diversity of communities affected by AIDS and the global AIDS crisis. "More irascible critics have found it "whitebread," maudlin, ingratiating, expensive." (Hawkins, 149–150)
4. Hawkins, 152.
5. Cleve Jones qtd. in Hawkins, 138.
6. Hawkins, 138–142.

7. Kim Berman's Paper Prayer's project has involved thousands of arts and crafts practitioners, activists, service providers and people living with or suffering losses from HIV/AIDS in South Africa in collective printmaking, papermaking, and embroidery workshops and creation of 'paper prayers.' This contemporary example of Craftivism was inspired by the Names Project AIDS Quilt and has incorporated collective methods, information sharing, skills teaching and sustainable methods of generating income for local communities to instigate broad social change. Kim Berman, "Paper Prayers—A strategy for AIDS Action in South Africa," paper presented at the *New Craft Future Voices International Conference* (University of Dundee, Scotland: Duncan of Jordanstone College of Art and Design, July 4–6, 2007).

8. Grant Neufeld, "The Revolutionary Knitting Circle Proclamation of Constructive Revolution," *Revolutionary Knitting Circle Website*, http://knitting.activist.ca/manifesto.html (accessed January 23, 2007).

9. Anthea Black, *exhibition text for SUPERSTRING* (Calgary: Stride Gallery, 2006).

10. The project is one of many international craftivist efforts to protest the current war in Iraq. Some, such as Lisa Auerbach's *Body Count Mittens The Workmanship of Risk* curated by Sabrina Gschwandtner, are ongoing personal projects by one person that blur the line between craftivism and a professional art practice. Others, like Sherri Wood's "Prayer Banner: REPENT" that depicts both the U.S. soldiers who have died in Iraq and every Iraqi citizen killed using tiny coffins and stitches alongside one another, are collaborative and participation-based projects like many of the examples that we have cited above. Sabrina Gschwandtner, "The Workmanship of Risk: the Performance of Craft," *KnitKnit Website*, http://www.knitknit.net/sabrina/_include/imgs/curatorial/pdfs/workmanship_of_risk.pdf (accessed December 2006).

11. Marianne Jørgensen, "Pink M.24 Chaffee: A tank wrapped in pink," *Pink M.24 Chaffee Website,* http://www.marianneart.dk/ (accessed January 2007).

12. Jørgensen.

13. Jørgensen.

14. Barb Hunt, "Re: antipersonnel story," email message to Anthea Black, January 2007.

15. Hunt.

16. Hunt.

17. These practices range from independent studios or boutiques (like Smoking Lily), blogs, artist/crafter websites (like anti-factory.com), and home sales, to larger web-shops (like Etsy.com), online listings (like craftrevolution.com), and independent craft fairs (like the Renegade Craft Fair). Etsy.com *http://www.etsy.com/* (accessed December 2006); Blatt, S. & Habbley, K. "The Renegade Craft Fair" Chicago. *http://www.renegadecraft.com/holiday/about.html* (accessed January 2007); Stephanie Syjuco, *Anti-Factory* http://www.stephaniesyjuco.com/antifactory/> (accessed January 2007); Smoking Lily, Victoria, http://www.smokinglily.com/ (accessed January 2007).

18. Wednesday Lupypciw, "Re: article for FFWD," email to Nicole Burisch, September 11, 2006.

19. Andrea Fraser, "A museum is not a business. It is run in a businesslike fashion," in *Beyond the Box: diverging curatorial practices*, ed. Melanie Townsend (Banff: Banff Centre Press, 2003), 115.

20. Alanna Heiss, qtd. in Clive Roberston, "Artist-run culture: Locating a history of the present," in *Policy Matters—Administrations of Art and Culture* (Toronto, ON: YYZ Books, 2006), 13.

21. Fraser, 117.

22. Recent exhibitions of politically engaged craft have foregrounded the agency of the artist and the audience to different degrees. Some examples include: *Radical Lace & Subversive Knitting* at the Museum of Art and Design in New York, *Common Threads* at the Confederation

Centre of the Arts in Charlottetown, and *At Home* at the Rugby Art Gallery and Museum in England.

23. While the corporatization of public arts venues has progressed at particularly alarming speeds in the last 15 years, the inclusion and appropriation of avant-garde, experimental or politically engaged works into institutional spaces and the curatorial conundrums that they present are nothing new. Fraser's essay also charts the development of cultural activism citing various media such as performance, site-specific installation, and relational practices as evidence of its tenuous relationship with institutional spaces (Fraser, 118–119).

24. Black.

25. Even the ability of the *SUPERSTRING* exhibition to make room for institutional change within the artist-run community in Calgary was limited by the dominant economies in which it exists: the gallery is a non-profit organization that relies on government grants, public funding and a small number of business sponsorships available to arts and culture organizations. When introduced with the idea of involvement with the Calgary Dollars alternate economy (which several Revolutionary Knitting Circle members also use), as part of the gallery's membership structure and fundraising efforts, the limited financial agency of the artist-run centre and non-profit culture made it difficult to use this grassroots currency within the smooth functioning of the organization. We see here that while artist-run centres are invested in and mandated to program politically charged artworks, dominant economies may also prevent them from incorporating activist ideals into their operations structures.

26. Chris Creighton-Kelly, "Bleeding the Memory Membrane: Arts Activism and Cultural Institutions," in *Questions of Community*, ed. Daina Augaitis et al. (Banff: Banff Centre Press, 1995), 96.

27. Barb Hunt, "*antipersonnel* booklet," information book exhibited in *Museopathy* at The Royal Military College of Canada Museum (Kingston, ON: self published, 2001-ongoing).

FURTHER READING

Eithne Farry, "¡Viva las craftivistas!," *The Guardian* (26 May 2006).

Stacey K. Sowards and Valerie R. Renegar, 'The Rhetorical Functions of Consciousness Raising in Third-Wave Feminism', *Communication Studies* 55 (2004).

THE POLITICS OF CRAFT: A ROUNDTABLE

Julia Bryan-Wilson, Liz Collins, Sabrina Gschwandtner, Cat Mazza and Allison Smith

This reader ends with the following transcribed conversation, featuring an all-star lineup of key players within the emerging crafter movement. One of the participants, Julia Bryan-Wilson, is an art historian; the others are all artists involved with textile media such as knitting, sewing and crochet. Already in Bryan-Wilson's opening line, 'Craft can be a moving target', one gets a sense of the openness and sense of purpose that mark these women's approach. Interweaving ideas from Feminism, political protest, performance art, industrial production, community-building programs and good old-fashioned craft theory, this group of women discuss the present state of affairs, and a future that they themselves bear the responsibility of crafting.

'The Politics of Craft', a roundtable with Julia Bryan-Wilson, Liz Collins, Sabrina Gschwandtner, Cat Mazza and Allison Smith. Edited and introduced by Julia Bryan-Wilson. This is a revised version of a text published in *Modern Painters* (February 2008), pp. 78–83.

INTRODUCTION

With bars hosting weekly crochet nights and knitting cafés proliferating, the current popularity of textile handicrafts in the US is undeniable. A growing number of artists, many of them women, are also producing critical, socially committed, conceptually oriented, collaborative craft-based work—so many that it could be called an emerging genre. For example, Cat Mazza uses her website, microRevolt.org, to solicit knitted and crocheted squares from crafters around the world for her *Nike Blanket Petition*, which protests sweatshop labor. Sabrina Gschwandtner engaged audiences in discussions about the Iraq war in her 2007 installation *Wartime Knitting Circle*, commenting on how knitting has historically been mobilized as a form of civic participation as well as protest. Allison Smith creates work that examines how handmade objects are embedded in political narratives, and in 2005 she orchestrated a public art project on Governor's Island, "The Muster," that deployed the rhetoric of Civil War–era "mustering" to explore the role of historical reenactment in a time of national crisis.

One of the more than 50 projects at "The Muster" was Liz Collins's *Knitting Nation: Knitting During Wartime*. For this work, Collins, a knitwear designer who teaches textiles at the Rhode Island School of Design, assembled a small army of women who used hand-powered knitting machines to churn out a massive abstract flag in red, white, and blue. As the Stars and Stripes grew to cover the ground, an orator clad in a dress that recalled a tattered US flag read texts about women knitting during wartime and about textile trade policies. (Full disclosure: that orator was me.)

Collins, Gschwandtner, Mazza, Smith, and I met in New York in fall 2007 to reflect on the nascent movement of crafted critique as well as to discuss the wider implications of handmaking today. The following conversation was distilled from many hours of talk.

—Julia Bryan-Wilson

JBW: Craft can be a moving target. I was part of a conference at the Getty last spring, "Craft at the Limits," that brought together studio craft artists and contemporary art historians who were not necessarily decorative-arts scholars. At some points it felt like we were on different planets—the art historians were discussing how craft was degraded or devalued before it was resurrected by '70s feminism. There was some talk, even, of it as embarrassing.

LC: Craft shame.

JBW: Exactly. You can see echoes of it in the way that the word has been dropped from the California College of Arts and Crafts [now the California College of the Arts] and the American Craft Museum [now Museum of Arts & Design]. Neither of those has craft in its name anymore. But the studio craft artists of an older generation at this conference were completely puzzled by the idea that craft has been marginalized: "What are you talking about? There's a stigma associated with craft?"

LC: I wonder if the fine-art versus craft split matters anymore. I teach a new generation that doesn't care about old art-craft hierarchies. The DIY movement doesn't think of craft as a dirty word.

AS: Even though the work each of us makes can be categorized as craft, we are all trying to stretch the boundaries of conceptual and collaborative art as well. We are

conceptual artists whose subject is craft. That's the difference between our work and studio craft artists: we are working within a theoretical framework. And yet all of us are actively courted by the craft arena—often more courted by that world than by contemporary art audiences.

JBW: One striking thing about you all as artists is how you explore the relationship of craft to politics. If craft implies utility, for you four, craft's "function" is to generate political dialogues.

AS: Well, American culture—a lot of cultures, actually—foregrounds craft as a mode of expression for defining national identity. In living-history reenactments and open-air museums, which are pedagogical sites, the focus is always on two things: craft and war. Obviously our master narratives of history are about war more than anything else, but at these sites it's craft that meticulously re-creates war and makes it visible. Handicraft becomes a metaphor through which to think about the terrain of war.

JBW: Allison, your project *Notion Nanny* (2005–), a community-based performance and installation in which you work with local craftspeople, is a good example of how you explore the relation between the two.

AS: In *Notion Nanny*, one sculptural element is a doll wearing a handmade Revolutionary War–era costume. I was trying to think about what can be considered revolutionary or politically relevant within very traditional craft practices.

JBW: *Notion Nanny* has a significant interactive/collaborative component. All of you take this approach in your work, and I'm curious how that gets elaborated for each of you in different institutional contexts.

SG: My *Wartime Knitting Circle* in the "Radical Lace and Subversive Knitting" show at New York's Museum of Arts & Design [2007] worked really well, because the museum was very generous in allowing people to knit at my table without paying an entry fee. But other times I run into installation problems, because curators don't always understand that this kind of work is not made through a studio practice that leads to discrete objects. The works are collaborative, and can be made outside the exhibition site or at the site itself, or some mixture of both. It's not the same as when artists come in and bring their materials and create an installation that somebody walks around in and then leaves. The audience is part of creating the work. So the exhibition site becomes a participatory space, an activist space, an education space, and a tactical space. It's also a working space.

AS: When *Notion Nanny* was exhibited at the Berkeley Art Museum in 2007, there was a similar institutional confusion about what the art actually was. It was a living, active process that would continue throughout the show. The public is coming in and doing things; I'm there doing things. Things are being added; things are disappearing. The show is accumulating objects and ideas as it goes on over time. I had apprenticeships with craftspeople in the area, and crafts we made collaboratively were integrated into the exhibit. You just said a great string of words, Sabrina, about how an activist, participatory, tactical space is part of the work. It's something that isn't always clear, I think, for the institution.

There were a lot of interesting moments trying to install *Notion Nanny*, because boxes were arriving from my studio and from a lot of different locations. Some of them had objects that I had made. Some of them had objects that people had given to me. Some of them had pencils and paper and art supplies that were there for the audience to use.

When these boxes came in, there was a registrar with a clipboard—

LC: She must have been going crazy.

AS: She was like, "What's this ball of string? What's this pencil?" And when I got everything back at the end of it, things like

Figure 55 Liz Collins, *Black Curtain Dress*, 2005.

pencils were wrapped in bubble wrap, preciously packaged as if they were artworks. Of course I love working in institutions, because there's sustained dialogue that's generated around the work that you don't always get in galleries.

SG: One problem I had with the *Wartime Knitting Circle* had to do with maintenance. It took me a while to figure out how to display it so that yarn wasn't always getting tangled. I was there all the time just to clean it up. It was like a full-time job, but I wasn't getting paid to do it. It would be great if museums could realize that support might be needed and budget for that.

What was wonderful was that two museum staff members really helped out, including the traveling exhibitions coordinator. He was really into the piece. He had a relative on duty in Iraq, and it was very therapeutic for him to be knitting. He learned to knit there. He would go almost every day on his lunch hour and tell people, "You can also sit and knit here with me."

CM: I had a different experience with the "Radical Lace" show. The piece I was showing is a 14-foot-wide blanket of the Nike swoosh. The swoosh is made up of four-by-four-inch squares that serve as petitions for fair-labor policies for Nike garment workers. It was created by knit hobbyists from over 25 countries who visited the microRevolt website. The interface [microrevolt.org/petition_overview.htm] allowed users to virtually sign their names and mail in their hand-stitched squares, instead of the traditional signing of a petition. And the public is supposed to add to it during the exhibition. But MAD decided not to show the Nike blanket; they decided to show *documentation* of it. Maybe the political rhetoric made the institution concerned; there were also

copyright questions because of the logo. But there was another issue, too.

AS: What was it?

CM: Aesthetics.

SG: That was the big one, I think. More so than the political issue.

LC: When was this decision made?

CM: They decided that the Nike blanket would be in the exhibition. Then, three days before the piece was supposed to go up, I got an e-mail from the curator saying, "We're so sorry; we just can't hang it. It's too difficult to hang, and it looks too 'funky' among the other work." I thought maybe someone from Nike was on their board. But in fact the aesthetic issues are just as interesting. The banner is made from iridescent orange yarn—acrylic, synthetic material—and the squares are made by hobbyists, so they are sometimes a bit amateur looking. The museum assumed that the banner aspired to some high-art quality, which was never the intention. The show was about radical and subversive artworks using knitting or lace, and while the piece fit into their title, it didn't fit into their aesthetic.

AS: All of us have tried to reach out to this incredible crafter energy that's out there waiting to be engaged, which is somewhat separate from the concern to reach a museum audience. There was something so exciting for me about the way "The Muster" brought together so many different groups of people beyond the contemporary art world.

CM: Looking back on the Nike piece's appearance in "Radical Lace," I was really disappointed, but on the other hand, it made sense. We negotiate these different venues, and our expectations have to change based on those venues. I use Web media to reach audiences beyond the museum. In fact this connects back to craft, because craft, like the Internet, is also seen as a democratizing medium, a social network that operates outside the institution.

SG: What's fascinating, too, is that people pick up knitting needles as an escape from the computer. In the face of everything fast and glinting, they want something real—a reinjection of the artisanal. But handicraft often brings them back online, because they go searching for instructions or tips, and they discover there's this whole online community of blogging crafters. So it feeds back into the digital environment.

LC: Maybe putting together a MySpace page is not that different from collaging or quilting. You're using different materials, to different ends, but along the way you're starting with matter and transforming it into something else, using your hands and your brain.

CM: There's a rich interface between textiles, technology, and labor issues. Theorists have written about the link between textiles and digital media, because the computer is made up of a binary machine code, ones and zeros, while knitting is also based on a two-digit system: the stitches, knit and purl.

JBW: Your point about labor, Cat, is crucial. The ways you all treat craft underscore the role of the hand in making art—which has been somewhat effaced since the Conceptual turn—as well as larger economic issues.

AS: Somebody who saw my project at Berkeley titled their blog entry "The Handmade vs. the Brain-made (Idea)." They were frustrated because the label of a piece in my show, which is a coverlet, read, "Coverlet by Allison Smith, woven by Leigh Alexander of Charleston, South Carolina." I take a lot of pride in foregrounding all of the people who are making this work with me, but for this person, the issue was that the weaver was secondarily mentioned. It's an interesting dilemma. This person was struggling with what that meant in terms of legitimate labor, as if to say, "It's her idea, but somebody else made it. Is that OK?" I find all of this, this exact conversation, to be the meat of the work.

LC: If at the top of the hierarchy is the artist, and underneath that is the crafter, there's a huge, important population of handmakers even lower than that. Art historian Glenn Adamson called our attention to this at the 2006 American Craft Council conference in Houston. If you want to talk about what's going on in craft today, let's look at the kids making soccer balls overseas.

AS: Exactly. We forget that even today Nikes are made by workers. We tend to think of mass-produced "machine-made" things as if they're totally devoid of human hands and workmanship, but machines aren't making all these things; people are making these things.

LC: The harsh reality is that a lot of US textile mills are still in business only because they have government contracts making uniforms for soldiers—all uniforms are supposed to be American made, which is why they're so expensive. There is a glove factory in North Carolina that makes trigger-finger gloves with antimicrobial liners for the soldiers in Iraq, so that their hands don't fall off from sweating so much. That's the kind of thing keeping domestic factories alive. But I think we're on a cusp of a new shift, perhaps, because of the anti-China backlash.

JBW: There's also the reinvigorated "made in the USA" movement that stems not from xenophobia but from localism and sustainability, as consumers become more aware of carbon footprints and the resources wasted in shipping.

LC: In *Knitting Nation* at "The Muster," I was trying to address this tension: "I hate patriotism; let's put the flag on the ground and walk all over it." But at the same time, I love where I live, and I want people here to have jobs. It's complicated; if we take back manufacturing and pull out of other countries, suddenly those people don't have any jobs, and then they're starving. There are no easy answers.

CM: Liz, maybe out of all of us, you have your finger on the pulse of manufacturing because of your work as a clothing designer.

LC: When I started out as a designer, I didn't know much about production and manufacturing. That's part of why I created a niche for my work as different and handmade. At a certain point, though, I knew I had to shift to outsourcing; my business could not survive if I continued to make clothes in my studio, paying the labor wages I had to pay. I think Natalie Chanin from Project Alabama, a fashion line that featured garments hand-sewn in Alabama, sold her company for this very reason. It was sad; she had this fantastic mission to have her stuff made in the US, and she really revived a community and gave them a lot of work. But the garments were too expensive. Another company bought her out, and now all of those pieces are made in India.

For me, trying to survive by outsourcing took an ironic twist: Barneys wasn't interested in my work when I had it made by a Chinese factory. They said, "We have other designers who do that. We want the handmade pieces from you." But if you do go handmade, you have to find a factory that can manage that quality of labor and be humane—which raises a lot of questions.

JBW: Ethical questions.

LC: Absolutely. I've been working with a factory in Peru that can maintain that level of integrity, but it required research on my part to discover that resource. The biggest problem for me in having a business was that I became so disconnected from my creative process. I wasn't getting to make things anymore; it was all about management and outsourcing. It was a real spiritual crisis for me—the commodification of something that had been about love and connection. It sounds hokey, but my work is so emotional. That's part of why I started doing fashion in the first place, because I used the knitting process as a way to do self-surgery. I was going through a lot of emotional pain when I started making knitwear, and I was also coming out, discovering my sexuality. Creating garments that are connected to the body was a way to connect to myself and describe my emotional landscape. To go from that to making expensive clothes for Barneys was a difficult transition, and I missed the hands-on aspect.

SG: That kind of alienation is one of the things that draw people to craft as a hobby right now. People don't see the end result of their labor in their jobs. You do a part of the administration or a part of the physical assembly, but your work goes out into the world and doesn't have your stamp on it. So these projects at home are something people can do from start to finish. And part of the pleasure of the knitting or sewing circle is that it doesn't happen in isolation. You're sitting with other people, talking and making together, so craft also implies a social space. Not only that, but while you're doing physical, repetitive gestures, it frees your brain to think about other things.

AS: A lot of what we've been talking about is the mind-body split—how intellectual labor is valued over manual labor. Craft, as a physical, performative act, is still considered to be at odds with the intellectual labor that has fed conceptual art and a lot of politically motivated art. For many of my political artist role models, the critique of the commodification of art entailed a rejection of hands-on making—and *craft* became a bad word. So bringing back the political, activist spirit to something interactive and bodily is really important.

LC: Part of my mission with *Knitting Nation*, which has taken various permutations since its first appearance at Allison's "The Muster," was to give people an inkling of the human effort involved in the process of making something. My intentions were multifaceted. It's a public art experience that lays bare the process of machine knitting and textile and garment construction. It's also a celebration of movement and the physicality of a manual process. And it's usually collaborative, with many people working together to build one thing; in this it's a commentary about labor.

CM: As we're talking about garment labor issues, I'm reminded, too, of the unjust conditions in computer manufacturing. This is something that I come up against when I'm discussing sweatshop labor using my mass-produced laptop. But, then, we all coexist in corporate culture, and we try to do what we can within it. What inspired me to call my website

Figure 56 Sabrina Gschwandtner, *Wartime Knitting Circle*, 2007.

microRevolt was Guattari's idea of molecular revolutions—the idea that social and cultural change can occur from small acts of resistance, that change is not simply a consequence of a governing or economic policy.

JBW: A café in my neighborhood in Long Beach, California, recently held a craft fair called "Handmade Revolution," and it made me think about the desire to connect craft, William Morris style, with utopia or radical politics. Yet as much as this event couched itself as resisting capitalist culture, it just featured a lot of tables with people selling stuff. It didn't, for instance, sponsor workshops on how to knit or crochet. While they could be part of an alternative microeconomy, more often these craft fairs have become about hipster shopping.

AS: *Craft* has become a trendy buzzword in the artworld, too. In so much contemporary art, though, when a work is supposed to be about craft, what that means is it's got some big, sloppy stitches on it. It's a kind of disrespect to craft traditions and the deep history of hand-making.

SG: Well, I'm a pretty bad crafter, I have to say. I took some sewing lessons and just learned how to do an invisible hem. I work with film, video, photography, sewing, crochet, and knitting, and I'm pretty much just technically proficient with all of those mediums.

LC: But that's not your process.

SG: No, it's not. I make installations, events, and publications, among other things, that challenge the boundaries between artist and curator, or archivist, and between art and craft. I think about audience and distribution a lot. When I wrote my book *KnitKnit*, for example, I chose a publisher, Stewart, Tabori & Chang, that specializes in craft books. It's a division of Abrams, the art book publisher, so

Abrams sells *KnitKnit* to yarn stores and also to museum books stores. It was really important for me to write with handicraft hobbyists and fine art readers in mind, because I wanted to mix perspectives on what art is and can be; I've often used handicraft as a site to engage ideas about what constitutes art.

AS: Even though crafted polish is not what you're aiming for, Sabrina, because your work is more conceptual, I've found that if you're using craft, the question of skill is going to enter the critique no matter what. Especially as women, we're expected to have a level of high expertise in handmaking.

SG: On that note, I'd like to hear some people's thoughts on their relationship to feminism.

LC: Some of my garments address female sexuality and ideas about erogenous zones, or areas of comfort, discomfort, bondage, and release. And that ties into queerness, but I don't know where it falls within feminism. *Knitting Nation* in "The Muster" was, importantly, all female, and it was about women working in a time of war and a tradition of women getting together for a cause. But I've moved on from that and thought about configurations that would be more about labor that's not gendered.

SG: Historically, hasn't textile labor been mostly done by women?

LC: Well, when you look around the manufacturing landscape globally, there are some areas where it's women's work, but there are others where it's men's work, and there are others where it's both. I was really jazzed by being in the factory in Peru and seeing men who were knitters. The men are the ones running the industrial machines and doing the programming, to be sure; but where people are doing things by hand, it's a real mix.

AS: The '60s and '70s feminist approach revealed the historically gendered nature of craft and tied it to domesticity. I grew up in a Martha Stewart–type household, with all the straight white rituals of suburban America that are embodied in craft projects. Feminism helped me be critical of those things while also implicating myself in the critique, and it also gave me a way to think about performance and queer identity. I've had a lot of students—women students—who don't want to have anything to do with feminism. Up until LTTR [the queer/feminist artist collective], really, the predominant attitude was, "It's done, it's in the past, feminism's outmoded." Feminism has been just as messy and unwieldy, and stigmatized, as craft.

CM: I have a mentor, Faith Wilding, who worked with Judy Chicago on *Womanhouse* and made influential craft-based artworks in the '70s. Wilding is also now involved in new media and cyberfeminism, and she made me aware of those overlaps. Third-wave feminist theory has introduced other aspects of identity—sexuality, race, and class. Feminism is still relevant; race, class, and gender are also played out in the global economy, as advanced countries subjugate an entire third-world workforce to create products. I try to consider this more global critique in my work.

JBW: Perhaps *craft* is a useful term today for the way it allows us to see those overlaps, to make connections between such different subjects: globalized labor, war, digital culture, feminism, collaboration, queer identity. Maybe precisely because it is so slippery and unfixed, it can encompass a broad spectrum of issues.

FURTHER READING

Ednie Kaeh Garrison, 'U.S. Feminism-Grrrl Style! Youth (Sub)Cultures and the Technologics of the Third Wave', *Feminist Studies* 26/1 (Spring 2000), pp. 141–70.

Sabrina Gschwandtner, *KnitKnit: Profiles and Products From Knitting's New Wave* (New York: Stewart, Tabori and Chang, 2007).

Sabrina Gschwandtner, "Knitting Is . . ." *Journal of Modern Craft* 1/2 (July 2008).

RECOMMENDED BIBLIOGRAPHY

While this reader includes indicative references for further reading throughout, a few general sources on craft history and theory deserve special mention. For the most part, these have not been cited elsewhere in the text. With a few exceptions, medium-specific texts, exhibition catalogues and monographic studies have not been included. Several books listed here were forthcoming as this anthology went to press.

Glenn Adamson, *Thinking through Craft* (Oxford: Berg/V&A Publications, 2007). Emphasizes craft's relations to modern and contemporary art, within the themes of supplementarity, materiality, skill, the pastoral, and the amateur. The book also includes some reference to educational theory and architecture.

Sandra Alfoldy, *Crafting Identity: The Development of Professional Fine Craft in Canada* (Toronto: McGill/Queen's University Press, 2005). History of craft in Canada, with particular attention to the role of institutions and exhibitions.

Sandra Alfoldy, ed., *Neocraft: Modernity and the Crafts* (Halifax: NSCAD Press, 2007). Collection of essays, most of which are historical case studies, with some theoretical contributions and discussions of contemporary 'crafter' activity.

Sandra Alfoldy and Janice Helland, eds, *Craft, Space and Interior Design, 1855–2005* (Burlington: Ashgate, 2008). Examination of the relations between craft and architecture, with case studies from a wide geographical range.

Elissa Auther, *The Material Boundaries of Modernism* (Minneapolis: University of Minnesota Press, forthcoming). An interdisciplinary analysis of fiber arts during the late 1960s and 1970s. Important for its use of materiality as a means of articulating the formation and maintenance of hierarchies within fine art, craft and folk production.

Kim Brandt, *The Kingdom of Beauty* (Durham, NC: Duke University Press, 2007). Brandt's history of the *mingei* movement adds significantly to the postcolonialist critique offered by Yuko Kikuchi. It is particularly strong in its coverage of the commercialization of folk crafts, and the political use of *mingei* during World War II.

Julia Bryan-Wilson, *Art Workers: Radical Practice in the Vietnam War Era* (Berkeley: University of California Press, 2009). Analysis of the postwar American avant-garde from the perspective of production politics, looking particularly at issues of labor.

Maria Elena Buszek, ed., *Extra/ordinary: Craft Culture and Contemporary Art* (Durham, NC: Duke University Press, forthcoming). Essays about the intersection between subculture, craft and contemporary art practice. The best scholarly book thus far on the new 'crafter' movement.

Louise Allison Cort and Bert Winther-Tamaki, *Isamu Noguchi and Modern Japanese Ceramics* (Washington, DC: Arthur M. Sackler Gallery, 2003). A fine study of the impact of the Japanese-American sculptor on potters in Japan, which provides a model for the examination of a traditional craft transformed by avant-garde ideas.

Peter Dormer, *The Art of the Maker* (London: Thames and Hudson, 1994). Searching interrogation of questions of skill by the key British writer on craft from the 1980s and 1990s, design historian Peter Dormer.

Peter Dormer, ed., *The Culture of Craft: Status and Future* (Manchester: Manchester University Press). Book of essays with an emphasis on relations between craft and technology.

Arindam Dutta, *The Bureaucracy of Beauty: Design in the Age of Its Global Reproducibility* (New York: Routledge, 2007). A theoretically informed history of the activities of the Department of Science and Arts, the British governmental body responsible for design reform in the late nineteenth century. Dutta's book is notable both for its attention to the Indian context, and its arguments about the construction of the 'traditional' artisan as a lynchpin of imperialist aesthetics.

Isabelle Frank, ed., *The Theory of Decorative Art: An Anthology of European and American Writings, 1750–1940* (New Haven, CT: Yale University Press, 2003). Frank's excellent anthology of writings about decorative art and ornament includes a section on 'materials and techniques of the decorative arts' that is especially relevant to craft history and discourse.

Paul Greenhalgh, ed., *The Persistence of Craft: The Applied Arts Today* (London: A & C Black, 2002). Book of essays looking at the art/craft debate; also includes focused analyses of glass, ceramics, jewelry, wood, textiles and metal.

Tanya Harrod, *The Crafts in Britain in the 20th Century* (New Haven, CT: Yale University Press, 1999). An idea-packed, definitive history of one nation's experience of modern craft. Harrod's approach combines social and political context, examinations of individual media, 'set piece' studies of particular commissions and organizations, and profiles of significant individuals.

Tanya Harrod, ed., *Obscure Objects of Desire: Reviewing the Crafts in the Twentieth Century* (London: Crafts Council, 1997). The best of a series of conference proceedings published by the British Crafts Council in the 1990s.

Janice Helland, *British and Irish Home Arts and Industries, 1880–1914* (Dublin: Irish Academic Press, 2007). Examination of the work of three women craft entrepreneurs in Ireland around the turn of the century, focusing on the sale of cottage crafts as a means of ethical uplift for both producers and consumers.

John Houston, ed., *Craft Classics: An Anthology of Belief and Comment* (London: Crafts Council, 1988). A compact, well-chosen selection of writings from the British craft movement.

Wendy Kaplan, ed., *The Arts and Crafts Movement in Europe and America: Design for the Modern World 1880–1920* (Los Angeles: Los Angeles County Museum of Art/Thames and Hudson, 2004). Catalogue, accompanying an exhibition of the same name, surveying craft reform and revival movements at the turn of the century. Includes a number of in-depth essays by key writers in the field.

Janet Kardon, *The Ideal Home, 1900–1920* (New York: H. N. Abrams/American Craft Museum, 1993); *Revivals! Diverse Traditions, 1920–1945* (New York: H. N. Abrams/American Craft Museum, 1994); *Craft in the Machine Age, 1920–1945* (New York: H. N. Abrams/American Craft Museum, 1995). The incomplete 'centenary' project was originally meant to document the twentieth-century history of American craft but was able to cover only the period prior to World War II. The volumes that resulted are uneven; the best is *Craft in the Machine Age*, an overview of American Art Deco. *Revivals!* also has its strong points, particularly in its coverage of regionalism. At the time of writing the Museum of Arts and Design (formerly the American Craft Museum) is currently planning its own revival, as it hopes to bring the centenary project to completion over the next few years.

Janet Koplos and Bruce Metcalf, *Makers: A History of American Studio Craft* (Chapel Hill: University of North Carolina Press, forthcoming). An indispensable grand tour of American craft from the Arts and Crafts movement on, organized as a textbook and emphasizing individual makers' biographies. Completed just as this reader was going to press, it promises to serve as a foundation for future studies of American craft history, much as Tanya Harrod's book does for Britain.

Bruno Latour and Peter Weibel, eds, *Making Things Public: Atmospheres of Democracy* (Cambridge, MA: MIT Press, 2005). A 100-author book of essays on the subject of concretizing politics. The key figure is Latour, whose pioneering efforts in "network theory" resonate in many of the essays. Though it takes digesting, the project suggests a theoretical environment in which craft theory (and other thing-specific thought) is part of a fluid mix of ideas rather than a separate set of concerns.

Karen Livingstone and Linda Parry, *International Arts and Crafts* (London: V&A Publications, 2005). Similar to Kaplan's volume above, this exhibition catalogue makes the case for the Arts and Crafts Movement as a pan-Euro-American phenomenon rather than a specifically British one.

Saloni Mathur, *India By Design: Colonial History and Cultural Display* (Berkeley: University of California Press, 2007). Important study of the exhibitionary complex surrounding Indian craft in the late nineteenth century, perhaps most rewardingly read alongside Arindam Dutta's text (see above). Mathur also examines contemporary postcolonial museum politics.

Howard Risatti, *A Theory of Craft: Function and Aesthetic Expression* (Durham: University of North Carolina Press, 2007). A recent book espousing the position that craft should be defined as a distinct field of practice, separate from art and design.

Sue Rowley, ed., *Craft and Contemporary Theory* (St. Leonards, Australia: Allen & Unwin, 1997). Book of essays by Australian authors, including practitioners as well as scholars.

Wendy R. Salmond, *Arts and Crafts in Late Imperial Russia: Reviving the Kustar Art Industries 1870–1917* (Cambridge: Cambridge University Press, 1996). Examination of the Russian experience of craft and design reform in the pre-Soviet era, with particular attention to the rhetoric surrounding traditional 'folk' crafts and architecture.

Frederic J. Schwartz, *The Werkbund: Design Theory and Mass Culture Before the First World War* (New Haven, CT: Yale University Press, 2005). Outstanding intellectual history of craft and design theory in Germany around the turn of the century.

David Summers, *Real Spaces: World Art History and the Rise of Western Modernism* (London: Phaidon, 2003). An ambitious attempt to ground a 'world art history' in terms of facture—the object conceived as a record of its own making. Though many readers will find the book unmanageably diverse, it offers useful models for interpreting artifacts of all kinds in terms of craft process.

Jorunn Veiteberg, *Craft in Transition* (Bergen: Bergen National Academy of the Arts, 2005). Theoretical and critical writings by a leading curator in Scandinavia.

ILLUSTRATIONS

Figure 12 John Ruskin, *Sketch of Gothic Tracery in Venice*, 1845. Watercolor. Victoria and Albert Museum.

Figure 13 William Morris and Philip Webb, '*Trellis*' *Wallpaper*, designed 1862. Block-printed in distemper colors on paper. Victoria and Albert Museum.

Figure 14 William Morris, *Sussex Chair*, designed ca. 1860. Ebonised beech, rush seat. Victoria and Albert Museum.

Figure 15 William Morris, '*Iris*' *Furnishing Fabric*, 1876. Retailed by Morris & Co. Victoria and Albert Museum.

Figure 16 *Ainu Robe*, mid-nineteenth century. Elm bark fiber (*ohyô*), cotton appliqué and cotton thread embroidery. Made in Hokkaidô, Japan. Victoria and Albert Museum.

Figure 17 Bernard Leach's work bench, showing pots, some of his working sketches, and his seals. Courtesy of Crafts Study Centre, Farnham.

Figure 18 *Architectural Model of the Jagannatha Temple in Puri*, nineteenth century. Silver filigree. Made in Cuttack, India. Victoria and Albert Museum.

Figure 19 Margaret Bourke-White, *Mohandas K. Gandhi, India's Leader in the Struggle for Independence from Great Britain, Reading near a Spinning Wheel at Home,* 1946. Time & Life Pictures.

Figure 20 Maija Grotell, *Self-Portrait Vase*, 1937. Courtesy Cranbrook Art Museum.

Figure 21 Alexandra Jacopetti, macramé playpen, erected at a sale of the Baulines Craftsmen's Guild in Bolinas, California, ca. 1973–4.

Figure 22 George Nakashima, *Conoid Bench with Back*, 1961. American black walnut, with hickory spindles and rosewood butterflies. Courtesy George Nakashima Woodworker, S.A.

Figure 23 Shop mark used by Sam Maloof from the late 1950s through 1971. Photograph by Jonathan Pollock.

Figure 24 Maloof, *Rocking Chair*, 1993. Walnut and ebony. Photograph by Jonathan Pollock.

Walter Nurnberg, *Female Machinist at United Steel, Sheffield*, 1947. This British lathe operator is shown cutting the frames for umbrellas. Science Museum, London.

Figure 25 Wooden shoe forms of famous women at shoemaker Ferragamo. Photo by David Lees/Time Life Pictures/Getty Images.

Figure 26a, b, c, d, e, f. Photographs of African textiles by David Doris.

Figure 27 Martin Bodilsen Kaldahl, *Nurbs and Loop 1*, 2007. Low-fired stoneware. 65 cm h. A Danish ceramist trained at the Royal College of Art, Kaldahl has conducted various experiments exploring the relationship between digital and physical objects. This piece was first conceived on the computer and later modelled by hand as it cannot be fabricated on a 3D printer (like many of his other designs). Courtesy of the artist.

Figure 28 Martin Bodilsen Kaldahl, digital rendering, 2007.

Figure 29 Luke Limner, *Artist and Artisan*; detail from the frontispiece for *Suggestions in Design* (1853).

Cover of "Do You Know Our Name?," pamphlet, Handcraft Cooperative League of America, November 1941. This was the first publication of the organization that was to become the American Craft Council.

Figure 30   David Pye, *Small Circular Box*, no date. Collection of Regency House, Tidebrook, Sussex. Courtesy of the Crafts Study Centre, Farnham.

Figure 31   "A Ming Vase of Kiang Ting Ware." As pictured in Elsie Fogerty, *Rhythm*.

Figure 32   "Irish Embroidery, 1863." As pictured in Elsie Fogerty, *Rhythm*.

Figure 33   *Bell*, collected in 1920. Bronze, iron. Made by the Edo ethnic group, Benin. British Museum.

Figure 34   *Amuletic Necklace*, 1880–1920. Leather. Made in Northern Nigeria. Science Museum, London.

Figure 35   *Jug*, ca. 1560–75. Tin glazed earthenware, painted. Victoria and Albert Museum.

Figure 36   Harry Bertoia, *Tea and Coffee Service*, 1940. Silver and wood. Courtesy Cranbrook Art Museum.

Figure 37   Alison Britton, *Pair with Black Lines*, 1981. Collection Ed Wolf. Photo by David Cripps.

Otto Hagel, *Marguerite Wildenhain Showing The Motion of Hands Making a Pot*, ca. 1945. Marguerite Wildenhain papers. Archives of American Art.

Figure 38   Elizabeth Parker, *Sampler*, 1830. Linen embroidered with red silk in cross stitch. Victoria and Albert Museum. This extraordinary sewn text, made by Elizabeth Parker when she was about 17 years old, documents the story of her life. Though it mentions her thoughts of suicide, Parker in fact lived to the age of 76.

Figure 39   Anella James in her handmade dress. Anella James (b. 1930) migrated to Britain from Jamaica in 1961. This studio photograph was taken at Griffith & Sons, East London, in the mid-1960s. The dress was designed and produced by Anella through the 'freehand' method of dressmaking, originally for a wedding.

Figure 40   Peter Voulkos, *Soleares II*, 1958. Ceramic. Photo by Anthony Cuñha. Collection of Pauline Annon. Courtesy Frank Lloyd Gallery, Los Angeles.

Figure 41   Ken Shores, *Little Red i*, 1962. Acrylic paint on clay. Museum of Contemporary Craft, Portland, Oregon. The work of the Oregon-based artist Ken Shores, less well-known than that of Peter Voulkos and his colleagues in California, attests to the spread of radical ceramics across the West Coast.

Figure 42   Lee Ufan, *Relatum*, 1979–1996. Iron and stone. Photo by Ken Adlard. Courtesy of the artist and Lisson Gallery.

Figure 43   Theodor Bogler, *Combination Teapot and Sugar Bowl*, 1923. Ceramic. Made at the Bauhaus. Victoria and Albert Museum.

Figure 44   Gunta Stölzl, *Wall Hanging*, 1926–7. Made at the Bauhaus. Victoria and Albert Museum.

Figure 45   Don Wallance, studies for *Design One* Cutlery, 1956, as depicted in *Shaping America's Products*.

Figure 46   Marianne Strengel, *Taj Mahal* automobile upholstery fabric, 1959. Courtesy Cranbrook Art Museum.

Figure 47   The fiber panel at the American Craftsmen's Council, Asilomar, 1957. Left to right: Jack Lenor Larsen, Anni Albers, Roy Ginstrom, Lenore Tawney. Courtesy, American Craft Council.

Figure 48   Otto Hagel, *Marguerite Wildenhain throwing at the wheel*, ca. 1945. Marguerite Wildenhain papers. Archives of American Art.

Figure 49   Charles Eames and Eero Saarinen, *Side chair*, 1940. Exhibited in the "Organic Design" exhibition at the Museum of Modern Art. Courtesy Cranbrook Art Museum.

Figure 50   Ettore Sottsass, *drawing for the Carlton bookcase*, 1981. Ink and colors on paper. Victoria and Albert Museum.

Liz Collins and collaborators, *Knitting Nation Phase 1: Knitting During Wartime,* part of "The Muster" organized by artist Allison Smith on Governors Island in 2005, sponsored by the Public Art Fund.

Figure 51   Kristian Kozul, *Wheelchair I*, 2003. From the series *Discoware*. Wheelchair, chrome beads, mirrors, rhinestones, feathers, rotating platform. Courtesy of the artist and Filip Trade Collection, Croatia.

Figure 52   Haimi Fenichel, *Passive Aggressive*, 2007. Carved Ytong construction block. Courtesy of the artist.

Figure 53   Zandra Ahl with her work *Autoshape*, 2008. Found object, foam rubber, and black pantyhose.

Figure 54   Marianne Jørgenson, *Pink M.24 Chaffee*, 2006. A World War II combat tank was covered with 4,000 knitted and crocheted squares in pink yarn, contributed by volunteers. For five days the tank and its crafted cozy were on view at the Nikolaj Contemporary Art Center, Copenhagen.

Figure 55   Liz Collins, *Black Curtain Dress*, 2005. Hand loomed (knit) cotton with vintage curtains. Photo by Karen Philippi. Courtesy of the artist.

Figure 56   Sabrina Gschwandtner, *Wartime Knitting Circle*, 2007. Machine knit cotton, cotton tablecloth, wooden table and chairs, wool yarn, knitting needles, tape measure, scissors, stitch markers and other knitting notions. As installed at the Museum of Arts and Design. Image courtesy of the artist and the Museum of Arts & Design, New York. Photo: Alan Klein.

INDEX

Italic numbers denote references to illustrations.